MW00838234

HEAT CONDUCTION

HEAT CONDUCTION

Third Edition

DAVID W. HAHN
M. NECATI ÖZIŞIK

JOHN WILEY & SONS, INC.

Cover image: Courtesy of author

This book is printed on acid-free paper. ♾

Published by John Wiley & Sons, Inc., Hoboken, New Jersey

Published simultaneously in Canada

Library of Congress Cataloging-in-Publication Data:

Hahn, David W., 1964-
Heat conduction.—3rd ed. / David W. Hahn.
p. cm.
Rev. ed. of: Heat conduction / M. Necati Özişik. 2nd ed. c1993.
Includes bibliographical references and index.
ISBN 978-0-470-90293-6 (hardback : acid-free paper); ISBN 978-1-118-32197-3 (ebk); ISBN 978-1-118-32198-0 (ebk); ISBN 978-1-118-33011-1 (ebk); ISBN 978-1-118-33285-6 (ebk); ISBN 978-1-118-33450-8 (ebk); ISBN 978-1-118-41128-5
1. Heat–Conduction. I. Title.
QC321.O34 2012
621.402′2–dc23

2011052322

Printed in the United States of America
10 9 8 7 6 5 4 3 2 1

To Allison
-DWH

To Gül
-MNO

CONTENTS

PREFACE

The decision to take on the third edition of Professor Özişik's book was not one that I considered lightly. Having taught from the second edition for more than a dozen years and to nearly 500 graduate students, I was intimately familiar with the text. For the last few years I had considered approaching Professor Özişik with the idea for a co-authored third edition. However, with his passing in October 2008 at the age of 85, before any contact between us, I was faced with the decision of moving forward with a revision on my own. Given my deep familiarity and appreciation for Professor Özişik's book, it was ultimately an easy decision to proceed with the third edition.

My guiding philosophy to the third edition was twofold: first, to preserve the spirit of the second edition as the standard contemporary analytic treatment of conduction heat transfer, and, second, to write a truly major revision with goals to improve and advance the presentation of the covered material. The feedback from literally hundreds of my students over the years combined with my own pedagogical ideas served to shape my overall approach to the revision. At the end of this effort, I sincerely feel that the result is a genuine collaboration between me and Professor Özişik, equally combining our approaches to conduction heat transfer. As noted in the second edition, this book is meant to serve as a graduate-level textbook on conduction, as well as a comprehensive reference for practicing engineers and scientists. The third edition remains true to these goals.

I will now attempt to summarize the changes and additions to the third edition, providing some commentary and motivating thoughts. Chapter 1 is very much in the spirit of a collaborative effort, combining the framework from the second edition with significant revision. Chapter 1 presents the concepts of conduction starting with the work of Fourier and providing a detailed derivation of the heat equation. We now present both differential and integral derivations, which together present additional insight into the governing heat equation, notably with regard to conservation of energy. Extension of the heat equation to other coordinate systems, partial lumping of the heat equation, and the detailed treatment of relevant boundary conditions complete the chapter.

Chapter 2 is completely new in the third edition, presenting the characteristic value problem (the Sturm–Liouville problem) and the concept of orthogonal functions in considerable detail. We then develop the trigonometric orthogonal functions, Bessel functions, and the Legendre polynomials, followed by a rigorous presentation of the Fourier series, and finally the Fourier integrals. This material was dispersed in Chapters 2–4 in the second edition. The current unified treatment is envisioned to provide a more comprehensive foundation for the following chapters, while avoiding discontinuity through dispersion of the material over many chapters.

Chapters 3, 4, and 5 present the separation of variables method for Cartesian, cylindrical, and spherical coordinate systems, respectively. The organization differs from the second edition in that we first emphasize the steady-state solutions and then the transient solutions, with the concept of superposition presented in Chapter 3. In particular, the superposition schemes are presented using a more systematic approach and added illustrative figures. The many tables of characteristic value problems and resulting eigenfunctions and eigenvalues are retained in this edition in Chapter 2, although the approach from the second edition of developing more generic solutions in conjunction with these tables was deemphasized. Finally, each of these chapters now ends with a capstone example problem, attempting to illustrate the numerical implementation of representative analytic solutions, with goals of discussing the numerical convergence of realization of our analytic solutions, as well as emphasizing the concepts of energy conservation. A significant change in the third edition was the removal of all semi-infinite and infinite domain material from these three chapters and consolidating this material into Chapter 6, a new chapter dedicated to the semi-infinite and infinite domain problems in the context of the Fourier integral. Pedagogically, it was felt that this topic was better suited to a dedicated chapter, rather than treated along with separation of variables. Overall, the revised treatment in Chapters 2–6 will hopefully assist students in learning the material, while also greatly improving the utility of the third edition as a reference book. Additional homework problems have been added throughout.

Chapters 7, 8, and 9 focus on the treatment of nonhomogeneities in heat conduction problems, notably time-dependent nonhomogeneities, using Duhamel's theorem, the Green's function approach, and the Laplace transform method, respectively. A new derivation of Duhamel's theorem is presented along with a new presentation of Duhamel's various solutions that is intended to clarify the use and limitations of this method. Also included in Chapter 7 is a new table giving closed-form solutions for various surface temperature functions. The Green's function chapter contains new sample problems with a greater variety of boundary condition types. The Laplace transform chapter contains additional problems focusing on the general solution method (i.e., not limited to the small-time approximations), and the Laplace transform tables are greatly expanded to facilitate such solutions. Together, Chapters 1–9 are considered the backbone of a graduate course on conduction heat transfer, with the remainder of the text providing additional topics to pursue depending on the scope of the class and the interests of the instructor and students.

Chapters 10, 11, and 12 cover the one-dimensional composite medium, the moving heat source problem, and phase-change heat transfer, respectively, much along the lines of the original treatment by Professor Özişik, although efforts were made to homogenize the style with the overall revision. Chapter 13 is on approximate analytic methods and takes a departure from the exact analytic treatment of conduction presented in the first dozen chapters. The emphasis on the integral method and method of residuals has applications to broader techniques (e.g., finite-element methods), although efforts are made in the current text to emphasize conservation of energy and formulation in the context of these methods. Chapter 14 presents the integral transform technique as a means for solution of the heat equation under a variety of conditions, setting the foundation for use of the integral transform technique in Chapter 15, which focuses on heat conduction in anisotropic solids.

Chapter 16 is a totally new contribution to the text and presents an introduction to microscale heat conduction. There were several motivating factors for the inclusion of Chapter 16. First, engineering and science fields are increasingly concerned with the micro- and nanoscales, including with regard to energy transfer. Second, given that the solution of the heat equation in conjunction with Fourier's law remains at the core of this book, it is useful to provide a succinct treatment of the limitations of these equations.

The above material considerably lengthened the manuscript with regard to the corresponding treatment in the second edition, and therefore, some material was omitted from the revised edition. It is here that I greatly missed the opportunity to converse with Professor Özişik, but ultimately, such decisions were mine alone. The lengthy treatment of finite-difference methods and the chapter on inverse heat conduction from the second edition were omitted from the third edition. With regard to the former, the logic was that the strength of this text is the analytic treatment of heat conduction, noting that many texts on numerical methods are available, including treatment of the heat equation. The latter topic has been the subject of entire monographs, and, therefore, the brief treatment of inverse conduction was essentially replaced with the introduction to microscale heat transfer.

I would like to express my sincere gratitude to the many heat transfer students that have provided me with their insight into a student's view of heat conduction and shared their thoughts, both good and bad, on the second edition. You remain my primary motivation for taking on this project. A few of my former students have reviewed chapters of the third edition, and I would like to express my thanks to Fotouh Al-Ragom, Roni Plachta, Lawrence Stratton and Nadim Zgheib for their feedback. I would also like to thank my graduate student Philip Jackson for his invaluable help with the numerical implementation of the Legendre polynomials for the capstone problem in Chapter 5. I would like to acknowledge several of my colleagues at the University of Florida, including Andreas Haselbacher for useful discussions and feedback regarding Chapter 1, Renwei Mei for his reviews of several chapters and his very generous contribution of Table 1 in Chapter 7, Greg Sawyer for his suggestion of the capstone problem in Chapter 5, as well as for many motivating discussions, and Simon Phillpot for his very insightful review of Chapter 16. I would like to thank my editors at John Wiley & Sons, Daniel Magers and Bob Argentieri, for their support and enthusiasm from the very beginning, as well as for their generous patience as this project neared completion.

On a more personal note, I want to thank my wonderful children, Katherine, William, and Mary-Margaret, for their support throughout this process. Many of the hours that I dedicated to this manuscript came at their expense, and I will never forget their patience with me to the very end. I also thank Mary-Margaret for her admirable proof-reading skills when needed. The three of you never let me down. Finally, I thank the person that gave me the utmost support throughout this project, my wife and dearest partner in life, Allison. This effort would never have been completed without your unwavering support, and it is to you that I dedicate this book.

DAVID W. HAHN

Gainesville, Florida

PREFACE TO SECOND EDITION

In preparing the second edition of this book, the changes have been motivated by the desire to make this edition a more application-oriented book than the first one in order to better address the needs of the readers seeking solutions to heat conduction problems without going through the details of various mathematical proofs. Therefore, emphasis is placed on the understanding and use of various mathematical techniques needed to develop exact, approximate, and numerical solutions for a broad class of heat conduction problems. Every effort has been made to present the material in a clear, systematic, and readily understandable fashion. The book is intended as a graduate-level textbook for use in engineering schools and a reference book for practicing engineers, scientists and researchers. To achieve such objectives, lengthy mathematical proofs and developments have been omitted, instead examples are used to illustrate the applications of various solution methodologies.

During the twelve years since the publication of the first edition of this book, changes have occurred in the relative importance of some of the application areas and the solution methodologies of heat conduction problems. For example, in recent years, the area of inverse heat conduction problems (IHCP) associated with the estimation of unknown thermophysical properties of solids, surface heat transfer rates, or energy sources within the medium has gained significant importance in many engineering applications. To answer the needs in such emerging application areas, two new chapters are added, one on the theory and application of IHCP and the other on the formulation and solution of moving heat source problems. In addition, the use of enthalpy method in the solution of phase-change problems has been expanded by broadening its scope of applications. Also, the chapters on the use of Duhamel's method, Green's function, and finite-difference methods have been revised in order to make them application-oriented. Green's function formalism provides an efficient, straightforward approach for developing exact analytic solutions to a broad class of heat conduction problems in the rectangular, cylindrical, and spherical coordinate systems, provided that appropriate Green's functions are available. Green's functions needed for use in such formal solutions are constructed by utilizing the tabulated eigenfunctions, eigenvalues and the normalization integrals presented in the tables in Chapters 2 and 3.

Chapter 1 reviews the pertinent background material related to the heat conduction equation, boundary conditions, and important system parameters. Chapters 2, 3, and 4 are devoted to the solution of time-dependent homogeneous heat conduction problems in the rectangular, cylindrical, and spherical coordinates, respectively, by the application of the classical method of separation of variables and orthogonal expansion technique. The resulting eigenfunctions, eigenconditions, and the normalization integrals are systematically tabulated for various combinations of the boundary conditions in Tables 2-2, 2-3, 3-1, 3-2, and 3-3. The results from such tables are used to construct the Green functions needed in solutions utilizing Green's function formalism.

Chapters 5 and 6 are devoted to the use of Duhamel's method and Green's function, respectively. Chapter 7 presents the use of Laplace transform technique in the solution of one-dimensional transient heat conduction problems.

Chapter 8 is devoted to the solution of one-dimensional, time-dependent heat conduction problems in parallel layers of slabs and concentric cylinders and spheres. A generalized orthogonal expansion technique is used to solve the homogeneous problems, and Green's function approach is used to generalize the analysis to the solution of problems involving energy generation.

Chapter 9 presents approximate analytical methods of solving heat conduction problems by the integral and Galerkin methods. The accuracy of approximate results are illustrated by comparing with the exact solutions. Chapter 10 is devoted to the formulation and the solution of moving heat source problems, while Chapter 11 is concerned with the exact, approximate, and numerical methods of solution of phase-change problems.

Chapter 12 presents the use of finite difference methods for solving the steady-state and time-dependent heat conduction problems. Chapter 13 introduces the use of integral transform technique in the solution of general time-dependent heat conduction equations. The application of this technique for the solution of heat conduction problems in rectangular, cylindrical, and spherical coordinates requires no additional background, since all basic relationships needed for constructing the integral transform pairs have already been developed and systematically tabulated in Chapters 2 to 4. Chapter 14 presents the formulation and methods of solution of inverse heat conduction problems and some background information on statistical material needed in the inverse analysis. Finally, Chapter 15 presents the analysis of heat conduction in anisotropic solids. A host of useful information, such as the roots of transcendental equations, some properties of Bessel functions, and the numerical values of Bessel functions and Legendre polynomials are included in Appendixes IV and V for ready reference.

I would like to express my thanks to Professors J. P. Bardon and Y. Jarny of University of Nantes, France, J. V. Beck of Michigan State University, and Woo Seung Kim of Hanyang University, Korea, for valuable discussions and suggestions in the preparation of the second edition.

M. NECATI ÖZIŞIK

Raleigh, North Carolina
December 1992

HEAT CONDUCTION

1

HEAT CONDUCTION FUNDAMENTALS

> No subject has more extensive relations with the progress of industry and the natural sciences; for the action of heat is always present, it penetrates all bodies and spaces, it influences the processes of the arts, and occurs in all the phenomena of the universe.
>
> —Joseph Fourier, *Théorie Analytique de la Chaleur,* 1822 [1]

All matter when considered at the macroscopic level has a definite and precise energy. Such a state of energy may be quantified in terms of a thermodynamic energy function, which partitions energy at the atomic level among, for example, electronic, vibrational, and rotational states. Under local equilibrium, the energy function may be characterized by a measurable scalar quantity called *temperature*. The energy exchanged by the constituent particles (e.g., atoms, molecules, or free electrons) from a region with a greater local temperature (i.e., greater thermodynamic energy function) to a region with a lower local temperature is called *heat*. The transfer of heat is classically considered to take place by conduction, convection, and radiation, and although it cannot be measured directly, the concept has physical meaning because of the direct relationship to temperature. *Conduction* is a specific mode of heat transfer in which this energy exchange takes place in solids or quiescent fluids (i.e., no convective motion resulting from the macroscopic displacement of the medium) from the region of high temperature to the region of low temperature due to the presence of a *temperature gradient* within the system. Once the temperature distribution $T(\hat{r}, t)$ is known within the medium as a function of space (defined by the position vector $\hat{r}$) and time (defined by scalar t), the flow of heat is then prescribed from the governing heat transfer laws. The study of heat conduction provides an enriching combination of fundamental science and mathematics. As the prominent thermodynamicist H. Callen wrote: "The history of the concept of heat as a form of energy transfer is unsurpassed as a case study in the tortuous development of scientific theory, as an illustration of the almost insuperable inertia presented by accepted physical doctrine, and as a superb tale of human ingenuity applied to a subtle and abstract problem" [2]. The science of heat conduction is principally concerned with the determination of the temperature distribution and flow of energy within solids. In this chapter, we present the basic laws relating the heat flux to the temperature gradient in the medium, the governing differential equation of heat conduction, the boundary conditions appropriate for the analysis of heat conduction problems, the rules of coordinate transformation needed for working in different orthogonal coordinate systems, and a general discussion of the various solution methods applicable to the heat conduction equation.

1-1 THE HEAT FLUX

Laws of nature provide accepted descriptions of natural phenomena based on observed behavior. Such laws are generally formulated based on a large body of empirical evidence accepted within the scientific community, although they usually can be neither proven nor disproven. To quote Joseph Fourier from the opening sentence of his *Analytical Theory of Heat*: "Primary causes are unknown to us; but are subject to simple and constant laws, which may be discovered by observation" [1]. These laws are considered *general laws*, as their application is independent of the medium. Well-known examples include Newton's laws of motion and the laws of thermodynamics. Problems that can be solved using only general laws of nature are referred to as deterministic and include, for example, simple projectile motion.

Other problems may require supplemental relationships in addition to the general laws. Such problems may be referred to as nondeterministic, and their solution requires laws that apply to the specific medium in question. These additional laws are referred to as *particular laws* or *constitutive relations*. Well-known examples include the ideal gas law, the relationship between shear stress and the velocity gradient for a Newtonian fluid, and the relationship between stress and strain for a linear-elastic material (Hooke's law).

The particular law that governs the relationship between the flow of heat and the temperature field is named after Joseph Fourier. For a homogeneous, isotropic solid (i.e., material in which thermal conductivity is independent of direction), *Fourier's law* may be given in the form

$$q''(\hat{r}, t) = -k\nabla T(\hat{r}, t) \quad \text{W/m}^2 \tag{1-1}$$

where the temperature gradient $\nabla T(\hat{r}, t)$ is a vector normal to the isothermal surface, the heat flux vector $q''(\hat{r}, t)$ represents the heat flow per unit time, per unit area of the isothermal surface in the direction of decreasing temperature gradient, and k is the *thermal conductivity* of the material. The thermal conductivity is a positive, scalar quantity for a homogeneous, isotropic material. The minus sign is introduced in equation (1-1) to make the heat flow a positive quantity in the positive coordinate direction (i.e., opposite of the temperature gradient), as described below. This text will consider the heat flux in the SI units W/m^2 and the temperature gradient in K/m (equivalent to the unit °C/m), giving the thermal conductivity the units of W/(m · K). In the Cartesian coordinate system (i.e., rectangular system), equation (1-1) is written as

$$q''(x, y, z, t) = -\hat{i}k\frac{\partial T}{\partial x} - \hat{j}k\frac{\partial T}{\partial y} - \hat{k}k\frac{\partial T}{\partial z} \tag{1-2}$$

where $\hat{i}, \hat{j}$, and $\hat{k}$ are the unit direction vectors along the x, y, and z directions, respectively. One may consider the three components of the heat flux vector in the x, y, and z directions, respectively, as given by

$$q''_x = -k\frac{\partial T}{\partial x} \quad q''_y = -k\frac{\partial T}{\partial y} \quad \text{and} \quad q''_z = -k\frac{\partial T}{\partial z} \tag{1-3a,b,c}$$

Clearly, the flow of heat for a given temperature gradient is directly proportional to the thermal conductivity of the material. Equation (1-3a) is generally used for one-dimensional (1-D) heat transfer in a rectangular coordinate system. Figure 1-1 illustrates the sign convention of Fourier's law for the 1-D Cartesian coordinate system. Both plots depict the heat flux (W/m^2) through the plane at $x = x_0$ based on the local temperature gradient. In Figure 1-1(*a*), the gradient *dT/dx* is negative with regard to the Cartesian coordinate system; hence the resulting flux is mathematically positive, and by convention is in the *positive x direction*, as shown in the figure. In contrast, Figure 1-1(*b*) depicts a positive gradient *dT/dx*. This yields a mathematically negative heat flux, which by convention

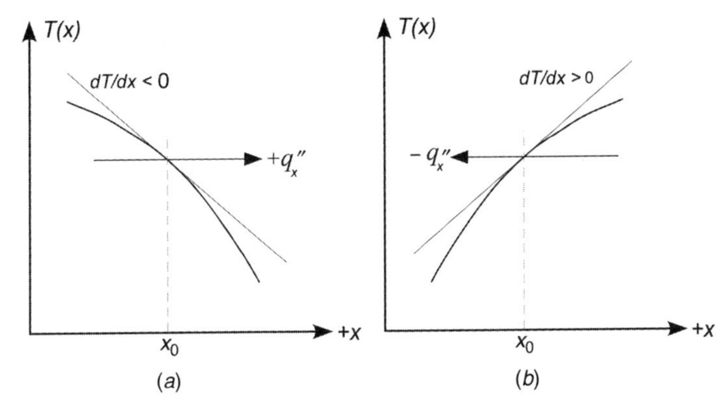

Figure 1-1 Fourier's law illustrated for a (a) positive heat flux and (b) a negative heat flux.

is in the negative x direction, as indicated in the figure. As defined, Fourier's law is directly tied to the coordinate system, with positive heat flux always flowing in the positive coordinate direction. While determining the actual direction of heat flow is often trivial for 1-D problems, multidimensional problems, and notably transient problems, can present considerable difficulty in determining the direction of the local heat flux terms. Adherence to the sign convention of Fourier's law will avoid any such difficulties of flux determination, which is useful in the context of overall energy conservation for a given heat transfer problem.

In addition to the heat flux, which is the flow of heat *per unit area* normal to the direction of flow (e.g., a plane perpendicular to the page in Fig. 1-1), one may define the total heat flow, often called the *heat rate*, in the unit of watts (W). The heat rate is calculated by multiplying the heat flux by the *total cross-sectional area* through which the heat flows for a 1-D problem or by integrating over the area of flow for a multidimensional problem. The heat rate in the x direction for one-, two-, and three-dimensional (1-D, 2-D, and 3-D) Cartesian problems is given by

$$q_x = -kA_x \frac{dT}{dx} \quad \text{W} \tag{1-4}$$

$$q_x = -kH \int_{y=0}^{L} \frac{\partial T(x, y)}{\partial x} dy \quad \text{W} \tag{1-5}$$

$$q_x = -k \int_{y=0}^{L} \int_{z=0}^{H} \frac{\partial T(x, y, z)}{\partial x} dz\, dy \quad \text{W} \tag{1-6}$$

where A_x is the total cross-sectional area for the 1-D problem in equation (1-4). The total cross-sectional area for the 2-D problem in equation (1-5) is defined by the surface from $y = 0$ to L in the second spatial dimension and by the length H in the z direction, for which there is no temperature dependence [i.e., $T \neq f(z)$]. The total cross-sectional area for the 3-D problem in equation (1-6) is defined by the surface from $y = 0$ to L and $z = 0$ to H in the second and third spatial dimensions, noting that $T = f(x, y, z)$.

1-2 THERMAL CONDUCTIVITY

Given the direct dependency of the heat flux on the thermal conductivity via Fourier's law, the thermal conductivity is an important parameter in the analysis of heat conduction. There is a wide range in the thermal conductivities of various engineering materials. Generally, the highest values are observed for pure metals and the lowest value by gases and vapors, with the amorphous insulating materials and inorganic liquids having thermal conductivities that lie in between. There are important exceptions. For example, natural type IIa diamond (nitrogen free) has the highest thermal conductivity of any bulk material (~2300 W/m · K at ambient temperature), due to the ability of the well-ordered crystal lattice to transmit thermal energy via vibrational quanta called *phonons*. In Chapter 16, we will explore in depth the physics of energy carriers to gain further insight into this important material property.

To give some idea of the order of magnitude of thermal conductivity for various materials, Figure 1-2 illustrates the typical range for various material classes. Thermal conductivity also varies with temperature and may change with orientation for nonisotropic materials. For most pure metals the thermal conductivity decreases with increasing temperature, whereas for gases it increases with increasing temperature. For most insulating materials it increases with increasing temperature. Figure 1-3 provides the effect of temperature on the thermal conductivity for a range of materials. At very low temperatures, thermal conductivity increases rapidly and then exhibits a sharp decrease as temperatures approach absolute zero, as shown in Figure 1-4, due to the dominance of energy carrier scattering from defects at extreme low temperatures. A comprehensive compilation of thermal conductivities of materials may be found in references 3–6. We present in Appendix I the thermal conductivity of typical engineering materials together with the specific heat c_p, density ρ, and the thermal diffusivity α. These latter properties are discussed in more detail in the following section.

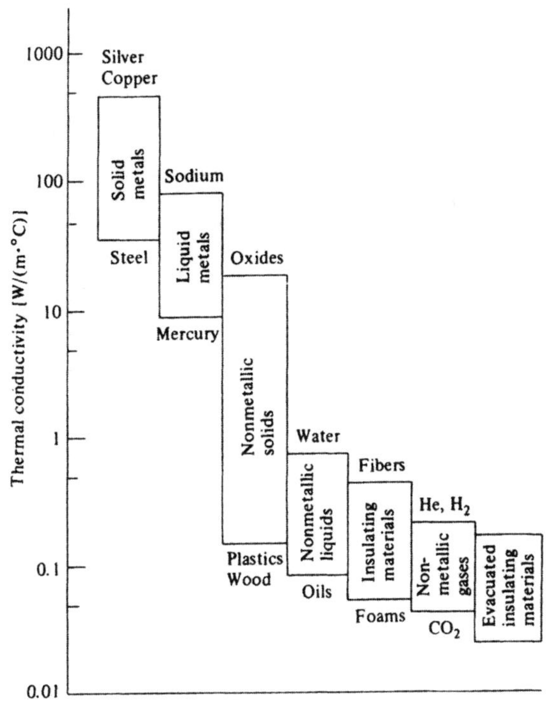

Figure 1-2 Typical range of thermal conductivity of various material classes.

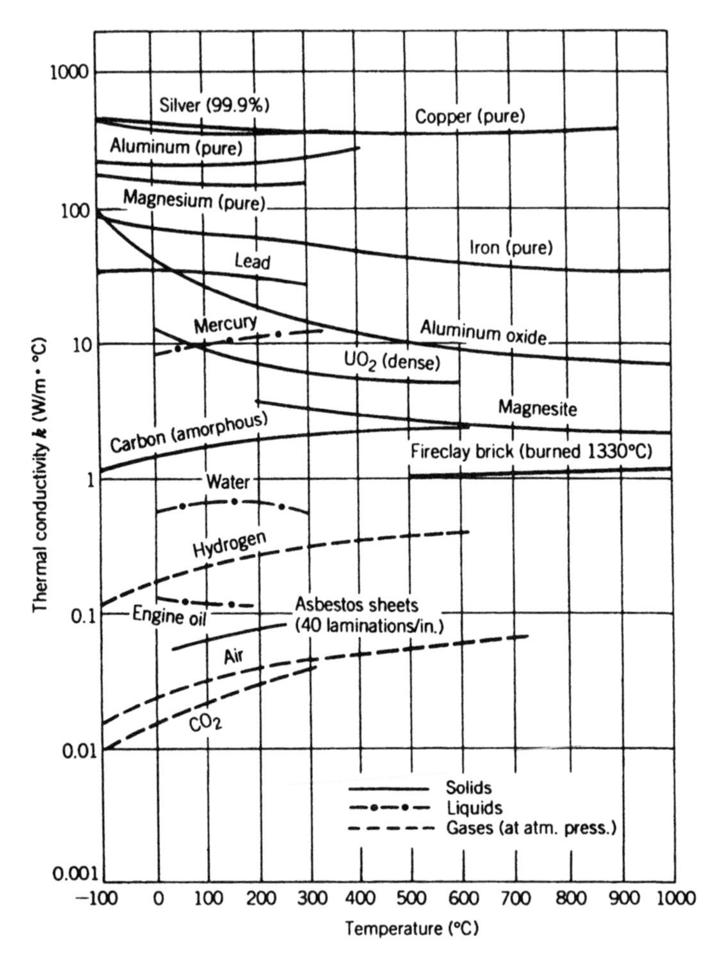

Figure 1-3 Effect of temperature on thermal conductivity.

1-3 DIFFERENTIAL EQUATION OF HEAT CONDUCTION

We now derive the differential equation of heat conduction, often called the *heat equation*, for a stationary, homogeneous, isotropic solid with heat generation within the body. *Internal heat generation* may be due to nuclear or chemical reactivity, electrical current (i.e., Joule heating), absorption of laser light, or other sources that may in general be a function of time and/or position. The heat equation may be derived using either a differential control volume approach or an integral approach. The former is perhaps more intuitive and will be presented first, while the latter approach is more general and readily extends the derivation to moving solids.

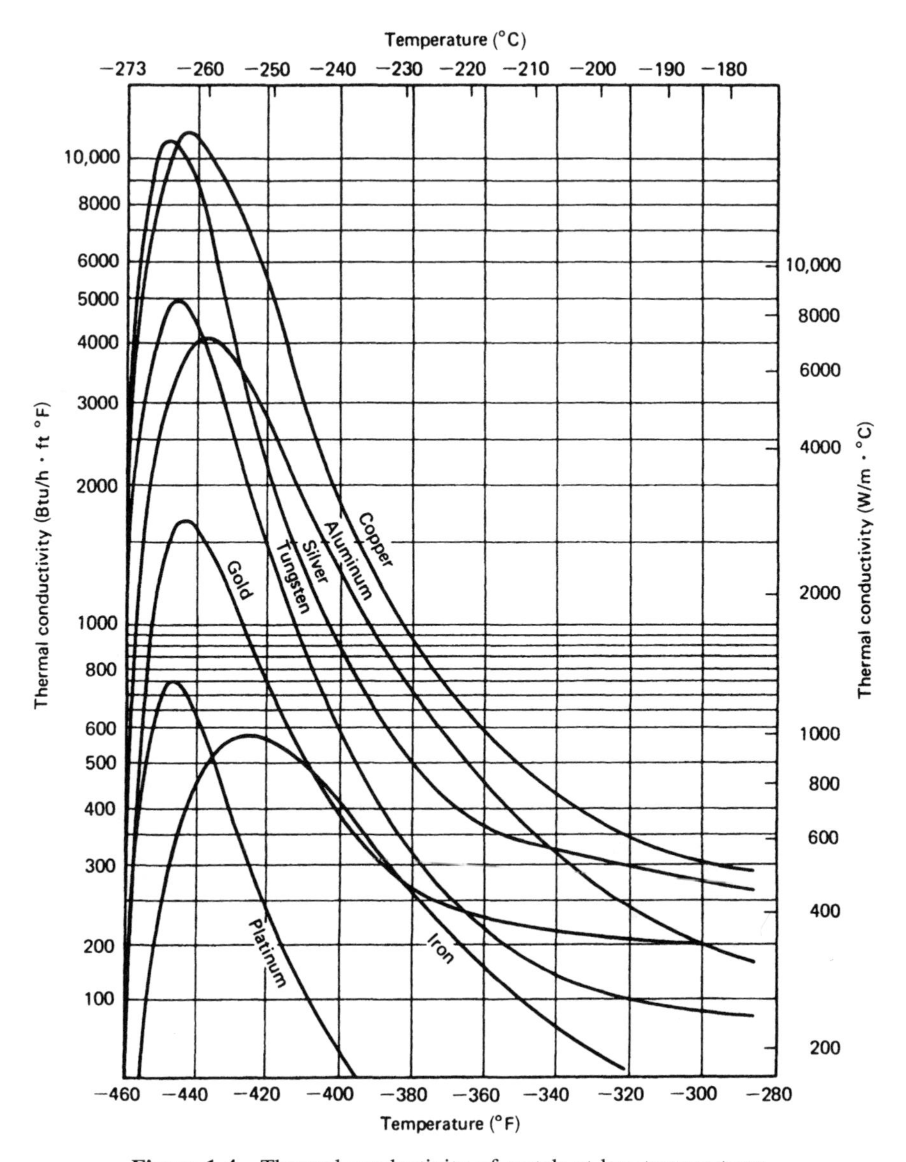

Figure 1-4 Thermal conductivity of metals at low temperatures.

The *differential control volume* is defined in Figure 1-5 for the Cartesian coordinate system. The corresponding volume and mass of the differential control volume are defined, respectively, as

$$dv = dx\,dy\,dz \quad \text{and} \quad dm = \rho dx\,dy\,dz \tag{1-7}$$

where ρ is the mass density (kg/m^3) of the control volume. The differential approach will assume a *continuum* such that all properties do not vary

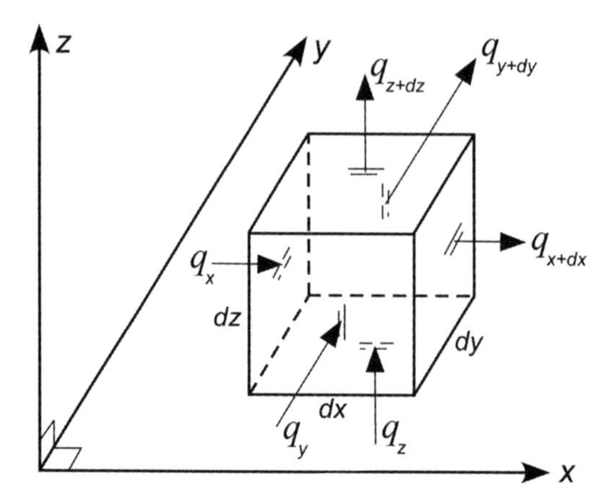

Figure 1-5 Differential control volume for derivation of the heat equation in Cartesian coordinates.

microscopically. The continuum assumption may be considered in terms of ε, a volume much larger than individual atoms. If we let L_c be the smallest length scale of interest for the heat transfer problem, then the continuum assumption is considered justified for the condition

$$\varepsilon \ll L_c^3 \tag{1-8}$$

Limitations on the continuum approach for the heat equation and Fourier's law are discussed in Chapter 16.

We begin with a general statement of *conservation of energy* based on the first law of thermodynamics, namely,

$$\left(h + \frac{1}{2}\overline{V}^2 + gz\right)_{\text{in}} \delta\dot{m}_{\text{in}} - \left(h + \frac{1}{2}\overline{V}^2 + gz\right)_{\text{out}} \delta\dot{m}_{\text{out}} + \delta\dot{Q} + \delta\dot{E}_{\text{gen}} - \delta\dot{W} = \frac{dE_{\text{cv}}}{dt} \tag{1-9}$$

where $\delta\dot{m}_{\text{in}}$ and $\delta\dot{m}_{\text{out}}$ represent the mass flow rates in and out of the differential control volume, respectively. We will derive the heat equation for a *quiescent medium*, hence the mass flow rates are zero, and assume that the rate of work done by the control volume is zero ($\delta\dot{W} = 0$). The rate of change of energy within the control volume may be expanded as

$$\frac{dE_{\text{cv}}}{dt} = \frac{d}{dt}\left[\left(u + \frac{1}{2}\overline{V}^2 + gz\right)_{\text{cv}} dm\right] \tag{1-10}$$

where u is the internal energy (J/kg), an intensive, scalar property associated with the thermodynamic state of the system. Neglecting any *changes* in the kinetic and potential energy of the control volume, and applying the above assumptions, conservation of energy becomes

$$\delta \dot{Q} + \delta \dot{E}_{\text{gen}} = \frac{d(u\ dm)}{dt} \tag{1-11}$$

where $\delta \dot{E}_{\text{gen}}$ (W) represents the rate at which energy is generated within the control volume due to internal energy generation as described above, and $\delta \dot{Q}$ (W) represents the net rate of heat transfer into the control volume, with positive $\delta \dot{Q}$ representing heat transfer into the system. We may now consider equation (1-11) term by term.

The net rate of heat transfer is given in terms of the heat rate in and out of the control volume, namely,

$$\delta \dot{Q} = \left(q_x - q_{x+dx}\right) + \left(q_y - q_{y+dy}\right) + \left(q_z - q_{z+dz}\right) \tag{1-12}$$

where the individual, entering heat rate terms may be defined using Fourier's law and the respective cross-sectional areas, as given by equation (1-4) for the x direction:

$$q_x = -kA_x \frac{\partial T}{\partial x} \quad \text{where} \quad A_x = dy\,dz \tag{1-13}$$

$$q_y = -kA_y \frac{\partial T}{\partial y} \quad \text{where} \quad A_y = dx\,dz \tag{1-14}$$

$$q_z = -kA_z \frac{\partial T}{\partial z} \quad \text{where} \quad A_z = dx\,dy \tag{1-15}$$

The individual, exiting heat rate terms may be defined using a Taylor series expansion of the entering terms. Neglecting higher-order terms, for the x direction this term becomes

$$q_{x+dx} = q_x + \frac{\partial q_x}{\partial x} dx = -kA_x \frac{\partial T}{\partial x} + \frac{\partial}{\partial x}\left(-kA_x \frac{\partial T}{\partial x}\right) dx \tag{1-16}$$

Using equations (1-13) and (1-16), the net heat rate entering the differential control volume from the x direction becomes

$$q_x - q_{x+dx} = \frac{\partial}{\partial x}\left(k \frac{\partial T}{\partial x}\right) dx\,dy\,dz \tag{1-17}$$

Similarly, the net heat rate in the y and z directions becomes

$$q_y - q_{y+dy} = \frac{\partial}{\partial y}\left(k \frac{\partial T}{\partial y}\right) dx\,dy\,dz \tag{1-18}$$

$$q_z - q_{z+dz} = \frac{\partial}{\partial z}\left(k \frac{\partial T}{\partial z}\right) dx\,dy\,dz \tag{1-19}$$

Equations (1-17)–(1-19) may now be substituted in equation (1-12).

The rate of internal energy generation (W) may be directly calculated from the *volumetric rate* of internal energy generation g (W/m^3), noting that in general $g = g(\hat{r}, t)$, and the control volume, namely,

$$\delta \dot{E}_{\text{gen}} = g \; dx \; dy \; dz \tag{1-20}$$

Finally, the rate of change of energy within the control volume may be defined by introducing the *constant volume specific heat* c_v (J/ kg · K), namely,

$$c_v \equiv \left. \frac{\partial u}{\partial T} \right|_v \quad \rightarrow \quad u = c_v T + u_{\text{ref}} \tag{1-21}$$

noting that for an incompressible solid or fluid, $c_v = c_p = c$, with the middle quantity defined as the constant pressure specific heat. Equation (1-21) may be substituted into the right-hand side of equation (1-11), which along with the assumption of constant properties ρ and c yields the net rate of change of energy within the control volume as

$$\frac{d(u \; dm)}{dt} = \rho c \frac{\partial T}{\partial t} dx \; dy \;\; dx \tag{1-22}$$

The above expressions may now be introduced into equation (1-11) to provide the *general heat equation* for the Cartesian coordinate system, which after cancelation of the $dx \; dy \; dz$ terms yields

$$\boxed{\frac{\partial}{\partial x}\left(k \frac{\partial T}{\partial x}\right) + \frac{\partial}{\partial y}\left(k \frac{\partial T}{\partial y}\right) + \frac{\partial}{\partial z}\left(k \frac{\partial T}{\partial z}\right) + g = \rho c \frac{\partial T}{\partial t}} \tag{1-23}$$

where each term has the units W/m^3. In simple terms, the heat equation expresses that the net rate of heat conducted per differential volume plus the rate of energy generated internally per volume is equal to the net rate of energy stored per differential volume. The heat equation may be expressed in several additional forms, including using vector notation:

$$\nabla \cdot (k \nabla T) + g = \rho c \frac{\partial T}{\partial t} \tag{1-24}$$

where ∇ is the vector differential operator $[\nabla = \hat{i}(\partial/\partial x) + \hat{j}(\partial/\partial y) + \hat{k}(\partial/\partial z)]$. When the thermal conductivity is a constant, equation (1-23) may be written in the form

$$\boxed{\frac{\partial^2 T}{\partial x^2} + \frac{\partial^2 T}{\partial y^2} + \frac{\partial^2 T}{\partial z^2} + \frac{g}{k} = \frac{1}{\alpha} \frac{\partial T}{\partial t}} \tag{1-25}$$

where each term now has the units K/m^2. The *thermal diffusivity* (m^2/s), which appears on the right-hand side, is defined as

$$\alpha = \frac{k}{\rho c} \tag{1-26}$$

and represents a thermal-physical property of the medium. The physical significance of thermal diffusivity is associated with the speed of propagation of heat into the solid during changes of temperature. In other words, the thermal diffusivity represents the flow of heat (i.e., conduction of heat) compared to the storage of energy (i.e., heat capacity). The higher the thermal diffusivity, the faster is the response of a medium to thermal perturbations, and the faster such changes propagate throughout the medium. This statement is better understood by referring to the following heat conduction problem: Consider a semi-infinite medium, $x \geq 0$, initially at a uniform temperature T_0. For times $t > 0$, the boundary surface at $x = 0$ is suddenly reduced to and kept at zero temperature. Clearly, the temperature within the medium will now vary with position and time. Suppose we are interested in the time required for the temperature to decrease from its initial value T_0 to half of this value, $\frac{1}{2}T_0$, at a position, say, 30 cm from the boundary surface. Table 1-1 gives the time required for this change for several different materials. It is apparent from these results that the greater the thermal diffusivity, the shorter is the time required for the boundary perturbation to penetrate into the depth of the solid. It is important to note, therefore, that the thermal response of a material depends not only on the thermal conductivity but also on the density and specific heat.

For a medium with constant thermal conductivity, no internal heat generation, and under steady-state conditions (i.e., $\partial T/\partial t = 0$), the heat equation takes the form

$$\frac{\partial^2 T}{\partial x^2} + \frac{\partial^2 T}{\partial y^2} + \frac{\partial^2 T}{\partial z^2} = 0 \quad \text{or} \quad \nabla^2 T = 0 \tag{1-27}$$

which is known as *Laplace's equation*, after the French mathematician Pierre-Simon Laplace.

We now present the *integral formulation* of the heat equation by considering the energy balance for a small control volume V, as illustrated in Figure 1-6. Conversation of energy may be stated as

$$\begin{bmatrix}\text{Rate of heat entering through} \\ \text{the bounding surface of } V\end{bmatrix} + \begin{bmatrix}\text{Rate of energy} \\ \text{generation in } V\end{bmatrix} = \begin{bmatrix}\text{Rate of energy} \\ \text{storage in } V\end{bmatrix} \tag{1-28}$$

Each term in equation (1-28) may now be evaluated individually, beginning with the rate of heat entering through the boundary. This can be calculated by

TABLE 1-1 Effect of Thermal Diffusivity on Rate of Heat Propagation

Material	Silver	Copper	Steel	Glass	Cork
$\alpha \times 10^6$ m^2/s	170	103	12.9	0.59	0.16
Time	9.5 min	16.5 min	2.2 h	2.0 days	7.7 days

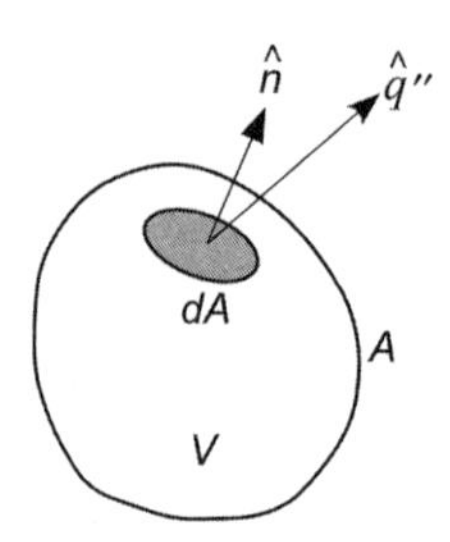

Figure 1-6 Control volume for integral formulation of the heat equation.

integrating the surface heat flux normal to the surface over the entire surface area of the control volume, as given by

$$\begin{bmatrix}\text{Rate of heat entering through}\\ \text{the bounding surface of } V\end{bmatrix} = -\int_A q'' \cdot \hat{n} dA = -\int_V \nabla \cdot q'' dv \tag{1-29}$$

where A is the surface area of the control volume V, $\hat{n}$ is the outward-drawn normal vector to the surface element dA, and $q''(\hat{r}, t)$ is the heat flux vector at the surface element dA. The minus sign is introduced to ensure that the heat flow is positive into the control volume in consideration of the negative sign in Fourier's law per equation (1-1). The *divergence theorem*, also known as Gauss' theorem, is then used to transform the surface integral into a volume integral, yielding the final form above.

We next consider the rate of energy generation within the control volume, which is readily evaluated by integrating the volumetric energy generation, as defined above, over the control volume

$$\begin{bmatrix}\text{Rate of energy}\\ \text{generation in } V\end{bmatrix} = \int_V g(\hat{r}, t) dv \tag{1-30}$$

For the rate of storage within the control volume, it is first useful to define the *material* or *total derivative* in terms of Eulerian derivatives for a generic property $\lambda(\hat{r}, t)$, namely,

$$\frac{D\lambda}{Dt} \equiv \frac{\partial \lambda}{\partial t} + u\frac{\partial \lambda}{\partial x} + v\frac{\partial \lambda}{\partial y} + w\frac{\partial \lambda}{\partial z} \tag{1-31}$$

where the velocity vector $\hat{u}$ has the components

$$\hat{u} = u\hat{i} + v\hat{j} + w\hat{k} \tag{1-32}$$

We now introduce *Reynolds transport theorem*, which allows one to readily calculate the material derivative of a volume integral. Again using our generic property

λ, Reynolds transport theorem may be expressed as

$$\frac{D}{Dt}\int_V \lambda(\hat{r},t)dv = \int_V \left[\frac{\partial \lambda}{\partial t} + \nabla \cdot (\lambda \hat{u})\right] dv \tag{1-33}$$

For the rate of change of energy within the control volume, we want our generic property λ to specifically equal the energy per unit volume (J/m^3). Using our definition of constant volume specific heat, this is accomplished by letting $\lambda(\hat{r},t) = \rho c T(\hat{r},t)$. Substituting into equation (1-33) yields

$$\begin{bmatrix}\text{Rate of energy}\\ \text{storage in } V\end{bmatrix} = \frac{D}{Dt}\int_V \rho c T(\hat{r},t)dv = \int_V \rho c\left[\frac{\partial T}{\partial t} + \nabla \cdot (T\hat{u})\right] dv \tag{1-34}$$

Now all three rate terms of equation (1-28) are expressed as volume integrals, namely equations (1-29), (1-30), and (1-34), which may be brought into a common integral, yielding

$$\int_V \left\{-\nabla \cdot q'' + g - \rho c\left[\frac{\partial T}{\partial t} + \nabla \cdot (T\hat{u})\right]\right\} dv = 0 \tag{1-35}$$

Because equation (1-35) is derived for an arbitrary control volume V, the only way it is satisfied for all choices of V is if the integrand itself is zero. With this condition, equation (1-35) becomes

$$-\nabla \cdot q'' + g = \rho c\left[\frac{\partial T}{\partial t} + \nabla \cdot (T\hat{u})\right] \tag{1-36}$$

This equation can now be simplified further by expanding the rightmost term,

$$\nabla \cdot (T\hat{u}) = T(\nabla \cdot \hat{u}) + \hat{u} \cdot \nabla T \tag{1-37}$$

and noting that $\nabla \cdot \hat{u} = 0$ per continuity for an incompressible medium. We then insert equation (1-1) for the heat flux vector in the left-hand side (LHS). Making these substitutions yields the desired final form of the conduction heat equation,

$$\nabla \cdot (k\nabla T) + g = \rho c\left[\frac{\partial T}{\partial t} + u\frac{\partial T}{\partial x} + v\frac{\partial T}{\partial y} + w\frac{\partial T}{\partial z}\right] \tag{1-38}$$

Equation (1-38) is valid for an *incompressible, moving solid*, assuming constant ρc. Overall, the bulk motion of the solid is regarded to give rise to convective or enthalpy fluxes, namely, in the x, y, and z directions, in addition to the conduction fluxes in these same directions. With these considerations, the components of the heat flux vector $q''(\hat{r},t)$ are taken as

$$q''_x = -k\frac{\partial T}{\partial x} + \rho c T\, u \tag{1-39a}$$

$$q_y'' = -k\frac{\partial T}{\partial y} + \rho c T\, v \tag{1-39b}$$

$$q_z'' = -k\frac{\partial T}{\partial z} + \rho c T\, w \tag{1-39c}$$

Clearly, on the right-hand sides (RHS) of the above three equations, the first term is the conduction heat flux and the second term is the convective heat flux due to the bulk motion of the solid. For the case of no motion (i.e., quiescent medium), the terms $u = v = w = 0$ and equation (1-38) reduces exactly to equation (1-24).

1-4 FOURIER'S LAW AND THE HEAT EQUATION IN CYLINDRICAL AND SPHERICAL COORDINATE SYSTEMS

Here we present Fourier's law and the heat conduction equation for other *orthogonal curvilinear coordinate systems*, namely, cylindrical and spherical coordinates. The heat equations may be directly derived using a differential control volume in the respective coordinate systems, following the approach of Section 1-3, or they may be obtained using the appropriate coordinate transformation into cylindrical or spherical coordinates. The results are presented here without derivation, although the respective differential control volumes are defined.

The expression for the heat flux vector (i.e., Fourier's law) in each new coordinate system may be given by the three principal components

$$q_i'' = -k\frac{1}{a_i}\frac{\partial T}{\partial u_i} \quad \text{for } i = 1, 2, 3, \ldots \tag{1-40}$$

where u_1, u_2, and u_3 are the curvilinear coordinates, and the coefficients a_1, a_2, and a_3 are the *coordinate scale factors*, which may be constants or functions of the coordinates. The expressions for the scale factors are derived for a general orthogonal curvilinear system by Özisik [7].

We will first consider the *cylindrical coordinate system*, as shown in Figure 1-7 along with a representative differential control volume. Using the appropriate scale factors, the three components of heat flux in the r, ϕ and z directions, respectively, become

$$q_r'' = -k\frac{\partial T}{\partial r} \quad q_\phi'' = -\frac{k}{r}\frac{\partial T}{\partial \phi} \quad \text{and} \quad q_z'' = -k\frac{\partial T}{\partial z} \tag{1-41a,b,c}$$

By inspection, it is seen that the scale factors a_r and a_z are unity, while the scale factor $a_\phi = r$. This scale factor also provides the correct units for flux, as the gradient term, $\partial T/\partial \phi$, in the ϕ direction (K/rad) is missing the spatial dimension. In the cylindrical coordinate system, the heat equations (1-23) and (1-25) become, respectively,

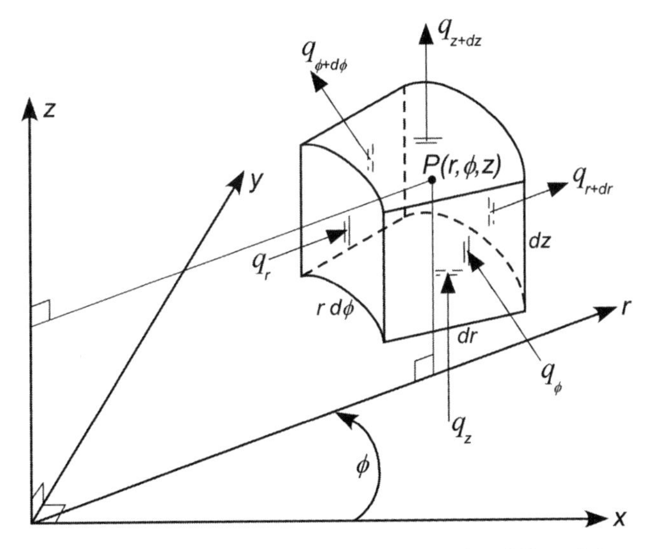

Figure 1-7 Cylindrical coordinate system and representative differential control volume.

$$\boxed{\frac{1}{r}\frac{\partial}{\partial r}\left(kr\frac{\partial T}{\partial r}\right)+\frac{1}{r^2}\frac{\partial}{\partial \phi}\left(k\frac{\partial T}{\partial \phi}\right)+\frac{\partial}{\partial z}\left(k\frac{\partial T}{\partial z}\right)+g=\rho c\frac{\partial T}{\partial t}} \quad (1\text{-}42)$$

$$\boxed{\frac{1}{r}\frac{\partial}{\partial r}\left(r\frac{\partial T}{\partial r}\right)+\frac{1}{r^2}\frac{\partial^2 T}{\partial \phi^2}+\frac{\partial^2 T}{\partial z^2}+\frac{g}{k}=\frac{1}{\alpha}\frac{\partial T}{\partial t}} \quad (1\text{-}43)$$

We now consider the *spherical coordinate system*, as shown in Figure 1-8 along with a representative differential control volume. Using the appropriate scale factors, the three components of heat flux in the r, ϕ and θ directions become, respectively,

$$q_r''=-k\frac{\partial T}{\partial r} \quad q_\phi''=-\frac{k}{r\ \sin\theta}\frac{\partial T}{\partial \phi} \quad \text{and} \quad q_\theta''=-\frac{k}{r}\frac{\partial T}{\partial \theta} \quad (1\text{-}44\text{a,b,c})$$

By inspection, it is seen now that only the scale factor a_r is unity, while the scale factors $a_\phi = r\sin\theta$ and $a_\theta = r$. As before, these scale factors also provide the correct units for flux, as the gradient terms in both the ϕ and θ directions (K/rad) are missing the spatial dimension. In the spherical coordinate system, the heat equations (1-23) and (1-25) become, respectively,

$$\boxed{\frac{1}{r^2}\frac{\partial}{\partial r}\left(kr^2\frac{\partial T}{\partial r}\right)+\frac{1}{r^2\sin\theta}\frac{\partial}{\partial \theta}\left(k\sin\theta\frac{\partial T}{\partial \theta}\right)+\frac{1}{r^2\sin^2\theta}\frac{\partial}{\partial \phi}\left(k\frac{\partial T}{\partial \phi}\right)+g=\rho c\frac{\partial T}{\partial t}} \quad (1\text{-}45)$$

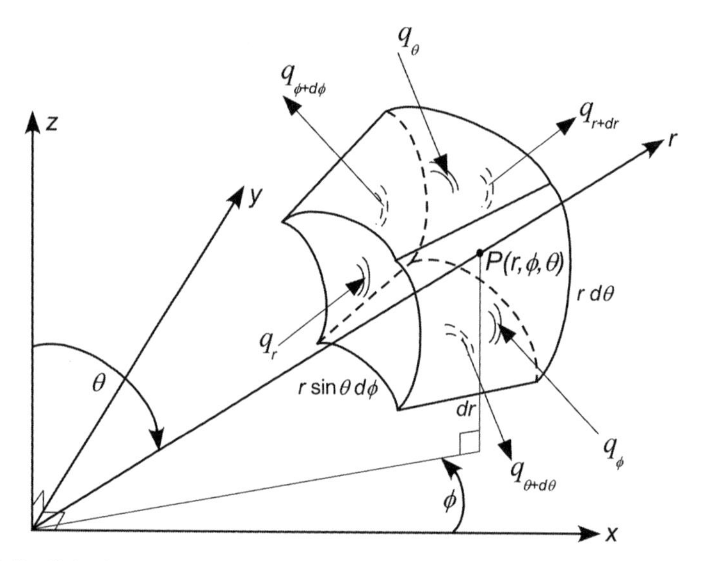

Figure 1-8 Spherical coordinate system and representative differential control volume.

$$\boxed{\frac{1}{r^2}\frac{\partial}{\partial r}\left(r^2\frac{\partial T}{\partial r}\right)+\frac{1}{r^2\sin\theta}\frac{\partial}{\partial\theta}\left(\sin\theta\frac{\partial T}{\partial\theta}\right)+\frac{1}{r^2\sin^2\theta}\frac{\partial^2 T}{\partial\phi^2}+\frac{g}{k}=\frac{1}{\alpha}\frac{\partial T}{\partial t}} \tag{1-46}$$

1-5 GENERAL BOUNDARY CONDITIONS AND INITIAL CONDITION FOR THE HEAT EQUATION

The differential equation of heat conduction, see, for example, equation (1-25), will require two boundary conditions for each spatial dimension, as well as one initial condition for the non–steady-state problem. The initial condition specifies the temperature distribution in the medium at the origin of the time coordinate, that is, $T(\hat{r}, t = 0)$, and the boundary conditions specify the temperature or the heat flux at the boundaries of the region. For example, at a given boundary surface, the temperature distribution may be prescribed, or the heat flux distribution may be prescribed, or there may be heat exchange by convection and/or radiation with an environment at a prescribed temperature. The boundary condition can be derived by writing an energy balance equation at the surface of the solid. Prior to considering formal boundary conditions, it is useful to define two additional particular laws for heat transfer, namely, for radiation and convection heat transfer.

The *Stefan–Boltzmann law* [8, 9] describes the heat flux emitted from a surface by radiation heat transfer

$$q''_{\text{rad}} = \varepsilon\sigma T^4 \quad \text{W/m}^2 \tag{1-47}$$

where ε is the *total, hemispherical emissivity* of the surface, and σ is the Stefan–Boltzmann constant, given as $\sigma = 5.670 \times 10^{-8}$ W/(m$^2 \cdot$ K^4). As presented, equation (1-47) represents energy radiated into all directions and over all wavelengths. The total, hemispherical emissivity represents an integration of the spectral, directional emissivity over all directions and wavelengths. Because the weighting of the spectral emissivity is by the Planck distribution, ε is in general a function of the surface temperature for a *nongray surface* (i.e., a surface for which the properties vary with wavelength).

A commonly used approximation for the *net* radiation heat flux between a surface and a surrounding medium is given as

$$q''_{\text{rad}} = \varepsilon\sigma(T^4 - T_\infty^4) \quad \text{W/m}^2 \tag{1-48}$$

where T_∞ is the temperature of the ambient, surrounding medium. Equation (1-48) assumes that the ambient surroundings form an *ideal enclosure*, which is satisfied if the surroundings are isothermal and of a much larger surface area, and assumes that the emitting surface is a *gray body* (i.e., neglect the wavelength dependency of the surface's emissivity and absorptivity).

Newton's law of cooling describes the heat flux to or from a surface by convection heat transfer

$$q''_{\text{conv}} = h(T - T_\infty) \quad \text{W/m}^2 \tag{1-49}$$

where T_∞ is the reference temperature of the surrounding ambient fluid (e.g., liquid or gas) and h is the *convection heat transfer coefficient* of units W/(m$^2 \cdot$ K). Equation (1-49) is not tied to the overall coordinate system; hence positive heat flux is considered in the direction of the surface normal (i.e., away from the surface). Unlike Fourier's law and the Stefan–Boltzmann law, Newton's law is not so much a particular law as it is the *definition* of the heat transfer coefficient, namely,

$$h = \frac{q''_{\text{conv}}}{T - T_\infty} \quad \text{W/(m}^2 \cdot \text{K)} \tag{1-50}$$

which reflects the dependency of h on the actual heat flux, and the difference between the surface temperature and a suitable *reference temperature*. Here the heat transfer coefficient h varies with the type of flow (laminar, turbulent, etc.), the geometry of the body and flow passage area, the physical properties of the fluid, the average surface and fluid temperatures, and many other parameters. As a result, there is a wide difference in the range of values of the heat transfer coefficient for various applications. Table 1-2 lists the typical values of h, in our units W/(m$^2 \cdot$ K), encountered in some applications.

To now develop the general boundary conditions, we consider conservation of energy at the surface, assumed to be stationary, noting that no energy can be accumulated (i.e., stored) at an infinitely thin surface. Figure 1-9 depicts a

TABLE 1-2 Typical Values of Convective Heat Transfer Coefficient

Type of Flow[a]	h, W/(m$^2 \cdot$ K)
Free Convection, $\Delta T = 25$ K	
0.25-m vertical plate in:	
Atmospheric air	5
Engine oil	40
Water	440
0.02-m (OD) horizontal cylinder in:	
Atmospheric air	10
Engine oil	60
Water	740
Forced Convection	
Atmospheric air at 25 K with $U_\infty = 10$ m/s over $L = 0.1$-m flat plate	40
Flow at 5 m/s across 1-cm (OD) cylinder of:	
Atmospheric air	90
Engine oil	1,800
Water flow at 1 kg/s inside 2.5-cm (ID) tube	10,500
Boiling of Water at 1 atm	
Pool boiling in a container	3,000
Pool boiling at peak heat flux	35,000
Film boiling	300
Condensation of Steam at 1 atm	
Film condensation on horizontal tubes	9,000–25,000
Film condensation on vertical surfaces	4,000–11,000
Dropwise condensation	60,000–120,000

[a] OD = outer diameter and ID = inner diameter.

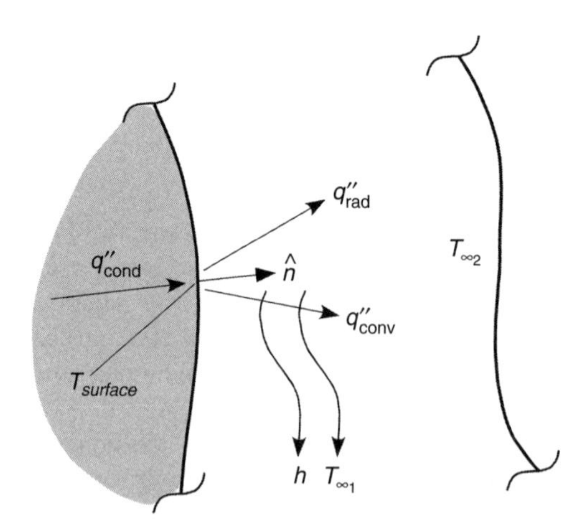

Figure 1-9 Energy balance at surface of a solid.

surface having an outward-drawn unit normal vector, $\hat{n}$, that is in the positive coordinate direction, subjected to convection heat transfer with some fluid, and to radiation heat transfer with an ideal surrounding. Conservation of energy at the surface boundary takes the form

$$q''_{\text{in}} = q''_{\text{out}} \tag{1-51}$$

or

$$-k \left.\frac{\partial T}{\partial n}\right|_{\text{surface}} = h(T|_{\text{surface}} - T_{\infty_1}) + \varepsilon\sigma(T^4|_{\text{surface}} - T^4_{\infty_2}) \tag{1-52}$$

In equation (1-52), Fourier's law follows our sign convention of positive flux in the positive coordinate direction, while Newton's law follows our convention of positive flux in the direction of the surface normal. The Stefan–Boltzmann law yields a positive flux away from the surface when the surface temperature is greater than the surrounding medium. Note also that the dependent variable T on the right-hand side is considered the value of T *at the surface*.

It is also useful to classify a given boundary or initial condition as either *homogeneous* or *nonhomogeneous*. A homogeneous condition is one in which all nonzero terms in the expression contain the dependent variable, $T(\hat{r}, t)$ in our case, or its derivative. The concept of homogeneous and nonhomogeneous boundary and initial conditions lies at the very core of the *method of separation of variables* that will be considered in the following chapters. In our treatment, for the analytic solution of linear heat conduction problems, we shall consider the following three types of linear boundary conditions.

1. ***Boundary Condition of the First Type (Prescribed Temperature)***. This is the situation when the temperature is prescribed at the boundary surface, that is,

$$T|_{\text{surface}} = T_0 \tag{1-53a}$$

or

$$T|_{\text{surface}} = f(\hat{r}, t) \tag{1-53b}$$

where T_0 is a prescribed constant temperature, and where $f(\hat{r}, t)$ is the prescribed surface temperature distribution that is, in general, a function of position and time. The special case of zero temperature on the boundary

$$T|_{\text{surface}} = 0 \tag{1-54}$$

is called the *homogeneous boundary condition of the first type*. In mathematics, boundary conditions of the first type are called *Dirichlet boundary conditions*.

2. ***Boundary Condition of the Second Type (Prescribed Heat Flux)***. This is the situation in which the heat flux is prescribed at the boundary surface, that is,

$$-k\left.\frac{\partial T}{\partial n}\right|_{\text{surface}} = q''_0 \tag{1-55a}$$

or

$$-k\frac{\partial T}{\partial n}\bigg|_{\text{surface}} = f(\hat{r},t) \tag{1-55b}$$

where $\partial T/\partial n$ is the derivative along the outward-drawn normal to the surface, q_0'' is a prescribed constant heat flux (W/m^2), and $f(\hat{r},t)$ is the prescribed surface heat flux distribution that is, in general, a function of position and time. The special case of zero heat flux at the boundary

$$\frac{\partial T}{\partial n}\bigg|_{\text{surface}} = 0 \quad \text{(perfectly insulated or adiabatic)} \tag{1-56}$$

is called the *homogeneous boundary condition of the second type*. In mathematics, boundary conditions of the second type (i.e., prescribed derivative values) are called *Neumann boundary conditions*.

3. ***Boundary Condition of the Third Type (Convection)***. This is the pure convection boundary condition, which is readily obtained from equation (1-52) by setting the radiation term to zero, that is,

$$-k\,\frac{\partial T}{\partial n}\bigg|_{\text{surface}} = h\left[T|_{\text{surface}} - T_\infty\right] \tag{1-57a}$$

For generality, the ambient fluid temperature T_∞ may assumed to be a function of position and time, yielding

$$-k\,\frac{\partial T}{\partial n}\bigg|_{\text{surface}} = h\left[T|_{\text{surface}} - T_\infty(\hat{r},t)\right] \tag{1-57b}$$

The special case of zero fluid temperature ($T_\infty = 0$), as given by

$$-k\,\frac{\partial T}{\partial n}\bigg|_{\text{surface}} = h\;T|_{\text{surface}} \tag{1-58}$$

is called the *homogeneous boundary condition of the third type*, since the dependent variable or its derivative now appears in all nonzero terms. This represents convection into a fluid medium at zero temperature, noting that a common practice is to redefine or shift the temperature scale such that the fluid temperature is now zero, as discussed in more detail in Section 1-8. A convection boundary condition is physically different than type 1 (prescribed temperature) or type 2 (prescribed flux) boundary conditions in that the temperature gradient within the solid at the surface is now *coupled* to the convective flux at the solid–fluid interface. Neither the flux nor the temperature are prescribed, but rather, a balance between conduction and convection is forced, see equation (1-51), with the exact surface temperature and surface heat flux determined by the *combination* of convection coefficient, thermal conductivity, and ambient fluid temperature. Clearly, the boundary conditions of the first and second type can be obtained from the type 3 boundary condition as special cases if k and h are treated as coefficients. For example, by setting $hT_\infty(\hat{r}, t) \equiv f(\hat{r}, t)$ and then letting $h = 0$ in the first term of the right-hand side, equation (1-57b) reduces to equation (1-55b).

A few final words are offered with regard to these three important boundary conditions. Mathematically speaking, convection boundary conditions provide the greatest complexity; however, from a physical point of view they are the simplest to realize in that many actual systems are governed by a natural energy balance between conduction and convection; hence no active control is necessary. In contrast, prescribed temperature boundary conditions, while mathematically simple, are actually rather difficult to realize in practice in that they are nearly always associated with surface heat flux. Therefore, for a transient problem, a constant temperature boundary condition necessitates a controlled, time-dependent surface heat flux to maintain the prescribed temperature. This is often difficult to achieve in practice. A prescribed temperature boundary condition is perhaps best realized when a physical phase change (e.g., evaporation/boiling) occurs on the surface. Such is the case of spray cooling with phase change heat transfer in which the surface will remain constant at the boiling point (i.e., saturation temperature) of the coolant fluid provided that sufficient coolant is applied to maintain a wetted surface and sufficient heat flux is present. Alternatively, the constant temperature boundary condition may be thought of as the limiting case of a convective boundary condition as $h \to \infty$, yielding $T_{\text{surface}} = T_\infty =$ constant. Boundary conditions of the second type may physically correspond to heaters (e.g., thin electric strip heaters) attached to the surface, which with low contact resistance and proper control can provide a prescribed heat flux condition.

In addition to the three linear boundary conditions discussed above, other boundary conditions are now considered here.

4. ***Interface Boundary Conditions***. When two materials having different thermal conductivities k_1 and k_2 are in *imperfect thermal contact* and have a common boundary as illustrated in Figure 1-10, the temperature profile through the solids experiences a sudden drop across the interface between the two materials. The physical significance of this temperature drop is envisioned better if we consider an enlarged view of the interface as shown in this figure, and note that actual solid-to-solid contact takes place at a limited number of spots, and that the void between them is filled with air (or other interfacial fluid), which is the surrounding fluid. As the thermal conductivity of air is much smaller than that of many solids (e.g., metals), a steep temperature drop occurs across the gap.

To develop the boundary condition for such an interface, we write the energy balance as

$$\begin{pmatrix}\text{Heat conduction}\\ \text{in solid 1}\end{pmatrix} = \begin{pmatrix}\text{Heat transfer}\\ \text{across the gap}\end{pmatrix} = \begin{pmatrix}\text{Heat conduction}\\ \text{in solid 2}\end{pmatrix} \tag{1-59a}$$

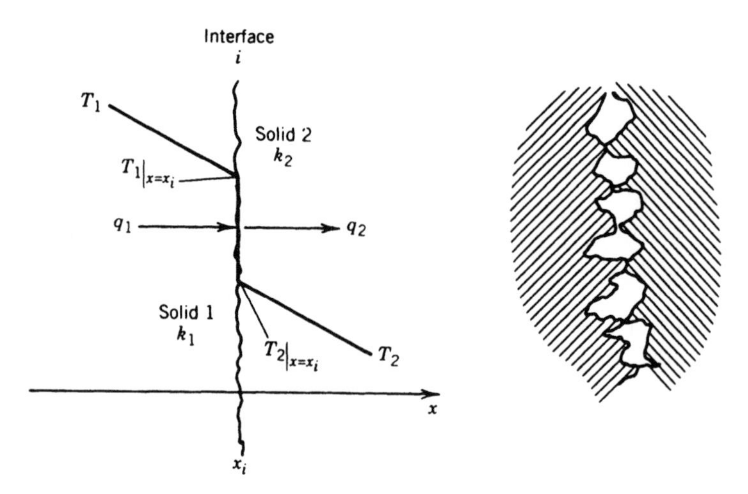

Figure 1-10 Boundary conditions at interface of two contacting solid surfaces.

$$q_i'' = -k_1 \left. \frac{\partial T_1}{\partial x} \right|_i = h_c(T_1 - T_2)_i = -k_2 \left. \frac{\partial T_2}{\partial x} \right|_i \tag{1-59b}$$

where subscript i denotes the interface, and h_c, in units W/(m$^2 \cdot$ K), is called the *contact conductance* for the interface. Equation (1-59b) provides two expressions for the boundary condition at the interface of two contacting solids, which together are generally called the *interface boundary conditions*. It is also common to consider the reciprocal of the contact conductance as the *thermal contact resistance*, R_c'', in units of (m$^2 \cdot$ K)/W.

For the special case of *perfect thermal contact* between the surfaces, we have $h_c \to \infty$, and equation (1-59b) is replaced with the following:

$$T_1|_i = T_2|_i \quad \text{at the surface interface} \tag{1-60a}$$

$$-k_1 \left. \frac{\partial T_1}{\partial x} \right|_i = -k_2 \left. \frac{\partial T_2}{\partial x} \right|_i \quad \text{at the surface interface} \tag{1-60b}$$

where equation (1-60a) is the continuity of temperature, and equation (1-60b) is the continuity of heat flux at the interface (i.e., conservation of energy).

Overall, the surface roughness, the interface contact pressure and temperature, thermal conductivities of the contacting solids, and the type of fluid in the gap are the principal factors that affect contact conductance. The experimentally determined values of contact conductance for typical materials in contact can be found in references 10–12.

To illustrate the effects of various parameters such as the surface roughness, the interface temperature, the interface pressure, and the type of material, we present in Table 1-3 the *interface thermal contact conductance* h_c for various material combinations. The results show that interface conductance increases with

TABLE 1-3 Interface Contact Conductance for Representative Solid–Solid Interfaces

Interface	Contact Pressure, atm	Interfacial Fluid	h_c, W/(m$^2\cdot$K)
Stainless steel to stainless steel	10	Air	9,000–11,500
[10] (0.76 μm roughness)	20	Air	10,000–12,000
Stainless steel to stainless steel	10	Air	2,800–4,000
[10] (2.5 μm roughness)	20	Air	3,100–4,200
Aluminum to aluminum [10]	10	Air	6,000–15,000
(3 μm roughness)	20	Air	10,500–28,000
Stainless steel to stainless steel	1	Vacuum	400–1,600
[13]	100	Vacuum	2,500–14,000
Copper to copper [13]	1	Vacuum	1,000–10,000
	100	Vacuum	20,000–100,000
Aluminum to aluminum [13]	1	Vacuum	2,000–6,600
	100	Vacuum	25,000–50,000
Aluminum to aluminum [13]	1	Air	3,600
(10 μm roughness)	1	Helium	10,000
	1	Silicone oil	19,000
Aluminum to aluminum [13, 14]	1	Dow Corning 340 grease	140,000
Stainless steel to stainless steel [13, 14]	35	Dow Corning 340 grease	250,000

increasing interface pressure, increasing interface temperature, and decreasing surface roughness. As might be expected, the interface conductance is higher with a softer material (e.g., aluminum) than with a harder material (e.g., stainless steel). The smoothness of the surface is another factor that affects contact conductance; a joint with a superior surface finish may exhibit lower contact conductance owing to waviness. The adverse effect of waviness can be overcome by introducing between the surfaces an interface shim from a soft material such as lead. Contact conductance also is reduced with a decrease in the ambient air pressure because the effective thermal conductance of the gas entrapped in the interface is lowered.

5. ***Other Boundary Conditions and Relations***. Two additional boundary conditions are frequently used during the solution of the heat conduction equation. When symmetry is present in a given coordinate direction, it is often desirable to limit the domain to one-half of the problem and use the *line of symmetry* as an alternative boundary condition. Since the net heat flux is zero across a line of symmetry, the boundary condition becomes

$$\left.\frac{\partial T}{\partial n}\right|_{\text{boundary}} = 0 \quad \text{(symmetry condition)} \tag{1-61}$$

which acts like an adiabatic (i.e., perfectly insulated) boundary in keeping with thermodynamic equilibrium. Care should be taken, however, when imposing symmetry as a boundary condition. For example, the initial condition $T(\hat{r}, t = 0)$ or nonuniform internal energy generation $g(\hat{r})$ may break the symmetry, even if the outer boundary conditions appear symmetric.

A second condition, more of a pseudoboundary condition, to consider here concerns the necessity for *finite temperature* throughout the domain of the problem. With curvilinear coordinate systems, as will be seen in later chapters, the solution of the heat equation often contains functions that tend to infinity as their argument approaches zero. Because such behavior violates the condition of finite temperature, these functions are eliminated from the general solution if zero is within the spatial variable domain of the problem. Under this scenario, the equivalent boundary condition at the coordinate origin ($r = 0$) may be stated as

$$\lim_{r \to 0} T(r) \neq \pm\infty \quad \text{(finite temperature condition)} \tag{1-62}$$

which implies that a finite temperature limit exists at the origin.

Example 1-1 Problem Formulation for 1-D Cylinder
Consider a hollow cylinder (i.e., thick-walled pipe) subjected to convection boundary conditions at the inner $r = a$ and outer $r = b$ surfaces into ambient fluids at constant temperatures $T_{\infty 1}$ and $T_{\infty 2}$, with heat transfer coefficients $h_{\infty 1}$ and $h_{\infty 2}$, respectively, as illustrated in Fig. 1-11. Write the boundary conditions.

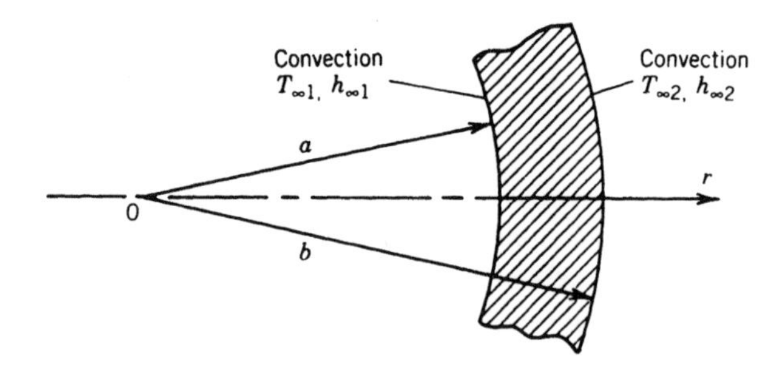

Figure 1-11 Boundary conditions for cylinder in Example 1-1.

The domain of the problem is $a \leq r \leq b$; hence boundary conditions are required at $r = a$ and $r = b$. The appropriate convection boundary condition is given by equation (1-57a), which is written here in the general cylindrical form

$$-k\frac{\partial T}{\partial r} = \pm h(T - T_\infty) \tag{1-63}$$

The positive conductive heat flux is always in the positive r direction per Fourier's law, while the outward-drawn surface normal at the boundary surfaces

$r = a$ and $r = b$ are in the negative r and positive r directions, respectively. Hence *positive* convection at the inner surface ($r = a$) is in the opposite direction of positive conduction, while positive conduction and convection are both in the same direction at the outer surface ($r = b$). Accordingly, we have to introduce a sign change to Newton's law for the inner surface, what we consider to be convection on the *back side* with respect to the coordinate direction. With these considerations in mind, the two boundary conditions (BC) become

BC1: $$-k\left.\frac{\partial T}{\partial r}\right|_{r=a} = -h_{\infty 1}\left(T|_{r=a} - T_{\infty 1}\right) \tag{1-64a}$$

BC2: $$-k\left.\frac{\partial T}{\partial r}\right|_{r=b} = h_{\infty 2}\left(T|_{r=b} - T_{\infty 2}\right) \tag{1-64b}$$

1-6 NONDIMENSIONAL ANALYSIS OF THE HEAT CONDUCTION EQUATION

In general, the solution and analysis of engineering problems benefit by first nondimensionalizing the governing equations. This process often yields important nondimensional groups, such as the Fourier number and Biot number, and reduces the dependency of the solution from a potentially large number of dimensional parameters. We consider the general 1-D Cartesian coordinate system, which from equation (1-25) is written here as

$$\frac{\partial^2 T}{\partial x^2} + \frac{g}{k} = \frac{1}{\alpha}\frac{\partial T}{\partial t} \tag{1-65}$$

over the domain $0 \leq x \leq L$, with the following initial (IC) and boundary conditions:

IC: $$T(x, t = 0) = T_0 \tag{1-66a}$$

BC1: $$\left.\frac{\partial T}{\partial x}\right|_{x=0} = 0 \tag{1-66b}$$

BC2: $$-k\left.\frac{\partial T}{\partial x}\right|_{x=L} = h(T|_{x=L} - T_\infty) \tag{1-66c}$$

It is now possible to define the *nondimensional independent variables*, denoted with an asterisk, using the available dimensional parameters of the problem. The independent variables become

$$x^* = \frac{x}{L} \tag{1-67a}$$

$$t^* = \frac{\alpha t}{L^2} \tag{1-67b}$$

where the nondimensional time is known as the *Fourier number*. The most common approach to define the *nondimensional temperature* is to use the reference temperature (e.g., the fluid temperature) in combination with the temperature difference between the initial and fluid temperatures, namely,

$$T^* = \frac{T - T_\infty}{T_0 - T_\infty} \tag{1-68}$$

A formal change of variables is now done via the chain rule, namely,

$$\frac{\partial T}{\partial x} = \frac{\partial T}{\partial T^*}\frac{\partial T^*}{\partial x} = \left(T_0 - T_\infty\right)\frac{\partial T^*}{\partial x} \tag{1-69a}$$

and

$$\frac{\partial T^*}{\partial x} = \frac{\partial T^*}{\partial x^*}\frac{\partial x^*}{\partial x} = \frac{\partial T^*}{\partial x^*}\frac{1}{L} \tag{1-69b}$$

Combining equations (1-69a) and (1-69b) yields

$$\frac{\partial T}{\partial x} = \frac{T_0 - T_\infty}{L}\frac{\partial T^*}{\partial x^*} \tag{1-70}$$

Further differentiating equation (1-70) yields

$$\frac{\partial^2 T}{\partial x^2} = \frac{\partial}{\partial x}\left(\frac{T_0 - T_\infty}{L}\frac{\partial T^*}{\partial x^*}\right) = \frac{T_0 - T_\infty}{L}\frac{\partial^2 T^*}{\partial x^{*2}}\frac{\partial x^*}{\partial x} = \frac{T_0 - T_\infty}{L^2}\frac{\partial^2 T^*}{\partial x^{*2}} \tag{1-71}$$

It is seen that the necessary dimension of the second derivative on the left-hand side of equation (1-71), namely K/m^2, is now supplied by the scaling factor of the right-hand side, since the right-hand side second derivative is dimensionless. In a similar manner,

$$\frac{\partial T}{\partial t} = \frac{\partial T}{\partial T^*}\frac{\partial T^*}{\partial t} = T_0 - T_\infty\frac{\partial T^*}{\partial t} \tag{1-72a}$$

$$\frac{\partial T^*}{\partial t} = \frac{\partial T^*}{\partial t^*}\frac{\partial t^*}{\partial t} = \frac{\partial T^*}{\partial t^*}\frac{\alpha}{L^2} \tag{1-72b}$$

which together yield

$$\frac{\partial T}{\partial t} = \frac{\alpha\left(T_0 - T_\infty\right)}{L^2}\frac{\partial T^*}{\partial t^*} \tag{1-73}$$

Inserting equations (1-71) and (1-73) into the heat equation yields

$$\boxed{\frac{\partial^2 T^*}{\partial x^{*2}} + \frac{gL^2}{k\left(T_0 - T_\infty\right)} = \frac{\partial T^*}{\partial t^*}} \tag{1-74}$$

All terms in equation (1-74) are now without dimension; hence it represents the *nondimensional form of the* 1*-D heat equation*. In a similar manner, the initial and boundary conditions are readily transformed:

IC: $$T^*(x^*, t^* = 0) = \frac{T_0 - T_\infty}{T_0 - T_\infty} = 1 \qquad (1\text{-}75a)$$

BC1: $$\left.\frac{\partial T^*}{\partial x^*}\right|_{x^*=0} = 0 \qquad (1\text{-}75b)$$

BC2: $$-\left.\frac{\partial T^*}{\partial x^*}\right|_{x^*=1} = \text{Bi}\ T^*\big|_{x^*=1} \qquad (1\text{-}75c)$$

where $\text{Bi} = hL/k$ is defined as the nondimensional *Biot number*, named after the physicist and mathematician Jean-Baptiste Biot. The Biot number is an important heat transfer parameter relating the conduction of heat within a solid to the convection of heat across the boundary, and is discussed in greater detail in Section 1-8. Examination of equations (1-74) and (1-75) reveals that the nondimensional temperature T^* depends on only two nondimensional parameters. If the internal energy generation is set to zero ($g = 0$), then

$$T^*(x^*, t^*) = f(\text{Bi}) \qquad (1\text{-}76)$$

such that the nondimensional temperature profile depends only on the Biot number. Clearly nondimensionalization is a powerful tool for engineering analysis.

1-7 HEAT CONDUCTION EQUATION FOR ANISOTROPIC MEDIUM

So far we considered the heat flux law for isotropic media, namely, the thermal conductivity k is independent of direction, and developed the heat conduction equation accordingly. However, there are natural as well as synthetic materials in which thermal conductivity varies with direction. For example, in a tree trunk the thermal conductivity may vary with direction; specifically, the thermal conductivities along the grain and across the grain are different. In lamellar materials, such as graphite and molybdenum disulfide, the thermal conductivity along and across the laminations may differ significantly. For example, in graphite, the thermal conductivity varies by about two orders of magnitude between the two principal orientations. Other examples include sedimentary rocks, fibrous reinforced structures, cables, heat shielding for space vehicles, and many others.

Orthotropic Medium

First we consider a situation in the rectangular coordinates in which the thermal conductivities k_x, k_y, and k_z in the x, y, and z dimensions, respectively, are different. Then the heat flux vector $q''(\hat{r}, t)$ given by equation (1-2) is modified as

$$q''(x, y, z, t) = -\left(\hat{i}k_x \frac{\partial T}{\partial x} + \hat{j}k_y \frac{\partial T}{\partial y} + \hat{k}k_z \frac{\partial T}{\partial z}\right) \tag{1-77}$$

and the three components of the heat flux vector in the x, y, and z directions, respectively, become

$$q''_x = -k_x \frac{\partial T}{\partial x} \qquad q''_y = -k_y \frac{\partial T}{\partial y} \qquad \text{and} \qquad q''_z = -k_z \frac{\partial T}{\partial z} \tag{1-78a,b,c}$$

Similar relations can be written for the heat flux components in the cylindrical and spherical coordinates. The materials in which thermal conductivity varies in the (x, y, z), (r, θ, z), or (r, θ, ϕ) directions are called *orthotropic materials*. The heat conduction equation for an orthotropic medium in the rectangular coordinate system is obtained by introducing the heat flux vector given by equation (1-77) into equation (1-23), which for a quiescent medium yields

$$\frac{\partial}{\partial x}\left(k_x \frac{\partial T}{\partial x}\right) + \frac{\partial}{\partial y}\left(k_y \frac{\partial T}{\partial y}\right) + \frac{\partial}{\partial z}\left(k_z \frac{\partial T}{\partial z}\right) + g = \rho c \frac{\partial T}{\partial t} \tag{1-79}$$

Thus the thermal conductivity has three distinct components.

Anisotropic Medium

In a more general situation encountered in heat flow through *crystals*, at any point in the medium, each component q''_x, q''_y, and q''_z of the heat flux vector is considered a linear combination of the temperature gradients $\partial T/dx$, $\partial T/dy$, and $\partial T/dz$, that is,

$$q''_x = -\left(k_{11} \frac{\partial T}{\partial x} + k_{12} \frac{\partial T}{\partial y} + k_{13} \frac{\partial T}{\partial z}\right) \tag{1-80a}$$

$$q''_y = -\left(k_{21} \frac{\partial T}{\partial x} + k_{22} \frac{\partial T}{\partial y} + k_{23} \frac{\partial T}{\partial z}\right) \tag{1-80b}$$

$$q''_z = -\left(k_{31} \frac{\partial T}{\partial x} + k_{32} \frac{\partial T}{\partial y} + k_{33} \frac{\partial T}{\partial z}\right) \tag{1-80c}$$

Such a medium is called an *anisotropic medium*, and the thermal conductivity for such a medium has nine components, k_{ij}, called the *conductivity coefficients*, that are considered to be the components of a second-order tensor $\bar{\bar{k}}$:

$$\bar{\bar{k}} \equiv \begin{vmatrix} k_{11} & k_{12} & k_{13} \\ k_{21} & k_{22} & k_{23} \\ k_{31} & k_{32} & k_{33} \end{vmatrix} \tag{1-81}$$

Crystals are typical examples of anisotropic materials involving nine conductivity coefficients [15]. The heat conduction equation for anisotropic solids in the rectangular coordinate system is obtained by introducing the expressions for the three components of heat flux given by equations (1-80) into the energy equation (1-23). Again for a quiescent medium, we find

$$k_{11}\frac{\partial^2 T}{\partial x^2} + k_{22}\frac{\partial^2 T}{\partial y^2} + k_{33}\frac{\partial^2 T}{\partial z^2} + \left(k_{12} + k_{21}\right)\frac{\partial^2 T}{\partial x\,\partial y} + \left(k_{13} + k_{31}\right)\frac{\partial^2 T}{\partial x\,\partial z}$$

$$+ \left(k_{23} + k_{32}\right)\frac{\partial^2 T}{\partial y\,\partial z} + g\left(x, y, z, t\right) = \rho c \frac{\partial T\left(x, y, z, t\right)}{\partial t} \qquad (1\text{-}82)$$

where $k_{12} = k_{21}$, $k_{13} = k_{31}$, and $k_{23} = k_{32}$ by the reciprocity relation. This matter will be discussed further in Chapter 15.

1-8 LUMPED AND PARTIALLY LUMPED FORMULATION

The transient heat conduction formulations considered up to this point assume a general temperature distribution varying both with time and position. There are many engineering applications in which the spatial variation of temperature within the medium can be neglected, and temperature is considered to be a function of time only. Such formulations, called *lumped system formulation* or *lumped capacitance method*, provide a great simplification in the analysis of transient heat conduction; but their range of applicability is very restricted. Here we illustrate the concept of the lumped formulation approach and examine its range of validity in terms of the dimensionless Biot number.

Consider a small, high-conductivity material, such as a metal, initially at a uniform temperature T_0, and then suddenly immersed into a well-stirred hot bath maintained at a uniform temperature T_∞. Let V be the volume, A the surface area, ρ the density, c the specific heat of the solid, and h the convection heat transfer coefficient between the solid's surface and the fluid. We assume that the temperature distribution within the solid remains sufficiently uniform for all times due to its small size and high thermal conductivity. Under such an assumption, the uniform temperature $T(t)$ of the solid can be considered to be a function of time only. The energy balance equation, taking the entire solid as the control volume, is stated as

$$\begin{pmatrix} \text{Rate of heat flow from the} \\ \text{solid through its boundaries} \end{pmatrix} = \begin{pmatrix} \text{Rate of change of the} \\ \text{internal energy of the solid} \end{pmatrix} \qquad (1\text{-}83)$$

Considering convection as the only means for heat to enter or leave the control volume, the energy equation (1-83) takes the form

$$-hA[T(t) - T_\infty] = \rho V c \frac{dT(t)}{dt} \qquad (1\text{-}84)$$

which is rearranged to yield

$$\frac{dT(t)}{dt} + \frac{hA}{\rho Vc}[T(t) - T_\infty] = 0 \qquad \text{for} \qquad t > 0 \tag{1-85a}$$

IC:

$$T(t = 0) = T_0 \tag{1-85b}$$

Equation (1-85a) is a nonhomogeneous ordinary differential equation, which is readily solved using the sum of the homogeneous and particular solutions. However, it is useful to remove the nonhomogeneity by defining the excess temperature $\theta(t)$ as

$$\theta(t) = T(t) - T_\infty \tag{1-86}$$

With this substitution, the lumped formulation becomes

$$\frac{d\theta(t)}{dt} + m\theta(t) = 0 \qquad \text{for} \qquad t > 0 \tag{1-87a}$$

IC:

$$\theta(t = 0) = T_0 - T_\infty = \theta_0 \tag{1-87b}$$

where

$$m = \frac{hA}{\rho Vc} \tag{1-87c}$$

The solution of equations (1-87) becomes

$$\theta(t) = \theta_0 e^{-mt} \tag{1-88}$$

This is a very simple expression for the temperature of the solid as a function of time, noting that the parameter m has the unit of s^{-1} and may be thought of as the inverse of the *thermal time constant*. The physical significance of the parameter m is better envisioned if its definition is rearranged in the form

$$\frac{1}{m} = (\rho cV)\left(\frac{1}{hA}\right) \tag{1-89}$$

which is the product of the *thermal heat capacitance* and the resistance to convection heat transfer. It follows that the smaller the thermal capacitance and/or the convective resistance, the larger is the value of m, and hence the faster is the rate of change of temperature $\theta(t)$ of the solid according to equation (1-88).

In order to establish some criterion for the range of validity of such a straightforward method for the analysis of transient heat conduction, we consider the definition of the Biot number, and rearrange it in the form

$$\text{Bi} \equiv \frac{hL_c}{k_s} = \frac{L_c/k_s A}{1/hA} = \frac{\text{internal conductive resistance}}{\text{external convective resistance}} \tag{1-90}$$

where k_s is the thermal conductivity of the solid. L_c is the *characteristic length* of the solid and is generally defined as $L_c = V/A$.

We recall that the lumped system analysis is applicable if the temperature distribution within the solid remains sufficiently uniform during the transients.

This may be interpreted as the condition in which the internal resistance to conduction within the solid is negligible as compared to the external resistance to convection heat transfer at the solid–fluid boundary. Now we refer to the above definition of the Biot number and note that the internal conductive resistance of the solid is very small in comparison to the external convective resistance if the Biot number is *much less than unity*, say one order of magnitude smaller. We therefore conclude that the lumped system analysis is valid only for small values of the Biot number, namely

$$\text{Bi} = \frac{hL_c}{k_s} < 0.1 \quad \text{(lumped analysis criterion)} \tag{1-91}$$

We note that equation (1-91) is a *guideline* for validity of lumped analysis, but must emphasize that spatial gradients *gradually diminish* with decreasing Bi number for transient problems. Hence a Bi $= 0.1$ does not represent an abrupt transition from the presence of spatial gradients to the absence of spatial gradients, but rather is considered a reasonable interpretation of Bi $\ll 1$. For example, exact analytic solutions of transient heat conduction for solids in the form of a slab, cylinder or sphere, subjected to convective cooling show that for a Bi < 0.1, the variation of temperature within the solid during transients is less than about 5%. Hence it may be concluded that the lumped system analysis may be applicable for most engineering applications if the Biot number is less than about 0.1.

It is useful here to examine graphically the behavior of the transient temperature profile for three different values of the Biot number, namely Bi $\ll 1$, Bi ≈ 1, and Bi $\gg 1$. This is done in Figure 1-12 for a symmetric plane wall with a

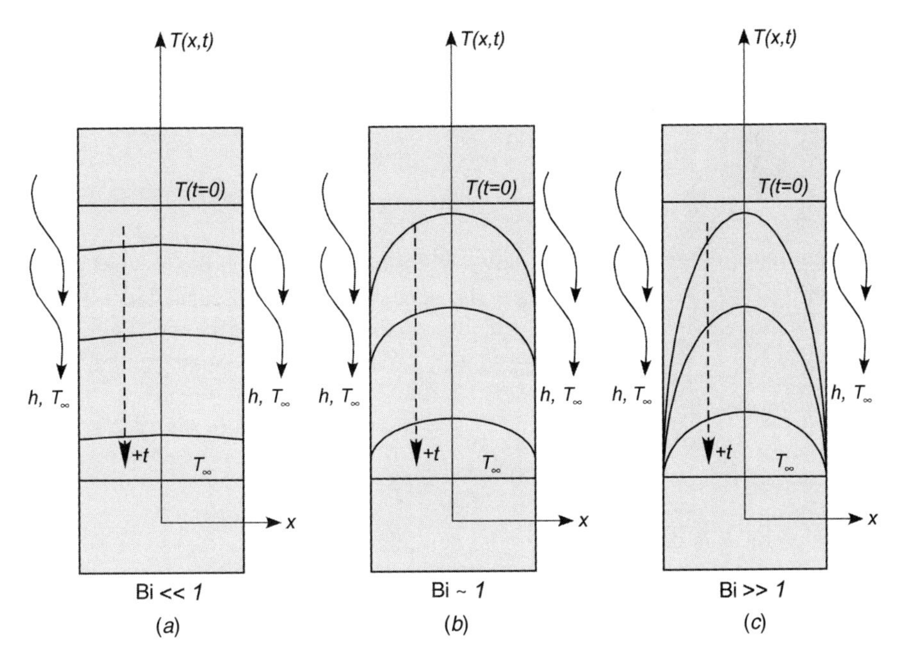

Figure 1-12 Temperature distribution *T(x,t)* for a symmetric plane wall cooled by convection heat transfer for various Biot numbers.

uniform initial temperature of T_0, which is then subjected to cooling via convection heat transfer at both surfaces from a fluid at T_∞.

For the case of $\mathrm{Bi} \ll 1$, Figure 1-12(*a*), the wall is observed to cool *uniformly* from the initial temperature T_0 to the steady-state temperature, which would be T_∞. This is consistent with spatial gradients being dominated by convective resistance, hence $T \simeq T(t)$. Figure 1-12(*b*) shows the case of $\mathrm{Bi} \approx 1$; hence comparable conductive and convective resistances lead to temperature gradients within the solid and temperature differences between the solid and fluid. Finally, Figure 1-12(*c*) depicts the $\mathrm{Bi} \gg 1$, which in the limiting case corresponds to $h \to \infty$, or essentially the case of prescribed surface temperature. Under this condition, the spatial temperature gradients dominate the problem. In particular, the gradient near the surfaces at $t \simeq 0$ is very steep.

Example 1-2 Lumped Analysis for a Solid Sphere
The temperature of a gas stream is to be measured with a thermocouple. The junction may be approximated as a sphere of diameter and properties: $D = \frac{3}{4}$ mm, $k = 30$ W/(m · K), $\rho = 8400$ kg/m^3, and $c = 400$ J/(kg · K). If the heat transfer coefficient between the junction and the gas stream is $h = 600$ W/(m^2 · K), how long does it take for the thermocouple to record 99% of the temperature difference between the gas temperature and the initial temperature of the thermocouple? Here we neglect any radiation losses.

The characteristic length L_c is

$$L_c = \frac{V}{A} = \frac{(\frac{4}{3})\pi r^3}{4\pi r^2} = \frac{r}{3} = \frac{D}{6} = \frac{\frac{3}{4}}{6} = \frac{1}{8} \text{ mm} = 1.25 \times 10^{-4} \text{ m}$$

The Biot number becomes

$$\mathrm{Bi} = \frac{hL_c}{k} = \frac{600}{30} 1.25 \times 10^{-4} = 2.5 \times 10^{-3}$$

The lumped system analysis is applicable since $\mathrm{Bi} < 0.1$. From equation (1-88) we have

$$\frac{T(t) - T_\infty}{T_0 - T_\infty} = \frac{1}{100} = e^{-mt}$$

or

$$e^{mt} = 100 \quad \to \quad mt = 4.605$$

The value of m is determined from its definition:

$$m = \frac{hA}{\rho Vc} = \frac{h}{\rho c L_c} = \frac{600}{8400 \times 400 \times 0.000125} = 1.429 \text{ s}^{-1}$$

Then

$$t = \frac{4.605}{m} = \frac{4.605}{1.429} \cong 3.2 \text{ s}$$

That is, about 3.2 s is needed for the thermocouple to record 99% of the applied temperature difference.

Partial Lumping

In the lumped system analysis described above, we considered a total lumping in all the space variables; as a result, the temperature for the fully lumped system became a function of the time variable only.

It is also possible to perform a *partial lumped analysis*, such that the temperature variation is retained in one of the space variables but lumped in the others. For example, if the temperature gradient in a solid is steep in the x direction and very small in the y and z directions, then it is possible to lump the system with regard to the y and z variables and let $T = T(x)$. To illustrate this matter we consider a solid with thermal conductivity k as shown in Figure 1-13 in which temperature gradients are assumed to be large along the x direction, but small over the y–z plane perpendicular to the x axis. This would be valid by considering the length scale δ as the width of the solid in the y and z directions, such that $\text{Bi}_y = \text{Bi}_z = {}^{h\delta}/_k < 0.1$. Let the solid dissipate heat by convection from its lateral surfaces into an ambient fluid at a constant temperature T_∞ with a uniform heat transfer coefficient h over the entire exposed surface.

To develop the steady-state heat conduction equation with lumping over the plane perpendicular to the x axis, we consider an energy balance for a differential disk of thickness dx located about the axial location x given by

$$\begin{pmatrix} \text{Net rate of heat} \\ \text{gain by conduction} \\ \text{in the } x \text{ direction} \end{pmatrix} - \begin{pmatrix} \text{Rate of heat loss} \\ \text{by convection from} \\ \text{the lateral surfaces} \end{pmatrix} = 0 \qquad (1\text{-}92)$$

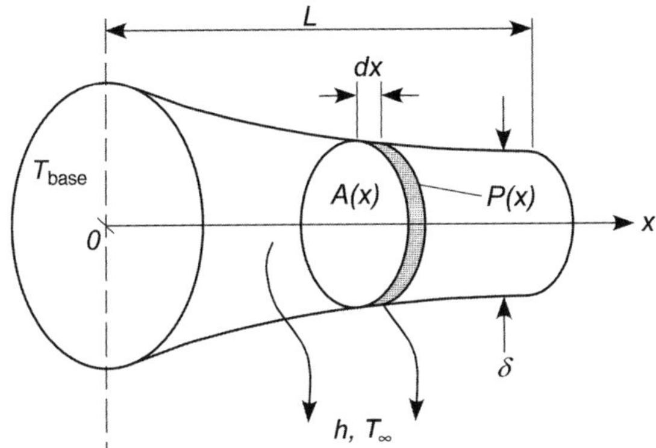

Figure 1-13 Nomenclature for derivation of partially lumped heat conduction equation.

When the appropriate rate expressions are introduced for each of these terms, we obtain

$$\left(q_x - q_{x+dx}\right) - hP(x)dx\left(T(x) - T_\infty\right) = 0 \tag{1-93}$$

where the heat rates q_x and q_{x+dx} are given by

$$q_x = -kA(x)\frac{\partial T}{\partial x} \tag{1-94a}$$

$$q_{x+dx} = q_x + \frac{\partial q_x}{\partial x}dx = -kA(x)\frac{\partial T}{\partial x} + \frac{\partial}{\partial x}\left[-kA(x)\frac{\partial T}{\partial x}\right]dx \tag{1-94b}$$

where we have again used a Taylor series expansion for q_{x+dx}, and where the other quantities are defined as

$A(x)$ = cross-sectional area of the disk

$P(x)$ = perimeter of the disk

h = convection heat transfer coefficient

k = thermal conductivity of the solid

T_∞ = ambient fluid temperature

We now introduce equations (1-94) into (1-93) to yield

$$\frac{d}{dx}\left[A(x)\frac{dT}{dx}\right] - \frac{hP(x)}{k}\left(T - T_\infty\right) = 0 \tag{1-95}$$

We have left equation (1-95) to reflect a variable cross-sectional area; hence $A(x)$ and $P(x)$ both are prescribed functions of x. If we further assume that the cross-sectional area is constant, namely, $A(x) = A_0$ = constant and $P(x) = P_0$ = constant, then equation (1-95) reduces to

$$\frac{d^2T}{dx^2} - \frac{hP_0}{kA_0}\left(T - T_\infty\right) = 0 \tag{1-96}$$

which is the *fin equation* for fins of uniform cross section. Rather than solving this second-order ordinary differential equation (ODE), it is useful to once again define the excess temperature $\theta(x)$:

$$\theta(x) = T(x) - T_\infty \tag{1-97}$$

which yields the final form of the 1-D constant-area fin equation, namely

$$\frac{d^2\theta}{dx^2} - \frac{hP_0}{kA_0}\theta = 0 \tag{1-98a}$$

or

$$\frac{d^2\theta}{dx^2} - m^2\theta = 0 \tag{1-99}$$

where the parameter m is defined as

$$m^2 = \frac{hP_0}{kA_0} \tag{1-99}$$

The solution to the fin equation (1-98b) can be expressed in the form

$$\theta(x) = C_1 e^{-mx} + C_2 e^{mx} \tag{1-100a}$$

or

$$\theta(x) = C_1 \cosh(mx) + C_2 \sinh(mx) \tag{1-100b}$$

We note here that the general solution form of the second-order ode in equation (1-98) is determined by the roots of the *auxiliary or characteristic equation*

$$\lambda^2 - m^2 = 0 \tag{1-101}$$

The present case of two real roots (i.e., $\lambda_{1,2} = \pm m$) yields the exponential solution given by equation (1-100a). However, for the case of a *conjugate pair* of real roots, the solution may also be formed using the *hyperbolic functions* as given by equation (1-100b). For this case, it is readily seen that a linear combination of the two exponential solutions yields the hyperbolics. The two unknown coefficients C_1 and C_2 are determined by the application of boundary conditions at $x = 0$ and $x = L$, which correspond to the base and tip of the fin, respectively. Typically, the temperature at the base of the fin is prescribed

$$\theta(x = 0) = T_b - T_\infty = \theta_b \tag{1-102}$$

However, there are several choices of boundary conditions at the tip of the fin, including the following four cases:

1. $-k \left.\frac{\partial T}{\partial x}\right|_{x=L} = h(T|_{x=L} - T_\infty)$ (convective tip) (1-103a)

   which yields $$-k \left.\frac{\partial \theta}{\partial x}\right|_{x=L} = h\, \theta|_{x=L} \tag{1-103b}$$

2. $\left.\frac{\partial T}{\partial x}\right|_{x=L} = 0$ (insulated/symmetric tip) (1-104a)

   which yields $$\left.\frac{\partial \theta}{\partial x}\right|_{x=L} = 0 \tag{1-104b}$$

3. $T(x = L) = T_{tip}$ (prescribed tip) (1-105a)

which yields $\theta(x = L) = T_{tip} - T_{\infty} = \theta_{tip}$ (1-105b)

4. $T(x \to \infty) = T_{\infty}$ (long/infinite fin) (1-106a)

which yields $\theta(x \to \infty) = 0$ (1-106b)

Generally, it is preferable to use the hyperbolic solution, equation (1-100b), for the finite domain problems, that is, boundary equations (1-103) – (1-105), and to use the exponential solution, equation (1-100a), for the infinite domain problem, namely boundary equation (1-106).

The solution of equation (1-95) for fins of a variable cross section is more involved, as the resulting ode has nonconstant coefficients. Analytic solutions of fins of various cross sections can be found in references 16 and 17.

REFERENCES

1. J. F., *Théorie Analytique de la Chaleur*, Paris, 1822. (English trans. by Al. Freeman, Cambridge University Press, 1878; reprinted by Dover, New York, 1955).
2. H. B. Callen, *Thermodynamics*, Wiley, New York 1960.
3. R. W. Powell, C. Y. Ho and P. E. Liley, *Thermal Conductivity of Selected Materials*, NSRDS-NBS 8, U.S. Department of Commerce, National Bureau of Standards, U.S. Government Printing Offer, Washington, DC, 1966.
4. Y. S. Touloukian, R. W. Powell, C. Y. Ho and P. G. Klemens, *Thermophysical Properties of Matter*, Vols. 1–3, IFI/Plenum, New York, 1970.
5. C. Y. Ho, R. W. Powell, and P. E. Liley, *Thermal Conductivity of Elements*, Vol. **1**, first supplement to J. Phys. Chem. Ref. Data, 1972.
6. National Institute of Standards and Technology (NIST), *NIST Data Gateway*, available at: http://srdata.nist.gov/gateway/gateway; accessed November 9, 2001.
7. M. N. Özisik, *Heat Conduction*, 2nd ed., Wiley, New York, 1993.
8. J. Stefan, Über die Beziehung zwischen der Wärmestrahlung und der Temperatur, Sitzber. Akad. Wiss. Wien, **79**, (2), 391–428, 1879.
9. L. Boltzmann, Ableitung des Stefan'schen Gesetzes, betreffend die Abhängigkeit der Wärmestrahlung von der Temperatur aus der electromagnetischen Lichttheorie, *Ann. Phys., Ser. 6,* **258**, 291–294, 1884.
10. M. E. Barzelay, K. N. Tong, and G. F. Holloway, Effects of pressure on thermal conductance of contact joints, NACA Tech. Note 3295, May 1955.
11. E. Fried and F. A. Costello, ARS J., **32**, 237–243, 1962.
12. H. L. Atkins and E. Fried, *J. Spacecraft Rockets*, **2**, 591–593, 1965.
13. E. Fried, Thermal Conduction Contribution to Heat Transfer at Contacts, in *Thermal Conductivity*, Vol. 2, R. P. Tye (Ed.), Academic, London, 1969.
14. B. Snaith, P. W. O'Callaghan, and S. D. Probert, *Appl. Energy*, **16**, 175–191 1984.

15. J. F. Nye, *Physical Properties of Crystals: Their Representation by Tensors and Matrices*, Oxford University Press, Oxford, 1985.
16. D. A. Kern and A. D. Kraus, *Extended Surface Heat Transfer*, McGraw-Hill, New York, 1972.
17. M. D. Mikhailov and M. N. Özisik, *Unified Analysis and Solutions of Heat and Mass Diffusion*, Wiley, New York, 1984.

PROBLEMS

1-1 Derive the heat conduction equation (1-43) in cylindrical coordinates using the differential control approach beginning with the general statement of conservation of energy. Show all steps and list all assumptions. Consider Figure 1-7.

1-2 Derive the heat conduction equation (1-46) in spherical coordinates using the differential control approach beginning with the general statement of conservation of energy. Show all steps and list all assumptions. Consider Figure 1-8.

1-3 Show that the following two forms of the differential operator in the cylindrical coordinate system are equivalent:

$$\frac{1}{r}\frac{d}{dr}\left(r\frac{dT}{dr}\right) = \frac{d^2T}{dr^2} + \frac{1}{r}\frac{dT}{dr}$$

1-4 Show that the following three different forms of the differential operator in the spherical coordinate system are equivalent:

$$\frac{1}{r^2}\frac{d}{dr}\left(r^2\frac{dT}{dr}\right) = \frac{1}{r}\frac{d^2}{dr^2}(rT) = \frac{d^2T}{dr^2} + \frac{2}{r}\frac{dT}{dr}$$

1-5 Set up the mathematical formulation of the following heat conduction problems. Formulation includes the simplified differential heat equation along with boundary and initial conditions. Do not solve the problems.

a. A slab in $0 \le x \le L$ is initially at a temperature $F(x)$. For times $t > 0$, the boundary at $x = 0$ is kept insulated, and the boundary at $x = L$ dissipates heat by convection into a medium at zero temperature.

b. A semi-infinite region $0 \le x < \infty$ is initially at a temperature $F(x)$. For times $t > 0$, heat is generated in the medium at a constant, uniform rate of g_0 (W/m^3), while the boundary at $x = 0$ is kept at zero temperature.

c. A hollow cylinder $a \leq r \leq b$ is initially at a temperature $F(r)$. For times $t > 0$, heat is generated within the medium at a rate of $g(r)$, (W/m^3), while both the inner boundary at $r = a$ and outer boundary $r = b$ dissipate heat by convection into mediums at fluid temperature T_∞.

d. A solid sphere $0 \leq r \leq b$ is initially at temperature $F(r)$. For times $t > 0$, heat is generated in the medium at a rate of $g(r)$ (W/m^3), while the boundary at $r = b$ is kept at a uniform temperature T_0.

1-6 A solid cube of dimension L is originally at a uniform temperature T_0. The cube is then dropped into a large bath where the cube rapidly settles flat on the bottom. The fluid in the bath provides convection heat transfer with coefficient h (W/m^2 K) from the fluid at constant temperature T_∞. Formulate the heat conduction problem. Formulation includes the simplified differential heat equation along with appropriate boundary and initial conditions. Include a sketch with your coordinate axis position. Do not solve the problem.

1-7 For an anisotropic solid, the three components of the heat conduction vector q_x, q_y, and q_z are given by equations (1-80). Write the similar expressions in the cylindrical coordinates for q_r, q_ϕ, q_z and in the spherical coordinates for q_r, q_ϕ, q_θ.

1-8 An infinitely long, solid cylinder (D = diameter) has the ability for uniform internal energy generation given by the rate g_0 (W/m^3) by passing a current through the cylinder. Initially ($t = 0$), the cylinder is at a uniform temperature T_0. The internal energy generation is then turned on (i.e., current passed) and maintained at a constant rate g_0, and at the same moment the cylinder is exposed to convection heat transfer with coefficient h (W/m^2 K) from a fluid at constant temperature T_∞, noting that $T_\infty > T_0$. The cylinder has uniform and constant thermal conductivity k (W/m K). The Biot number $hD/k \ll 1$. Solve for time t at which point the surface heat flux is exactly zero. Present your answer in variable form.

1-9 A long cylindrical iron bar of diameter $D = 5$ cm, initially at temperature $T_0 = 650°C$, is exposed to an air stream at $T_\infty = 50°C$. The heat transfer coefficient between the air stream and the surface of the bar is $h = 80\ W/(m^2 \cdot K)$. Thermophysical properties are constant: $\rho = 7800\ kg/m^3$, $c = 460\ J/(kg \cdot K)$, and $k = 60\ W/(m \cdot K)$. Determine the time required for the temperature of the bar to reach 250°C by using the lumped system analysis.

1-10 A thermocouple is to be used to measure the temperature in a gas stream. The junction may be approximated as a sphere having thermal conductivity $k = 25\ W/(m \cdot K)$, $\rho = 8400\ kg/m^3$, and $c = 0.4\ kJ/(kg \cdot K)$. The heat transfer coefficient between the junction and the gas stream is

$h = 560$ W/(m$^2 \cdot$ K). Calculate the diameter of the junction if the thermocouple should measure 95% of the applied temperature difference in 3 s.

1-11 Determine the constants C_1 and C_2 for the constant area fin solution of equations (1-100) for the case of the prescribed base temperature of equation (1-102), and the following tip conditions:

a. Convective tip per equation (1-103)

b. Insulated or symmetric tip per equation (1-104)

c. Prescribed temperature tip per equation (1-105)

d. Infinitely long fin per equation (1-106)

2

ORTHOGONAL FUNCTIONS, BOUNDARY VALUE PROBLEMS, AND THE FOURIER SERIES

In the preceding chapter, we developed the governing equation for conduction heat transfer, the heat equation, along with the appropriate boundary conditions. In the following chapters, we will explore in great detail various solution schemes, including separation of variables, Duhamel's theorem, and the Green's function method. However, it is now useful to develop a set of mathematical tools that will be used extensively in our subsequent analytical solutions. As Joseph Fourier wrote in his *Analytical Theory of Heat*: "The theory of heat will always attract the attention of mathematicians, by the rigorous exactness of its elements and the analytical difficulties peculiar to it, and above all by the extent and usefulness of its applications." We address here the former part of his statement, with the goal of providing a robust foundation upon which to build our analytical solutions. More details of the topics covered in this chapter are available in references 1–9.

2-1 ORTHOGONAL FUNCTIONS

Consider two functions $\phi_1(x)$and $\phi_2(x)$that are both defined for all values of x on an interval $a \le x \le b$. These two functions are said to be *orthogonal functions* if

$$\int_{x=a}^{b} \phi_1(x)\phi_2(x)\, dx = 0 \tag{2-1}$$

The property of orthogonality of two functions $\phi_1(x)$and $\phi_2(x)$ does not denote anything about their being perpendicular, although equation (2-1) is somewhat analogous to the scalar product of two vectors, which is zero when the vectors are perpendicular or orthogonal. In fact, two orthogonal functions have the unique property that the product $\phi_1(x)\phi_2(x)$assumes both positive and negative values in a way that satisfies equation (2-1), namely an integral value of exactly zero over the specified interval. We note here that functions are always orthogonal *over a defined interval*, called the *fundamental interval*.

The definition of orthogonality can be extended to a general set of orthogonal functions $\phi_n(x)$for $n = 1, 2, 3, \ldots$ by the following expression:

$$\int_{x=a}^{b} \phi_n(x)\phi_m(x)\, dx = \begin{cases} 0 & \text{if } n \neq m \\ N(\lambda) & \text{if } n = m \end{cases} \tag{2-2}$$

where once again the interval of orthogonality is $a \leq x \leq b$, and where the term $N(\lambda)$, also called the *norm* or the *normalization integral*, is defined as

$$N(\lambda) = \int_{x=a}^{b} \left[\phi_n(x)\right]^2 dx \tag{2-3}$$

The λ symbol in $N(\lambda)$ is related to the nature of the specific orthogonal function, as determined by the general boundary value problem, and is discussed in the next section. For the special case of $N(\lambda) = 1$, the function is both orthogonal and normalized and is said to be *orthonormal*.

Finally, we may extend the definition of an orthogonal set of functions to a more general case, namely,

$$\int_{x=a}^{b} w(x)\phi_n(x)\phi_m(x)\, dx = \begin{cases} 0 & \text{if } n \neq m \\ N(\lambda) & \text{if } n = m \end{cases} \tag{2-4}$$

where $w(x)$ is the *weighting function* $[w(x) \geq 0]$, as discussed below, and again $N(\lambda)$ is the norm, defined now as

$$N(\lambda) = \int_{x=a}^{b} w(x) \left[\phi_n(x)\right]^2 dx \tag{2-5}$$

Equation (2-4) is interpreted as a set of functions $\phi_n(x)$ that is orthogonal over the interval $a \leq x \leq b$ with weighting function $w(x)$. Equations (2-4) and (2-5) reduce to equations (2-2) and (2-3), respectively, for a weighting function of unity.

Orthogonal functions may take many forms, including trigonometric functions, Bessel functions, Legendre polynomials, Hermite polynomials, Tchebysheff polynomials, and Laguerre polynomials.

2-2 BOUNDARY VALUE PROBLEMS

In the preceding section, we introduced the concept and properties of an orthogonal function. Here we present the means to generate sets of orthogonal functions in terms of the *boundary value problem* or what is called the *Sturm–Liouville problem*, named after the mathematicians J.C.F. Sturm and J. Liouville who published their studies on this topic in *Journal de Mathématique Pures et Appliquées*, between 1836 and 1838.

The general Sturm–Liouville problem takes the form of the homogeneous ordinary differential equation

$$\boxed{\frac{d}{dx}\left[p(x)\frac{dX(x)}{dx}\right]+[q(x)+\lambda w(x)]X(x)=0} \tag{2-6}$$

along with the homogeneous boundary conditions

$$A_1\frac{dX(x)}{dx}+A_2X(x)=0 \qquad \text{at } x=a \tag{2-7a}$$

$$B_1\frac{dX(x)}{dx}+B_2X(x)=0 \qquad \text{at } x=b \tag{2-7b}$$

where the functions $p(x)$, $q(x)$, $w(x)$, and $dp(x)/dx$ are assumed to be real valued and continuous over the interval $a \le x \le b$, and $p(x) > 0$ and $w(x) > 0$ over the interval $a < x < b$. The constants A_1, A_2, B_1, and B_2 are real and independent of the parameter λ. For the rather general conditions of equations (2-6) and (2-7) it can be shown that there is an infinite set $(\lambda_1, \lambda_2, \ldots)$ of values for the parameter λ, which are called the *eigenvalues* or *characteristic numbers*. The corresponding solutions $X_n(x, \lambda_n)$ of the ODE in equation (2-6) and corresponding boundary conditions are called *eigenfunctions* or *characteristic functions*. It is readily shown [3] that the eigenfunctions are in fact orthogonal functions over the interval $a \le x \le b$ with weighting function $w(x)$, namely,

$$\int_{x=a}^{b} w(x)X_nX_m\,dx=\begin{cases}0 & \text{if } n\neq m\\ N(\lambda_n) & \text{if } n=m\end{cases} \tag{2-8}$$

where the norm is defined as

$$N(\lambda_n)=\int_{x=a}^{b} w(x)\left[X_n(x,\lambda_n)\right]^2dx \tag{2-9}$$

provided that both X_n and its derivative are continuous (implies being bounded) on the interval $a \le x \le b$. Orthogonality also holds for the following three cases: (1) For the special case of $p(a) = 0$, boundary condition (2-7a) may be dropped from the problem; (2) for the special case of $p(b) = 0$, boundary condition (2-7b) may be dropped from the problem; and (3) for the special case of $p(a) = p(b)$, boundary conditions (2-7a) and (2-7b) may be replaced by the periodic boundary conditions, namely, $X(a) = X(b)$ and $X'(a) = X'(b)$. We will now solve equations (2-6) and (2-7) for three specific cases that we will use extensively in the following chapters for solution of the heat equation under Cartesian, cylindrical, and spherical coordinate systems.

Trigonometric Functions

Consider equation (2-6) for the special case of $q(x) = 0$, $p(x) = 1$, and $w(x) = 1$:

$$\frac{d^2X}{dx^2} + \lambda X = 0 \tag{2-10}$$

over the interval $0 \leq x \leq L$, with the general boundary conditions

$$A_1 \frac{dX}{dx} + A_2 X = 0 \qquad \text{at } x = 0 \tag{2-11a}$$

$$B_1 \frac{dX}{dx} + B_2 X = 0 \qquad \text{at } x = L \tag{2-11b}$$

The boundary value problem and eigenfunctions are only realized for the condition of *positive* λ. Letting $\lambda = \beta^2$ for convenience, the general solution of equation (2-10) becomes

$$X(x) = C_1 \cos \beta x + C_2 \sin \beta x \tag{2-12}$$

where C_1 and C_2 are now considered along with β and the boundary conditions (2-11) to complete the boundary value problem.

We will first consider equations (2-11) in terms of the heat transfer homogeneous boundary conditions, namely, equations (1-54), (1-56), and (1-58). It is readily observed that equation (2-11a) reduces to these three equations for the case of (i) $A_1 = 0$ and $A_2 = 1$, (ii) $A_1 = 1$ and $A_2 = 0$, and (iii) $A_1 = -k$ and $A_2 = \pm h$, respectively. We will now continue to solve equation (2-12) for the specific case (i); hence with $A_1 = 0$, $A_2 = 1$, $B_1 = 0$, and $B_2 = 1$:

$$X = 0 \qquad \text{at } x = 0 \tag{2-13a}$$

$$X = 0 \qquad \text{at } x = L \tag{2-13b}$$

Applying equation (2-13a) to equation (2-12) yields

$$X(x = 0) = C_1 = 0 \quad \rightarrow \quad C_1 = 0 \tag{2-14}$$

Now applying equation (2-13b) yields

$$X(x = L) = C_2 \sin \beta L = 0 \tag{2-15}$$

where letting $C_2 = 0$ yields only the trivial solution of (2-10), namely, $X(x) = 0$. Therefore equation (2-15) is now satisfied by

$$\beta L = n\pi \quad \rightarrow \quad \beta_n = \frac{n\pi}{L} \quad \text{for } n = 0, 1, 2, \cdots \tag{2-16}$$

where the set of parameters β_n are the eigenvalues of the boundary value problem, and the eigenfunctions become

$$X_n(x) = \sin \beta_n x \tag{2-17}$$

where we have dropped the remaining constant C_2 from the actual eigenfunction, recognizing that an undetermined constant C_n still exists with each eigenfunction X_n. By our statement of the general Sturm–Liouville problem, X_n is orthogonal, with

$$\int_{x=0}^{L} X_n(\beta_n, x) X_m(\beta_m, x)\, dx = \begin{cases} 0 & \text{if } n \neq m \\ N(\beta_n) & \text{if } n = m \end{cases} \tag{2-18}$$

over the interval $0 \leq x \leq L$ with a weighing function of unity [i.e.,$w(x) = 1$]. The norm of equation (2-18) becomes

$$N(\beta_n) = \int_{x=0}^{L} \left(\sin \beta_n x\right)^2 dx = \frac{L}{2} \quad \text{for } n \geq 1 \tag{2-19}$$

We may now generalize the above procedure for the trigonometric sine and cosine functions. For the boundary equation (2-11a), we have a choice of three unique boundary condition types at each boundary by letting either $A_1 = 0$ or $A_2 = 0$ or letting both A_1 and A_2 be nonzero. These correspond to heat transfer boundary conditions of the first, second, and third type, respectively, as defined by equations (1-54), (1-56), and (1-58). The same situation applies to the boundary equation (2-11b). Together, these yield a total of *nine unique combinations of homogeneous boundary conditions* for the boundary value problem given by equations (2-10) and (2-11). For all nine cases, the above procedure holds, namely, that one of the constants C_1 and C_2 are eliminated using one of the boundary conditions, and that the second boundary condition is used to define the eigenvalues. The result is a unique set of orthogonal eigenfunctions over the interval $0 \leq x \leq L$.

The procedure is identical for all nine cases; however, the algebraic complexity grows as the type of boundary condition progresses from the first type to the third type. In addition, when type 3 boundary conditions (i.e., convective boundary conditions) are present, the eigenvalues are no longer explicitly defined, as they are, for example, by equation (2-16), but are now implicitly defined by a transcendental equation (i.e., an equation containing transcendental functions with no explicit algebraic solution). We have summarized the nine cases in Table 2-1, showing the resulting eigenvalues, eigenfunctions, and the norm of the eigenfunctions. As examples, we present here cases 2 and 7 from Table 2-1.

Example 2-1 Boundary Value Problem for Cartesian Coordinates
Consider the boundary value problem of equations (2-10) and (2-11) for the case of a type 1 boundary (i.e., prescribed value) at $x = 0$ and a type 3 boundary condition (i.e., convective) at $x = L$:

$$X = 0 \qquad \text{at } x = 0 \tag{2-20a}$$

$$\frac{dX}{dx} + H_2 X = 0 \qquad \text{at } x = L \tag{2-20b}$$

where we have defined $H_2 = B_2/B_1$ $(\approx h_2/k)$ for the homogeneous type 3 boundary condition (see equation 1-58). We eliminate C_1 in the identical manner as equation (2-14). We then plug $X(x) = C_2 \sin \beta x$ into equation (2-20b), giving

$$C_2 \beta \cos \beta L + H_2 C_2 \sin \beta L = 0 \tag{2-21}$$

which yields the following transcendental equation for eigenvalues β_n :

$$\beta_n \cot \beta_n L = -H_2 \quad \rightarrow \quad \beta_1, \beta_2, \ldots \tag{2-22}$$

For the case of eigenvalues as defined by equation (2-22), the norm becomes

$$N(\beta_n) = \int_{x=0}^{L} \left(\sin \beta_n x\right)^2 dx = \frac{L\left(\beta_n^2 + H_2^2\right) + H_2}{2\left(\beta_n^2 + H_2^2\right)} \tag{2-23}$$

where the eigenfunctions, dropping the constant C_n, are defined as

$$X_n(x) = \sin \beta_n x \tag{2-24}$$

These values agree with the results of case 7 in Table 2-1.

Example 2-2 Boundary Value Problem for Cartesian Coordinates
Consider the boundary value problem of equations (2-10) and (2-11) for the case of a type 3 boundary (i.e., convective) at $x = 0$ and a type 2 boundary condition (i.e., insulated) at $x = L$:

$$-\frac{dX}{dx} + H_1 X = 0 \qquad \text{at } x = 0 \tag{2-25a}$$

$$\frac{dX}{dx} = 0 \qquad \text{at } x = L \tag{2-25b}$$

where we have defined $H_1 = A_2/A_1$ $(\approx h_1/k)$ for the homogeneous type 3 boundary condition (see equation 1-58). Plugging equation (2-12) into (2-25a) gives

$$-C_2\beta + H_1 C_1 = 0 \quad \rightarrow \quad C_2 = C_1 \frac{H_1}{\beta} \tag{2-26}$$

which allows elimination of one constant, yielding

$$X(x) = C_1 \left[\cos \beta x + \frac{H_1}{\beta} \sin \beta x\right] \tag{2-27}$$

We now plug equation (2-27) into the boundary condition at $x = L$, giving

$$C_1 \left[-\beta \sin \beta L + H_1 \cos \beta L\right] = 0 \tag{2-28}$$

TABLE 2-1 Solution $X(\lambda_n, x)$, the Norm $N(\lambda_n)$, and the Eigenvalues λ_n of the Differential Equation

$$\frac{d^2X(x)}{dx^2} + \lambda^2 X(x) = 0 \quad \text{in} \quad 0 < x < L$$

Case No.	Boundary Condition at $x = 0$	Boundary Condition at $x = L$	$X(\lambda_n, x)$	$1/N(\lambda_n)$	Eigenvalues λ_n Are Positive Roots of
1	$-\frac{dX}{dx} + H_1 X = 0$	$\frac{dX}{dx} + H_2 X = 0$	$\lambda_n \cos\lambda_n x + H_1 \sin\lambda_n x$	$2\left[(\lambda_n^2 + H_1^2)\left(L + \frac{H}{\lambda_n^2 + H_2^2}\right) + H_1\right]^{-1}$	$\tan\lambda_n L = \frac{\lambda_n(H_1 + H_2)}{\lambda_n^2 - H_1 H_2}$
2	$-\frac{dX}{dx} + H_1 X = 0$	$\frac{dX}{dx} = 0$	$\cos\lambda_n(L - x)$	$2\frac{\lambda_n^2 + H_1^2}{L(\lambda_n^2 + H_1^2) + H_1}$	$\lambda_n \tan\lambda_n L = H_1$
3	$-\frac{dX}{dx} + H_1 X = 0$	$X = 0$	$\sin\lambda_n(L - x)$	$2\frac{\lambda_n^2 + H_1^2}{L(\lambda_n^2 + H_1^2) + H_1}$	$\lambda_n \cot\lambda_n L = -H_1$
4	$\frac{dX}{dx} = 0$	$\frac{dX}{dx} + H_2 X = 0$	$\cos\lambda_n x$	$2\frac{\lambda_n^2 + H_2^2}{L(\lambda_n^2 + H_2^2) + H_2}$	$\lambda_n \tan\lambda_n L = H_2$
5	$\frac{dX}{dx} = 0$	$\frac{dX}{dx} = 0$	$\cos\lambda_n x$[a]	$\frac{2}{L}$ for $\lambda_n \neq 0$; $\frac{1}{L}$ for $\lambda_0 = 0$[a]	$\sin\lambda_n L = 0$[a]
6	$\frac{dX}{dx} = 0$	$X = 0$	$\cos\lambda_n x$	$\frac{2}{L}$	$\cos\lambda_n L = 0$
7	$X = 0$	$\frac{dX}{dx} + H_2 X = 0$	$\sin\lambda_n x$	$2\frac{\lambda_n^2 + H_2^2}{L(\lambda_n^2 + H_2^2) + H_2}$	$\lambda_n \cot\lambda_n L = -H_2$
8	$X = 0$	$\frac{dX}{dx} = 0$	$\sin\lambda_n x$	$\frac{2}{L}$	$\cos\lambda_n L = 0$
9	$X = 0$	$X = 0$	$\sin\lambda_n x$	$\frac{2}{L}$	$\sin\lambda_n L = 0$

[a] For this particular case $\lambda_0 = 0$ is also an eigenvalue corresponding to $X = 1$. $H_1 = h_1/k$ and $H_2 = h_2/k$.

which simplifies to the transcendental equation for eigenvalues β_n :

$$\beta_n \tan \beta_n L = H_1 \quad \rightarrow \quad \beta_1, \beta_2, \ldots \tag{2-29}$$

Together, equations (2-27) and (2-29) complete the boundary value problem for this example. However, while the eigenfunctions in equation (2-27) are orthogonal as expressed for the general β_n, they are often recast by making the following substitution, per equation (2-29):

$$\frac{H_1}{\beta_n} = \tan \beta_n L \tag{2-30}$$

which produces the eigenvalue expression

$$X_n(x) = C_n \left[\cos \beta_n x + \tan \beta_n L \sin \beta x\right] \tag{2-31}$$

We can then replace $\tan \beta_n L$ with the $\sin \beta_n L / \cos \beta_n L$ and factor the $\cos \beta_n L$ term out of the brackets, yielding

$$X_n(x) = \frac{C_n}{\cos \beta_n L} \left[\cos \beta_n L \cos \beta_n x + \sin \beta_n L \sin \beta x\right] \tag{2-32}$$

We now make two final simplifications to equation (2-32). The term in the brackets is replaced using the trigonometric *addition formula identity*, and the undetermined constant term in front of the brackets, namely, $C_n / \cos \beta_n L$, may be replaced with a new constant, which may be called simply C_n'. This gives the final form of the eigenvalues, omitting here the new constant C_n', as

$$X_n(x) = \cos \beta_n (L - x) \tag{2-33}$$

along with norm, for eigenvalues per equation (2-29), given by

$$N(\beta_n) = \int_{x=0}^{L} \left[\cos \beta_n (L - x)\right]^2 dx = \frac{L\left(\beta_n^2 + H_1^2\right) + H_1}{2\left(\beta_n^2 + H_1^2\right)} \tag{2-34}$$

which together agree with the values presented in Table 2-1 for case 2.

Bessel's Equation and Bessel Functions

We now consider the Sturm–Liouville equation (2-6) in cylindrical or spherical coordinates *R(r)*:

$$\frac{d}{dr}\left[p(r)\frac{dR(r)}{dr}\right] + [q(r) + \lambda w(r)]R(r) = 0 \tag{2-35}$$

which for the special case of $p(r) = r, q(r) = -\nu^2/r$, and $w(r) = r$, which yields

$$\frac{d}{dr}\left[r\frac{dR}{dr}\right] + \left[\frac{-\nu^2}{r} + \lambda^2 r\right] R = 0 \tag{2-36a}$$

or

$$\frac{d^2R}{dr^2} + \frac{1}{r}\frac{dR}{dr} + \left[\lambda^2 - \frac{\nu^2}{r^2}\right] R = 0 \tag{2-36b}$$

where we have replaced λ with λ^2, and where the parameter ν will be limited generally in our treatment to integer values ($\nu = 0, 1, 2, \ldots$) or to half-integer values ($\nu = \frac{1}{2}, \frac{3}{2}, \frac{5}{2}, \ldots$), but need not be so limited. For the general parameter ν, equations (2-36a) and (2-36b) are known as *Bessel's equation*, named after the mathematician Friedrich Bessel. If we consider the interval $a \le r \le b$ along with the boundary conditions

$$A_1\frac{dR}{dr} + A_2 R = 0 \qquad \text{at } r = a \tag{2-37a}$$

$$B_1\frac{dR}{dr} + B_2 R = 0 \qquad \text{at } r = b \tag{2-37b}$$

then the solution of Bessel's equation will yield orthogonal eigenfunctions over the interval of orthogonality $a \le r \le b$ with weighting function $w(r) = r$.

The solution of equation (2-36) is by the *extended power series method*, or the *method of Frobenius*, and takes the most general form

$$R(r) = C_1 J_\nu(\lambda r) + C_2 Y_\nu(\lambda r) \tag{2-38}$$

where $J_\nu(\lambda r)$ is known as a *Bessel function of the first kind* of order ν, and $Y_\nu(\lambda r)$ is known as a *Bessel function of the second kind* of order ν or as a Neumann function of order ν. The eigenvalues λ then come from the specific boundary conditions, generating the final eigenfunctions.

Before further discussing the boundary value problem for Bessel's equation, it is useful to explore the Bessel functions in greater detail. References 3 and 6–9, and notably 10–11 provide additional information on Bessel's equation and Bessel functions, while Appendix IV provides the actual series form of the functions, as well as useful properties. Figure 2-1 shows the Bessel functions $J_0(z)$, $J_1(z)$, $Y_0(z)$, and $Y_1(z)$ over the interval $0 \le z \le 10$.

Several important features are noted with regard to the Bessel function behavior in Figure 2-1. Both $J_\nu(z)$ and $Y_\nu(z)$ functions have oscillatory behavior, which is characteristic of all orthogonal functions; see for example, the discussion regarding equation (2-1), although the period is not necessarily constant as with the trigonometric functions. The following hold for Bessel functions of the first kind for an argument of zero:

$$J_0(0) = 1 \tag{2-39a}$$

$$J_\nu(0) = 0 \quad \text{for } \nu \ne 0 \tag{2-39b}$$

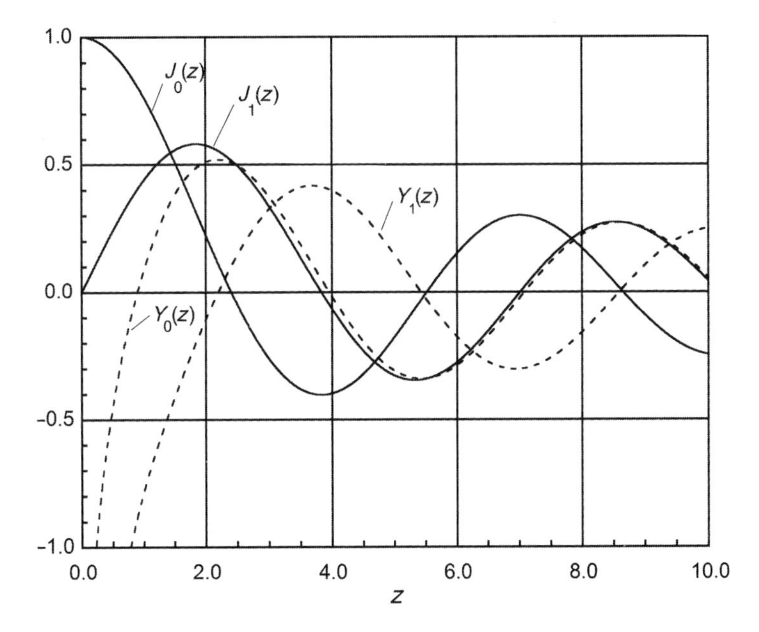

Figure 2-1 Bessel functions of the first and second kind.

Equations (2-39) are important when considering eigenvalues, notably for boundary conditions of the second type (e.g., insulated). The first 10 zeros for a representative set of integer-order Bessel functions are tabulated in Appendix IV. Between any two consecutive positive zeros of $J_n(z)$, there is precisely one zero of $J_{n+1}(z)$; hence the Bessel functions of the first kind are interwoven as observed in Figure 2-1. A second important feature concerns the behavior of Bessel functions of the second kind as the argument approaches zero, namely

$$Y_\nu(z \to 0) \to -\infty \quad \text{for } \nu \geq 0 \tag{2-40}$$

Equation (2-40) becomes very important when the domain of the boundary value problem contains the origin, such as the case of a solid cylinder or sphere.

It is also useful here to list the first derivatives of the Bessel functions $J_\nu(z)$ and $Y_\nu(z)$, as well as of the Bessel functions $J_\nu(\lambda z)$ and $Y_\nu(\lambda z)$:

$$\frac{d}{dz}\left[J_0(z)\right] = -J_1(z) \tag{2-41a}$$

$$\frac{d}{dz}\left[J_0(\lambda z)\right] = -\lambda J_1(\lambda z) \tag{2-41b}$$

$$\frac{d}{dz}\left[Y_0(z)\right] = -Y_1(z) \tag{2-42a}$$

$$\frac{d}{dz}\left[Y_0(\lambda z)\right] = -\lambda Y_1(\lambda z) \tag{2-42b}$$

$$\frac{d}{dz}\left[J_\nu(z)\right] = \frac{\nu}{z}J_\nu(z) - J_{\nu+1}(z) \tag{2-43a}$$

$$\frac{d}{dz}\left[J_\nu(\lambda z)\right] = \frac{\nu}{z}J_\nu(\lambda z) - \lambda J_{\nu+1}(\lambda z) \tag{2-43b}$$

$$\frac{d}{dz}\left[Y_\nu(z)\right] = \frac{\nu}{z}Y_\nu(z) - Y_{\nu+1}(z) \tag{2-44a}$$

$$\frac{d}{dz}\left[Y_\nu(\lambda z)\right] = \frac{\nu}{z}Y_\nu(\lambda z) - \lambda Y_{\nu+1}(\lambda z) \tag{2-44b}$$

It is readily seen that equations (2-41) and (2-42) follow directly from (2-43) and (2-44), respectively, for the special case of $\nu = 0$. See Appendix IV for additional formulas related to the Bessel functions.

We stated above that the solution of the boundary value problem for Bessel's equations yields orthogonal eigenfunctions. We now present two examples for representative boundary conditions.

Example 2-3 Boundary Value Problem for Solid Cylinder
Consider the boundary value problem of equations (2-36b) and (2-37) over the interval $0 \le r \le b$ with a type 1 boundary (i.e., prescribed value) at $r = b$. Considering the general boundary value problem per equations (2-6) and (2-7), this is the case of $a = 0$ and $p(a) = p(0) = 0$; hence equation (2-7a) or (2-37a) is dropped from consideration. However, if we consider an actual cylindrical or spherical problem domain, while there is no physical boundary at $r = 0$, we still need to impose the condition of finiteness. The boundary conditions become

$$R(r \to 0) \neq \pm\infty \qquad \text{(requirement for finiteness)} \tag{2-45a}$$

$$R(r = b) = 0 \tag{2-45b}$$

The solution of equation (2-36b) takes the general form

$$R(r) = C_1 J_\nu(\lambda r) + C_2 Y_\nu(\lambda r) \tag{2-46}$$

where we can immediately drop $Y_\nu(\lambda r)$ from further consideration (i.e., $C_2 = 0$) per boundary condition (2-45a) in view of equation (2-40). We now apply the remaining boundary condition (2-45b):

$$C_1 J_\nu(\lambda b) = 0 \quad \rightarrow \quad J_\nu(\lambda b) = 0 \tag{2-47}$$

For *each* value of ν, there is a unique set of eigenvalues $\lambda_1, \lambda_2, \ldots,$ which are related to the zeros of the $J_\nu(z)$ as follows. Let the set of positive zeros of $J_\nu(z)$ be $\alpha_1, \alpha_2, \ldots,$ where α_n represents the nth root. We note here that the first root $\alpha_1 = 0$ is valid for the case of $\nu \neq 0$, while zero is not a root for $\nu = 0$. The set of eigenvalues then becomes

$$\lambda_{\nu n} = \frac{\alpha_n}{b} \qquad n = 1, 2, 3, \ldots, \qquad \text{for each } \nu \tag{2-48}$$

where we use the double subscript to denote a set of eigenvalues indexed $n = 1, 2, \ldots$ for *each value* of ν. The eigenfunctions become

$$R_\nu(r, \lambda_{\nu n}) = J_\nu(\lambda_{\nu n} r) \tag{2-49}$$

which are orthogonal with weighting function $w(r) = r$, namely,

$$\int_{r=0}^{b} r J_\nu(\lambda_{\nu n} r) J_\nu(\lambda_{\nu m} r) dr = \begin{cases} 0 & \text{if } n \neq m \\ N(\lambda_{\nu n}) & \text{if } n = m \end{cases} \tag{2-50}$$

We note here that the set of eigenfunctions are only orthogonal for each value of ν; hence $J_1(\lambda_{1n} r)$ and $J_1(\lambda_{1m} r)$ are orthogonal, while $J_1(\lambda_{1n} r)$ and $J_2(\lambda_{2m} r)$ are not. The norm of equation (2-50) is given as

$$N(\lambda_{\nu n}) = \int_{r=0}^{b} r J_\nu^2(\lambda_{\nu n} r) dr = \frac{b^2}{2} \left[J_{\nu+1}(\lambda_{\nu n} b) \right]^2 \tag{2-51}$$

where the definite integral is dependent on the specific nature of $\lambda_{\nu n}$; hence the final value of equation (2-51) is valid only for the eigenvalues of equation (2-47).

Example 2-4 Boundary Value Problem for Hollow Cylinder

We now examine a case where $r = 0$ is excluded from the domain (e.g., hollow sphere or cylinder). Consider the boundary value problem of equations (2-36b) and (2-37) over the interval $a \leq r \leq b$ with a type 1 boundary condition (i.e., prescribed value) at both $r = a$ and $r = b$. Because $p(a) \neq 0$, equation (2-7a) or (2-37a) is included in the formulation:

$$R(r = a) = 0 \tag{2-52a}$$

$$R(r = b) = 0 \tag{2-52b}$$

We consider the general solution of equation (2-38) and the first boundary condition given by equation (2-52a), which yields

$$C_1 J_\nu(\lambda a) + C_2 Y_\nu(\lambda a) = 0 \quad \rightarrow \quad C_2 = -C_1 \frac{J_\nu(\lambda a)}{Y_\nu(\lambda a)} \tag{2-53a}$$

which upon substitution of equation (2-53a) into equation (2-38), yields

$$R(r) = C_1 \left[J_\nu(\lambda r) - \frac{J_\nu(\lambda a)}{Y_\nu(\lambda a)} Y_\nu(\lambda r) \right] \tag{2-54a}$$

or

$$R(r) = \frac{C_1}{Y_\nu(\lambda a)} \left[J_\nu(\lambda r) Y_\nu(\lambda a) - J_\nu(\lambda a) Y_\nu(\lambda r) \right] \tag{2-54b}$$

As we have done before, we may replace the undetermined constant in front of the bracket with a new, single constant C_1'. The second boundary condition equation (2-52b) now yields

$$C_1' \left[J_\nu(\lambda b) Y_\nu(\lambda a) - J_\nu(\lambda a) Y_\nu(\lambda b) \right] = 0 \tag{2-55a}$$

The term in the brackets now defines the set of eigenvalues $\lambda_{\nu n}$, for each value of ν, as the roots of the transcendental equation:

$$J_\nu(\lambda b) Y_\nu(\lambda a) - J_\nu(\lambda a) Y_\nu(\lambda b) = 0 \quad \rightarrow \quad \lambda_{\nu 1}, \lambda_{\nu 2}, \ldots \text{for each } \nu \tag{2-55b}$$

The corresponding eigenfunctions become

$$R_\nu(\lambda_{\nu n}, r) = J_\nu(\lambda_{\nu n} r) Y_\nu(\lambda_{\nu n} a) - J_\nu(\lambda_{\nu n} a) Y_\nu(\lambda_{\nu n} r) \tag{2-56}$$

with the property of orthogonality over the interval $a \leq r \leq b$ expressed as

$$\int_{r=a}^{b} r R_\nu(\lambda_{\nu n}, r) R_\nu(\lambda_{\nu m}, r) dr = \begin{cases} 0 & \text{if } n \neq m \\ N(\lambda_{\nu n}) & \text{if } n = m \end{cases} \tag{2-57}$$

For the eigenfunctions of equation (2-56), the norm of equation (2-57) is

$$N(\lambda_{\nu n}) = \int_{r=a}^{b} r R_\nu^2(\lambda_{\nu n}, r) dr = \frac{2}{\pi^2} \frac{J_\nu^2(\lambda_{\nu n} a) - J_\nu^2(\lambda_{\nu n} b)}{\lambda_{\nu n}^2 J_\nu^2(\lambda_{\nu n} b)} \tag{2-58}$$

as evaluated for the specific case of eigenvalues defined per equation (2-55b) for each value ν.

We conclude this treatment of Bessel's equation with a discussion directed toward the context of heat transfer problems. For the solid cylinder or sphere, the condition of finiteness will apply at the origin $r = 0$, which will *always exclude* the Bessel function of the second kind, namely, $Y_\nu(\lambda r)$, from the solution. This leaves only a single boundary condition at $r = b$ to complete the boundary value problem. Hence there are only three distinct problems that we will encounter, namely, for boundary conditions of the first, second, and third type. The three cases are presented in Table 2-2, including the resulting transcendental equations for determination of the eigenvalues, the eigenfunctions, and the norm of the eigenfunctions. In a similar manner, for the case of the hollow cylinder defined over the domain $a \leq r \leq b$, there are nine distinct cases similar to the outcome for the trigonometric functions discussed in the previous section. We present in Table 2-3 the resulting transcendental equations for determination of the eigenvalues, the eigenfunctions, and the norm of the eigenfunctions for the four different combinations of the boundary conditions of the first and second type. The additional cases for boundary conditions of the third type (i.e., convective) follow the same procedure but are of considerable algebraic complexity and are not included in this table. Additional cases that include the convective boundary condition are included in Appendix 3 of reference 12 for all but the convective–convective boundary case.

TABLE 2-2 Eigenfunctions $R_\nu(\lambda_n, r)$, the Norm $N(\lambda_n)$, and the Eigenvalues λ_n of the Differential Equation

$$\frac{d^2 R_\nu}{dr^2} + \frac{1}{r}\frac{dR_\nu}{dr} + \left(\lambda^2 - \frac{\nu^2}{r^2}\right) R_\nu = 0 \text{ in } 0 \leq r < b$$

Subject to the Boundary Conditions Shown

Case No.	Boundary Condition at $r = b$	$R_\nu(\lambda_n, r)$	$\frac{1}{N(\lambda_n)}$	Eigenvalues λ_n Are the Positive Roots of
1	$\frac{dR_\nu}{dr} + HR_\nu = 0$	$J_\nu(\lambda_n r)$	$\frac{2}{J_\nu^2(\lambda_n b)} \cdot \frac{\lambda_n^2}{b^2(H^2 + \lambda_n^2) - \nu^2}$	$\left.\frac{dJ_\nu(\lambda_n r)}{dr}\right\rvert_{r=b} + HJ_\nu(\lambda_n b) = 0$
2	$\frac{dR_\nu}{dr} = 0$	$J_\nu(\lambda_n r)^a$	$\frac{2}{J_\nu^2(\lambda_n b)} \cdot \frac{\lambda_n^2}{b^2\lambda_n^2 - \nu^2}$ [a]	$\left.\frac{dJ_\nu(\lambda_n r)}{dr}\right\rvert_{r=b} = 0^a$
3	$R_\nu = 0$	$J_\nu(\lambda_n r)$	$\frac{2}{b^2 J_{\nu+1}^2(\lambda_n b)}$	$J_\nu(\lambda_n b) = 0$

[a] For this particular case $\lambda_0 = 0$ is also an eigenvalue with $\nu = 0$; then the corresponding eigenfunction is $R_0 = 1$ and the norm $1/N(\lambda_0) = 2/b^2$. $H = h/k$.

TABLE 2-3 Solution $R_\nu(\lambda_n, r)$, the Norm $N(\lambda_n)$, and the Eigenvalues λ_n of the Differential Equation

$$\frac{d^2R_\nu(r)}{dr^2} + \frac{1}{r}\frac{dR_\nu(r)}{dr} + \left(\lambda^2 - \frac{\nu^2}{r^2}\right)R_\nu(r) = 0 \text{ in } a < r < b$$

Case No.	Boundary Condition at $r = a$	Boundary Condition at $r = b$	$R_\nu(\lambda_n, r)^a$ and $\frac{1}{N(\lambda_n)}$	λ_n Values Are the Positive Roots of[a]
1	$\frac{dR_\nu}{dr} = 0$	$\frac{dR_\nu}{dr} = 0$	$R_\nu(\lambda_n, r) = J_\nu(\lambda_n r)Y'_\nu(\lambda_n a) - J'_\nu(\lambda_n a)Y_\nu(\lambda_n r)$ $\frac{1}{N(\lambda_n)} = \frac{\pi^2}{2}\frac{\lambda_n^2 J'^2_\nu(\lambda_n b)}{\left[1-\left(\frac{\nu}{\lambda_n b}\right)^2\right]J'^2_\nu(\lambda_n a) - \left[1-\left(\frac{\nu}{\lambda_n a}\right)^2\right]J'^2_\nu(\lambda_n b)}$	$J'_\nu(\lambda_n a)Y'_\nu(\lambda_n b)$ $-J'_\nu(\lambda_n b)Y'_\nu(\lambda_n a) = 0^b$
2	$\frac{dR_\nu}{dr} = 0$	$R_\nu = 0$	$R_\nu(\lambda_n, r) = J_\nu(\lambda_n r)Y'_\nu(\lambda_n a) - J'_\nu(\lambda_n a)Y_\nu(\lambda_n r)$ $\frac{1}{N(\lambda_n)} = \frac{\pi^2}{2}\frac{\lambda_n^2 J^2_\nu(\lambda_n b)}{J'^2_\nu(\lambda_n a) - \left[1-\left(\frac{\nu}{\lambda_n a}\right)^2\right]J_\nu{}^2(\lambda_n b)}$	$J_\nu(\lambda_n b)Y'_\nu(\lambda_n a)$ $-J'_\nu(\lambda_n a)Y_\nu(\lambda_n b) = 0$
3	$R_\nu = 0$	$\frac{dR_\nu}{dr} = 0$	$R_\nu(\lambda_n, r) = J_\nu(\lambda_n r)Y_\nu(\lambda_n a) - J_\nu(\lambda_n a)Y_\nu(\lambda_n r)$ $\frac{1}{N(\lambda_n)} = \frac{\pi^2}{2}\frac{\lambda_n^2 J'^2_\nu(\lambda_n b)}{\left[1-\left(\frac{\nu}{\lambda_n b}\right)^2\right]J_\nu{}^2(\lambda_n a) - J'^2_\nu(\lambda_n b)}$	$J'_\nu(\lambda_n b)Y_\nu(\lambda_n a)$ $-J_\nu(\lambda_n a)Y'_\nu(\lambda_n b) = 0$
4	$R_\nu = 0$	$R_\nu = 0$	$R_\nu(\lambda_n, r) = J_\nu(\lambda_n r)Y_\nu(\lambda_n a) - J_\nu(\lambda_n a)Y_\nu(\lambda_n r)$ $\frac{1}{N(\lambda_n)} = \frac{\pi^2}{2}\frac{\lambda_n^2 J^2_\nu(\lambda_n b)}{J^2_\nu(\lambda_n a) - J^2_\nu(\lambda_n b)}$	$J_\nu(\lambda_n a)Y_\nu(\lambda_n b)$ $-J_\nu(\lambda_n b)Y_\nu(\lambda_n a) = 0$

[a] We note here that the prime notation denotes differentiation with respect to the entire argument. For example, $J'_\nu(\lambda b) = dJ_\nu(\lambda b)/d(\lambda b)$.
[b] For this particular case $\lambda_0 = 0$ is also an eigenvalue with $\nu = 0$; the corresponding eigenfunction is $R_0(\lambda_0, r) = 1$ and the norm $1/N(\lambda_0) = 2/(b^2 - a^2)$.
Source: From reference 12.

Modified Bessel's Equation

We now consider a modification to Bessel's equation (2-36) to yield the form of

$$\frac{d^2R}{dr^2} + \frac{1}{r}\frac{dR}{dr} - \left[\eta^2 + \frac{\nu^2}{r^2}\right] R = 0 \tag{2-59}$$

which corresponds to the replacement of λ^2 with $-\eta^2$. This may be alternatively interpreted as the case of weighting function $w(r) = -r$ in equation (2-35). Equation (2-59) is known as the *modified Bessel's equation*, which yields the most general solution

$$R(r) = C_1 I_\nu(\eta r) + C_2 K_\nu(\eta r) \tag{2-60}$$

where $I_\nu(\eta r)$ is known as a *modified Bessel function of the first kind* of order ν, and $K_\nu(\eta r)$ is known as a *modified Bessel function of the second kind* of order ν. In consideration of the complete Sturm–Liouville boundary value problem, $w(r) > 0$ must hold over the interval of interest. This is clearly not the case for equation (2-59), with $w(r) = -r$; therefore, the modified Bessel functions do not satisfy the criteria for the Sturm–Liouville problem. As a result, *modified Bessel functions are not orthogonal* under any conditions. In fact, the modified Bessel functions are analogous to the nonorthogonal hyperbolic trigonometric functions. The exact series form for the modified Bessel functions are presented in Appendix IV along with a detailed treatment of useful properties. Figure 2-2 presents the Bessel functions $I_0(z)$, $I_1(z)$, $K_0(z)$, and $K_1(z)$ over the interval $0 \le z \le 2.5$.

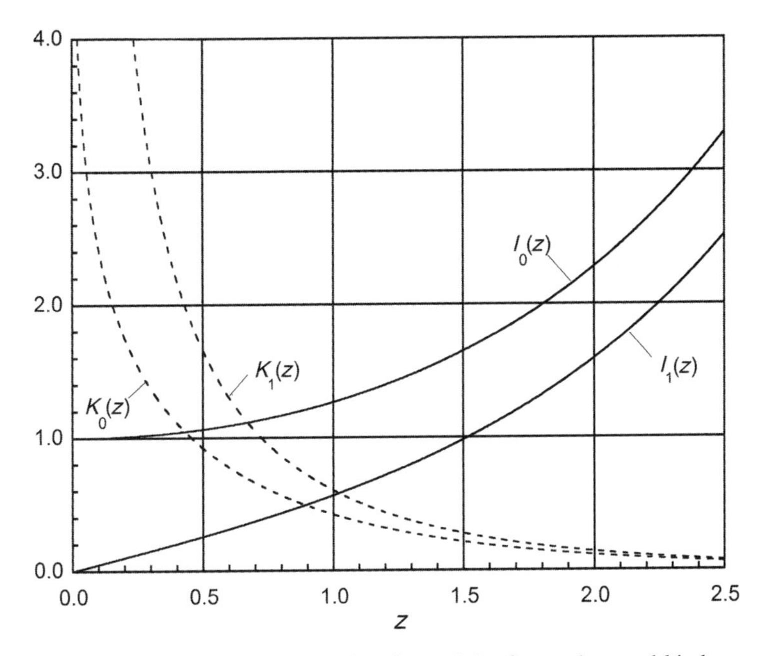

Figure 2-2 Modified Bessel functions of the first and second kind.

Several important features are noted with regard to the modified Bessel function behavior in Figure 2-2. Both $I_\nu(z)$ and $K_\nu(z)$ exhibit nonoscillatory behavior, which is consistent with their hyperbolic trigonometric analogs. The following hold for modified Bessel functions of the first kind for an argument of zero

$$I_0(0) = 1 \tag{2-61a}$$

$$I_\nu(0) = 0 \quad \text{for } \nu \neq 0 \tag{2-61b}$$

Another important feature concerns the behavior of modified Bessel functions of the second kind as the argument approaches zero, namely,

$$K_\nu(z \to 0) \to +\infty \quad \text{for } \nu \geq 0 \tag{2-62}$$

As with Bessel functions of the second kind, equation (2-62) becomes very important when the domain of the heat transfer problem contains the origin, such as the case of a solid cylinder or sphere.

It is also useful here to list the first derivatives of the modified Bessel functions $I_\nu(z)$ and $K_\nu(z)$:

$$\frac{d}{dz}\left[I_0(\eta z)\right] = \eta I_1(\eta z) \tag{2-63}$$

$$\frac{d}{dz}\left[K_0(\eta z)\right] = -\eta K_1(\eta z) \tag{2-64}$$

$$\frac{d}{dz}\left[I_\nu(\eta z)\right] = \frac{\nu}{z} I_\nu(\eta z) + \eta I_{\nu+1}(\eta z) \tag{2-65}$$

$$\frac{d}{dz}\left[K_\nu(\eta z)\right] = \frac{\nu}{z} K_\nu(\eta z) - \eta K_{\nu+1}(\eta z) \tag{2-66}$$

It is readily seen that equations (2-63) and (2-64) follow directly from (2-65) and (2-66), respectively, for the special case of $\nu = 0$.

Legendre's Equation and Legendre Polynomials

We consider here our last set of orthogonal eigenfunctions. We begin with the Sturm–Liouville equation in the form of *Legendre's equation*:

$$\frac{d}{d\mu}\left[(1-\mu^2)\frac{dM}{d\mu}\right] + n(n+1)M = 0 \tag{2-67}$$

where we now use $M = f(\mu)$ to reflect our future choice of independent variable, and where we have set $p(\mu) = 1 - \mu^2$, $q(\mu) = 0$, $w(\mu) = 1$, and $\lambda = n(n+1)$ in equation (2-6). By defining the eigenvalues for *integer values* of n

$$\lambda_n = n(n+1) \quad \text{for } n = 0, 1, 2, \ldots \tag{2-68}$$

the Legendre polynomials are generated as the solution of the Sturm–Liouville problem, with the most general solution of the form

$$M(\mu) = C_1 P_n(\mu) + C_2 Q_n(\mu) \tag{2-69}$$

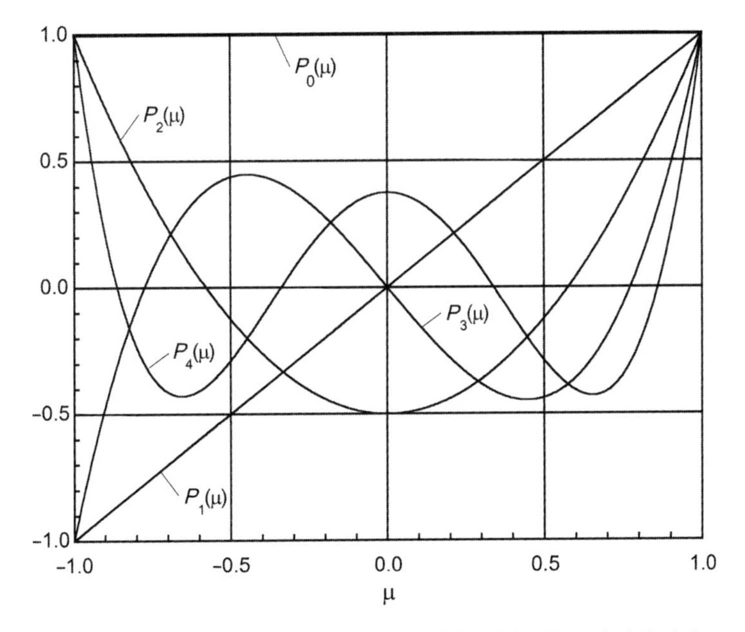

Figure 2-3 First five Legendre polynomials of the first kind $P_n(\mu)$.

where $P_n(\mu)$ and $Q_n(\mu)$ are *Legendre polynomials of the first and second kind*, respectively, of *degree n*. The first few Legendre polynomials of the first kind are given here [3], while we provide a plot of the first five in Figure 2-3:

$$\begin{aligned} &P_0(\mu) = 1 && P_1(\mu) = \mu \\ &P_2(\mu) = \tfrac{1}{2}\left(3\mu^2 - 1\right) && P_3(\mu) = \tfrac{1}{2}\left(5\mu^3 - 3\mu\right) \\ &P_4(\mu) = \tfrac{1}{8}\left(35\mu^4 - 30\mu^2 + 3\right) && P_5(\mu) = \tfrac{1}{8}\left(63\mu^5 - 70\mu^3 + 15\mu\right) \end{aligned} \tag{2-70}$$

Using the above expressions, any other $P_n(\mu)$, when n is a positive integer, is obtainable from the following recurrence relation:

$$(n+1)\,P_{n+1}(\mu) - (2n+1)\,\mu P_n(\mu) + nP_{n-1}(\mu) = 0 \tag{2-71}$$

The Legendre polynomials $P_n(\mu)$ are also obtainable from the *Rodrigues formula:*

$$P_n(\mu) = \frac{1}{2^n n!}\frac{d^n}{d\mu^n}\left(\mu^2 - 1\right)^n \tag{2-72}$$

We note here that with $p(\mu) = 1 - \mu^2$, both $p(-1) = 0$ and $p(1) = 0$. Therefore, no boundary conditions are needed to complete the Sturm–Liouville problem on the interval $-1 \leq \mu \leq 1$ for the Legendre polynomials of the first kind. However, the Legendre polynomials of the second kind are not continuous at the end points of the interval $-1 \leq \mu \leq 1$, but rather the Legendre polynomial $Q_n(\mu)$ is infinite at $\mu = \pm 1$ for all values of n. We therefore exclude $Q_n(\mu)$ from our general solution, equation (2-69), on physical grounds as well as the requirements of the Sturm–Liouville boundary value problem, namely that the solution be continuous on the entire interval of orthogonality.

In view of the above, the eigenfunctions become

$$M_n(\mu) = P_n(\mu) \quad \text{for } n = 0, 1, 2, \ldots \tag{2-73}$$

which are orthogonal over $-1 \leq \mu \leq 1$ with unity weighting function, namely

$$\int_{\mu=-1}^{1} P_n(\mu) P_k(\mu) d\mu = \begin{cases} 0 & \text{if } n \neq k \\ N(n) & \text{if } n = k \end{cases} \tag{2-74}$$

and where the norm of equation (2-74) is equal to

$$N(n) = \int_{\mu=-1}^{1} \left[P_n(\mu) \right]^2 d\mu = \frac{2}{2n+1} \tag{2-75}$$

We can also limit the interval of orthogonality for the Legendre polynomials to the reduced interval $0 \leq \mu \leq 1$; but because $p(0) \neq 0$, we now require a true boundary condition at $\mu = 0$. We consider two specific cases, which correspond to type 1 or type 2 boundary conditions. The first case (type 1) becomes

$$P_n(\mu = 0) = 0 \quad \rightarrow \quad n = 1, 3, 5, \ldots \tag{2-76}$$

with solutions restricted only to *odd Legendre polynomials*, which is readily seen in Figure 2-3 to satisfy equation (2-76). The second case (type 2) becomes

$$\left. \frac{dP_n(\mu)}{d\mu} \right|_{\mu=0} = 0 \quad \rightarrow \quad n = 0, 2, 4, \ldots \tag{2-77}$$

with solutions restricted only to *even Legendre polynomials*. This corresponds to a tangent of zero slope at the origin ($\mu = 0$), as seen to be satisfied in Figure 2-3 for the even Legendre polynomials. We now have orthogonality over $0 \leq \mu \leq 1$ given by

$$\int_{\mu=0}^{1} P_n(\mu) P_k(\mu) d\mu = \begin{cases} 0 & \text{if } n \neq k\text{: both even or both odd} \\ N(n) & \text{if } n = k \end{cases} \tag{2-78}$$

and where the norm of equation (2-78) is equal to

$$N(n) = \int_{\mu=0}^{1} \left[P_n(\mu) \right]^2 d\mu = \frac{1}{2n+1} \tag{2-79}$$

Associated Legendre's Equation and Associated Legendre Polynomials

We finally consider a modification to Legendre's equation, which is known as the *associated Legendre's equation:*

$$\frac{d}{d\mu}\left[(1-\mu^2)\frac{dM}{d\mu}\right]+\left[n(n+1)-\frac{m^2}{1-\mu^2}\right]M=0 \tag{2-80}$$

where we have now set $p(\mu)=1-\mu^2$, $q(\mu)=-m^2/(1-\mu^2)$, $w(\mu)=1$, and $\lambda=n(n+1)$ in equation (2-6). By considering *positive integer values* for both n and m, along with the requirement that $m\leq n$, the eigenvalues again become

$$\lambda_n=n(n+1) \quad \text{for } n=0,1,2,\ldots \tag{2-81}$$

and the associated Legendre polynomials are generated as the solution of the Sturm–Liouville problem, with the most general solution of the form

$$M(\mu)=C_1P_n^m(\mu)+C_2Q_n^m(\mu) \tag{2-82}$$

where $P_n^m(\mu)$ and $Q_n^m(\mu)$ are *associated Legendre polynomials of the first and second kind*, respectively, of *degree n* and *order m*.

The $P_n^m(\mu)$ functions can also be expressed in terms of the $P_n(\mu)$ functions by the following differential relation [7, 8]:

$$P_n^m(\mu)=(-1)^m\left(1-\mu^2\right)^{m/2}\frac{d^m}{d\mu^m}P_n(\mu) \tag{2-83}$$

The first few of $P_n^m(\mu)$ functions for integer values of n and m over the interval of $-1\leq\mu\leq1$ are given by

$$\begin{aligned}
&P_1^1(\mu)=-\left(1-\mu^2\right)^{1/2} && P_2^1(\mu)=-3\left(1-\mu^2\right)^{1/2}\mu \\
&P_2^2(\mu)=3\left(1-\mu^2\right) && P_3^1(\mu)=-\tfrac{3}{2}\left(1-\mu^2\right)^{1/2}\left(5\mu^2-1\right) \\
&P_3^2(\mu)=15\mu\left(1-\mu^2\right) && P_3^3(\mu)=-15\left(1-\mu^2\right)^{3/2}
\end{aligned} \tag{2-84}$$

The recurrence formula among $P_n^m(\mu)$ functions for integer values of m and n is given by

$$(n-m+1)P_{n+1}^m(\mu)-(2n+1)\mu P_n^m(\mu)+(n+m)P_{n-1}^m(\mu)=0 \tag{2-85}$$

For $m=0$, this expression reduces to the recurrence relation (2-71) for the Legendre polynomials, noting that the associated Legendre polynomials of order $m=0$ reduce directly to the Legendre polynomials, namely, $P_n^0(\mu)\equiv P_n(\mu)$.

As with the Legendre polynomials discussed above, the associated Legendre polynomials of the second kind are not continuous at the end points of the interval $-1\leq\mu\leq1$, but rather the associated Legendre polynomial $Q_n^m(\mu)$ is infinite at

$\mu = \pm 1$ for all values of n and m. We therefore exclude $Q_n^m(\mu)$ from our general solution, equation (2-82), on physical grounds, as well as the requirements of the Sturm–Liouville boundary value problem, namely that the solution be continuous on the entire interval of orthogonality.

In view of the above, the eigenfunctions become

$$M_n(\mu) = P_n^m(\mu) \quad \text{for } n, m = 0, 1, 2, \ldots \text{ and } m \leq n \tag{2-86}$$

which are orthogonal on the interval $-1 \leq \mu \leq 1$, with unity weighting function,

$$\int_{\mu=-1}^{1} P_n^m(\mu) P_k^m(\mu) d\mu = \begin{cases} 0 & \text{if } n \neq k \\ N(n, m) & \text{if } n = k \end{cases} \tag{2-87}$$

and where the norm of equation (2-87) is equal to

$$N(n, m) = \int_{\mu=-1}^{1} \left[P_n^m(\mu)\right]^2 d\mu = \frac{2}{2n+1} \frac{(n+m)!}{(n-m)!} \tag{2-88}$$

As with the Bessel functions, we note that the set of associated Legendre polynomials are only orthogonal for each given order m; hence $P_n^3(\mu)$ and $P_k^3(\mu)$ are orthogonal, while $P_n^3(\mu)$ and $P_k^5(\mu)$ are not.

Finally, one may extend the orthogonality of the Legendre functions to the interval $\mu_0 \leq \mu \leq 1$, with $\mu_0 > -1$. This is consistent with the top part of a sphere cut out by the cone defined by θ_0 (i.e., $\mu_0 = \cos\theta_0$). Then we have

$$\int_{\mu=\mu_0}^{1} P_n^{-m}(\mu) P_k^{-m}(\mu) d\mu = \begin{cases} 0 & \text{if } n \neq k \\ N(n, m) & \text{if } n = k \end{cases} \tag{2-89}$$

where now $P_n^{-m}(\mu)$ is the *generalized Legendre function* [13], where m is zero or a positive integer, but with n (or k) no longer restricted to integer values. Rather, for each value of m, n are the roots greater than $-\frac{1}{2}$ of the equation

$$P_n^{-m}(\mu_0) = 0 \qquad n > -\tfrac{1}{2} \tag{2-90}$$

We see that equation (2-90) corresponds to a type 1 boundary condition on the surface of the cone. We note that the generalized Legendre function $P_n^{-m}(\mu)$ has a singularity at $\mu_0 = -1$ for nonintegral values of n; hence the generalized Legendre functions are not useful for the whole sphere problem.

2-3 THE FOURIER SERIES

The treatment of *nonhomogeneous* boundary conditions or initial conditions during our solution of the heat conduction equation will generally lead to the expression of these conditions in the form of a series expansion composed of a set of orthogonal eigenfunctions. Such a treatment is known as a generalized Fourier series expansion. Quoting from Fourier's *Analytical Theory of Heat*: "The examination of this condition shows that we may develop in convergent series, or express by definite integrals, functions which are not subject to a constant law... and extends the employment of arbitrary functions by submitting them to the ordinary processes of analysis." While Fourier significantly advanced the representation of arbitrary functions in what we now call a *Fourier series expansion*, the rigorous treatment of the necessary conditions (e.g. series convergence) was later provided by a number of other mathematicians, notably P.G. Dirichlet. The reader is referred to references 1–4 for an in depth analysis of the Fourier series and to reference 9 for an expansive discussion of the historical development. Here we provide a concise treatment of the Fourier series that is consistent both with the solution of many classes of partial differential equations and with our above presentation of orthogonal functions and the general Sturm–Liouville boundary value problem.

Consider a set of orthogonal eigenfunctions $\phi_n(x)$ on the interval $a \le x \le b$. Now let $f(x)$ be a function that is integrable (see comments below) on the same interval $a \le x \le b$. We then define the *general Fourier series expansion* of $f(x)$ as

$$f(x) = \sum_{n=0}^{\infty} C_n \phi_n(x) \tag{2-91}$$

where the coefficients C_n are known as the *Fourier constants* or *Fourier coefficients*.

Convergence of the Fourier series is a topic of considerable mathematical depth. Here we use H. Lebeque's definition of an *integrable function* (see reference 1) along with our treatment of the Sturm–Liouville boundary value problem for eigenfunction generation, as sufficient conditions for the general Fourier series expansion of equation (2-91) to converge in the mean to $f(x)$ [2].

Determination of the Fourier constants is straightforward using the property of orthogonality. Multiplying both sides of equation (2-91) by $w(x)\phi_m(x)$, with m a fixed arbitrary value, and then integrating on the interval $a \le x \le b$, yields

$$\int_{x=a}^{b} w(x) f(x) \phi_m(x)\, dx = \int_{x=a}^{b} \sum_{n=0}^{\infty} C_n w(x) \phi_n(x) \phi_m(x)\, dx \tag{2-92}$$

where $w(x)$ is the weighting function for the orthogonal functions. Since the integral of a sum is the sum of the integrals, equation (2-92) becomes

$$\int_{x=a}^{b} w(x) f(x) \phi_m(x)\, dx = \sum_{n=0}^{\infty} \left[C_n \int_{x=a}^{b} w(x) \phi_n(x) \phi_m(x)\, dx \right] \tag{2-93}$$

However, by equation (2-4), the integral on the right-hand side is equal to zero whenever $n \ne m$ and equal to the norm when $n = m$; hence only a single term of the summation remains, namely $C_m \int_{x=a}^{b} w(x)\phi_m(x)\phi_m(x)\,dx$. Solving for the Fourier coefficients yields

$$C_n = \frac{\displaystyle\int_{x=a}^{b} w(x) f(x)\phi_n(x)\,dx}{\displaystyle\int_{x=a}^{b} w(x)\left[\phi_n(x)\right]^2\,dx} = \frac{\displaystyle\int_{x=a}^{b} w(x) f(x)\phi_n(x)\,dx}{N(\lambda)} \tag{2-94}$$

where we have replaced C_m with C_n in equation (2-94), since m was arbitrary. Recall here that the norm, $N(\lambda)$, will depend on the exact nature of the eigenvalues of the boundary value problem and may therefore change with changing boundary conditions even for a given orthogonal function.

We present three examples of the Fourier series expansion for our three general sets of orthogonal functions developed in the preceding sections, namely, the trigonometric functions, Bessel functions, and Legendre polynomials.

Example 2-5 Fourier Series for Cartesian Coordinates
Consider the set of orthogonal trigonometric functions generated with Example 2-2, namely $\phi_n(x) = \cos\beta_n\,(L - x)$, per equation (2-33), as defined on the interval $0 \le x \le L$. The eigenvalues are given by the transcendental equation (2-29). We wish to express the arbitrary function $f(x)$ as a *Fourier cosine series*. Equation (2-91) becomes

$$f(x) = \sum_{n=1}^{\infty} C_n \cos\beta_n(L - x) \tag{2-95}$$

Multiplying by $\phi_m(x) = \cos\beta_m\,(L - x)$ and integrating yields the Fourier coefficients

$$C_n = \frac{\displaystyle\int_{x=0}^{L} f(x)\cos\beta_n(L - x)\,dx}{\displaystyle\int_{x=0}^{L} \cos^2\beta_n(L - x)\,dx} = \frac{\displaystyle\int_{x=0}^{L} f(x)\cos\beta_n(L - x)\,dx}{\dfrac{L\left(\beta_n^2 + H_1^2\right) + H_1}{2\left(\beta_n^2 + H_1^2\right)}} \tag{2-96}$$

where we have used the norm defined by equation (2-34) for this specific set of eigenvalues.

Example 2-6 Fourier Series for Cylindrical Coordinates
Consider the set of orthogonal functions generated with Example 2-3, namely, $\phi_n(r) = J_\nu(\lambda_{\nu n} r)$, per equation (2-49), as defined on the interval $0 \le r \le b$. The eigenvalues are given by the transcendental equation (2-47). We wish to express the arbitrary function $f(r)$ as a *Fourier–Bessel series*. Equation (2-91) becomes

$$f(r) = \sum_{n=1}^{\infty} C_n J_\nu(\lambda_{\nu n} r) \tag{2-97}$$

Multiplying by $\phi_m(r) = J_\nu(\lambda_{\nu m} r)$ and by the necessary weighting function $w(r) = r$, and then integrating over $0 \le r \le b$ yields the Fourier coefficients

$$C_n = \frac{\int_{r=0}^{b} r f(r) J_\nu(\lambda_{\nu n} r) dr}{\int_{r=0}^{b} r J_\nu^2(\lambda_{\nu n} r) dr} = \frac{\int_{r=0}^{b} r f(r) J_\nu(\lambda_{\nu n} r) dr}{(b^2/2)\left[J_{\nu+1}(\lambda_{\nu n} b)\right]^2} \tag{2-98}$$

where we have used the norm defined by equation (2-51), Table 2-2 case 3, for this set of eigenvalues. We note that there is a unique Fourier–Bessel series expansion for each value of ν and related eigenvalues $\lambda_{\nu n}$.

Example 2-7 Fourier Series for Spherical Coordinates
Consider the set of orthogonal Legendre polynomials of the first kind $P_n(\mu)$ limited to even integer values, as defined over the interval $0 \le \mu \le 1$. Recall that the Legendre polynomials are orthogonal on this limited interval when restricted to either all even or all odd integers. We wish to express the arbitrary function $f(\mu)$ as a *Fourier–Legendre series*. Equation (2-91) then becomes

$$f(\mu) = \sum_{n=0}^{\infty} C_n P_n(\mu) \qquad \text{for even } n \tag{2-99}$$

Multiplying by $\phi_m(\mu) = P_m(\mu)$ and then integrating over $0 \le \mu \le 1$ yields the Fourier coefficients

$$C_n = \frac{\int_{\mu=0}^{1} f(\mu) P_n(\mu)\, d\mu}{\int_{\mu=0}^{1} \left[P_n(\mu)\right]^2 d\mu} = \frac{\int_{\mu=0}^{1} f(\mu) P_n(\mu) d\mu}{1/(2n+1)} \tag{2-100}$$

where we have used the norm defined by equation (2-79) for this set of even integer-order Legendre polynomials.

2-4 COMPUTATION OF EIGENVALUES

Once the eigenvalues λ_n are computed from the solution of the boundary value problem, the eigenfunctions $X(\lambda_n)$ and the normalization integral $N(\lambda_n)$ become known. Some of the eigenvalue equations, such as $\sin \lambda_n L = 0$ or $\cos \lambda_n L = 0$, are simple expressions; hence the eigenvalues λ_n are readily evaluated in explicit form. Now consider, for example, the transcendental equation for the case 1 in Table 2-1 with $H_1 = H_2 = H$ ($\approx h/k$). The resulting expression is written as

$$\tan \lambda L = 2 \frac{(\lambda L)(HL)}{(\lambda L)^2 - (HL)^2} \tag{2-101a}$$

For convenience this result is rearranged as

$$\cot \xi = \frac{1}{2}\left(\frac{\xi}{B} - \frac{B}{\xi}\right) \equiv Z \tag{2-101b}$$

where $\xi = \lambda L$ and $B = HL$. Clearly, the solution of this transcendental equation is not so easy. First we present a graphical interpretation of the roots of this transcendental equation before discussing its numerical solution.

Graphical Representation

The result given by equation (2-101b) represents the following two curves:

$$Z = \frac{1}{2}\left(\frac{\xi}{B} - \frac{B}{\xi}\right) \quad \text{and} \quad Z = \cot \xi \tag{2-101c, d}$$

The first of these curves represents a hyperbola whose center is at the origin and its asymptotes are $\xi = 0$ and $Z = \xi/2B$, while the second represents a set of cotangent curves as illustrated in Figure 2-4. The ξ values corresponding to the intersections of the hyperbola with the cotangent curves are the roots of the above transcendental equation. Clearly there are an infinite number of such points, each successively located in intervals $(0 - \pi)$, $(\pi - 2\pi)$, $(2\pi - 3\pi)$, and so forth. Because of symmetry, the negative roots are equal in absolute value to the positive ones; therefore, only the positive roots need to be considered in

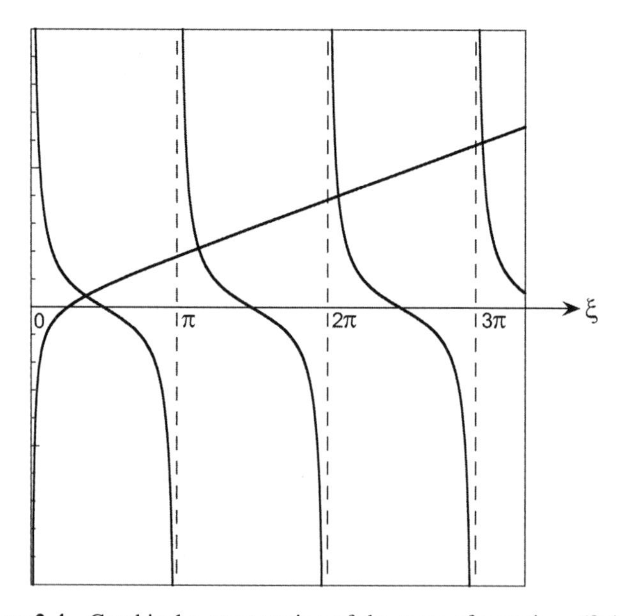

Figure 2-4 Graphical representation of the roots of equations (2-101).

the solution since the solution remains unaffected by the sign of the root. The graphical representation of the roots shown in Figure 2-4 is useful to establish the regions where the roots exist; but accurate values of the roots are determined by numerical solution of the transcendental equations as described next.

Numerical Solutions

Various methods are available for solving transcendental equations numerically [14, 15]. Here we consider the bisection, Newton–Raphson, and secant methods for the determination of the roots of transcendental equations.

Bisection Method: Consider a transcendental equation written compactly in the form

$$F(\xi) = 0 \tag{2-102}$$

and suppose it has only one root in the region $\xi_i \leq \xi \leq \xi_{i+1}$ as illustrated in Figure 2-5. We wish to determine this root by the bisection method. The interval $\xi_i \leq \xi \leq \xi_{i+1}$ is divided into two subintervals by a point $\xi_{i+(1/2)}$ defined by

$$\xi_{i+(1/2)} = \frac{1}{2}(\xi_i + \xi_{i+1}) \tag{2-103a}$$

and the sign of the product $F(\xi_i) \cdot F(\xi_{i+(1/2)})$ is examined. If the product

$$F(\xi_i)F(\xi_{i+(1/2)}) < 0 \tag{2-103b}$$

then the root lies in the first subinterval $\xi_i \leq \xi \leq \xi_{i+(1/2)}$, since the sign change occurs in this interval. If the product is positive, the root must lie in the second subinterval. If the product is exactly zero, $\xi_{i+(1/2)}$ is the exact root. The procedure for determining the root is now apparent. The subinterval containing the root is bisected, and the bisection procedure is continued until the change in the value of root from one bisection to the next becomes less than a specified tolerance ε.

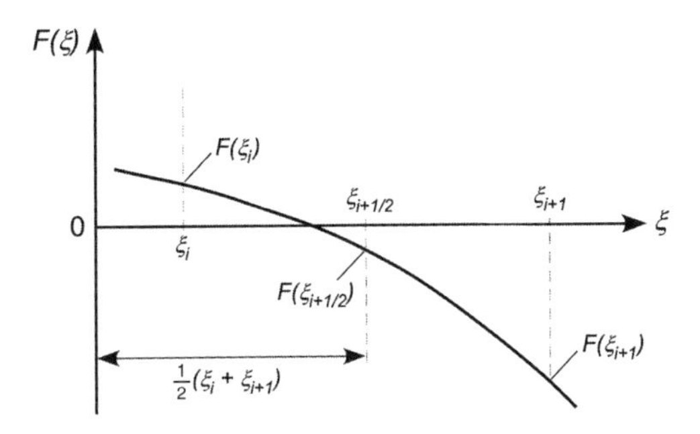

Figure 2-5 Bisection method.

The bisection procedure always yields a root if a region is found over which $F(\xi)$ changes sign and has only one root. Therefore, the graphical interpretation of roots as illustrated in Figure 2-4 is useful to locate the regions where the roots exist. In the absence of graphical representation, one starts with $\xi = 0$ and evaluates $F(\xi)$ for each small increment of ξ until $F(\xi)$ changes sign. Then, a root must lie in that interval, and the bisection procedure is applied for its determination.

In each bisection, the interval is reduced by half; therefore, after n bisections the original interval is reduced by a factor 2^n. For example, 10 bisections reduce the original interval by a factor more than 1000, and 20 bisections reduce more than one million.

Newton–Raphson Method: Consider a function $F(\xi) = 0$ plotted against ξ as illustrated in Figure 2-6. Let the tangent drawn to this curve at $\xi = \xi_i$ intersect the ξ axis at $\xi = \xi_{i+1}$. The slope of this tangent is given by

$$F'(\xi_i) = \frac{F(\xi_i)}{\xi_i - \xi_{i+1}} \tag{2-104a}$$

where prime denotes the derivative with respect to ξ. Solving this equation for ξ_{i+1} we obtain

$$\xi_{i+1} = \xi_i - \frac{F(\xi_i)}{F'(\xi_i)} \tag{2-104b}$$

Equation (2-104b) provides an expression for calculating ξ_{i+1} from the knowledge of $F(\xi_i)$ and $F'(\xi_i)$ by iteration, until the change in the value of ξ_{i+1} from one iteration to the next is less than a specified convergence criteria ε. The method is widely used in practice because of its rapid convergence; however, there are situations that may give rise to convergence difficulties. For example, if the initial approximation to the root is not sufficiently close to the exact value of the root or the second derivative $F''(\xi_i)$ changes sign near the root, convergence difficulties arise.

Secant Method: The Newton–Raphson method requires the derivative of the function for each iteration. However, if the function is difficult to differentiate,

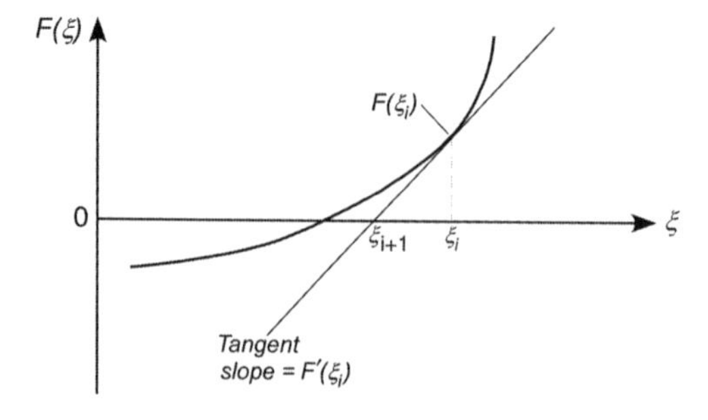

Figure 2-6 Newton–Raphson method.

the derivative is approximated by a difference approximation; hence equation (2-104b) takes the form

$$\xi_{i+1} = \xi_i - \frac{F(\xi_i)}{[F(\xi_i) - F(\xi_{i-1})]/(\xi_i - \xi_{i-1})} \tag{2-105}$$

The secant method may not be as rapidly convergent as the Newton–Raphson method; but if the evaluation of $F'(\xi_i)$ is time consuming, then the secant method may require less computational time than Newton's method.

Tabulated Eigenvalues: In Appendix II we tabulate the first six roots of the transcendental equations

$$F(\beta) \equiv \beta \tan \beta - C = 0 \tag{2-106a}$$

$$F(\beta) \equiv \beta \cot \beta + C = 0 \tag{2-106b}$$

for several different values of C. These transcendental equations are associated with cases 2 and 3 in Table 2-1, respectively.

When using the secant or Newton–Raphson method for solving such transcendental equations, it is preferable to establish the region where a root lies by a bisection method or a graphical approach and then apply the secant or Newton–Raphson method. An examination of the roots listed in the Appendix II reveals that the roots lie in the intervals, which are multiples of π. Consider, for example, the transcendental equation (2-106a). For large values of C, the roots lie in the regions where the slope of the tangent curve is very steep; similarly, difficulty is experienced in the determination of roots from Eq. (2-106b) when the roots lie in the regions where the slope of the cotangent curve is very steep. In such situations, the convergence difficulty is alleviated if equations (2-106a) and (2-106b) are rearranged, respectively, in the forms

$$F(\beta) \equiv \beta \sin \beta - C \cos \beta = 0 \tag{2-107a}$$

$$F(\beta) = \beta \cos \beta + C \sin \beta = 0 \tag{2-107b}$$

2-5 FOURIER INTEGRALS

In the final section of this chapter, we now consider the expansion of an arbitrary function over an unlimited interval, both semi-infinite and infinite. While such a treatment does not directly make use of orthogonality as with the case of the Fourier series expansion, the results are useful for the solution of the heat equation over semi-infinite and infinite domains, as will be seen in Chapter 6.

Trigonometric Functions

Consider a function $f(x)$ that is sectionally continuous for $x > 0$, that $f(x)$ is defined at each point of discontinuity by the mean value theorem, and that the integral

$\int_{x=0}^{\infty} |f(x)|\,dx$ exists. One may then express *f(x)* in terms of the *Fourier sine integral* as [1–3]

$$f(x) = \int_{\lambda=0}^{\infty} c(\lambda) \sin \lambda x d\lambda \tag{2-109a}$$

where the unknown coefficients *c(λ)* are no longer discrete and are defined by

$$c(\lambda) = \frac{2}{\pi} \int_{x'=0}^{\infty} f(x') \sin \lambda x' \, dx' \tag{2-109b}$$

Equations (2-109a) and (2-109b) may be combined to state the Fourier integral as a single expression

$$f(x) = \frac{2}{\pi} \int_{\lambda=0}^{\infty} \sin \lambda x \int_{x'=0}^{\infty} f(x') \sin \lambda x' \, dx' d\lambda \tag{2-110}$$

Under the same necessary conditions described above, one may also express *f(x)* in terms of the *Fourier cosine integral* as

$$f(x) = \int_{\lambda=0}^{\infty} c(\lambda) \cos \lambda x d\lambda \tag{2-111a}$$

where the unknown coefficients *c(λ)* are now defined by

$$c(\lambda) = \frac{2}{\pi} \int_{x'=0}^{\infty} f(x') \cos \lambda x' \, dx' \tag{2-111b}$$

Equations (2-111a) and (2-111b) may likewise be combined to state the Fourier cosine integral as a single expression

$$f(x) = \frac{2}{\pi} \int_{\lambda=0}^{\infty} \cos \lambda x \int_{x'=0}^{\infty} f(x') \cos \lambda x' \, dx' d\lambda \tag{2-112}$$

As we will see in Chapter 6, the Fourier sine and cosine integrals will enable us to treat the 1-D conduction problem for homogeneous conditions at the $x = 0$ boundary for the *semi-infinite domain conduction problem*. For example, the Fourier sine integral will be used to treat the homogeneous prescribed temperature boundary condition [i.e., $\sin(0) = 0$], while the Fourier cosine integral will be used to treat the homogeneous prescribed flux boundary condition, that is,

$$\left. \frac{d \cos(\lambda x)}{dx} \right|_{x=0} = 0$$

In addition, we can treat the homogeneous convective boundary condition as follows [16], defining the boundary condition as

$$-\left.\frac{dX}{dx}\right|_{x=0} + A\, X|_{x=0} = 0 \tag{2-113}$$

where $A\ (= h/k)$ is a constant. For our same conditions on *f(x)* defined above, with the additional requirement that *df/dx* be sectionally continuous on each finite interval, we can define the Fourier integral expansion as

$$f(x) = \int_{\lambda=0}^{\infty} c(\lambda)\,[\lambda \cos \lambda x + A \sin \lambda x]\, d\lambda \tag{2-114}$$

where $X(\lambda, x) = \lambda \cos \lambda x + A \sin \lambda x$, which satisfies equation (2-113). The parameter *c(λ)* is given by

$$c(\lambda) = \frac{2}{\pi(\lambda^2 + A^2)} \int_{x'=0}^{\infty} f(x')\left[\lambda \cos \lambda x' + A \sin \lambda x'\right] dx' \tag{2-115}$$

Equation (2-115) may be substituted into equation (2-114) to yield

$$f(x) = \int_{\lambda=0}^{\infty} \frac{2\,[\lambda \cos \lambda x + A \sin \lambda x]}{\pi(\lambda^2 + A^2)} \int_{x'=0}^{\infty} f(x')\left[\lambda \cos \lambda x' + A \sin \lambda x'\right] dx'\, d\lambda \tag{2-116}$$

We now consider a function *f(x)* that is sectionally continuous over $-\infty < x < \infty$, that *f(x)* is defined at each point of discontinuity by the mean value theorem, and that the integral $\int_{x=-\infty}^{\infty} |f(x)|\, dx$ exists. Under these conditions one may define the *general Fourier integral* expansion of *f(x)* as

$$f(x) = \int_{\lambda=0}^{\infty} [a(\lambda) \cos \lambda x + b(\lambda) \sin \lambda x] d\lambda \tag{2-117}$$

where the unknown coefficients *a(λ)* and *b(λ)* are defined by the expressions

$$a(\lambda) = \frac{1}{\pi} \int_{x'=-\infty}^{\infty} f(x') \cos \lambda x'\, dx' \tag{2-118a}$$

$$b(\lambda) = \frac{1}{\pi} \int_{x'=-\infty}^{\infty} f(x') \sin \lambda x'\, dx' \tag{2-118b}$$

Equations (2-118a) and (2-118b) may be substituted into equation (2-117) and the trigonometric terms combined to yield

$$f(x) = \int_{\lambda=0}^{\infty} \frac{1}{\pi} \int_{x'=-\infty}^{\infty} f(x') \cos \lambda(x - x')\, dx'\, d\lambda \tag{2-119}$$

The representation given by equation (2-119) is valid for the conditions outlined above for *f(x)*, with additional requirement that *df/dx* is also sectionally continuous over $-\infty < x < \infty$ [3, p.115].

Bessel Functions

Finally, we can extend the general Fourier integral approach to the Bessel functions of the first kind. Consider a function *f(r)* over the interval $0 \leq r < \infty$ such that the integral $\int_{r=0}^{\infty} f(r)dr$ is absolutely convergent, and that the function *f(r)* is of bounded variation in the neighborhood of each point r. Under these conditions, we may express *f(r)* as a Fourier–Bessel integral of the form [17–19]

$$f(r) = \int_{\lambda=0}^{\infty} c(\lambda) J_\nu(\lambda r) d\lambda \tag{2-120}$$

for $\nu \geq -\frac{1}{2}$, where the coefficients $c(\lambda)$ are defined by

$$c(\lambda) = \lambda \int_{r'=0}^{\infty} r' J_\nu(\lambda r') f(r') dr' \tag{2-121}$$

Combining equations (2-120) and (2-121) yields the following expression:

$$f(r) = \int_{\lambda=0}^{\infty} \lambda J_\nu(\lambda r) \int_{r'=0}^{\infty} r' J_\nu(\lambda r') f(r') dr' d\lambda \tag{2-122}$$

where the inner integral $\int_{r'=0}^{\infty} r' J_\nu(\lambda r') f(r') dr'$ is called the *Hankel transform of order* ν *of the function f(r).*

A similar integral representation is needed for the solution of heat conduction problems in the semi-infinite region $a < r < \infty$ in the cylindrical coordinate system for an azimuthally symmetric geometry (i.e. no ϕ dependency). The representation of an arbitrary function $f(r)$ in the region $a < r < \infty$ in terms of the function $R_0(\lambda, r)$ is considered in reference 20, and the result can be written in the form

$$f(r) = \int_{\lambda=0}^{\infty} \frac{1}{N(\lambda)} \lambda R_0(\lambda, r) \int_{r'=a}^{\infty} r' R_0(\lambda, r') f(r')\, dr' d\lambda \quad \text{in} \quad a < r < \infty \tag{2-123}$$

Equation (2-123) may also be considered in the form

$$f(r) = \int_{\lambda=0}^{\infty} c(\lambda) R_0(\lambda, r) d\lambda \quad \text{in} \quad a < r < \infty \tag{2-124}$$

which reveals the coefficients to be defined as

$$c(\lambda) \equiv \frac{1}{N(\lambda)} \lambda \int_{r'=a}^{\infty} r' R_0(\lambda, r') f(r') dr' \tag{2-125}$$

Here, the norm $N(\lambda)$ and the function $R_0(\lambda, r)$ depend on the type of the boundary condition at $r = a$; that is, whether it is of the first, second, or the third type. We now present the expressions for $R_0(\lambda, r)$ and $N(\lambda)$ for these three different types of boundary conditions at $r = a$, for the specific case of $\nu = 0$.

The Boundary Condition at r = a of the Third Type The solution of Bessel's equation of order $\nu = 0$ satisfying the boundary condition

$$-\left.\frac{dR_0}{dr}\right|_{r=a} + H\left.R_0\right|_{r=a} = 0 \tag{2-126}$$

with $H = h/k$, is taken as

$$R_0(\lambda, r) = J_0(\lambda r)[\lambda Y_1(\lambda a) + HY_0(\lambda a)] - Y_0(\lambda r)[\lambda J_1(\lambda a) + HJ_0(\lambda a)] \tag{2-127}$$

and the norm $N(\lambda)$ is given by

$$N(\lambda) = [\lambda J_1(\lambda a) + HJ_0(\lambda a)]^2 + [\lambda Y_1(\lambda a) + HY_0(\lambda a)]^2 \tag{2-128}$$

Boundary Condition at r = a of the Second Type: For this special case we have $H = 0$ in equation (2-126). The solution of Bessel's equation of order $\nu = 0$ satisfying this boundary condition is taken as

$$R_0(\lambda, r) = J_0(\lambda r)Y_1(\lambda a) - Y_0(\lambda r)J_1(\lambda a) \tag{2-129}$$

and the norm becomes

$$N(\lambda) = J_1^2(\lambda a) + Y_1^2(\lambda a) \tag{2-130}$$

Boundary Condition at r = a of the First Type: For this special case we have $H \to \infty$ (i.e., $k = 0$) in equation (2-126), which yields the boundary condition

$$\left.R_0\right|_{r=a} = 0 \tag{2-131}$$

The solution of Bessel's equation of order $\nu = 0$ satisfying this boundary condition is taken as

$$R_0(\lambda, r) = J_0(\lambda r)Y_0(\lambda a) - Y_0(\lambda r)J_0(\lambda a) \tag{2-132}$$

and the corresponding norm becomes

$$N(\lambda) = J_0^2(\lambda a) + Y_0^2(\lambda a) \tag{2-133}$$

We summarize in Table 2-4 the above results for $R_0(\lambda, r)$ and $N(\lambda)$ for the boundary conditions of the first, second, and third types at $r = a$.

TABLE 2-4 Solution $R_0(\lambda, r)$ and the Norm $N(\lambda)$ of the Differential Equation

$$\frac{dR_0^2(r)}{dr^2} + \frac{1}{r}\frac{dR_0(r)}{dr} + \lambda^2 R_0(r) = 0 \text{ in } a < r < \infty$$

Subject to the Boundary Conditions Shown

Case No.	Boundary Condition at $r = a$	$R_\nu(\lambda, r)$ and $\frac{1}{N(\lambda)}$
1	$-\frac{dR_0}{dr} + HR = 0$	$R_0(\lambda, r) = J_0(\lambda r)[\lambda Y_1(\lambda a) + HY_0(\lambda a)] - Y_0(\lambda r)[\lambda J_1(\lambda a) + HJ_0(\lambda a)]$ $\frac{1}{N(\lambda)} = \left\{[\lambda J_1(\lambda a) + HJ_0(\lambda a)]^2 + [\lambda Y_1(\lambda a) + HY_0(\lambda a)]^2\right\}^{-1}$
2	$\frac{dR_0}{dr} = 0$	$R_0(\lambda, r) = J_0(\lambda r)Y_1(\lambda a) - Y_0(\lambda r)J_1(\lambda a)$ $\frac{1}{N(\lambda)} = \left\{J_1^2(\lambda a) + Y_1^2(\lambda a)\right\}^{-1}$
3	$R_0 = 0$	$R_0(\lambda, r) = J_0(\lambda r)Y_0(\lambda a) - Y_0(\lambda r)J_0(\lambda a)$ $\frac{1}{N(\lambda)} = \left\{J_0^2(\lambda a) + Y_0^2(\lambda a)\right\}^{-1}$

REFERENCES

1. H. S. Carslaw, *Introduction to the Theory of Fourier's Series and Integrals*, 3rd ed., Dover, New York, 1930.
2. H. F. Davis, *Fourier Series and Orthogonal Functions*, Allyn and Bacon, Boston, 1963.
3. R. V. Churchill, *Fourier Series and Boundary Value Problems*, 2nd ed., McGraw-Hill, New York, 1963.
4. G. H. Hardy and W. W. Rogosinski, *Fourier Series*, Cambridge University Press, New York, 1944.
5. M. N. Özişik, *Boundary Value Problems of Heat Conduction*, International Textbook, Scranton, PA, 1968. also Dover, New York, 1989.
6. M. L. Boas, *Mathematical Methods in the Physical Sciences*, 2nd ed., Wiley, New York, 1983.
7. T. M. MacRobert, *Spherical Harmonics*, 3rd ed., Pergamon Oxford, England, 1967.
8. W. Magnus and F. Oberhettinger, *Formulas and Theorems for the Spherical Functions of Mathematical Physics*, Chelsea, New York, 1949.
9. R. E. Langer, Fourier's Series: The Genesis and Evolution of a Theory, *Am. Math. Monthly*, **54**, (7, Part 2), 1–86, 1947.
10. G. N. Watson, *A Treatise on the Theory of Bessel Functions*, 2nd ed., Cambridge University Press, London, 1966.
11. N. W. McLachlan, *Bessel Functions for Engineers*, 2nd ed., Clarendon, London, 1961.
12. G. Cinelli, An Extension of the Finite Hankel Transform and Applications, Int. *J. Eng. Sci.*, **3**, 539–559, 1965.
13. H. S. Carslaw and J. C. Jaeger, *Conduction of Heat in Solids*, Oxford University Press, London, 1947.
14. M. L. James, G. M. Smith, and J. C. Wolford, *Applied Numerical Methods for Digital Computations with Fortran and CSMP*, 2nd ed., IEP, New York, 1977.
15. Y. Jaluria, *Computer Methods for Engineering*, Allyn and Bacon, London, 1988.
16. A. Erdelyi, W. Magnus, F. Oberhettinger, and F. G. Tricomi, *Higher Transcendental Functions*, McGraw-Hill, New York, 1953.
17. I. N. Sneddon, *Fourier Transforms*, McGraw-Hill, New York, 1951.
18. E. C. Titchmarsh, *Introduction to the Theory of Fourier Integrals*, 2nd ed., Oxford University Press, Oxford, England, 1948.
19. E. C. Titchmarsh, *Eigenfunction Expansions, Part I*, Clarendon Press, London, 1962.
20. S. Goldstein, *Proc. Lond. Math. Soc.*, S2, **34**, 51–88, 1932.

PROBLEMS

2-1 Let $\phi_n(x) = \cos \lambda_n x$, with eigenvalues $\lambda_n = (2n+1)\pi/2L$, for $n = 0, 1, 2 \ldots$ over the interval $0 \le x \le L$, noting that the weighting function is 1.

a. Show that the integral $\int_{x=0}^{L} \phi_n(x)\phi_m(x)\,dx = 0$ for $n \ne m$.

b. Evaluate the norm of the eigenfunction, namely, $\int_{x=0}^{L} \left[\phi_n(x)\right]^2 dx$.

c. Determine the general Fourier series expansion coefficients (C_n) for the following, simplifying all integrals as much as possible.

$$2x^2 = \sum_{n=0}^{\infty} C_n \cos(\lambda_n x)$$

2-2 Let $\phi_n(x) = \sin \lambda_n x$, with eigenvalues $\lambda_n = n\pi/L$, for $n = 1, 2, 3 \ldots$ over the interval $0 \leq x \leq L$, noting that the weighting function is 1.

a. Show that the integral $\int_{x=0}^{L} \phi_n(x)\phi_m(x)\,dx = 0$ for $n \neq m$.

b. Evaluate the norm of the eigenfunction, namely, $\int_{x=0}^{L} [\phi_n(x)]^2\,dx$.

c. Determine the general Fourier series expansion coefficients (C_n) for the following, simplifying all integrals as much as possible.

$$4x = \sum_{n=1}^{\infty} C_n \sin(\lambda_n x)$$

2-3 Derive the eigenvalue expression and the norm for case 3 in Table 2-2.

2-4 Derive the eigenvalue expression and the norm for case 4 in Table 2-2.

2-5 Derive the eigenvalue expression and the norm for case 5 in Table 2-2.

2-6 Derive the eigenvalue expression and the norm for case 6 in Table 2-2.

3

SEPARATION OF VARIABLES IN THE RECTANGULAR COORDINATE SYSTEM

The method of separation of variables has been widely used in the solution of heat conduction problems and will be the basis for this chapter. Homogeneous problems, as defined below, are readily handled with this method. The multidimensional steady-state heat conduction problems with no generation can also be solved with this method if only one of the boundary conditions is nonhomogeneous; problems involving more than one nonhomogeneous boundary condition can be split up into simpler problems using the principle of superposition. In this chapter, we discuss the general problem of the separability of the heat conduction equation; examine the separation in the rectangular coordinate system; determine the elementary solutions and the eigenvalues of the resulting separated equations for different combinations of boundary conditions; examine the solution of one- and multidimensional (spatial and temporal) homogeneous problems by the method of separation of variables; examine the solution of multidimensional steady-state heat conduction problems with and without heat generation; and describe the splitting up of a nonhomogeneous problem into a set of simpler problems that can be solved by the separation of variables technique. The reader should consult references 1–4 for a discussion of the mathematical aspects of the method of separation of variables and references 5–8 for additional applications on the solution of heat conduction problems.

3-1 BASIC CONCEPTS IN THE SEPARATION OF VARIABLES METHOD

The concept of a homogeneous boundary condition or homogeneous differential equation is fundamental to understanding and implementing the method of separation of variables. As introduced in Chapter 1, see, for example, equations (1-54) and (1-56), an equation is *homogeneous* if each nonzero term contains the dependent variable (in our case T) or its derivatives (e.g., T' or T''). We now state the requirements for solution of a partial differential equation (PDE), specifically the heat equation, by the *method of separation of variables*:

1. Homogeneous PDE
2. For steady-state problems, all homogeneous boundary conditions with the exception of a single nonhomogeneous boundary condition
3. For transient problems, all homogeneous boundary conditions and a nonhomogeneous initial condition
4. Linear PDE and linear boundary conditions

When the above conditions are not satisfied, we will attempt to use the principle of superposition to realize the necessary conditions, along with other approaches, such as shifting of the temperature scale or making use of symmetry.

Procedure for Steady-State Problems

To illustrate the basic procedures associated with the method of separation of variables, we consider a homogeneous, steady-state boundary value problem of heat conduction for a two-dimensional rectangular solid over the domain $0 \le x \le L$ and $0 \le y \le W$, as shown in Figure 3-1.

We first formulate the problem, which includes a statement of the simplified governing equation, namely, steady state (i.e., $\partial T/\partial t = 0$) and with no energy generation (i.e., $g = 0$), and with the boundary conditions:

$$\frac{\partial^2 T}{\partial x^2} + \frac{\partial^2 T}{\partial y^2} = 0 \quad \text{in} \quad 0 < x < L, \quad 0 < y < W \tag{3-1}$$

$$\text{BC1:} \quad T(x = 0, y) = T_1 \qquad \text{BC2:} \quad T(x = L, y) = T_1 \tag{3-2a}$$

$$\text{BC3:} \quad T(x, y = 0) = T_1 \qquad \text{BC4:} \quad T(x, y = W) = T_2 \tag{3-2b}$$

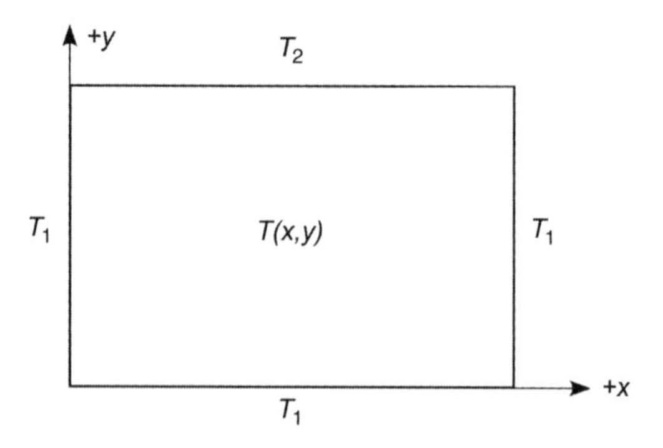

Figure 3-1 Heat conduction in a rectangular solid.

The requirement of a homogeneous PDE is satisfied by equation (3-1), but the requirement for a steady-state problem of all homogeneous boundary conditions with the exception of a single nonhomogenous boundary is *not satisfied*, as all four boundary conditions are nonhomogeneous by the presence of the prescribed, nonzero temperature (T_1 or T_2). This type of nonhomogeneity is readily removed by *linearly shifting the temperature scale*, namely defining a new temperature:

$$\Theta(x, y) = T(x, y) - T_1 \tag{3-3}$$

which here has the effect of removing three of the four nonhomogeneities. We now reformulate the problem in terms of our new dependent variable $\Theta(x, y)$:

$$\frac{\partial^2 \Theta}{\partial x^2} + \frac{\partial^2 \Theta}{\partial y^2} = 0 \quad \text{in} \quad 0 < x < L, \quad 0 < y < W \tag{3-4}$$

$$\text{BC1:} \quad \Theta(x = 0, y) = 0 \qquad \text{BC2:} \quad \Theta(x = L, y) = 0 \tag{3-5a}$$

$$\text{BC3:} \quad \Theta(x, y = 0) = 0 \qquad \text{BC4:} \quad \Theta(x, y = W) = T_2 - T_1 = T_3 \tag{3-5b}$$

With this temperature shift, all requirements for separation of variables are met. We now proceed with our solution by assuming a separation of $\Theta(x, y)$ into two spatially dependent functions of a single variable each in the form

$$\Theta(x, y) = X(x)Y(y) \tag{3-6}$$

Substituting equation (3-6) into the PDE of equation (3-4) and dividing both sides by XY yields

$$\frac{1}{X(x)} \frac{d^2 X(x)}{dx^2} + \frac{1}{Y(y)} \frac{d^2 Y(y)}{dy^2} = 0 \tag{3-7}$$

Equation (3-7) may be *separated* as follows:

$$\frac{1}{Y} \frac{d^2 Y}{dy^2} = -\frac{1}{X} \frac{d^2 X}{dx^2} \tag{3-8}$$

where the left-hand side is only a function of y, while the right-hand side is only a function of x. Such an equality can only hold if both sides are equal to the same constant value, say λ^2:

$$\frac{1}{Y} \frac{d^2 Y}{dy^2} = -\frac{1}{X} \frac{d^2 X}{dx^2} = \lambda^2 \tag{3-9}$$

In equation (3-9), λ^2 is known as the *separation constant*, and we emphasize here that the sign of λ^2 becomes very important in our solution. We had the choice of introducing $\pm\lambda^2$ as the separation constant, and specifically selected $+\lambda^2$ to force a desired ordinary differential equation in the x direction, as summarized here.

Selection Rule for the Separation Constant: Always select the sign of the separation constant to produce a boundary value problem (i.e., Sturm–Liouville problem) in the dimension that corresponds to all homogenous boundary conditions.

Recalling the general Sturm–Liouville problem, see Chapter 2, of equations (2-6) and (2-7), we note that the solution generates orthogonal functions, which are an essential component of the separation of variables approach. Keeping this in mind, equation (3-9) produces the following two ordinary differential equations:

$$\frac{d^2X}{dx^2} + \lambda^2 X = 0 \tag{3-10a}$$

$$\text{BC1:} \quad X(x=0) = 0 \qquad \text{BC2:} \quad X(x=L) = 0 \tag{3-10b}$$

and

$$\frac{d^2Y}{dy^2} - \lambda^2 Y = 0 \tag{3-11a}$$

$$\text{BC3:} \quad Y(y=0) = 0 \tag{3-11b}$$

noting the corresponding separation of all the *homogeneous* boundary conditions. We see now that equations (3-10) do in fact define the Sturm–Liouville problem in Cartesian coordinates as given by equations (2-10) and (2-13).

Under these conditions, we consider the x direction to be the *homogeneous dimension* and will always solve the ODE from the homogeneous dimension(s) first. Equation (3-10a) yields

$$X(x) = C_1 \cos \lambda x + C_2 \sin \lambda x \tag{3-12}$$

Boundary condition 1 yields $C_1 = 0$, while boundary condition 2 yields

$$X(x=L) = 0 = C_2 \sin \lambda L \tag{3-13}$$

where $C_2 = 0$ gives the trivial solution $X(x) = 0$; hence we define the *eigenvalues*

$$\lambda L = n\pi \quad \rightarrow \quad \lambda_n = \frac{n\pi}{L} \quad \text{for} \quad n = 0, 1, 2, \ldots \tag{3-14}$$

to satisfy BC2, which then yields the desired, *orthogonal eigenfunctions*

$$X_n(x) = C_2 \sin \lambda_n x \tag{3-15}$$

Only after solving the boundary value problem corresponding to the homogeneous dimension (noting that in general there may be multiple homogenous dimensions) does one turn to the nonhomogeneous dimension, equation (3-11a). This ODE yields a *conjugate pair of real roots* $\pm\lambda$ for which case the solution may be written in two forms, noting their equivalence:

$$Y(y) = C_3 e^{\lambda y} + C_4 e^{-\lambda y} \tag{3-16a}$$

$$Y(y) = C_3 \cosh \lambda y + C_4 \sinh \lambda y \tag{3-16b}$$

While both of the above are valid forms of the solution, we shall follow a general rule that serves to simplify the algebra. For the case of a *finite domain*, as we have here, the hyperbolic solutions will always be used. For the case of a *semi-infinite domain*, we will generally use the exponential form. While each solution is valid, our approach will generally simplify the algebra of the problem. The homogeneous boundary condition (BC3) is then applied to determine or eliminate one of the two constants. This process yields

$$Y(y=0) = 0 = C_3 \cosh 0 + C_4 \sinh 0 \quad \rightarrow \quad C_3 = 0 \tag{3-17}$$

At this point, all the homogeneous boundary conditions have been satisfied, and it is time to consider the product solution of the separated functions:

$$\Theta(x, y) = C_2 \sin \lambda_n x C_4 \sinh \lambda_n y \tag{3-18}$$

Equation (3-18) may be simplified by defining a new constant $C_n = C_2 C_4$, yielding for each eigenvalue *a solution* of the original PDE of the form

$$\Theta_n(x, y) = C_n \sin \lambda_n x \sinh \lambda_n y \tag{3-19}$$

The most general solution then becomes the sum of all solutions, as given by

$$\Theta(x, y) = \sum_{n=0}^{\infty} C_n \sin \lambda_n x \sinh \lambda_n y \tag{3-20}$$

The only step that remains is to define the constant C_n. This process will always make use of the final, nonhomogeneous boundary condition. Applying here BC4, as defined by equation (3-5b), yields

$$\Theta(x, y = W) = T_3 = \sum_{n=0}^{\infty} C_n \sin \lambda_n x \sinh \lambda_n W \tag{3-21}$$

If we momentarily consider $C_n \sinh \lambda_n W$ as a single constant, say a_n, this becomes

$$T_3 = \sum_{n=0}^{\infty} a_n \sin \lambda_n x \tag{3-22}$$

which is readily seen as a Fourier series expansion of the constant T_3 in terms of an orthogonal function over the interval $0 \le x \le L$, noting the weighting function of unity. As detailed in Chapter 2, we then multiply both sides by the orthogonal function for an arbitrary eigenvalue, say $\sin \lambda_m x$, and integrate from 0 to L, which we will signify with the following operator:

$$* \int_{x=0}^{L} \sin \lambda_m x \, dx$$

This operation yields

$$\int_{x=0}^{L} T_3 \sin \lambda_m x \, dx = \int_{x=0}^{L} \sum_{n=0}^{\infty} a_n \sin \lambda_n x \sin \lambda_m x \, dx = a_m \int_{x=0}^{L} \sin^2 \lambda_m x \, dx \tag{3-23}$$

where we made use of the property of orthogonality to eliminate all terms of the series except for the term with $m = n$. Solving for a_m, replacing arbitrary index m with n, and substituting back to $a_n = C_n \sinh \lambda_n W$ yields the coefficients of equation (3-20):

$$C_n = \frac{\int_{x=0}^{L} T_3 \sin \lambda_n x \, dx}{\sinh \lambda_n W \int_{x=0}^{L} \sin^2 \lambda_n x \, dx} = \frac{T_3 \int_{x=0}^{L} \sin \lambda_n x \, dx}{\sinh \lambda_n W N(\lambda_n)} \tag{3-24}$$

where $N(\lambda_n)$ is the norm of the eigenfunction and is dependent on the specific nature of the eigenvalues, as given here by equation (3-14), and summarized in Table 2-1 for the nine general cases related to equation (3-10a). As a general practice, we will leave C_n in the form of definite integrals. However, here we simplify equation (3-24) to yield the following for $n > 0$:

$$C_n = \frac{T_3 \dfrac{L}{\pi} \left[\dfrac{(-1)^{n+1} + 1}{n} \right]}{\sinh \lambda_n W \cdot (L/2)} \tag{3-25}$$

The solution of$\Theta(x, y)$ may now be simplified as

$$\Theta(x, y) = \frac{2T_3}{\pi} \sum_{n=1}^{\infty} \left[\frac{(-1)^{n+1} + 1}{n} \right] \frac{\sin \lambda_n x \sinh \lambda_n y}{\sinh \lambda_n W} \tag{3-26}$$

We finally define the desired solution $T(x, y)$ as

$$T(x, y) = \frac{2T_3}{\pi} \sum_{n=1}^{\infty} \left[\frac{(-1)^{n+1} + 1}{n} \right] \frac{\sin \lambda_n x \sinh \lambda_n y}{\sinh \lambda_n W} + T_1 \tag{3-27}$$

Procedure for Transient Problems

We now illustrate the basic concepts of separation of variables for a transient, boundary value problem of heat conduction for a 1-D slab (i.e., plane wall) over the domain $0 \le x \le L$. Initially the slab is at a temperature $T = F(x)$, and for times $t > 0$, the boundary surface at $x = 0$ is kept insulated while the boundary at $x = L$ dissipates heat by convection with a heat transfer coefficient h into a

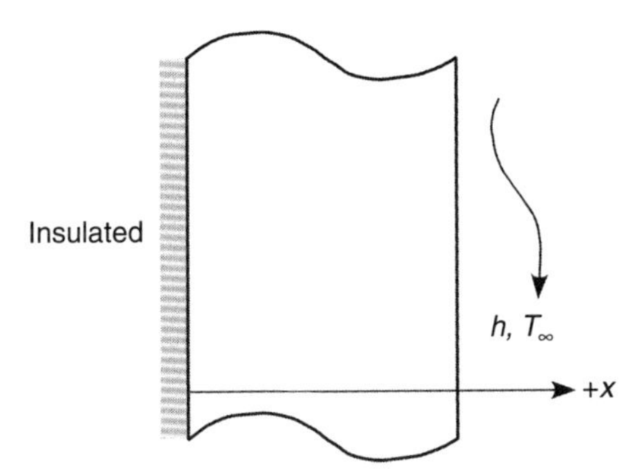

Figure 3-2 Transient heat conduction in a slab.

fluid of temperature T_∞, as depicted in Figure 3-2. There is no heat generation in the medium. The mathematical formulation of this problem is given as

$$\frac{\partial^2 T(x,t)}{\partial x^2} = \frac{1}{\alpha}\frac{\partial T(x,t)}{\partial t} \quad \text{in} \quad 0 < x < L, \quad t > 0 \tag{3-28}$$

$$\text{BC1:} \quad \left.\frac{\partial T}{\partial x}\right|_{x=0} = 0 \qquad \text{BC2:} \quad -k\left.\frac{\partial T}{\partial x}\right|_{x=L} = h\left[T|_{x=L} - T_\infty\right] \tag{3-29a}$$

$$\text{IC:} \quad T(x, t=0) = F(x) \tag{3-29b}$$

The requirement of a homogeneous PDE is satisfied by equation (3-28), but the requirement for a transient problem of *all homogeneous boundary conditions* is *not satisfied*, as the boundary condition at $x = L$ contains the nonhomogeneous term T_∞. This type of nonhomogeneous convective BC is readily removed by *linearly shifting the temperature scale*, namely defining a new temperature

$$\Theta(x,t) = T(x,t) - T_\infty \tag{3-30}$$

We now reformulate the problem in terms of the new dependent variable $\Theta(x,t)$:

$$\frac{\partial^2 \Theta(x,t)}{\partial x^2} = \frac{1}{\alpha}\frac{\partial \Theta(x,t)}{\partial t} \quad \text{in} \quad 0 < x < L, \quad t > 0 \tag{3-31}$$

$$\text{BC1:} \quad \left.\frac{\partial \Theta}{\partial x}\right|_{x=0} = 0 \qquad \text{BC2:} \quad -k\left.\frac{\partial \Theta}{\partial x}\right|_{x=L} = h\,\Theta|_{x=L} \tag{3-32a}$$

$$\text{IC:} \quad \Theta(x, t=0) = F(x) - T_\infty = G(x) \tag{3-32b}$$

With this temperature shift, all requirements for separation of variables are met. We now proceed with our solution by assuming a separation of $\Theta(x,t)$ into space-dependent and time-dependent functions of a single variable each, namely,

$$\Theta(x,t) = X(x)\Gamma(t) \tag{3-33}$$

We now substitute equation (3-33) into (3-31) and introduce a separation constant

$$\frac{1}{X}\frac{d^2X}{dx^2} = \frac{1}{\alpha\Gamma}\frac{d\Gamma}{dt} = -\lambda^2 \tag{3-34}$$

In keeping with our guideline established above, we selected the sign of the separation constant to force the boundary value problem in the homogenous x dimension. Equation (3-34) produces the following two ordinary differential equations:

$$\frac{d^2X}{dx^2} + \lambda^2 X = 0 \tag{3-35a}$$

$$\text{BC1:} \quad \left.\frac{dX}{dx}\right|_{x=0} = 0 \qquad \text{BC2:} \quad -k\left.\frac{dX}{dx}\right|_{x=L} = h\, X|_{x=L} \tag{3-35b}$$

and

$$\frac{d\Gamma}{dt} + \alpha\lambda^2\Gamma = 0 \tag{3-36}$$

Beginning with the homogeneous dimension, Equation (3-35a) yields

$$X(x) = C_1 \cos\lambda x + C_2 \sin\lambda x \tag{3-37}$$

Boundary condition 1 yields $C_2 = 0$, while boundary condition 2 yields

$$-k\left[-C_1\lambda \sin\lambda L\right] = h\left[C_1 \cos\lambda L\right] \tag{3-38}$$

which is readily rearranged to the following *transcendental equation:*

$$\lambda_n \tan\lambda_n L = \frac{h}{k} \quad \rightarrow \quad \lambda_n \qquad \text{for } n = 1, 2, 3, \ldots \tag{3-39}$$

where the roots λ_n are the eigenvalues, as discussed in Example 2-2. The desired, orthogonal eigenfunctions become

$$X_n(x) = C_1 \cos\lambda_n x \tag{3-40}$$

When solving equation (3-39), it is useful to multiple by L and define a new parameter $\beta_n = L\lambda_n$, yielding the transcendental equation

$$\beta_n \tan\beta_n = \frac{hL}{k} = \text{Bi} \quad \rightarrow \quad \beta_n \quad \text{for } n = 1, 2, 3, \ldots \tag{3-41}$$

The transcendental equation is now cast in terms of a single parameter, which is recognized as the nondimensional Biot number, see equation (1-90). We note here that for nonzero Bi numbers, $\lambda_0 = 0$ is not an eigenvalue; hence we begin our index at $n = 1$ rather than $n = 0$ for emphasis of this point.

As before, after solving the boundary value problem corresponding to the homogeneous dimension, we now turn to the nonhomogeneous dimension, equation (3-36). This ODE yields the solution

$$\Gamma(t) = C_3 e^{-\alpha\lambda^2 t} \tag{3-42}$$

We note here that transient problems will always produce a decaying exponential function from the time-dependent ODE. This is consistent with the physics of heat transfer, namely, that for constant boundary conditions as considered here, the transient component of the solution decays to zero, giving way to the steady-state temperature profile. For a transient problem, there is only the nonhomogeneous initial condition, which we always consider *after* recombining the functions.

Equations (3-40) and (3-42) are now combined, introducing a new constant $C_n = C_1 C_3$, to produce a solution of the original PDE of the form

$$\Theta_n(x,t) = C_n \cos\lambda_n x e^{-\alpha\lambda_n^2 t} \tag{3-43}$$

We again construct the most general solution by a summation of all solutions:

$$\Theta(x,t) = \sum_{n=1}^{\infty} C_n \cos\lambda_n x e^{-\alpha\lambda_n^2 t} \tag{3-44}$$

The final step is to define the constant C_n using the nonhomogeneous initial condition. Applying the initial condition, equation (3-32b), yields

$$G(x) = \sum_{n=1}^{\infty} C_n \cos\lambda_n x \tag{3-45}$$

which is a Fourier series expansion of $G(x)$ in terms of an orthogonal function over the interval $0 \le x \le L$. As detailed above, as well as in Chapter 2, we multiply both sides by the orthogonal function for an arbitrary eigenvalue, say $\cos\lambda_m x$, and integrate from 0 to L, which we will denote by the following operator:

$$* \int_{x=0}^{L} \cos\lambda_m x \, dx$$

This operation eliminates all terms of the summation except one, yielding

$$C_n = \frac{\int_{x=0}^{L} G(x)\cos\lambda_n x \, dx}{\int_{x=0}^{L} \cos^2\lambda_n x \, dx} = \frac{\int_{x=0}^{L} G(x)\cos\lambda_n x \, dx}{N(\lambda_n)} \tag{3-46}$$

where $N(\lambda_n)$ is the norm of the eigenfunction and is dependent on the specific nature of the eigenvalues, as given here by the transcendental equation (3-39), and summarized in Table 2-1. We finally define $T(x, t)$, the desired solution, as

$$T(x,t) = \sum_{n=1}^{\infty} C_n \cos \lambda_n x e^{-\alpha \lambda_n^2 t} + T_\infty \tag{3-47}$$

We may substitute equation (3-46) into (3-47), yielding

$$T(x,t) = \sum_{n=1}^{\infty} \left[\frac{\cos \lambda_n x e^{-\alpha \lambda_n^2 t}}{N(\lambda_n)} \int_{x'=0}^{L} G(x') \cos \lambda_n x' \, dx' \right] + T_\infty \tag{3-48}$$

where we have introduced *prime notation* to distinguish the variable of integration within the definite integral from our independent variable x.

General Procedure

We have provided the above two solutions in considerable detail, pointing out the important elements with regard to the general separation of variables method for the heat equation. Here we summarize the key steps for the separation of variables:

1. Identify the homogeneous dimensions (i.e., all homogeneous boundary conditions) and the nonhomogeneous dimension. For transient problems, the nonhomogeneous dimension will always be time.
2. Express the temperature as a product of separated functions, one function for each dimension.
3. Isolate the first separated function that corresponds to a homogeneous dimension, and introduce a separation constant to force a boundary value problem (Sturm–Liouville problem) for this isolated function.
4. Solve the boundary value problem, yielding eigenvalues for the separation constant, corresponding orthogonal eigenfunctions, and elimination of one of the two solution constants.
5. Repeat steps 3 and 4 for each homogeneous dimension, yielding new eigenvalues and new orthogonal functions in each homogeneous dimension.
6. Solve the ODE in the nonhomogeneous dimension, yielding a nonorthogonal function. For a steady-state problem, use the single homogeneous boundary condition to eliminate one of the two solution constants. For a transient problem, simply retain the solution constant.
7. Assemble the general solution as the sum of all product solutions, yielding, in general, a single undetermined constant.
8. Apply the final nonhomogeneous boundary or initial condition, which will be satisfied as a Fourier series expansion. The undetermined constant(s) will be solved as the Fourier coefficient(s).

3-2 GENERALIZATION TO MULTIDIMENSIONAL PROBLEMS

The method of separation of variables as illustrated above for the solution of the heat conduction problem is now formally generalized to the solution of the following multidimensional, transient homogeneous problem:

$$\nabla^2 T(\hat{r}, t) = \frac{1}{\alpha}\frac{\partial T(\hat{r}, t)}{\partial t} \qquad \text{in domain R} \qquad t > 0 \tag{3-49}$$

$$k_i \frac{\partial T}{\partial n_i} + h_i T = 0 \qquad \text{on boundary } S_i \quad i = 1 \ to \ N \qquad t > 0 \tag{3-50a}$$

$$T(\hat{r}, t = 0) = F(\hat{r}) \qquad \text{in domain } R \tag{3-50b}$$

where $\partial/\partial n_i$ denotes differentiation along the outward-drawn normal to the boundary surface S_i, and $\hat{r}$ denotes the general space coordinates. It is assumed that the domain R has a number of continuous boundary surfaces $S_i, i = 1, 2, \ldots, N$ in number, such that each boundary surface S_i fits the coordinate surface of the chosen orthogonal coordinate system. The 1-D slab problem considered above is obtainable as a special case from this more general problem; that is, the slab has two continuous boundary surfaces, one at $x = 0$ and the other at $x = L$. The boundary conditions for the slab problem are readily obtainable from the general boundary condition (3-50a) by choosing the coefficients h_i and k_i, accordingly.

To solve the above general problem we assume a separation in the form

$$T(\hat{r}, t) = \Psi(\hat{r})\Gamma(t) \tag{3-51}$$

where the function $\Psi(\hat{r})$ in general depends on three spatial variables. We substitute equation (3-51) into equation (3-49) and carry out the analysis with a similar argument as discussed above to obtain

$$\frac{1}{\Psi(\hat{r})}\nabla^2\Psi(\hat{r}) = \frac{1}{\alpha\Gamma(t)}\frac{d\Gamma(t)}{dt} = -\lambda^2 \tag{3-52}$$

where λ^2 is the first separation constant. Clearly, the function $\Gamma(t)$ satisfies an ordinary differential equation of the same form as equation (3-36) and its solution is taken as $\exp(-\alpha\lambda^2 t)$. The spatial variable function $\Psi(\hat{r})$ satisfies the following auxiliary problem:

$$\nabla^2\Psi(\hat{r}) + \lambda^2\Psi(\hat{r}) = 0 \qquad \text{in domain } R \tag{3-53a}$$

$$k_i \frac{\partial \Psi}{\partial n_i} + h_i \Psi = 0 \qquad \text{on boundary } S_i \tag{3-53b}$$

TABLE 3-1 Orthogonal Coordinate Systems Allowing Simple Separation of the Helmholtz and Laplace Equations

Coordinate System	Functions That Appear in Solution
1 Rectangular	Exponential, circular, hyperbolic
2 Circular cylinder	Bessel, exponential, circular
3 Elliptic cylinder	Mathieu, circular
4 Parabolic cylinder	Weber, circular
5 Spherical	Legendre, power, circular
6 Prolate spheroidal	Legendre, circular
7 Oblate spheroidal	Legendre, circular
8 Parabolic	Bessel, circular
9 Conical	Lamé, power
10 Ellipsoidal	Lamé
11 Paraboloidal	Baer

Sources: From references 1, 3, and 9.

where again $i = 1, 2, \ldots, N$. The differential equation (3-53a) is called the *Helmholtz equation* and is a partial differential equation, in general, in the three spatial variables. The solution of this PDE is essential for the solution of the above heat conduction problem. The Helmholtz equation can be solved by the method of separation of variables, as outlined in the previous section, provided that its separation into a set of ordinary differential equations is possible. The separability of the Helmholtz equation has been studied and it has been shown that a simple separation of the Helmholtz equation (also of the *Laplace equation*) into ordinary differential equations is possible in 11 orthogonal coordinate systems. We list in Table 3-1 these 11 orthogonal coordinate systems and also indicate the type of functions that may appear as solutions of the separated functions [1, 3, 9]. A discussion of the separation of the Helmholtz equation will be presented in this chapter for the rectangular coordinate system and in the following two chapters for the cylindrical and spherical coordinate systems. The reader should consult references 9 and 10 for the definition of various functions listed in Table 3-1.

3-3 SOLUTION OF MULTIDIMENSIONAL HOMOGENOUS PROBLEMS

Having established the basic solution methodology for the separation of variables technique, we are now in a position to apply this approach to the solution of multidimensional homogenous heat conduction problems under steady-state and transient conditions in the Cartesian coordinate system.

Steady-State Problems

Example 3-1 Three-Dimensional Geometry

A semi-infinite, 3-D rectangular solid of dimensions L × W is maintained with a constant temperature on the left end, $T(z = 0) = T_0$, has an insulated base ($y = 0$), while all other surfaces are maintained at zero temperature, as illustrated in Figure 3-3.

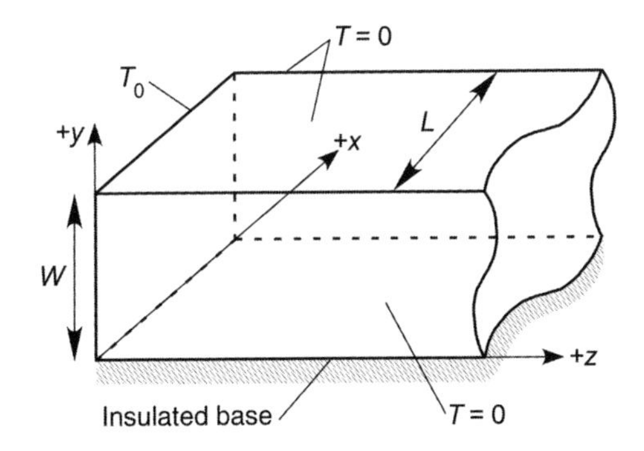

Figure 3-3 Problem description for Example 3-1.

The mathematical formulation of the problem is given as

$$\frac{\partial^2 T}{\partial x^2} + \frac{\partial^2 T}{\partial y^2} + \frac{\partial^2 T}{\partial z^2} = 0 \quad \text{in} \quad 0 < x < L \quad 0 < y < W \quad 0 < z < \infty \tag{3-54}$$

$$\text{BC1:} \quad T(x = 0) = 0 \qquad \text{BC2:} \quad T(x = L) = 0 \tag{3-55a}$$

$$\text{BC3:} \quad \left.\frac{\partial T}{\partial y}\right|_{y=0} = 0 \qquad \text{BC4:} \quad T(y = W) = 0 \tag{3-55b}$$

$$\text{BC5:} \quad T(z = 0) = T_0 \qquad \text{BC6:} \quad T(z \to \infty) = 0 \tag{3-55c}$$

The problem formulation contains five homogenous boundary conditions and a single nonhomogeneity at BC5; hence separation of variables is appropriate. We assume separation of the form

$$T(x, y, z) = X(x)Y(y)Z(z) \tag{3-56}$$

Substituting equation (3-56) into the PDE of equation (3-54) yields

$$\frac{1}{X}\frac{d^2X}{dx^2} + \frac{1}{Y}\frac{d^2Y}{dy^2} + \frac{1}{Z}\frac{d^2Z}{dz^2} = 0 \tag{3-57}$$

We have two homogenous dimensions for this problem, so we elect to separate the homogeneous x dimension first:

$$\frac{1}{Y}\frac{d^2Y}{dy^2} + \frac{1}{Z}\frac{d^2Z}{dz^2} = -\frac{1}{X}\frac{d^2X}{dx^2} = \lambda^2 \tag{3-58}$$

where λ^2 is the first separation constant, selected to yield a boundary value problem in the x dimension, as formulated here (See Table 2-1, case 9):

$$\frac{d^2X}{dx^2} + \lambda^2 X = 0 \tag{3-59a}$$

$$\text{BC1:} \quad X(x=0) = 0 \qquad \text{BC2:} \quad X(x=L) = 0 \tag{3-59b}$$

Solution of equation (3-59a) yields

$$X(x) = C_1 \cos\lambda x + C_2 \sin\lambda x \tag{3-60}$$

where BC1 then eliminates one constant ($C_1 = 0$), while BC2 yields

$$\sin\lambda L = 0 \quad \rightarrow \quad \lambda_n = \frac{n\pi}{L} \quad \text{for } n = 0, 1, 2, \ldots \tag{3-61}$$

to define the eigenvalues from the x dimension, noting that $\lambda_0 = 0$ is a trivial eigenvalue, and the orthogonal eigenfunctions

$$X_n(x) = C_2 \sin\lambda_n x \tag{3-62}$$

We now separate the remaining homogeneous y dimension from equation (3-58):

$$\frac{1}{Z}\frac{d^2Z}{dz^2} - \lambda^2 = -\frac{1}{Y}\frac{d^2Y}{dy^2} = \beta^2 \tag{3-63}$$

where β^2 is the second separation constant, selected to yield a boundary value problem in the y dimension, as formulated here (See Table 2-1, case 6):

$$\frac{d^2Y}{dy^2} + \beta^2 Y = 0 \tag{3-64a}$$

$$\text{BC3:} \quad \left.\frac{dY}{dy}\right|_{y=0} = 0 \qquad \text{BC4:} \quad Y(y=W) = 0 \tag{3-64b}$$

Solution of equation (3-64) follows the $X(x)$ solution given by equation (3-60), with BC3 eliminating the $\sin\beta y$ term, and with BC4 then yielding

$$\cos\beta W = 0 \quad \rightarrow \quad \beta_m = \frac{(2m+1)\pi}{2W} \quad \text{for } m = 0, 1, 2, \ldots \tag{3-65}$$

to define the eigenvalues from the y dimension, and the orthogonal eigenfunctions

$$Y_m(y) = C_3 \cos \beta_m y \tag{3-66}$$

noting that we have specifically used a different index (m) to avoid confusion with the x-dimension index (n).

Finally, we consider the nonhomogenous z dimension as given by

$$\frac{d^2 Z}{dz^2} - (\lambda_n^2 + \beta_m^2) Z = 0 \tag{3-67a}$$

$$\text{BC6:} \qquad Z(z \to \infty) = 0 \tag{3-67b}$$

where we can only separate the homogenous boundary condition. Because the eigenvalues λ_n^2 and β_m^2 are defined, it is convenient to introduce the parameter

$$\mu_{nm}^2 = \lambda_n^2 + \beta_m^2 \tag{3-68}$$

With this substitution, solution of equation (3-67a) yields

$$Z(z) = C_5 e^{\mu_{nm} z} + C_6 e^{-\mu_{nm} z} \tag{3-69}$$

where we have selected the exponential form for the semi-infinite domain. Satisfaction of BC6 requires that $C_5 = 0$. With all homogeneous boundary conditions satisfied, we combine the separated functions:

$$T_{nm}(x, y, z) = C_{nm} \sin \lambda_n x \cos \beta_m y \, e^{-\mu_{nm} z} \tag{3-70}$$

where we have introduced the new constant $C_{nm} = C_2 C_3 C_6$. The most general solution is now constructed by a summation of all nontrivial solutions:

$$T(x, y, z) = \sum_{n=1}^{\infty} \sum_{m=0}^{\infty} C_{nm} \sin \lambda_n x \cos \beta_m y \, e^{-\mu_{nm} z} \tag{3-71}$$

The final step is to determine the constant C_{nm} using the nonhomogeneous BC5 at $z = 0$ given by equation (3-55c), which yields

$$T_0 = \sum_{n=1}^{\infty} \sum_{m=0}^{\infty} C_{nm} \sin \lambda_n x \cos \beta_m y \tag{3-72}$$

Equation (3-72) is a double Fourier series expansion of T_0. We use the property of orthogonality of our two eigenfunctions over their respective intervals by applying successively the following operators to both sides of equation (3-72), namely,

$$* \int_{x=0}^{L} \sin \lambda_k x \, dx \qquad \text{and} \qquad * \int_{y=0}^{W} \cos \beta_p y \, dy$$

which yields the desired expression:

$$C_{nm} = \frac{\int_{x=0}^{L} \int_{y=0}^{W} T_0 \sin \lambda_n x \cos \beta_m y dy\, dx}{\int_{x=0}^{L} \sin^2 \lambda_n x\, dx \int_{y=0}^{W} \cos^2 \beta_m y dy} = \frac{T_0 4\,[1-(-1)^n]\,(-1)^m}{\lambda_n \beta_m L W} \tag{3-73}$$

Example 3-2 Two Dimensional: Temperature Shift and Use of Symmetry A semi-infinite, 2-D rectangular solid of width $2L$ is maintained with a constant base temperature $T(x=0)=T_0$, while the upper and lower surfaces dissipate heat by convection with a heat transfer coefficient h into a fluid of temperature T_∞, as illustrated in Figure 3-4.

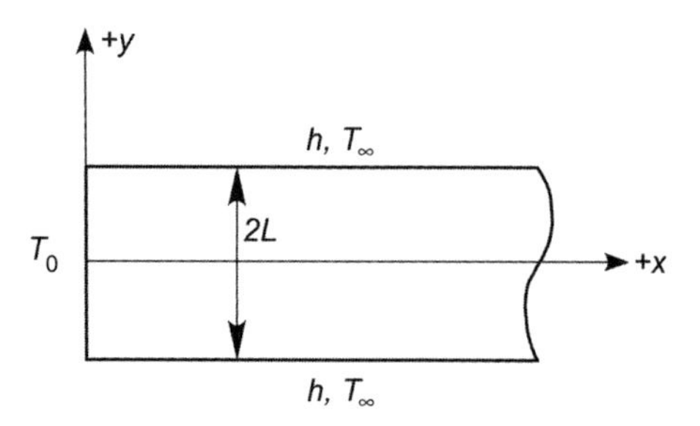

Figure 3-4 Problem description for Example 3-2.

The mathematical formulation of the problem is given as

$$\frac{\partial^2 T}{\partial x^2} + \frac{\partial^2 T}{\partial y^2} = 0 \quad \text{in} \quad 0 < x < \infty \quad 0 < y < L \tag{3-74}$$

$$\text{BC1:} \quad T(x=0) = T_0 \qquad \text{BC2:} \quad T(x \to \infty) = T_\infty \tag{3-75a}$$

$$\text{BC3:} \quad \left.\frac{\partial T}{\partial y}\right|_{y=0} = 0 \qquad \text{BC4:} \quad -k \left.\frac{\partial T}{\partial y}\right|_{y=L} = h\left[T|_{y=L} - T_\infty\right] \tag{3-75b}$$

where we have made use of *symmetry in the y direction*, which effectively removes one nonhomogeneity, and solved the problem over the reduced domain $0 \le y \le L$. The problem formulation still contains three nonhomogenous boundary conditions, which necessitates reformulation. Because we *cannot have a boundary value problem over a semi-infinite domain*, per the general Sturm–Liouville problem, we shift by the fluid temperature, letting $\Theta(x,y) = T(x,y) - T_\infty$, and giving the new formulation:

$$\frac{\partial^2 \Theta}{\partial x^2} + \frac{\partial^2 \Theta}{\partial y^2} = 0 \quad \text{in} \quad 0 < x < \infty \quad 0 < y < L \tag{3-76}$$

$$\text{BC1:} \quad \Theta(x=0) = T_0 - T_\infty = \Theta_0 \qquad \text{BC2:} \quad \Theta(x \to \infty) = 0 \quad \text{(3-77a)}$$

$$\text{BC3:} \quad \left.\frac{\partial \Theta}{\partial y}\right|_{y=0} = 0 \qquad \text{BC4:} \quad -k \left.\frac{\partial \Theta}{\partial y}\right|_{y=L} = h\,\Theta|_{y=L} \quad \text{(3-77b)}$$

With only a single nonhomogeneity, we now assume separation of the form

$$\Theta(x, y) = X(x)Y(y) \quad \text{(3-78)}$$

Substituting the above into the PDE of equation (3-76) and separating yields

$$\frac{1}{X}\frac{d^2X}{dx^2} = -\frac{1}{Y}\frac{d^2Y}{dy^2} = \lambda^2 \quad \text{(3-79)}$$

where the separation constant λ^2 produces the boundary value problem in the homogeneous y dimension, as formulated here (See Table 2-1, case 4):

$$\frac{d^2Y}{dy^2} + \lambda^2 Y = 0 \quad \text{(3-80a)}$$

$$\text{BC3:} \quad \left.\frac{dY}{dy}\right|_{y=0} = 0 \qquad \text{BC4:} \quad -k \left.\frac{dY}{dy}\right|_{y=L} = h\,Y|_{y=L} \quad \text{(3-80b)}$$

Solution of equation (3-80a) yields

$$Y(y) = C_1 \cos \lambda y + C_2 \sin \lambda y \quad \text{(3-81)}$$

where BC3 then eliminates one constant $(C_2 = 0)$, while BC4 yields the transcendental equation

$$\lambda_n \tan \lambda_n L = \frac{h}{k} \quad \to \quad \lambda_n \quad \text{for } n = 1, 2, 3, \ldots \quad \text{(3-82)}$$

where the roots λ_n are the eigenvalues, as discussed above, and the orthogonal eigenfunctions become

$$Y_n(y) = C_1 \cos \lambda_n y \quad \text{(3-83)}$$

We now consider the nonhomogenous x dimension as given by

$$\frac{d^2X}{dx^2} - \lambda_n^2 X = 0 \quad \text{(3-84a)}$$

$$\text{BC2:} \quad X(x \to \infty) = 0 \quad \text{(3-84b)}$$

where we only consider the homogenous boundary condition. Solution of equation (3-84a) yields

$$X(x) = C_3 e^{\lambda_n x} + C_4 e^{-\lambda_n x} \quad \text{(3-85)}$$

having selected the exponential form for the semi-infinite domain. Satisfaction of BC2 requires that $C_3 = 0$. With all homogeneous boundary conditions satisfied, we combine and sum the separated functions to yield

$$\Theta(x, y) = \sum_{n=1}^{\infty} C_n \cos \lambda_n y \, e^{-\lambda_n x} \tag{3-86}$$

where we have introduced the new constant $C_n = C_1 C_4$. We finally apply the nonhomogeneous BC1, equation (3-77a), to determine C_n, which yields

$$\Theta_0 = \sum_{n=1}^{\infty} C_n \cos \lambda_n y \tag{3-87}$$

Equation (3-87) is a Fourier series expansion of Θ_0. We use the property of orthogonality by applying the operator to both sides of equation (3-87):

$$* \int_{y=0}^{L} \cos \lambda_m y \, dy$$

which yields the desired expression

$$C_n = \frac{\int_{y=0}^{L} \Theta_0 \cos \lambda_n y \, dy}{\int_{y=0}^{L} \cos^2 \lambda_n y \, dy} \tag{3-88}$$

The denominator of equation (3-88) is recognized as the norm $N(\lambda_n)$ and is readily evaluated for the eigenvalues of equation (3-82) per Table 2-1. Now $T(x, y)$ is defined to complete the problem:

$$T(x, y) = \sum_{n=1}^{\infty} C_n \cos \lambda_n y \, e^{-\lambda_n x} + T_\infty \tag{3-89}$$

Example 3-3 Two Dimensional: Convective Nonhomogeneous BC
A 2-D rectangular solid of dimensions $L \times W$ is maintained at a constant temperature of zero on the left, right, and lower surfaces, while the upper surface dissipates heat by convection with heat transfer coefficient h into a fluid of temperature T_∞, as illustrated in Figure 3-5.

The mathematical formulation of the problem is given as

$$\frac{\partial^2 T}{\partial x^2} + \frac{\partial^2 T}{\partial y^2} = 0 \qquad \text{in} \qquad 0 < x < L \quad 0 < y < W \tag{3-90}$$

$$\text{BC1:} \quad T(x=0)=0 \qquad \text{BC2:} \quad T(x=L)=0 \tag{3-91a}$$

$$\text{BC3:} \quad T(y=0)=0 \qquad \text{BC4:} \quad -k\left.\frac{\partial T}{\partial y}\right|_{y=W} = h\left[T|_{y=W} - T_\infty\right] \tag{3-91b}$$

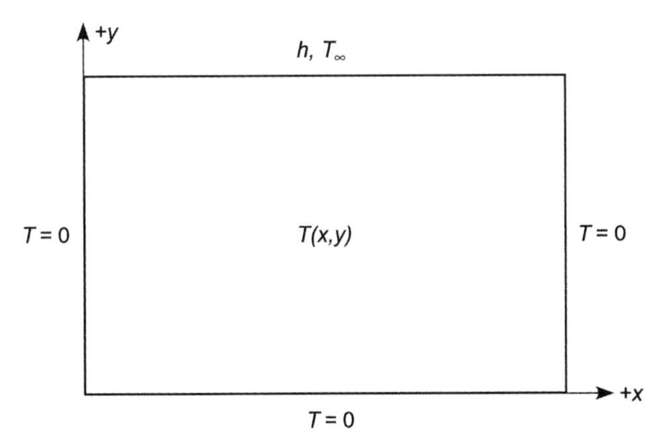

Figure 3-5 Problem description for Example 3-3.

With a single nonhomogeneity, we are ready to assume separation of the form

$$T(x,y) = X(x)Y(y) \tag{3-92}$$

Substituting the above into the PDE of equation (3-90) and separating yields

$$\frac{1}{Y}\frac{d^2Y}{dy^2} = -\frac{1}{X}\frac{d^2X}{dx^2} = \lambda^2 \tag{3-93}$$

where the separation constant λ^2 produces the boundary value problem in the homogeneous x dimension, as formulated here (See Table 2-1, case 9):

$$\frac{d^2X}{dx^2} + \lambda^2 X = 0 \tag{3-94a}$$

$$\text{BC1:} \quad X(x=0)=0 \qquad \text{BC2:} \quad X(x=L)=0 \tag{3-94b}$$

Solution of equation (3-94a) yields

$$X(x) = C_1 \cos\lambda x + C_2 \sin\lambda x \tag{3-95}$$

where BC1 then eliminates one constant ($C_1 = 0$), while BC2 yields the eigenvalues

$$\lambda L = n\pi \quad \rightarrow \quad \lambda_n = \frac{n\pi}{L} \quad \text{for } n = 0, 1, 2, \ldots \tag{3-96}$$

noting $\lambda_0 = 0$ is a trivial eigenvalue, and the orthogonal eigenfunctions become

$$X_n(x) = C_2 \sin \lambda_n x \tag{3-97}$$

We now consider the nonhomogenous y dimension as given by

$$\frac{d^2Y}{dy^2} - \lambda_n^2 Y = 0 \tag{3-98a}$$

$$\text{BC3:} \qquad Y(y = 0) = 0 \tag{3-98b}$$

where we only consider the homogenous boundary condition. Solution of equation (3-98a) yields

$$Y(y) = C_3 \cosh \lambda y + C_4 \sinh \lambda y \tag{3-99}$$

having selected the hyperbolic form for the finite domain. Satisfaction of BC3 requires that $C_3 = 0$. With all homogeneous boundary conditions satisfied, we combine and sum the separated functions, omitting the trivial case, to yield

$$T(x, y) = \sum_{n=1}^{\infty} C_n \sin \lambda_n x \sinh \lambda_n y \tag{3-100}$$

where we have introduced the new constant $C_n = C_2 C_4$. We lastly apply the nonhomogeneous BC4, equation (3-91b), to determine C_n, which yields

$$-k\left[\sum_{n=1}^{\infty} C_n \sin \lambda_n x \, \lambda_n \cosh \lambda_n W\right] = h\left[\sum_{n=1}^{\infty} C_n \sin \lambda_n x \sinh \lambda_n W\right] - hT_\infty \tag{3-101}$$

The two summations in equation (3-101) may be combined to a single series as

$$hT_\infty = \sum_{n=1}^{\infty} C_n \sin \lambda_n x \left\{k\lambda_n \cosh \lambda_n W + h \sinh \lambda_n W\right\} \tag{3-102}$$

which is now simply a Fourier series expansion of hT_∞. We use the property of orthogonality by applying the operator to both sides of equation (3-102):

$$* \int_{x=0}^{L} \sin \lambda_m x \, dx$$

which yields the desired expression for C_n to complete the problem:

$$C_n = \frac{hT_\infty \int_{x=0}^{L} \sin \lambda_n x \, dx}{\left\{k\lambda_n \cosh \lambda_n W + h \sinh \lambda_n W\right\} \int_{x=0}^{L} \sin^2 \lambda_n x \, dx} \tag{3-103}$$

The integral in the denominator of equation (3-103) is recognized as the norm $N(\lambda_n)$ and is equal to $L/2$ for the eigenvalues of equation (3-96).

Transient Problems

Example 3-4 Three-Dimensional Geometry
A 3-D, rectangular solid of dimensions $L \times W \times H$ is initially characterized by a temperature distribution $T(t = 0) = F(x, y, z)$. For $t>0$, four of the boundary surfaces are maintained at zero temperature, while the back surface is insulated and the top surface dissipates heat by convection with heat transfer coefficient h into a zero temperature fluid ($T_\infty = 0$), as illustrated in Figure 3-6.

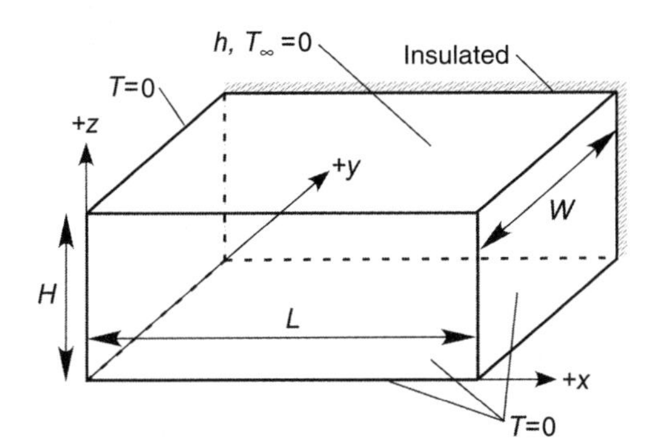

Figure 3-6 Problem description for Example 3-4.

The mathematical formulation of the problem is given as

$$\frac{\partial^2 T}{\partial x^2} + \frac{\partial^2 T}{\partial y^2} + \frac{\partial^2 T}{\partial z^2} = \frac{1}{\alpha}\frac{\partial T}{\partial t} \quad \text{in} \quad 0 < x < L \quad 0 < y < W \quad 0 < z < H \tag{3-104}$$

$$\text{BC1:} \quad T(x = 0) = 0 \qquad \text{BC2:} \quad T(x = L) = 0 \tag{3-105a}$$

$$\text{BC3:} \quad T(y = 0) = 0 \qquad \text{BC4:} \quad \left.\frac{\partial T}{\partial y}\right|_{y=W} = 0 \tag{3-105b}$$

$$\text{BC5:} \quad T(z = 0) = 0 \qquad \text{BC6:} \quad -k\left.\frac{\partial T}{\partial z}\right|_{z=H} = h\, T|_{z=H} \tag{3-105c}$$

$$\text{IC:} \quad T(x, y, z, t = 0) = F(x, y, z) \tag{3-105d}$$

The problem formulation contains all homogenous boundary conditions and a single nonhomogeneous initial condition; hence separation of variables is appropriate. We note that symmetry in the x direction is *not a valid assumption* given the *unspecified* initial condition. We assume separation of the form

$$T(x, y, z, t) = X(x)Y(y)Z(z)\Gamma(t) \tag{3-106}$$

Substituting equation (3-106) into the PDE of equation (3-104) yields

$$\frac{1}{X}\frac{d^2X}{dx^2} + \frac{1}{Y}\frac{d^2Y}{dy^2} + \frac{1}{Z}\frac{d^2Z}{dz^2} = \frac{1}{\alpha\Gamma}\frac{d\Gamma}{dt} = -\lambda^2 \tag{3-107}$$

We consider first the time-dependent ODE:

$$\frac{d\Gamma}{dt} + \alpha\lambda^2\Gamma = 0 \tag{3-108}$$

which yields the expected decaying exponential solution

$$\Gamma(t) = C_1 e^{-\alpha\lambda^2 t} \tag{3-109}$$

where λ will be determined by the homogeneous spatial dimensions. We now consider the spatial variables of equation (3-107), rearranging as

$$\frac{1}{Y}\frac{d^2Y}{dy^2} + \frac{1}{Z}\frac{d^2Z}{dz^2} + \lambda^2 = -\frac{1}{X}\frac{d^2X}{dx^2} = \beta^2 \tag{3-110}$$

Separating the x dimension as our first boundary value problem yields

$$X(x) = C_2 \cos\beta x + C_3 \sin\beta x \tag{3-111}$$

where separated BC1 then eliminates one constant ($C_2 = 0$), while separated BC2 yields the eigenvalues

$$\beta L = n\pi \quad \rightarrow \quad \beta_n = \frac{n\pi}{L} \quad \text{for } n = 0, 1, 2, \ldots \tag{3-112}$$

noting $\beta_0 = 0$ is a trivial eigenvalue. The orthogonal eigenfunctions become

$$X_n(x) = C_3 \sin\beta_n x \tag{3-113}$$

We now rearrange equation (3-110) to separate the next homogeneous dimension:

$$\frac{1}{Z}\frac{d^2Z}{dz^2} + \lambda^2 - \beta^2 = -\frac{1}{Y}\frac{d^2Y}{dy^2} = \eta^2 \tag{3-114}$$

Having forced the y dimension as the next boundary value problem yields

$$Y(y) = C_4 \cos \eta y + C_5 \sin \eta y \tag{3-115}$$

where separated BC3 then eliminates one constant ($C_4 = 0$), while separated BC4 yields the eigenvalues

$$\eta_m = \frac{(2m+1)\pi}{2W} \quad \text{for} \quad m = 0, 1, 2, \ldots \tag{3-116}$$

and the orthogonal eigenfunctions become

$$Y_m(y) = C_5 \sin \eta_m y \tag{3-117}$$

We now consider the last homogeneous dimension from equation (3-114):

$$\frac{1}{Z}\frac{d^2 Z}{dz^2} + \lambda^2 - \beta^2 - \eta^2 = 0 \tag{3-118}$$

Letting $\mu^2 = \lambda^2 - \beta^2 - \eta^2$ yields the final boundary value problem, giving

$$Z(z) = C_6 \cos \mu z + C_7 \sin \mu z \tag{3-119}$$

Applying BC5 eliminates one constant ($C_6 = 0$), and BC6 yields the following transcendental equation:

$$\mu_p \tan \mu_p H = \frac{h}{k} \quad \rightarrow \quad \mu_p \quad \text{for} \quad p = 1, 2, 3, \ldots \tag{3-120}$$

where the roots μ_p are the eigenvalues. The desired, orthogonal eigenfunctions become

$$Z_p(z) = C_7 \sin \mu_p z \tag{3-121}$$

With all homogeneous boundary conditions satisfied, we combine the separated functions:

$$T_{nmp}(x, y, z, t) = C_{nmp} \sin \beta_n x \sin \eta_m y \sin \mu_p z \, e^{-\alpha \lambda_{nmp}^2 t} \tag{3-122}$$

where we have introduced the new constant $C_{nmp} = C_1 C_3 C_5 C_7$, and where we now define λ in terms of the three sets of eigenvalues:

$$\lambda_{nmp}^2 = \beta_n^2 + \eta_m^2 + \mu_p^2 \tag{3-123}$$

The most general solution is now constructed by a summation of all solutions

$$T(x, y, z, t) = \sum_{n=1}^{\infty} \sum_{m=0}^{\infty} \sum_{p=1}^{\infty} C_{nmp} \sin \beta_n x \sin \eta_m y \sin \mu_p z \, e^{-\alpha \lambda_{nmp}^2 t} \tag{3-124}$$

The final step is to determine the constant C_{nmp} using the nonhomogeneous initial condition of equation (3-105d). This process yields

$$F(x, y, z) = \sum_{n=1}^{\infty}\sum_{m=0}^{\infty}\sum_{p=1}^{\infty} C_{nmp} \sin \beta_n x \sin \eta_m y \sin \mu_p z \tag{3-125}$$

which is a triple Fourier series expansion of $F(x,y,z)$. We use the property of orthogonality of our three eigenfunctions over their respective intervals by applying successively the following operators to both sides of equation (3-125):

$$*\int_{x=0}^{L} \sin \beta_i x \, dx \qquad *\int_{y=0}^{W} \sin \eta_j y \, dy \qquad \text{and} \qquad *\int_{z=0}^{H} \sin \mu_k z \, dz$$

which yields the desired expression

$$C_{nmp} = \frac{\int_{x=0}^{L}\int_{y=0}^{W}\int_{z=0}^{H} F(x, y, z) \sin \beta_n x \sin \eta_m y \sin \mu_p z \, dz \, dy \, dx}{\int_{x=0}^{L} \sin^2 \beta_n x \, dx \int_{y=0}^{W} \sin^2 \eta_m y \, dy \int_{z=0}^{H} \sin^2 \mu_p z \, dz} \tag{3-126}$$

The three integrals in the denominator correspond to the norms of the respective eigenfunctions and are defined in Table 2-1.

3-4 MULTIDIMENSIONAL NONHOMOGENEOUS PROBLEMS: METHOD OF SUPERPOSITION

In the above section, we have considered both steady-state and transient homogeneous problems with the separation of variables technique. We now introduce the *method of superposition* as a means to treat nonhomogeneous problems, meaning problems with more than the single nonhomogeneity as required for a solution with separation of variables. We present the method of superposition by beginning with the most general problem, which then is divided into problems of decreasing complexity, which are either ready for solution by separation of variables or further divided into problems more suitable for direct solution.

Consider a general nonhomogeneous problem in which the generation term and the nonhomogeneous boundary conditions are steady with time:

$$\nabla^2 T(\hat{r}, t) + \frac{g(\hat{r})}{k} = \frac{1}{\alpha}\frac{\partial T(\hat{r}, t)}{\partial t} \qquad \text{in domain } R \qquad t > 0 \tag{3-127}$$

$$k_i \frac{\partial T}{\partial n_i} + h_i T = f_i(\hat{r}) \qquad \text{on boundary } S_i \qquad i = 1 - N \qquad t > 0 \tag{3-128a}$$

$$T(\hat{r}, t = 0) = F(\hat{r}) \qquad \text{in domain } R \tag{3-128b}$$

where $\partial/\partial n_i$ denotes differentiation along the outward-drawn normal to the boundary surface S_i, and $\hat{r}$ denotes the general space coordinates. It is assumed that the domain R has a number of continuous boundary surfaces $S_i, i = 1, 2, \ldots, N$ in number, such that each boundary surface S_i fits the coordinate surface of the chosen orthogonal coordinate system. In general, the nonhomogeneous energy generation term $g(\hat{r})$ may depend on the spatial variables. The nonhomogeneity at each boundary condition is represented by $f_i(\hat{r})$, which readily describes a prescribed temperature, prescribed flux, or prescribed convective fluid temperature by the adjustment of constants k_i and h_i. Here we note that $g(\hat{r})$ and $f_i(\hat{r})$ *may not depend on time*. Time-dependent nonhomogeneities may not be treated with separation of variables and are handled in Chapters 7–9. This general problem is now split into two simpler problems, one steady state and one transient, with the solution achieved by superposition of the two simpler problem solutions. This approach is shown schematically in Fig. 3-7 for a representative 2-D problem and summarized as follows:

$$T(\hat{r}, t) = T_H(\hat{r}, t) + T_{SS}(\hat{r}) \qquad \text{in domain } R \tag{3-129}$$

where $T_H(\hat{r}, t)$ is the *homogeneous problem* that takes the homogenous form of all the boundary conditions, as well as the homogeneous PDE (i.e., no generation), and where $T_{SS}(\hat{r})$ is the *steady-state problem* that takes all the nonhomogeneous boundary conditions and the nonhomogeneous generation term in the PDE.

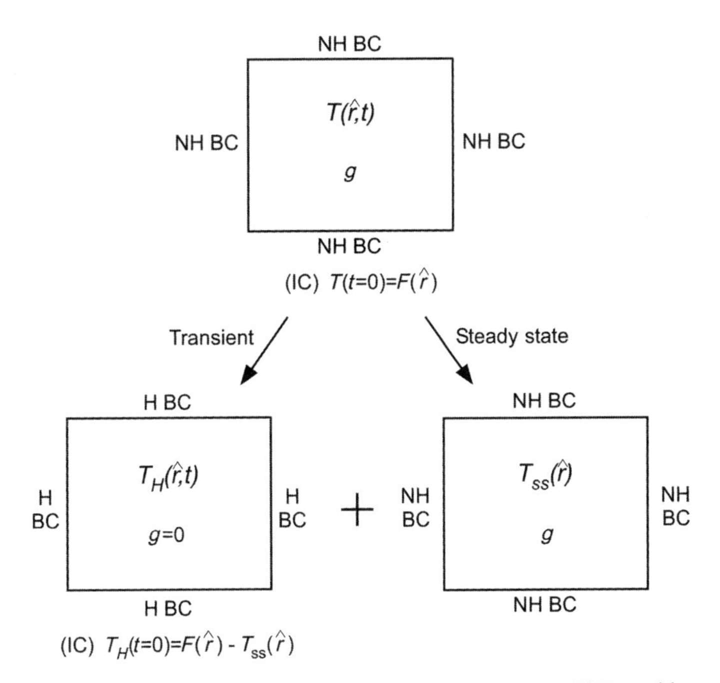

Figure 3-7 Superposition for general nonhomogeneous (NH) problem.

The formulations of these two reduced problems are as follows. Starting with the homogeneous, transient problem, we have

$$\nabla^2 T_H(\hat{r}, t) = \frac{1}{\alpha}\frac{\partial T_H(\hat{r}, t)}{\partial t} \qquad \text{in domain } R \qquad t > 0 \tag{3-130}$$

$$k_i \frac{\partial T_H}{\partial n_i} + h_i T_H = 0 \qquad \text{on boundary } S_i \qquad i = 1 - N \qquad t > 0 \tag{3-131a}$$

$$T_H(\hat{r}, t = 0) = F(\hat{r}) - T_{SS}(\hat{r}) \qquad \text{in domain } R \tag{3-131b}$$

and for the steady-state problem, we have

$$\nabla^2 T_{SS}(\hat{r}) + \frac{g}{k} = 0 \qquad \text{in domain } R \tag{3-132}$$

$$k_i \frac{\partial T_{SS}}{\partial n_i} + h_i T_{SS} = f_i(\hat{r}) \qquad \text{on boundary } S_i \qquad i = 1 - N \tag{3-133}$$

The homogeneous problem given by equations (3-130) and (3-131) is the homogeneous version of the original problem (3-127), except the *initial condition is modified* by subtracting from it the solution of the steady-state problem as defined by equations (3-132) and (3-133). This *coupling of the steady-state and the transient solutions* is necessary for superposition involving the time-dependent heat equation.

The transient problem is now readily solved with separation of variables given all the homogenous boundary conditions and homogeneous PDE. However, the steady-state problem does not yet meet the requirements for direct separation of variables; hence we must further simplify this problem. We present now the approach for handling a *nonhomogeneous, steady-state problem* as given by $T_{SS}(\hat{r})$, which is split into a homogeneous PDE with nonhomogeneous boundary conditions, and a nonhomogenous PDE with homogeneous boundary conditions. This approach is shown schematically in Figure 3-8 for our continuing representative 2-D problem and summarized as follows:

$$T_{SS}(\hat{r}) = T_A(\hat{r}) + T_B(\hat{r}) \qquad \text{in domain } R \tag{3-134}$$

where $T_A(\hat{r})$ is defined by the *homogeneous* PDE that takes the homogenous form of the differential equation (i.e., no generation), but with all of the nonhomogeneous boundary conditions, and where $T_B(\hat{r})$ is defined by the *nonhomogeneous* PDE that takes the homogeneous form of all boundary conditions.

The formulations of these two reduced problems are given as

$$\nabla^2 T_A(\hat{r}) = 0 \qquad \text{in domain } R \tag{3-135}$$

$$k_i \frac{\partial T_A}{\partial n_i} + h_i T_A = f_i(\hat{r}) \qquad \text{on boundary } S_i \qquad i = 1 - N \tag{3-136}$$

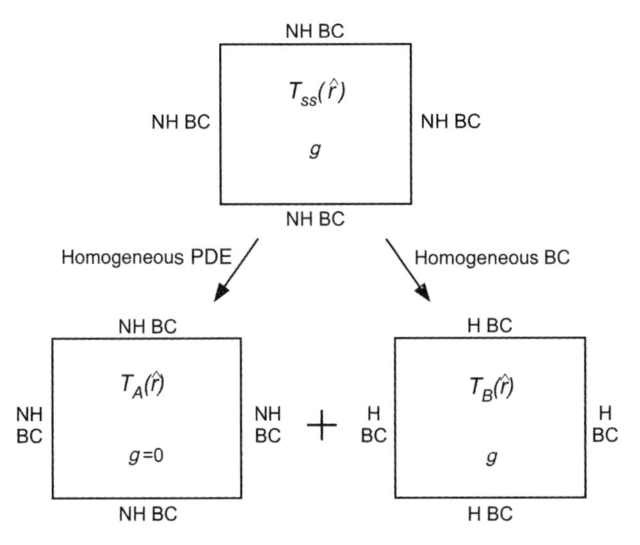

Figure 3-8 Superposition for general steady-state problem.

and

$$\nabla^2 T_B(\hat{r}) + \frac{g}{k} = 0 \qquad \text{in domain } R \tag{3-137}$$

$$k_i \frac{\partial T_B}{\partial n_i} + h_i T_B = 0 \qquad \text{on boundary } S_i \qquad i = 1 - N \tag{3-138}$$

However, it is readily observed that neither of these problems meets the requirements for direct separation of variables; hence we must further simplify each problem. We will first present the approach for handling the homogeneous PDE with multiple nonhomogeneous boundary conditions, as given by $T_A(\hat{r})$, and will then present the approach for handling the nonhomogeneous PDE with homogeneous boundary conditions, as given by $T_B(\hat{r})$.

Our first approach is shown schematically in Figure 3-9 for equations (3-135) and (3-136) and is summarized as follows:

$$T_A(\hat{r}) = \sum_{i=1}^{N} T_i(\hat{r}) \qquad \text{in domain } R \qquad \text{for} \qquad i = 1 - N \tag{3-139}$$

where N is the number of nonhomogeneous boundary conditions, and $T_i(\hat{r})$ is the solution of the identical homogeneous PDE as equation (3-135), but with only a *single nonhomogeneous boundary condition*. All other boundary conditions take their homogeneous form [i.e., $f_i(\hat{r}) = 0$]. It is now apparent that each problem $T_i(\hat{r})$ is readily solved using separation of variables, as the conditions for a steady-state problem are satisfied given the homogeneous PDE and only a single nonhomogeneous boundary condition.

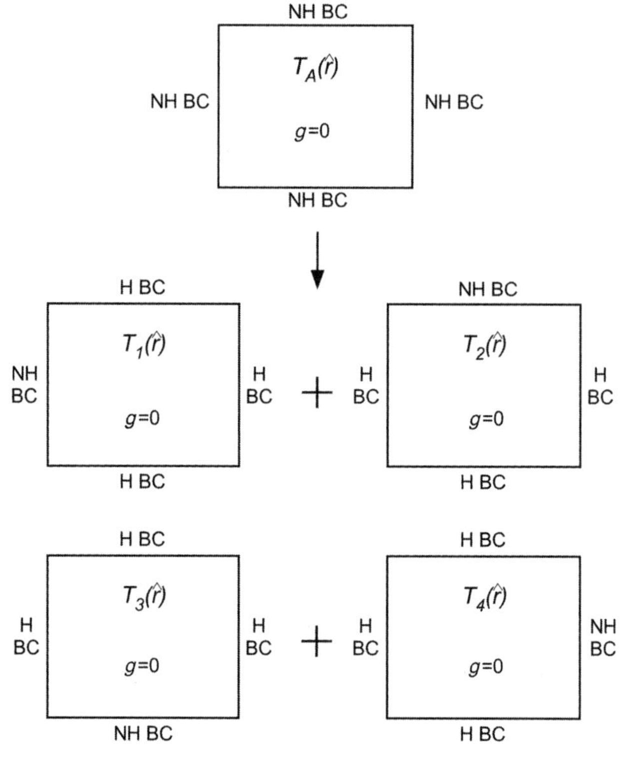

Figure 3-9 Superposition for nonhomogeneous BCs.

Before we formulate the nonhomogeneous PDE for solution, it is useful to clarify what we mean by the *homogeneous form* of the nonhomogeneous boundary condition. Superposition requires that the nature of the boundary condition be preserved, or with regard to the heat transfer problem that the *physics of the boundary condition be preserved*. For example, a nonhomogeneous prescribed temperature must be transformed to a homogeneous prescribed temperature ($T = 0$), and a nonhomogeneous convective boundary condition must be transformed to a homogeneous convective boundary condition ($T_\infty = 0$). In other words, the physics of a convective boundary condition requires that the surface temperature must remain coupled to the surface heat flux. One may also make the same argument from a purely mathematical perspective. The corresponding homogeneous forms of our three nonhomogeneous boundary conditions are summarized here:

$$T = f_i(\hat{r}) \quad \rightarrow \quad T = 0 \quad \text{on boundary } S_i \tag{3-140a}$$

$$k\frac{\partial T}{\partial n_i} = f_i(\hat{r}) \quad \rightarrow \quad \frac{\partial T}{\partial n_i} = 0 \quad \text{on boundary } S_i \tag{3-140b}$$

$$k\frac{\partial T}{\partial n_i} + hT = f_i(\hat{r}) \quad \rightarrow \quad k\frac{\partial T}{\partial n_i} + hT = 0 \quad \text{on boundary } S_i \tag{3-140c}$$

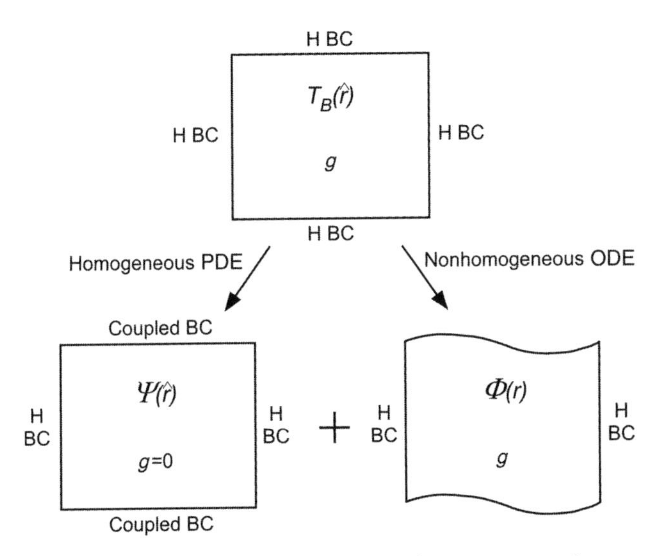

Figure 3-10 Superposition for nonhomogeneous pde.

Finally, we consider the nonhomogenous PDE of equation (3-137), which contains all homogeneous boundary conditions. Our approach is to divide this problem into a homogeneous PDE and a nonhomogeneous ODE, as shown schematically in Figure 3-10, and summarized as follows:

$$T_B(\hat{r}) = \Psi(\hat{r}) + \Phi(r) \qquad \text{in domain } R \tag{3-141}$$

where $\Psi(\hat{r})$ is the solution of the identical *homogeneous* PDE (i.e., $g = 0$) as equation (3-137), and where $\Phi(r)$ is the solution of a 1-D nonhomogeneous ODE of a single spatial dimension, containing the energy generation term, and with all homogeneous boundary conditions. Importantly, $\Psi(\hat{r})$ will have the identical homogeneous boundary conditions as equation (3-138) in the same spatial dimension as $\Phi(r)$ but will now have *one or more nonhomogeneous boundary conditions coupled* to $\Phi(r)$ in the remaining spatial dimensions.

The solution of the homogeneous PDE is now readily treated with separation of variables for the case of a single, nonhomogeneous coupled boundary condition. If the 1-D solution couples into the homogeneous PDE solution at more than one boundary, creating more than one nonhomogeneity, then simple superposition as given by Figure 3-9 is used to solve the homogeneous PDE problem. The nonhomogeneous ODE is then readily solved using traditional ODE solution techniques, namely, as

$$\Phi(r) = \Phi_H(r) + \Phi_P(r) \qquad \text{in domain } R \tag{3-142}$$

where we have used the sum of the homogeneous and particular ODE solutions. We leave it to Example 3-7 to provide further details about the selection of the spatial dimension for the 1-D ODE problem. While any spatial dimension will generally work, careful selection can minimize the coupling into the homogeneous PDE as nonhomogeneous boundary conditions, thereby simplifying the solution.

Before we present a series of examples to illustrate the concepts outlined above, we end our general discussion of the solution of the heat equation by separation of variables and the method of superposition by noting that no attempt has been made to prove the uniqueness of a solution or to establish that the general equations indeed have a solution. Instead we take the approach of Carslaw and Jaeger, who nicely summarized such thoughts in their classic treatise on heat transfer [5]: "The physical interpretation of the equations requires that there be a solution: The mathematical demonstration of such existence theorems belongs to pure analysis."

Example 3-5 Superposition; Steady State

A 2-D rectangular solid of dimensions $L \times W$ is maintained at steady-state conditions with all four boundary conditions being nonhomogeneous, as illustrated in Figure 3-11.

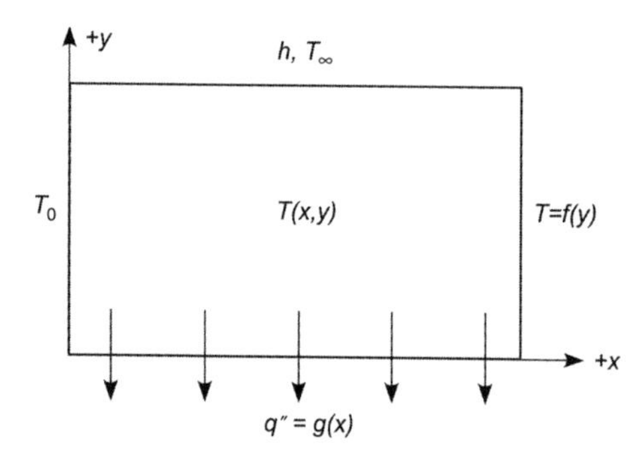

Figure 3-11 Problem description for Example 3-5.

The mathematical formulation of the problem is given as

$$\frac{\partial^2 T}{\partial x^2} + \frac{\partial^2 T}{\partial y^2} = 0 \quad \text{in} \quad 0 < x < L \quad 0 < y < W \tag{3-143}$$

$$\text{BC1:} \quad T(x=0) = T_0 \qquad \text{BC2:} \quad T(x=L) = f(y) \tag{3-144a}$$

$$\text{BC3:} \quad -k\left.\frac{\partial T}{\partial y}\right|_{y=0} = -g(x) \qquad \text{BC4:} \quad -k\left.\frac{\partial T}{\partial y}\right|_{y=W} = h\left[T|_{y=W} - T_\infty\right] \tag{3-144b}$$

Clearly the problem requires superposition given the four nonhomogeneous boundary conditions. The nonhomogeneity at $x = 0$ is readily removed by shifting the temperature by T_0, or the nonhomogeneity at $y = W$ is readily removed by shifting the temperature by T_∞. Either of these temperature shifts is recommended to simplify the problem prior to superposition, thereby eliminating one nonhomogeneity; however, for illustrative purposes, we formulate the original problem as a superposition of four problems, each with the required single nonhomogeneous boundary condition. Using equation (3-139), our superposition solution takes the form

$$T(x, y) = T_1(x, y) + T_2(x, y) + T_3(x, y) + T_4(x, y) \tag{3-145}$$

where we now formulate each of the four reduced solutions as

$$\frac{\partial^2 T_1}{\partial x^2} + \frac{\partial^2 T_1}{\partial y^2} = 0 \quad \text{in} \quad 0 < x < L \quad 0 < y < W \tag{3-146}$$

$$\text{BC1:} \quad T_1(x = 0) = T_0 \qquad \text{BC2:} \quad T_1(x = L) = 0 \tag{3-147a}$$

$$\text{BC3:} \quad \left.\frac{\partial T_1}{\partial y}\right|_{y=0} = 0 \qquad \text{BC4:} \quad -k \left.\frac{\partial T_1}{\partial y}\right|_{y=W} = h\, T_1\big|_{y=W} \tag{3-147b}$$

and

$$\frac{\partial^2 T_2}{\partial x^2} + \frac{\partial^2 T_2}{\partial y^2} = 0 \quad \text{in} \quad 0 < x < L \quad 0 < y < W \tag{3-148}$$

$$\text{BC1:} \quad T_2(x = 0) = 0 \qquad \text{BC2:} \quad T_2(x = L) = f(y) \tag{3-149a}$$

$$\text{BC3:} \quad \left.\frac{\partial T_2}{\partial y}\right|_{y=0} = 0 \qquad \text{BC4:} \quad -k \left.\frac{\partial T_2}{\partial y}\right|_{y=W} = h\, T_2\big|_{y=W} \tag{3-149b}$$

and

$$\frac{\partial^2 T_3}{\partial x^2} + \frac{\partial^2 T_3}{\partial y^2} = 0 \quad \text{in} \quad 0 < x < L \quad 0 < y < W \tag{3-150}$$

$$\text{BC1:} \quad T_3(x = 0) = 0 \qquad \text{BC2:} \quad T_3(x = L) = 0 \tag{3-151a}$$

$$\text{BC3:} \quad -k \left.\frac{\partial T_3}{\partial y}\right|_{y=0} = -g(x) \qquad \text{BC4:} \quad -k \left.\frac{\partial T_3}{\partial y}\right|_{y=W} = h\, T_3\big|_{y=W} \tag{3-151b}$$

and

$$\frac{\partial^2 T_4}{\partial x^2} + \frac{\partial^2 T_4}{\partial y^2} = 0 \quad \text{in} \quad 0 < x < L \quad 0 < y < W \tag{3-152}$$

$$\text{BC1:} \quad T_4(x = 0) = 0 \qquad \text{BC2:} \quad T_4(x = L) = 0 \tag{3-153a}$$

$$\text{BC3:} \quad \left.\frac{\partial T_4}{\partial y}\right|_{y=0} = 0 \qquad \text{BC4:} \quad -k \left.\frac{\partial T_4}{\partial y}\right|_{y=W} = h\left[T_4\big|_{y=W} - T_\infty \right] \tag{3-153b}$$

Each of the above four problems contains only a single nonhomogeneous boundary condition and is therefore readily solved using separation of variables. We have used equations (3-140) for the superposition process, maintaining the original form of each boundary condition. It is important to note that each nonhomogeneous boundary is *included once and only once*, which ensures that the original boundary conditions of equations (3-144) are indeed satisfied with the summation of equation (3-145). For example, we see that the original BC1 is readily satisfied by the superposition of all four solutions, namely,

$$T(x = 0, y) = T_1(x = 0) + T_2(x = 0) + T_3(x = 0) + T_4(x = 0) \qquad \text{(3-154a)}$$

which gives

$$T(x = 0, y) = T_0 + 0 + 0 + 0 = T_0 \qquad \text{(3-154b)}$$

Similarly, it is readily shown that the other three boundary conditions are satisfied by superposition. We present the superposition scheme schematically in Figure 3-12.

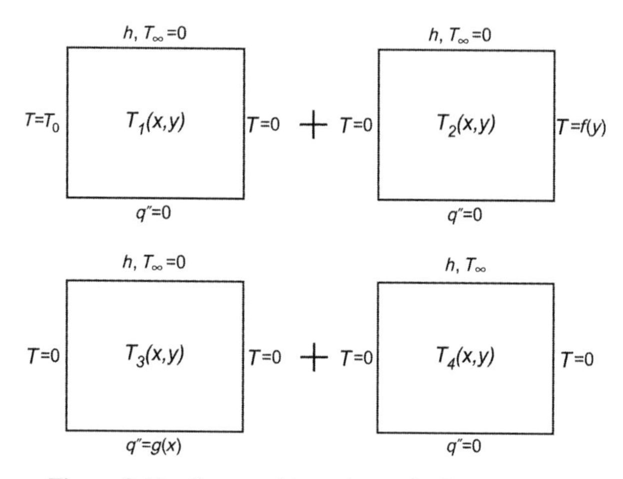

Figure 3-12 Superposition scheme for Example 3-5.

Example 3-6 Superposition; Transient

Consider a 1-D slab (i.e., plane wall) over the domain $0 \le x \le L$. Initially the slab is at a temperature $T = F(x)$, and for times $t > 0$ the boundary surface at $x = 0$ is exposed to an incident heat flux, while the boundary at $x = L$ dissipates heat by convection with a heat transfer coefficient h into a zero temperature fluid ($T_\infty = 0$), as illustrated in Figure 3-13. The mathematical formulation of this problem is given as

$$\frac{\partial^2 T(x,t)}{\partial x^2} = \frac{1}{\alpha}\frac{\partial T(x,t)}{\partial t} \quad \text{in} \quad 0 < x < L \quad t < 0 \qquad \text{(3-155)}$$

$$\text{BC1:} \quad -k\frac{\partial T}{\partial x}\bigg|_{x=0} = q_0'' \qquad \text{BC2:} \quad -k\frac{\partial T}{\partial x}\bigg|_{x=L} = h\,T|_{x=L} \tag{3-156a}$$

$$\text{IC:} \quad T(x, t=0) = F(x) \tag{3-156b}$$

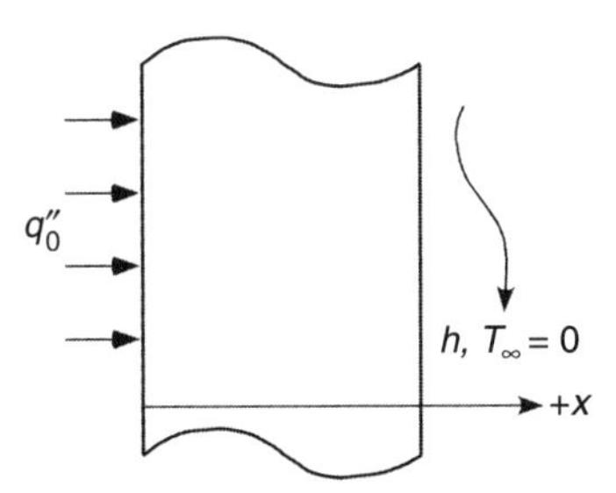

Figure 3-13 Problem description for Example 3-6.

The requirement of a homogeneous PDE is satisfied by equation (3-155), but the requirement for a transient problem of all homogeneous boundary conditions is *not satisfied*, as the boundary condition at $x = 0$ contains the nonhomogeneous prescribed heat flux. The problem is solved by a superposition as detailed in Figure 3-3, namely,

$$T(x, t) = T_H(x, t) + T_{SS}(x) \tag{3-157}$$

where $T_H(x, t)$ is the transient, homogeneous problem that takes the homogenous form of the boundary conditions, and where $T_{SS}(x)$ is the steady-state problem that takes the nonhomogeneous boundary condition. This formulation yields

$$\frac{\partial^2 T_H(x,t)}{\partial x^2} = \frac{1}{\alpha}\frac{\partial T_H(x,t)}{\partial t} \quad \text{in} \quad 0 > x > L \quad t < 0 \tag{3-158}$$

$$\text{BC1:} \quad \frac{\partial T_H}{\partial x}\bigg|_{x=0} = 0 \qquad \text{BC2:} \quad -k\frac{\partial T_H}{\partial x}\bigg|_{x=L} = h\,T_H\big|_{x=L} \tag{3-159a}$$

$$\text{IC:} \quad T_H(x, t=0) = F(x) - T_{SS}(x) = G(x) \tag{3-159b}$$

for the transient problem, noting the adjustment of the initial condition, and

$$\frac{d^2 T_{SS}(x)}{dx^2} = 0 \quad \text{in} \quad 0 > x > L \tag{3-160}$$

$$\text{BC1:} \quad -k\frac{dT_{SS}}{dx}\bigg|_{x=0} = q_0'' \qquad \text{BC2:} \quad -k\frac{dT_{SS}}{dx}\bigg|_{x=L} = h\,T_{SS}\big|_{x=L} \tag{3-161}$$

for the steady-state solution. The ODE of equation (3-160) yields

$$T_{SS}(x) = C_1 x + C_2 \tag{3-162a}$$

which after the application of BC1 and BC2 yields

$$T_{SS}(x) = \frac{q_0''}{k}(L - x) + \frac{q_0''}{h} \tag{3-162b}$$

The transient problem of equations (3-158) and (3-159) is now identical to the problem given by equations (3-31) and (3-32a); hence we have

$$T_H(x, t) = \sum_{n=1}^{\infty} C_n \cos \lambda_n x \, e^{-\alpha \lambda_n^2 t} \tag{3-163}$$

with eigenvalues given by

$$\lambda_n \tan \lambda_n L = \frac{h}{k} \quad \rightarrow \quad \lambda_n \quad \text{for} \quad n = 1, 2, 3, \ldots \tag{3-164}$$

We now apply the modified initial condition equation (3-159b), yielding

$$G(x) = \sum_{n=1}^{\infty} C_n \cos \lambda_n x \tag{3-165}$$

The following operator is applied to both sides:

$$* \int_{x=0}^{L} \cos \lambda_m x \, dx$$

which yields the Fourier coefficients:

$$C_n = \frac{\int_{x=0}^{L} G(x) \cos \lambda_n x \, dx}{\int_{x=0}^{L} \cos^2 \lambda_n x \, dx} = \frac{\int_{x=0}^{L} G(x) \cos \lambda_n x \, dx}{N(\lambda_n)} \tag{3-166}$$

We finally define $T(x, t)$, the desired solution per equation (3-157), as

$$T(x, t) = \sum_{n=1}^{\infty} C_n \cos \lambda_n x \, e^{-\alpha \lambda_n^2 t} + \frac{q_0''}{k}(L - x) + \frac{q_0''}{h} \tag{3-167}$$

Example 3-7 Superposition; Steady-State with Generation
A 2-D rectangular solid of dimensions $L \times 2W$ is maintained at steady-state conditions with all four boundary conditions at zero temperature, and with internal energy generation, as illustrated in Figure 3-14.

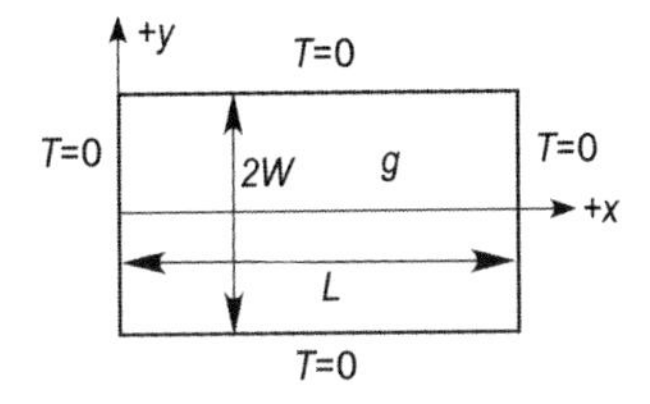

Figure 3-14 Problem description for Example 3-7.

The mathematical formulation of the problem is given as

$$\frac{\partial^2 T}{\partial x^2} + \frac{\partial^2 T}{\partial y^2} + \frac{g}{k} = 0 \quad \text{in} \quad 0 < x < L \quad 0 < y < W \tag{3-168}$$

$$\text{BC1:} \quad T(x = 0) = 0 \qquad \text{BC2:} \quad T(x = L) = 0 \tag{3-169a}$$

$$\text{BC3:} \quad \left.\frac{\partial T}{\partial y}\right|_{y=0} = 0 \qquad \text{BC4:} \quad T(y = W) = 0 \tag{3-169b}$$

where we have made use of symmetry in the y dimension and solved the problem over the half-domain. Our approach is to now divide this problem into a homogeneous PDE and a nonhomogeneous ODE, as detailed above in Figure 3-10:

$$T(x, y) = \Psi(x, y) + \Phi(x) \tag{3-170}$$

where $\Psi(x, y)$ takes the homogeneous form of the PDE of equation (3-168), and where $\Phi(x)$ corresponds to the 1-D nonhomogeneous ODE of the x dimension, containing the energy generation term. Because we have selected $\Phi(x)$ to correspond to the x dimension, we are able to simply *superimpose* the two homogeneous boundary conditions in the x dimension. In other words, using BC1 from equation (3-169a) yields

$$\text{BC1:} \quad T(x = 0) = \Psi(x = 0) + \Phi(x = 0) = 0 \tag{3-171}$$

which is readily satisfied by setting each of the respective terms equal to zero. Similar logic holds for BC2 at $x = L$. In general, however, careful attention must be given to the boundary conditions that correspond to the spatial dimension (or dimensions) opposite of the Φ problem, in our case, the y dimension. Let us first consider BC3 of equation (3-169b):

$$\text{BC3:} \quad \left.\frac{\partial T}{\partial y}\right|_{y=0} = \left.\frac{\partial \Psi}{\partial y}\right|_{y=0} + \left.\frac{\partial \Phi}{\partial y}\right|_{y=0} = 0 \tag{3-172}$$

Because the derivative of $\Phi(x)$ with respect to y is identically *zero*, we readily yield the boundary condition $\partial\Psi/\partial y = 0$ at $y = 0$. Now we consider BC4 of equation (3-169b), which yields

$$\text{BC4:} \qquad T(y = W) = \Psi(y = W) + \Phi(x) = 0 \tag{3-173}$$

and is now only satisfied for a nontrivial $\Phi(x)$ by letting $\Psi(y = W) = -\Phi(x)$.

With these comments in mind, we formulate our two problems as

$$\frac{d^2\Phi}{dx^2} + \frac{g}{k} = 0 \qquad \text{in} \qquad 0 < x < L \tag{3-174}$$

$$\text{BC1:} \qquad \Phi(x = 0) = 0 \qquad \text{BC2:} \qquad \Phi(x = L) = 0 \tag{3-175}$$

and

$$\frac{\partial^2\Psi}{\partial x^2} + \frac{\partial^2\Psi}{\partial y^2} = 0 \qquad \text{in} \qquad 0 < x < L \quad 0 < y < W \tag{3-176}$$

$$\text{BC1:} \qquad \Psi(x = 0) = 0 \qquad \text{BC2:} \qquad \Psi(x = L) = 0 \tag{3-177a}$$

$$\text{BC3:} \qquad \left.\frac{\partial\Psi}{\partial y}\right|_{y=0} = 0 \qquad \text{BC4:} \qquad \Psi(y = W) = -\Phi(x) \tag{3-177b}$$

The above process for treating BC4, equation (3-177b), is what we referred to above as a *coupled boundary condition* in terms of the 2-D homogeneous PDE and the 1-D nonhomogeneous ODE problems. Our choice of the x dimension for the ODE problem $\Phi(x)$ was done intentionally, however, to make use of the derivative boundary condition at $y = 0$. This had the effect of uncoupling $\Phi(x)$ from the 2-D boundary condition at $y = 0$. If we had instead selected $\Phi(y)$ for the ODE problem, then $\Phi(y)$ would have coupled to the 2-D problem as a nonhomogeneity at both $x = 0$ and $x = L$. That problem would still have been readily solved, but superposition of the 2-D problem would have been necessary to treat each nonhomogenous boundary condition separately, thereby doubling the required work. This brings us to our general rule of thumb when separating out the energy generation term: W*henever possible select the spatial dimension of the 1-D generation problem to be orthogonal to a spatial dimension containing a homogeneous, first-derivative boundary condition as given by an insulated surface or symmetry condition*. We therefore took advantage of the symmetry boundary condition at $y = 0$ and deliberately let $\Phi = \Phi(x)$.

We are now ready to solve the two superimposed problems, starting with equation (3-174), which yields

$$\Phi(x) = C_1 x + C_2 - \frac{g}{2k}x^2 \tag{3-178}$$

which we recognize as the classic parabolic temperature profile for 1-D, steady-state heat generation in the Cartesian system. Applying BC1 and BC2 of equation (3-175) yields

$$\Phi(x) = \frac{gL^2}{2k}\left[\frac{x}{L} - \left(\frac{x}{L}\right)^2\right] \tag{3-179}$$

We now solve the homogenous PDE of equation (3-176) by assuming

$$\Psi(x, y) = X(x)Y(y) \tag{3-180}$$

Substituting the above into the PDE and separating yields

$$\frac{d^2X}{dx^2} + \lambda^2 X = 0 \tag{3-181}$$

and

$$\frac{d^2Y}{dy^2} - \lambda^2 Y = 0 \tag{3-182}$$

where we have forced the characteristic value problem in the homogeneous x dimension. The solution of equation (3-181) yields

$$X(x) = C_1 \cos \lambda x + C_2 \sin \lambda x \tag{3-183}$$

where BC1 then eliminates one constant ($C_1 = 0$), while BC2 yields the eigenvalues

$$\lambda L = n\pi \quad \rightarrow \quad \lambda_n = \frac{n\pi}{L} \quad \text{for} \quad n = 0, 1, 2, \ldots \tag{3-184}$$

noting $\lambda_0 = 0$ is a trivial eigenvalue. Solution of equation (3-182) yields

$$Y(y) = C_3 \cosh \lambda y + C_4 \sinh \lambda y \tag{3-185}$$

having selected the hyperbolics. Satisfaction of BC3 in equation (3-177b) requires that $C_4 = 0$. With all homogeneous boundary conditions satisfied, we combine and sum the separated functions, omitting the trivial case, to yield

$$\Psi(x, y) = \sum_{n=1}^{\infty} C_n \sin \lambda_n x \cosh \lambda_n y \tag{3-186}$$

where we have introduced the new constant $C_n = C_2 C_3$. We lastly apply the nonhomogeneous BC4, equation (3-177b), to determine C_n, which yields

$$-\Phi(x) = \sum_{n=1}^{\infty} C_n \sin \lambda_n x \cosh \lambda_n W \tag{3-187}$$

which is now a Fourier series expansion of $-\Phi(x)$. We apply the following operator to both sides of equation (3-187):

$$* \int_{x=0}^{L} \sin \lambda_m x \, dx$$

which yields the desired expression for C_n:

$$C_n = \frac{-\int_{x=0}^{L} \Phi(x) \sin \lambda_n x \, dx}{\cosh \lambda_n W \int_{x=0}^{L} \sin^2 \lambda_n x \, dx} \tag{3-188}$$

We can now superimpose the two solutions to generate the desired solution:

$$T(x, y) = \sum_{n=1}^{\infty} \left[C_n \sin \lambda_n x \cosh \lambda_n y \right] + \frac{gL^2}{2k} \left[\frac{x}{L} - \left(\frac{x}{L} \right)^2 \right] \tag{3-189}$$

It is useful to note here that the generation term is present in the $\Phi(x)$ expression, as readily seen in the rightmost term. However, generation is also coupled into the 2-D expression of the series summation by the Fourier constant, as observed in the numerator of equation (3-188); hence the generation term is very much driving the temperature distribution in both the x and y dimensions. Accordingly, failure to couple the 1-D generation-containing Φ solution to *at least one* of the boundary conditions of the 2-D problem will lead to a physically unrealistic solution.

3-5 PRODUCT SOLUTION

In the rectangular coordinate system, the solution of multidimensional, transient homogeneous heat conduction problems can be written down very simply as the product of the solutions of one-dimensional problems if the initial temperature distribution in the medium is expressible *as a product of single-space variable functions*. For example, for a two-dimensional problem the initial condition may be expressed in the form $T(x, y, t = 0) = F(x, y) = F_1(x)F_2(y)$, or for a three-dimensional problem in the form $T(x, y, z, t = 0) = F(x, y, z) = F_1(x)F_2(y)F_3(z)$. Clearly, the case of uniform temperature initial condition is always expressible in the product form.

To illustrate this method, we consider the following two-dimensional homogeneous heat conduction problem for a rectangular solid over the domain given by $0 \le x \le L$, $0 \le y \le W$:

$$\frac{\partial^2 T}{\partial x^2} + \frac{\partial^2 T}{\partial y^2} = \frac{1}{\alpha} \frac{\partial T}{\partial t} \quad \text{in} \quad 0 < x < L \quad 0 < y < W \quad t > 0 \tag{3-190}$$

$$\text{BC1:} \qquad -k\left.\frac{\partial T}{\partial x}\right|_{x=0} = -h_1 \left. T\right|_{x=0} \tag{3-191a}$$

$$\text{BC2:} \qquad -k\left.\frac{\partial T}{\partial x}\right|_{x=L} = h_2 \left. T\right|_{x=L} \tag{3-191b}$$

$$\text{BC3:} \qquad -k\left.\frac{\partial T}{\partial y}\right|_{y=0} = -h_3 \left. T\right|_{y=0} \tag{3-191c}$$

$$\text{BC4:} \qquad -k\left.\frac{\partial T}{\partial y}\right|_{y=W} = h_4 \left. T\right|_{y=W} \tag{3-191d}$$

$$\text{IC:} \qquad T(t=0) = F_1(x)F_2(y) \quad \text{in the domain} \tag{3-191e}$$

where

$$T \equiv T(x, y, t)$$

To solve this problem, we consider the following two 1-D homogeneous heat conduction problems for slabs $0 \leq x \leq L$ and $0 \leq y \leq W$, given as

$$\frac{\partial^2 T_1}{\partial x^2} = \frac{1}{\alpha}\frac{\partial T_1}{\partial t} \qquad \text{in} \qquad 0 < x < L \quad t > 0 \tag{3-192}$$

$$\text{BC1:} \qquad -k\left.\frac{\partial T_1}{\partial x}\right|_{x=0} = -h_1 \left. T_1\right|_{x=0} \tag{3-193a}$$

$$\text{BC2:} \qquad -k\left.\frac{\partial T_1}{\partial x}\right|_{x=L} = h_2 \left. T_1\right|_{x=L} \tag{3-193b}$$

$$\text{IC:} \qquad T_1(x, t=0) = F_1(x) \qquad \text{in} \qquad 0 \leq x \leq L \tag{3-193c}$$

and

$$\frac{\partial^2 T_2}{\partial y^2} = \frac{1}{\alpha}\frac{\partial T_2}{\partial t} \qquad \text{in} \qquad 0 < y < W \quad t > 0 \tag{3-194}$$

$$\text{BC3:} \qquad -k\left.\frac{\partial T_2}{\partial y}\right|_{y=0} = -h_3 \left. T_2\right|_{y=0} \tag{3-195a}$$

$$\text{BC4:} \qquad -k\left.\frac{\partial T_2}{\partial y}\right|_{y=W} = h_4 \left. T_2\right|_{y=W} \tag{3-195b}$$

$$\text{IC:} \qquad T_2(y, t=0) = F_2(y) \qquad \text{in} \qquad 0 \leq y \leq W \tag{3-195c}$$

Here we note that the boundary conditions for problem (3-192) are the same as those given by equations (3-191a,b), and those for problem (3-194) are the same as those given by equations (3-191c,d). Then the solution of the original two-dimensional problem (3-190) is given as the product solution of the above one-dimensional problems, namely, as

$$T(x, y, t) = T_1(x, t)T_2(y, t) \tag{3-196}$$

To prove the validity of this result, we substitute equation (3-196) into equation (3-190) and utilize equations (3-192) and (3-194), which gives

$$T_2\frac{\partial^2 T_1}{\partial x^2} + T_1\frac{\partial^2 T_2}{\partial y^2} = \frac{1}{\alpha}T_2\frac{\partial T_1}{\partial t} + \frac{1}{\alpha}T_1\frac{\partial T_2}{\partial t} \tag{3-197a}$$

or

$$T_2\left(\frac{\partial^2 T_1}{\partial x^2} - \frac{1}{\alpha}\frac{\partial T_1}{\partial t}\right) + T_1\left(\frac{\partial^2 T_2}{\partial y^2} - \frac{1}{\alpha}\frac{\partial T_2}{\partial t}\right) = 0 \tag{3-197b}$$

Thus, the differential equation is satisfied in view of equations (3-192) and (3-194), given that each of the terms in parenthesis is identically zero.

Similarly, the substitution of equation (3-196) into the boundary conditions (3-191) shows that they are also satisfied. Hence, equation (3-196) is the solution of the problem defined by equations (3-190) and (3-191). We emphasize that for solution of all transient problems, the boundary conditions must be homogeneous. This condition also holds for the use of a product solution, namely, both boundary conditions of the 1-D problems must be homogeneous.

Example 3-8 Product Solution

A 2-D rectangular solid of dimensions $L \times W$ is initially at a uniform temperature of T_0. For t>0, all four boundary conditions are homogeneous, as illustrated in Figure 3-15.

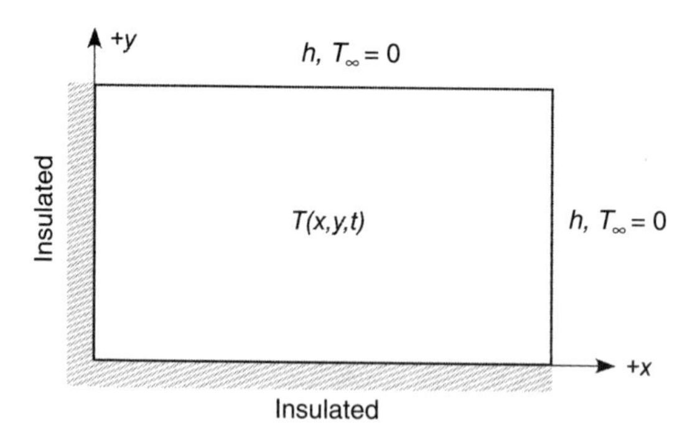

Figure 3-15 Problem description for Example 3-8.

The mathematical formulation of the problem is given as

$$\frac{\partial^2 T}{\partial x^2} + \frac{\partial^2 T}{\partial y^2} = \frac{1}{\alpha}\frac{\partial T}{\partial t} \qquad \text{in} \qquad 0 < x < L \quad 0 < y < W \tag{3-198}$$

$$\text{BC1:} \quad \left.\frac{\partial T}{\partial x}\right|_{x=0} = 0 \qquad \text{BC2:} \quad -k\left.\frac{\partial T}{\partial x}\right|_{x=L} = h\,T|_{x=L} \tag{3-199a}$$

$$\text{BC3:} \quad \left.\frac{\partial T}{\partial y}\right|_{y=0} = 0 \qquad \text{BC4:} \quad -k\left.\frac{\partial T}{\partial y}\right|_{y=W} = h\,T|_{y=W} \tag{3-199b}$$

$$\text{IC:} \quad T(x, y, t = 0) = T_0 \tag{3-199c}$$

The solution of this 2-D problem can be expressed as the product of the solutions of the following two 1-D slab problems: (1) $T_1(x, t)$, for a slab, $0 \le x \le L$, initially at a temperature $F(x) = 1$ and for times $t > 0$ the boundary at $x = 0$ is insulated and the boundary at $x = L$ dissipates heat by convection into an environment at zero temperature with a heat transfer coefficient h (or $H = h/k$); and (2) $T_2(y, t)$, for a slab, $0 \le y \le W$, initially at a temperature $F(y) = T_0$ and for times $t > 0$ the boundary at $y = 0$ is insulated and the boundary at $y = W$ dissipates heat by convection into an environment at zero temperature with a heat transfer coefficient h (or $H = h/k$).

Both of these 1-D problems are identical with regard to formulation and solution and in fact, are identical to the problem given by equations (3-158) and (3-159), with the appropriate adjustment of the initial condition $G(x)$. For $T_1(x, t)$, we consider equations (3-163) and (3-166), and make the substitution $G(x) = 1$:

$$T_1(x, t) = \sum_{n=1}^{\infty} C_n \cos \lambda_n x e^{-\alpha\lambda_n^2 t} \tag{3-200}$$

with eigenvalues given by

$$\lambda_n \tan \lambda_n L = \frac{h}{k} \quad \rightarrow \quad \lambda_n \quad \text{for} \quad n = 1, 2, 3, \ldots \tag{3-201}$$

and with the Fourier coefficients given by

$$C_n = \frac{\int_{x=0}^{L} \cos \lambda_n x\, dx}{\int_{x=0}^{L} \cos^2 \lambda_n x\, dx} = \frac{2(\lambda_n^2 + H^2)}{L(\lambda_n^2 + H^2) + H} \frac{\sin \lambda_n L}{\lambda_n}$$

$$= \frac{2H}{L(\lambda_n^2 + H^2) + H} \frac{1}{\cos \lambda_n L} \tag{3-202}$$

where $H = h/k$, and where we have made use of equation (3-201) in simplifying the middle term to yield the right-hand term. Using equation (3-202),

$$T_1(x, t) = \sum_{n=1}^{\infty} \frac{2H}{L(\lambda_n^2 + H^2) + H} \frac{\cos \lambda_n x}{\cos \lambda_n L} e^{-\alpha\lambda_n^2 t} \tag{3-203}$$

In a similar manner, we solve for $T_2(y, t)$ after making the substitution $G(y) = T_0$:

$$T_2(y,t) = T_0 \sum_{m=1}^{\infty} \frac{2H}{W(\beta_m^2 + H^2) + H} \frac{\cos \beta_m y}{\cos \beta_m W} e^{-\alpha \beta_m^2 t} \tag{3-204}$$

with eigenvalues now given by

$$\beta_m \tan \beta_m W = \frac{h}{k} \quad \rightarrow \quad \beta_m \quad \text{for} \quad m = 1, 2, 3, \ldots \tag{3-205}$$

We finally construct the solution for the 2-D rectangular region as

$$T(x, y, t) = T_1(x, t) \cdot T_2(y, t) \tag{3-206}$$

which upon substitution of equations (3-203) and (3-204) becomes

$$T(x, y, t) = \sum_{n=1}^{\infty} \sum_{m=1}^{\infty} \frac{4T_0 H^2 e^{-\alpha(\beta_m^2 + \lambda_n^2)t}}{\left[W(\beta_m^2 + H^2) + H\right]\left[L(\lambda_n^2 + H^2) + H\right]} \frac{\cos \lambda_n x}{\cos \lambda_n L} \frac{\cos \beta_m y}{\cos \beta_m W} \tag{3-207}$$

The same solution could have been obtained by directly solving the 2-D equation using separation of variables.

3-6 CAPSTONE PROBLEM

In this chapter we have developed in considerable detail the methods of separation of variables and superposition to solve the heat conduction equation in Cartesian coordinates. The focus has been to develop the analytical solutions, which always entail a series summation, eigenvalues, and the necessary Fourier coefficients. Our general approach has been to leave the Fourier coefficients in integral form, although at times we have performed the integrals and substituted the expression back into the series summation. It is useful, however, to consider the implementation of our analytic solutions with regard to convergence, boundary conditions, and initial conditions. Furthermore, knowledge of the temperature distribution also provides knowledge of the heat flux via Fourier's law, which then allows for additional calculations, for example, of the total energy flow across a boundary. Here we consider a final problem in depth, addressing the issues outlined above, in an attempt to integrate and illustrate the many aspects of our temperature field solutions presented in this chapter. Along the way, we have provided comments regarding both the physics and mathematics of our solution.

We consider a large sheet of plate glass with a total thickness of L that is initially at a uniform temperature T_0. At $t = 0$, both surfaces of the glass are rapidly quenched to a surface temperature of T_1. Physically, let us assume that phase change heat transfer is used to provide such a large heat transfer coefficient

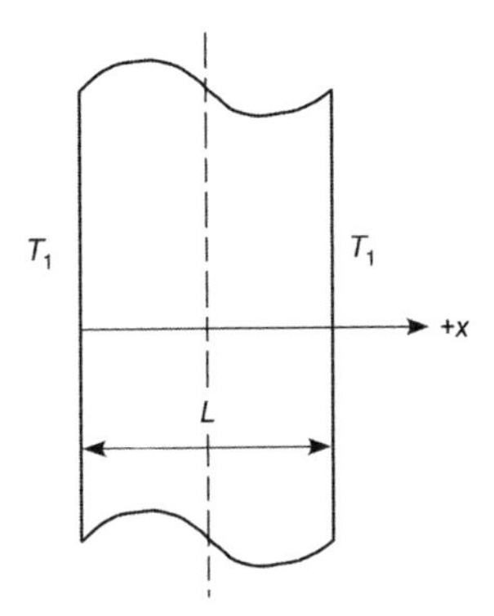

Figure 3-16 Description for Capstone Problem.

that we may assume infinite convection (i.e., Bi $\rightarrow \infty$); hence the imposition of a prescribed temperature boundary condition $T = T_1$ for $t > 0$. We want to find $T(x,t)$ for this problem, which is illustrated in Figure 3-16.

The formulation of this problem becomes

$$\frac{\partial^2 T(x,t)}{\partial x^2} = \frac{1}{\alpha}\frac{\partial T(x,t)}{\partial t} \quad \text{in} \quad 0 < x < L \quad t > 0 \tag{3-208}$$

$$\text{BC1:} \quad T(x = 0) = T_1 \qquad \text{BC2:} \quad T(x = L) = T_1 \tag{3-209a}$$

$$\text{IC:} \quad T(x, t = 0) = T_0 \tag{3-209b}$$

Alternatively, we could have made use of symmetry to solve the problem over the half-domain, namely, $0 \le x \le L/2$, using a perfectly insulated boundary condition at the center ($x = 0$). The requirement that a transient problem must have all homogeneous boundary conditions is *not satisfied*, as the boundary conditions at $x = 0$ and $x = L$ contain the nonhomogeneous prescribed temperature T_1. We therefore shift the temperature scale, letting $\Theta(x, t) = T(x, t) - T_1$, which yields the new formulation:

$$\frac{\partial^2 \Theta(x,t)}{\partial x^2} = \frac{1}{\alpha}\frac{\partial \Theta(x,t)}{\partial t} \quad \text{in} \quad 0 < x < L \quad t > 0 \tag{3-210}$$

$$\text{BC1:} \quad \Theta(x = 0) = 0 \qquad \text{BC2:} \quad \Theta(x = L) = 0 \tag{3-211a}$$

$$\text{IC:} \quad \Theta(x, t = 0) = T_0 - T_1 = \Theta_0 \tag{3-211b}$$

We now assume a separation of the form $\Theta(x, t) = X(x)\Gamma(t)$, which yields

$$\frac{1}{X}\frac{d^2 X}{dx^2} = \frac{1}{\alpha\Gamma}\frac{d\Gamma}{dt} = -\lambda^2 \tag{3-212}$$

As with all transient problems, we have forced the boundary value problem in the homogenous x dimension. Equation (3-212) produces two ordinary differential equations, with solutions:

$$\Gamma(t) = C_1 e^{-\alpha\lambda^2 t} \tag{3-213}$$

and

$$X(x) = C_2 \cos\lambda x + C_3 \sin\lambda x \tag{3-214}$$

Boundary condition 1 yields $C_2 = 0$, while boundary condition 2 yields

$$\lambda L = n\pi \quad \rightarrow \quad \lambda_n = \frac{n\pi}{L} \quad \text{for} \quad n = 0, 1, 2, \ldots \tag{3-215}$$

noting $\lambda_0 = 0$ is a trivial eigenvalue. We now combine the separated functions, introducing a new constant $C_n = C_1 C_3$, and sum over all solutions to yield

$$\Theta(x, t) = \sum_{n=1}^{\infty} C_n \sin\lambda_n x e^{-\alpha\lambda_n^2 t} \tag{3-216}$$

The initial condition is now applied to determine the Fourier constants, yielding

$$\Theta_0 = \sum_{n=1}^{\infty} C_n \sin\lambda_n x \tag{3-217}$$

to which we apply the following operator to both sides:

$$* \int_{x=0}^{L} \sin\lambda_m x \, dx$$

which yields the desired expression for C_n as follows

$$C_n = \frac{\Theta_0 \int_{x=0}^{L} \sin\lambda_n x \, dx}{\int_{x=0}^{L} \sin^2\lambda_n x \, dx} = \frac{2\Theta_0}{\pi} \frac{1 - (-1)^n}{n} \tag{3-218}$$

To complete our solution, we now substitute equation (3-218) into (3-217) and shift back to $T(x,t)$, yielding

$$\boxed{T(x, t) = \frac{2(T_0 - T_1)}{\pi} \sum_{n=1}^{\infty} \left[\frac{1 - (-1)^n}{n} \sin\lambda_n x \, e^{-\alpha\lambda_n^2 t} \right] + T_1} \tag{3-219}$$

To further explore this problem, let the glass thickness $L = 0.008$ m, $T_0 = 700$ K, and $T_1 = 373$ K. In addition, let us use representative properties for glass: Thermal conductivity $k = 1.4$ W/m K and thermal diffusivity $\alpha = 5.2 \times 10^{-7}$ m^2/s.

We will now use the first 200 terms ($n = 1$–200) of the series solution given by equation (3-219) to calculate and plot the temperature at the center of the glass ($x = 0.004$) and half-way between the center and the surface ($x = 0.002$) for the first 80 s. We note here that we will simply use the initial condition, $T_0 =$ 700 K, at $t = 0$, but below we will also discuss recovery of the initial condition using the series solution. Figure 3-17 shows these two temperature profiles over the specified time domain.

We observe in Figure 3-17 that the temperature at these two locations decays smoothly toward the steady-state value of 373 K within the 80 s covered by this plot. The center temperature at *L*/2 is observed to lag behind the temperature value at *L*/4, which makes physical sense in that heat is always flowing outward, hence a temperature gradient between these two points. In addition, it can be seen that the center temperature remains at the initial condition of 700 K for nearly 2 s before feeling the influence of the altered boundary conditions. As discussed in Chapter 1, the response time to the boundary change is strongly influenced by the *thermal diffusivity*, noting that the influence of the thermal conductivity on the temperature profile is only within the thermal diffusivity (i.e., $\alpha = k/\rho c$).

As mentioned above, it is useful to explore the convergence of the series summation. For the data in Figure 3-17, 200 terms were always sufficient for convergence, but we note that the initial condition was used at $t = 0$ rather than the series summation. In contrast, we plot in Figure 3-18 the calculated center

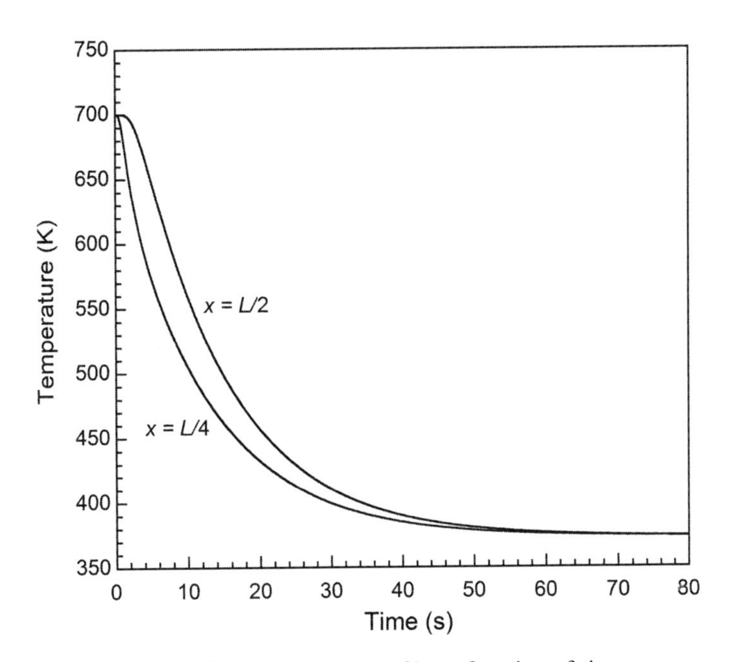

Figure 3-17 Temperature profile as function of time.

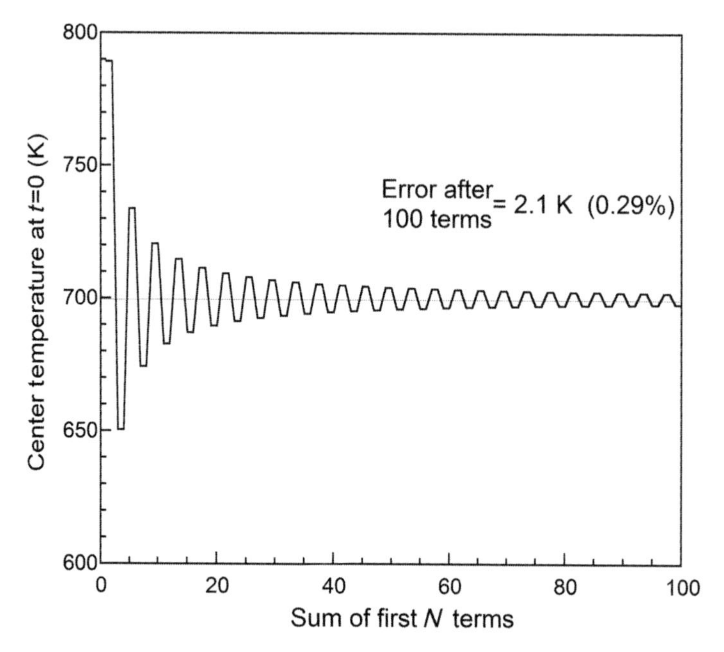

Figure 3-18 Convergence of the center temperature at $t = 0$.

temperature ($x = L/2$) at $t = 0$ using the series summation as a function of the number of terms summed, over the range from 1 to 100 terms.

It can be seen in Figure 3-18 that the convergence at $t = 0$ is rather slow, with accuracy of 2 K reached after 100 terms. The error after summing only the first 10 terms is greater than 20 K. We note that every other term, namely, the even n terms, are identically zero. In contrast, at $t = 10$ s, the center temperature converges to within 0.001 K of the exact value after summing only the first 5 terms of the series. In general, as time increases, the convergence of the series summation becomes much more rapid. This is readily explained by the exponential term of equation (3-219), which depends on the power of $-\alpha\lambda_n^2 t$. Clearly this term decreases exponentially with increasing time and increasing eigenvalues, hence rapid convergence of the series. A useful nondimensional parameter to explore this point is the *Fourier number*

$$\text{Fo} = \frac{\alpha t}{L^2} \tag{3-220}$$

which may also be interpreted as a ratio of heat conduction to heat storage within a given volume element. Using our eigenvalues of equation (3-215), the exponential term of equation (3-219) may be recast in terms of the Fourier number as

$$e^{-\alpha\lambda_n^2 t} = e^{-n^2\pi^2 \text{Fo}} \tag{3-221}$$

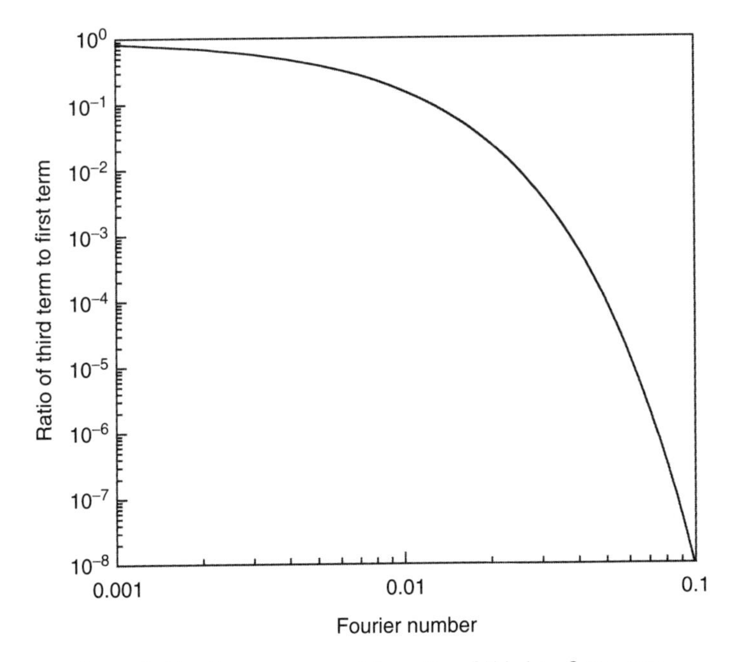

Figure 3-19 Convergence of the ratio of third-to-first terms.

For our temperature solution, we can explore the convergence of the series by comparing the third term of the series ($n = 3$) to the first term of the series ($n = 1$), noting that the second term is zero and contributes nothing to the solution. In Figure 3-9 we plot the ratio of the third-to-first terms as a function of Fourier number. This figure reveals the very rapid convergence of the series with increasing Fourier number. In fact, for a Fourier number of 0.1, the third term of the series is a factor of 10^8 less than the first term. For our problem, a Fourier number of 0.1 corresponds to 12.3 s. This rapid convergence is in perfect agreement with the findings noted above at 10 s. The phenomenon of rapid convergence with increasing Fourier number is well known, and, in fact, a simple *one-term approximation* (i.e., only the *first term* of the series summation) is often used to calculate the temperature for Fo > 0.2, which is the basis for the *Heisler charts* [11].

Knowledge of the temperature profile enables calculation of the heat flux as a function of space and time by applying Fourier's law to equation (3-219):

$$q''(x,t) = -k\frac{\partial T}{\partial x} = \frac{-2k\left(T_0 - T_1\right)}{\pi}\sum_{n=1}^{\infty}\left[\frac{1-(-1)^n}{n}\lambda_n \cos\lambda_n x\, e^{-\alpha\lambda_n^2 t}\right] \tag{3-222}$$

The heat flux is the instantaneous rate of energy per unit area (W/m^2) crossing a plane at any given position. Let us now consider the heat flux specifically crossing the outer surface at $x = 0$, given as

$$q''(x=0,t) = -k\frac{\partial T}{\partial x}\bigg|_{x=0} = \frac{-2k\left(T_0 - T_1\right)}{\pi}\sum_{n=1}^{\infty}\left[\frac{1-(-1)^n}{n}\lambda_n e^{-\alpha\lambda_n^2 t}\right] \tag{3-223}$$

where we note that this is equivalent in value but opposite in sign to the heat flux crossing the surface at $x = L$ (i.e., symmetry). From the surface heat flux, we can calculate the total amount of energy per unit area (J/m^2) that has been removed from the glass surface starting from $t = 0$ up to any time t_1 by integrating the surface heat flux in equation (3-223), namely,

$$E''_{\text{out}} = \int_{t=0}^{t_1} q''(x=0,t)\,dt = \frac{-2k\left(T_0 - T_1\right)}{\pi}\sum_{n=1}^{\infty}\left[\frac{1-(-1)^n}{n}\lambda_n \int_{t=0}^{t_1} e^{-\alpha\lambda_n^2 t}\,dt\right] \tag{3-224}$$

which yields

$$E''_{\text{out}} = \frac{-2k\left(T_0 - T_1\right)}{\pi}\sum_{n=1}^{\infty}\left[\frac{1-(-1)^n}{n}\frac{1}{\alpha\lambda_n}\left(1 - e^{-\alpha\lambda_n^2 t_1}\right)\right] \tag{3-225}$$

We note that this quantity is negative, which reflects energy transferred in the negative x direction, as appropriate for the left surface. In Figure 3-20, we plot the *total energy* removed from the glass per unit area as a function of time, which is directly calculated from equation (3-225) after multiplying by 2 to account for both sides, and introducing a sign change to report this as a positive quantity.

The behavior observed in Figure 3-20 is consistent with the physics of the problem, namely, a very rapid removal of energy from the glass at initial times when the temperature gradients are steepest, followed by a steady decline in rate of removal as the quantity of integrated energy asymptotically approaches a final value. At $t = 0$, the rate of heat flux at the surface is actually infinite, which corresponds to the instantaneous change in the boundary condition from T_0 to T_1. This is also consistent with surface heat flux as given by equation (3-223), in which the series is nonconvergent at $t = 0$. The final value in Figure 3-20 ($t_1 \to \infty$) corresponds to the maximum amount of energy that is possible to remove from the glass, which is readily evaluated from simple conservation of energy using the heat capacity and maximum ΔT,

$$E''_{\text{out}}\Big|_{\max} = \rho c L(T_0 - T_1) = \frac{k}{\alpha}L(T_0 - T_1) \tag{3-226}$$

The maximum energy removed should also follow directly from (3-225), after multiplying by -2 and by letting $t_1 \to \infty$, which yields

$$E''_{\text{out}}\Big|_{\max} = \frac{4kL\left(T_0 - T_1\right)}{\alpha\pi^2}\sum_{n=1}^{\infty}\left[\frac{1-(-1)^n}{n^2}\right] \tag{3-227}$$

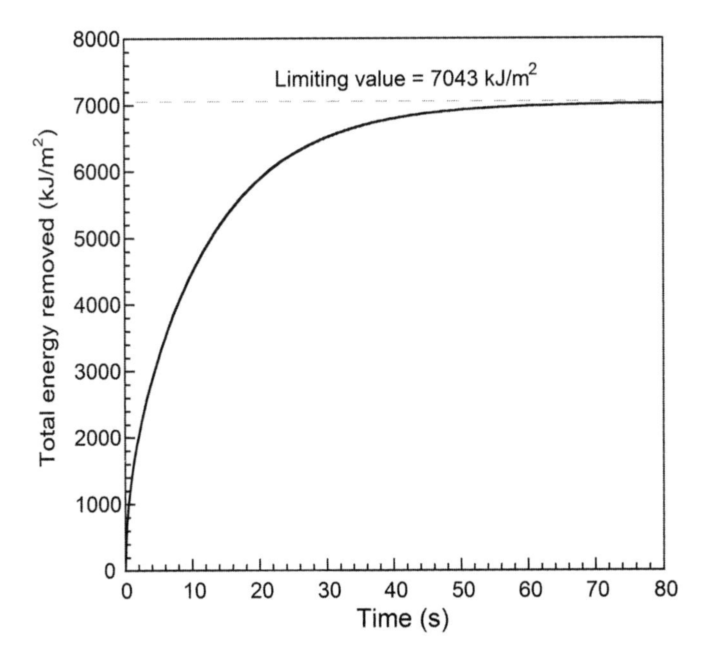

Figure 3-20 Energy removed from the glass vs. time.

where the series summation converges exactly (but slowly) to the sum of $\pi^2/4$, which gives perfect agreement with equation (3-226). Equation (3-226) yields a total amount of energy (i.e., at steady state) removed per unit surface area equal to 7043 kJ/m^2, while by 80 s, we have removed a total of 7019 kJ/m^2, within 99.7% of the maximum value. This is consistent with the temperature profiles in Figure 3-17, which show that the steady-state temperature is nearly reached by 80 s in the center of the glass plate [i.e., $T(x = L/2, t = 80 \text{ s}) = 373.7\ K$].

In summary, this capstone problem is intended to explore the details and nuances associated with solutions of the heat equation using separation of variables, including evaluation of the Fourier series coefficients, summation and convergence of the series summation and the influence of the Fourier number, as well as calculation of the heat flux, convergence toward steady-state conditions, and overall conservation of energy within the domain of the problem.

REFERENCES

1. M. P. Morse and H. Feshbach, *Methods of Theoretical Physics*, Part I, McGraw-Hill, New York, 1953.
2. J. W. Dettman, *Mathematical Methods in Physics and Engineering*, McGraw-Hill, New York, 1962.

3. P. Moon and D. E. Spencer, *Field Theory for Engineers*, Van Nostrand, Princeton, NJ, 1961.
4. R. V. Churchill, *Fourier Series and Boundary Value Problems*, McGraw-Hill, New York, 1963.
5. H. S. Carslaw and J. C. Jaeger, *Conduction of Heat in Solids*, Oxford, Clarendon, London, 1959.
6. V. S. Arpaci, *Conduction Heat Transfer*, Addison-Wesley, Reading, MA, 1966.
7. M. N. Özişik, *Boundary Value Problems of Heat Conduction*, International Textbook, Scranton, PA, 1968; also Dover, New York, 1989.
8. M. N. Özişik, *Heat Conduction*, Wiley, New York, 1980.
9. M. D. Mikhailov and M. N. Özişik, *Unified Analysis and Solutions of Heat and Mass Diffusion*, Wiley, New York, 1984.
10. P. Moon and D. E. Spencer, *Q. Appl. Math.* **16**, 1–10, 1956.
11. M. P. Heisler, *Trans. ASME*, **69**, 227–236, 1947.

PROBLEMS

3-1 A 1-D slab, $0 \leq x \leq L$, is initially at a temperature $F(x)$. For times $t > 0$, the boundary surface at $x = 0$ is kept at zero temperature, and the boundary surface at $x = L$ dissipates heat by convection into a medium with fluid temperature of zero and with a heat transfer coefficient h. Obtain an expression for the temperature distribution $T(x, t)$ in the slab for times $t > 0$, and for the heat flux at the boundary surface $x = L$.

3-2 A 1-D slab, $0 \leq x \leq L$, is initially at a temperature $F(x)$. For times $t > 0$, both of the boundary surfaces are perfectly insulated. Obtain an expression for the temperature $T(x, t)$ in the slab. Clearly show the steady-state temperature from your expression. Independently derive the steady-state temperature from simple conservation of energy and show that this agrees with your above expression.

3-3 A 2-D rectangular region $0 \leq x \leq a$, $0 \leq y \leq b$ is initially at a temperature $F(x, y)$. For times $t > 0$, it dissipates heat by convection from all its boundary surfaces into an environment at zero temperature and with convection coefficient h, which is the same for all the boundaries. Obtain an expression for the temperature distribution $T(x, y, t)$. Do not assume symmetry with regard to the initial condition.

3-4 A 2-D rectangular region $0 \leq x \leq a$, $0 \leq y \leq b$ is initially at a uniform temperature T_0. For times $t > 0$, the boundaries at $x = 0$ and $y = 0$ are kept at zero temperature and the boundaries at $x = a$ and $y = b$ dissipate heat by convection into an environment at zero temperature. The heat transfer coefficients h are the same for both of these boundaries. Using the product solution, obtain an expression for the temperature distribution $T(x, y, t)$.

3-5 A 3-D rectangular parallelepiped $0 \le x \le a, 0 \le y \le b, 0 \le z \le c$ is initially at temperature $F(x, y, z)$. For times $t > 0$, the boundaries at $x = 0$, $y = 0$, and $z = 0$ are insulated, and the boundaries at $x = a$, $y = b$, and $z = c$ are kept at a constant temperature of T_1. Obtain an expression for the temperature distribution $T(x, y, z, t)$ in the region.

3-6 A 3-D rectangular parallelepiped $0 \le x \le a,\ 0 \le y \le b,\ 0 \le z \le c$ is initially at temperature $F(x, y, z)$. For times $t > 0$, the boundaries at $x = 0$, $y = 0$, and $z = 0$ are insulated, and the boundaries at $x = a$, $y = b$, and $z = c$ all dissipate heat by convection into an environment with fluid temperature T_∞ and with identical heat transfer coefficients h at all three boundaries. Obtain an expression for the temperature distribution $T(x, y, z, t)$ in the region.

3-7 Obtain an expression for the steady-state temperature distribution $T(x, y)$ in a 2-D rectangular region $0 \le x \le a, 0 \le y \le b$ for the following boundary conditions: The boundary at $x = 0$ is kept insulated, the boundary at $y =$ b is kept at a temperature $f(x)$, and the boundaries at $x = a$ and $y = 0$ dissipate heat by convection into an environment at zero temperature. Assume the heat transfer coefficient h to be the same for both convective boundaries.

3-8 Obtain an expression for the steady-state temperature distribution $T(x, y, z)$ in a 3-D rectangular parallelepiped $0 \le x \le a,\ 0 \le y \le b,\ 0 \le z \le c$ for the following boundary conditions: The boundary surface at $x = 0$ is kept at temperature T_0, the boundaries at $y = 0$ and $z = 0$ are kept insulated, the boundary at $x = a$ is kept at zero temperature, and the boundaries at $y = b$ and $z = c$ dissipate heat by convection into an environment at zero temperature. The heat transfer coefficients h are the same for the two convective surfaces.

3-9 Consider the 2-D, steady-state rectangular region $0 \le x \le a,\ 0 \le y \le b$ in which internal energy is generated at a constant rate g_0 (W/m^3) and subjected to the following boundary conditions: The boundaries at $x = 0$ and $y = 0$ are kept insulated, whereas the boundaries at $x = a$ and $y = b$ are kept at zero temperature. Calculate the temperature distribution $T(x, y)$.

3-10 A 1-D slab, $0 \le x \le L$, is initially at zero temperature. For times $t > 0$, the boundary at $x = 0$ is kept insulated, the boundary at $x = L$ is kept at zero temperature, and there is internal energy generation within the solid at a constant rate of g_0 (W/m^3). Obtain an expression for the temperature distribution $T(x, t)$ in the slab for times $t > 0$.

3-11 Consider the 2-D, steady-state rectangular region $0 \le x \le a,\ 0 \le y \le b$ subjected to the following boundary conditions: The boundary at $x = 0$ is maintained at a prescribed temperature $f(y)$, the boundary at $x = a$ is maintained at zero temperature, the boundary at $y = 0$ is exposed to an incident heat flux $q''(x)$, and the boundary at $y = b$ is perfectly insulated. Calculate the temperature distribution $T(x, y)$.

3-12 Consider the 2-D, steady-state rectangular region $0 \leq x \leq a$, $0 \leq y \leq b$ subjected to the following boundary conditions: The boundary at $x = 0$ is maintained at a prescribed temperature T_1, the boundary at $x = a$ dissipates heat by convection into a medium with fluid temperature of T_∞ and with a heat transfer coefficient h, the boundary at $y = 0$ is exposed to an incident heat flux $q''(x)$, and the boundary at $y = b$ is perfectly insulated. Calculate the temperature distribution $T(x, y)$.

3-13 Consider the 2-D, steady-state rectangular region $0 \leq x \leq a$, $0 \leq y \leq b$ subjected to the following boundary conditions: The boundary at $x = 0$, and the boundary conditions at $y = 0$ and $y = b$ are all maintained at a prescribed temperature T_1, while the boundary at $x = a$ dissipates heat by convection into a medium with fluid temperature of T_∞ and with a heat transfer coefficient h. Calculate the temperature distribution $T(x, y)$.

3-14 Consider the 2-D, steady-state rectangular region $0 \leq x \leq a$, $0 \leq y \leq b$ in which internal energy is generated at a constant rate g_0 (W/m^3) and subjected to the following boundary conditions: The boundaries at $x = 0$ and $y = 0$ are maintained at zero temperature, while the boundary at $x = a$ dissipates heat by convection into a medium with fluid temperature of zero and with a heat transfer coefficient h, and the boundary condition at $y = b$ is perfectly insulated. Calculate the temperature distribution $T(x, y)$.

3-15 Consider the 2-D rectangular region $0 \leq x \leq a$, $0 \leq y \leq b$, which is initially at temperature $F(x, y)$. For $t > 0$, internal energy is generated at a constant rate g_0 (W/m^3) and the region is subjected to the following boundary conditions: The boundaries at $x = 0$ are maintained at a prescribed temperature $f(y)$, the boundary at $x = a$ is subjected to a prescribed heat flux $q''(y)$ out of the solid, the boundary at $y = 0$ is maintained at constant temperature T_1, and the boundary at $y = b$ dissipates heat by convection into a medium with fluid temperature T_∞ and with a heat transfer coefficient h. Formulate the problem using superposition for solution of $T(x, y, t)$, such that each subproblem is in a solvable form. Formulation includes the simplified PDE or ODE and the corresponding boundary and/or initial conditions. Do not shift the temperature scale for any problems, and do not actually solve any of the subproblems, only formulate.

3-16 Obtain an expression for the steady-state temperature distribution $T(x, y, z)$ in a 3-D rectangular parallelepiped $0 \leq x \leq a$, $0 \leq y \leq b$, $0 \leq z \leq c$ for the following boundary conditions: The boundary surfaces at $x = 0$, $x = a$, and the bottom surface at $z = 0$ are all perfectly insulated; the boundary at $y = 0$ is kept at zero temperature, the boundary at $y = b$ is subjected to a prescribed heat flux $q''(x, z)$ into the solid, and the boundary at $z = c$ dissipates heat by convection into an environment at zero temperature with heat transfer coefficient h.

3-17 A 3-D rectangular parallelepiped $0 \leq x \leq a$, $0 \leq y \leq b$, $0 \leq z \leq c$ is initially at a temperature given by $F(x, y, z)$. For $t > 0$, the following boundary conditions are imposed: the boundary surfaces at $x = 0$ and $y = b$ are kept at zero temperature, the boundaries at $x = a$ and $z = 0$ are perfectly insulated, the boundary surfaces at $y = 0$ and $z = c$ both dissipate heat by convection into an environment at zero temperature with heat transfer coefficient h. Calculate the temperature distribution $T(x, y, z, t)$.

3-18 Consider the 2-D rectangular region $0 \leq x \leq a$, $0 \leq y \leq b$ that has an initial temperature profile $F(x)$. For $t > 0$, the region is subjected to the following boundary conditions: The boundary surfaces at $x = 0$ and $x = a$ are maintained at a prescribed temperature T_1, the boundary at $y = 0$ is perfectly insulated, and the boundary at $y = b$ is exposed to an incident heat flux $q''(x)$. Calculate the temperature distribution $T(x, y, t)$.

3-19 Obtain an expression for the steady-state temperature distribution $T(x, y, z)$ in a 3-D rectangular parallelepiped $0 \leq x \leq a$, $0 \leq y \leq b$, $0 \leq z \leq c$ for the following boundary conditions: The boundary surfaces at $x = 0$ and $x = a$ are maintained at a prescribed temperature T_1, the boundary surfaces at $y = 0$, $z = 0$, and $z = c$ are all perfectly insulated, and the boundary at $y = b$ is maintained at a prescribed temperature $F(x, z)$.

3-20 Consider the 2-D rectangular region $0 \leq x \leq a$, $0 \leq y \leq b$ that has an initial uniform temperature $F(x, y)$. For $t > 0$, the region is subjected to the following boundary conditions: The boundary surfaces at $y = 0$ and $y = b$ are maintained at a prescribed temperature T_∞, the boundary at $x = 0$ dissipates heat by convection into a medium with fluid temperature T_∞ and with a heat transfer coefficient h, and the boundary surface at $x = a$ is exposed to constant incident heat flux q''_0. Calculate the temperature $T(x, y, t)$.

4

SEPARATION OF VARIABLES IN THE CYLINDRICAL COORDINATE SYSTEM

In this chapter we examine the separation of the heat conduction equation in the cylindrical coordinate system; determine the elementary solutions, the norms, and the eigenvalues of the separated problems for different combinations of boundary conditions; discuss the solution of the one- and multidimensional homogeneous problems by the method of separation of variables; examine the solutions of steady-state and transient multidimensional problems with and without the heat generation in the medium; and illustrate the splitting up of nonhomogeneous problems into a set of simpler problems. The reader should consult references 1–4 for additional applications on the solution of heat conduction problems in the cylindrical coordinate system.

4-1 SEPARATION OF HEAT CONDUCTION EQUATION IN THE CYLINDRICAL COORDINATE SYSTEM

We will first review the cylindrical coordinate system, as shown in Figure 4-1, along with the corresponding components of heat flux in the r, ϕ and z directions (i.e., Fourier's law), respectively, which are given by

$$q_r'' = -k\frac{\partial T}{\partial r}, \quad q_\phi'' = -\frac{k}{r}\frac{\partial T}{\partial \phi}, \quad \text{and} \quad q_z'' = -k\frac{\partial T}{\partial z} \tag{4-1}$$

Now we consider the three-dimensional, homogeneous differential equation of heat conduction in the cylindrical coordinate system:

$$\frac{1}{r}\frac{\partial}{\partial r}\left(r\frac{\partial T}{\partial r}\right) + \frac{1}{r^2}\frac{\partial^2 T}{\partial \phi^2} + \frac{\partial^2 T}{\partial z^2} = \frac{1}{\alpha}\frac{\partial T}{\partial t} \tag{4-2}$$

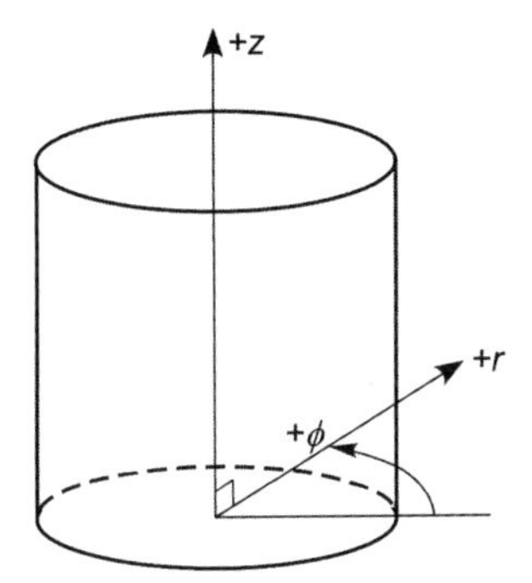

Figure 4-1 Cylindrical coordinate system.

or alternatively

$$\frac{\partial^2 T}{\partial r^2} + \frac{1}{r}\frac{\partial T}{\partial r} + \frac{1}{r^2}\frac{\partial^2 T}{\partial \phi^2} + \frac{\partial^2 T}{\partial z^2} = \frac{1}{\alpha}\frac{\partial T}{\partial t} \tag{4-3}$$

where $T \equiv T(r, \phi, z, t)$. Because the differential equation is homogeneous (i.e., no internal energy generation), we can assume a separation of variables in the form

$$T(r, \phi, z, t) = \Psi(r, \phi, z)\Gamma(t) \tag{4-4}$$

Substitution of equation (4-4) into the heat equation of (4-3) yields

$$\frac{1}{\Psi}\left(\frac{\partial^2 \Psi}{\partial r^2} + \frac{1}{r}\frac{\partial \Psi}{\partial r} + \frac{1}{r^2}\frac{\partial^2 \Psi}{\partial \phi^2} + \frac{\partial^2 \Psi}{\partial z^2}\right) = \frac{1}{\alpha\Gamma(t)}\frac{d\Gamma(t)}{dt} = -\lambda^2 \tag{4-5}$$

where we have introduced the separation constant $-\lambda^2$ to produce a decaying exponential function in time, as we discussed in Chapter 3. Equation (4-5) leads to the separated differential equations

$$\frac{d\Gamma}{dt} + \alpha\lambda^2\Gamma(t) = 0 \tag{4-6}$$

and

$$\frac{\partial^2 \Psi}{\partial r^2} + \frac{1}{r}\frac{\partial \Psi}{\partial r} + \frac{1}{r^2}\frac{\partial^2 \Psi}{\partial \phi^2} + \frac{\partial^2 \Psi}{\partial z^2} + \lambda^2\Psi = 0 \tag{4-7}$$

The solution of equation (4-6) yields

$$\Gamma(t) = C_1 e^{-\alpha\lambda^2 t} \tag{4-8}$$

Equation (4-7) is known as the *Helmholtz equation*, for which we assume a separation of the form

$$\Psi(r, \phi, z) = R(r)\Phi(\phi)Z(z) \tag{4-9}$$

We now substitute equation (4-9) into the Helmholtz equation and divide both sides by $R(r)\Phi(\phi)Z(z)$, which yields

$$\frac{1}{R}\left(\frac{d^2R}{dr^2} + \frac{1}{r}\frac{dR}{dr}\right) + \frac{1}{r^2}\frac{1}{\Phi}\frac{d^2\Phi}{d\phi^2} + \frac{1}{Z}\frac{d^2Z}{dz^2} + \lambda^2 = 0 \tag{4-10}$$

Because the initial condition will always provide the nonhomogeneity for transient problems, here we expect all three spatial dimensions to have corresponding homogeneous boundary conditions. Therefore, *we seek to force a characteristic boundary value problem in each spatial dimension*. Because the z dimension is readily isolated, we begin by first separating the z-variable term:

$$\frac{1}{R}\left(\frac{d^2R}{dr^2} + \frac{1}{r}\frac{dR}{dr}\right) + \frac{1}{r^2}\frac{1}{\Phi}\frac{d^2\Phi}{d\phi^2} + \lambda^2 = -\frac{1}{Z}\frac{d^2Z}{dz^2} = \eta^2 \tag{4-11}$$

where we have introduced a new separation constant $+\eta^2$ to force the ODE

$$\frac{d^2Z}{dz^2} + \eta^2 Z = 0 \tag{4-12}$$

Solution of equation (4-12) yields the desired solution of the form

$$Z(z) = C_2 \cos \eta z + C_3 \sin \eta z \tag{4-13}$$

which will produce orthogonal functions and eigenvalues η for the complete characteristic value problem in the z dimension. We now consider the remaining terms of equation (4-11), namely,

$$\frac{1}{R}\left(\frac{d^2R}{dr^2} + \frac{1}{r}\frac{dR}{dr}\right) + \frac{1}{r^2}\frac{1}{\Phi}\frac{d^2\Phi}{d\phi^2} + \lambda^2 = \eta^2 \tag{4-14}$$

To isolate the ϕ term, we multiply both sides by r^2, which yields

$$\frac{r^2}{R}\left(\frac{d^2R}{dr^2} + \frac{1}{r}\frac{dR}{dr}\right) + \left(\lambda^2 - \eta^2\right) r^2 = -\frac{1}{\Phi}\frac{d^2\Phi}{d\phi^2} = \nu^2 \tag{4-15}$$

We have now introduced another separation constant $+\nu^2$ to force the ODE

$$\frac{d^2\Phi}{d\phi^2} + \nu^2\Phi = 0 \tag{4-16}$$

which yields the desired solution form in the ϕ dimension, namely,

$$\Phi(\phi) = C_4 \cos \nu\phi + C_5 \sin \nu\phi \tag{4-17}$$

which will produce orthogonal functions and eigenvalues ν for the complete characteristic value problem in the ϕ dimension. Finally, we consider the remaining r terms of equation (4-15):

$$\frac{r^2}{R}\left(\frac{d^2R}{dr^2} + \frac{1}{r}\frac{dR}{dr}\right) + \left(\lambda^2 - \eta^2\right) r^2 = \nu^2 \tag{4-18}$$

We will let $\beta^2 = \lambda^2 - \eta^2$ and then multiply both sides by R/r^2, yielding

$$\frac{d^2R}{dr^2} + \frac{1}{r}\frac{dR}{dr} + \left(\beta^2 - \frac{\nu^2}{r^2}\right) R = 0 \tag{4-19}$$

which we recognize as Bessel's equation, recalling equation (2-36b). The solution of equation (4-19) yields

$$R(r) = C_6 J_\nu(\beta r) + C_7 Y_\nu(\beta r) \tag{4-20}$$

which will produce orthogonal functions and eigenvalues β for the complete characteristic value problem in the r dimension. We note that given the eigenvalues β and η as defined from the boundary value problems in the r dimension and z dimension, respectively, that λ is now defined by

$$\lambda^2 = \beta^2 + \eta^2 \tag{4-21}$$

An additional comment is offered with regard to equation (4-21) for the general case of ϕ dependency. The eigenvalues ν originating from the ϕ dimension do not appear explicitly in this expression for λ; however, the eigenvalues β from the r dimension will be indexed with regard to ν as we demonstrated in Chapter 2 (see Example 2-3). As a result, λ will also be indexed to the ϕ dimension, and therefore to all three spatial dimensions, as we expect for a multidimensional, transient problem.

4-2 SOLUTION OF STEADY-STATE PROBLEMS

In this section we will solve a range of steady-state problems in the cylindrical coordinate system. Our general guidelines established for separation of variables in the rectangular coordinate system equally apply to the cylindrical coordinate problems. Therefore, steady-state problems with a homogeneous PDE must always be solved with only a single nonhomogeneous boundary condition. If more than one boundary condition is nonhomogeneous, the principle of superposition is used to treat each nonhomogeneity individually. For a nonhomogeneous PDE (i.e., when heat generation is present), superposition is used to treat the heat generation within a 1-D ODE, which is coupled to a homogeneous PDE at one or more boundary conditions. Prior to solving a range of problems to illustrate the above procedures, a few additional guidelines are offered.

For two-dimensional, steady-state *axisymmetric* conduction problems, namely, $T = T(r, z)$, two distinct cases are possible. First, the characteristic value problem arises from homogeneous boundary conditions in the r dimension, for which we expect the orthogonal Bessel functions $J_0(\beta r)$ and $Y_0(\beta r)$, where we note that these are of order zero for the specific case of no ϕ dependency. The z dimension will generate nonorthogonal functions, either the exponentials or hyperbolics. In the second case, the characteristic value problem arises from homogeneous boundary conditions in the z dimension, for which we expect the orthogonal trigonometric functions $\cos \eta z$ and $\sin \eta z$. The r dimension will likewise generate nonorthogonal functions, namely, the modified Bessel functions of order zero.

For two or three-dimensional, steady-state problems that contain ϕ dependency, the ϕ dimension *must be homogeneous*. This is satisfied for a full cylinder $(0 \le \phi \le 2\pi)$ by the condition of 2π periodicity or for the case of a partial cylinder $(0 \le \phi \le \phi_0$ with $\phi_0 < 2\pi)$ by homogeneous boundary conditions at the cylinder surfaces defined by $\phi = 0$ and $\phi = \phi_0$. The nonhomogeneous boundary may then arise from the r dimension or the z dimension. For the 2-D steady-state problem with ϕ dependency, only the case of $T = T(r, \phi)$ makes physical sense, what we consider the very long cylinder, as it is difficult to imagine on physical grounds temperature dependency of a cylinder in the z and ϕ dimensions that does not likewise have r dependency.

Example 4-1 $T = T(r, \phi)$ Solid Cylinder

A 2-D, long solid cylinder of radius b is maintained at steady-state conditions with a prescribed surface temperature $T(r = b) = f(\phi)$. The mathematical formulation of the problem is given as

$$\frac{\partial^2 T}{\partial r^2} + \frac{1}{r}\frac{\partial T}{\partial r} + \frac{1}{r^2}\frac{\partial^2 T}{\partial \phi^2} = 0 \quad \text{in} \quad 0 \le r < b, \quad 0 \le \phi \le 2\pi \tag{4-22}$$

$$\text{BC1:} \quad T(r \to 0) \Rightarrow \text{finite} \qquad \text{BC2:} \quad T(r = b) = f(\phi) \tag{4-23a}$$

$$\text{BC3:} \quad T(\phi) = T(\phi + 2\pi) \quad \to \quad 2\pi\text{-periodicity} \tag{4-23b}$$

$$\text{BC5:} \quad \left.\frac{\partial T(\phi)}{\partial \phi}\right|_{\phi} = \left.\frac{\partial T(\phi)}{\partial \phi}\right|_{\phi+2\pi} \quad \to \quad 2\pi\text{-periodicity} \tag{4-23c}$$

In general, equations (4-23b) and (4-23c) are not considered true boundary conditions, as no physical boundary is present at any particular slice of the full cylinder $(0 \le \phi \le 2\pi)$. Physically, however, we must maintain *continuity* of both temperature and heat flux when traveling one full revolution (2π radians) around the cylinder. Therefore, we will always impose the requirement that our solution to a full cylinder have 2π periodicity. The condition of 2π periodicity is considered a *homogeneous boundary condition*, which is consistent with equations (4-23b) and (4-23c). The problem formulation contains only a single nonhomogeneous boundary condition, namely, BC2; hence the problem is ready for solution using separation of variables. We note that the condition of finiteness imposed as $r \to 0$ may be treated as a homogeneous boundary condition in the context of separation of variables in that it has the effect of eliminating one of the solution constants. We now assume separation of the form

$$T(r, \phi) = R(r)\Phi(\phi) \tag{4-24}$$

Substituting equation (4-24) into the PDE of equation (4-22) yields

$$\frac{r^2}{R}\left(\frac{d^2R}{dr^2} + \frac{1}{r}\frac{dR}{dr}\right) = \frac{-1}{\Phi}\frac{d^2\Phi}{d\phi^2} = \lambda^2 \tag{4-25}$$

after multiplying both sides by $r^2/R(r)\Phi(\phi)$ and selecting the separation constant λ^2 to force the characteristic value problem in the homogeneous ϕ dimension. Solution of the separated ODE in the ϕ dimension yields

$$\Phi(\phi) = C_1 \cos \lambda\phi + C_2 \sin \lambda\phi \tag{4-26}$$

We note here that the general trigonometric functions $\cos (n\pi/L)\phi$ and $\sin (n\pi/L)\phi$ have $2L$ periodicity for integer values of n. Therefore, the requirement of 2π periodicity for our solution of equation (4-26) is satisfied here by considering $L = \pi$, and letting

$$\lambda_n = n = 0, 1, 2, 3, \ldots$$

We note that we retain both constants C_1 and C_2, which will always be the case for the 2π-periodicity requirement. The remaining r terms in equation (4-25) yield

$$\frac{d^2R}{dr^2} + \frac{1}{r}\frac{dR}{dr} - \frac{\lambda^2}{r^2}R = 0 \tag{4-27}$$

which may be rearranged to the form

$$r^2\frac{d^2R}{dr^2} + r\frac{dR}{dr} - \lambda^2 R = 0 \tag{4-28}$$

Equation (4-28) is a special case of the more general ODE given by

$$\boxed{r^2\frac{d^2R}{dr^2} + a_0 r\frac{dR}{dr} + b_0 R = 0} \tag{4-29}$$

for the specific values of $a_0 = 1$ and $b_0 = -\lambda^2$. Equation (4-29) is known as the *Cauchy–Euler equation* (or Cauchy's equation) and is readily solved by making a change of variables $z = \ln(r)$ and solving the resulting auxiliary equation

$$\gamma^2 + (a_0 - 1)\gamma + b_0 = 0 \tag{4-30}$$

For our case given by equation (4-28), this becomes

$$\gamma^2 - \lambda_n^2 = 0 \quad \rightarrow \quad \gamma_{1,2} = \pm\lambda_n = \pm n \tag{4-31}$$

For the case of *two real roots* (i.e., $\pm n$), the ODE solution becomes

$$R(r) = C_3 r^{\gamma_1} + C_4 r^{\gamma_2} = C_3 r^n + C_4 r^{-n} \tag{4-32}$$

The two solutions in equation (4-32) are not orthogonal functions, as expected for the nonhomogeneous r dimension. We apply what we consider to be the homogeneous boundary condition BC1, namely, the requirement of finiteness at the origin, which eliminates C_4 given that this term goes to infinity as $r \rightarrow 0$ for $n > 0$. Having considered all but the final, nonhomogeneous boundary condition, we now recombine our separated solutions as a product:

$$T_n(r, \phi) = C_3 r^n \left(C_1 \cos n\phi + C_2 \sin n\phi\right) \tag{4-33}$$

The general solution is formed by summing over all solutions, yielding

$$T(r, \phi) = \sum_{n=0}^{\infty} r^n \left(a_n \cos n\phi + b_n \sin n\phi\right) \tag{4-34}$$

where we have introduced the two new constants, $a_n = C_1 C_3$ and $b_n = C_2 C_3$. We lastly apply the nonhomogeneous BC2, equation 4-23a, to determine a_n and b_n, which yields

$$T(r = b) = f(\phi) = \sum_{n=0}^{\infty} b^n \left(a_n \cos n\phi + b_n \sin n\phi\right) \tag{4-35}$$

Equation (4-35) is recognized as a general Fourier series expansion. We first solve for the Fourier coefficients a_n by applying to both sides the operator

$$* \int_{\phi=0}^{2\pi} \cos m\phi \, d\phi$$

noting that the integral

$$\int_{\phi=0}^{2\pi} \sin n\phi \, \cos m\phi \, d\phi = 0 \quad \text{for all integer values } n \text{ and } m \tag{4-36}$$

This yields the coefficients a_n as

$$a_n = \frac{\int_{\phi=0}^{2\pi} f(\phi)\cos n\phi \, d\phi}{b^n \int_{\phi=0}^{2\pi} \cos^2 n\phi \, d\phi} \tag{4-37}$$

The integral in the denominator of equation (4-37) is recognized as the norm $N(\lambda_n)$, and is equal to 2π for $n = 0$ and to π for $n \geq 1$. The Fourier coefficients b_n are now found by applying to both sides the operator

$$* \int_{\phi=0}^{2\pi} \sin m\phi \, d\phi$$

which yields the coefficients b_n as

$$b_n = \frac{\int_{\phi=0}^{2\pi} f(\phi)\sin n\phi \, d\phi}{b^n \int_{\phi=0}^{2\pi} \sin^2 n\phi \, d\phi} \tag{4-38}$$

The integral in the denominator of equation (4-38) is recognized as the norm $N(\lambda_n)$ and is equal to π for $n \geq 1$, noting that the sin $n\phi$ term drops from the solution for $n = 0$. The solution is now complete with equations (4-34), (4-37), and (4-38). We may simplify the $n = 0$ term, yielding the following solution:

$$T(r,\phi) = \frac{1}{2\pi}\int_{\phi=0}^{2\pi} f(\phi)\, d\phi + \sum_{n=1}^{\infty} r^n \left(a_n \cos n\phi + b_n \sin n\phi\right) \tag{4-39}$$

Example 4-2 $T = T(r, z)$ with Homogeneous z Dimension
A 2-D, solid cylinder of radius b and length L is maintained at steady-state conditions with a prescribed surface temperature $T(r = b) = f(z)$, and with the two ends maintained at zero temperature. The mathematical formulation of the problem is given as

$$\frac{\partial^2 T}{\partial r^2} + \frac{1}{r}\frac{\partial T}{\partial r} + \frac{\partial^2 T}{\partial z^2} = 0 \quad \text{in} \quad 0 \leq r < b, \quad 0 < z < L \tag{4-40}$$

$$\text{BC1:} \quad T(r \to 0) \Rightarrow \text{finite} \quad \text{or} \quad \left.\frac{\partial T}{\partial r}\right|_{r=0} = 0 \quad \text{(symmetry)} \tag{4-41a}$$

$$\text{BC2:} \quad T(r = b) = f(z) \tag{4-41b}$$

$$\text{BC3:} \quad T(z = 0) = 0 \qquad \text{BC4:} \quad T(z = L) = 0 \tag{4-41c}$$

where we may impose finiteness as $r \to 0$, or for this *axisymmetric case*, we may impose the symmetry condition at $r = 0$. We have a single nonhomogeneous boundary condition, so we proceed with separation of variables in the form

$$T(r, z) = R(r)Z(z) \tag{4-42}$$

Substituting equation (4-42) into the PDE of equation (4-40) yields

$$\frac{1}{R}\left(\frac{d^2R}{dr^2}+\frac{1}{r}\frac{dR}{dr}\right)=\frac{-1}{Z}\frac{d^2Z}{dz^2}=\lambda^2 \tag{4-43}$$

after multiplying both sides by $1/R(r)Z(z)$, and selecting the separation constant to force the characteristic value problem in the homogeneous z dimension. Solution of the separated ODE in the z dimension yields

$$Z(z)=C_1\cos\lambda z+C_2\sin\lambda z \tag{4-44}$$

This characteristic value problem is identical to the problems encountered with rectangular coordinate systems (see case 9 in Table 2-1). Applying BC3 yields constant $C_1=0$, while applying BC4 yields the eigenvalues

$$\lambda L=n\pi \quad\rightarrow\quad \lambda_n=\frac{n\pi}{L} \quad \text{for} \quad n=0,1,2,\ldots \tag{4-45}$$

Considering the r terms from equation (4-43) yields the ODE

$$\frac{d^2R}{dr^2}+\frac{1}{r}\frac{dR}{dr}-\lambda^2R=0 \tag{4-46}$$

which is the modified Bessel equation (see equation 2-59) for the specific case of $\nu=0$. The solution of equation (4-46) yields

$$R(r)=C_3I_0(\lambda r)+C_4K_0(\lambda r) \tag{4-47}$$

Considering first BC1, we can readily apply the requirement for finiteness as $r\rightarrow 0$ of equation (4-41a) to eliminate $K_0(\lambda r)$, that is, $C_4=0$, given the behavior of the modified Bessel functions of the second kind for argument zero (see Fig. 2-2). However, we can also consider the requirement of symmetry at $r=0$:

$$\frac{dR}{dr}=C_3\lambda I_1(\lambda r)-C_4\lambda K_1(\lambda r) \tag{4-48}$$

which yields

$$\left.\frac{dR}{dr}\right|_{r=0}=0=C_3\lambda I_1(0)-C_4\lambda K_1(r\rightarrow 0) \quad\rightarrow\quad C_4=0 \tag{4-49}$$

We see that both symmetry and finiteness at the origin provide the same outcome, namely, the elimination of the $K_0(\lambda r)$ term. We now form a product solution and sum over all product functions to give the general solution

$$T(r,z)=\sum_{n=1}^{\infty}C_nI_0(\lambda_n r)\sin\lambda_n z \tag{4-50}$$

where we have dropped the trivial solution corresponding to $n = 0$, and where we have introduced $C_n = C_2 C_3$. Finally, we apply the nonhomogeneous BC2:

$$T(r = b) = f(z) = \sum_{n=1}^{\infty} C_n I_0(\lambda_n b) \sin \lambda_n z \tag{4-51}$$

to which we apply the operator

$$* \int_{z=0}^{L} \sin \lambda_m z \, dz$$

which yields the coefficients C_n as

$$C_n = \frac{\int_{z=0}^{L} f(z) \sin \lambda_n z \, dz}{I_0(\lambda_n b) \int_{z=0}^{L} \sin^2 \lambda_n z \, dz} \tag{4-52}$$

The norm $N(\lambda_n)$ is equal to $L/2$ for the nontrivial eigenvalues of equation (4-45).

Example 4-3 $T = T(r, z)$ with Homogeneous r Dimension
A 2-D, semi-infinite, solid cylinder of radius b is maintained at steady-state conditions with a convective boundary condition at $r = b$ with convection coefficient h and fluid temperature T_∞, and with the end face at $z = 0$ maintained at the prescribed temperature distribution $T(z = 0) = f(r)$. The problem is illustrated in Figure 4-2. We note here that the Bi $= hb/k$ is not $\ll 1$; hence we maintain the r dependency and have $T = T(r, z)$.

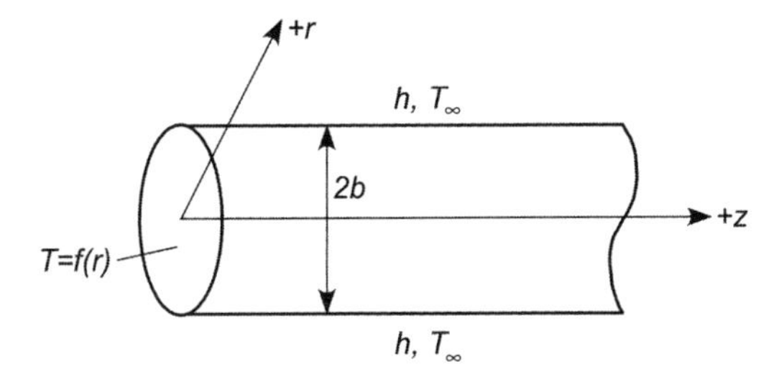

Figure 4-2 Problem description for Example 4-3.

The mathematical formulation of the problem is given as

$$\frac{\partial^2 T}{\partial r^2} + \frac{1}{r}\frac{\partial T}{\partial r} + \frac{\partial^2 T}{\partial z^2} = 0 \quad \text{in} \quad 0 \le r < b, \quad 0 < z < \infty \tag{4-53}$$

$$\text{BC1:} \quad T(r \to 0) \Rightarrow \text{finite} \quad \text{or} \quad \left.\frac{\partial T}{\partial r}\right|_{r=0} = 0 \quad \text{(symmetry)} \tag{4-54a}$$

$$\text{BC2:} \quad -k\left.\frac{\partial T}{\partial r}\right|_{r=b} = h\left[T|_{r=b} - T_\infty\right] \tag{4-54b}$$

$$\text{BC3:} \quad T(z=0) = f(r) \quad \text{BC4: } T(z \to \infty) = T_\infty \tag{4-54c}$$

As formulated, there are too many nonhomogeneities for separation of variables. Because a characteristic value problem is not suited to the semi-infinite z dimension, we will shift to $\Theta(r, z) = T(r, z) - T_\infty$, which yields

$$\frac{\partial^2 \Theta}{\partial r^2} + \frac{1}{r}\frac{\partial \Theta}{\partial r} + \frac{\partial^2 \Theta}{\partial z^2} = 0 \quad \text{in} \quad 0 \le r < b, \quad 0 < z < \infty \tag{4-55}$$

$$\text{BC1:} \quad \Theta(r \to 0) \Rightarrow \text{finite} \quad \text{or} \quad \left.\frac{\partial \Theta}{\partial r}\right|_{r=0} = 0 \quad \text{(symmetry)} \tag{4-56a}$$

$$\text{BC2:} \quad -k\left.\frac{\partial \Theta}{\partial r}\right|_{r=b} = h\,\Theta|_{r=b} \tag{4-56b}$$

$$\text{BC3:} \quad \Theta(z=0) = f(r) - T_\infty = g(r) \qquad \text{BC4:} \quad \Theta(z \to \infty) = 0 \tag{4-56c}$$

The problem is now set for separation of variables, assuming a form

$$\Theta(r, z) = R(r)Z(z) \tag{4-57}$$

Substituting equation (4-57) into the PDE of equation (4-55) yields

$$\frac{1}{Z}\frac{d^2 Z}{dz^2} = \frac{-1}{R}\left(\frac{d^2 R}{dr^2} + \frac{1}{r}\frac{dR}{dr}\right) = \lambda^2 \tag{4-58}$$

where we have selected the separation constant to force the characteristic value problem in the homogeneous r dimension. The corresponding ODE yields

$$\frac{d^2 R}{dr^2} + \frac{1}{r}\frac{dR}{dr} + \lambda^2 R = 0 \tag{4-59}$$

which is Bessel's equation (see equation 2-36b) for the specific case of $\nu = 0$. The solution of equation (4-59) yields

$$R(r) = C_1 J_0(\lambda r) + C_2 Y_0(\lambda r) \tag{4-60}$$

In a similar manner as detailed above in Example 4-2, we can readily apply the requirement for finiteness as $r \to 0$ of equation (4-56a) to eliminate $Y_0(\lambda r)$, that is, $C_2 = 0$, given the behavior of the Bessel functions of the second kind for argument zero (see Fig. 2-1). We can similarly eliminate $Y_0(\lambda r)$ in consideration of the symmetry boundary condition:

$$\left.\frac{dR}{dr}\right|_{r=0} = 0 = -C_1\lambda J_1(0) - C_2\lambda Y_1(r \to 0) \quad \to \quad C_2 = 0 \tag{4-61}$$

Application of BC2 yields the equation

$$-k\left(C_1 \left.\frac{dJ_0(\lambda r)}{dr}\right|_{r=b}\right) = hC_1 J_0(\lambda b) \tag{4-62}$$

which provides the following transcendental equation for eigenvalues λ_n:

$$J_0'(\lambda b) + \frac{h}{k} J_0(\lambda b) = 0 \quad \to \quad \lambda_n \quad \text{for} \quad n = 1, 2, 3, \ldots \tag{4-63}$$

We note here that the prime nomenclature represents integration with respect to r,

$$\left.\frac{dJ_0(\lambda r)}{dr}\right|_{r=b} \equiv J_0'(\lambda b) = -\lambda J_1(\lambda b) \tag{4-64}$$

Substitution of equation (4-64) into (4-63) and multiplication by b yields

$$\beta J_1(\beta) = Bi\, J_0(\beta) \quad \to \quad \beta_n \quad \text{for} \quad n = 1, 2, 3, \ldots \tag{4-65}$$

after making the substitution $\beta_n = \lambda_n b$, where the transcendental equation is now parameterized in terms of the Biot number (Bi $= hb/k$), and the roots β_n are related to the actual eigenvalues λ_n by $\lambda_n = \beta_n/b$. Table IV-3 in Appendix IV presents the first six roots of equation (4-65) for various values of Bi.

The remaining z terms of equation (4-58) give the ODE

$$\frac{d^2Z}{dz^2} - \lambda^2 Z = 0 \tag{4-66}$$

which yields a solution of the form

$$Z(z) = C_3 e^{\lambda z} + C_4 e^{-\lambda z} \tag{4-67}$$

We have selected the exponential form for the ODE solution given the semi-infinite domain in the z dimension. Application of BC4, equation (4-56c), eliminates constant C_3. We now recombine the separated functions to form a product solution and sum over all possible solutions:

$$\Theta(r, z) = \sum_{n=1}^{\infty} C_n J_0(\lambda_n r) e^{-\lambda_n z} \tag{4-68}$$

where we have introduced $C_n = C_1C_4$. We may now apply the nonhomogeneous BC3, given by equation (4-56c), to yield the Fourier–Bessel series expansion

$$\Theta(z=0) = g(r) = \sum_{n=1}^{\infty} C_n J_0(\lambda_n r) \tag{4-69}$$

We apply the following operator to both sides, noting the weighting factor r:

$$* \int_{r=0}^{b} r J_0(\lambda_m r) dr$$

which yields the coefficients C_n as

$$C_n = \frac{\int_{r=0}^{b} r g(r) J_0(\lambda_n r)\, dr}{\int_{r=0}^{b} r J_0^2(\lambda_n r)\, dr} \tag{4-70}$$

The norm $N(\lambda_n)$ in the denominator is given by case 1 in Table 2-2 for the specific eigenvalues of equation (4-63). Finally, we shift back to $T(r, z)$:

$$T(r,z) = \sum_{n=1}^{\infty} C_n J_0(\lambda_n r) e^{-\lambda_n z} + T_\infty \tag{4-71}$$

Example 4-4 $T = T(r, \phi)$ with Cylindrical Shell
A 2-D, long cylindrical wedge defined by angle ϕ_0, inner radius a, and outer radius b is maintained at steady-state conditions with a angularly varying incident heat flux on the outer surface, with prescribed temperature of zero on the inner surface, and insulated on the two sides of the wedge. The problem is illustrated in Figure 4-3, with $T = T(r, \phi)$.

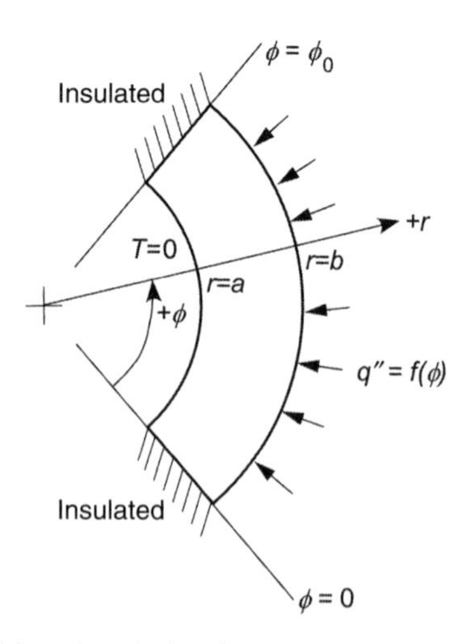

Figure 4-3 Problem description for Example 4-4.

The mathematical formulation of the problem is given as

$$\frac{\partial^2 T}{\partial r^2} + \frac{1}{r}\frac{\partial T}{\partial r} + \frac{1}{r^2}\frac{\partial^2 T}{\partial \phi^2} = 0 \quad \text{in} \quad a < r < b, \quad 0 < \phi < \phi_0 \tag{4-72}$$

$$\text{BC1:} \quad T(r = a) = 0 \qquad \text{BC2:} \quad -k\left.\frac{\partial T}{\partial r}\right|_{r=b} = -f(\phi) \tag{4-73a}$$

$$\text{BC3:} \quad \left.\frac{\partial T}{\partial \phi}\right|_{\phi=0} = 0 \qquad \text{BC4:} \quad \left.\frac{\partial T}{\partial \phi}\right|_{\phi=\phi_0} = 0 \tag{4-73b}$$

With a single nonhomogeneous boundary condition, we assume separation of the form

$$T(r, \phi) = R(r)\Phi(\phi) \tag{4-74}$$

Substituting equation (4-74) into the PDE of equation (4-72) yields

$$\frac{r^2}{R}\left(\frac{d^2 R}{dr^2} + \frac{1}{r}\frac{dR}{dr}\right) = \frac{-1}{\Phi}\frac{d^2\Phi}{d\phi^2} = \lambda^2 \tag{4-75}$$

where we have selected the separation constant to force the characteristic value problem in the homogeneous ϕ dimension. Solution of separated ODE in the ϕ dimension yields

$$\Phi(\phi) = C_1 \cos \lambda\phi + C_2 \sin \lambda\phi \tag{4-76}$$

We no longer have 2π periodicity but in fact have a traditional characteristic value problem in the ϕ dimension (case 5 in Table 2-1). Boundary condition BC3 eliminates the constant C_2, and BC4 determines the eigenvalues as

$$\lambda\phi_0 = n\pi \quad \rightarrow \quad \lambda_n = \frac{n\pi}{\phi_0} \quad \text{for} \quad n = 0, 1, 2, \ldots \tag{4-77}$$

The remaining r terms in equation (4-75) yield Cauchy's equation:

$$r^2\frac{d^2 R}{dr^2} + r\frac{dR}{dr} - \lambda^2 R = 0 \tag{4-78}$$

For *nonzero* values of λ_n, we again have the case of *two real roots* ($\pm\lambda$) for the auxiliary equation, and the ODE solution becomes

$$R_n(r) = C_3 r^{\lambda_n} + C_4 r^{-\lambda_n} \tag{4-79}$$

However, $\lambda_0 = 0$ is a nontrivial eigenvalue, which changes the nature of the roots of the auxiliary equation for Cauchy's equation. For this case, $a_0 = 1$

and $b_0 = -\lambda_0^2 = 0$, with the resulting auxiliary equation [see equation (4-30)] given by

$$\gamma^2 = 0 \quad \rightarrow \quad \gamma_{1,2} = 0 \quad \text{a double root} \tag{4-80}$$

For the case of a *real double root* (i.e., $\gamma_1 = \gamma_2 = \gamma = 0$), the ODE solution becomes

$$R_0(r) = C_3 r^{\gamma} + C_4 \ln(r) r^{\gamma} = C_3 + C_4 \ln(r) \tag{4-81}$$

We now use the homogeneous boundary condition BC1 in the r dimension to eliminate one of the constants for both $R_0(r)$ and $R_n(r)$:

$$R_0(r = a) = 0 = C_3 + C_4 \ln(a) \quad \rightarrow \quad C_3 = -C_4 \ln(a) \tag{4-82}$$

which upon substitution into equation (4-81) yields

$$R_0(r) = C_4 \ln\left(\frac{r}{a}\right) \tag{4-83}$$

Similarly for $R_n(r)$,

$$R_n(r = a) = 0 = C_3 a^{\lambda_n} + C_4 a^{-\lambda_n} \quad \rightarrow \quad C_4 = -C_3 a^{2\lambda_n} \tag{4-84}$$

which upon substitution into equation (4-79) yields

$$R_n(r) = C_5 \left[\left(\frac{r}{a}\right)^{\lambda_n} - \left(\frac{r}{a}\right)^{-\lambda_n}\right] \tag{4-85}$$

where we have factored out an a^{λ_n} term and introduced a new constant C_5. The general solution is then formed by summation of all product solutions, which after combining the constants, becomes

$$T(r, \phi) = \sum_{n=0}^{\infty} C_n R_n(r) \cos \lambda_n \phi \tag{4-86}$$

where $R_0(r)$ and $R_n(r)$ are defined by equations (4-83) and (4-85), respectively. We now apply the nonhomogeneous boundary condition BC2, equation (4-73a), which yields

$$f(\phi) = k \left.\frac{\partial T}{\partial r}\right|_{r=b} = k \sum_{n=0}^{\infty} C_n \left.\frac{dR_n}{dr}\right|_{r=b} \cos \lambda_n \phi \tag{4-87}$$

where we define the two derivative terms as

$$\left.\frac{dR_n}{dr}\right|_{r=b} = \begin{cases} \dfrac{1}{b} & \text{for} \quad n = 0 \\ \dfrac{\lambda_n}{b}\left[\left(\dfrac{b}{a}\right)^{\lambda_n} + \left(\dfrac{b}{a}\right)^{-\lambda_n}\right] & \text{for} \quad n \geq 1 \end{cases} \tag{4-88}$$

Equation (4-87) is a straightforward Fourier cosine series expansion of $f(\phi)$, so we apply the following operator to both sides:

$$*\int_{\phi=0}^{\phi_0} \cos\lambda_m\phi \, d\phi$$

which yields the constants C_0 and C_n as follows:

$$C_0 = \frac{(1/k)\int_{\phi=0}^{\phi_0} f(\phi)\, d\phi}{(1/b)\int_{\phi=0}^{\phi_0} d\phi} = \frac{b}{k\phi_0}\int_{\phi=0}^{\phi_0} f(\phi)\, d\phi \tag{4-89}$$

and

$$\begin{aligned} C_n &= \frac{(1/k)\int_{\phi=0}^{\phi_0} f(\phi)\cos\lambda_n\phi\, d\phi}{(\lambda_n/b)\left[(b/a)^{\lambda_n} + (b/a)^{-\lambda_n}\right]\int_{\phi=0}^{\phi_0}\cos^2\lambda_n\phi\, d\phi} \\ &= \frac{(2b/k\lambda_n\phi_0)\int_{\phi=0}^{\phi_0} f(\phi)\cos\lambda_n\phi\, d\phi}{\left[(b/a)^{\lambda_n} + (b/a)^{-\lambda_n}\right]} \end{aligned} \tag{4-90}$$

Finally, we define the solution as

$$T(r,\phi) = C_0\ln(r/a) + \sum_{n=1}^{\infty} C_n\left[\left(\frac{r}{a}\right)^{\lambda_n} - \left(\frac{r}{a}\right)^{-\lambda_n}\right]\cos\lambda_n\phi \tag{4-91}$$

While this problem is somewhat unique in that the nonhomogeneous problem was split into two functions, the overall solution process was a classical separation of variables. The split into two functions for the nonorthogonal r-dimension solution was purely due to the nature of the solution of Cauchy's equation for the situations of both two real roots and a double root.

Example 4-5 $T = T(r, \phi)$ with Convective Boundary Condition

A 2-D, long solid cylinder of radius b is maintained at steady-state condition with convection heat transfer at the outer surface characterized by convection coefficient h and by a fluid temperature that varies along the cylinder with temperature distribution $f(\phi)$. The mathematical formulation of the problem is given as:

$$\frac{\partial^2 T}{\partial r^2} + \frac{1}{r}\frac{\partial T}{\partial r} + \frac{1}{r^2}\frac{\partial^2 T}{\partial \phi^2} = 0 \quad \text{in} \quad 0 \le r < b \quad 0 \le \phi \le 2\pi \tag{4-92}$$

$$\text{BC1:} \quad T(r \to 0) \Rightarrow \text{finite} \tag{4-93a}$$

$$\text{BC2:} \quad -k\left.\frac{\partial T}{\partial r}\right|_{r=b} = h\left[T|_{r=b} - f(\phi)\right] \tag{4-93b}$$

$$\text{BC3:} \quad T(\phi) = T(\phi + 2\pi) \quad \rightarrow \quad 2\pi\text{-periodicity} \tag{4-93c}$$

$$\text{BC5:} \quad \left.\frac{\partial T(\phi)}{\partial \phi}\right|_{\phi} = \left.\frac{\partial T(\phi)}{\partial \phi}\right|_{\phi+2\pi} \quad \rightarrow \quad 2\pi\text{-periodicity} \tag{4-93d}$$

We note that symmetry about the origin does not apply given the ϕ dependency of the fluid temperature. We now assume separation of the form

$$T(r, \phi) = R(r)\Phi(\phi) \tag{4-94}$$

which upon substitution into the PDE of equation (4-92) yields

$$\frac{r^2}{R}\left(\frac{d^2R}{dr^2} + \frac{1}{r}\frac{dR}{dr}\right) = \frac{-1}{\Phi}\frac{d^2\Phi}{d\phi^2} = \lambda^2 \tag{4-95}$$

We have forced the characteristic value problem in the homogeneous ϕ dimension, which yields the desired solution

$$\Phi(\phi) = C_1 \cos \lambda\phi + C_2 \sin \lambda\phi \tag{4-96}$$

The requirement of 2π periodicity is satisfied by letting

$$\lambda_n = n = 0, 1, 2, 3, \ldots$$

where both constants C_1 and C_2 are retained. The remaining r terms in equation (4-95) yield Cauchy's equation

$$r^2\frac{d^2R}{dr^2} + r\frac{dR}{dr} - \lambda^2 R = 0 \tag{4-97}$$

with, for $\lambda_n = n$, the solution

$$R_n(r) = C_3 r^n + C_4 r^{-n} \tag{4-98}$$

As with Example 4-4, $\lambda_0 = 0$ yields a double root for the solution of Cauchy's equation, as given by equation (4-81). However, both the $\ln(r)$ and r^{-n} terms of the $\lambda_0 = 0$ and $\lambda_n \neq 0$ solutions, respectively, are eliminated due to the condition of finiteness as $r \rightarrow 0$, as given by boundary condition BC1. In particular, this leaves the solution $R_0(r) = C_3$. Therefore, equation (4-98) reduces to the correct solution of equation (4-97) for all values of λ_n defined for this problem, and we need not further divide the r dimension into two distinct solutions. We now form the general solution as a sum of all solutions:

$$T(r, \phi) = \sum_{n=0}^{\infty} r^n \left(a_n \cos n\phi + b_n \sin n\phi\right) \tag{4-99}$$

where we have introduced the two new constants, $a_n = C_1C_3$ and $b_n = C_2C_3$. We finally apply the nonhomogeneous BC2, equation (3-93b), to determine a_n and b_n. First, restating the boundary condition,

$$f(\phi) = \frac{h}{k} T|_{r=b} + \left.\frac{\partial T}{\partial r}\right|_{r=b} \tag{4-100}$$

this process yields

$$f(\phi) = \sum_{n=0}^{\infty} b^{n-1}\left(\frac{hb}{k} + n\right)\left(a_n \cos n\phi + b_n \sin n\phi\right) \tag{4-101}$$

where we have combined the two series terms on the right-hand side into a single summation. Equation (4-101) is a general Fourier series expansion of $f(\phi)$. We solve for the Fourier coefficients a_n and b_n by applying the operators

$$* \int_{\phi=0}^{2\pi} \cos m\phi \, d\phi \quad \text{and} \quad * \int_{\phi=0}^{2\pi} \sin m\phi \, d\phi$$

respectively, to both sides. This yields the coefficients a_n and b_n:

$$a_n = \frac{\int_{\phi=0}^{2\pi} f(\phi) \cos n\phi \, d\phi}{b^{n-1}\left[(hb/k) + n\right] \int_{\phi=0}^{2\pi} \cos^2 n\phi \, d\phi} \tag{4-102}$$

and

$$b_n = \frac{\int_{\phi=0}^{2\pi} f(\phi) \sin n\phi \, d\phi}{b^{n-1}\left[(hb/k) + n\right] \int_{\phi=0}^{2\pi} \sin^2 n\phi \, d\phi} \tag{4-103}$$

While equations (4-102) and (4-103) complete the problem, we may substitute these two equations back into the solution of equation (4-100), which after a trigonometric substitution, yields

$$T(r, \phi) = \frac{b}{\pi} \sum_{n=0}^{\infty} \left(\frac{r}{b}\right)^n \frac{1}{hb/k + n} \int_{\phi'=0}^{2\pi} f(\phi') \cos n(\phi - \phi') d\phi' \tag{4-104}$$

where the term π is replaced by 2π for the first term of $n = 0$.

Example 4-6 $T = T(r, z)$ with Internal Energy Generation

A 2-D, solid cylinder of radius b and length L is maintained at steady-state conditions with a prescribed surface temperature $T(r = b) = f(z)$, with the two ends maintained at zero temperature, and with internal energy generation g_0(W/m^3).

The mathematical formulation of this problem, using the general cylindrical heat equation from equation (1-43), is given as

$$\frac{\partial^2 T}{\partial r^2} + \frac{1}{r}\frac{\partial T}{\partial r} + \frac{\partial^2 T}{\partial z^2} + \frac{g_0}{k} = 0 \quad \text{in} \quad 0 \leq r < b, \quad 0 < z < L \tag{4-105}$$

BC1: $$\left.\frac{\partial T}{\partial r}\right|_{r=0} = 0 \quad \text{(symmetry)} \tag{4-106a}$$

BC2: $$T(r = b) = f(z) \tag{4-106b}$$

BC3: $$T(z = 0) = 0 \qquad \text{BC4:} \qquad T(z = L) = 0 \tag{4-106c}$$

where, in anticipation of superposition, we impose the symmetry condition at $r = 0$. We have a nonhomogeneous PDE, so we proceed with *superposition* of the form

$$T(r, z) = \Psi(r, z) + \Phi(z) \tag{4-107}$$

where $\Psi(r, z)$ will take the homogenous PDE and $\Phi(z)$ will take the nonhomogeneous ODE. As we developed in Chapter 3, for treatment of internal energy generation, it is useful to take advantage of the derivative boundary condition in the r dimension (i.e., type 2 boundary) to uncouple the $\Phi(z)$ solution from the derivative boundary condition of the homogeneous PDE problem. The formulation of the nonhomogeneous $\Phi(z)$ problem becomes

$$\frac{d^2\Phi}{dz^2} + \frac{g_0}{k} = 0 \quad \text{in} \quad 0 < z < L \tag{4-108}$$

BC3: $$\Phi(z = 0) = 0 \qquad \text{BC4:} \qquad \Phi(z = L) = 0 \tag{4-109}$$

The solution of equations (4-108) and (4-109) is identical to the solution in equation (3-179) of Example 3-7, which we repeat here as

$$\Phi(z) = \frac{g_0 L^2}{2k}\left[\frac{z}{L} - \left(\frac{z}{L}\right)^2\right] \tag{4-110}$$

We now formulate the 2-D homogeneous PDE, namely, the $\Psi(r, z)$ problem:

$$\frac{\partial^2 \Psi}{\partial r^2} + \frac{1}{r}\frac{\partial \Psi}{\partial r} + \frac{\partial^2 \Psi}{\partial z^2} = 0 \quad \text{in} \quad 0 \leq r < b \quad 0 < z < L \tag{4-111}$$

BC1: $$\left.\frac{\partial \Psi}{\partial r}\right|_{r=0} = 0 \quad \text{(symmetry)} \tag{4-112a}$$

BC2: $$\Psi(r = b) = f(z) - \Phi(z) = g(z) \tag{4-112b}$$

BC3: $$\Psi(z = 0) = 0 \qquad \text{BC4:} \qquad \Psi(z = L) = 0 \tag{4-112c}$$

where we have coupled the $\Phi(z)$ problem to the boundary condition at $r = b$, as we developed in Chapter 3. As mentioned above, because of the derivative boundary condition at $r = 0$, we avoided coupling the $\Phi(z)$ problem a second time, which would have necessitated the solution of the $\Psi(r, z)$ problem with an *additional* superposition step (i.e., two nonhomogeneous boundary conditions). Had we selected the 1-D nonhomogeneous problem as $\Phi(r)$, that is exactly the outcome that would have occurred, namely, a nonhomogeneous boundary condition of $-\Phi(r)$ at both the $z = 0$ and the $z = L$ boundaries, requiring superposition for solution. With a single nonhomogeneous boundary condition, we now proceed with separation of variables in the form

$$\Psi(r, z) = R(r)Z(z) \tag{4-113}$$

Substituting equation (4-113) into the PDE of equation (4-111) yields

$$\frac{1}{R}\left(\frac{d^2R}{dr^2} + \frac{1}{r}\frac{dR}{dr}\right) = \frac{-1}{Z}\frac{d^2Z}{dz^2} = \lambda^2 \tag{4-114}$$

after selecting the separation constant to force the characteristic value problem in the homogeneous z dimension. Solution of separated ODE in the z dimension yields

$$Z(z) = C_1 \cos \lambda z + C_2 \sin \lambda z \tag{4-115}$$

This characteristic value problem is identical to the problem in Example 4-2 (see case 9 in Table 2-1). Applying BC3 yields constant $C_1 = 0$, while applying BC4 yields the eigenvalues

$$\lambda L = n\pi \quad \rightarrow \quad \lambda_n = \frac{n\pi}{L} \quad \text{for} \quad n = 0, 1, 2, \ldots \tag{4-116}$$

Considering the r terms from equation (4-114) yields the ODE

$$\frac{d^2R}{dr^2} + \frac{1}{r}\frac{dR}{dr} - \lambda^2 R = 0 \tag{4-117}$$

and the corresponding modified Bessel functions

$$R(r) = C_3 I_0(\lambda r) + C_4 K_0(\lambda r) \tag{4-118}$$

By applying BC1, equation (4-112a), we let $C_4 = 0$, thereby eliminating $K_0(\lambda r)$, given the behavior of the modified Bessel functions of the second kind for argument zero (see Fig. 2-2):

$$\left.\frac{dR}{dr}\right|_{r=0} = 0 = C_3 \lambda I_1(0) - C_4 \lambda K_1(r \rightarrow 0) \quad \rightarrow \quad C_4 = 0 \tag{4-119}$$

We now form a product solution and sum over all possible products:

$$\Psi(r, z) = \sum_{n=1}^{\infty} C_n I_0(\lambda_n r) \sin \lambda_n z \tag{4-120}$$

where we have dropped the trivial solution corresponding to $n = 0$, and where we have introduced $C_n = C_2 C_3$. Finally, we apply the nonhomogeneous BC2:

$$\Psi(r = b) = g(z) = \sum_{n=1}^{\infty} C_n I_0(\lambda_n b) \sin \lambda_n z \tag{4-121}$$

to which we apply the operator

$$* \int_{z=0}^{L} \sin \lambda_m z \, dz$$

which yields the Fourier coefficients C_n as

$$C_n = \frac{\int_{z=0}^{L} g(z) \sin \lambda_n z \, dz}{I_0(\lambda_n b) \int_{z=0}^{L} \sin^2 \lambda_n z \, dz} \tag{4-122}$$

The norm $N(\lambda_n)$ is equal to $L/2$ for the nontrivial eigenvalues of equation (4-116). The complete solution is now given by

$$T(r, z) = \sum_{n=1}^{\infty} C_n I_0(\lambda_n r) \sin \lambda_n z + \frac{gL^2}{2k} \left[\frac{z}{L} - \left(\frac{z}{L} \right)^2 \right] \tag{4-123}$$

Example 4-7 $\boldsymbol{T = T(r, \phi, z)}$

A 3-D, solid cylinder of radius b and length L is maintained at steady-state conditions with a prescribed surface heat flux $f(z,\phi)$ incident at the surface $r = b$, with the end at $z = 0$ maintained at zero temperature, and with the end at $z = L$ perfectly insulated. The mathematical formulation of this problem is given as

$$\frac{\partial^2 T}{\partial r^2} + \frac{1}{r}\frac{\partial T}{\partial r} + \frac{1}{r^2}\frac{\partial^2 T}{\partial \phi^2} + \frac{\partial^2 T}{\partial z^2} = 0 \quad \text{in} \quad 0 \le r < b, \quad 0 < z < L \quad 0 \le \phi \le 2\pi \tag{4-124}$$

BC1: $\quad T(r \to 0) \Rightarrow \text{finite}$ (4-125a)

BC2: $\quad -k \left. \dfrac{\partial T}{\partial r} \right|_{r=b} = -f(z, \phi)$ (4-125b)

BC3: $\quad T(z = 0) = 0$ BC4: $\left. \dfrac{\partial T}{\partial z} \right|_{z=L} = 0$ (4-125c)

BC5: $T(\phi) = T(\phi + 2\pi) \quad \rightarrow \quad 2\pi\text{-periodicity}$ (4-125d)

BC6: $\left.\dfrac{\partial T(\phi)}{\partial \phi}\right|_{\phi} = \left.\dfrac{\partial T(\phi)}{\partial \phi}\right|_{\phi+2\pi} \quad \rightarrow \quad 2\pi\text{-periodicity}$ (4-125e)

Because we have only a single nonhomogeneous boundary condition, we can assume a separation of the form

$$T(r, \phi, z) = R(r)\Phi(\phi)Z(z) \tag{4-126}$$

which upon substitution into equation (4-124) yields

$$\frac{1}{R}\left(\frac{d^2R}{dr^2} + \frac{1}{r}\frac{dR}{dr}\right) + \frac{1}{r^2}\frac{1}{\Phi}\frac{d^2\Phi}{d\phi^2} = -\frac{1}{Z}\frac{d^2Z}{dz^2} = \lambda^2 \tag{4-127}$$

where we have introduced a separation constant $+\lambda^2$ to force the ODE

$$\frac{d^2Z}{dz^2} + \lambda^2 Z = 0 \tag{4-128}$$

along with the desired trigonometric functions in the z dimension, namely,

$$Z(z) = C_1 \cos \lambda z + C_2 \sin \lambda z \tag{4-129}$$

Boundary condition BC3 eliminates constant C_1, while boundary condition BC4 yields the eigenvalues (case 8 in Table 2-1):

$$\lambda L = \frac{(2m+1)\pi}{2} \quad \rightarrow \quad \lambda_m = \frac{(2m+1)\pi}{2L} \quad \text{for} \quad m = 0, 1, 2 \ldots \tag{4-130}$$

The remaining terms of equation (4-127) are rearranged as

$$\frac{r^2}{R}\left(\frac{d^2R}{dr^2} + \frac{1}{r}\frac{dR}{dr}\right) - \lambda^2 r^2 = -\frac{1}{\Phi}\frac{d^2\Phi}{d\phi^2} = n^2 \tag{4-131}$$

where we have now introduced a new separation constant $+n^2$ to force the ODE:

$$\frac{d^2\Phi}{d\phi^2} + n^2\Phi = 0 \tag{4-132}$$

which yields the desired solution form in the ϕ dimension as

$$\Phi(\phi) = C_3 \cos n\phi + C_4 \sin n\phi \tag{4-133}$$

Here we let $n = 0, 1, 2, 3 \ldots$ to satisfy 2π periodicity. Finally, we consider the remaining r terms of equation (4-131), which are rearranged to

$$\frac{d^2R}{dr^2} + \frac{1}{r}\frac{dR}{dr} - \left(\lambda^2 + \frac{n^2}{r^2}\right)R = 0 \tag{4-134}$$

which we recognize as modified Bessel's equation of order n. The solution of equation (4-134) yields

$$R(r) = C_5 I_n(\lambda r) + C_6 K_n(\lambda r) \tag{4-135}$$

Boundary condition BC1 eliminates $K_n(\lambda r)$ from further consideration. Having considered all the homogeneous boundary conditions, we form the general solution by summing over all possible solutions:

$$T(r, \phi, z) = \sum_{m=0}^{\infty} \sum_{n=0}^{\infty} I_n(\lambda_m r) \sin \lambda_m z \left[a_{nm} \cos n\phi + b_{nm} \sin n\phi\right] \tag{4-136}$$

where we have defined new constants $a_{nm} = C_2 C_3 C_5$ and $b_{nm} = C_2 C_4 C_5$. Finally, we apply the nonhomogeneous boundary condition given by equation (4-125b),

$$f(z, \phi) = k \sum_{m=0}^{\infty} \sum_{n=0}^{\infty} \left. \frac{dI_n(\lambda_m r)}{dr} \right|_{r=b} \sin \lambda_m z \left[a_{nm} \cos n\phi + b_{nm} \sin n\phi\right] \tag{4-137}$$

which is a Fourier–Bessel–trigonometric series expansion of the known function $f(z,\phi)$. To solve for the coefficients a_{nm}, we apply the two operators successively to both sides

$$* \int_{\phi=0}^{2\pi} \cos p\phi \, d\phi \quad \text{and} \quad * \int_{z=0}^{L} \sin \lambda_k z \, dz$$

which yields the result

$$a_{nm} = \frac{\int_{\phi=0}^{2\pi} \int_{z=0}^{L} f(z, \phi) \sin \lambda_m z \cos n\phi \, dz \, d\phi}{k I_n'(\lambda_m b) \int_{z=0}^{L} \sin^2 \lambda_m z \, dz \int_{\phi=0}^{2\pi} \cos^2 n\phi \, d\phi} \tag{4-138}$$

where we define the prime notation to denote the derivative with respect to r (see equation 4-137). In a similar manner, we solve for the coefficients b_{nm} using the following two operators successively:

$$* \int_{\phi=0}^{2\pi} \sin p\phi \, d\phi \quad \text{and} \quad * \int_{z=0}^{L} \sin \lambda_k z \, dz$$

which yields the result

$$b_{nm} = \frac{\int_{\phi=0}^{2\pi} \int_{z=0}^{L} f(z, \phi) \sin \lambda_m z \sin n\phi \, dz \, d\phi}{k I_n'(\lambda_m b) \int_{z=0}^{L} \sin^2 \lambda_m z \, dz \int_{\phi=0}^{2\pi} \sin^2 n\phi \, d\phi} \tag{4-139}$$

4-3 SOLUTION OF TRANSIENT PROBLEMS

In this section we will solve a range of transient problems in the cylindrical coordinate system. Our general guidelines established for separation of variables in the rectangular coordinate system equally apply to the cylindrical coordinate problems. Therefore, transient problems with a homogeneous PDE must always be solved with all homogeneous boundary conditions and with the single nonhomogeneity being the initial condition. If one or more boundary conditions are nonhomogeneous, the principle of superposition is used to treat the nonhomogeneous boundary conditions separately in a steady-state problem. For a nonhomogeneous PDE (i.e., when heat generation is present), superposition is used to treat the heat generation within a homogeneous steady-state problem that is coupled to a homogeneous transient problem at the initial condition. Prior to solving a range of problems to illustrate the above procedures, a few additional guidelines are offered for the transient heat equation in cylindrical coordinates.

For two-dimensional, transient conduction problems two distinct cases are physically possible, namely, $T = T(r, z, t)$ and $T = T(r, \phi, t)$. For both instances we expect the orthogonal Bessel functions, with order zero for the case of no ϕ dependency. We expect the orthogonal trigonometric functions $\cos \eta z$ and $\sin \eta z$ from the z dimension and once again state that the ϕ dimension must always be homogeneous, generating the trigonometric functions $\cos n\phi$ and $\sin n\phi$ for the condition of 2π periodicity for the full cylinder ($0 \le \phi \le 2\pi$).

For three-dimensional, transient problems, all three spatial dimensions must be homogeneous; hence we expect orthogonal Bessel functions of order ν since we have ϕ dependency. We expect the orthogonal trigonometric functions $\cos \eta z$ and $\sin \eta z$ from the z dimension, and the trigonometric functions $\cos \nu\phi$ and $\sin \nu\phi$ from the ϕ dimension, where ν is an integer for the case of 2π periodicity for the full cylinder ($0 \le \phi \le 2\pi$).

Example 4-8 $\boldsymbol{T = T(r, t)}$ **with Insulated Boundary**

A long solid cylinder of radius b is initially at temperature $F(r)$. For $t>0$, the boundary condition at $r = b$ is perfectly insulated. The mathematical formulation of the problem is given as

$$\frac{\partial^2 T}{\partial r^2} + \frac{1}{r}\frac{\partial T}{\partial r} = \frac{1}{\alpha}\frac{\partial T(r,t)}{\partial t} \quad \text{in} \quad 0 \le r < b, \quad t > 0 \tag{4-140}$$

$$\text{BC1:} \quad T(r \to 0) \Rightarrow \text{finite} \qquad \text{BC2:} \quad \left.\frac{\partial T}{\partial r}\right|_{r=b} = 0 \tag{4-141a}$$

$$\text{IC:} \quad T(t = 0) = F(r) \tag{4-141b}$$

With all homogeneous boundary conditions, we assume a form

$$T(r, t) = R(r)\Gamma(t) \tag{4-142}$$

which after substitution into equation (4-140) yields

$$\frac{1}{R}\left(\frac{d^2R}{dr^2}+\frac{1}{r}\frac{dR}{dr}\right)=\frac{1}{\alpha\Gamma}\frac{d\Gamma}{dt}=-\lambda^2 \tag{4-143}$$

Solution of separated ODE in the t dimension yields the desired solution

$$\Gamma(t)=C_1e^{-\alpha\lambda^2t} \tag{4-144}$$

while the r terms yield Bessel's equation of order zero, with the solution

$$R(r)=C_2J_0(\lambda r)+C_3Y_0(\lambda r) \tag{4-145}$$

The requirement for finiteness as $r\to 0$ stated by BC1 eliminates the $Y_0(\lambda r)$ term, while BC2 yields

$$\left.\frac{dR}{dr}\right|_{r=b}=0=-C_2\lambda J_1(\lambda b) \tag{4-146}$$

which produces eigenvalues from the corresponding transcendental equation

$$J_1(\lambda_n b)=0 \quad\rightarrow\quad \lambda_n \quad \text{for} \quad n=0,1,2,3,\ldots \tag{4-147}$$

We note that the eigenvalue $\lambda_0=0$ is a nontrivial eigenvalue given that $J_1(0)=0$ and $J_0(0)=1$. The general solution is formed by the summation over all products

$$T(r,t)=\sum_{n=0}^{\infty}C_nJ_0(\lambda_n r)e^{-\alpha\lambda_n^2t} \tag{4-148}$$

with $C_n=C_1C_2$. The initial condition is now applied, yielding

$$T(t=0)=F(r)=\sum_{n=0}^{\infty}C_nJ_0(\lambda_n r) \tag{4-149}$$

where we solve for the Fourier–Bessel coefficients by applying the operator

$$*\int_{r=0}^{b} r J_0(\lambda_m r)\,dr$$

noting the inclusion of the weighting factor r. This process yields the result

$$C_n=\frac{\int_{r=0}^{b} rF(r)J_0(\lambda_n r)\,dr}{\int_{r=0}^{b} rJ_0^2(\lambda_n r)\,dr} \tag{4-150}$$

The denominator is the norm $N(\lambda_n)$ and may be simplified by case 2 in Table 2-2. Let us now consider the eigenvalue $\lambda_0 = 0$, which reduces equation (4-150) to

$$C_0 = \frac{\int_{r=0}^{b} rF(r)\,dr}{\int_{r=0}^{b} r\,dr} = \frac{2}{b^2}\int_{r=0}^{b} rF(r)\,dr \tag{4-151}$$

Using equation (4-151), we rewrite our solution in the form

$$T(r,t) = \frac{2}{b^2}\int_{r=0}^{b} rF(r)\,dr + \sum_{n=1}^{\infty} C_n J_0(\lambda_n r)e^{-\alpha\lambda_n^2 t} \tag{4-152}$$

We can now readily examine the steady-state solution as $t \to \infty$ from equation (4-152),

$$T(t \to \infty) \equiv T_{ss} = \frac{2}{b^2}\int_{r=0}^{b} rF(r)\,dr \tag{4-153}$$

We see that the first eigenvalue $\lambda_0 = 0$ actually yields the important steady-state solution. It is useful to examine the steady-state solution given by equation (4-153) in the context of conservation of energy. For the perfectly insulated cylinder, we must conserve total energy per unit length, namely,

$$E'_{\text{initial}} \equiv E'_{\text{final}} \quad \text{J/m} \tag{4-154}$$

We may evaluate the left-side term by integrating the initial energy over the cross-sectional area, while the right-side term follows from $E'_{\text{final}} = mcT_{ss}/L$, yielding

$$\int_{r=0}^{b} (\rho c)2\pi r F(r)\,dr = \frac{(\rho c)\left(\pi b^2 L\right) T_{ss}}{L} \tag{4-155}$$

Simplification of equation (4-155) directly yields equation (4-153); hence our solution of the heat equation, including the steady-state solution, correctly reflects conservation of energy.

Example 4-9 $T = T(r, \phi, t)$ with Convective Boundary

A long solid cylinder of radius b is initially at a temperature $F(r, \phi)$. For $t>0$, the boundary condition at $r = b$ is exposed to convection heat transfer with convection coefficient h and fluid temperature of zero. The mathematical formulation of the problem is given as

$$\frac{\partial^2 T}{\partial r^2} + \frac{1}{r}\frac{\partial T}{\partial r} + \frac{1}{r^2}\frac{\partial^2 T}{\partial \phi^2} = \frac{1}{\alpha}\frac{\partial T}{\partial t} \quad \text{in} \quad 0 \le r < b, \quad 0 \le \phi \le 2\pi, \quad t > 0 \tag{4-156}$$

$$\text{BC1:} \quad T(r \to 0) \Rightarrow \text{finite} \qquad \text{BC2:} \quad -k\frac{\partial T}{\partial r}\bigg|_{r=b} = h\ T|_{r=b} \tag{4-157a}$$

$$\text{BC3:} \quad T(\phi) = T(\phi + 2\pi) \quad \to \quad 2\pi\text{-periodicity} \tag{4-157b}$$

$$\text{BC4:} \quad \frac{\partial T(\phi)}{\partial \phi}\bigg|_{\phi} = \frac{\partial T(\phi)}{\partial \phi}\bigg|_{\phi+2\pi} \quad \to \quad 2\pi\text{-periodicity} \tag{4-157c}$$

$$\text{IC:} \quad T(t = 0) = F(r, \phi) \tag{4-157d}$$

We note that we do not have symmetry in the r dimension given the initial condition; hence we use finiteness. With all homogeneous boundary conditions, we assume separation of a form

$$T(r, \phi, t) = R(r)\Phi(\phi)\Gamma(t) \tag{4-158}$$

which after substitution into equation (4-156) yields

$$\frac{1}{R}\left(\frac{d^2R}{dr^2} + \frac{1}{r}\frac{dR}{dr}\right) + \frac{1}{r^2\Phi}\frac{d^2\Phi}{d\phi^2} = \frac{1}{\alpha\Gamma}\frac{d\Gamma}{dt} = -\lambda^2 \tag{4-159}$$

Solution of a separated ODE in the t dimension yields

$$\Gamma(t) = C_1 e^{-\alpha\lambda^2 t} \tag{4-160}$$

while the r terms and ϕ term of equation (4-159) are now separated as

$$\frac{r^2}{R}\left[\frac{d^2R}{dr^2} + \frac{1}{r}\frac{dR}{dr}\right] + \lambda^2 r^2 = -\frac{1}{\Phi}\frac{d^2\Phi}{d\phi^2} = n^2 \tag{4-161}$$

Solution of the ϕ dimension ODE yields

$$\Phi(\phi) = C_2 \cos n\phi + C_3 \sin n\phi \tag{4-162}$$

where we let $n = 0, 1, 2, 3 \ldots$ to satisfy 2π periodicity. Finally, we consider the remaining r terms of equation (4-161), which are rearranged as

$$\frac{d^2R}{dr^2} + \frac{1}{r}\frac{dR}{dr} + \left(\lambda^2 - \frac{n^2}{r^2}\right)R = 0 \tag{4-163}$$

which we recognize as Bessel's equation of order n, with the solution

$$R(r) = C_4 J_n(\lambda r) + C_5 Y_n(\lambda r) \tag{4-164}$$

The requirement for finiteness as $r \to 0$ stated by BC1 eliminates the $Y_n(\lambda r)$ term, while BC2 yields the transcendental equation

$$\frac{dJ_n(\lambda r)}{dr}\bigg|_{r=b} + \frac{h}{k}J_n(\lambda b) = 0 \quad \to \quad \lambda_{nm} \quad \text{for} \quad m = 1, 2, 3, \ldots \tag{4-165}$$

which generates eigenvalues λ_{nm}. We use a double subscript for the eigenvalues to denote that *for each value of* n, there is a full set of eigenvalues λ_m. For example, with $n = 0$, equation (4-165) becomes

$$-\lambda_m J_1(\lambda_m b) + \frac{h}{k} J_0(\lambda_m b) = 0 \quad \rightarrow \quad \lambda_m \quad \text{for} \quad m = 1, 2, 3, \ldots \tag{4-166}$$

while for $n = 1$, equation (4-165) becomes

$$\left(\frac{1}{b} + \frac{h}{k}\right) J_1(\lambda_m b) - \lambda_m J_2(\lambda_m b) = 0 \quad \rightarrow \quad \lambda_m \quad \text{for} \quad m = 1, 2, 3, \ldots \tag{4-167}$$

Clearly the set of roots are different for these two equations. In addition, we note that the eigenvalue $\lambda = 0$ is a trivial eigenvalue for $n \geq 1$, while zero is not an eigenvalue for the case of $n = 0$. We now form the general solution by summing the product solutions over all eigenvalues n and λ_{nm}, yielding

$$T(r, \phi, t) = \sum_{n=0}^{\infty} \sum_{m=1}^{\infty} J_n(\lambda_{nm} r) \left(a_{nm} \cos n\phi + b_{nm} \sin n\phi\right) e^{-\alpha \lambda_{nm}^2 t} \tag{4-168}$$

with constants $a_{nm} = C_1 C_2 C_4$ and $b_{nm} = C_1 C_3 C_4$. The order of summation is intentional, such that the inner series will converge while summing over the eigenvalues λ_{nm} for each value n of the outer series. With increasing values of n, the *sum of the entire inner series* is found to converge. We now apply the initial condition of equation (4-157d), yielding

$$F(r, \phi) = \sum_{n=0}^{\infty} \sum_{m=1}^{\infty} J_n(\lambda_{nm} r) \left(a_{nm} \cos n\phi + b_{nm} \sin n\phi\right) \tag{4-169}$$

to which we apply successively the two operators to both sides

$$* \int_{r=0}^{b} r J_n(\lambda_{nk} r)\, dr \quad \text{and} \quad * \int_{\phi=0}^{2\pi} \cos p\phi \, d\phi$$

to yield the coefficients a_{nm} as follows:

$$a_{nm} = \frac{\int_{\phi=0}^{2\pi} \int_{r=0}^{b} r F(r, \phi) J_n(\lambda_{nm} r) \cos n\phi \, dr \, d\phi}{\int_{r=0}^{b} r J_n^2(\lambda_{nm} r)\, dr \int_{\phi=0}^{2\pi} \cos^2 n\phi \, d\phi} \tag{4-170}$$

Similarly, we apply successively the two operators to both sides

$$* \int_{r=0}^{b} r J_n(\lambda_{nk} r)\, dr \quad \text{and} \quad * \int_{\phi=0}^{2\pi} \sin p\phi \, d\phi$$

to yield the coefficients b_{nm} as follows:

$$b_{nm} = \frac{\int_{\phi=0}^{2\pi} \int_{r=0}^{b} r F(r,\phi) J_n(\lambda_{nm} r) \sin n\phi \, dr \, d\phi}{\int_{r=0}^{b} r J_n^2(\lambda_{nm} r) dr \int_{\phi=0}^{2\pi} \sin^2 n\phi \, d\phi} \tag{4-171}$$

The steady-state temperature for this problem is the fluid temperature, namely, zero, which is readily achieved as $t \to \infty$ in equation (4-168). The norms for the integrals of the Bessel functions in the denominator of equations (4-170) and (4-171) are given by case 1 in Table 2-2 corresponding the specific eigenvalues of equation (4-165).

Example 4-10 $T = T(r, t)$ for a Hollow Cylinder

A long, hollow cylinder of inner radius a and outer radius b is initially at temperature $F(r)$. For $t>0$, the inner and outer surfaces are maintained at a constant temperature of zero. The mathematical formulation of the problem is given as

$$\frac{\partial^2 T}{\partial r^2} + \frac{1}{r}\frac{\partial T}{\partial r} = \frac{1}{\alpha}\frac{\partial T(r,t)}{\partial t} \quad \text{in} \quad a < r < b, \quad t > 0 \tag{4-172}$$

$$\text{BC1:} \quad T(r = a) = 0 \qquad \text{BC2:} \quad T(r = b) = 0 \tag{4-173a}$$

$$\text{IC:} \quad T(t = 0) = F(r) \tag{4-173b}$$

With all homogeneous boundary conditions, we assume a form

$$T(r,t) = R(r)\Gamma(t) \tag{4-174}$$

Substitution into equation (4-172) yields

$$\frac{1}{R}\left(\frac{d^2 R}{dr^2} + \frac{1}{r}\frac{dR}{dr}\right) = \frac{1}{\alpha\Gamma}\frac{d\Gamma}{dt} = -\lambda^2 \tag{4-175}$$

with the solution of the separated ODE in the t dimension yielding

$$\Gamma(t) = C_1 e^{-\alpha\lambda^2 t} \tag{4-176}$$

while the solution of the separated ODE in the r dimension yields

$$R(r) = C_2 J_0(\lambda r) + C_3 Y_0(\lambda r) \tag{4-177}$$

Applying BC1 allows elimination of constant C_3, namely,

$$R(r = a) = 0 = C_2 J_0(\lambda a) + C_3 Y_0(\lambda a) \quad \to \quad C_3 = -C_2 \frac{J_0(\lambda a)}{Y_0(\lambda a)} \tag{4-178}$$

which, upon substitution into equation (4-177), and after factoring out the $Y_0(\lambda a)$ term, gives the eigenfunctions

$$R(r) = \frac{C_2}{Y_0(\lambda a)} \left[J_0(\lambda r) Y_0(\lambda a) - J_0(\lambda a) Y_0(\lambda r) \right] \tag{4-179}$$

where the constant in front of the bracket can be replaced by a new constant C_4. Applying BC2 gives

$$R(r = b) = 0 = C_4 \left[J_0(\lambda b) Y_0(\lambda a) - J_0(\lambda a) Y_0(\lambda b) \right] \tag{4-180}$$

which gives the following transcendental equation for eigenvalues λ_n:

$$J_0(\lambda b) Y_0(\lambda a) - J_0(\lambda a) Y_0(\lambda b) = 0 \quad \rightarrow \quad \lambda_n \quad \text{for} \quad n = 1, 2, 3, \ldots \tag{4-181}$$

where we note that $\lambda_0 = 0$ is not an eigenvalue since the Bessel functions of the second kind are undefined for an argument of zero. The general solution is now formed by summing all of the individual product solutions

$$T(r, t) = \sum_{n=1}^{\infty} C_n \left[J_0(\lambda_n r) Y_0(\lambda_n a) - J_0(\lambda_n a) Y_0(\lambda_n r) \right] e^{-\alpha \lambda_n^2 t} \tag{4-182}$$

with $C_n = C_1 C_4$. Applying the initial condition yields

$$T(t = 0) = F(r) = \sum_{n=1}^{\infty} C_n \left[J_0(\lambda_n r) Y_0(\lambda_n a) - J_0(\lambda_n a) Y_0(\lambda_n r) \right] \tag{4-183}$$

The entire term in brackets is an orthogonal function over the interval $a \leq r \leq b$ with weighting function r. To solve for the Fourier coefficients, we apply the following operator to both sides:

$$* \int_{r=a}^{b} r \left[J_0(\lambda_p r) Y_0(\lambda_p a) - J_0(\lambda_p a) Y_0(\lambda_p r) \right] dr$$

which yields the expression

$$C_n = \frac{\int_{r=a}^{b} r F(r) \left[J_0(\lambda_n r) Y_0(\lambda_n a) - J_0(\lambda_n a) Y_0(\lambda_n r) \right] dr}{\int_{r=a}^{b} r \left[J_0(\lambda_n r) Y_0(\lambda_n a) - J_0(\lambda_n a) Y_0(\lambda_n r) \right]^2 dr} \tag{4-184}$$

The integral in the denominator is recognized as the norm, and is given by case 4 in Table 2-3 for the eigenvalues of equation (4-181).

Example 4-11 $T = T(r, \phi, t)$ **with Cylindrical Wedge**
A very long cylindrical wedge defined by angle ϕ_0, with radius b, is initially maintained at a temperature $F(r, \phi)$. For $t>0$, all three boundary surfaces are maintained at a prescribed temperature of zero, as illustrated in Figure 4-4.

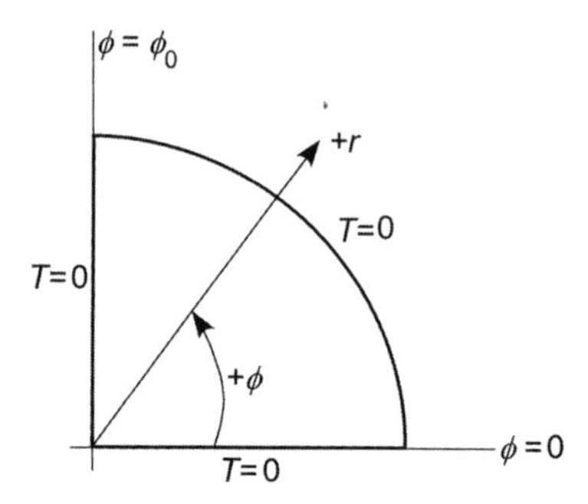

Figure 4-4 Problem description for Example 4-11.

The mathematical formulation of the problem is given as

$$\frac{\partial^2 T}{\partial r^2} + \frac{1}{r}\frac{\partial T}{\partial r} + \frac{1}{r^2}\frac{\partial^2 T}{\partial \phi^2} = \frac{1}{\alpha}\frac{\partial T(r,t)}{\partial t} \quad \text{in} \quad 0 \le r < b, \quad 0 < \phi < \phi_0 \tag{4-185}$$

BC1: $T(r \to 0) \Rightarrow$ finite BC2: $T(r = b) = 0$ (4-186a)

BC3: $T(\phi = 0) = 0$ BC4: $T(\phi = \phi_0) = 0$ (4-186b)

IC: $T(t = 0) = F(r, \phi)$ (4-186c)

With all homogeneous boundary conditions, we assume separation of the form

$$T(r, \phi, t) = R(r)\Phi(\phi)\Gamma(t) \tag{4-187}$$

Substituting equation (4-187) into equation (4-185) yields

$$\frac{1}{R}\left(\frac{d^2 R}{dr^2} + \frac{1}{r}\frac{dR}{dr}\right) + \frac{1}{r^2 \Phi}\frac{d^2 \Phi}{d\phi^2} = \frac{1}{\alpha \Gamma}\frac{d\Gamma}{dt} = -\lambda^2 \tag{4-188}$$

Solution of separated ODE in the t dimension yields the solution

$$\Gamma(t) = C_1 e^{-\alpha \lambda^2 t} \tag{4-189}$$

with the r terms and ϕ term of equation (4-188) yielding

$$\frac{r^2}{R}\left(\frac{d^2 R}{dr^2} + \frac{1}{r}\frac{dR}{dr}\right) + \lambda^2 r^2 = -\frac{1}{\Phi}\frac{d^2 \Phi}{d\phi^2} = \nu^2 \tag{4-190}$$

Solution of the ϕ dimension ODE yields

$$\Phi(\phi) = C_2 \cos \nu\phi + C_3 \sin \nu\phi \tag{4-191}$$

where we no longer have 2π periodicity but, in fact, have a traditional characteristic value problem in the ϕ dimension (case 9 in Table 2-1). Boundary condition BC3 eliminates the constant C_2, and BC4 determines the eigenvalues as

$$\nu\phi_0 = n\pi \quad \rightarrow \quad \nu_n = \frac{n\pi}{\phi_0} \quad \text{for} \quad n = 0, 1, 2, \ldots \tag{4-192}$$

We next consider the remaining r terms of equation (4-190), which we recognize as Bessel's equation of order ν, with the solution

$$R(r) = C_4 J_\nu(\lambda r) + C_5 Y_\nu(\lambda r) \tag{4-193}$$

The requirement for finiteness as $r \rightarrow 0$ stated by BC1 eliminates the $Y_\nu(\lambda r)$ term, while BC2 yields the transcendental equation

$$J_\nu(\lambda b) = 0 \quad \rightarrow \quad \lambda_{nm} \quad \text{for} \quad m = 1, 2, 3, \ldots \tag{4-194}$$

which generates eigenvalues λ_{nm}. As described with Example 4-9, we use a double subscript for the eigenvalues to denote that for each value of ν_n, there is a full set of eigenvalues λ_m. We now form the general solution by summing the product solutions over all eigenvalues ν_n and λ_{nm}, yielding

$$T(r, \phi, t) = \sum_{n=1}^{\infty} \sum_{m=1}^{\infty} C_{nm} J_\nu(\lambda_{nm} r) \sin \nu_n \phi \, e^{-\alpha\lambda_{nm}^2 t} \tag{4-195}$$

with constant $C_{nm} = C_1 C_3 C_4$. We have omitted the eigenvalue $\nu_0 = 0$, which yields a trivial solution, from the summation, and note that for nonzero values of ν_n, $\lambda = 0$ is a trivial eigenvalue. Application of the initial condition of equation (4-157d) yields

$$F(r, \phi) = \sum_{n=1}^{\infty} \sum_{m=1}^{\infty} C_{nm} J_\nu(\lambda_{nm} r) \sin \nu_n \phi \tag{4-196}$$

to which we apply successively the two operators

$$* \int_{r=0}^{b} r J_\nu(\lambda_{nk} r)\, dr \quad \text{and} \quad * \int_{\phi=0}^{\phi_0} \sin \nu_p \phi \, d\phi$$

to yield the Fourier–Bessel–sine coefficients C_{nm} as

$$C_{nm} = \frac{\int_{\phi=0}^{\phi_0} \int_{r=0}^{b} r F(r, \phi) J_\nu(\lambda_{nm} r) \sin \nu_n \phi \, dr \, d\phi}{\int_{r=0}^{b} r J_\nu^2(\lambda_{nm} r)\, dr \int_{\phi=0}^{\phi_0} \sin^2 \nu_n \phi \, d\phi} \tag{4-197}$$

The norms of the denominator are given by $\phi_0/2$ for the sine functions and given by case 3 in Table 2-2 for the Bessel functions. By substituting the two norms, as well as the Fourier coefficients, into equation (4-195), we may express the solution in a combined form as

$$T(r,\phi,t) = \frac{4}{b^2\phi_0}\sum_{n=1}^{\infty}\sum_{m=1}^{\infty}\left\{\frac{J_\nu(\lambda_{nm}r)\sin\nu_n\phi}{J_{\nu+1}^2(\lambda_{nm}b)}e^{-\alpha\lambda_{nm}^2 t} \times \int_{\phi'=0}^{\phi_0}\int_{r'=0}^{b} r'F(r',\phi')J_\nu(\lambda_{nm}r')\sin\nu_n\phi'\,dr'\,d\phi'\right\} \tag{4-198}$$

Example 4-12 $T = T(r, z, t)$ for a Hollow Cylinder

A hollow cylinder of length L with inner radius a and outer radius b is initially at temperature $F(r,z)$. For $t>0$, the inner and outer surfaces are maintained at a constant temperature of zero, the boundary at $z = 0$ is insulated, and the boundary at $z = L$ dissipates heat by convection with convection coefficient h into a fluid at zero temperature. The mathematical formulation of the problem is given as

$$\frac{\partial^2 T}{\partial r^2} + \frac{1}{r}\frac{\partial T}{\partial r} + \frac{\partial^2 T}{\partial z^2} = \frac{1}{\alpha}\frac{\partial T(r,t)}{\partial t} \quad \text{in} \quad a < r < b, \quad 0 < z < L, \quad t > 0 \tag{4-199}$$

$$\text{BC1:} \quad T(r=a) = 0 \qquad \text{BC2:} \quad T(r=b) = 0 \tag{4-200a}$$

$$\text{BC3:} \quad \left.\frac{\partial T}{\partial z}\right|_{z=0} = 0 \qquad \text{BC4:} \quad -k\left.\frac{\partial T}{\partial z}\right|_{z=L} = h\,T|_{z=L} \tag{4-200b}$$

$$\text{IC:} \quad T(t=0) = F(r,z) \tag{4-200c}$$

With all homogeneous boundary conditions, we assume a form

$$T(r,z,t) = R(r)Z(z)\Gamma(t) \tag{4-201}$$

Substitution into equation (4-199) yields

$$\frac{1}{R}\left(\frac{d^2R}{dr^2} + \frac{1}{r}\frac{dR}{dr}\right) + \frac{1}{Z}\frac{d^2Z}{dz^2} = \frac{1}{\alpha\Gamma}\frac{d\Gamma}{dt} = -\lambda^2 \tag{4-202}$$

with the solution of the separated ODE in the t dimension yielding

$$\Gamma(t) = C_1 e^{-\alpha\lambda^2 t} \tag{4-203}$$

The remaining r terms and z term of equation (4-202) yield

$$\frac{1}{R}\left(\frac{d^2R}{dr^2} + \frac{1}{r}\frac{dR}{dr}\right) + \lambda^2 = -\frac{1}{Z}\frac{d^2Z}{dz^2} = \beta^2 \tag{4-204}$$

where we have forced the characteristic value problem in the z dimension, with separation constant β^2, yielding the functions

$$Z(z) = C_2 \cos \beta z + C_3 \sin \beta z \tag{4-205}$$

Boundary condition BC3 eliminates the constant C_3, and BC4 yields

$$-k\left(-C_2\beta \sin \beta L\right) = h\left(C_2 \cos \beta L\right) \tag{4-206}$$

which is readily rearranged to the following transcendental equation:

$$\beta_n \tan \beta_n L = \frac{h}{k} \quad \rightarrow \quad \beta_n \quad \text{for} \quad n = 1, 2, 3, \ldots \tag{4-207}$$

where the roots β_n are the eigenvalues, as seen in case 4 of Table 2-1. The r terms of equation (4-204) yield Bessel's equation of order zero, giving the solution

$$R(r) = C_4 J_0(\eta r) + C_5 Y_0(\eta r) \tag{4-208}$$

where we have made the substitution $\eta^2 = \lambda^2 - \beta^2$. Applying BC1 allows elimination of constant C_5, namely,

$$R(r = a) = 0 = C_4 J_0(\eta a) + C_5 Y_0(\eta a) \quad \rightarrow \quad C_5 = -C_4 \frac{J_0(\eta a)}{Y_0(\eta a)} \tag{4-209}$$

which, upon substitution into equation (4-207) and after factoring out the $Y_0(\eta a)$ term, gives the eigenfunctions

$$R(r) = C_6\left[J_0(\eta r)Y_0(\eta a) - J_0(\eta a)Y_0(\eta r)\right] \tag{4-210}$$

As we did with Example 4-10, we introduced a new constant C_6 in front of the brackets. Applying BC2 yields

$$R(r = b) = 0 = C_6\left[J_0(\eta b)Y_0(\eta a) - J_0(\eta a)Y_0(\eta b)\right] \tag{4-211}$$

which gives the following transcendental equation for eigenvalues η_m:

$$J_0(\eta b)Y_0(\eta a) - J_0(\eta a)Y_0(\eta b) = 0 \quad \rightarrow \quad \eta_m \quad \text{for} \quad m = 1, 2, 3, \ldots \tag{4-212}$$

noting here again that $\eta_0 = 0$ is not an eigenvalue. The general solution is now formed by summing all of the individual product solutions

$$T(r, z, t) = \sum_{n=1}^{\infty}\sum_{m=1}^{\infty} C_{nm}\left[J_0(\eta_m r)Y_0(\eta_m a) - J_0(\eta_m a)Y_0(\eta_m r)\right] \cos \beta_n z\, e^{-\alpha\lambda_{nm}^2 t} \tag{4-213}$$

with $C_{nm} = C_1 C_2 C_6$. With both eigenvalues defined, λ^2 is now given by

$$\lambda_{nm}^2 = \beta_n^2 + \eta_m^2 \tag{4-214}$$

We apply the initial condition given by equation (4-200c), which yields

$$F(r,z) = \sum_{n=1}^{\infty}\sum_{m=1}^{\infty} C_{nm}\left[J_0(\eta_m r)Y_0(\eta_m a) - J_0(\eta_m a)Y_0(\eta_m r)\right]\cos\beta_n z \tag{4-215}$$

To calculate the Fourier coefficients, we apply the two operators to both sides:

$$*\int_{r=a}^{b} r\left[J_0(\eta_k r)Y_0(\eta_k a) - J_0(\eta_k a)Y_0(\eta_k r)\right]dr \quad \text{and} \quad *\int_{z=0}^{L} \sin\beta_p z\, dz$$

which yields the final result

$$C_{nm} = \frac{\int_{z=0}^{L}\int_{r=a}^{b} rF(r,z)\left[J_0(\eta_m r)Y_0(\eta_m a) - J_0(\eta_m a)Y_0(\eta_m r)\right]\sin\beta_n z\, dr\, dz}{\int_{r=a}^{b} r\left[J_0(\eta_m r)Y_0(\eta_m a) - J_0(\eta_m a)Y_0(\eta_m r)\right]^2 dr \int_{z=0}^{L}\sin^2\beta_n z\, dz} \tag{4-216}$$

The norms of equation (4-216) are given by case 4 of Table 2-1 for the sine functions, and by case 4 of Table 2-3 for the Bessel functions.

Example 4-13 $T = T(r,t)$ with Internal Energy Generation

A long solid cylinder of radius b is initially at temperature $F(r)$. For $t>0$, the boundary condition at $r = b$ is maintained at zero temperature, while the cylinder is subjected to uniform internal energy generation g_0 (W/m^3). The mathematical formulation of the problem is given as:

$$\frac{\partial^2 T}{\partial r^2} + \frac{1}{r}\frac{\partial T}{\partial r} + \frac{g_0}{k} = \frac{1}{\alpha}\frac{\partial T(r,t)}{\partial t} \quad \text{in} \quad 0 \le r < b, \quad t>0 \tag{4-217}$$

$$\text{BC1:} \quad T(r\to 0) \Rightarrow \text{finite} \qquad \text{BC2:} \quad T(r=b) = 0 \tag{4-218a}$$

$$\text{IC:} \quad T(t=0) = F(r) \tag{4-218b}$$

We have a nonhomogeneous PDE, so we proceed with *superposition* of the form

$$T(r,t) = \Psi(r,t) + \Phi(r) \tag{4-219}$$

where $\Psi(r,t)$ will take the homogenous PDE and $\Phi(r)$ will take the nonhomogeneous ODE. As we developed in Chapter 3, the solution of the nonhomogeneous

ODE must couple to the homogeneous PDE problem at the initial condition. The formulation of the nonhomogeneous $\Phi(r)$ problem becomes

$$\frac{d^2\Phi}{dr^2} + \frac{1}{r}\frac{d\Phi}{dr} + \frac{g_0}{k} = 0 \quad \text{in} \quad 0 \le r < b \tag{4-220}$$

$$\text{BC1:} \quad \Phi(r \to 0) \Rightarrow \text{finite} \qquad \text{BC2:} \quad \Phi(r = b) = 0 \tag{4-221}$$

The solution of equation (4-220) yields

$$\Phi(r) = C_1 + C_2 \ln(r) - \frac{g_0}{4k} r^2 \tag{4-222}$$

where BC1 eliminates the ln(r) term, and BC2 yields the constant C_1. With these substitutions, the solution becomes

$$\Phi(r) = \frac{g_0}{4k}\left(b^2 - r^2\right) \tag{4-223}$$

We now formulate the homogeneous problem $\Psi(r, t)$:

$$\frac{\partial^2\Psi}{\partial r^2} + \frac{1}{r}\frac{\partial\Psi}{\partial r} = \frac{1}{\alpha}\frac{\partial\Psi(r,t)}{\partial t} \quad \text{in} \quad 0 \le r < b \quad t > 0 \tag{4-224}$$

$$\text{BC1:} \quad \Psi(r \to 0) \Rightarrow \text{finite} \qquad \text{BC2:} \quad \Psi(r = b) = 0 \tag{4-225a}$$

$$\text{IC:} \quad \Psi(t = 0) = F(r) - \Phi(r) = G(r) \tag{4-225b}$$

where we have coupled the solution of $\Phi(r)$ to the initial condition. With all homogeneous boundary conditions, the $\Psi(r, t)$ problem is ready for separation of variables. We assume a solution of the form

$$\Psi(r, t) = R(r)\Gamma(t) \tag{4-226}$$

which after substitution into equation (4-224) yields

$$\frac{1}{R}\left(\frac{d^2R}{dr^2} + \frac{1}{r}\frac{dR}{dr}\right) = \frac{1}{\alpha\Gamma}\frac{d\Gamma}{dt} = -\lambda^2 \tag{4-227}$$

Solution of the separated ODE in the t dimension yields the desired solution

$$\Gamma(t) = C_1 e^{-\alpha\lambda^2 t} \tag{4-228}$$

while the r terms yield Bessel's equation of order zero, with solution

$$R(r) = C_2 J_0(\lambda r) + C_3 Y_0(\lambda r) \tag{4-229}$$

The requirement for finiteness as $r \to 0$ stated by BC1 eliminates the $Y_0(\lambda r)$ term, while BC2 yields the following transcendental equation

$$J_0(\lambda_n b) = 0 \quad \to \quad \lambda_n \quad \text{for} \quad n = 1, 2, 3, \ldots \tag{4-230}$$

for eigenvalues λ_n. We note that $\lambda_0 = 0$ is not an eigenvalue. The general solution is now formed by the summation over all products

$$\Psi(r,t) = \sum_{n=1}^{\infty} C_n J_0(\lambda_n r) e^{-\alpha\lambda_n^2 t} \tag{4-231}$$

with $C_n = C_1 C_2$. The initial condition equation (4-225b) is now applied, yielding

$$\Psi(t=0) = G(r) = \sum_{n=1}^{\infty} C_n J_0(\lambda_n r) \tag{4-232}$$

where we solve for the Fourier–Bessel coefficients by applying the operator

$$* \int_{r=0}^{b} r J_0(\lambda_m r)\, dr$$

noting the inclusion of the weighting factor r. This process yields the result

$$C_n = \frac{\int_{r=0}^{b} r\,[F(r) - \Phi(r)]\, J_0(\lambda_n r)\, dr}{\int_{r=0}^{b} r J_0^2(\lambda_n r)\, dr} \tag{4-233}$$

where we have substituted for $G(r)$. The denominator is the norm $N(\lambda_n)$ and may be simplified by case 3 in Table 2-2. The overall solution then becomes

$$T(r,t) = \sum_{n=1}^{\infty} C_n J_0(\lambda_n r) e^{-\alpha\lambda_n^2 t} + \frac{g_0}{4k}\left(b^2 - r^2\right) \tag{4-234}$$

The steady-state solution is realized as $t \to \infty$ in equation (4-234), which yields the expected result for a 1-D cylinder with generation and outer temperature of zero:

$$T_{ss}(r) = \frac{g_0}{4k}\left(b^2 - r^2\right) \tag{4-235}$$

Example 4-14 $T = T(r, \phi, z, t)$ for a Solid Cylinder

A solid cylinder of radius b and length L is initially at a temperature $F(r, \phi, z)$. For $t > 0$, the boundary condition at $r = b$ is exposed to convection heat transfer with convection coefficient h and fluid temperature of zero, the boundary at $z = 0$ is insulated, and the boundary at $z = L$ is maintained at zero temperature. The mathematical formulation of the problem is given as

$$\frac{\partial^2 T}{\partial r^2} + \frac{1}{r}\frac{\partial T}{\partial r} + \frac{1}{r^2}\frac{\partial^2 T}{\partial \phi^2} + \frac{\partial^2 T}{\partial z^2} = \frac{1}{\alpha}\frac{\partial T}{\partial t} \quad \text{in} \quad \begin{array}{ll} 0 \le r < b & 0 < z < L \\ 0 \le \phi \le 2\pi & \text{for} \quad t > 0 \end{array} \tag{4-236}$$

$$\text{BC1:} \quad T(r \to 0) \Rightarrow \text{finite} \qquad \text{BC2:} \quad -k\left.\frac{\partial T}{\partial r}\right|_{r=b} = h\, T|_{r=b} \tag{4-237a}$$

$$\text{BC3:} \quad T(\phi) = T(\phi + 2\pi) \quad \rightarrow \quad 2\pi - \text{periodicity} \tag{4-237b}$$

$$\text{BC4:} \quad \left.\frac{\partial T(\phi)}{\partial \phi}\right|_{\phi} = \left.\frac{\partial T(\phi)}{\partial \phi}\right|_{\phi+2\pi} \quad \rightarrow \quad 2\pi - \text{periodicity} \tag{4-237c}$$

$$\text{BC5:} \quad \left.\frac{\partial T}{\partial z}\right|_{z=0} = 0 \qquad \text{BC6:} \quad T(z = L) = 0 \tag{4-237d}$$

$$\text{IC:} \quad T(t = 0) = F(r, \phi, z) \tag{4-237e}$$

We note that we do not have symmetry in the r dimension given the initial condition; hence we use finiteness at the origin. With all homogeneous boundary conditions, we assume separation of the form

$$T(r, \phi, z, t) = R(r)\Phi(\phi)Z(z)\Gamma(t) \tag{4-238}$$

which after substitution into equation (4-236) yields

$$\frac{1}{R}\left(\frac{d^2R}{dr^2} + \frac{1}{r}\frac{dR}{dr}\right) + \frac{1}{r^2\Phi}\frac{d^2\Phi}{d\phi^2} + \frac{1}{Z}\frac{d^2Z}{dz^2} = \frac{1}{\alpha\Gamma}\frac{d\Gamma}{dt} = -\lambda^2 \tag{4-239}$$

Solution of the separated ODE in the t dimension yields the solution

$$\Gamma(t) = C_1 e^{-\alpha\lambda^2 t} \tag{4-240}$$

with the remaining terms of equation (4-239) giving

$$\frac{1}{R}\left(\frac{d^2R}{dr^2} + \frac{1}{r}\frac{dR}{dr}\right) + \frac{1}{r^2}\frac{1}{\Phi}\frac{d^2\Phi}{d\phi^2} + \lambda^2 = -\frac{1}{Z}\frac{d^2Z}{dz^2} = \eta^2 \tag{4-241}$$

Solution of the z-dimension ODE yields the desired solution form

$$Z(z) = C_2 \cos \eta z + C_3 \sin \eta z \tag{4-242}$$

with the characteristic value problem corresponding to case 6 in Table 2-1. Applying BC5 yields constant $C_3 = 0$, while applying BC6 yields the eigenvalues

$$\cos \eta L = 0 \quad \rightarrow \quad \eta_m = \frac{(2m+1)\,\pi}{2L} \quad \text{for} \quad m = 0, 1, 2, \ldots \tag{4-243}$$

We now consider the remaining terms of equation (4-241), namely,

$$\frac{r^2}{R}\left(\frac{d^2R}{dr^2} + \frac{1}{r}\frac{dR}{dr}\right) + \left(\lambda^2 - \eta^2\right) r^2 = -\frac{1}{\Phi}\frac{d^2\Phi}{d\phi^2} = n^2 \tag{4-244}$$

where we have introduced another separation constant $+n^2$ to force the ODE

$$\frac{d^2\Phi}{d\phi^2} + n^2\Phi = 0 \tag{4-245}$$

which yields the desired solution in the ϕ dimension

$$\Phi(\phi) = C_4 \cos n\phi + C_5 \sin n\phi \tag{4-246}$$

We let $n = 0, 1, 2, 3, \ldots$ to satisfy 2π periodicity in the ϕ dimension as stated by BC3 and BC4. Finally, we consider the remaining r terms of equation (4-244), where we will let $\beta^2 = \lambda^2 - \eta^2$ and then multiply both sides by R/r^2, yielding

$$\frac{d^2R}{dr^2} + \frac{1}{r}\frac{dR}{dr} + \left(\beta^2 - \frac{n^2}{r^2}\right)R = 0 \tag{4-247}$$

We recognize equation (4-247) as Bessel's equation of integer order n, with the solution yielding

$$R(r) = C_6 J_n(\beta r) + C_7 Y_n(\beta r) \tag{4-248}$$

The requirement for finiteness as $r \to 0$ stated by BC1 eliminates the $Y_n(\beta r)$ term, while BC2 yields the transcendental equation

$$\left.\frac{dJ_n(\beta r)}{dr}\right|_{r=b} + \frac{h}{k}J_n(\beta b) = 0 \quad \to \quad \beta_{np} \quad \text{for} \quad p = 1, 2, 3, \ldots \tag{4-249}$$

which generates the eigenvalues β_{np}. As we discussed in Example 4-9, we use a double subscript for the eigenvalues to denote that *for each value of* n, there is a full set of eigenvalues β_p. Having defined the eigenvalues for all three spatial dimensions, we form a product solution of the separated functions and sum over all possible solutions, yielding

$$T(r, \phi, z, t) = \sum_{m=0}^{\infty}\sum_{n=0}^{\infty}\sum_{p=1}^{\infty} J_n(\beta_{np}r)\cos \eta_m z\left(a_{nmp}\cos n\phi + b_{nmp}\sin n\phi\right)e^{-\alpha\lambda_{nmp}^2 t} \tag{4-250}$$

with constants $a_{nmp} = C_1C_2C_4C_6$ and $b_{nmp} = C_1C_2C_5C_6$. The order of summation is intentional, such that the inner series over p will converge while summing over the eigenvalues β_{np} for each value n. We define λ_{nmp} as

$$\lambda_{nmp}^2 = \beta_{np}^2 + \eta_m^2 \tag{4-251}$$

We now apply the initial condition of equation (4-237e), yielding

$$F(r, \phi, z) = \sum_{m=0}^{\infty}\sum_{n=0}^{\infty}\sum_{p=1}^{\infty} J_n(\beta_{np}r)\cos \eta_m z\left(a_{nmp}\cos n\phi + b_{nmp}\sin n\phi\right) \tag{4-252}$$

to which we apply successively the three operators to both sides

$$* \int_{r=0}^{b} r J_n(\beta_{nl} r)\, dr \quad * \int_{\phi=0}^{2\pi} \cos k\phi\, d\phi \quad \text{and} \quad * \int_{z=0}^{L} \cos \eta_q z\, dz$$

to yield the coefficients a_{nmp} as follows:

$$a_{nmp} = \frac{\int_{z=0}^{L} \int_{\phi=0}^{2\pi} \int_{r=0}^{b} r F(r,\phi,z) J_n(\beta_{np} r) \cos \eta_m z \cos n\phi\, dr\, d\phi\, dz}{\int_{z=0}^{L} \cos^2 \eta_m z\, dz \int_{r=0}^{b} r J_n^2(\beta_{np} r)\, dr \int_{\phi=0}^{2\pi} \cos^2 n\phi\, d\phi} \tag{4-253}$$

We then apply successively the three operators to both sides:

$$* \int_{r=0}^{b} r J_n(\beta_{nl} r)\, dr \quad * \int_{\phi=0}^{2\pi} \sin k\phi\, d\phi \quad \text{and} \quad * \int_{z=0}^{L} \cos \eta_q z\, dz$$

to yield the coefficients b_{nmp} as follows:

$$b_{nmp} = \frac{\int_{z=0}^{L} \int_{\phi=0}^{2\pi} \int_{r=0}^{b} r F(r,\phi,z) J_n(\beta_{np} r) \cos \eta_m z \sin n\phi\, dr\, d\phi\, dz}{\int_{z=0}^{L} \cos^2 \eta_m z\, dz \int_{r=0}^{b} r J_n^2(\beta_{np} r)\, dr \int_{\phi=0}^{2\pi} \sin^2 n\phi\, d\phi} \tag{4-254}$$

The norms in the denominators of equations (4-253) and (4-254) are equal to $L/2$ for the z-dimension eigenvalues, to π for the ϕ-dimension eigenvalues except for the case of $n = 0$, for which the ϕ-integral norm of equation (4-253) is then 2π, and, finally, for the Bessel functions as given by case 1 in Table 2-2 for the specific eigenvalues of equation (4-249).

4-4 CAPSTONE PROBLEM

In this chapter we have developed the solution schemes for the solution of the cylindrical heat equation using both the method of separation of variables and superposition. Our focus has been to develop the analytical solutions, with emphasis on the resulting eigenfunctions and eigenvalues, and the necessary Fourier coefficients. The general approach has been to leave the Fourier coefficients in integral form, although at times we have performed the integrals and substituted back into the series summation. Here we consider a final problem in detail in an attempt to integrate and illustrate the many aspects of our temperature field solutions presented in this chapter, including the transient behavior of the heat flux and temperature field in the context of conservation of energy.

We consider here a solid cylinder as shown in Figure 4-5. The cylinder is of radius b and length L and may be considered perfectly insulated at the ends $z = 0$ and $z = L$. Initially, the cylinder is characterized by a parabolic temperature distribution along the z direction given by $T = F(z)$. For $t > 0$, the surface $r = b$

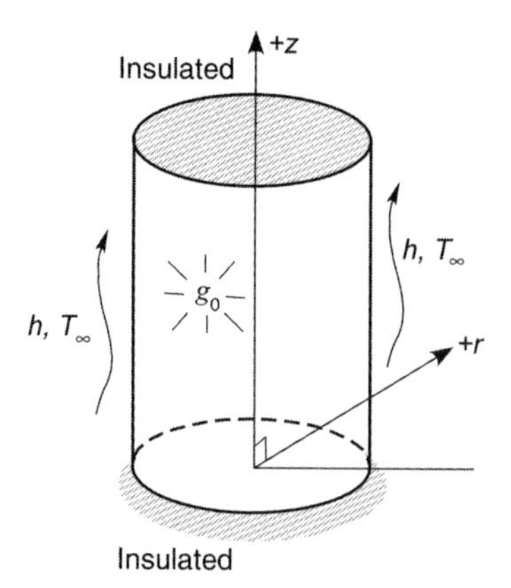

Figure 4-5 Description for capstone problem.

is subjected to convection heat transfer with convection coefficient h and fluid temperature T_∞, while the cylinder is also subjected to uniform internal energy generation g_0 (W/m^3).

The mathematical formulation of the problem is

$$\frac{\partial^2 T}{\partial r^2} + \frac{1}{r}\frac{\partial T}{\partial r} + \frac{\partial^2 T}{\partial z^2} + \frac{g_0}{k} = \frac{1}{\alpha}\frac{\partial T(r,t)}{\partial t} \quad \text{in} \quad 0 \le r < b \quad 0 < z < L, \quad t > 0 \tag{4-255}$$

$$\text{BC1:} \quad T(r \to 0) \Rightarrow \text{finite} \quad \text{BC2:} \quad -k\left.\frac{\partial T}{\partial r}\right|_{r=b} = h\left[T|_{r=b} - T_\infty\right] \tag{4-256a}$$

$$\text{BC3:} \quad \left.\frac{\partial T}{\partial z}\right|_{z=0} = 0 \qquad \text{BC4:} \quad \left.\frac{\partial T}{\partial z}\right|_{z=L} = 0 \tag{4-256b}$$

$$\text{IC:} \quad T(t=0) = F(z) \tag{4-256c}$$

We have a nonhomogeneous PDE and a nonhomogeneous boundary condition in a transient problem. We first treat the nonhomogeneous boundary condition by shifting the temperature to $\Theta(r, z, t) = T(r, z, t) - T_\infty$, which yields

$$\frac{\partial^2 \Theta}{\partial r^2} + \frac{1}{r}\frac{\partial \Theta}{\partial r} + \frac{\partial^2 \Theta}{\partial z^2} + \frac{g_0}{k} = \frac{1}{\alpha}\frac{\partial \Theta(r,t)}{\partial t} \quad \text{in} \quad 0 \le r < b \quad 0 < z < L, \quad t > 0 \tag{4-257}$$

$$\text{BC1:} \quad \Theta(r \to 0) \Rightarrow \text{finite} \quad \text{BC2:} \quad -k\left.\frac{\partial \Theta}{\partial r}\right|_{r=b} = h\,\Theta|_{r=b} \tag{4-258a}$$

$$\text{BC3:} \quad \left.\frac{\partial \Theta}{\partial z}\right|_{z=0} = 0 \qquad \text{BC4:} \quad \left.\frac{\partial \Theta}{\partial z}\right|_{z=L} = 0 \tag{4-258b}$$

$$\text{IC:} \quad \Theta(t=0) = F(z) - T_\infty = G(z) \tag{4-258c}$$

We now treat the nonhomogeneous PDE using superposition of the form

$$\Theta(r, z, t) = \Psi(r, z, t) + \Phi(r) \tag{4-259}$$

where $\Psi(r, z, t)$ will take the homogenous form of the *PDE*, and the steady-state problem $\Phi(r)$ will take the generation term. As we developed in Chapter 3, the steady-state problem would normally be of the same spatial dimensions as the transient problem. However, given the perfectly insulated boundaries at both $z = 0$ and $z = L$, the steady-state problem has no z dependency; hence we may directly consider a one-dimensional problem of r-dimension.

The formulation of the nonhomogeneous $\Phi(r)$ problem becomes

$$\frac{d^2\Phi}{dr^2} + \frac{1}{r}\frac{d\Phi}{dr} + \frac{g_0}{k} = 0 \quad \text{in} \quad 0 \le r < b$$

$$\text{BC1:} \quad \Phi(r \to 0) \Rightarrow \text{finite} \qquad \text{BC2:} \quad -k\left.\frac{d\Phi}{dr}\right|_{r=b} = h\, \Phi|_{r=b} \tag{4-261}$$

The solution of equation (4-260) yields

$$\Phi(r) = C_1 + C_2 \ln(r) - \frac{g_0}{4k} r^2 \tag{4-262}$$

where BC1 eliminates the ln(*r*) term, and BC2 yields the constant C_1. With these substitutions, the solution becomes

$$\Phi(r) = \frac{bg_0}{2h} + \frac{g_0}{4k}\left(b^2 - r^2\right) \tag{4-263}$$

The formulation of the transient problem now becomes

$$\frac{\partial^2\Psi}{\partial r^2} + \frac{1}{r}\frac{\partial\Psi}{\partial r} + \frac{\partial^2\Psi}{\partial z^2} = \frac{1}{\alpha}\frac{\partial\Psi(r,t)}{\partial t} \quad \text{in} \quad 0 \le r < b \quad 0 < z < L, \quad t > 0 \tag{4-264}$$

$$\text{BC1:} \quad \Psi(r \to 0) \Rightarrow \text{finite} \quad \text{BC2:} \quad -k\left.\frac{\partial\Psi}{\partial r}\right|_{r=b} = h\, \Psi|_{r=b} \tag{4-265a}$$

$$\text{BC3:} \quad \left.\frac{\partial\Psi}{\partial z}\right|_{z=0} = 0 \qquad \text{BC4:} \quad \left.\frac{\partial\Psi}{\partial z}\right|_{z=L} = 0 \tag{4-256b}$$

$$\text{IC:} \quad \Psi(t = 0) = G(z) - \Phi(r) = H(r, z) \tag{4-265c}$$

With all homogeneous boundary conditions, we can assume a form

$$\Psi(r, z, t) = R(r)Z(z)\Gamma(t) \tag{4-266}$$

which after substitution into equation (4-264) yields

$$\frac{1}{R}\left(\frac{d^2R}{dr^2}+\frac{1}{r}\frac{dR}{dr}\right)+\frac{1}{Z}\frac{d^2Z}{dz^2}=\frac{1}{\alpha\Gamma}\frac{d\Gamma}{dt}=-\lambda^2 \tag{4-267}$$

Solution of separated ODE in the t dimension yields the expected form

$$\Gamma(t)=C_1e^{-\alpha\lambda^2 t} \tag{4-268}$$

with the remaining terms of equation (4-267) yielding

$$\frac{1}{R}\left(\frac{d^2R}{dr^2}+\frac{1}{r}\frac{dR}{dr}\right)+\lambda^2=-\frac{1}{Z}\frac{d^2Z}{dz^2}=\eta^2 \tag{4-269}$$

Solution of the z-dimension ODE yields the desired solution form

$$Z(z)=C_2\cos\eta z+C_3\sin\eta z \tag{4-270}$$

with the characteristic value problem corresponding to case 5 in Table 2-1. Applying BC3 yields constant $C_3=0$, while applying BC4 yields the eigenvalues

$$\eta L=n\pi \quad\rightarrow\quad \eta_n=\frac{n\pi}{L} \quad \text{for} \quad n=0,1,2,\ldots \tag{4-271}$$

The eigenvalue $\eta_0=0$ is a nontrivial eigenvalue and must be included in the overall summation. We now consider the remaining r terms of equation (4-269), where we first let $\beta^2=\lambda^2-\eta^2$, and then multiply both sides by R, yielding

$$\frac{d^2R}{dr^2}+\frac{1}{r}\frac{dR}{dr}+\beta^2R=0 \tag{4-272}$$

We recognize equation (4-272) as Bessel's equation of order zero, with the solution yielding

$$R(r)=C_4J_0(\beta r)+C_5Y_0(\beta r) \tag{4-273}$$

The requirement for finiteness as $r\rightarrow 0$ stated by BC1 eliminates the $Y_0(\beta r)$ term, while BC2 then yields

$$-k\left[C_4\left.\frac{dJ_0(\beta r)}{dr}\right|_{r=b}\right]=hC_4J_0(\beta b) \tag{4-274}$$

Simplification of the derivative term using equation (2-41), multiplication by b, and making the substitution $\gamma=\beta b$, yields the transcendental equation

$$\gamma J_1(\gamma)=\text{Bi}J_0(\gamma) \quad\rightarrow\quad \gamma_m \quad \text{for} \quad m=1,2,3,\ldots \tag{4-275}$$

where equation (4-275) is now parameterized in terms of the Biot number ($\text{Bi} \equiv hb/k$). The roots γ_m are related to the actual eigenvalues β_m by $\beta_m = \gamma_m/b$. We now define λ_{nm} as

$$\lambda_{nm}^2 = \beta_m^2 + \eta_n^2 \tag{4-276}$$

Having defined the eigenvalues and eigenfunctions for both spatial dimensions, we form a product solution of the separated functions and sum over all possible solutions, yielding

$$\Psi(r, z, t) = \sum_{n=0}^{\infty} \sum_{m=1}^{\infty} C_{nm} J_0(\beta_m r) \cos \eta_n z \, e^{-\alpha \lambda_{nm}^2 t} \tag{4-277}$$

We apply the initial condition given by equation (4-265c), which yields

$$H(r, z) = \sum_{n=0}^{\infty} \sum_{m=1}^{\infty} C_{nm} J_0(\beta_m r) \cos \eta_n z \tag{4-278}$$

To find the Fourier coefficients, we apply the two operators to both sides

$$* \int_{z=0}^{L} \cos \eta_q z \, dz \quad \text{and} \quad * \int_{r=0}^{b} r J_0(\beta_k r) \, dr$$

which yields

$$C_{nm} = \frac{\int_{z=0}^{L} \int_{r=0}^{b} r H(r, z) J_0(\beta_m r) \cos \eta_n z \, dr \, dz}{\int_{r=0}^{b} r J_0^2(\beta_m r) \, dr \int_{z=0}^{L} \cos^2 \eta_n z \, dz} \tag{4-279}$$

The norms in the denominators of equation (4-279) are equal to $L/2$ for the z-dimension eigenvalues with $\eta_n \neq 0$, and to L with $\eta_0 = 0$, and for the Bessel functions as given by case 1 in Table 2-2 for the specific eigenvalues of equation (4-275). The overall temperature solution is now the sum of the homogenous, transient solution $\Psi(r, z, t)$ and the nonhomogeneous, steady-state solution $\Phi(r)$, with the additional reshifting of temperature by T_∞. This process yields

$$T(r, z, t) = \sum_{n=0}^{\infty} \sum_{m=1}^{\infty} \left[C_{nm} J_0(\beta_m r) \cos \eta_n z \, e^{-\alpha \lambda_{nm}^2 t} \right] + \frac{b g_0}{2h} + \frac{g_0}{4k} \left(b^2 - r^2 \right) + T_\infty \tag{4-280}$$

The steady-state solution is realized as $t \to \infty$ in equation (4-280), which yields the expected one-dimensional profile:

$$T_{ss}(r) = \frac{b g_0}{2h} + \frac{g_0}{4k} \left(b^2 - r^2 \right) + T_\infty \tag{4-281}$$

Let the initial parabolic temperature distribution $F(z)$ be given by the temperature distribution

$$F(z) = \frac{-4T_1}{L^2}z^2 + \frac{4T_1}{L}z + T_2 \tag{4-282}$$

where the initial temperature at the cylinder ends ($z = 0$ and $z = $ L) is given by T_2, and the initial temperature at the center ($z = L/2$) is given by $T_1 + T_2$. With $F(z)$ and $\Phi(r)$ now defined by equations (4-282) and (4-263), respectively, $H(r,z)$ becomes

$$H(r, z) = \frac{-4T_1}{L^2}z^2 + \frac{4T_1}{L}z + T_2 - T_\infty - \frac{bg_0}{2h} - \frac{g_0}{4k}\left(b^2 - r^2\right) \tag{4-283}$$

Substitution of equation (4-283) into equation (4-279) and evaluation of the integrals yields the following expressions for the Fourier coefficients:

$$C_{om} = \left(\frac{1}{L}\right)\left\{\frac{2\beta_m^2}{b^2 J_0^2(\beta_m b)\left[\beta_m^2 + \left(\frac{h}{k}\right)^2\right]}\right\} \times \left[\frac{bLJ_1(\beta_m b)}{\beta_m}\left(T_2 - T_\infty + \frac{2T_1}{3} - \overline{A} + b^2\overline{B}\right) - \frac{2Lb^2\overline{B}J_2(\beta_m b)}{\beta_m^2}\right] \tag{4-284}$$

which corresponds to the $n = 0$ case, and

$$C_{nm} = \left(\frac{2}{L}\right)\left\{\frac{2\beta_m^2}{b^2 J_0^2(\beta_m b)\left[\beta_m^2 + \left(\frac{h}{k}\right)^2\right]}\right\}\left\{\frac{-4bT_1 J_1(\beta_m b)}{\beta_m \eta_n^2 L}\left[(-1)^2 + 1\right]\right\} \tag{4-285}$$

for the case of $n>0$, where we have made the following two substitutions in equation (2-284) for simplification,

$$\overline{A} = \frac{bg_0}{2h} + \frac{g_0 b^2}{4k} \tag{4-286}$$

and

$$\overline{B} = \frac{g_0}{4k}. \tag{4-287}$$

We note that the units of C_{nm} are K, which follows directly from equations (4-284)–(4-287) when using SI units for all expressions. In addition, we note that the true eigenvalues β_m have the unit m^{-1}, while the roots of equation (4-275), namely, γ_m, are dimensionless.

The above expressions may now be used to explore the behavior of the 2-D transient response for a set of parameters corresponding to a stainless steel cylindrical solid. For these calculations, let the cylinder have the following dimensions: radius $b = 0.04$ m and length $L = 0.2$ m, which gives an aspect ratio $L/b = 5$. Let us use representative properties for stainless steel: Thermal conductivity $k = 15$ W/m K, and thermal diffusivity $\alpha = 4 \times 10^{-6}$ m^2/s. The convection heat transfer is characterized by a fluid temperature $T_\infty = 500$ K and with convection coefficient $h = 375$ W/m^2 K. This yields a Biot number of unity ($\text{Bi} = hb/k = 1.0$); hence the thermal resistances to conduction within the cylinder and to convection at the cylinder–fluid boundary are comparable. Finally, we define the initial temperature distribution, equation (4-282), with the two parameters $T_1 = 200$ K and $T_2 = 200$ K, while the rate of internal energy generation is $g_0 = 1 \times 10^6$ W/m^3.

We have used the first 6 terms ($m = 1$–6) for the inner series summation, as given by equation (2-280), and have used the first 51 terms ($n = 0$–50) of the outer series summation. For the roots of equation (4-275), namely, γ_m, the first 6 roots as given by Table IV-3 in Appendix IV for the unity Biot number were used for calculation of the eigenvalues β_m. Over the range of temporal and spatial variables explored below, these parameters were found to yield excellent convergence (less than 0.005%) for both the inner and outer series. However, for very small times ($t \lesssim 1$s), the convergence of the inner series, particularly at $r \simeq 0$, was closer to 2%; hence exploration of very small times would require additional terms for the inner series summation (i.e., $m > 6$) to achieve greater accuracy.

Figure 4-6 presents the radial temperature profiles for a cross section located halfway along the length of the cylinder ($z = L/2$) as a function of time. Included are the initial temperature (uniform at $T = 400$ K for $z = L/2$) per equation (4-282) and the steady-state temperature per equation (4-281). The series summation as given by equation (2-280) converges nicely to the exact steady-state solution, equation (2-281), as expected, with agreement to within 0.1 K at $t = 2100$ s. At $t = 3500$ s, agreement is within 0.001 K. Clearly, the convergence of the series summation is rapid for large time.

The temperature profiles in Figure 4-6 reveal several interesting trends that are rooted in the physics of the boundary condition and the presence of internal energy generation. For relatively small times (e.g., first 100 s), the temperature profiles reveal a greater temperature near the outer surface with a distinct positive slope at $r = b$. Simultaneously, the inner temperatures are rising, but with zero slope at $r = 0$ due to the symmetry (i.e., zero flux across the radial centerline). The temperature increase and positive gradient near the outer surface are due to heat flowing *into the cylinder* via convection heat transfer because the temperatures of the solid at this location (initially at 400 K) are less than the fluid temperature $T_\infty = 500$ K. Near the centerline ($r \sim 0$), the temperature is increasing primarily by the internal energy generation, although some contribution of the convected heat is being conducted toward the center as the temperature gradient develops. In particular, at $t = 10$ s, the gradient near the centerline is still negligible, meaning the influence of the convective heating at the outer surface has not yet

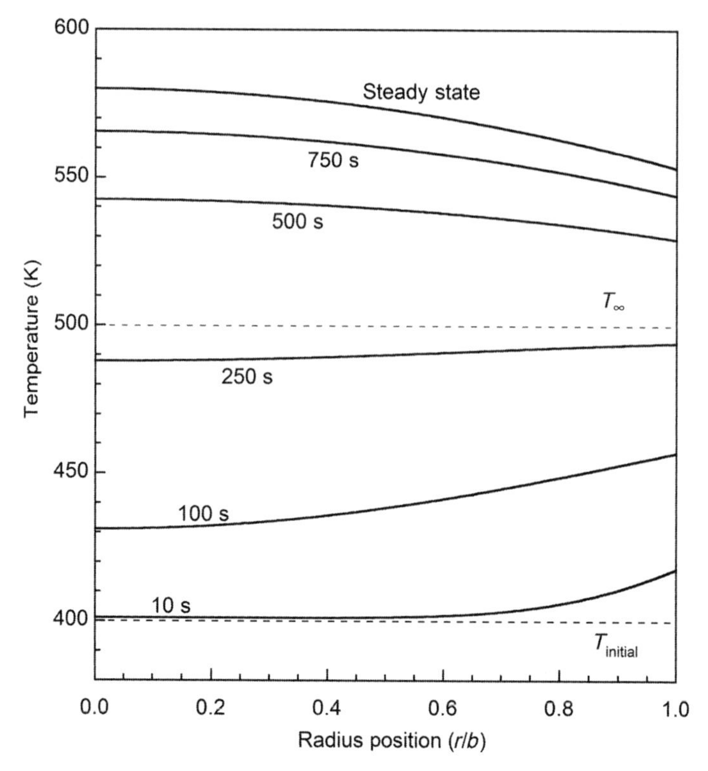

Figure 4-6 Radial temperature profiles at $z = L/2$ for various times, including the initial and steady-state profiles. The dashed line at 500 K represents the convective fluid temperature.

reached the cylinder's center. Accordingly, the rate of temperature increase near $r = 0$ at very early times is due to internal energy generation. However, as the temperature increases on the outer surface and approaches the fluid temperature, the temperature gradient is decreased as the rate of convection heat transfer decreases. This is the classic case of a convection boundary condition in that the gradient at the surface $(\partial T/\partial r)$ is proportional to the temperature difference between the surface and fluid $(T_{r=b} - T_\infty)$. For an instant in time, sometime soon after $t = 250$ s, the surface temperature and the fluid temperature are identical at 500 K, and the slope of the temperature profile at $r = b$ is identically zero. Beyond this moment, the surface temperature will always be greater than the fluid temperature, with heat flowing *out of the cylinder* via convection heat transfer. The temperature within the cylinder continues to rise due to the internal energy generation, although the gradient is now negative and increasing toward the outer surface. This behavior is demonstrated by the temperature profiles at 500 and 750 s. As noted above, the gradient at $r = 0$ remains zero at all times due to symmetry across the radial centerline. Finally, the expected parabolic temperature profile, see equation (4-281), is achieved for steady-state conditions. The gradient of the steady-state temperature profile becomes steeper with increasing radius, which reflects the fact that the accumulated internal energy generation within each radial location must be conducted outward, although one must also keep in mind that the circumferential area is increasing with radius as well.

The steady-state temperature at the outer surface ($r = b$) is readily calculated from the series summation as $t \to \infty$, which directly yields equation (4-281), giving

$$T_{ss}(r = b) = \frac{bg_0}{2h} + T_\infty \tag{4-288}$$

It is also useful to consider conservation of energy under steady-state conditions, which may be expressed as

$$\dot{Q}_{\text{generated}} = \dot{Q}_{\text{out}} = hA_{\text{sur}}(T_{r=b} - T_\infty) \tag{4-289}$$

where the rate of total energy generated (W) within the cylinder is simply

$$\dot{Q}_{\text{generated}} = \text{Volume} \cdot g_0 = \pi b^2 L g_0 \tag{4-290}$$

and where the appropriate surface area is the outer curved surface at $r = b$, as given by $A_{\text{sur}} = 2\pi bL$, since the cylinder's ends are perfectly insulated and do not contribute to heat transfer out of the cylinder. Substitution of the above yields

$$\pi b^2 L g_0 = h(2\pi bL)(T_{r=b} - T_\infty) \tag{4-291}$$

which may be readily solved for the surface temperature, giving

$$T_{r=b} = \frac{bg_0}{2h} + T_\infty \tag{4-292}$$

which is in perfect agreement with equation (4-288).

Figure 4-7 presents the axial temperature profiles along the centerline ($r = 0$) and the outer surface ($r = b$) as a function of time. Included are the initial temperature profile per equation (4-282), and the steady-state temperatures per equation (4-281). While the axial plots reveal a different spatial perspective to the temperature gradients as compared to Figure 4-6, the physics is perfectly consistent with the discussion presented above regarding the radial profiles. The initial parabolic temperature profile presents strong gradients in the z direction; hence for the first few hundred seconds, considerable axial gradients are present with heat flowing from the cylinder's center toward the ends. Note that the gradient remains zero at the center ($z = L/2$) due to symmetry, and for $t > 0$ remains zero

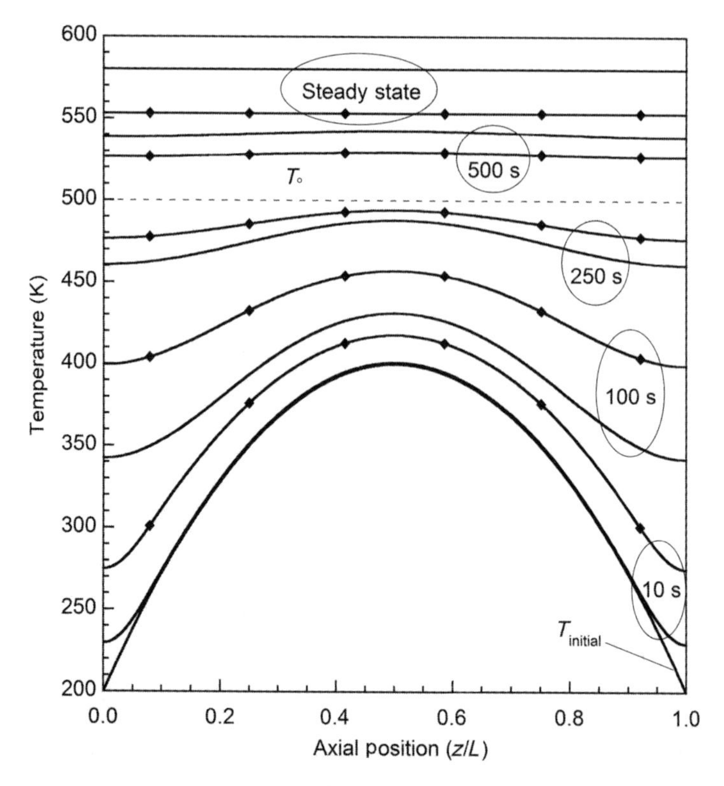

Figure 4-7 Axial temperature profiles for various times, including the initial and steady-state profiles. The dashed line at 500 K represents the convective fluid temperature. For each time, two profiles are presented that correspond to the centerline (r=0) and the outer surface (r=b), with the latter marked by the diamonds.

at the ends due to the insulated boundary conditions. Note, however, that when the temperature is less than the fluid temperature (<500 K), which is approximately within the first 250 s, the outer surface ($r = b$) temperatures are greater than the centerline ($r = 0$) temperatures. This agrees with Figure 4-6. However, once the temperature profiles cross over the fluid temperature (>500 K), then the centerline temperature exceeds the surface temperature. This condition exists for $t = 500$ s, as shown in Figure 4-7, and persists at steady-state conditions, as it must. In addition, Figure 4-7 nicely illustrates the comments made during the problem formulation, namely, that the steady-state solution has no z dependency, as there are no physical drivers of heat in the axial direction once the initial driver (i.e., the axially parabolic initial profile) has decayed to zero. Therefore, the z dependency of the temperature profiles are observed to steadily decay away in Figure 4-7, with only a slight axial gradient observed at $t = 500$ s and no gradient for the steady-state profiles. One also notices a greater rate of temperature increase near the ends ($z = 0$ and $z = L$) at $r = b$ during the first 250 s, as compared to the center of cylinder ($z = L/2$). This is due to the greater temperature difference between the cylinder and fluid at these locations. Specifically, at $t = 0$, the value of $T_{r=b} - T_\infty = 300\,\text{K}$ at the ends, while $T_{r=b} - T_\infty = 100\,\text{K}$ at $z = L/2$.

In addition to exploring the temperature profiles, knowledge of the temperature solution allows calculation of the heat flux per Fourier's law. The radial heat flux is therefore given by

$$q_r'' = -k\frac{\partial T}{\partial r} \tag{4-293}$$

which when applied to equation (4-280) yields

$$q_r''(r, z, t) = -k\sum_{n=0}^{\infty}\sum_{m=1}^{\infty}\left(C_{nm}(-\beta_m)J_1(\beta_m r)\cos\,\eta_n z\, e^{-\alpha\lambda_{nm}^2 t}\right) + \frac{g_0}{2}r \tag{4-294}$$

Using the same number of summation terms as used for the temperature profiles, as discussed above, the radial heat flux at the outer surface $r = b$ was explored as a function of time. Figure 4-8 presents the surface heat flux as a function of axial position for times from 10 s to steady-state conditions.

Once again, the physics of heat transfer is nicely illustrated by the trends observed in Figure 4-8. For times less than 200 s, the radial surface heat flux is negative at all axial locations. Given the sign convention of Fourier's law, a negative flux is in the negative r direction, hence into the cylinder. This agrees with our discussions above, namely, that for these times, the outer surface of our cylinder is cooler than the fluid temperature, and therefore heat transfer is convected *into the cylinder*. Furthermore, as noted above, for relatively small times, the temperature difference between the surface and fluid is much greater near the cylinder ends; hence the convective heat flux is much greater. For example, at $t = 10$ s, the radial heat flux at $z = 0$ is $-84{,}415$ W/m^2 as compared to the radial heat flux of $-30{,}854$ W/m^2 at $z = L/2$. The axial variation in heat flux steadily diminishes with time, as the temperature profile decays from a 2-D profile to a 1-D profile, as discussed above. The heat flux profile corresponding to $t = 300$ s is an interesting case, as the heat flux is still negative at the end points (i.e., into the cylinder), while the heat flux at $z = L/2$ is now positive (i.e., out of the cylinder). Because the center of the initial temperature distribution was so much hotter than the end points (i.e., 400 K vs. 200 K), the surface temperature at $z = L/2$ reaches the fluid temperature much faster, crossing over to a local net heat flux out of the cylinder, while heat is still flowing into the cylinder nearer to the two ends. Clearly the transient behavior is two dimensional due to the strong z dependency of the initial condition. However, as described above, the radial heat flux at the surface loses its axial dependency with time, showing a slight z dependency at 400 s, essentially no such dependency at 600 s, and zero such dependency at steady-state conditions.

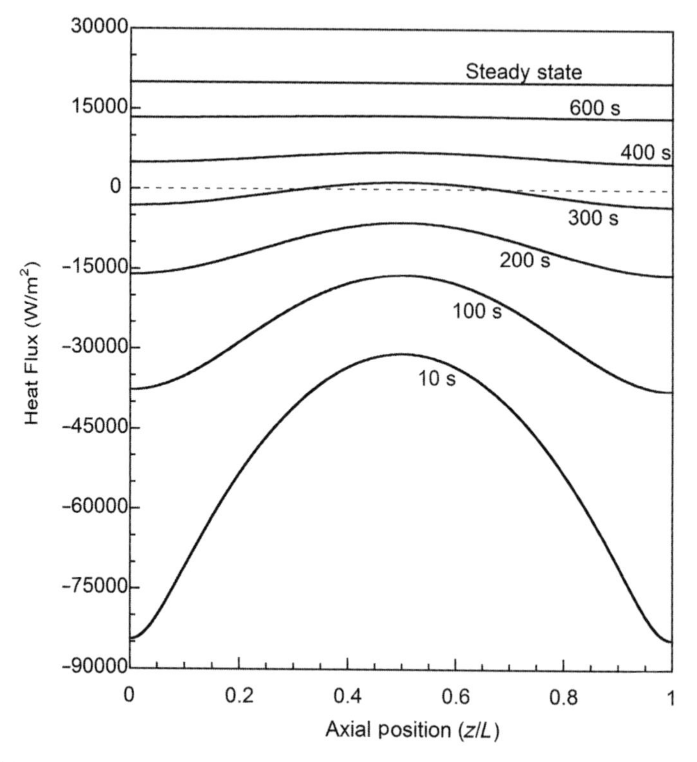

Figure 4-8 Radial heat flux at the surface $r = b$ as a function of axial position for various times and at steady-state conditions.

The steady-state radial heat flux at $r = b$ is 20,000 W/m^2, as calculated from the steady-state solution of equation (4-294), namely,

$$q_r''(r, z, t \to \infty) = \frac{g_0}{2} r \tag{4-295}$$

which also follows directly from equation (4-281):

$$-k\frac{dT_{ss}(r)}{dr} = -k\frac{d}{dr}\left[\frac{bg_0}{2h} + \frac{g_0}{4k}\left(b^2 - r^2\right) + T_\infty\right] = \frac{g_0}{2} r \tag{4-296}$$

The steady-state surface heat flux is therefore given by

$$q_r''(r = b) = \frac{bg_0}{2} \tag{4-297}$$

One might also calculate the steady-state surface heat flux directly from Newton's law, using the steady-state surface temperature, as described by equation (4-292).

This approach yields

$$q_r''(r=b) = h(T_{r=b} - T_\infty) = h\left(\frac{bg_0}{2h} + T_\infty - T_\infty\right) = \frac{bg_0}{2} \tag{4-298}$$

which is in perfect agreement with equation (4-297).

REFERENCES

1. H. S. Carslaw and J. C. Jaeger, *Conduction of Heat in Solids*, Clarendon, London, 1959.
2. M. N. Özişik, *Heat Conduction*, Wiley, New York, 1980.
3. M. D. Mikhailov and M. N. Özişik, *Unified Analysis and Solutions of Heat and Mass Diffusion*, Wiley, New York, 1984.
4. P. Moon and D. E. Spencer, *Field Theory for Engineers*, Van Nostrand, Princeton, NJ, 1961.

PROBLEMS

4-1 A long, hollow cylinder, $a \le r \le b$, is initially at a temperature $T = F(r)$. For times $t > 0$ the boundaries at $r = a$ and $r = b$ are kept insulated. Obtain an expression for temperature distribution $T(r, t)$ in the solid for times $t > 0$.

4-2 A solid cylinder, $0 \le r \le b$, $0 \le z \le L$, is initially at temperature $T = F(r, z)$. For times $t > 0$, the boundary at $z = 0$ is insulated, the boundary at $z = L$ is dissipating heat by convection into a medium at zero temperature, and the boundary at $r = b$ is kept at zero temperature. Obtain an expression for the temperature distribution $T(r, z, t)$ in the solid for times $t > 0$.

4-3 A long, solid cylinder, $0 \le r \le b$, $0 \le \phi \le 2\pi$, is initially at temperature $T = F(r, \phi)$. For times $t > 0$, the boundary at $r = b$ is kept insulated. Obtain an expression for the temperature distribution $T(r, \phi, t)$ in the solid for times $t > 0$. Independently derive the steady-state temperature from simple conservation of energy and show that this agrees with your above expression.

4-4 A long, hollow cylinder, $a \le r \le b$, $0 \le \phi \le 2\pi$, is initially at temperature $T = F(r, \phi)$. For times $t > 0$, the boundaries at $r = a$ and $r = b$ are kept insulated. Obtain an expression for the temperature distribution $T(r, \phi, t)$ in the region for times $t > 0$. Independently derive the steady-state temperature from simple conservation of energy, and show that this agrees with your above expression.

4-5 A portion of a long, solid cylinder $0 \leq r \leq b$, $0 \leq \phi \leq \phi_0 \, (< 2\pi)$ is initially at temperature $T = F(r, \phi)$. For times $t > 0$, the boundary at $r = b$ dissipates heat by convection into a medium at zero temperature, the boundaries at $\phi = 0$ and $\phi = \phi_0$ are kept at zero temperature. Obtain an expression for the temperature distribution $T(r, \phi, t)$ in the solid for times $t > 0$.

4-6 A portion of a long, hollow cylinder $a \leq r \leq b$, $0 \leq \phi \leq \phi_0 \, (< 2\pi)$ is initially at temperature $T = F(r, \phi)$. For times $t > 0$, the boundaries at $r = a$, $r = b$, $\phi = 0$, and $\phi = \phi_0$ are all kept at zero temperature. Obtain an expression for the temperature distribution $T(r, \phi, t)$ in the solid for times $t > 0$.

4-7 A long, solid cylinder, $0 \leq r \leq b$, $0 \leq \phi \leq 2\pi$, is initially at temperature $T = F(r, \phi)$. For times $t > 0$, the boundary at $r = b$ is kept at constant temperature T_0. Obtain an expression for the temperature distribution $T(r, \phi, t)$ in the solid for times $t > 0$.

4-8 A solid cylinder $0 \leq r \leq b$, $0 \leq z \leq L$, $0 \leq \phi \leq 2\pi$, is initially at temperature $T = F(r, \phi, z)$. For times $t > 0$, the boundary at $z = 0$ is kept insulated, the boundaries at $z = L$ and $r = b$ are kept at zero temperature. Obtain an expression for the temperature distribution $T(r, z, \phi, t)$ in the solid for times $t > 0$.

4-9 A portion of a solid cylinder, $0 \leq r \leq b$, $0 \leq \phi \leq \phi_0 \, (< 2\pi)$, $0 \leq z \leq L$, is initially at temperature $T = F(r, \phi, z)$. For times $t > 0$, the boundary surface at $z = 0$ is kept insulated, the boundary at $z = L$ dissipates heat by convection into an environment at zero temperature, and the remaining boundaries are kept at zero temperature. Obtain an expression for temperature distribution $T(r, \phi, z, t)$ in the solid for times $t > 0$.

4-10 Obtain an expression for the steady-state temperature distribution $T(r, z)$ in a 2-D solid cylinder, $0 \leq r \leq b$, $0 \leq z \leq L$ for the following boundary conditions: The boundary at $z = 0$ is kept at temperature $F(r)$, and the boundaries at surfaces $r = b$ and $z = L$ dissipate heat by convection into a medium at zero temperature. Assume the heat transfer coefficients h to be the same for two convective surfaces.

4-11 Obtain an expression for the steady-state temperature distribution $T(r, z)$ in a hollow cylinder $a \leq r \leq b$, $0 \leq z \leq L$ for the following boundary conditions: The boundary at $r = b$ is kept at temperature $F(z)$, and other boundaries at $r = a$, $z = 0$, and $z = L$ are kept at zero temperature.

4-12 Obtain an expression for the steady-state temperature $T(r, z)$ in a hollow cylinder $a \leq r \leq b$, $0 \leq z \leq L$ for the following boundary conditions: The heat flux into the surface at $r = a$ is given by $q_r''(r = a) = f(z)$, and the other boundaries at $r = b$, $z = 0$, and $z = L$ are kept at zero temperature.

4-13 Obtain an expression for the steady-state temperature distribution $T(r, \phi)$ in a long, solid cylinder $0 \le r \le b$, $0 \le \phi \le 2\pi$ for the following boundary conditions: The boundary at $r = b$ is subjected to a prescribed temperature distribution $f(\phi)$.

4-14 Obtain an expression for the steady-state temperature distribution $T(r, z)$ in a cylinder $0 \le r \le b$, $0 \le z \le L$ for the following boundary conditions: The cylinder ends at $z = 0$ and $z = L$ are both maintained at temperature T_1, while the surface at $r = b$ dissipates heat by convection into a fluid at T_∞ with convection coefficient h. Force the z direction to be the homogeneous direction for the solution.

4-15 Obtain an expression for the steady-state temperature distribution $T(r, z)$ in a cylinder $0 \le r \le b$, $0 \le z \le L$ for the following boundary conditions: The cylinder ends at $z = 0$ and $z = L$ are both maintained at temperature T_1, while the surface at $r = b$ dissipates heat by convection into a fluid at T_∞ with convection coefficient h. Force the r direction to be the homogeneous direction for the solution. Hint: Consider symmetry in the z direction and work the half-domain problem.

4-16 Obtain an expression for the steady-state temperature distribution $T(r, z)$ in a cylinder $0 \le r \le b$, $0 \le z \le L$ for the following boundary conditions: The cylinder end at $z = 0$ is maintained at temperature T_1, the cylinder end at $z = L$ is perfectly insulated, and the surface at $r = b$ dissipates heat by convection into a fluid at T_∞ with convection coefficient h.

4-17 A solid cylinder $0 \le r \le b$, $0 \le z \le L$ is initially at a temperature given by $T{=}F(r, z)$. For times $t > 0$, the boundary at $r = b$ is kept perfectly insulated, the boundary at $z = 0$ is maintained at temperature T_1, and the boundary at $z = L$ is subjected to a uniform incident heat flux $q_z''(z = L) = -q_0''$. Obtain an expression for the temperature distribution $T(r, z, t)$ in the solid cylinder for times $t > 0$. What is the steady-state temperature distribution?

4-18 A hollow cylinder, $a \le r \le b$, $0 \le z \le L$, is initially at temperature $T = F(r, z)$. For times $t > 0$, the boundary surfaces at $z = 0$, $z = L$, and at $r = b$ are all insulated, while the boundary at $r = a$ is dissipating heat by convection into a fluid at T_∞ with convection coefficient h. Obtain an expression for the temperature distribution $T(r, z, t)$ in the solid for times $t > 0$. Obtain an expression for the heat flux at the inner surface $r = a$ for times $t > 0$.

4-19 Consider a solid cylinder $0 \le r \le b$, $0 \le z \le 2L$ with the following boundary conditions: The boundary at $r = b$ is maintained at temperature T_1, and boundary surfaces at $z = 0$ and $z = 2L$ are subjected to a radially dependent, incident heat flux given by the expression $q_z''(r) = Q_0 \cos(\pi r/2b)$. Obtain an expression for the steady-state temperature distribution $T(r, z)$. Provide an expression for conservation of energy with regard to the total heat rate (W) into the cylinder as compared to the total heat rate out of the cylinder. You may leave integrals in your answer.

4-20 Consider a solid cylinder $0 \le r \le b$, $0 \le z \le L$ with the following boundary conditions: The boundary at $r = b$ is subjected to constant incident heat flux q_0'', the boundary surface at $z = 0$ maintained at a constant temperature T_∞, and the boundary at $z = L$ dissipates heat by convection into a fluid at temperature T_∞ with convection coefficient h. Obtain an expression for the steady-state temperature distribution $T(r, z)$. Provide an expression for conservation of energy with regard to the total heat rate (W) into the cylinder from the curved surface as compared to the total heat rate dissipated through the end surfaces. You may leave integrals in your answer.

5

SEPARATION OF VARIABLES IN THE SPHERICAL COORDINATE SYSTEM

In this chapter, we examine the separation of the heat conduction equation in our last coordinate system of interest, namely, the spherical coordinate system. We determine the elementary solutions, the norms, and the eigenvalues of the separated problems for different combinations of boundary conditions; discuss the solution of the one- and multidimensional homogeneous problems by the method of separation of variables for spheres; examine the solutions of steady-state and transient multidimensional problems with and without the heat generation in the medium; and illustrate the splitting up of nonhomogeneous problems into a set of simpler problems. The reader should consult references 1–7 for additional applications on the solution of heat conduction problems in the spherical coordinate system.

5-1 SEPARATION OF HEAT CONDUCTION EQUATION IN THE SPHERICAL COORDINATE SYSTEM

We will first review the spherical coordinate system, as shown in Figure 5-1, along with the corresponding components of heat flux in the r, ϕ, and θ directions (i.e., Fourier's law), which are given, respectively, by

$$q_r'' = -k\frac{\partial T}{\partial r}, \qquad q_\phi'' = -\frac{k}{r\sin\theta}\frac{\partial T}{\partial \phi} \qquad \text{and} \qquad q_\theta'' = -\frac{k}{r}\frac{\partial T}{\partial \theta} \tag{5-1}$$

Now we consider the three-dimensional, differential equation of heat conduction in the spherical coordinate system, given as

$$\frac{\partial^2 T}{\partial r^2} + \frac{2}{r}\frac{\partial T}{\partial r} + \frac{1}{r^2\sin\theta}\frac{\partial}{\partial \theta}\left(\sin\theta\frac{\partial T}{\partial \theta}\right) + \frac{1}{r^2\sin^2\theta}\frac{\partial^2 T}{\partial \phi^2} + \frac{g_0}{k} = \frac{1}{\alpha}\frac{\partial T}{\partial t} \tag{5-2}$$

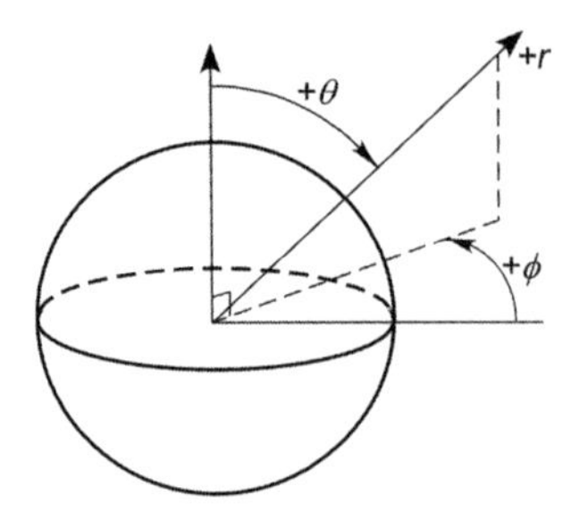

Figure 5-1 Spherical coordinate system.

To aid our solution of equation (5-2), it is convenient to define a new independent variable μ as

$$\boxed{\mu = \cos\theta} \tag{5-3}$$

where the domain of μ is given by $1 \geq \mu \geq -1$, which maps to the variable θ over the corresponding domain $0 \leq \theta \leq \pi$. With this change, equation (5-2) becomes

$$\frac{\partial^2 T}{\partial r^2} + \frac{2}{r}\frac{\partial T}{\partial r} + \frac{1}{r^2}\frac{\partial}{\partial \mu}\left[\left(1-\mu^2\right)\frac{\partial T}{\partial \mu}\right] + \frac{1}{r^2\left(1-\mu^2\right)}\frac{\partial^2 T}{\partial \phi^2} + \frac{g_0}{k} = \frac{1}{\alpha}\frac{\partial T}{\partial t} \tag{5-4}$$

where $T \equiv T(r, \mu, \phi, t)$. Finally, we may define a new dependent variable V,

$$\boxed{V(r, \mu, \phi, t) = r^{1/2}T(r, \mu, \phi, t)} \tag{5-5}$$

which upon substitution into equation (5-4) yields

$$\frac{\partial^2 V}{\partial r^2} + \frac{1}{r}\frac{\partial V}{\partial r} - \frac{1}{4}\frac{V}{r^2} + \frac{1}{r^2}\frac{\partial}{\partial \mu}\left[\left(1-\mu^2\right)\frac{\partial V}{\partial \mu}\right] + \frac{1}{r^2\left(1-\mu^2\right)}\frac{\partial^2 V}{\partial \phi^2} + \frac{g_0 r^{1/2}}{k} = \frac{1}{\alpha}\frac{\partial V}{\partial t} \tag{5-6}$$

Equations (5-4) and (5-6) are both alternative forms of the spherical coordinate heat equation that will be considered in this chapter. Equation (5-6) will be used only when temperature depends on the (r, μ, t) or (r, μ, ϕ, t) variables. The reason for this is that when equation (5-6) is used in such situations, the elementary solutions of the separated differential equation for $R(r)$ become the Bessel functions, which have already been developed in Chapter 2. However, if equation (5-4) is used for such a problem, the elementary solutions are the spherical Bessel functions. For all other cases, including the problems involving (r), (r, t), (r, μ), and (r, μ, ϕ) variables, the governing equation will be obtained

TABLE 5-1 Solution Schemes for the Spherical Heat Equation

Variables	Equation to Solve	Transforms	Expected Equations
$T(r)$	(5-2) or (5-4)	None	Cauchy equation
$T(r, \mu)$	(5-4)	None	Cauchy equation in r Legendre equation in μ
$T(r, \mu, \phi)$	(5-4)	None	Cauchy equation in r Associated Legendre in μ
$T(r, t)$	(5-2) or (5-4)	$U(r, t) = rT(r, t)$	$\frac{\partial^2 U}{\partial r^2} + \frac{rg}{k} = \frac{1}{\alpha}\frac{\partial U}{\partial t}$
$T(r, \mu, t)$	(5-6)	$V = r^{1/2}T$	Bessel equation in r Legendre equation in μ
$T(r, \mu, \phi, t)$	(5-6)	$V = r^{1/2}T$	Bessel equation in r Associated Legendre in μ

from equation (5-2) or (5-4). In Table 5-1, we summarize the various solution schemes for the spherical heat equation along with the expected ODE equations.

We now begin with the general separation of the spherical heat equation for the case of $T \equiv T(r, \mu, \phi, t)$, for which we make the transformation to $V(r, \mu, \phi, t)$ and subsequently solve equation (5-6). Here we will consider the homogeneous version of equation (5-6), that is, no internal energy generation, recalling that the generation term is always handled in a steady-state solution via superposition as we developed in Chapter 3. For the homogeneous version of equation (5-6), we can assume a separation of variables in the form

$$V(r, \mu, \phi, t) = R(r)M(\mu)\Phi(\phi)\Gamma(t) \tag{5-7}$$

Substitution of equation (5-7) into the heat equation of (5-6) yields

$$\begin{aligned} \frac{1}{R}\left(\frac{d^2R}{dr^2} + \frac{1}{r}\frac{dR}{dr} - \frac{1}{4}\frac{R}{r^2}\right) + \frac{1}{r^2 M}\frac{d}{d\mu}\left[\left(1-\mu^2\right)\frac{dM}{d\mu}\right] \\ + \frac{1}{r^2\left(1-\mu^2\right)}\frac{1}{\Phi}\frac{d^2\Phi}{d\phi^2} = \frac{1}{\alpha\Gamma}\frac{d\Gamma}{dt} = -\lambda^2 \end{aligned} \tag{5-8}$$

where we have introduced the separation constant $-\lambda^2$ to produce a decaying exponential function in time, as we discussed in Chapter 3. Equation (5-8) leads to the separated ordinary differential equation in the t dimension:

$$\frac{d\Gamma}{dt} + \alpha\lambda^2\Gamma(t) = 0 \tag{5-9}$$

with corresponding solution given by

$$\Gamma(t) = C_1 e^{-\alpha\lambda^2 t} \tag{5-10}$$

We now consider the remaining spatial variables of equation (5-8), namely,

$$\frac{1}{R}\left(\frac{d^2R}{dr^2}+\frac{1}{r}\frac{dR}{dr}-\frac{1}{4}\frac{R}{r^2}\right)+\frac{1}{r^2M}\frac{d}{d\mu}\left[\left(1-\mu^2\right)\frac{dM}{d\mu}\right]$$
$$+\frac{1}{r^2\left(1-\mu^2\right)}\frac{1}{\Phi}\frac{d^2\Phi}{d\phi^2}+\lambda^2=0 \tag{5-11}$$

which is a separated form of the Helmholtz equation for spherical coordinates. Because the initial condition will always provide the nonhomogeneity for transient problems, here we expect all three spatial dimensions to have corresponding homogeneous boundary conditions. Therefore, we seek to force a characteristic value problem in each spatial dimension. Because none of the three spatial dimensions are currently separated, we first multiply by r^2 and then isolate the r terms, yielding

$$\frac{-1}{M}\frac{d}{d\mu}\left[\left(1-\mu^2\right)\frac{dM}{d\mu}\right]-\frac{1}{\left(1-\mu^2\right)}\frac{1}{\Phi}\frac{d^2\Phi}{d\phi^2}$$
$$=\frac{r^2}{R}\left(\frac{d^2R}{dr^2}+\frac{1}{r}\frac{dR}{dr}-\frac{1}{4}\frac{R}{r^2}\right)+\lambda^2r^2=n(n+1) \tag{5-12}$$

where we have introduced a new separation constant $n(n+1)$, where n will ultimately be limited to integer values as discussed below, to force the following ODE in the r dimension:

$$\frac{d^2R}{dr^2}+\frac{1}{r}\frac{dR}{dr}-\frac{1}{4}\frac{R}{r^2}+\lambda^2R-\frac{n(n+1)}{r^2}R=0 \tag{5-13}$$

Equation (5-13) is readily rearranged to yield

$$\frac{d^2R}{dr^2}+\frac{1}{r}\frac{dR}{dr}+\left[\lambda^2-\frac{\left(n+\frac{1}{2}\right)^2}{r^2}\right]R=0 \tag{5-14}$$

which is recognized as Bessel's equation of order $n+\frac{1}{2}$. The solution of equation (5-14) for integer n yields half-integer order Bessel functions

$$R(r)=C_2J_{n+1/2}(\lambda r)+C_3Y_{n+1/2}(\lambda r) \tag{5-15}$$

We now consider the left-hand side of equation (5-12), which after multiplication by $1-\mu^2$, yields the following:

$$\frac{1-\mu^2}{M}\frac{d}{d\mu}\left[\left(1-\mu^2\right)\frac{dM}{d\mu}\right]+\left(1-\mu^2\right)n(n+1)=\frac{-1}{\Phi}\frac{d^2\Phi}{d\phi^2}=m^2 \tag{5-16}$$

We have introduced a new separation constant m^2, which yields the desired ODE and corresponding solution for the ϕ dimension, as given by

$$\Phi(\phi) = C_4 \cos m\phi + C_5 \sin m\phi \tag{5-17}$$

The requirement for 2π periodicity, as discussed in detail in the previous chapter, see equation (4-26), is satisfied for integer m, namely,

$$m = 0, 1, 2, 3, \ldots \tag{5-18}$$

and where both constants C_4 and C_5 are retained. Finally, we consider the remaining μ terms of equation (5-16), which yield

$$\frac{d}{d\mu}\left[\left(1-\mu^2\right)\frac{dM}{d\mu}\right] + \left[n(n+1) - \frac{m^2}{1-\mu^2}\right]M = 0 \tag{5-19}$$

We recognize equation (5-19) as the associated Legendre equation, as discussed in Chapter 2 in reference to equation (2-80). As noted above, we now limit n to *integer values*, which along with integer values of m as defined by equation (5-18), yields the orthogonal associated Legendre polynomials as the solution of equation (5-19),

$$M(\mu) = C_6 P_n^m(\mu) + C_7 Q_n^m(\mu) \tag{5-20}$$

As defined for the general characteristic value problem (i.e., with corresponding homogeneous boundary conditions), the functions given by equations (5-15), (5-17), and (5-20) are all orthogonal functions. The homogeneous boundary conditions in the respective spatial dimensions would define the eigenvalues and eigenfunctions, and the general solution of V would then be formed by summation over all eigenfunctions.

An additional comment is offered with regard to the general case of ϕ dependency. As noted in Table 5-1, the presence of ϕ dependency will always yield the associated Legendre functions in the μ dimension as coupled to the integer-order eigenvalues that arise from the conditions of 2π periodicity. When no ϕ dependency is present, this coupling does not exist and Legendre's equation, rather than the associated Legendre equation, is then generated during separation. The integer eigenvalues n originating from the μ dimension will always couple to the order of Bessel's equation, specifically half-integer order, whenever there is μ dependency for a transient problem. Finally, we note that ϕ dependency in the absence of μ dependency is not covered in Table 5-1, as such a scenario is difficult to envision from a physical point of view. We first explore various steady-state problems and then extend our treatment to transient problems of various spatial dimensions.

5-2 SOLUTION OF STEADY-STATE PROBLEMS

In this section we will solve a range of steady-state problems in the spherical coordinate system. Our general guidelines established for separation of variables in the rectangular coordinate system, equally apply to the spherical coordinate problems. Therefore, steady-state problems with a homogeneous PDE must always be solved with only a single nonhomogeneous boundary condition. If more than one boundary condition is nonhomogeneous, the principle of superposition is used to treat each nonhomogeneity individually. For a nonhomogeneous PDE (i.e., when heat generation is present), superposition is used to treat the heat generation within a 1-D ODE, which is coupled to a homogeneous PDE at one or more boundary conditions. Prior to solving a range of problems to illustrate the above procedures, a few additional guidelines are offered with regard to the solution of steady-state problems in spherical coordinates.

For steady-state 1-D problems, $T = T(r)$ only and the solution scheme reduces to the solution of an ODE, which turns out to be the Cauchy equation. For steady-state 2-D and 3-D problems, namely, (r, μ) or (r, μ, ϕ) dependency, respectively, the r dimension must always be the nonhomogeneous dimension and again the ODE in the r dimension will be the Cauchy equation. The orthogonal functions from the μ and ϕ dimensions will then take the form of the Legendre or associated Legendre polynomials and the trigonometric sin and cos functions, respectively.

Example 5-1 $T = T(r, \mu)$ for Solid Hemisphere
A solid hemisphere of radius b is maintained at steady-state conditions with a prescribed surface temperature $T(r = b) = f(\mu)$, and with a perfectly insulated base at $\mu = 0$ (Fig. 5-2). We now calculate the temperature distribution.

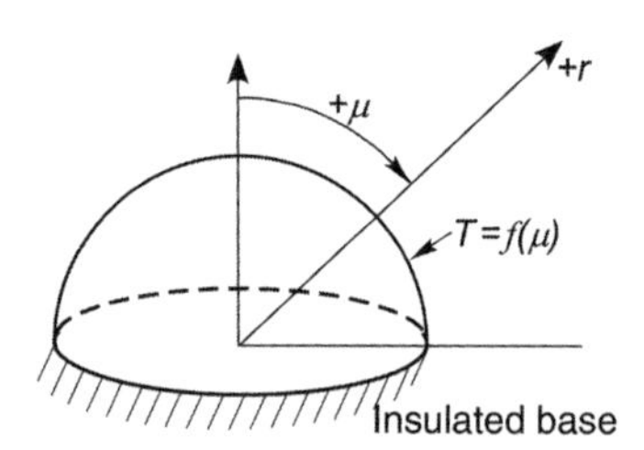

Figure 5-2 Problem description for Example 5-1.

The mathematical formulation of the problem is given as

$$\frac{\partial^2 T}{\partial r^2} + \frac{2}{r}\frac{\partial T}{\partial r} + \frac{1}{r^2}\frac{\partial}{\partial \mu}\left[\left(1-\mu^2\right)\frac{\partial T}{\partial \mu}\right] = 0 \quad \text{in} \quad 0 \le r < b \quad 0 < \mu \le 1 \tag{5-21}$$

BC1: $T(r \to 0) \Rightarrow$ finite BC2: $T(r = b) = f(\mu)$ (5-22a)

BC3: $T(\mu \to +1) \Rightarrow$ finite BC4: $\left.\dfrac{\partial T}{\partial \mu}\right|_{\mu=0} = 0$ (5-22b)

The problem formulation contains only a single nonhomogenous boundary condition, namely, BC2; hence the problem is ready for solution using separation of variables. We note that the condition of finiteness imposed as $r \to 0$ may be treated as a homogeneous boundary condition in the context of separation of variables in that it has the effect of eliminating one of the solution constants in the nonhomogeneous r dimension. Similarly so for the condition of finiteness as $\mu \to +1$. We now assume separation of the form

$$T(r, \mu) = R(r)M(\mu) \tag{5-23}$$

Substituting equation (5-23) into the PDE of equation (5-21), multiplying both sides by r^2, and separating yields

$$\frac{r^2}{R}\left[\frac{d^2R}{dr^2} + \frac{2}{r}\frac{dR}{dr}\right] = \frac{-1}{M}\frac{d}{d\mu}\left[\left(1-\mu^2\right)\frac{dM}{d\mu}\right] = n(n+1) \tag{5-24}$$

where we have introduced the separation constant $n(n+1)$ to force the characteristic value problem in the homogeneous μ dimension. Considering the μ terms, the following ODE, recognized as the Legendre equation, is realized:

$$\frac{d}{d\mu}\left[\left(1-\mu^2\right)\frac{dM}{d\mu}\right] + n(n+1)M = 0 \tag{5-25}$$

The orthogonal Legendre polynomials are generated for integer values of n; hence we define $n = 0, 1, 2, 3, \ldots$ and the solution of equation (5-25) becomes

$$M(\mu) = C_1 P_n(\mu) + C_2 Q_n(\mu) \tag{5-26}$$

We may eliminate the Legendre polynomials of the second kind ($C_2 = 0$) because $Q_n(\mu)$ is infinite at $\mu = \pm 1$, noting that $\mu = 1$(i.e., $\theta = 0$) is in the domain of the problem. For the hemisphere, we have an actual boundary condition (BC4) to consider corresponding to the insulated base of the sphere at $\mu = 0$ (i.e., $\theta = \pi/2$), namely,

$$\left.\frac{dP_n(\mu)}{d\mu}\right|_{\mu=0} = 0 \qquad \text{yielding} \qquad n = 0, 2, 4, \ldots \text{ (even)} \tag{5-27}$$

Examination of the first few Legendre polynomials, see equation (2-70), reveals that the derivatives are all zero at $\mu = 0$ only for the even Legendre polynomials, and we therefore limit n to the even integer values. The remaining r terms in equation (5-24) yield

$$r^2\frac{d^2R}{dr^2} + 2r\frac{dR}{dr} - n(n+1)R = 0 \tag{5-28}$$

which is recognized as Cauchy's equation, as developed in equation (4-29). Here we have the specific values of $a_0 = 2$ and $b_0 = -n(n+1)$, which yield the auxiliary equation

$$\gamma^2 + \gamma - n(n+1) = 0 \tag{5-29}$$

Equation (5-29) is readily factored to yields roots $\gamma_1 = n$ and $\gamma_2 = -(n+1)$, giving

$$R(r) = C_3 r^n + C_4 r^{-(n+1)} \tag{5-30}$$

We now apply what we consider to be the homogeneous boundary condition (BC1), namely, the requirement of finiteness at the origin, which eliminates C_4 given that this term goes to infinity as $r \to 0$ for n equal to even positive integers. Having considered all but the final, nonhomogeneous boundary condition, we now recombine our separated solutions as products and sum over all possible solutions:

$$T(r, \mu) = \sum_{\substack{n=0 \\ \text{even}}}^{\infty} C_n r^n P_n(\mu) \tag{5-31}$$

where we have introduced the new constant $C_n = C_1 C_3$. We lastly apply the nonhomogeneous BC2, equation (5-22a), which yields

$$T(r = b) = f(\mu) = \sum_{\substack{n=0 \\ \text{even}}}^{\infty} C_n b^n P_n(\mu) \tag{5-32}$$

Equation (5-32) is recognized as a Fourier–Legendre series expansion of the function $f(\mu)$. We apply the following operator to both sides:

$$* \int_{\mu=0}^{1} P_q(\mu)\, d\mu$$

noting that the arbitrary constant q must be an *even integer*, recalling that

$$\int_{\mu=0}^{1} P_q(\mu) P_n(\mu)\, d\mu = 0 \qquad \text{for} \qquad n \neq q \tag{5-33}$$

provided that n and q are both even or both odd. This yields the coefficients C_n as

$$C_n = \frac{\int_{\mu=0}^{1} f(\mu) P_n(\mu)\, d\mu}{b^n \int_{\mu=0}^{1} \left[P_n(\mu)\right]^2 d\mu} \tag{5-34}$$

The integral in the denominator of equation (5-34) is recognized as the norm and is equal to $1/(2n+1)$. The solution is now complete with equations (5-31) and

(5-34). We may combine both expressions and simplify, yielding the following expression:

$$T(r,\mu) = \sum_{\substack{n=0\\ \text{even}}}^{\infty} (2n+1)\left(\frac{r}{b}\right)^n P_n(\mu)\int_{\mu'=0}^{1} f(\mu')\,P_n(\mu')\,d\mu' \tag{5-35}$$

Example 5-2 $T = T(r, \mu, \phi)$ for Solid Sphere

A solid sphere of radius b is maintained at steady-state conditions with a prescribed surface temperature $T(r=b) = f(\mu,\phi)$. The mathematical formulation of the problem is given as

$$\frac{\partial^2 T}{\partial r^2} + \frac{2}{r}\frac{\partial T}{\partial r} + \frac{1}{r^2}\frac{\partial}{\partial \mu}\left[\left(1-\mu^2\right)\frac{\partial T}{\partial \mu}\right] + \frac{1}{r^2\left(1-\mu^2\right)}\frac{\partial^2 T}{\partial \phi^2} = 0 \tag{5-36}$$

$$\text{in} \quad 0 \le r < b \qquad -1 \le \mu \le 1 \qquad 0 \le \phi \le 2\pi$$

$$\text{BC1:} \quad T(r \to 0) \Rightarrow \text{finite} \qquad \text{BC2:} \quad T(r=b) = f(\mu,\phi) \tag{5-37a}$$

$$\text{BC3:} \quad T(\mu \to \pm 1) \Rightarrow \text{finite} \tag{5-37b}$$

$$\text{BC4:} \quad T(\phi) = T(\phi + 2\pi) \quad \rightarrow \quad 2\pi - \text{periodicity} \tag{5-37c}$$

$$\text{BC5:} \quad \left.\frac{\partial T(\phi)}{\partial \phi}\right|_{\phi} = \left.\frac{\partial T(\phi)}{\partial \phi}\right|_{\phi+2\pi} \quad \rightarrow \quad 2\pi - \text{periodicity} \tag{5-37d}$$

where we have imposed finiteness as $r \to 0$ and as $\mu \to \pm 1$. Boundary conditions BC4 and BC5 are recognized as the requirement of 2π periodicity in the ϕ dimension. With only a single nonhomogeneous boundary condition, we proceed with separation of variables in the form

$$T(r,\mu,\phi) = R(r)M(\mu)\Phi(\phi) \tag{5-38}$$

Substituting equation (5-38) into the PDE of equation (5-36) yields

$$\frac{-1}{M}\frac{d}{d\mu}\left[\left(1-\mu^2\right)\frac{dM}{d\mu}\right] - \frac{1}{1-\mu^2}\frac{1}{\Phi}\frac{d^2\Phi}{d\phi^2} = \frac{r^2}{R}\left(\frac{d^2R}{dr^2} + \frac{2}{r}\frac{dR}{dr}\right) = n(n+1) \tag{5-39}$$

after multiplying both sides by $r^2/R(r)M(\mu)\Phi(\phi)$. The separation constant is selected here to ultimately yield the associated Legendre equation in the μ dimension. The r dimension is now separated to yield the Cauchy equation as the ODE, as detailed in the previous problem, giving the solution

$$R(r) = C_1 r^n + C_2 r^{-(n+1)} \tag{5-40}$$

The requirement of finiteness at the origin eliminates C_2 given that this term goes to infinity as $r \to 0$ for $n > -1$. The term associated with C_1 is finite for

$n \geq 0$. We now consider the left-hand side of equation (5-39), multiplying first by $1 - \mu^2$ and separating, to yield

$$\frac{1-\mu^2}{M}\frac{d}{d\mu}\left[\left(1-\mu^2\right)\frac{dM}{d\mu}\right] + n(n+1)\left(1-\mu^2\right) = \frac{-1}{\Phi}\frac{d^2\Phi}{d\phi^2} = m^2 \qquad (5\text{-}41)$$

We have forced a characteristic value problem in the homogeneous ϕ dimension with the introduction of the separation constant m^2, which yields the desired solution

$$\Phi(\phi) = C_3 \cos m\phi + C_4 \sin m\phi \qquad (5\text{-}42)$$

The requirement of 2π periodicity is satisfied for positive integers, namely,

$$m = 0, 1, 2, 3, \ldots$$

and both constants C_3 and C_4 are retained. We finally consider the remaining μ terms in equation (5-41), rearranging as

$$\frac{d}{d\mu}\left[\left(1-\mu^2\right)\frac{dM}{d\mu}\right] + \left[n(n+1) - \frac{m^2}{1-\mu^2}\right]M = 0 \qquad (5\text{-}43)$$

Equation (5-43) is recognized as the associated Legendre equation. With m set to positive integers to satisfy 2π periodicity in the ϕ dimension, we now limit n to the positive integers,

$$n = 0, 1, 2, 3, \ldots$$

to generate the orthogonal associated Legendre polynomials as the solution of equation (5-43). This provides the solution

$$M(\mu) = C_5 P_n^m(\mu) + C_6 Q_n^m(\mu) \qquad (5\text{-}44)$$

We may now eliminate the associated Legendre polynomials of the second kind ($C_6 = 0$) in keeping with BC3 because $Q_n^m(\mu)$ is infinite at $\mu = \pm 1$. The complete solution may now be formed by summing over all possible product solutions,

$$T(r, \mu, \phi) = \sum_{n=0}^{\infty}\sum_{m=0}^{n} r^n P_n^m(\mu)\left(a_{nm}\cos m\phi + b_{nm}\sin m\phi\right) \qquad (5\text{-}45)$$

where we have defined new constants $a_{nm} = C_1C_3C_5$ and $b_{nm} = C_1C_4C_5$. The inner summation is terminated at $m = n$, noting that the associated Legendre polynomials are zero for $m > n$, as seen by equation (2-83). The nonhomogeneous boundary condition, equation (5-37a), may now be applied:

$$T(r = b) = f(\mu, \phi) = \sum_{n=0}^{\infty}\sum_{m=0}^{n} b^n P_n^m(\mu)\left(a_{nm}\cos m\phi + b_{nm}\sin m\phi\right) \qquad (5\text{-}46)$$

Equation (5-46) is a double Fourier series expansion in terms of the orthogonal associated Legendre polynomials and trigonometric functions. We first find the Fourier coefficients a_{nm} by applying the following operators to both sides:

$$* \int_{\mu=-1}^{1} P_k^m(\mu)\, d\mu \qquad \text{and} \qquad * \int_{\phi=0}^{2\pi} \cos q\phi\, d\phi$$

which yields the expression

$$a_{nm} = \frac{\displaystyle\int_{\phi=0}^{2\pi} \int_{\mu=-1}^{1} f(\mu, \phi) P_n^m(\mu) \cos m\phi\, d\mu\, d\phi}{\displaystyle b^n \int_{\mu=-1}^{1} [P_n^m(\mu)]^2\, d\mu \int_{\phi=0}^{2\pi} \cos^2 m\phi\, d\phi} \tag{5-47}$$

In a similar manner, the Fourier coefficients b_{nm} are found by applying the following operators to both sides:

$$* \int_{\mu=-1}^{1} P_k^m(\mu)\, d\mu \qquad \text{and} \qquad * \int_{\phi=0}^{2\pi} \sin q\phi\, d\phi$$

to yield the expression

$$b_{nm} = \frac{\displaystyle\int_{\phi=0}^{2\pi} \int_{\mu=-1}^{1} f(\mu, \phi) P_n^m(\mu) \sin m\phi\, d\mu\, d\phi}{\displaystyle b^n \int_{\mu=-1}^{1} [P_n^m(\mu)]^2\, d\mu \int_{\phi=0}^{2\pi} \sin^2 m\phi\, d\phi} \tag{5-48}$$

The norms $N(\mathrm{m})$ of the two trigonometric functions in the denominators of equations (5-47) and (5-48) are equal to π except for the special case of $m = 0$ in equation (5-47), for which the norm is equal to 2π. The norm $N(n,m)$ of the associated Legendre polynomials was previously defined, see equation (2-88), as

$$N(n, m) = \int_{\mu=-1}^{1} \left[P_n^m(\mu)\right]^2 d\mu = \frac{2}{2n+1} \frac{(n+m)!}{(n-m)!} \tag{5-49}$$

We may introduce the Fourier coefficients into equation (5-45) to yield

$$T(r, \mu, \phi) = \frac{1}{\pi} \sum_{n=0}^{\infty} \sum_{m=0}^{n} \left[\frac{2n+1}{2} \frac{(n-m)!}{(n+m)!} \left(\frac{r}{b}\right)^n P_n^m(\mu) \cdot \int_{\phi'=0}^{2\pi} \int_{\mu'=-1}^{1} \right.$$

$$\left. \times\, f(\mu', \phi')\, P_n^m(\mu') \cos m(\phi - \phi')\, d\mu'\, d\phi' \right] \tag{5-50}$$

where π should be replaced by 2π for the case of $m = 0$, and noting that we have introduced a trigonometric substitution to combine the sin and cos terms into a single expression.

5-3 SOLUTION OF TRANSIENT PROBLEMS

In this section we will solve a range of transient problems in the spherical coordinate system. Our general guidelines established for separation of variables in the rectangular coordinate system equally apply to the spherical coordinate problems. Therefore, transient problems with a homogeneous PDE must always be solved with all homogeneous boundary conditions and with the single nonhomogeneity being the initial condition. If one or more boundary conditions are nonhomogeneous, removal of the nonhomogeneities via a temperature shift may be used, or the principle of superposition must be used to treat the nonhomogeneous boundary conditions separately in a steady-state problem. For a nonhomogeneous PDE (i.e., when heat generation is present), superposition is used to treat the heat generation within a steady-state problem, which is then coupled to a homogeneous transient problem at the initial condition. Prior to solving a range of problems to illustrate the above procedures, a few additional guidelines are offered for the transient heat equation in spherical coordinates.

As introduced in Section 5-1, the last three entries of Table 5-1 concern the transient spherical problems. For the 1-D transient problem, namely, $T = T(r, t)$, the transformation $U = rT$ is used to remove the nonconstant coefficients from the PDE, effectively reducing the spherical heat equation to the form of the Cartesian heat equation, for which we have the necessary tools to solve. The 2-D and 3-D transient problems involve $T = T(r, \mu, t)$ and $T = T(r, \mu, \phi, t)$, respectively. For both instances, we utilize the $V = r^{1/2}T$ transformation, and expect orthogonal functions in all spatial dimensions. For the r dimension, half-integer-order Bessel functions will always arise due to the separation constant introduced to force the Legendre polynomials (or associated Legendre polynomials) in the μ dimension. For the 3-D problem, as with cylindrical systems, the ϕ dimension must always be homogeneous, generating the trigonometric functions $\cos m\phi$ and $\sin m\phi$ for the condition of 2π periodicity for the full sphere ($0 \le \phi \le 2\pi$). The necessity for 2π periodicity will limit the separation constants m to integer values, which will then couple to the order of the associated Legendre polynomials with ϕ dependency. For the spherical coordinate system, the ϕ dimension will always correspond to the full sphere ($0 \le \phi \le 2\pi$), with a hemisphere defined by limiting the domain of the μ dimension, namely, to $0 \le \mu \le 1$. Finally, as described above, ϕ dependency in the absence of μ dependency is not covered from a physical point of view. Similarly, any combination of ϕ or μ dependency in the absence of r dependency is not considered (i.e., physically unrealistic).

Example 5-3 $T = T(r, t)$ for Solid Sphere with Convective Boundary
A solid sphere of radius b is initially at temperature $F(r)$. For $t > 0$, the boundary condition at $r = b$ dissipates heat by convection, with convection coefficient h into a fluid at zero temperature. The mathematical formulation of the problem is given as

$$\frac{\partial^2 T}{\partial r^2} + \frac{2}{r}\frac{\partial T}{\partial r} = \frac{1}{\alpha}\frac{\partial T}{\partial t} \qquad \text{in} \qquad 0 \le r < b \qquad t > 0 \tag{5-51}$$

$$\text{BC1:} \quad T(r \to 0) \Rightarrow \text{finite} \qquad \text{BC2:} \quad -k\left.\frac{\partial T}{\partial r}\right|_{r=b} = h\, T|_{r=b} \tag{5-52a}$$

$$\text{IC:} \quad T(t=0) = F(r) \tag{5-52b}$$

Equation (5-51) may be recast in the equivalent form

$$\frac{1}{r}\frac{\partial^2}{\partial r^2}(rT) = \frac{1}{\alpha}\frac{\partial T}{\partial t} \tag{5-53}$$

We now introduce a new dependent variable $U(r,t)$, using the transform

$$\boxed{U(r,t) = rT(r,t)} \tag{5-54}$$

With this substitution, the PDE of equations (5-51) and (5-53) becomes

$$\frac{\partial^2 U}{\partial r^2} = \frac{1}{\alpha}\frac{\partial U}{\partial t} \tag{5-55}$$

which is equivalent to the 1-D Cartesian formulation. It is also necessary to transform the boundary conditions. We first consider a change of dependent variable for the spatial derivative of T, namely,

$$\frac{\partial T}{\partial r} = \frac{\partial}{\partial r}\left(\frac{U}{r}\right) = \frac{1}{r}\frac{\partial U}{\partial r} - \frac{1}{r^2}U \tag{5-56}$$

Equation (5-56) has the result of adding an additional term, $1/r^2 U$, to each derivative term in a boundary condition; hence both insulated and convective boundary conditions each obtain an additional term. One must also consider the condition of finiteness as $r \to 0$, which now takes the form

$$T(r \to 0) \quad \Rightarrow \quad \lim_{r \to 0} \frac{U(r)}{r} \tag{5-57}$$

which is only finite for the expected trigonometric functions if the numerator is zero in the limit, which then enables the application of L'Hôpital's rule. Thus finiteness at the origin requires $U(r=0) = 0$. We now summarize the complete transformation of the general $U(r,t) = rT(r,t)$ formulation in Table 5-2.

In view of the above discussion and Table 5-2, we now reformulate the problem, including both the boundary conditions and initial condition:

$$\frac{\partial^2 U}{\partial r^2} = \frac{1}{\alpha}\frac{\partial U}{\partial t} \quad \text{in} \quad 0 \le r < b \qquad t > 0 \tag{5-58}$$

TABLE 5-2 PDE and Boundary Condition Transformations for $U = rT$

Equation	$T(r,t)$	$U(r,t)$
PDE	$\frac{\partial^2 T}{\partial r^2} + \frac{2}{r}\frac{\partial T}{\partial r} + \frac{g_0}{k} = \frac{1}{\alpha}\frac{\partial T(r,t)}{\partial t}$	$\frac{\partial^2 U}{\partial r^2} + \frac{g_0 r}{k} = \frac{1}{\alpha}\frac{\partial U(r,t)}{\partial t}$
BC1	$T(r=b) = 0$	$U(r=b) = 0$
BC2	$\left.\frac{\partial T}{\partial r}\right\vert_{r=b} = 0$	$\left.\frac{\partial U}{\partial r}\right\vert_{r=b} - \frac{1}{b} U\vert_{r=b} = 0$
BC3	$\left.\frac{\partial T}{\partial r}\right\vert_{r=b} + \frac{h}{k} T\vert_{r=b} = 0$	$\left.\frac{\partial U}{\partial r}\right\vert_{r=b} + \left(\frac{h}{k} - \frac{1}{b}\right) U\vert_{r=b} = 0$
BC4	$T(r \to 0) \Rightarrow$ finite	$U(r=0) = 0$

$$\text{BC1:} \quad U(r=0) = 0 \qquad \text{BC2:} \quad \left.\frac{\partial U}{\partial r}\right|_{r=b} + \left(\frac{h}{k} - \frac{1}{b}\right) U|_{r=b} = 0 \tag{5-59a}$$

$$\text{IC:} \quad U(t=0) = rF(r) \tag{5-59b}$$

For convenience, we will introduce the constant $K = (h/k - 1/b)$ into equation (5-59a). With the successful transformation and with all homogeneous boundary conditions, we are now ready to proceed with separation of the form

$$U(r,t) = R(r)\Gamma(t) \tag{5-60}$$

which after substitution into equation (5-58) yields

$$\frac{1}{R}\frac{d^2 R}{dr^2} = \frac{1}{\alpha\Gamma}\frac{d\Gamma}{dt} = -\lambda^2 \tag{5-61}$$

Solution of separated ODE in the t dimension yields

$$\Gamma(t) = C_1 e^{-\alpha\lambda^2 t} \tag{5-62}$$

while solution of the separated ODE in the r dimension gives

$$R(r) = C_2 \cos\lambda r + C_3 \sin\lambda r \tag{5-63}$$

Boundary condition BC1 yields $C_2 = 0$, while BC2 yields the following transcendental equation:

$$\lambda_n \cot\lambda_n b = -K \quad \rightarrow \quad \lambda_n \quad \text{for} \quad n = 1, 2, 3, \ldots \tag{5-64}$$

The transcendental equation (5-64) may be recast in the form $b\lambda_n \cot \lambda_n b + bK = 0$, which has all real roots if $bK > -1$, and noting that $\lambda_0 = 0$ is an eigenvalue only for the special case of $bK = -1$. When the value of K as defined above is introduced into the inequality $bK > -1$, we find the necessary condition for all real roots given as $bh/k > 0$ (i.e., Bi >0). This result is satisfied with the physical requirements of a positive radius, convection coefficient, and thermal conductivity. We now sum over all possible solutions, yielding the general solution

$$U(r, t) = \sum_{n=1}^{\infty} C_n \sin(\lambda_n r) e^{-\alpha \lambda_n^2 t} \tag{5-65}$$

with $C_n = C_1 C_3$. The initial condition, equation (5-29b), is now applied, yielding

$$U(t = 0) = rF(r) = \sum_{n=1}^{\infty} C_n \sin(\lambda_n r) \tag{5-66}$$

We solve for the Fourier coefficients by applying the operator

$$* \int_{r=0}^{b} \sin(\lambda_q r)\, dr$$

which yields the result

$$C_n = \frac{\int_{r=0}^{b} rF(r) \sin(\lambda_n r)\, dr}{\int_{r=0}^{b} \sin^2(\lambda_n r)\, dr} \tag{5-67}$$

The denominator is the norm $N(\lambda_n)$ and may be simplified by case 7 in Table 2-1, with the replacement of H_2 with K. We finally transform the problem back to $T(r,t)$, yielding the result

$$T(r, t) = \frac{U(r, t)}{r} = \sum_{n=1}^{\infty} C_n \frac{\sin(\lambda_n r)}{r} e^{-\alpha \lambda_n^2 t} \tag{5-68}$$

We know that the limit as $r \to 0$ exists and is finite per L'Hôpital's rule. In fact, in this limit, the temperature at the center of the sphere becomes

$$T(r = 0, t) = \sum_{n=1}^{\infty} C_n \lambda_n e^{-\alpha \lambda_n^2 t} \tag{5-69}$$

The steady-state solution for the entire sphere is simply $T = 0$, which follows directly from equations (5-68) and (5-69) as $t \to \infty$, and is in agreement with the physics of the problem, namely, equilibration at the fluid temperature of zero.

Example 5-4 $T = T(r, t)$ for Solid Sphere with Insulated Boundary
A solid sphere of radius b is initially at temperature $F(r)$. For $t > 0$, the boundary condition at $r = b$ is perfectly insulated. The mathematical formulation of the problem is given as

$$\frac{\partial^2 T}{\partial r^2} + \frac{2}{r}\frac{\partial T}{\partial r} = \frac{1}{\alpha}\frac{\partial T}{\partial t} \quad \text{in} \quad 0 \le r < b \quad t > 0 \tag{5-70}$$

$$\text{BC1:} \quad T(r \to 0) \Rightarrow \text{finite} \qquad \text{BC2:} \quad \left.\frac{\partial T}{\partial r}\right|_{r=b} = 0 \tag{5-71a}$$

$$\text{IC:} \quad T(t = 0) = F(r) \tag{5-71b}$$

As in the previous problem, we introduce the $U(r, t) = rT(r, t)$ transformation of the PDE, boundary conditions, and initial condition. Using Table 5-2, this yields

$$\frac{\partial^2 U}{\partial r^2} = \frac{1}{\alpha}\frac{\partial U}{\partial t} \quad \text{in} \quad 0 \le r < b \quad t > 0 \tag{5-72}$$

$$\text{BC1:} \quad U(r = 0) = 0 \qquad \text{BC2:} \quad \left.\frac{\partial U}{\partial r}\right|_{r=b} - \frac{1}{b}U|_{r=b} = 0 \tag{5-73a}$$

$$\text{IC:} \quad U(t = 0) = rF(r) \tag{5-73b}$$

With the successful transformation, and with all homogeneous boundary conditions, we are now ready to proceed with separation of the form

$$U(r, t) = R(r)\Gamma(t) \tag{5-74}$$

which after substitution into equation (5-72) yields

$$\frac{1}{R}\frac{d^2 R}{dr^2} = \frac{1}{\alpha\Gamma}\frac{d\Gamma}{dt} = -\lambda^2 \tag{5-75}$$

Solution of separated ODE in the t dimension yields the desired solution

$$\Gamma(t) = C_1 e^{-\alpha\lambda^2 t} \tag{5-76}$$

while solution of the separated ODE in the r dimension gives

$$R(r) = C_2 \cos\lambda r + C_3 \sin\lambda r \tag{5-77}$$

Boundary condition BC1 yields $C_2 = 0$, while BC2 yields the following transcendental equation:

$$b\lambda_n \cot\lambda_n b = 1 \quad \to \quad \lambda_n \quad \text{for} \quad n = 0, 1, 2, 3, \ldots \tag{5-78}$$

noting that $\lambda_0 = 0$ is an eigenvalue for this special case of the right-hand side equal to unity (see Appendix II). The general solution is now formed by summing over all possible solutions, yielding

$$U(r,t) = \sum_{n=0}^{\infty} C_n \sin(\lambda_n r) e^{-\alpha\lambda_n^2 t} \tag{5-79}$$

with $C_n = C_1 C_3$. Given the physics of the problem with regard to perfect insulation, as discussed below, it is useful to transform the problem back to $T(r,t)$ prior to considering the initial condition, which gives

$$T(r,t) = \sum_{n=0}^{\infty} C_n \frac{\sin(\lambda_n r)}{r} e^{-\alpha\lambda_n^2 t} \tag{5-80}$$

Examination of equation (5-80) as $t \to \infty$ yields a steady-state temperature equal to zero, which is inconsistent with the physics given that finite energy is initially contained within the sphere, as given by the initial temperature $F(r)$. Unlike with the case of the perfectly insulated cylinder, see Example 4-8, the necessary steady-state temperature does not arise directly from the zero eigenvalue term. Nevertheless, as noted by Carslaw and Jaeger [1, p. 203], for the case of the perfectly insulated sphere, $\lambda_0 = 0$ is an eigenvalue, and an additional term must be added to the solution (see the note at the end of this chapter). With this in mind, we express our general solution in the form

$$T(r,t) = C_0 + \sum_{n=1}^{\infty} C_n \frac{\sin(\lambda_n r)}{r} e^{-\alpha\lambda_n^2 t} \tag{5-81}$$

Ultimately, we expect C_0 to be the steady-state solution, although we continue here using our standard approach to evaluate the constants. We now apply the initial condition as given by equation (5-71b), which yields

$$T(t=0) = F(r) = C_0 + \sum_{n=1}^{\infty} C_n \frac{\sin(\lambda_n r)}{r} \tag{5-82}$$

The orthogonality of the sin function is with respect to a weighting function of unity; hence we first multiply both sides by r, giving

$$rF(r) = C_0 r + \sum_{n=1}^{\infty} C_n \sin(\lambda_n r) \tag{5-83}$$

We now evaluate the Fourier coefficients C_n by applying the operator to both sides

$$* \int_{r=0}^{b} \sin(\lambda_m r)\, dr$$

which yields the following expression:

$$\int_{r=0}^{b} rF(r)\sin(\lambda_m r)\,dr = C_0 \int_{r=0}^{b} r\sin(\lambda_m r)\,dr + \sum_{n=1}^{\infty} C_n \int_{r=0}^{b} \sin(\lambda_n r)\sin(\lambda_m r)\,dr \tag{5-84}$$

We note that the integral $\int_{r=0}^{b} r\sin(\lambda_m r)\,dr$ is identically zero for the eigenvalues as defined by equation (5-78). Using the property of orthogonality for the remaining integral on the right-hand side yields the Fourier coefficients

$$C_n = \frac{\int_{r=0}^{b} rF(r)\sin(\lambda_n r)\,dr}{\int_{r=0}^{b} \sin^2(\lambda_n r)\,dr} \tag{5-85}$$

The norm $N(\lambda_n)$ corresponds to case 7 in Table 2-1 with the replacement of H_2 with $-1/b$; hence the norm becomes

$$\int_{r=0}^{b} \sin^2(\lambda_n r)\,dr = \frac{b\left(\lambda_n^2 + 1/b^2\right) - 1/b}{2\left(\lambda_n^2 + 1/b^2\right)} \tag{5-86}$$

To now evaluate the remaining coefficient C_0, we apply the following operator to both sides of equation (5-83)

$$* \int_{r=0}^{b} r\,dr$$

which yields the following expression:

$$\int_{r=0}^{b} r^2 F(r)\,dr = C_0 \int_{r=0}^{b} r^2\,dr + \sum_{n=1}^{\infty} C_n \int_{r=0}^{b} r\sin(\lambda_n r)\,dr \tag{5-87}$$

As described above, the rightmost integral is zero for all terms in the summation, yielding the expression for C_0 as

$$C_0 = \frac{\int_{r=0}^{b} r^2 F(r)\,dr}{\int_{r=0}^{b} r^2\,dr} = \frac{3}{b^3}\int_{r=0}^{b} r^2 F(r)\,dr \tag{5-88}$$

Equations (5-81), (5-85), and (5-88) complete the problem, where C_0 is now recognized as the steady-state temperature of the sphere. This is readily checked by considering conservation of energy between the initial and final states. For a perfectly insulated sphere, we must conserve total energy, namely,

$$E_{\text{initial}} \equiv E_{\text{final}} \qquad \text{(J)} \tag{5-89}$$

We may evaluate the left-hand term by integrating the initial energy over the full sphere, while the right-hand term follows from $E_{\text{final}} = mcT_{\text{ss}}$, yielding

$$\int_{r=0}^{b} (\rho c) 4\pi r^2 F(r)\, dr = (\rho c)\left(\frac{4}{3}\pi b^3\right) T_{\text{ss}} \tag{5-90}$$

Simplification of equation (5-90) directly yields equation (5-88) with $C_0 \equiv T_{\text{ss}}$; hence our solution of the heat equation, including the steady-state solution, correctly reflects conservation of energy.

Example 5-5 $T = T(r, t)$ for Hollow Sphere
A hollow sphere of inner radius a and outer radius b is initially at temperature $F(r)$ (Fig. 5-3). For $t > 0$, the inner and outer surfaces are maintained at zero temperature.

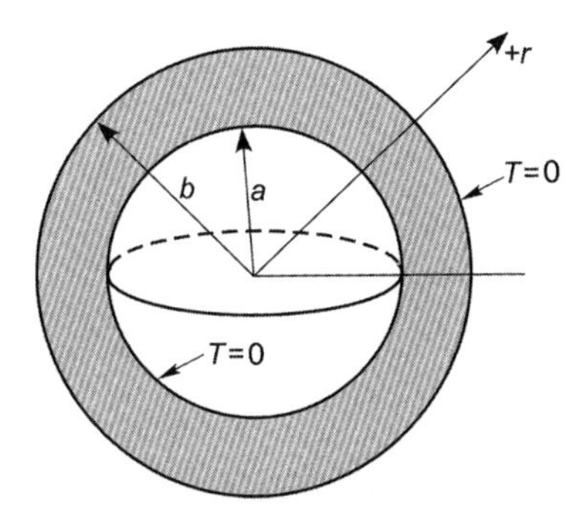

Figure 5-3 Problem description for Example 5-5.

The mathematical formulation of the problem is given as

$$\frac{\partial^2 T}{\partial r^2} + \frac{2}{r}\frac{\partial T}{\partial r} = \frac{1}{\alpha}\frac{\partial T}{\partial t} \quad \text{in} \quad a < r < b \quad t > 0 \tag{5-91}$$

BC1: $T(r = a) = 0$ BC2: $T(r = b) = 0$ (5-92a)

IC: $T(t = 0) = F(r)$ (5-92b)

We introduce the $U(r, t) = rT(r, t)$ transformation of the PDE, boundary conditions, and initial condition. Using Table 5-2, this yields

$$\frac{\partial^2 U}{\partial r^2} = \frac{1}{\alpha}\frac{\partial U}{\partial t} \quad \text{in} \quad a < r < b \quad t > 0 \tag{5-93}$$

BC1: $U(r = a) = 0$ BC2: $U(r = b) = 0$ (5-94a)

IC: $U(t = 0) = rF(r)$ (5-94b)

With all homogeneous boundary conditions, we proceed with separation of the form

$$U(r,t) = R(r)\Gamma(t) \tag{5-95}$$

which after substitution into equation (5-93) yields

$$\frac{1}{R}\frac{d^2R}{dr^2} = \frac{1}{\alpha\Gamma}\frac{d\Gamma}{dt} = -\lambda^2 \tag{5-96}$$

Solution of separated ODE in the t dimension yields

$$\Gamma(t) = C_1 e^{-\alpha\lambda^2 t} \tag{5-97}$$

while the solution of the separated ODE in the r dimension yields

$$R(r) = C_2\cos\lambda r + C_3\sin\lambda r \tag{5-98}$$

Boundary condition BC1 gives

$$C_3 = -C_2\frac{\cos\lambda a}{\sin\lambda a} \tag{5-99}$$

which upon substitution into equation (5-98) gives

$$R(r) = C_2\cos\lambda r - C_2\frac{\cos\lambda a}{\sin\lambda a}\sin\lambda r \tag{5-100}$$

We may now factor out the $1/\sin(\lambda a)$ term, yielding

$$R(r) = \frac{C_2}{\sin\lambda a}\left[\sin\lambda a\cos\lambda r - \cos\lambda a\sin\lambda r\right] \tag{5-101}$$

As we have done before, the term in front of the bracket is now defined as a new constant, say C_4, and a trigonometric substitution is applied to the bracket, giving

$$R(r) = C_4\sin\left[\lambda(r-a)\right] \tag{5-102}$$

We now apply the boundary condition BC2,

$$0 = C_4\sin\lambda\,(b-a) \quad \rightarrow \quad \lambda_n = \frac{n\pi}{b-a} \quad \text{for} \quad n = 0, 1, 2, \ldots \tag{5-103}$$

to yield the eigenvalues λ_n, noting that $\lambda_0 = 0$ is a trivial eigenvalue. We note that equation (5-103) is equivalent to the roots generated by the equation $\cot\lambda_n a = \cot\lambda_n b$. The general solution now becomes

$$U(r,t) = \sum_{n=1}^{\infty} C_n\sin\left[\lambda_n(r-a)\right]e^{-\alpha\lambda_n^2 t} \tag{5-104}$$

We now apply the initial condition as given by equation (5-94b), which yields

$$U(t=0) = rF(r) = \sum_{n=1}^{\infty} C_n \sin\left[\lambda_n(r-a)\right] \tag{5-105}$$

We evaluate the Fourier coefficients C_n by applying the operator to both sides

$$* \int_{r=a}^{b} \sin\left[\lambda_m(r-a)\right] dr$$

which yields the following expression:

$$C_n = \frac{\int_{r=a}^{b} rF(r)\sin\left[\lambda_n(r-a)\right] dr}{\int_{r=a}^{b} \sin^2\left[\lambda_n(r-a)\right] dr} \tag{5-106}$$

Transforming back to $T(r,t)$, the solution takes the form

$$T(r,t) = \sum_{n=1}^{\infty} C_n \frac{\sin\left[\lambda_n(r-a)\right]}{r} e^{-\alpha\lambda_n^2 t} \tag{5-107}$$

The norm of equation (5-106) is $(b-a)/2$ for the eigenvalues of equation (5-103). We may finally substitute the Fourier coefficients and the norm into equation (5-107), yielding the solution of the form

$$T(r,t) = \frac{2}{b-a} \sum_{n=1}^{\infty} \frac{\sin\left[\lambda_n(r-a)\right]}{r} e^{-\alpha\lambda_n^2 t} \int_{r'=a}^{b} r'F(r')\sin\left[\lambda_n(r'-a)\right] dr' \tag{5-108}$$

Example 5-6 $T = T(r,t)$ for Solid Sphere with Internal Energy Generation

A solid sphere of radius b is initially at temperature $F(r)$. For $t > 0$, the boundary condition at $r = b$ is subjected to convection heat transfer with convection coefficient h and zero fluid temperature. In addition, the sphere is subjected to uniform internal energy generation g_0 (W/m^3). The mathematical formulation of the problem is given as

$$\frac{\partial^2 T}{\partial r^2} + \frac{2}{r}\frac{\partial T}{\partial r} + \frac{g_0}{k} = \frac{1}{\alpha}\frac{\partial T}{\partial t} \quad \text{in} \quad 0 \le r < b \quad t > 0 \tag{5-109}$$

$$\text{BC1:} \quad T(r \to 0) \Rightarrow \text{finite} \qquad \text{BC2:} \quad -k\left.\frac{\partial T}{\partial r}\right|_{r=b} = h\, T|_{r=b} \tag{5-110a}$$

$$\text{IC:} \quad T(t=0) = F(r) \tag{5-110b}$$

We introduce the $U(r,t) = rT(r,t)$ transformation of the PDE, boundary conditions, and initial condition. Using Table 5-2, this yields

$$\frac{\partial^2 U}{\partial r^2} + \frac{g_o r}{k} = \frac{1}{\alpha}\frac{\partial U}{\partial t} \qquad \text{in} \qquad 0 \le r < b \quad t > 0 \tag{5-111}$$

$$\text{BC1:} \qquad U(r=0) = 0 \qquad \text{BC2:} \qquad \left.\frac{\partial U}{\partial r}\right|_{r=b} + \left(\frac{h}{k} - \frac{1}{b}\right) U|_{r=b} = 0 \tag{5-112a}$$

$$\text{IC:} \qquad U(t=0) = rF(r) \tag{5-112b}$$

The PDE of equation (5-111) is nonhomogeneous; hence we must seek superposition of the form $U(r,t) = \Psi(r,t) + \Phi(r)$, where $\Psi(r,t)$ takes the homogeneous form of the PDE and boundary conditions, while $\Phi(r)$ is a nonhomogeneous ODE. We first consider the 1-D problem of $\Phi(r)$:

$$\frac{d^2\Phi}{dr^2} + \frac{g_0 r}{k} = 0 \qquad \text{in} \qquad 0 \le r < b \tag{5-113}$$

$$\text{BC1:} \qquad \Phi(r=0) = 0 \qquad \text{BC2:} \qquad \left.\frac{d\Phi}{dr}\right|_{r=b} + K\,\Phi|_{r=b} = 0 \tag{5-114}$$

where we have introduced the constant $K = (h/k - 1/b)$. The solution of equation (5-113) takes the form of a homogeneous solution plus a particular solution, giving

$$\Phi(r) = C_1 + C_2 r - \frac{g_0}{6k} r^3 \tag{5-115}$$

Boundary condition BC1 eliminates constant C_1, while BC2 yields C_2,

$$C_2 = \left(\frac{bg_0}{6h}\right)\left(2 + \frac{hb}{k}\right) \tag{5-116}$$

which gives the final expression for $\Phi(r)$ as

$$\Phi(r) = r\left(C_2 - \frac{g_0}{6k} r^2\right) \tag{5-117}$$

We now define the homogeneous $\Psi(r,t)$ problem, noting that $\Phi(r)$ is coupled to the initial condition, as detailed in Chapter 3. The formulation becomes

$$\frac{\partial^2 \Psi}{\partial r^2} = \frac{1}{\alpha}\frac{\partial \Psi}{\partial t} \qquad \text{in} \qquad 0 \le r < b \quad t > 0 \tag{5-118}$$

$$\text{BC1:} \qquad \Psi(r=0) = 0 \qquad \text{BC2:} \qquad \left.\frac{\partial \Psi}{\partial r}\right|_{r=b} + K\,\Psi|_{r=b} = 0 \tag{5-119a}$$

$$\text{IC:} \qquad \Psi(t=0) = rF(r) - \Phi(r) \tag{5-119b}$$

With homogeneous PDE and boundary conditions, we are ready to separate using

$$\Psi(r,t) = R(r)\Gamma(t) \tag{5-120}$$

which after substitution into equation (5-118) yields

$$\frac{1}{R}\frac{d^2R}{dr^2} = \frac{1}{\alpha\Gamma}\frac{d\Gamma}{dt} = -\lambda^2 \tag{5-121}$$

Solution of the separated ODE in the t dimension yields the solution

$$\Gamma(t) = C_1 e^{-\alpha\lambda^2 t} \tag{5-122}$$

while solution of the separated ODE in the r dimension gives

$$R(r) = C_2 \cos\lambda r + C_3 \sin\lambda r \tag{5-123}$$

Boundary condition BC1 yields $C_2 = 0$, while BC2 yields the following transcendental equation:

$$\lambda_n \cot\lambda_n b = -K \quad \rightarrow \quad \lambda_n \quad \text{for} \quad n = 1, 2, 3, \ldots \tag{5-124}$$

We now sum over all possible solutions, yielding the general solution

$$\Psi(r,t) = \sum_{n=1}^{\infty} C_n \sin(\lambda_n r) e^{-\alpha\lambda_n^2 t} \tag{5-125}$$

with $C_n = C_1 C_3$. The initial condition, equation (5-119b), is now applied, yielding

$$\Psi(t=0) = rF(r) - \Phi(r) = \sum_{n=1}^{\infty} C_n \sin(\lambda_n r) \tag{5-126}$$

We solve for the Fourier coefficients by applying the operator

$$* \int_{r=0}^{b} \sin(\lambda_q r)\, dr$$

which yields the result

$$C_n = \frac{\int_{r=0}^{b} [rF(r) - \Phi(r)] \sin(\lambda_n r)\, dr}{\int_{r=0}^{b} \sin^2(\lambda_n r)\, dr} \tag{5-127}$$

The denominator is the norm $N(\lambda_n)$ and may be simplified by case 7 in Table 2-1, with the replacement of H_2 with K. We finally transform the problem back to $T(r,t)$, yielding the result

$$T(r,t) = \frac{\Psi(r,t)}{r} + \frac{\Phi(r)}{r} = \sum_{n=1}^{\infty} C_n \frac{\sin(\lambda_n r)}{r} e^{-\alpha\lambda_n^2 t} + \left(C_2 - \frac{g_0}{6k} r^2\right) \quad (5\text{-}128)$$

where the constant C_2 is defined by equation (5-116). We note that the rightmost term in equation (5-128) is the steady-state solution, giving the parabolic temperature distribution expected for a steady-state, 1-D sphere with generation.

Example 5-7 $T = T(r, \mu, t)$ for Solid Sphere

A solid sphere of radius b is initially at temperature $F(r,\mu)$. For $t > 0$, the boundary condition at $r = b$ is maintained at temperature T_1. Because all boundary conditions must be homogeneous for a transient problem, we first shift the temperature, defining $\Theta(r, \mu, t) = T(r, \mu, t) - T_1$. With this shift, the mathematical formulation of the problem is given as

$$\frac{\partial^2\Theta}{\partial r^2} + \frac{2}{r}\frac{\partial\Theta}{\partial r} + \frac{1}{r^2}\frac{\partial}{\partial\mu}\left[\left(1-\mu^2\right)\frac{\partial\Theta}{\partial\mu}\right] = \frac{1}{\alpha}\frac{\partial\Theta}{\partial t} \quad \text{in} \quad 0 \le r < b,\ -1 \le \mu \le 1,\ t > 0 \quad (5\text{-}129)$$

$$\text{BC1:} \quad \Theta(r \to 0) \Rightarrow \text{finite} \qquad \text{BC2:} \quad \Theta(r = b) = 0 \quad (5\text{-}130a)$$

$$\text{BC3:} \quad \Theta(\mu \to \pm 1) \Rightarrow \text{finite} \quad (5\text{-}130b)$$

$$\text{IC:} \quad \Theta(t = 0) = F(r,\mu) - T_1 = G(r,\mu) \quad (5\text{-}130c)$$

As discussed above and detailed in Table 5-1, transient problems with both r and μ dependency are solved using the $V(r,\mu,t) = r^{1/2}\Theta(r,\mu,t)$ transformation. Using such a transformation, the mathematical formulation becomes

$$\frac{\partial^2 V}{\partial r^2} + \frac{1}{r}\frac{\partial V}{\partial r} - \frac{1}{4}\frac{V}{r^2} + \frac{1}{r^2}\frac{\partial}{\partial\mu}\left[\left(1-\mu^2\right)\frac{\partial V}{\partial\mu}\right] = \frac{1}{\alpha}\frac{\partial V}{\partial t} \quad \text{in} \quad 0 \le r < b,\ -1 \le \mu \le 1,\ t > 0 \quad (5\text{-}131)$$

$$\text{BC1:} \quad V(r \to 0) \Rightarrow \text{finite} \qquad \text{BC2:} \quad V(r = b) = 0 \quad (5\text{-}132a)$$

$$\text{BC3:} \quad V(\mu \to \pm 1) \Rightarrow \text{finite} \quad (5\text{-}132b)$$

$$\text{IC:} \quad V(t = 0) = r^{1/2} G(r,\mu) \quad (5\text{-}132c)$$

In addition, it will be necessary to check that $V/r^{1/2}$ is finite as $r \to 0$ after transforming back to $\Theta(r,\mu,t)$. We now assume separation of the form

$$V(r,\mu,t) = R(r)M(\mu)\Gamma(t) \quad (5\text{-}133)$$

Substituting equation (5-133) into equation (5-131) yields

$$\frac{1}{R}\left[\frac{d^2R}{dr^2}+\frac{1}{r}\frac{dR}{dr}-\frac{1}{4}\frac{R}{r^2}\right]+\frac{1}{r^2M}\frac{d}{d\mu}\left[\left(1-\mu^2\right)\frac{dM}{d\mu}\right]=\frac{1}{\alpha\Gamma}\frac{d\Gamma}{dt}=-\lambda^2 \tag{5-134}$$

Solution of separated ODE in the t dimension yields the solution

$$\Gamma(t)=C_1e^{-\alpha\lambda^2t} \tag{5-135}$$

with the remaining r terms and μ term of equation (5-134) yielding

$$\frac{r^2}{R}\left[\frac{d^2R}{dr^2}+\frac{1}{r}\frac{dR}{dr}-\frac{1}{4}\frac{R}{r^2}\right]+\lambda^2r^2=-\frac{1}{M}\frac{d}{d\mu}\left[\left(1-\mu^2\right)\frac{dM}{d\mu}\right]=n(n+1) \tag{5-136}$$

after introduction of a new separation constant $n(n+1)$. Solution of the μ dimension ODE yields the orthogonal Legendre polynomials for integer values of n; hence we define $n=0,1,2,3,\ldots$, and the solution becomes

$$M(\mu)=C_2P_n(\mu)+C_3Q_n(\mu) \tag{5-137}$$

We may eliminate the Legendre polynomials of the second kind ($C_3=0$) because $Q_n(\mu)$ is infinite at $\mu=\pm1$, with both values within the domain of the problem. We next consider the remaining r terms of equation (5-136), which we reorganize to the form

$$\frac{d^2R}{dr^2}+\frac{1}{r}\frac{dR}{dr}+\left[\lambda^2-\frac{(n+1/2)^2}{r^2}\right]R=0 \tag{5-138}$$

Equation (5-138) is recognized as Bessel's equation of half-integer order, with the solution

$$R(r)=C_4J_{n+1/2}(\lambda r)+C_5Y_{n+1/2}(\lambda r) \tag{5-139}$$

The requirement for finiteness as $r\to0$ stated by BC1 eliminates the $Y_{n+1/2}(\lambda r)$ term from further consideration, while BC2 yields the equation

$$J_{n+1/2}(\lambda_{nm}b)=0 \quad\to\quad \lambda_{nm} \quad \text{for} \quad m=1,2,3,\ldots \quad \text{for each } n \tag{5-140}$$

which generates eigenvalues λ_{nm}. As discussed previously, we use a double subscript for the eigenvalues to denote that for each value of n, there is a *full set* of eigenvalues λ_m. We now form the general solution by summing the product solutions over all eigenvalues n and λ_{nm}, yielding

$$V(r,\mu,t)=\sum_{n=0}^{\infty}\sum_{m=1}^{\infty}C_{nm}J_{n+1/2}(\lambda_{nm}r)P_n(\mu)e^{-\alpha\lambda_{nm}^2t} \tag{5-141}$$

with constant $C_{nm} = C_1 C_2 C_4$. Application of the initial condition, equation (5-132c), yields

$$r^{1/2} G(r, \mu) = \sum_{n=0}^{\infty} \sum_{m=1}^{\infty} C_{nm} J_{n+1/2}(\lambda_{nm} r) P_n(\mu) \tag{5-142}$$

to which we apply successively the two operators

$$* \int_{r=0}^{b} r J_{n+1/2}(\lambda_{nk} r)\, dr \qquad \text{and} \qquad * \int_{\mu=-1}^{1} P_q(\mu)\, d\mu$$

to yield the Fourier–Bessel–Legendre coefficients C_{nm} as

$$C_{nm} = \frac{\int_{\mu=-1}^{1} \int_{r=0}^{b} r^{3/2} G(r, \mu) J_{n+1/2}(\lambda_{nm} r) P_n(\mu)\, dr\, d\mu}{\int_{r=0}^{b} r J_{n+1/2}^2(\lambda_{nm} r)\, dr \int_{\mu=-1}^{1} \left[P_n(\mu)\right]^2\, d\mu} \tag{5-143}$$

The norms are defined by equation (2-79) for the Legendre polynomials and by case 3 with $\nu = n + 1/2$ in Table 2-2 for the Bessel functions. Finally, we shift back to $\Theta\,(r, \mu, t)$, and then $T\,(r, \mu, t)$, giving

$$T(r, \mu, t) = \sum_{n=0}^{\infty} \sum_{m=1}^{\infty} C_{nm} \frac{J_{n+1/2}(\lambda_{nm} r)}{r^{1/2}} P_n(\mu) e^{-\alpha \lambda_{nm}^2 t} + T_1 \tag{5-144}$$

As a final step, it is necessary to explore the requirement for finiteness as $r \to 0$. Considering only the r dependence of the problem, we have

$$\lim_{r \to 0} \frac{J_{n+1/2}(\lambda r)}{r^{1/2}} = \lim_{r \to 0} \frac{(\text{constant}) r^{n+1/2}}{r^{1/2}} \sum_{k} \frac{(\lambda r)^{2k}}{f(n, k)} \tag{5-145}$$

where we have considered the series form of the Bessel functions, see Appendix IV. As seen in equation (5-145), the $r^{1/2}$ term cancels from the denominator, and the overall solution is finite as $r \to 0$ for all values of $n \geq 0$.

Example 5-8 $T = T(r, \mu, t)$ for a Hemisphere

A hemisphere of radius b is initially at temperature $F(r,\mu)$ (Fig. 5-4). For $t > 0$, the boundary condition at $r = b$ is maintained at temperature T_1, while the boundary condition at $\mu = 0$ (the base) is maintained at temperature T_2.

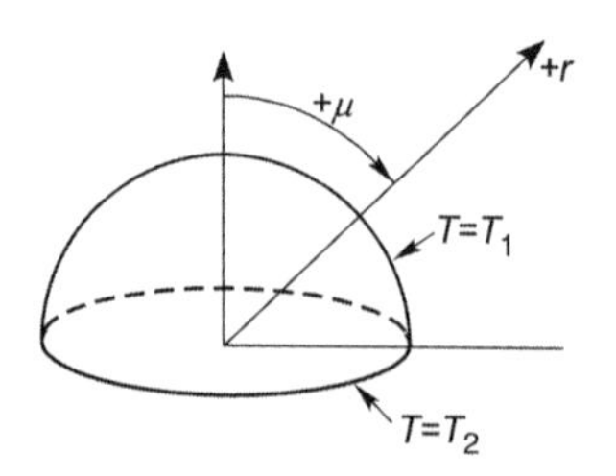

Figure 5-4 Problem description for Example 5-8.

The problem formulation becomes

$$\frac{\partial^2 T}{\partial r^2} + \frac{2}{r}\frac{\partial T}{\partial r} + \frac{1}{r^2}\frac{\partial}{\partial \mu}\left[\left(1-\mu^2\right)\frac{\partial T}{\partial \mu}\right] = \frac{1}{\alpha}\frac{\partial T}{\partial t} \quad \text{in} \quad 0 \le r < b,\ 0 < \mu \le 1,\ t > 0 \tag{5-146}$$

BC1: $T(r \to 0) \Rightarrow \text{finite}$ BC2: $T(r = b) = T_1$ (5-147a)

BC3: $T(\mu \to +1) \Rightarrow \text{finite}$ BC4: $T(\mu = 0) = T_2$ (5-147b)

IC: $T(t = 0) = F(r, \mu)$ (5-147c)

Because all boundary conditions must be homogeneous for a transient problem, and we are unable to remove both nonhomogeneous conditions by a simple temperature shift, we use superposition of the form

$$T(r, \mu, t) = \Psi(r, \mu, t) + \Phi(r, \mu) \tag{5-148}$$

The formulation for the steady-state problem becomes

$$\frac{\partial^2 \Phi}{\partial r^2} + \frac{2}{r}\frac{\partial \Phi}{\partial r} + \frac{1}{r^2}\frac{\partial}{\partial \mu}\left[\left(1-\mu^2\right)\frac{\partial \Phi}{\partial \mu}\right] = 0 \quad \text{in} \quad 0 \le r < b \quad 0 < \mu \le 1 \tag{5-149}$$

BC1: $\Phi(r \to 0) \Rightarrow \text{finite}$ BC2: $\Phi(r = b) = T_1$ (5-150a)

BC3: $\Phi(\mu \to +1) \Rightarrow \text{finite}$ BC4: $\Phi(\mu = 0) = T_2$ (5-150b)

where the steady-state solution takes both nonhomogeneous boundary conditions. However, the steady-state formulation must contain only a single nonhomogeneity for the solution, and since the μ dimension *must be homogeneous*, we introduce a temperature shift $\Theta(r, \mu) = \Phi(r, \mu) - T_2$, which leads to the new formulation:

$$\frac{\partial^2 \Theta}{\partial r^2} + \frac{2}{r}\frac{\partial \Theta}{\partial r} + \frac{1}{r^2}\frac{\partial}{\partial \mu}\left[\left(1-\mu^2\right)\frac{\partial \Theta}{\partial \mu}\right] = 0 \quad \text{in} \quad 0 \le r < b \quad 0 < \mu \le 1 \tag{5-151}$$

BC1: $\Theta(r \to 0) \Rightarrow \text{finite}$ BC2: $\Theta(r = b) = T_1 - T_2 = T_3$ (5-152a)

BC3: $\Theta(\mu \to +1) \Rightarrow \text{finite}$ BC4: $\Theta(\mu = 0) = 0$ (5-152b)

The problem is now correctly set for separation of the form

$$\Theta(r, \mu) = R(r)M(\mu) \tag{5-153}$$

Substituting equation (5-153) into the PDE of equation (5-151), multiplying both sides by r^2, and separating yields

$$\frac{r^2}{R}\left[\frac{d^2R}{dr^2}+\frac{2}{r}\frac{dR}{dr}\right]=\frac{-1}{M}\frac{d}{d\mu}\left[\left(1-\mu^2\right)\frac{dM}{d\mu}\right]=n(n+1) \tag{5-154}$$

where we have introduced the separation constant $n(n+1)$ to force the characteristic value problem in the homogeneous μ dimension. As with Example 5-1, the μ terms yield the orthogonal Legendre polynomials for integer values of n ($n = 0, 1, 2, 3, \ldots$), giving the solution

$$M(\mu) = C_1 P_n(\mu) + C_2 Q_n(\mu) \tag{5-155}$$

The Legendre polynomials of the second kind are eliminated ($C_2 = 0$) because $Q_n(\mu)$ is infinite at $\mu = +1$. The boundary condition at $\mu = 0$ is given by equation (5-152b), namely,

$$P_n(\mu = 0) = 0 \qquad \text{yielding} \qquad n = 1, 3, 5, \ldots \quad \text{(odd integers)} \tag{5-156}$$

Examination of the first few Legendre polynomials, see equation (2-70), reveals that these functions are zero at $\mu = 0$ only for the odd Legendre polynomials, and we therefore limit n to the odd integer values. The remaining r terms in equation (5-154) yield Cauchy's equation as described above, with the solution

$$R(r) = C_3 r^n + C_4 r^{-(n+1)} \tag{5-157}$$

where we eliminate the constant C_4 for the condition of finiteness as $r \to 0$ for n equal to odd positive integers. We now recombine our separated solutions as products, and sum over all possible solutions:

$$\Theta(r, \mu) = \sum_{\substack{n=1\\ \text{odd}}}^{\infty} C_n r^n P_n(\mu) \tag{5-158}$$

where we have introduced the new constant $C_n = C_1 C_3$. We lastly apply the nonhomogeneous BC2, equation (5-152a), which yields

$$\Theta(r = b) = T_3 = \sum_{\substack{n=1\\ \text{odd}}}^{\infty} C_n b^n P_n(\mu) \tag{5-159}$$

Equation (5-159) is recognized as a Fourier–Legendre series expansion of a constant. We apply the following operator to both sides:

$$* \int_{\mu=0}^{1} P_q(\mu)\, d\mu$$

noting that the arbitrary constant q must be an odd integer, recalling that

$$\int_{\mu=0}^{1} P_q(\mu) P_n(\mu)\, d\mu = 0 \qquad \text{for} \qquad n \neq q \tag{5-160}$$

provided that n and q are both even or both odd. This yields the coefficients C_n as

$$C_n = \frac{T_3 \int_{\mu=0}^{1} P_n(\mu)\, d\mu}{b^n \int_{\mu=0}^{1} \left[P_n(\mu)\right]^2 d\mu} \tag{5-161}$$

The integral in the denominator of equation (5-161) is recognized as the norm and is equal to $1/(2n+1)$. The solution is now complete and may be combined and the temperature shifted back, yielding the following expression:

$$\Phi(r, \mu) = \left(T_1 - T_2\right) \sum_{\substack{n=1 \\ \text{odd}}}^{\infty} (2n+1) \left(\frac{r}{b}\right)^n P_n(\mu) \int_{\mu'=0}^{1} P_n\left(\mu'\right) d\mu' + T_2 \tag{5-162}$$

With the steady-state solution now fully solved, we formulate the homogeneous transient problem as

$$\frac{\partial^2 \Psi}{\partial r^2} + \frac{2}{r}\frac{\partial \Psi}{\partial r} + \frac{1}{r^2}\frac{\partial}{\partial \mu}\left[\left(1-\mu^2\right)\frac{\partial \Psi}{\partial \mu}\right] = \frac{1}{\alpha}\frac{\partial \Psi}{\partial t} \quad \text{in} \quad 0 \le r < b,\ 0 < \mu \le 1,\ t > 0 \tag{5-163}$$

BC1:	$\Psi(r \to 0) \Rightarrow$ finite	BC2:	$\Psi(r = b) = 0$	(5-164a)
BC3:	$\Psi(\mu \to +1) \Rightarrow$ finite	BC4:	$\Psi(\mu = 0) = 0$	(5-164b)
IC:	$\Psi(t = 0) = F(r, \mu) - \Phi(r, \mu) = G(r, \mu)$			(5-164c)

where we have coupled the steady-state solution to the initial condition, as always required for superposition of the form given by equation (5-148). As detailed in Table 5-1, we now use the $V(r, \mu, t) = r^{1/2}\Psi(r, \mu, t)$ transformation, which gives

$$\frac{\partial^2 V}{\partial r^2} + \frac{1}{r}\frac{\partial V}{\partial r} - \frac{1}{4}\frac{V}{r^2} + \frac{1}{r^2}\frac{\partial}{\partial \mu}\left[\left(1-\mu^2\right)\frac{\partial V}{\partial \mu}\right] = \frac{1}{\alpha}\frac{\partial V}{\partial t} \quad \text{in} \quad 0 \le r < b,\ 0 < \mu \le 1,\ t > 0 \tag{5-165}$$

BC1:	$V(r \to 0) \Rightarrow$ finite	BC2:	$V(r = b) = 0$	(5-166a)
BC3:	$V(\mu \to +1) \Rightarrow$ finite	BC4:	$V(\mu = 0) = 0$	(5-166b)
IC:	$V(t = 0) = r^{1/2}\, G(r, \mu)$			(5-166c)

We now assume separation of the form

$$V(r, \mu, t) = R(r)M(\mu)\Gamma(t) \tag{5-167}$$

which upon substitution into equation (5-165) yields

$$\frac{1}{R}\left[\frac{d^2R}{dr^2} + \frac{1}{r}\frac{dR}{dr} - \frac{1}{4}\frac{R}{r^2}\right] + \frac{1}{r^2 M}\frac{d}{d\mu}\left[\left(1-\mu^2\right)\frac{dM}{d\mu}\right] = \frac{1}{\alpha\Gamma}\frac{d\Gamma}{dt} = -\lambda^2 \tag{5-168}$$

Solution of separated ODE in the t dimension yields the solution

$$\Gamma(t) = C_1 e^{-\alpha\lambda^2 t} \tag{5-169}$$

with the r terms and μ term of equation (5-168) yielding

$$\frac{r^2}{R}\left[\frac{d^2R}{dr^2} + \frac{1}{r}\frac{dR}{dr} - \frac{1}{4}\frac{R}{r^2}\right] + \lambda^2 r^2 = -\frac{1}{M}\frac{d}{d\mu}\left[\left(1-\mu^2\right)\frac{dM}{d\mu}\right] = n(n+1) \tag{5-170}$$

after introduction of a new separation constant $n(n+1)$. Solution of the μ-dimension ODE yields the orthogonal Legendre polynomials for integer values $n = 0, 1, 2, 3, \ldots$, and the solution becomes

$$M(\mu) = C_2 P_n(\mu) + C_3 Q_n(\mu) \tag{5-171}$$

We eliminate the Legendre polynomials of the second kind ($C_3 = 0$) for the requirement of finiteness at $\mu = +1$. The boundary condition at $\mu = 0$ as given by equation (5-166b) further limits the values of n, namely,

$$P_n(\mu = 0) = 0 \qquad \text{yielding} \qquad n = 1, 3, 5, \ldots \text{(odd integers)} \tag{5-172}$$

The remaining r terms of equation (5-170) yield Bessel's equation of half-integer order, with the solution

$$R(r) = C_4 J_{n+1/2}(\lambda r) + C_5 Y_{n+1/2}(\lambda r) \tag{5-173}$$

The requirement for finiteness as $r \to 0$ stated by BC1 eliminates the $Y_{n+1/2}(\lambda r)$ term, while BC2 yields the transcendental equation

$$J_{n+1/2}(\lambda_{nm} b) = 0 \quad \to \quad \lambda_{nm} \quad \text{for} \quad m = 1, 2, 3 \ldots \text{for each } n \tag{5-174}$$

which produces eigenvalues λ_{nm}, noting that there is a full set of eigenvalues λ_m for each value of n. We now form the general solution by summing the product solutions over all eigenvalues n and λ_{nm}, yielding

$$V(r, \mu, t) = \sum_{\substack{n=1 \\ \text{odd}}}^{\infty} \sum_{m=1}^{\infty} C_{nm} J_{n+1/2}(\lambda_{nm} r) P_n(\mu) e^{-\alpha\lambda_{nm}^2 t} \tag{5-175}$$

with constant $C_{nm} = C_1 C_2 C_4$. Application of the initial condition, equation (5-166c), yields

$$r^{1/2} G(r, \mu) = \sum_{\substack{n=1 \\ \text{odd}}}^{\infty} \sum_{m=1}^{\infty} C_{nm} J_{n+1/2}(\lambda_{nm} r) P_n(\mu) \tag{5-176}$$

to which we apply successively the two operators

$$* \int_{r=0}^{b} r J_{n+1/2}(\lambda_{nk} r)\, dr \qquad \text{and} \qquad * \int_{\mu=0}^{1} P_q(\mu)\, d\mu$$

to yield the Fourier–Bessel–Legendre coefficients C_{nm} as

$$C_{nm} = \frac{\int_{\mu=0}^{1} \int_{r=0}^{b} r^{3/2} G(r, \mu) J_{n+1/2}(\lambda_{nm} r) P_n(\mu)\, dr\, d\mu}{\int_{r=0}^{b} r J_{n+1/2}^2(\lambda_{nm} r)\, dr \int_{\mu=0}^{1} \left[P_n(\mu)\right]^2 d\mu} \tag{5-177}$$

The norms are defined by equation (2-79) for the Legendre polynomials, and by case 3 with $\nu = n + \frac{1}{2}$ in Table 2-2 for the Bessel functions. Finally, we shift back to $\Psi(r, \mu, t)$, giving

$$\Psi(r, \mu, t) = \sum_{\substack{n=1 \\ \text{odd}}}^{\infty} \sum_{m=1}^{\infty} C_{nm} \frac{J_{n+1/2}(\lambda_{nm} r)}{r^{1/2}} P_n(\mu) e^{-\alpha \lambda_{nm}^2 t} \tag{5-178}$$

The total solution is then formed by superposition of the transient and steady-state solutions,

$$\begin{aligned} T(r, \mu, t) = {} & \sum_{\substack{n=1 \\ \text{odd}}}^{\infty} \sum_{m=1}^{\infty} C_{nm} \frac{J_{n+1/2}(\lambda_{nm} r)}{r^{1/2}} P_n(\mu) e^{-\alpha \lambda_{nm}^2 t} \\ & + (T_1 - T_2) \sum_{\substack{n=1 \\ \text{odd}}}^{\infty} (2n+1) \left(\frac{r}{b}\right)^n P_n(\mu) \int_{\mu'=0}^{1} P_n(\mu')\, d\mu' + T_2 \end{aligned} \tag{5-179}$$

where C_{nm} is defined by equation (5-177). As explored in the previous example, the overall solution is indeed finite as $r \to 0$ for all values of $n \geq 0$.

Example 5-9 $T = T(r, \mu, t)$ for a Hemisphere

A hemisphere of radius b is initially at temperature $F(r,\mu)$. For $t > 0$, the boundary condition at $r = b$ dissipates heat by convection with convection coefficient

h into a fluid of zero temperature, while the boundary condition at $\mu = 0$ (the base) is perfectly insulated. The problem formulation becomes

$$\frac{\partial^2 T}{\partial r^2} + \frac{2}{r}\frac{\partial T}{\partial r} + \frac{1}{r^2}\frac{\partial}{\partial \mu}\left[\left(1 - \mu^2\right)\frac{\partial T}{\partial \mu}\right] = \frac{1}{\alpha}\frac{\partial T}{\partial t} \quad \text{in} \quad 0 \le r < b,\ 0 < \mu \le 1,\ t > 0 \tag{5-180}$$

BC1: $T(r \to 0) \Rightarrow \text{finite}$ BC2: $-k\left.\frac{\partial T}{\partial r}\right|_{r=b} = h\,T|_{r=b}$ (5-181a)

BC3: $T(\mu \to +1) \Rightarrow \text{finite}$ BC4: $\left.\frac{\partial T}{\partial \mu}\right|_{\mu=0} = 0$ (5-181b)

IC: $T(t = 0) = F(r, \mu)$ (5-181c)

As detailed in Table 5-1, we now use the $V(r, \mu, t) = r^{1/2}T(r, \mu, t)$ transformation, which gives

$$\frac{\partial^2 V}{\partial r^2} + \frac{1}{r}\frac{\partial V}{\partial r} - \frac{1}{4}\frac{V}{r^2} + \frac{1}{r^2}\frac{\partial}{\partial \mu}\left[\left(1 - \mu^2\right)\frac{\partial V}{\partial \mu}\right] = \frac{1}{\alpha}\frac{\partial V}{\partial t} \quad \text{in} \quad 0 \le r < b,\ 0 < \mu \le 1,\ t > 0 \tag{5-182}$$

BC1: $V(r \to 0) \Rightarrow \text{finite}$ BC2: $\left.\frac{\partial V}{\partial r}\right|_{r=b} + \left(\frac{h}{k} - \frac{1}{2b}\right)V|_{r=b} = 0$ (5-183a)

BC3: $V(\mu \to +1) \Rightarrow \text{finite}$ BC4: $\left.\frac{\partial V}{\partial \mu}\right|_{\mu=0} = 0$ (5-183b)

IC: $V(t = 0) = r^{1/2}F(r, \mu)$ (5-183c)

The derivative boundary condition BC2 of equation (5-181a) was modified using

$$\boxed{\frac{\partial T}{\partial r} = \frac{\partial}{\partial r}\left(\frac{V}{r^{1/2}}\right) = \frac{1}{r^{1/2}}\frac{\partial V}{\partial r} - \frac{1}{2r^{3/2}}V} \tag{5-184}$$

which has the effect of introducing an additional term to BC2, as seen in equation (5-183a). We list all such boundary condition transformations in Table 5-3 corresponding to the $V = r^{1/2}T$ transformation.

We now assume separation of the form

$$V(r, \mu, t) = R(r)M(\mu)\Gamma(t) \tag{5-185}$$

TABLE 5-3 Boundary Condition Transformations for $V = r^{1/2}T$

Boundary Equation	$T(r,\mu,t)$, $T(r,\mu,\phi,t)$	$V(r,\mu,t)$, $V(r,\mu,\phi,t)$
BC1	$T(r=b)=0$	$V(r=b)=0$
BC2	$\left.\dfrac{\partial T}{\partial r}\right\|_{r=b}=0$	$\left.\dfrac{\partial V}{\partial r}\right\|_{r=b}-\dfrac{1}{2b}V\|_{r=b}=0$
BC3	$\left.\dfrac{\partial T}{\partial r}\right\|_{r=b}+\dfrac{h}{k}T\|_{r=b}=0$	$\left.\dfrac{\partial V}{\partial r}\right\|_{r=b}+\left(\dfrac{h}{k}-\dfrac{1}{2b}\right)V\|_{r=b}=0$
BC4	$T(r\rightarrow 0)\Rightarrow$ finite	$V(r\rightarrow 0)\Rightarrow$ finite

which upon substitution into equation (5-182) yields

$$\frac{1}{R}\left(\frac{d^2R}{dr^2}+\frac{1}{r}\frac{dR}{dr}-\frac{1}{4}\frac{R}{r^2}\right)+\frac{1}{r^2M}\frac{d}{d\mu}\left[\left(1-\mu^2\right)\frac{dM}{d\mu}\right]=\frac{1}{\alpha\Gamma}\frac{d\Gamma}{dt}=-\lambda^2 \tag{5-186}$$

Solution of the separated ODE in the t dimension yields the solution

$$\Gamma(t)=C_1e^{-\alpha\lambda^2t} \tag{5-187}$$

with the remaining r terms and μ term of equation (5-186) yielding

$$\frac{r^2}{R}\left(\frac{d^2R}{dr^2}+\frac{1}{r}\frac{dR}{dr}-\frac{1}{4}\frac{R}{r^2}\right)+\lambda^2r^2=-\frac{1}{M}\frac{d}{d\mu}\left[\left(1-\mu^2\right)\frac{dM}{d\mu}\right]=n(n+1) \tag{5-188}$$

after introduction of a new separation constant $n(n+1)$. Solution of the μ dimension ODE yields the orthogonal Legendre polynomials for integer values $n=0,1,2,3,\ldots$, with the solution becoming

$$M(\mu)=C_2P_n(\mu)+C_3Q_n(\mu) \tag{5-189}$$

We eliminate the Legendre polynomials of the second kind ($C_3=0$) for the requirement of finiteness at $\mu=+1$. The boundary condition BC4 at $\mu=0$, as given by equation (5-183b), further limits the values of n, namely,

$$\left.\frac{dP_n(\mu)}{d\mu}\right|_{\mu=0}=0 \quad \text{yielding} \quad n=0,2,4,\ldots\text{(even integers)} \tag{5-190}$$

The remaining r terms of equation (5-188) yield Bessel's equation of half-integer order, with the solution

$$R(r)=C_4J_{n+1/2}(\lambda r)+C_5Y_{n+1/2}(\lambda r) \tag{5-191}$$

The requirement for finiteness as $r \to 0$ stated by BC1 eliminates the $Y_{n+1/2}(\lambda r)$ term, while BC2 yields the transcendental equation

$$\left.\frac{d}{dr} J_{n+1/2}(\lambda_{nm} r)\right|_{r=b} + K J_{n+1/2}(\lambda_{nm} b) = 0 \quad \to \quad \lambda_{nm} \quad \text{for} \quad m = 1, 2, 3, \ldots \tag{5-192}$$

where we have introduced the constant $K = (h/k - 1/2b)$. Equation (5-192) produces eigenvalues λ_{nm}, noting that there is a full set of eigenvalues λ_m for each value of n. We now form the general solution by summing the product solutions over all eigenvalues n and λ_{nm}, yielding

$$V(r, \mu, t) = \sum_{\substack{n=0 \\ \text{even}}}^{\infty} \sum_{m=1}^{\infty} C_{nm} J_{n+1/2}(\lambda_{nm} r) P_n(\mu) e^{-\alpha \lambda_{nm}^2 t} \tag{5-193}$$

with constant $C_{nm} = C_1 C_2 C_4$. Application of the initial condition, equation (5-183c), yields

$$V(t = 0) = r^{1/2} F(r, \mu) = \sum_{\substack{n=0 \\ \text{even}}}^{\infty} \sum_{m=1}^{\infty} C_{nm} J_{n+1/2}(\lambda_{nm} r) P_n(\mu) \tag{5-194}$$

to which we apply successively the two operators

$$* \int_{r=0}^{b} r J_{n+1/2}(\lambda_{nk} r)\, dr \qquad \text{and} \qquad * \int_{\mu=0}^{1} P_q(\mu)\, d\mu$$

to yield the Fourier–Bessel–Legendre coefficients C_{nm} as

$$C_{nm} = \frac{\int_{\mu=0}^{1} \int_{r=0}^{b} r^{3/2} F(r, \mu) J_{n+1/2}(\lambda_{nm} r) P_n(\mu)\, dr\, d\mu}{\int_{r=0}^{b} r J_{n+1/2}^2(\lambda_{nm} r)\, dr \int_{\mu=0}^{1} \left[P_n(\mu)\right]^2 d\mu} \tag{5-195}$$

The norms are defined by equation (2-79) for the Legendre polynomials and by case 1 with $\nu = n + 1/2$ and $H = K$ in Table 2-2 for the Bessel functions. Finally, we shift back to $T(r, \mu, t)$, giving

$$T(r, \mu, t) = \sum_{\substack{n=0 \\ \text{even}}}^{\infty} \sum_{m=1}^{\infty} C_{nm} \frac{J_{n+1/2}(\lambda_{nm} r)}{r^{1/2}} P_n(\mu) e^{-\alpha \lambda_{nm}^2 t} \tag{5-196}$$

As explored in the previous example, the overall solution is indeed finite as $r \to 0$ for all values of $n \geq 0$.

For the special case of the surface at $r = b$ now being perfectly insulated, several modifications to the above solution are made. The eigenvalues of equation (5-192) are adjusted by replacing K with $-1/2b$ (i.e., letting $h = 0$), as seen in Table 5-5. For the case of $n = 0$ only, there is a new eigenvalue $\lambda_{00} = 0$. This can be derived by expanding the derivative of the Bessel function using

$$\left.\frac{d}{dr}J_{1/2}(\lambda r)\right|_{r=b} = \lambda J_{-1/2}(\lambda b) - \frac{1}{2b}J_{1/2}(\lambda b) \tag{5-197}$$

in addition to the following substitutions for Bessel functions of order $\frac{1}{2}$, namely,

$$J_{1/2}(z) = \left(\frac{2}{\pi z}\right)^{1/2} \sin z \tag{5-198}$$

$$J_{-1/2}(z) = \left(\frac{2}{\pi z}\right)^{1/2} \cos z \tag{5-199}$$

The term that results from the zero eigenvalue solution should then be added to the solution, giving

$$T(r, \mu, t) = \frac{\int_{\mu=0}^{1}\int_{r=0}^{b} F(r, \mu) r^2\, dr\, d\mu}{\int_{\mu=0}^{1}\int_{r=0}^{b} r^2\, dr\, d\mu} + \sum_{\substack{n=0 \\ \text{even}}}^{\infty}\sum_{m=1}^{\infty} C_{nm}\frac{J_{n+1/2}(\lambda_{nm} r)}{r^{1/2}} P_n(\mu) e^{-\alpha\lambda_{nm}^2 t} \tag{5-200}$$

where the integral-containing expression is the steady-state temperature and is reflective of conservation of energy. Conservation of energy may be independently checked by considering the following:

$$E_{\text{initial}} \equiv E_{\text{final}} \qquad \text{(J)} \tag{5-201}$$

We may evaluate the left-hand term by integrating the initial energy over the hemisphere, while the right-hand term follows from $E_{\text{final}} = mcT_{\text{ss}}$, yielding

$$\int_{\mu=0}^{1}\int_{r=0}^{b} (\rho c) F(r, \mu)(2\pi r) r\, dr\, d\mu = (\rho c)\frac{1}{2}\left(\frac{4}{3}\pi b^3\right) T_{\text{ss}} \tag{5-202}$$

Simplification of equation (5-202) is in exact agreement with the steady-state solution of equation (5-200); hence the solution correctly reflects conservation of energy.

Example 5-10 $T = T(r, \mu, \phi, t)$ for a Sphere
A solid sphere of radius b is initially at temperature $F(r, \mu, \phi)$. For $t > 0$, the boundary condition at $r = b$ is maintained at zero temperature. The problem formulation becomes

$$\frac{\partial^2 T}{\partial r^2} + \frac{2}{r}\frac{\partial T}{\partial r} + \frac{1}{r^2}\frac{\partial}{\partial \mu}\left[\left(1-\mu^2\right)\frac{\partial T}{\partial \mu}\right] + \frac{1}{r^2\left(1-\mu^2\right)}\frac{\partial^2 T}{\partial \phi^2} = \frac{1}{\alpha}\frac{\partial T}{\partial t} \tag{5-203}$$

$$\text{in} \quad 0 \le r < b \quad -1 \le \mu \le 1 \quad 0 \le \phi \le 2\pi \quad t > 0$$

$$\text{BC1:} \quad T(r \to 0) \Rightarrow \text{finite} \qquad \text{BC2:} \quad T(r = b) = 0 \tag{5-204a}$$

$$\text{BC3:} \quad T(\mu \to \pm 1) \Rightarrow \text{finite} \tag{5-204b}$$

$$\text{BC4:} \quad T(\phi) = T(\phi + 2\pi) \quad \to \quad 2\pi\text{-periodicity} \tag{5-204c}$$

$$\text{IC:} \quad T(t = 0) = F(r, \mu, \phi) \tag{5-204d}$$

As detailed in Table 5-1, we now use the $V(r, \mu, \phi, t) = r^{1/2}T(r, \mu, \phi, t)$ transformation, which gives the mathematical formulation

$$\frac{\partial^2 V}{\partial r^2} + \frac{1}{r}\frac{\partial V}{\partial r} - \frac{1}{4}\frac{V}{r^2} + \frac{1}{r^2}\frac{\partial}{\partial \mu}\left[\left(1-\mu^2\right)\frac{\partial V}{\partial \mu}\right] + \frac{1}{r^2\left(1-\mu^2\right)}\frac{\partial^2 V}{\partial \phi^2} = \frac{1}{\alpha}\frac{\partial V}{\partial t} \tag{5-205}$$

$$\text{in} \quad 0 \le r < b \quad -1 \le \mu \le 1 \quad 0 \le \phi \le 2\pi \quad t > 0$$

$$\text{BC1:} \quad V(r \to 0) \Rightarrow \text{finite} \qquad \text{BC2:} \quad V(r = b) = 0 \tag{5-206a}$$

$$\text{BC3:} \quad V(\mu \to \pm 1) \Rightarrow \text{finite} \tag{5-206b}$$

$$\text{BC4:} \quad V(\phi) = V(\phi + 2\pi) \quad \to \quad 2\pi\text{-periodicity} \tag{5-206c}$$

$$\text{IC:} \quad V(t = 0) = r^{1/2}F(r, \mu, \phi) \tag{5-206d}$$

For all homogeneous boundary conditions, we can assume a separation of variables in the form

$$V(r, \mu, \phi, t) = R(r)M(\mu)\Phi(\phi)\Gamma(t) \tag{5-207}$$

Substitution of equation (5-207) into the heat equation of (5-205) yields

$$\frac{1}{R}\left(\frac{d^2 R}{dr^2} + \frac{1}{r}\frac{dR}{dr} - \frac{1}{4}\frac{R}{r^2}\right) + \frac{1}{r^2 M}\frac{d}{d\mu}\left[\left(1-\mu^2\right)\frac{dM}{d\mu}\right] + \frac{1}{r^2\left(1-\mu^2\right)}\frac{1}{\Phi}\frac{d^2\Phi}{d\phi^2} = \frac{1}{\alpha\Gamma}\frac{d\Gamma}{dt} = -\lambda^2 \tag{5-208}$$

where we have introduced the separation constant $-\lambda^2$. Solution of the separated ODE in the t dimension yields

$$\Gamma(t) = C_1 e^{-\alpha\lambda^2 t} \tag{5-209}$$

We now consider the remaining spatial variables of equation (5-208), first multiplying by r^2, and then isolating the r terms to yield

$$\frac{-1}{M}\frac{d}{d\mu}\left[\left(1-\mu^2\right)\frac{dM}{d\mu}\right] - \frac{1}{1-\mu^2}\frac{1}{\Phi}\frac{d^2\Phi}{d\phi^2}$$
$$= \frac{r^2}{R}\left(\frac{d^2R}{dr^2} + \frac{1}{r}\frac{dR}{dr} - \frac{1}{4}\frac{R}{r^2}\right) + \lambda^2 r^2 = n(n+1) \tag{5-210}$$

where we have introduced a new separation constant $n(n+1)$. Separation of the r terms produces Bessel's equation of order $\left(n+\frac{1}{2}\right)$,

$$\frac{d^2R}{dr^2} + \frac{1}{r}\frac{dR}{dr} + \left[\lambda^2 - \frac{\left(n+\frac{1}{2}\right)^2}{r^2}\right]R = 0 \tag{5-211}$$

The solution of equation (5-211) for integer n yields half-integer-order Bessel functions, namely,

$$R(r) = C_2 J_{n+1/2}(\lambda r) + C_3 Y_{n+1/2}(\lambda r) \tag{5-212}$$

Applying boundary condition BC1 eliminates the Bessel functions of the second kind ($C_3 = 0$), while boundary condition BC2 yields the transcendental equation

$$J_{n+1/2}(\lambda_{np} b) = 0 \quad \rightarrow \quad \lambda_{np} \quad \text{for} \quad p = 1, 2, 3, \ldots \quad \text{for each } n \tag{5-213}$$

which produces eigenvalues λ_{np}, noting that there is a full set of eigenvalues λ_p for each value of n. We now consider the left-hand side of equation (5-210), which after multiplication by $1-\mu^2$, yields the following separated equation:

$$\frac{1-\mu^2}{M}\frac{d}{d\mu}\left[\left(1-\mu^2\right)\frac{dM}{d\mu}\right] + \left(1-\mu^2\right)n(n+1) = \frac{-1}{\Phi}\frac{d^2\Phi}{d\phi^2} = m^2 \tag{5-214}$$

We have introduced a new separation constant m^2, which yields the desired ODE and corresponding solution for the ϕ dimension, as given by

$$\Phi(\phi) = C_4 \cos m\phi + C_5 \sin m\phi \tag{5-215}$$

The requirement for 2π periodicity is satisfied for integer m, namely,

$$m = 0, 1, 2, 3, \ldots \tag{5-216}$$

and both constants C_4 and C_5 are retained. Finally, we consider the remaining μ terms of equation (5-214), which yield

$$\frac{d}{d\mu}\left[\left(1-\mu^2\right)\frac{dM}{d\mu}\right]+\left[n(n+1)-\frac{m^2}{1-\mu^2}\right]M=0 \tag{5-217}$$

We recognize equation (5-217) as the associated Legendre equation, and now limit n to integer values ($n = 0, 1, 2, 3, \dots$), which along with integer values of m as defined by equation (5-216), yield the orthogonal associated Legendre polynomials as the solution, giving

$$M(\mu) = C_6 P_n^m(\mu) + C_7 Q_n^m(\mu) \tag{5-218}$$

We eliminate the Legendre polynomials of the second kind ($C_7 = 0$) for the requirement of finiteness at $\mu = \pm 1$. With all three boundary value problems considered for the three spatial dimensions, we now form a solution as the sum of all four separated functions, and sum over all possible solutions, giving

$$V(r,\mu,\phi,t) = \sum_{n=0}^{\infty}\sum_{p=1}^{\infty}\sum_{m=0}^{n} J_{n+1/2}(\lambda_{np}r)P_n^m(\mu)\left(a_{nmp}\cos m\phi + b_{nmp}\sin m\phi\right)e^{-\alpha\lambda_{np}^2 t} \tag{5-219}$$

with constants $a_{nmp} = C_1C_2C_4C_6$ and $b_{nmp} = C_1C_2C_5C_6$. Application of the initial condition, equation (5-206d), yields

$$r^{1/2}F(r,\mu,\phi) = \sum_{n=0}^{\infty}\sum_{p=1}^{\infty}\sum_{m=0}^{n} J_{n+1/2}(\lambda_{np}r)P_n^m(\mu)\left(a_{nmp}\cos m\phi + b_{nmp}\sin m\phi\right) \tag{5-220}$$

to which we apply successively the three operators

$$*\int_{r=0}^{b} rJ_{n+1/2}(\lambda_{nl}r)\,dr, \qquad *\int_{\mu=-1}^{1} P_q^m(\mu)\,d\mu \qquad \text{and} \qquad *\int_{\phi=0}^{2\pi}\cos k\phi\,d\phi$$

to yield the Fourier–Bessel–Legendre coefficients a_{nmp} as

$$a_{nmp} = \frac{\int_{\phi=0}^{2\pi}\int_{\mu=-1}^{1}\int_{r=0}^{b} r^{3/2}F(r,\mu,\phi)J_{n+1/2}(\lambda_{np}r)P_n^m(\mu)\cos m\phi\,dr\,d\mu\,d\phi}{\int_{r=0}^{b} rJ_{n+1/2}^2(\lambda_{np}r)\,dr\int_{\mu=-1}^{1}[P_n^m(\mu)]^2\,d\mu\int_{\phi=0}^{2\pi}\cos^2 m\phi\,d\phi} \tag{5-221}$$

In a similar manner, we apply successively the three operators

$$*\int_{r=0}^{b} rJ_{n+1/2}(\lambda_{nl}r)\,dr, \qquad *\int_{\mu=-1}^{1} P_q^m(\mu)\,d\mu \qquad \text{and} \qquad *\int_{\phi=0}^{2\pi}\sin k\phi\,d\phi$$

to yield the Fourier–Bessel–Legendre coefficients b_{nmp} as

$$b_{nmp} = \frac{\int_{\phi=0}^{2\pi} \int_{\mu=-1}^{1} \int_{r=0}^{b} r^{3/2} F(r, \mu, \phi) J_{n+1/2}(\lambda_{np} r) P_n^m(\mu) \sin m\phi \, dr \, d\mu \, d\phi}{\int_{r=0}^{b} r J_{n+1/2}^2(\lambda_{np} r) \, dr \int_{\mu=-1}^{1} [P_n^m(\mu)]^2 \, d\mu \int_{\phi=0}^{2\pi} \sin^2 m\phi \, d\phi} \tag{5-222}$$

The norms for the three orthogonal functions have all been previously defined. The associated Legendre polynomials, see equation (2-88), yield the norm

$$\int_{\mu=-1}^{1} \left[P_n^m(\mu)\right]^2 d\mu = \frac{2}{2n+1} \frac{(n+m)!}{(n-m)!} \tag{5-223}$$

while the Bessel functions, see case 3 of Table 2-2, yield the norm

$$\int_{r=0}^{b} r J_{n+1/2}^2(\lambda_{np} r) \, dr = \frac{b^2 J_{n+3/2}^2(\lambda_{np} b)}{2} \tag{5-224}$$

The trigonometric functions yield the norm π for $m > 0$, with the norm of the $\cos(m\phi)$ becoming 2π for the special case of $m = 0$. The Fourier coefficients and the norms may be substituted into equation (5-219), which yields the following overall solution after making a trigonometric substitution and transforming back to $T(r, \mu, \phi, t)$ from $V(r, \mu, \phi, t)$:

$$T(r, \mu, \phi, t) = \frac{1}{\pi} \sum_{n=0}^{\infty} \sum_{p=1}^{\infty} \sum_{m=0}^{n}$$

$$\left\{ \begin{aligned} & \frac{(2n+1)(n-m)!}{b^2 J_{n+3/2}^2(\lambda_{np} b)(n+m)!} \cdot \frac{J_{n+1/2}(\lambda_{np} r)}{r^{1/2}} P_n^m(\mu) e^{-\alpha \lambda_{np}^2 t} \\ & \cdot \int_{\phi'=0}^{2\pi} \int_{\mu'=-1}^{1} \int_{r'=0}^{b} \begin{bmatrix} r'^{3/2} J_{n+1/2}\left(\lambda_{np} r'\right) P_n^m\left(\mu'\right) \\ \cdot \cos m\left(\phi - \phi'\right) F(r', \mu', \phi') \end{bmatrix} dr' \, d\mu' \, d\phi' \end{aligned} \right\} \tag{5-225}$$

where the π term is replaced by 2π for the case of $m = 0$. As discussed previously, this solution is finite as $r \to 0$ for all values of $n \geq 0$.

5-4 CAPSTONE PROBLEM

In this chapter, we have developed the solution schemes for the solution of the spherical heat equation using both the method of separation of variables and superposition. Our focus has been to develop the analytical solutions, with emphasis on the resulting eigenfunctions and eigenvalues, and the necessary Fourier coefficients, while also considering conservation of energy. The general approach has been to leave the Fourier coefficients in integral form, although

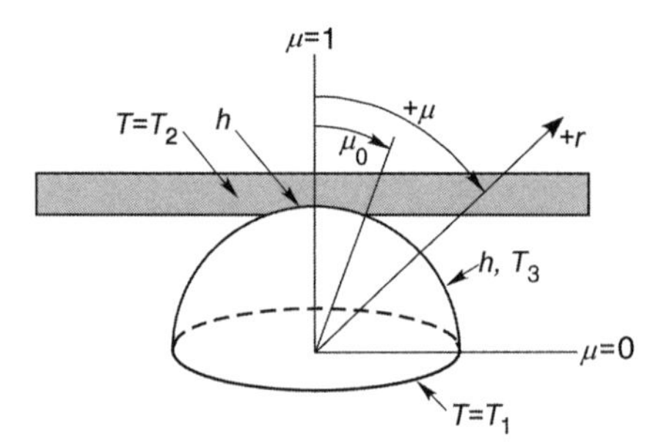

Figure 5-5 Description for capstone problem.

at times we have performed the integrals and substituted back into the series summation as in the above problem. Here we consider a final problem in detail in an attempt to integrate and illustrate the many aspects of our temperature field solutions presented in this chapter, including the temperature field in the context of conservation of energy.

We consider here a hemisphere as shown in Figure 5-5. The hemisphere is of radius b and at steady-state conditions. The base ($\mu = 0$) is maintained at a constant temperature T_1, while the surface at $r = b$ is exposed to convection heat transfer with convection coefficient h and with the reference fluid temperature $T_\infty = f(\mu)$. Specifically, for $0 < \mu \leq \mu_0$, the fluid temperature is equal to T_3, while for $\mu_0 < \mu \leq 1$, the reference temperature is equal to T_2. By letting the convection coefficient $h = h_c$, with h_c being the contact conductance, see Table 1-3, we approximate the problem of heat transfer through the hemisphere in thermal contact over $\mu_0 < \mu \leq 1$ with an isothermal reservoir at T_2, while the base is in perfect thermal contact with an isothermal reservoir at T_1. Ideally, different values of the convection coefficient should be assigned to regions in and out of the contact region, with an actual convection coefficient used over the region of fluid contact, namely, $0 < \mu \leq \mu_0$, and a true contact conductance used over the region $\mu_0 < \mu \leq 1$. However, formulation of the problem does not allow the convection coefficient to vary with μ; hence adjustment can be made to the value of T_3 as appropriate to best match the expected convective heat transfer through the exposed surface region outside of the zone of contact.

The mathematical formulation of the problem is given as

$$\frac{\partial^2 T}{\partial r^2} + \frac{2}{r}\frac{\partial T}{\partial r} + \frac{1}{r^2}\frac{\partial}{\partial \mu}\left[\left(1-\mu^2\right)\frac{\partial T}{\partial \mu}\right] = 0 \quad \text{in} \quad 0 \leq r < b \quad 0 < \mu \leq 1 \tag{5-225}$$

$$\text{BC1:} \quad T(r \to 0) \Rightarrow \text{finite} \qquad \text{BC2:} \quad -k\left.\frac{\partial T}{\partial r}\right|_{r=b} = h\left(T|_{r=b} - T_\infty\right) \tag{5-226a}$$

$$\text{where} \quad T_\infty = f(\mu) = \begin{cases} T_2 & \text{for } \mu_0 < \mu \leq 1 \\ T_3 & \text{for } 0 < \mu \leq \mu_0 \end{cases}$$

BC3: $T(\mu \to +1) \Rightarrow$ finite (5-227b)

BC4: $T(\mu = 0) = T_1$ (5-227c)

With two nonhomogeneous boundary conditions, one must be removed prior to separation of variables. Since the μ dimension must be homogeneous, we will introduce a shift of temperature as $\Theta(r, \mu) = T(r, \mu) - T_1$. With this shift, the problem formulation becomes

$$\frac{\partial^2 \Theta}{\partial r^2} + \frac{2}{r}\frac{\partial \Theta}{\partial r} + \frac{1}{r^2}\frac{\partial}{\partial \mu}\left[\left(1-\mu^2\right)\frac{\partial \Theta}{\partial \mu}\right] = 0 \quad \text{in} \quad 0 \le r < b \quad 0 < \mu \le 1 \tag{5-228}$$

BC1: $\Theta(r \to 0) \Rightarrow$ finite BC2: $-k\left.\frac{\partial \Theta}{\partial r}\right|_{r=b} = h\left[\Theta|_{r=b} - \left(T_\infty - T_1\right)\right]$ (5-229a)

$$\text{where} \quad T_\infty = f(\mu) = \begin{cases} T_2 & \text{for } \mu_0 < \mu \le 1 \\ T_3 & \text{for } 0 < \mu \le \mu_0 \end{cases}$$

BC3: $\Theta(\mu \to +1) \Rightarrow$ finite (5-229b)

BC4: $\Theta(\mu = 0) = 0$ (5-229c)

Now with only a single nonhomogeneous boundary condition, we proceed with separation of variables in the form

$$\Theta(r, \mu) = R(r)M(\mu) \tag{5-230}$$

which upon substitution into equation (5-228) yields

$$\frac{r^2}{R}\left(\frac{d^2R}{dr^2} + \frac{2}{r}\frac{dR}{dr}\right) = \frac{-1}{M}\frac{d}{d\mu}\left[\left(1-\mu^2\right)\frac{dM}{d\mu}\right] = n(n+1) \tag{5-231}$$

We have introduced the separation constant $n(n+1)$ to force the characteristic value problem in the homogeneous μ dimension. The resulting ODE is recognized as the Legendre equation, with solution

$$M(\mu) = C_1 P_n(\mu) + C_2 Q_n(\mu) \tag{5-232}$$

We may eliminate the Legendre polynomials of the second kind ($C_2 = 0$) because $Q_n(\mu)$ is infinite at $\mu = \pm 1$, noting that $\mu = 1$ (i.e., $\theta = 0$) is in the domain of the problem. We then consider the boundary condition BC4 at $\mu = 0$ (i.e., $\theta = \pi/2$), namely,

$$P_n(\mu = 0) = 0 \quad \text{yielding} \quad n = 1, 3, 5, \ldots \text{(odd integers)} \tag{5-233}$$

and therefore limit n to the odd integer values. The remaining r terms in equation (5-231) yield

$$r^2 \frac{d^2 R}{dr^2} + 2r \frac{dR}{dr} - n(n+1)R = 0 \tag{5-234}$$

which is recognized as Cauchy's equation, as developed in equation (4-29). As described in previous examples, the solution is

$$R(r) = C_3 r^n + C_4 r^{-(n+1)} \tag{5-235}$$

We now consider the requirement of finiteness at the origin, which eliminates C_4 given that this terms goes to infinity as $r \to 0$ for n equal to odd positive integers. Having considered all but the final, nonhomogeneous boundary condition, we now recombine our separated solutions as products, and sum over all possible solutions:

$$\Theta(r, \mu) = \sum_{\substack{n=1 \\ \text{odd}}}^{\infty} C_n r^n P_n(\mu) \tag{5-236}$$

We lastly apply the nonhomogeneous BC2, equation (5-229a), which yields

$$-k \frac{d}{dr} \left[\sum_{\substack{n=1 \\ \text{odd}}}^{\infty} C_n r^n P_n(\mu) \right]_{r=b} = h \left[\sum_{\substack{n=1 \\ \text{odd}}}^{\infty} C_n b^n P_n(\mu) - \left(T_\infty - T_1\right) \right] \tag{5-237}$$

The two summations may be combined into a single summation as follows:

$$h\left(T_\infty - T_1\right) = \sum_{\substack{n=1 \\ \text{odd}}}^{\infty} C_n \left(nkb^{n-1} + hb^n\right) P_n(\mu) \tag{5-238}$$

Equation (5-238) is recognized as a Fourier–Legendre series expansion of the function $h\left(T_\infty - T_1\right) \equiv h\left[f(\mu) - T_1\right]$. We apply the following operator to both sides:

$$* \int_{\mu=0}^{1} P_q(\mu)\, d\mu$$

noting that the arbitrary constant q must be an odd integer, recalling that

$$\int_{\mu=0}^{1} P_q(\mu) P_n(\mu)\, d\mu = 0 \qquad \text{for} \qquad n \neq q \tag{5-239}$$

provided that n and q are both even or both odd. This yields the coefficients C_n as

$$C_n = \frac{\int_{\mu=0}^{1} h\left[f(\mu) - T_1\right] P_n(\mu)\, d\mu}{\left(nkb^{n-1} + hb^n\right) \int_{\mu=0}^{1} \left[P_n(\mu)\right]^2 d\mu} \tag{5-240}$$

The integral in the denominator of equation (5-240) is recognized as the norm and is equal to $1/(2n+1)$. The solution is now complete with equations (5-236) and (5-240). We may combine both expressions, simplify, and shift back to $T(r, \mu)$, yielding the following expression:

$$T(r,\mu) = \sum_{\substack{n=1\\ \text{odd}}}^{\infty} (2n+1) \frac{r^n P_n(\mu)}{nkb^{n-1} + hb^n} \int_{\mu'=0}^{1} h\left[f(\mu') - T_1\right] P_n(\mu')\, d\mu' + T_1 \tag{5-241}$$

If we further define $f(\mu)$ as given by equation (5-229a), the above expression becomes

$$T(r,\mu) = \sum_{\substack{n=1\\ \text{odd}}}^{\infty} (2n+1) \frac{hr^n P_n(\mu)}{nkb^{n-1} + hb^n} \left[\begin{array}{l} (T_3 - T_1) \displaystyle\int_{\mu'=0}^{\mu_0} P_n(\mu')\, d\mu' \\ + (T_2 - T_1) \displaystyle\int_{\mu'=\mu_0}^{1} P_n(\mu')\, d\mu' \end{array} \right] + T_1 \tag{5-242}$$

The above expressions may now be used to explore the behavior of the 2-D temperature profiles and heat transfer for a set of parameters corresponding to a stainless steel hemisphere in contact with a large solid surface made of stainless steel. For these calculations, let the hemisphere have a radius of 1 cm ($b = 0.01$ m) and have properties for stainless steel: Thermal conductivity $k = 14$ W/m · K, and thermal diffusivity $\alpha = 3.5 \times 10^{-6}$ m^2/s. Let the contact interface be at 10 atm pressure, with air as the interfacial fluid, and with a mean surface roughness of the two surfaces equal to 2.5 μm. Using Table 1-3, a representative contact conductance (i.e., reciprocal of the thermal contact resistance) is taken as $h = h_c = 3400$ W/m^2 · K for this steel-on-steel interface. The upper and lower temperatures are maintained at $T_1 = 325$ K and $T_2 = 300$ K. To minimize actual convective heat transfer to the surrounding air outside of the contact zone, we will let $T_3 = T_1 = 325$ K. Finally, we will assume a contact angle of 5°; hence $\mu_0 = 0.9962$.

The first 150 nonzero terms ($n = 1 - 299$) were used for the series summation, as given by equation (5-242). The Legendre polynomials were generated using the recursion formula given by equation (2-71). Over the range of spatial variables explored below, these parameters were found to yield reasonable convergence for temperature and heat flux, although there was significant spatial dependence. In particular, convergence was always greater near the center ($r \sim 0$) and poorest near the edge ($r \sim b$). For example, convergence along the radial line corresponding to $\mu = 0.9659$ ($\theta = 15°$) for various values of r is explored in Figure 5-6, specifically at $r = 0.25b$, $r = 0.5b$, $r = 0.75b$, and at $r = b$. As observed in the figure, convergence is very rapid for the two innermost positions explored, with the first 5 terms of the series providing accuracy to within 0.001 K. At $r = 0.75b$, such accuracy is achieved with the first 9 terms. However, at the surface of the hemisphere, the first 19 terms achieve a convergence to less

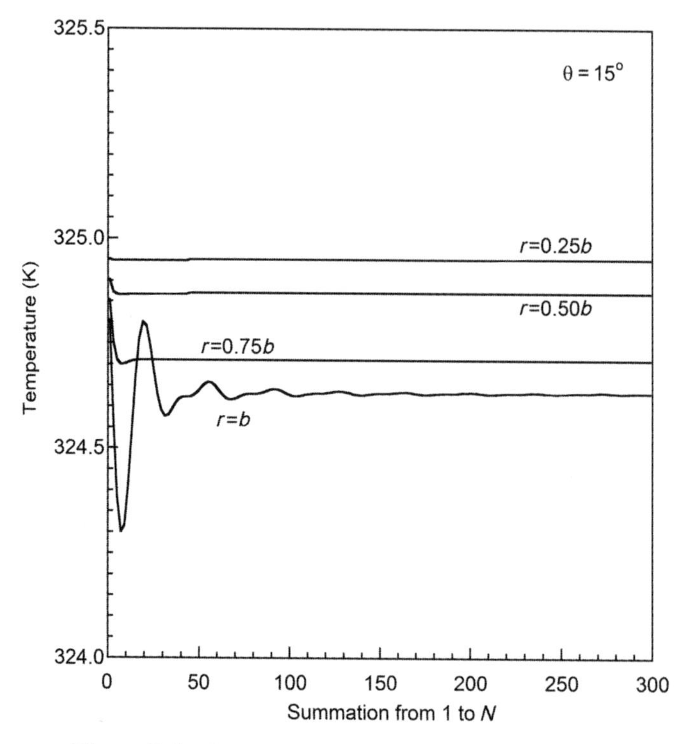

Figure 5-6 Convergence of the temperature solution.

than 0.1 K, while the first 150 terms are required to reach a convergence to less than 0.001 K. This behavior was consistent along the surface for all values of μ.

It is noted, however, that the coefficients of the Legendre polynomials grow very large with increasing degree, and for Legendre polynomials greater than a degree of about 37, round-off error becomes significant even with double-precision computation. Therefore, rather than calculate and store the Legendre polynomial coefficients for subsequent evaluation of the Legendre polynomials at each value of μ, the first two Legendre polynomials were evaluated at each value of μ, and the recurrence relations where then used to evaluate the Legendre polynomial value for each higher degree. This avoids the precision issues associated with very large coefficients, as the evaluated Legendre polynomials themselves over the range of $0 \leq \mu \leq 1$ present no computational challenge. For the Fourier coefficients, the integrals of the Legendre polynomials were done numerically using the trapezoid rule to once again avoid the calculation and storage of the Legendre coefficients for higher-order polynomials. This procedure was checked against the exact analytical integration for lower-order polynomials and found to be satisfactory within the precision of 0.001 K.

Convergence also tended to be worse as μ was varied from 1 to 0, although the boundary condition ($T_1 = 325$ K) is recovered exactly at $\mu = 0$ (i.e., along the base) for the odd Legendre polynomials. For calculation of the heat flux using Fourier's law (i.e., using the appropriate derivatives), convergence was again poor toward the outer surface. Therefore, for calculating the heat flux along the surface at $r = b$, the convective condition ($h\Delta T$) was used rather than Fourier's law ($-k\partial T/\partial r$), which provided sufficient convergence for the first 150 nonzero terms. In particular, with the heat flux calculations using Fourier's law, the terms in the series summation alternate signs, making convergence very slow, as the terms tend to cancel out in pairs. This was the case for the heat flux along the base of the sphere near the outer edge ($r \sim b$). For calculation of the derivatives, the same approach outlined above was used, namely, the recurrence relationships were used to calculate the exact values rather than the coefficients themselves. With these comments in mind, we now explore the behavior of our solution.

Figure 5-7 presents the radial temperature profile along the central axis of the hemisphere, namely, $\mu = 1$ (i.e., $\theta = 0$). As observed in the profile, the temperature decreases from the boundary condition of $T_1 = 325$ K at $r = 0$ to a value of 320.8 K at the point of contact $r = b$. This corresponds to a temperature drop of 20.8 K across the thermal contact resistance, with a corresponding heat flux of 70.9 kW/m^2. The influence of T_2 within the zone of contact is apparent in

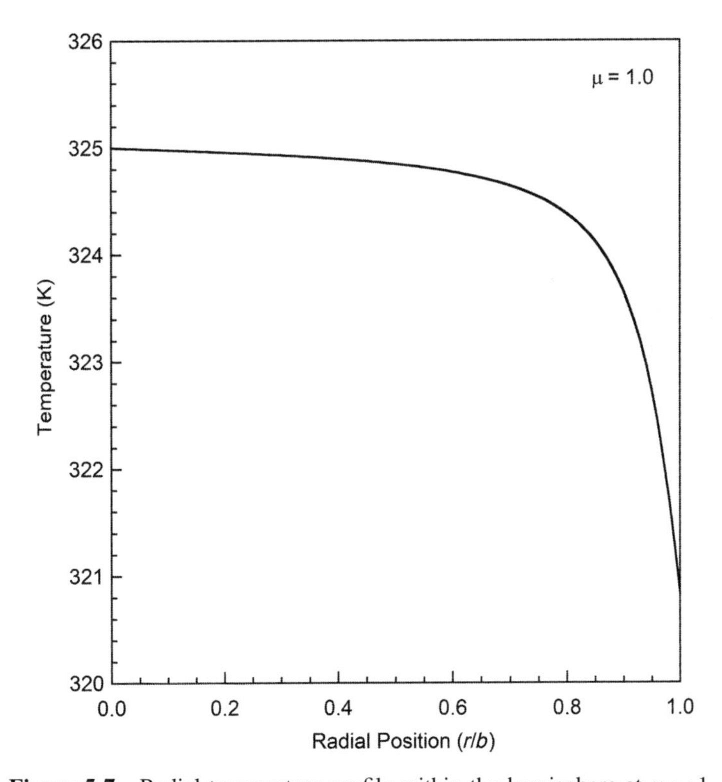

Figure 5-7 Radial temperature profile within the hemisphere at $\mu = 1$.

the temperature profile, with the pronounced increase in the temperature gradient when approaching the contact surface.

Figure 5-8 presents the radial temperature profiles for various values of μ, namely, for $\mu = 0.9848$, 0.9659, and 0.9397, corresponding to values of $\theta = 10^\circ$, 15° and 20°. As observed in the profile for $\theta = 10^\circ$, the temperature decreases monotonically to a minimum value just below the surface and then increases to a value of 324.3 K at the surface $r = b$. This is in contrast to the radial profiles that end within the zone of contact, as shown in Figure 5-4. The profiles for $\theta = 15^\circ$ and 20° reveal similar trends. Specifically, there is a slight increase in temperature nearing the very edge, reflecting the influence of $T_3 = 325$ K. In other words, while the central region of the hemisphere is suppressed in the temperature reflecting the temperature of the region in contact (i.e., a gradient toward the reservoir T_2), the outer region is actually being warmed by the convective fluid outside of the contact region; hence heat is flowing into the hemisphere from the surrounding convective fluid.

Another way to explore the temperature profile behavior, notably the decrease in temperature about the central axis region, is to make a horizontal cut through

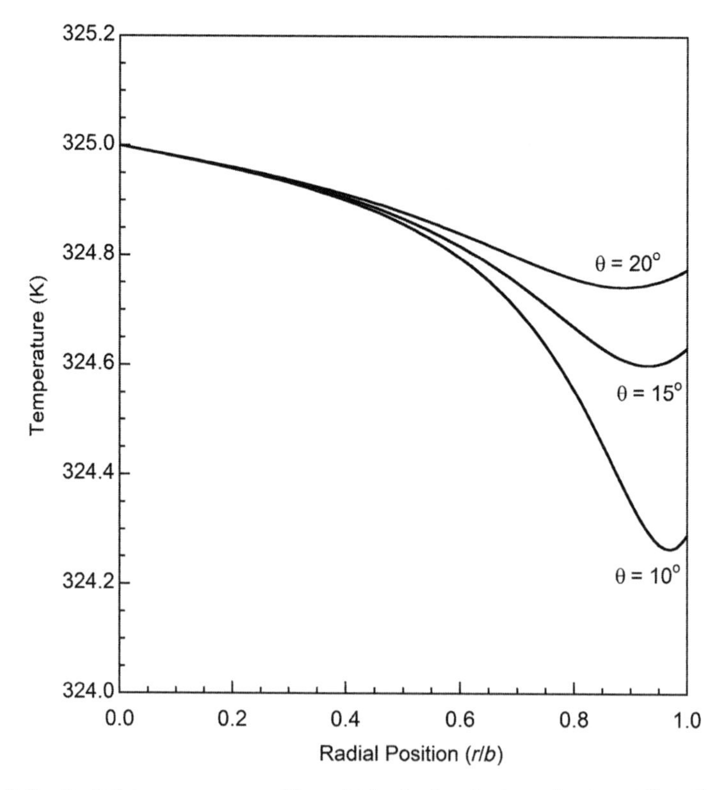

Figure 5-8 Radial temperature profiles within the hemisphere for $\theta = 10^\circ$, 15° and 20°, corresponding to $\mu = 0.9848$, 0.9659, and 0.9397.

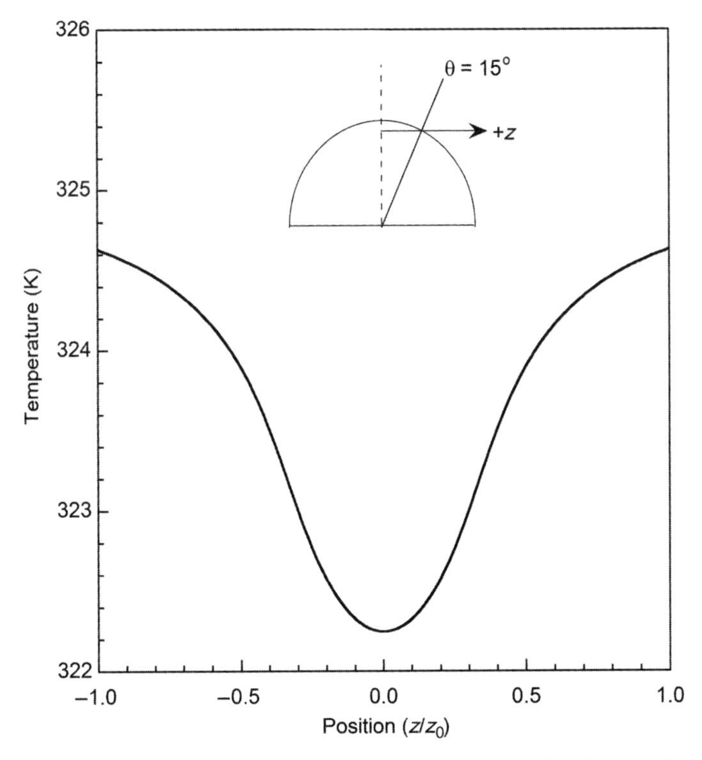

Figure 5-9 Temperature profile across the horizontal cross section intersecting the outer surface at $\theta = 15^{\circ}$. The z direction is defined as the radial distance perpendicular to the central axis of $\mu = 1$, and z_0 is the radius of this horizontal slice, as shown in the inset.

the hemisphere. This was done by taking a horizontal line from the center axis outward such that it intersects the outer surface at the same location as a radial line from the origin for $\theta = 15^{\circ}$. We show the temperature profile along such a slice in Figure 5-9, where we have passed fully from one side to the other to emphasize the symmetry across the centerline, as observed, for no ϕ dependence. We have normalized the spatial coordinate along this cut with respect to the outer radius of this circle, therefore the temperature at $z/z_0 = 1$ in Figure 5-9 is identical to the temperature at $r/b = 1$ in Figure 5-8 for the $\theta = 15^{\circ}$ trace. Clearly heat is flowing inward from the outer surface toward the central axis region and ultimately into the region of contact. The slope of this profile is identically zero across the center line for this horizontal slice, properly reflecting the physics of axial symmetry for this problem formulation. It is interesting to note that this behavior does not arise from any boundary conditions in this region. In fact, the spherical coordinate system has no orthogonal boundary for such a horizontal slice. Rather, such behavior arises from conservation of energy at each location inherent in the heat equation.

It is also useful to consider the heat flux entering and leaving the hemisphere. We first explore the heat flux crossing the outer surface at $r = b$, which as noted above, may be calculated by two different means, namely,

$$q_r''(r = b) = -k\frac{\partial T}{\partial r}\bigg|_{r=b} = h\left(T|_{r=b} - T_\infty\right) \qquad (5\text{-}243)$$

which simply reflects the boundary condition given by equation (2-226a). However, as discussed above, convergence was significantly better using the convective boundary condition; hence the surface temperature was used to calculate the surface heat flux. Figure 5-10 presents the surface heat flux at $r = b$ as a function of μ, over the entire surface of the hemisphere, namely, over the range $0 < \mu \leq 1$. As consistent with the physics of the problem, heat is flowing out of the hemisphere within the zone of contact with T_2, while heat is flowing into the sphere outside of the contact zone via convection heat transfer with the fluid at temperature T_3. There is a discontinuity in heat flux at $\theta = 5^\circ$ due to the step change in T_∞. Within the region of the convection heat transfer ($\theta > 5^\circ$), the heat flux is negative (i.e., into the hemisphere) as consistent with Newton's law of cooling,

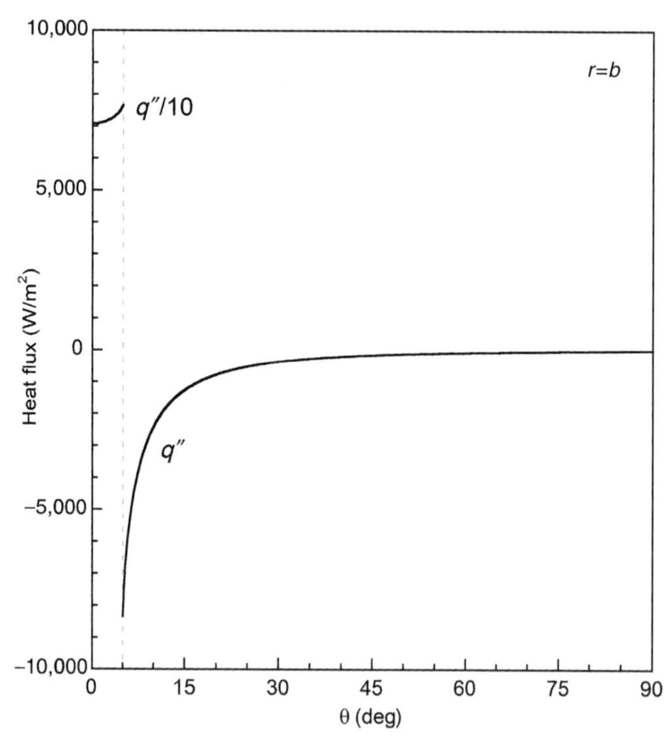

Figure 5-10 Surface heat flux ($r = b$) as a function of μ. Values within the zone of contact are scaled by a factor of 10.

where it decays monotonically from a value of -8335 W/m^2 at the boundary of the contact zone, to zero at the base of the hemisphere ($\theta = 90^{\text{o}}$, $\mu = 0$).

The heat flux entering the base is readily calculated using Fourier's law in the θ direction, given as

$$q''_\theta = -\frac{k}{r}\frac{\partial T}{\partial \theta} \tag{5-244}$$

With the change of variables to the μ coordinate, Fourier's law becomes

$$q''_\mu = \frac{k\left(1-\mu^2\right)^{1/2}}{r}\frac{\partial T}{\partial \mu} \tag{5-245}$$

Equation (5-245) is readily evaluated from equation (5-242) by simply differentiating the Legendre polynomials, noting that the integral terms (i.e., the Fourier coefficient terms) are unchanged. Evaluating equation (5-245) along the base of the hemisphere yields

$$q''_\mu(\mu = 0) = \sum_{\substack{n=1\\ \text{odd}}}^{\infty} \frac{(2n+1)\,khr^{n-1}}{nkb^{n-1}+hb^{n}} \left.\frac{dP_n}{d\mu}\right|_{\mu=0} \left[\begin{array}{l}\left(T_3 - T_1\right)\displaystyle\int_{\mu'=0}^{\mu_o} P_n\left(\mu'\right)d\mu' \\ +\left(T_2 - T_1\right)\displaystyle\int_{\mu'=\mu_o}^{1} P_n\left(\mu'\right)d\mu'\end{array}\right] \tag{5-246}$$

The surface heat flux along the base of the hemisphere is plotted in Figure 5-11. As discussed above, convergence of the series summation in equation (5-246) becomes an issue near the outer surface ($r \sim b$). In fact, convergence was sufficient for $r < 0.99b$ using the first 150 terms, although convergence directly on the surface ($r = b$) was insufficient for 150 nonzero terms; hence we have extrapolated the final value in Figure 5-11 at $r = b$. We note that the negative values in Figure 5-11 reflect heat entering the sphere (i.e., in the negative μ direction). Overall, the heat flux entering the base of the hemisphere is observed to decay monotonically from a maximum flux of 282.5 W/m^2 to a flux of 44 W/m^2 at the edge of the base. The heat flux at $r = 0$ as calculated with equation (2-246) is in exact agreement with the heat flux calculated using

$$q''_r(r = 0, \mu = 1) = -k\left.\frac{\partial T}{\partial r}\right|_{\substack{r=0\\ \mu=1}} \tag{5-247}$$

noting that only the first term of the series contributes at the origin (r=0).

Finally, it is useful to consider conservation of energy under steady-state conditions, which may be expressed as

$$\dot{Q}_{\text{in}} = \dot{Q}_{\text{out}} \tag{5-248}$$

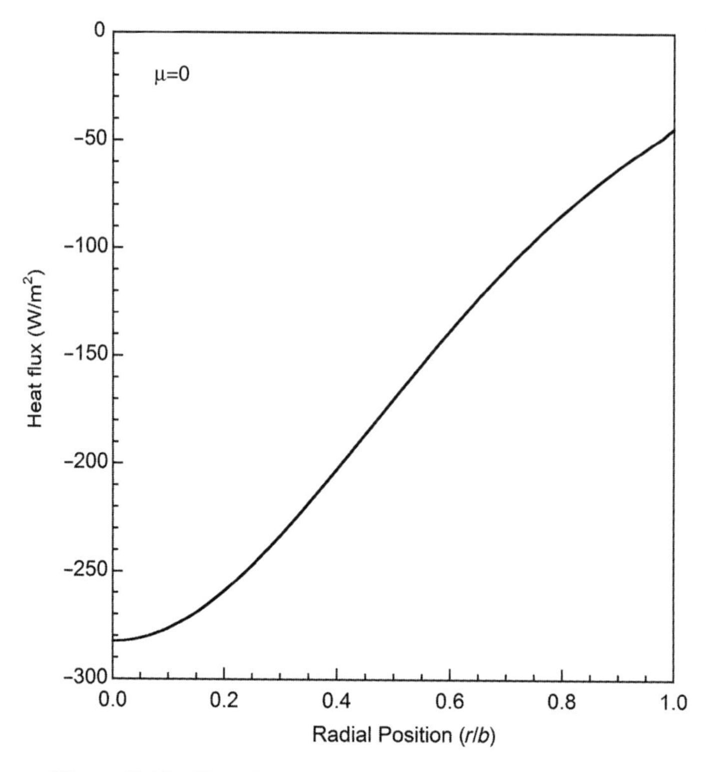

Figure 5-11 Heat flux entering the base of the hemisphere.

where the rate of energy entering the hemisphere must be balanced by the rate of energy leaving the sphere through the contact area with the solid at T_2. In view of the problem formulation, equation (5-248) may be expressed as

$$\int_{r=0}^{b} 2\pi r \, q''_\mu \big|_{\mu=0} \, dr + \int_{\theta=\theta_o}^{\pi/2} 2\pi b^2 \, q''_r \big|_{r=b} \sin\theta d\theta = \int_{\theta=0}^{\theta_o} 2\pi b^2 \, q''_r \big|_{r=b} \sin\theta d\theta \tag{5-249}$$

where we have set up the differential surface area over the surface $r = b$ in terms of the spherical coordinate θ. Rather than integrating the exact series summations analytically, the flux profiles of Figures 5-5 to 5-10 and 5-11 were integrated using the trapezoid rule, which yields

$$39.14 + 135.84 = 174.98 \quad \text{(mW)} \tag{5-250}$$

As observed with equation (5-250), conservation of energy is indeed realized, and it is observed that with respect to the ~175 mW of heat leaving the zone of contact, about 23.4% enters the sphere through the base, with the remaining 77.6% being transferred via convection heat transfer from the surrounding fluid.

REFERENCES

1. H. S. Carslaw and J. C. Jaeger, *Conduction of Heat in Solids*, Clarendon Press, London, 1959.
2. M. N. Özişik, *Boundary Value Problems of Heat Conduction*, International Textbook, Scranton, PA, 1968.
3. A. V. Luikov, *Analytical Heat Diffusion Theory*, Academic, New York, 1968.
4. J. Crank, *The Mathematics of Diffusion*, Clarendon Press, London, 1957.
5. T. M. MacRobert, *Spherical Harmonics*, 3rd ed., Pergamon, New York, 1967.
6. E. T. Whitaker and G. N. Watson, *A Course of Modern Analysis*, Cambridge University Press, London, 1965.
7. W. E. Byerly, *Fourier's Series and Spherical, Cylindrical and Ellipsoidal Harmonics*, Dover, New York, 1959.

PROBLEMS

5-1 A hollow sphere $a \leq r \leq b$ is initially at temperature $T = F(r)$. For times $t > 0$, the boundary surface at $r = a$ is kept insulated, and the boundary at $r = b$ dissipates heat by convection with convection coefficient h into a medium at zero temperature. Obtain an expression for the temperature distribution $T(r, t)$ in the sphere for times $t > 0$.

5-2 A hollow sphere $a \leq r \leq b$ is initially at temperature $T = F(r)$. For times $t > 0$, the boundary surface at $r = a$ is maintained at a constant temperature of zero, and the boundary at $r = b$ dissipates heat by convection with convection coefficient h into a medium at zero temperature. Obtain an expression for the temperature distribution $T(r, t)$ in the sphere for times $t > 0$.

5-3 A hollow sphere $a \leq r \leq b$ is initially at temperature $T = F(r)$. For times $t > 0$, the boundary surface at $r = a$ dissipates heat by convection with convection coefficient h_1 into a medium at temperature T_∞, and the boundary at $r = b$ dissipates heat by convection with convection coefficient h_2 into a medium at temperature T_∞. Obtain an expression for the temperature distribution $T(r, t)$ in the sphere for times $t > 0$.

5-4 A solid sphere $0 \leq r \leq b$ is initially at temperature $T = F(r)$. For times $t > 0$, the boundary surface at $r = b$ is maintained at a constant temperature of T_1. Obtain an expression for the temperature distribution $T(r, t)$ in the sphere for times $t > 0$.

5-5 A solid sphere $0 \leq r \leq b$ is initially at temperature $T = F(r)$. For times $t > 0$, the boundary surface at $r = b$ is maintained at a constant temperature of T_1. In addition, the sphere is subjected to uniform internal energy generation g_0 (W/m^3). Obtain an expression for the temperature distribution $T(r, t)$ in the sphere for times $t > 0$.

5-6 A hemisphere $0 \le r \le b$, $0 \le \mu \le 1$ is maintained at steady-state conditions with a prescribed surface temperature $T(r = b) = f(\mu)$, and with the base ($\mu = 0$) maintained at a constant temperature of T_1. Obtain an expression for the steady-state temperature distribution $T(r, \mu)$.

5-7 A solid sphere $0 \le r \le b$ is maintained at steady-state conditions with a prescribed surface temperature $T(r = b) = f(\mu)$. Obtain an expression for the steady-state temperature distribution $T(r, \mu)$.

5-8 A hollow sphere $a \le r \le b$ is initially at temperature $T = F(r)$. For times $t > 0$, the boundary surface at $r = a$ is maintained at temperature T_1, and the boundary surface at $r = b$ is maintained at temperature T_2. In addition, the sphere is subjected to uniform internal energy generation g_0 (W/m^3). Obtain an expression for the temperature distribution $T(r, t)$ in the sphere for times $t > 0$.

5-9 A solid sphere $0 \le r \le b$ is maintained at steady-state conditions. The surface at $r = b$ is subjected to convection heat transfer with convection coefficient h and with fluid temperature $T_\infty = f(\mu, \phi)$. Obtain an expression for the steady-state temperature distribution $T(r, \mu, \phi)$.

5-10 A hemisphere $0 \le r \le b$, $0 \le \mu \le 1$ is maintained at steady-state conditions. The surface at $r = b$ is maintained at a prescribed temperature $T(r = b) = f(\mu, \phi)$, while the base is maintained at zero temperature. Obtain an expression for the steady-state temperature distribution $T(r,\mu,\phi)$. See note 2.

5-11 A solid sphere $0 \le r \le b$, $-1 \le \mu \le 1$ is maintained at steady-state conditions. The surface at $r = b$ is subjected to a prescribed heat flux $f(\mu, \phi)$. Obtain an expression for the steady-state temperature distribution $T(r,\mu,\phi)$.

5-12 A hemisphere $0 \le r \le b$, $0 \le \mu \le 1$ is maintained at steady-state conditions. The surface at $r = b$ is subjected to convection heat transfer with convection coefficient h and with fluid temperature $T_\infty = f(\mu, \phi)$, while the base is maintained at zero temperature. Obtain an expression for the steady-state temperature distribution $T(r,\mu,\phi)$. See note 2.

5-13 A hemisphere $0 \le r \le b$, $0 \le \mu \le 1$ is maintained at steady-state conditions. The surface at $r = b$ is subjected to a prescribed heat flux $f(\mu, \phi)$, while the base is maintained at zero temperature. Obtain an expression for the steady-state temperature distribution $T(r,\mu,\phi)$. See note 2.

5-14 A hemisphere $0 \le r \le b$, $0 \le \mu \le 1$ is initially at temperature $T = F(r, \mu)$. For times $t > 0$, the boundary at the spherical surface $r = b$ is maintained at zero temperature, and at the base $\mu = 0$ is perfectly insulated. Obtain an expression for the temperature distribution $T(r,\mu,t)$ for times $t > 0$.

5-15 A solid sphere $0 \le r \le b$, $-1 \le \mu \le 1$ is initially at temperature $T = F(r, \mu)$. For times $t > 0$, the boundary at the surface $r = b$ dissipates heat by convection with convection coefficient h into a medium at temperature T_∞. Obtain an expression for the temperature distribution $T(r,\mu,t)$ for times $t > 0$.

5-16 A solid sphere $0 \le r \le b, -1 \le \mu \le 1, 0 \le \phi \le 2\pi$ is initially at temperature $T = F(r, \mu, \phi)$. For times $t > 0$, the boundary at the surface $r = b$ dissipates heat by convection with convection coefficient h into a medium at temperature T_∞. Obtain an expression for the temperature distribution $T(r,\mu ,\phi ,t)$ for times $t > 0$.

5-17 A solid sphere $0 \le r \le b, -1 \le \mu \le 1, 0 \le \phi \le 2\pi$ is initially at temperature $T = F(r, \mu, \phi)$. For times $t > 0$, the boundary at the surface $r = b$ is maintained at temperature T_1. In addition, the sphere is subjected to uniform internal energy generation g_0 (W/m^3). Obtain an expression for the temperature distribution $T(r, \mu, \phi, t)$ for times $t > 0$.

NOTES

1. For the fully insulated sphere in Example 5-4, we consider equation (5-75) for the special case of $\lambda = 0$, which yields in the r dimension

$$\frac{1}{R(r)}\frac{d^2 R(r)}{dr^2} = 0 \tag{1}$$

This ODE yields a solution of the form

$$R(r) = C_0 r + C_1 \tag{2}$$

which in view of BC1, namely,

$$\text{BC1:} \qquad U(r = 0) = 0 \quad \rightarrow \quad R(r = 0) = 0 \tag{3}$$

yields $C_1 = 0$. We then have $R_0(r) = C_0 r$ for the $U(r, t)$ problem, or simply $R_0(r) = C_0$ after transforming back to $T(r, t)$, which is consistent with our addition of C_0 to equation (5-81).

2. The associated Legendre polynomials, $P_n^m(\mu)$ and $P_k^m(\mu)$, are orthogonal over the interval $0 \le \mu \le 1$ for both n and k even integers, or for both n and k odd integers, for a given integer value of m. For integer values of m and n, $P_n^m(0) = 0$ for $m = n + 1$.

6

SOLUTION OF THE HEAT EQUATION FOR SEMI-INFINITE AND INFINITE DOMAINS

In the previous three chapters, we considered the method of separation of variables in our three principle coordinate systems. As developed in detail, the related boundary value problems and the resulting orthogonal functions are an integral part of our solution technique. During our treatment, it was mentioned numerous times the need for the boundary value problem to be limited to a dimension of finite domain. In this chapter, we now extend our solution schemes to include semi-infinite and infinite domain problems in our three principle coordinate systems, namely, Cartesian, cylindrical, and spherical coordinates. While the solution approach somewhat mirrors the approach of separation of variables, there is a clear distinction for the semi-infinite and infinite domain problems. Specifically, the solutions do not involve a true boundary value problem in their respective domains and, therefore, do not generate eigenvalues and eigenfunctions, nor do they yield Fourier series expansions of the nonhomogeneities. In contrast, the separation constants will be left unspecified, and, therefore, rather than summing over a discrete set of eigenvalues, the solutions will instead involve integration over all possible values of the separation constant. In an analogous manner, the nonhomogeneities will then be expressed in terms of a Fourier integral rather than a Fourier series.

6-1 ONE-DIMENSIONAL HOMOGENEOUS PROBLEMS IN A SEMI-INFINITE MEDIUM FOR THE CARTESIAN COORDINATE SYSTEM

We now consider the solution of a homogeneous 1-D heat conduction problem for a semi-infinite region. That is, a semi-infinite region, $0 \leq x < \infty$, is initially

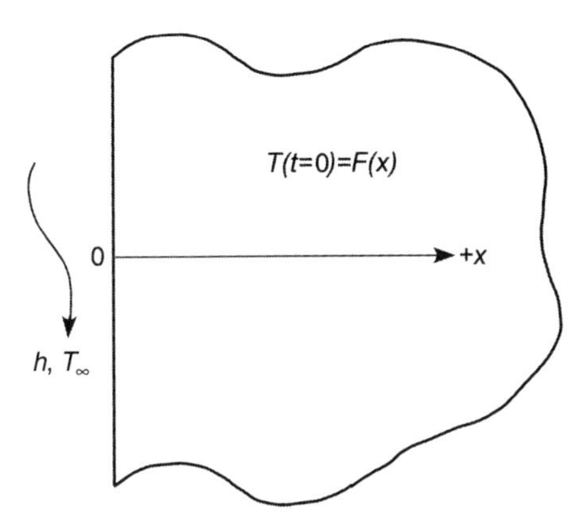

Figure 6-1 One-dimensional heat conduction in a semi-infinite medium.

at a temperature $F(x)$, and for times $t>0$ the boundary surface at $x=0$ dissipates heat by convection into a medium at zero temperature ($T_\infty = 0$) as illustrated in Figure 6-1.

The mathematical formulation of this problem is given as

$$\frac{\partial^2 T(x,t)}{\partial x^2} = \frac{1}{\alpha}\frac{\partial T(x,t)}{\partial t} \quad \text{in} \quad 0 < x < \infty, \quad t > 0 \tag{6-1}$$

$$\text{BC1:} \quad -k\left.\frac{\partial T}{\partial x}\right|_{x=0} = -hT|_{x=0} \tag{6-2a}$$

$$\text{IC:} \quad T(x, t=0) = F(x) \tag{6-2b}$$

With the boundary condition BC1 homogeneous, we assume a separation of the form

$$T(x,t) = X(x)\Gamma(t) \tag{6-3}$$

As we developed in Chapter 3, we now substitute equation (6-3) into (6-1) and introduce a separation constant

$$\frac{1}{X}\frac{d^2X}{dx^2} = \frac{1}{\alpha\Gamma}\frac{d\Gamma}{dt} = -\lambda^2 \tag{6-4}$$

The resulting ODE in the t dimension yields the desired solution of the form

$$\Gamma(t) = C_1 e^{-\alpha\lambda^2 t} \tag{6-5}$$

while the space-variable function produces the following ordinary differential equation:

$$\frac{d^2X}{dx^2} + \lambda^2 X = 0 \tag{6-6}$$

Equation (6-6) yields the solution

$$X(x) = C_2 \cos \lambda x + C_3 \sin \lambda x \tag{6-7}$$

with the application of boundary condition BC1 yielding

$$-k\left[C_3\lambda\right] = -h\left[C_2\right] \tag{6-8}$$

Elimination of C_3 and substitution into equation (6-7) yields

$$X(x) = \frac{C_2}{\lambda}\left[\lambda \cos \lambda x + \frac{h}{k} \sin \lambda x\right] \tag{6-9}$$

after factoring from within the brackets the term $1/\lambda$. The constant in front of the brackets may be replaced by a new unspecified constant, $c(\lambda)$, yielding the expression

$$X(x) = c(\lambda)\left[\lambda \cos \lambda x + \frac{h}{k} \sin \lambda x\right] \tag{6-10}$$

The functional form of the nomenclature used for $c(\lambda)$ implies that the constant will depend on all values of λ rather than discrete values as obtained with a true boundary value problem. The general solution for $T(x, t)$ is now constructed by the superposition of all these elementary solutions, which is accomplished for a continuous λ by integrating over the values of λ from zero to infinity, yielding

$$T(x, t) = \int_{\lambda=0}^{\infty} c(\lambda) e^{-\alpha\lambda^2 t} (\lambda \cos \lambda x + H \sin \lambda x) d\lambda \tag{6-11}$$

where we have replaced h/k with H. The application of the initial condition equation (6-2b) yields

$$F(x) = \int_{\lambda=0}^{\infty} c(\lambda)(\lambda \cos \lambda x + H \sin \lambda x) d\lambda \tag{6-12}$$

Equation (6-12) is recognized as a *Fourier integral expansion* of the function $F(x)$, as we developed in Chapter 2 with equation (2-114). Provided that $F(x)$ meets the functional requirements as outlined in the treatment accompanying equation (2-114), we may express the coefficients $c(\lambda)$ by

$$c(\lambda) = \frac{1}{N(\lambda)} \int_{x'=0}^{\infty} F(x')\left[\lambda \cos \lambda x' + H \sin \lambda x'\right] dx' \tag{6-13}$$

or

$$c(\lambda) = \frac{2}{\pi(\lambda^2 + H^2)} \int_{x'=0}^{\infty} F(x')\left[\lambda \cos \lambda x' + H \sin \lambda x'\right] dx' \tag{6-14}$$

Equations (6-11) and (6-14) may now be combined to produce the overall solution of the form

$$T(x,t) = \int_{\lambda=0}^{\infty} \left[\begin{array}{l} \dfrac{2\,(\lambda \cos \lambda x + H \sin \lambda x)}{\pi (\lambda^2 + H^2)} e^{-\alpha \lambda^2 t} \\ \cdot \displaystyle\int_{x'=0}^{\infty} F(x') \left(\lambda \cos \lambda x' + H \sin \lambda x' \right) dx' \end{array} \right] d\lambda \tag{6-15}$$

The functions $X(\lambda, x)$ and $N(\lambda)$ given by equations (6-10) and (6-14) are for a boundary condition of the third type (i.e., convection boundary) at x=0. The boundary condition at x=0 may also be of the first type (prescribed temperature) or of the second type (insulated). We list in Table 6-1 the functions $X(\lambda, x)$ and $N(\lambda)$ for these three different homogeneous boundary conditions at x=0. Thus, the solution of the 1-D homogeneous heat conduction problem of equations (6-1) and (6-2b) for a semi-infinite medium $0 \le x < \infty$ is given by the following equations:

$$T(x,t) = \int_{\lambda=0}^{\infty} c(\lambda) X(\lambda, x) e^{-\alpha \lambda^2 t} d\lambda \tag{6-16}$$

$$c(\lambda) = \frac{1}{N(\lambda)} \int_{x'=0}^{\infty} F(x') X(\lambda, x') dx' \tag{6-17}$$

which together yield

$$T(x,t) = \int_{\lambda=0}^{\infty} \left[\frac{1}{N(\lambda)} X(\lambda, x) e^{-\alpha \lambda^2 t} \int_{x'=0}^{\infty} X(\lambda, x') F(x') dx' \right] d\lambda \tag{6-18}$$

Equation (6-18) is obtainable for the three different homogeneous boundary conditions at $x = 0$, if $X(\lambda, x)$ and $N(\lambda)$ are taken from Table 6-1, accordingly.

With the general methodology established above for the 1-D semi-infinite problems, we now proceed to solve several related problems.

TABLE 6-1 Solution $X(\lambda, x)$ and the Norm $N(\lambda)$ for the 1-D Cartesian Semi-Infinite Domain Problem

No.	Boundary Condition at $x = 0$	$X(\lambda, x)$	$1/N(\lambda)$
1	$-\dfrac{dX}{dx} + HX = 0$	$\lambda \cos \lambda x + H \sin \lambda x$	$\dfrac{2}{\pi} \dfrac{1}{\lambda^2 + H^2}$
2	$\dfrac{dX}{dx} = 0$	$\cos \lambda x$	$\dfrac{2}{\pi}$
3	$X = 0$	$\sin \lambda x$	$\dfrac{2}{\pi}$

Example 6-1 Semi-infinite Region with Type I Boundary Condition Consider the solution of a homogeneous 1-D heat conduction problem for a semi-infinite region, $0 \leq x < \infty$, that is initially at a temperature $F(x)$. For times $t > 0$, the boundary surface at x=0 is maintained at a temperature of zero. Also, examine the case when the initial temperature distribution $F(x) = T_0 =$ constant. The mathematical formulation of this problem is given as

$$\frac{\partial^2 T(x,t)}{\partial x^2} = \frac{1}{\alpha}\frac{\partial T(x,t)}{\partial t} \quad \text{in} \quad 0 < x < \infty, \quad t > 0 \tag{6-19}$$

$$\text{BC1:} \qquad T(x = 0) = 0 \tag{6-20a}$$

$$\text{IC:} \qquad T(x, t = 0) = F(x) \tag{6-20b}$$

With the boundary condition BC1 homogeneous, we assume a separation of the form

$$T(x,t) = X(x)\Gamma(t) \tag{6-21}$$

Substitution of equation (6-21) into (6-19) yields

$$\frac{1}{X}\frac{d^2X}{dx^2} = \frac{1}{\alpha\Gamma}\frac{d\Gamma}{dt} = -\lambda^2 \tag{6-22}$$

with introduction of the separation constant λ. The resulting ODE in the t dimension yields the solution

$$\Gamma(t) = C_1 e^{-\alpha\lambda^2 t} \tag{6-23}$$

while the space-variable ODE produces the solution

$$X(x) = C_2 \cos\lambda x + C_3 \sin\lambda x \tag{6-24}$$

The application of boundary condition BC1 yields $C_2 = 0$, giving

$$X(x) = C_3 \sin\lambda x \tag{6-25}$$

as consistent with Table 6-1, case 3, for a type 1 boundary condition. Integrating over all possible values of λ gives the overall solution of the form

$$T(x,t) = \int_{\lambda=0}^{\infty} c(\lambda) \sin\lambda x\, e^{-\alpha\lambda^2 t} d\lambda \tag{6-26}$$

where $c(\lambda) = C_1 C_3$. Application of the initial condition yields the Fourier integral expansion of the initial temperature distribution $F(x)$, namely,

$$F(x) = \int_{\lambda=0}^{\infty} c(\lambda) \sin\lambda x\, d\lambda \tag{6-27}$$

Introducing the norm from Table 6-1, the solution of the coefficients yields

$$c(\lambda) = \frac{2}{\pi} \int_{x'=0}^{\infty} F(x') \sin \lambda x' \, dx' \tag{6-28}$$

Substitution of equation (6-28) into equation (6-26) yields the final expression

$$T(x,t) = \frac{2}{\pi} \int_{\lambda=0}^{\infty} \left(\sin \lambda x \, e^{-\alpha \lambda^2 t} \int_{x'=0}^{\infty} \sin \lambda x' F(x') dx' \right) d\lambda \tag{6-29}$$

We now consider the special case of a constant initial condition, namely, an initial temperature distribution $T(t=0) = T_0$. First, we rearrange the integrals in equation (6-29) to yield

$$T(x,t) = \frac{2}{\pi} \int_{x'=0}^{\infty} F(x') \int_{\lambda=0}^{\infty} e^{-\alpha \lambda^2 t} \sin \lambda x \sin \lambda x' \, d\lambda \, dx' \tag{6-30}$$

The integration with respect to λ is evaluated by making use of the following trigonometric relation:

$$2 \sin \lambda x \sin \lambda x' = \cos \lambda (x - x') - \cos \lambda (x + x') \tag{6-31}$$

and the two integrals from Dwight [1], namely,

$$\int_{\lambda=0}^{\infty} e^{-\alpha \lambda^2 t} \cos \lambda (x - x') \, d\lambda = \sqrt{\frac{\pi}{4\alpha t}} \exp \left[-\frac{(x - x')^2}{4\alpha t} \right] \tag{6-32}$$

and

$$\int_{\lambda=0}^{\infty} e^{-\alpha \lambda^2 t} \cos \lambda (x + x') \, d\lambda = \sqrt{\frac{\pi}{4\alpha t}} \exp \left[-\frac{(x + x')^2}{4\alpha t} \right] \tag{6-33}$$

Using equations (6-31)–(6-33), we can express the following integral as

$$\frac{2}{\pi} \int_{\lambda=0}^{\infty} e^{-\alpha \lambda^2 t} \sin \lambda x \, \sin \lambda x' \, d\lambda = \frac{1}{(4\pi \alpha t)^{1/2}} \left\{ \exp \left[-\frac{(x - x')^2}{4\alpha t} \right] - \exp \left[-\frac{(x + x')^2}{4\alpha t} \right] \right\} \tag{6-34}$$

which upon substitution into equation (6-30) yields

$$T(x,t) = \frac{1}{(4\pi \alpha t)^{1/2}} \int_{x'=0}^{\infty} F(x') \left\{ \exp \left[-\frac{(x - x')^2}{4\alpha t} \right] - \exp \left[-\frac{(x + x')^2}{4\alpha t} \right] \right\} dx' \tag{6-35}$$

For our case of constant initial temperature, we substitute $F(x) = T_0$ into equation (6-35), yielding

$$T(x,t) = \frac{T_0}{(4\pi\alpha t)^{1/2}} \left\{ \int_{x'=0}^{\infty} \exp\left[-\frac{(x-x')^2}{4\alpha t}\right] dx' - \int_{x'=0}^{\infty} \exp\left[-\frac{(x+x')^2}{4\alpha t}\right] dx' \right\} \tag{6-36}$$

The above integrals may be solved by introducing the following new variables:

$$-\eta = \frac{x-x'}{\sqrt{4\alpha t}} \qquad dx' = \sqrt{4\alpha t}\, d\eta \quad \text{for the first integral}$$

and

$$\eta = \frac{x+x'}{\sqrt{4\alpha t}} \qquad dx' = \sqrt{4\alpha t}\, d\eta \quad \text{for the second integral}$$

Substitution of the above into equation (6-36) yields

$$T(x,t) = \frac{T_0}{\sqrt{\pi}} \left(\int_{-x/\sqrt{4\alpha t}}^{\infty} e^{-\eta^2} d\eta - \int_{x/\sqrt{4\alpha t}}^{\infty} e^{-\eta^2} d\eta \right) \tag{6-37}$$

Because $e^{-\eta^2}$ is symmetrical about η=0, equation (6-37) may be written in the form

$$\frac{T(x,t)}{T_0} = \frac{2}{\sqrt{\pi}} \int_{\eta=0}^{x/\sqrt{4\alpha t}} e^{-\eta^2} d\eta \tag{6-38}$$

The right-hand side of equation (6-38) is recognized as the error function of argument $x/\sqrt{4\alpha t}$, and the solution may therefore be expressed in the form

$$T(x,t) = T_0 \operatorname{erf}\left(\frac{x}{\sqrt{4\alpha t}}\right) \tag{6-39}$$

The values of the error functions are tabulated in Appendix III, along with a brief discussion of the properties of the error function. We note two properties of the error function here, namely,

$$\boxed{\operatorname{erf}(0) = 0} \tag{6-40}$$

$$\boxed{\operatorname{erf}(z)|_{z\to\infty} = 1} \tag{6-41}$$

From equation (6-40), it is seen that the boundary condition at x=0 is always recovered for our solution of equation (6-39). Likewise, the argument goes to infinity at $t=0$; hence the initial condition of T_0 is recovered in view of equation (6-41). In addition, at any *finite value of time*, the initial condition is recovered as $x \to \infty$, which may be interpreted as the requirement of infinite time for the boundary condition to penetrate an infinite distance. We present the behavior of equation (6-39) in Figure 6-2 (letting $\alpha = 1 \times 10^{-6}$ m^2/s) for various times.

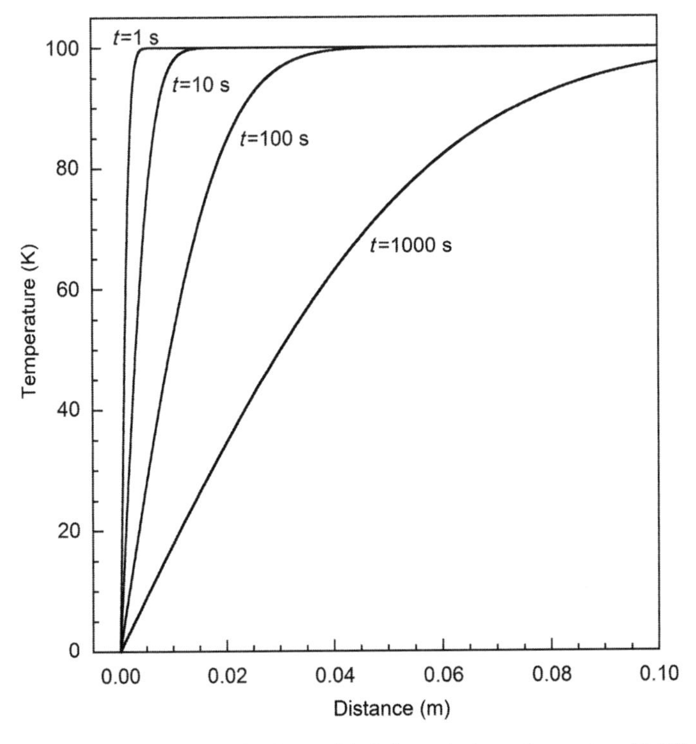

Figure 6-2 Behavior of 1-D semi-infinite solution of equation (6-39).

Examination of Figure 6-2 reveals several trends consistent with the physics of the problem. The temperature gradient at the surface decreases with time as the influence of the boundary condition change diminishes, noting that the surface gradient is essentially infinite at the moment of the boundary condition imposition. The penetration of the change in the surface boundary condition is also observed to increase with time, having penetrated to less than 0.015 m by 10 s and to less than 0.045 m by 100 s.

The heat flux at any location in space and time is readily calculated from Fourier's law and equation (6-39), noting the derivative of the error function is given by

$$\frac{d}{dz}\text{erf}(z) = \frac{2}{\sqrt{\pi}}e^{-z^2} \tag{6-42}$$

Applying equation (6-42) to equation (6-39), along with the chain rule, the heat flux becomes

$$q''(x,t) = \frac{-kT_0}{\sqrt{\pi\alpha t}} e^{-x^2/4\alpha t} \tag{6-43}$$

From equation (6-43), it is readily seen that the surface heat flux (x=0) is infinite at t=0 and that the heat flux decays with time at any fixed spatial location.

Heat Flux Formulation

The one-dimensional transient heat conduction equation, customarily given in terms of temperature $T(x,t)$, can be expressed in terms of the heat flux, $q''(x,t)$. Such a formulation is useful for solving heat conduction problems in a semi-infinite medium with a prescribed heat flux boundary condition. Consider the heat conduction equation

$$\frac{\partial^2 T(x,t)}{\partial x^2} = \frac{1}{\alpha}\frac{\partial T(x,t)}{\partial t} \tag{6-44}$$

and the definition of the heat flux (i.e., Fourier's law)

$$q''(x,t) = -k\frac{\partial T(x,t)}{\partial x} \tag{6-45}$$

Equation (6-44) is differentiated with respect to the space variable, while equation (6-45) is differentiated once with respect to time and separately differentiated twice with respect to the space variable. The results are then manipulated to obtain

$$\frac{\partial^2 q''(x,t)}{\partial x^2} = \frac{1}{\alpha}\frac{\partial q''(x,t)}{\partial t} \tag{6-46}$$

which is a differential equation in heat flux and is of the same form as equation (6-44).

To illustrate the application of equation (6-46), we consider 1-D heat conduction in a semi-infinite medium $0 \le x < \infty$, initially at zero temperature. For times $t > 0$, a constant heat flux q_0'' is applied at the boundary surface x=0. The mathematical formulation of this problem, in the flux formulation, is given by

$$\frac{\partial^2 q''(x,t)}{\partial x^2} = \frac{1}{\alpha}\frac{\partial q''(x,t)}{\partial t} \quad \text{in} \quad 0 < x < \infty, \quad t > 0 \tag{6-47}$$

$$\text{BC1:} \quad q''(x=0) = q_0'' = \text{constant} \tag{6-48a}$$

$$\text{IC:} \quad q''(t=0) = 0 \tag{6-48b}$$

To move the nonhomogeneity from the boundary condition to the initial condition, we introduce a new dependent variable $Q(x,t)$ defined as

$$Q(x,t) = q''(x,t) - q_0'' \tag{6-49}$$

With this shift, the problem is reformulated as

$$\frac{\partial^2 Q(x,t)}{\partial x^2} = \frac{1}{\alpha}\frac{\partial Q(x,t)}{\partial t} \quad \text{in} \quad 0 < x < \infty, \quad t > 0 \tag{6-50}$$

$$\text{BC1:} \qquad Q(x=0) = 0 \tag{6-51a}$$

$$\text{IC:} \qquad Q(t=0) = -q_0'' \tag{6-51b}$$

The solution is immediately obtained from equation (6-39), yielding

$$Q(x,t) = -q_0''\text{erf}\left(\frac{x}{\sqrt{4\alpha t}}\right) \tag{6-52}$$

Shifting back to $q''(x,t)$ yields

$$q''(x,t) = q_0'' + Q(x,t) = q_0''\left[1 - \text{erf}\left(\frac{x}{\sqrt{4\alpha t}}\right)\right] \tag{6-53}$$

or using the definition of the *complementary error function*,

$$q''(x,t) = q_0''\text{erfc}\left(\frac{x}{\sqrt{4\alpha t}}\right) \tag{6-54}$$

Once the function $q''(x,t)$ is known, the temperature distribution $T(x,t)$ is determined by the integration of equation (6-45). We then obtain

$$T(x,t) = \frac{q_0''}{k}\int_{x'=x}^{\infty} \text{erfc}(x'/\sqrt{4\alpha t})\, dx' \tag{6-55}$$

The integration is performed by utilizing the relationship given by equations (5) and (6) in Appendix III, which yields the result

$$\text{T}(x,t) = \frac{2q_0''}{k}\left[\left(\frac{\alpha t}{\pi}\right)^{1/2} e^{-x^2/4\alpha t} - \frac{x}{2}\text{erfc}(\frac{x}{\sqrt{4\alpha t}})\right] \tag{6-56}$$

The temperature at the surface x=0 becomes

$$T(x=0,t) = \frac{2q_0''}{k}\left(\frac{\alpha t}{\pi}\right)^{1/2} \tag{6-57}$$

The surface temperature continues to rise with the square root of time, consistent with the physics of a constant surface flux, as there is no steady-state solution for this formulation.

Example 6-2 Semi-infinite Region with Type 2 Boundary Condition Consider the solution of a homogeneous 1-D heat conduction problem for a semi-infinite region, $0 \leq x < \infty$, that is initially at a temperature $F(x)$. For times $t > 0$, the boundary surface at $x = 0$ is kept perfectly insulated. The mathematical formulation of this problem is given as

$$\frac{\partial^2 T(x,t)}{\partial x^2} = \frac{1}{\alpha} \frac{\partial T(x,t)}{\partial t} \quad \text{in} \quad 0 < x < \infty, \quad t > 0 \tag{6-58}$$

$$\text{BC1:} \quad \left. \frac{\partial T}{\partial x} \right|_{x=0} = 0 \tag{6-59a}$$

$$\text{IC:} \quad T(x, t = 0) = F(x) \tag{6-59b}$$

With the boundary condition BC1 homogeneous, we assume a separation of the form

$$T(x,t) = X(x)\Gamma(t) \tag{6-60}$$

Substitution of equation (6-60) into (6-58) yields

$$\frac{1}{X} \frac{d^2 X}{dx^2} = \frac{1}{\alpha \Gamma} \frac{d\Gamma}{dt} = -\lambda^2 \tag{6-61}$$

with introduction of the separation constant λ. The resulting ODE in the t dimension yields the solution

$$\Gamma(t) = C_1 e^{-\alpha \lambda^2 t} \tag{6-62}$$

while the space-variable ODE produces the solution

$$X(x) = C_2 \cos \lambda x + C_3 \sin \lambda x \tag{6-63}$$

The application of boundary condition BC1 yields $C_3 = 0$, giving

$$X(x) = C_2 \cos \lambda x \tag{6-64}$$

as consistent with Table 6-1, case 2, for a type 2 boundary condition. Integrating over all possible values of λ gives the overall solution of the form

$$T(x,t) = \int_{\lambda=0}^{\infty} c(\lambda) \cos \lambda x \, e^{-\alpha \lambda^2 t} \, d\lambda \tag{6-65}$$

where $c(\lambda) = C_1 C_2$. Application of the initial condition yields the Fourier integral expansion of the initial temperature distribution $F(x)$, namely,

$$F(x) = \int_{\lambda=0}^{\infty} c(\lambda) \cos \lambda x \, d\lambda \tag{6-66}$$

Introducing the norm from Table 6-1, case 2, the solution of the coefficients yields

$$c(\lambda) = \frac{2}{\pi} \int_{x'=0}^{\infty} F(x') \cos \lambda x' \, dx' \tag{6-68}$$

Substitution of equation (6-68) into equation (6-65) yields the expression

$$T(x,t) = \frac{2}{\pi} \int_{x'=0}^{\infty} F(x') \int_{\lambda=0}^{\infty} e^{-\alpha\lambda^2 t} \cos \lambda x \cos \lambda x' \, d\lambda \, dx' \tag{6-69}$$

We may now introduce the trigonometric substitution

$$2 \cos \lambda x \cos \lambda x' = \cos \lambda (x - x') + \cos \lambda (x + x') \tag{6-70}$$

With the substitution of equation (6-70) into equation (6-69), we may perform the integration using equations (6-32) and (6-33), which yields the solution

$$T(x,t) = \frac{1}{(4\pi\alpha t)^{1/2}} \int_{x'=0}^{\infty} F(x') \left\{ \exp\left[-\frac{(x-x')^2}{4\alpha t} \right] + \exp\left[-\frac{(x+x')^2}{4\alpha t} \right] \right\} dx' \tag{6-71}$$

A comparison of solutions (6-35) and (6-71) reveals that the two exponential terms are subtracted in the former and added in the latter.

6-2 MULTIDIMENSIONAL HOMOGENEOUS PROBLEMS IN A SEMI-INFINITE MEDIUM FOR THE CARTESIAN COORDINATE SYSTEM

We now consider the solution of homogeneous, multidimensional heat conduction problems that include at least one semi-infinite region. We will make use of the product solution technique developed in Chapter 3 as appropriate. We begin with a combination of finite and semi-infinite domains.

Example 6-3 Two-Dimensional Problem with a Finite Domain and Semi-infinite Domain

Consider a semi-infinite rectangular strip over the region $0 \le x < \infty$, $0 \le y \le b$ that is initially at a temperature $F(x,y)$. For times $t>0$, the boundary surfaces at $x=0$ and $y=b$ are maintained at zero temperature, while the boundary at $y=0$ dissipates heat by convection with convection coefficient h into a fluid at zero temperature, as illustrated in Figure 6-3. Obtain an expression for the temperature $T(x,y,t)$ for $t>0$.

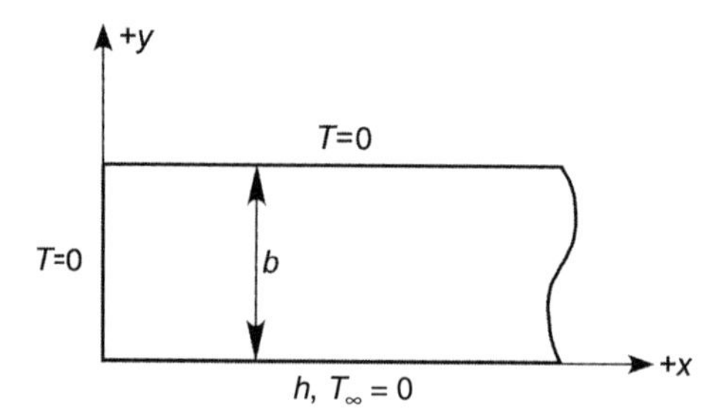

Figure 6-3 Problem description for Example 6-3.

The mathematical formulation of this problem is given as

$$\frac{\partial^2 T}{\partial x^2} + \frac{\partial^2 T}{\partial y^2} = \frac{1}{\alpha}\frac{\partial T}{\partial t} \quad \text{in} \quad 0 < x < \infty, \quad 0 < y < b, \quad t > 0 \tag{6-72}$$

$$\text{BC1:} \quad T(x = 0) = 0 \tag{6-73a}$$

$$\text{BC2:} \quad -k\left.\frac{\partial T}{\partial y}\right|_{y=0} = -h\,T|_{y=0} \qquad \text{BC3:} \quad T(y = b) = 0 \tag{6-73b}$$

$$\text{IC:} \quad T(x, y, t = 0) = F(x, y) \tag{6-73c}$$

With all boundary conditions homogeneous, we assume a separation of the form

$$T(x, y, t) = X(x)Y(y)\Gamma(t) \tag{6-74}$$

Substitution of equation (6-74) into (6-72) yields

$$\frac{1}{X}\frac{d^2X}{dx^2} + \frac{1}{Y}\frac{d^2Y}{dy^2} = \frac{1}{\alpha\Gamma}\frac{d\Gamma}{dt} = -\lambda^2 \tag{6-75}$$

The resulting ODE in the t dimension yields the solution

$$\Gamma(t) = C_1 e^{-\alpha\lambda^2 t} \tag{6-76}$$

while the spatial dimensions yield the following two auxiliary problems:

$$\frac{d^2X}{dx^2} + \beta^2 X(x) = 0 \quad \text{in} \quad 0 < x < \infty \tag{6-77}$$

$$\text{BC1:} \quad X(x = 0) = 0 \tag{6-78}$$

and

$$\frac{d^2Y}{dy^2} + \gamma^2 Y(y) = 0 \quad \text{in} \quad 0 < y < b \tag{6-79}$$

$$\text{BC2:} \quad -\left.\frac{dY}{\partial y}\right|_{y=0} + H\,Y|_{y=0} = 0 \qquad \text{BC3:} \quad Y(y = b) = 0 \tag{6-80}$$

where we have introduced the separation constants β and γ, with the relation

$$\lambda^2 = \beta^2 + \gamma^2 \tag{6-81}$$

and where we have made the substitution $H{=}h/k$.

We first consider the true boundary value problem of equations (6-79) and (6-80). This type of problem was considered in detail in Chapter 3 and corresponds to case 3 in Table 2-1. The solution becomes

$$Y(\gamma_n, y) = C_2 \sin \gamma_n (b - y) \tag{6-82}$$

with the eigenvalues coming from the transcendental equation

$$\gamma_n \cot \gamma_n b = -H \quad \rightarrow \quad \gamma_n \quad \text{for} \quad n = 1, 2, 3, \ldots \tag{6-83}$$

We now consider the semi-infinite domain problem defined by equations (6-77) and (6-78). This problem was considered above and corresponds to case 3 in Table 6-1. The solution becomes

$$X(\beta, x) = C_3 \sin \beta x \tag{6-84}$$

Now with the eigenvalues γ_n limited to discrete values per equation (6-83), while the separation constant β is still defined for all positive values, we form the general solution by *summing* over all values of γ_n and *integrating* overall all values of β, yielding

$$T(x, y, t) = \sum_{n=1}^{\infty} \int_{\beta=0}^{\infty} c_n(\beta) \sin \beta x \sin \gamma_n (b - y) e^{-\alpha(\beta^2+\gamma_n^2)t} \, d\beta \tag{6-85}$$

where we have introduced the new coefficient $c_n(\beta) = C_1 C_2 C_3$. The application of the initial condition equation (6-73c) yields

$$F(x, y) = \sum_{n=1}^{\infty} \int_{\beta=0}^{\infty} c_n(\beta) \sin \beta x \sin \gamma_n (b - y) \, d\beta \tag{6-86}$$

To determine the unknown coefficient, we first use the operator

$$* \int_{y=0}^{b} \sin \gamma_k (b - y) \, dy$$

and utilize the property of orthogonality in the y dimension. We obtain

$$f^*(x) = \int_{\beta=0}^{\infty} c_n(\beta) \sin \beta x \, d\beta \quad \text{in} \quad 0 < x < \infty \tag{6-87}$$

where

$$f^*(x) \equiv \frac{1}{N(\gamma_n)} \int_{y'=0}^{b} \sin \gamma_n (b - y') F(x, y')\, dy' \tag{6-88a}$$

and

$$N(\gamma_n) \equiv \int_{y=0}^{b} \sin^2 \gamma_n (b - y)\, dy = \frac{b(\gamma_n^2 + H^2) + H}{2\left(\gamma_n^2 + H^2\right)} \tag{6-88b}$$

for the eigenvalues as defined by equation (6-83). Equation (6-87) is now a Fourier integral representation of a prescribed function $f^*(x)$, defined on the interval $0 < x < \infty$, in terms of the function $\sin \beta x$. Such an expansion was developed in Chapter 2, equation (2-109a), and is summarized by case 3 in Table 6-1. This yields the following expression for $c_n(\beta)$:

$$c_n(\beta) = \frac{2}{\pi} \int_{x'=0}^{\infty} f^*(x') \sin \beta x' dx' \tag{6-89}$$

Substitution of equation (6-88a) into the above yields the final expression

$$c_n(\beta) = \frac{2}{\pi} \int_{x'=0}^{\infty} \sin \beta x' \left[\frac{1}{N(\gamma_n)} \int_{y'=0}^{b} \sin \gamma_n (b - y') F(x', y') dy' \right] dx' \tag{6-90}$$

Substitution of equation (6-90) into equation (6-85), with the additional substitution of equation (6-88b) and rearrangement of the order of integration, yields the final form of the temperature solution:

$$\begin{aligned} T(x, y, t) = \frac{4}{\pi} \sum_{n=1}^{\infty} \Bigg[& \frac{\gamma_n^2 + H^2}{b(\gamma_n^2 + H^2) + H} \sin \gamma_n (b - y) e^{-\alpha \gamma_n^2 t} \\ & \times \int_{x'=0}^{\infty} \int_{y'=0}^{b} F(x', y') \sin \gamma_n (b - y') \\ & \times \left(\int_{\beta=0}^{\infty} e^{-\alpha \beta^2 t} \sin \beta x' \sin \beta x d\beta \right) dy'\, dx' \Bigg] \end{aligned} \tag{6-91}$$

The inner integral over β was evaluated previously, see equation (6-34), which allows equation (6-91) to be written in the form

$$\begin{aligned} T(x, y, t) = \frac{1}{\sqrt{\pi \alpha t}} \sum_{n=1}^{\infty} \Bigg(& \frac{\gamma_n^2 + H^2}{b(\gamma_n^2 + H^2) + H} \sin \gamma_n (b - y) e^{-\alpha \gamma_n^2 t} \\ & \times \int_{x'=0}^{\infty} \int_{y'=0}^{b} F(x', y') \sin \gamma_n (b - y') \\ & \times \left\{ \exp\left[-\frac{(x - x')^2}{4\alpha t} \right] - \exp\left[-\frac{(x + x')^2}{4\alpha t} \right] \right\} dy'\, dx' \Bigg) \end{aligned} \tag{6-92}$$

Example 6-4 Two-Dimensional Problem for a Semi-infinite Corner Consider a semi-infinite corner over the region $0 \leq x < \infty$, $0 \leq y < \infty$ that is initially at a temperature T_0. For times $t>0$, the boundary surfaces at $x=0$ and $y=0$ are maintained at zero temperature, as illustrated in Figure 6-4. Obtain an expression for the temperature $T(x,y,t)$ in the region for $t>0$.

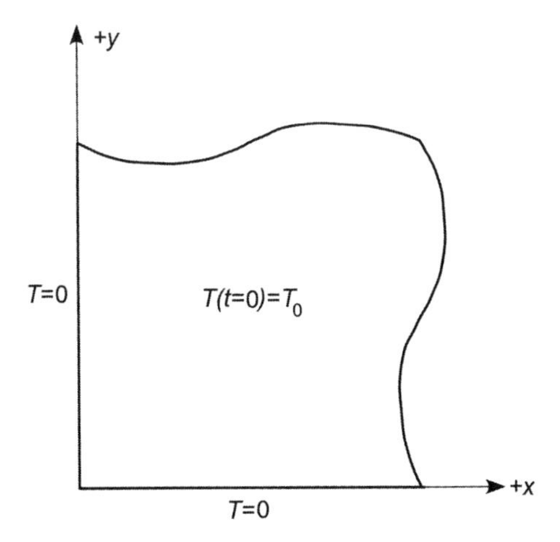

Figure 6-4 Problem description for Example 6-4.

The mathematical formulation of this problem is given as

$$\frac{\partial^2 T}{\partial x^2} + \frac{\partial^2 T}{\partial y^2} = \frac{1}{\alpha}\frac{\partial T}{\partial t} \quad \text{in} \quad 0 < x < \infty, \quad 0 < y < \infty, \quad t > 0 \quad \text{(6-93)}$$

BC1: $T(x = 0) = 0$ (6-94a)

BC2: $T(y = 0) = 0$ (6-94b)

IC: $T(x, y, t = 0) = T_0$ (6-94c)

The solution of this problem can be expressed as the product of the solutions of the following two one-dimensional problems: (1) $T_1(x, t)$, the solution for a semi-infinite region $0 \leq x < \infty$ initially at a temperature $F_1(x) = 1$, and for times $t > 0$ the boundary surface at $x=0$ is kept at zero temperature; (2) and $T_2(y, t)$, the solution for a semi-infinite region $0 \leq y < \infty$ initially at a temperature $F_2(y) = T_0$, and for times $t > 0$ the boundary surface at $y=0$ is kept at zero temperature. Clearly, the initial condition for the 2-D problem is expressible as a product, namely, $T_0 = 1 \cdot T_0$. In keeping with our guidelines for use of product solutions, as developed in Chapter 3, the 2-D solution is expressible as a product of 1-D solutions. The resulting two 1-D problems are formulated as follows:

$$\frac{\partial^2 T_1(x,t)}{\partial x^2} = \frac{1}{\alpha}\frac{\partial T_1(x,t)}{\partial t} \quad \text{in} \quad 0 < x < \infty, \quad t > 0 \quad \text{(6-95)}$$

$$\text{BC1:} \quad T_1(x=0) = 0 \tag{6-96a}$$

$$\text{IC:} \quad T_1(x, t=0) = 1 \tag{6-96b}$$

and

$$\frac{\partial^2 T_2(y,t)}{\partial y^2} = \frac{1}{\alpha}\frac{\partial T_2(y,t)}{\partial t} \quad \text{in} \quad 0 < y < \infty, \quad t > 0 \tag{6-97}$$

$$\text{BC1:} \quad T_2(y=0) = 0 \tag{6-98a}$$

$$\text{IC:} \quad T_2(y, t=0) = T_0 \tag{6-98b}$$

Both problems are nearly identical, with the only difference being the constant corresponding to the initial condition. Using the solution of Example 6-1, namely, equation (6-39), we write the two solutions as

$$T_1(x,t) = \text{erf}\left(\frac{x}{\sqrt{4\alpha t}}\right) \tag{6-99}$$

and

$$T_2(y,t) = T_0 \,\text{erf}\left(\frac{y}{\sqrt{4\alpha t}}\right) \tag{6-100}$$

Combining equations (6-99) and (6-100) as a product, the solution of the above 2-D problem becomes

$$T(x,y,t) = T_1(x,t)T_2(y,t) = T_0 \,\text{erf}\left(\frac{x}{\sqrt{4\alpha t}}\right)\text{erf}\left(\frac{y}{\sqrt{4\alpha t}}\right) \tag{6-101}$$

Example 6-5 Two-Dimensional Problem for a Steady-State, Semi-infinite Rectangular Strip

Consider a 2-D semi-infinite rectangular strip over the region $0 \le x < \infty$, $0 \le y \le b$ that is at steady-state conditions. The boundary surfaces at x=0 and y=b are maintained at zero temperature, while the boundary surface at y=0 is at a prescribed temperature given by T=$f(x)$, as illustrated in Figure 6-5. Obtain an expression for the temperature $T(x,y)$ in the region.

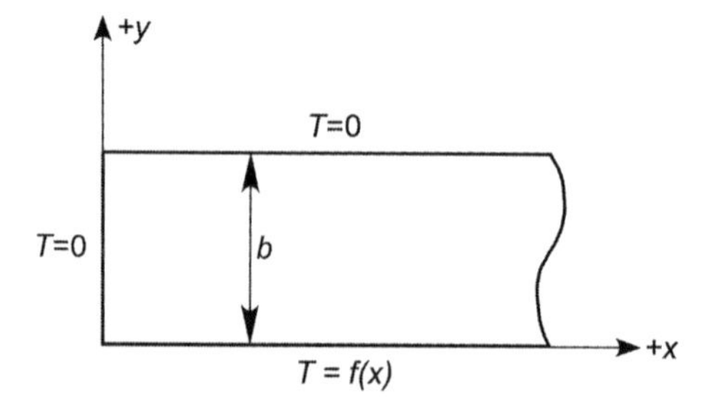

Figure 6-5 Problem description for Example 6-5.

The mathematical formulation of this problem is given as

$$\frac{\partial^2 T}{\partial x^2} + \frac{\partial^2 T}{\partial y^2} = 0 \quad \text{in} \quad 0 < x < \infty, \quad 0 < y < b \tag{6-102}$$

$$\text{BC1:} \qquad T(x = 0) = 0 \tag{6-103a}$$

$$\text{BC2:} \qquad T(y = 0) = f(x) \tag{6-103b}$$

$$\text{BC3:} \qquad T(y = b) = 0 \tag{6-103c}$$

As we discussed in previous chapters, a true characteristic boundary problem must be over a finite dimension, noting here, however, that the finite y dimension contains the nonhomogeneity. We therefore plan to force a Fourier integral expansion in the semi-infinite x dimension. With this in mind, we precede with a separation of the form

$$\frac{1}{Y}\frac{d^2 Y}{dy^2} = \frac{-1}{X}\frac{d^2 X}{dx^2} = \lambda^2 \tag{6-104}$$

The separated problems then become

$$\frac{d^2 X}{dx^2} + \lambda^2 X = 0 \quad \text{in} \quad 0 < x < \infty \tag{6-105}$$

$$\text{BC1:} \qquad X(x = 0) = 0 \tag{6-106}$$

and

$$\frac{d^2 Y}{dy^2} - \lambda^2 Y = 0 \quad \text{in} \quad 0 < y < b \tag{6-107}$$

$$\text{BC2:} \qquad Y(y = b) = 0 \tag{6-108}$$

where we have introduced the separation constant λ. The solution of equations (6-105) and (6-106) follows Example 6-1 and corresponds to case 3 of Table 6-1. After solution of the ODE, equation (6-106) eliminates the $\cos \lambda x$ term, yielding

$$X(x) = C_2 \sin \lambda x \tag{6-109}$$

where we retain all values of λ. The solution of equation (6-107) yields

$$Y(y) = C_3 \cosh \lambda y + C_4 \sinh \lambda y \tag{6-110}$$

with BC2 yielding

$$C_4 = -C_3 \frac{\cosh \lambda b}{\sinh \lambda b} \tag{6-111}$$

Substitution of equation (6-111) into equation (6-110), along with the introduction of a hyperbolic-trigonometric substitution, yields

$$Y(y) = C_5 \sinh \lambda \, (b - y) \tag{6-112}$$

after the introduction of a new constant C_5. The complete solution is now constructed as the product of equations (6-109) and (6-112), with integration over all λ,

$$T(x, y) = \int_{\lambda=0}^{\infty} c(\lambda) \sinh \lambda (b - y) \sin \lambda x \, d\lambda \tag{6-113}$$

Application of the nonhomogeneous boundary condition of equation (6-103b) yields

$$f(x) = \int_{\lambda=0}^{\infty} c(\lambda) \sinh \lambda b \, \sin \lambda x \, d\lambda \quad \text{in} \quad 0 < x < \infty \tag{6-114}$$

which, as expected, is a Fourier integral expansion of the boundary temperature distribution $f(x)$. Equation (6-114) is recognized as a special case of the Fourier integral representation as given by equations (2-109a) and (2-109b). By carrying along the $\sinh \lambda b$ term, the coefficients of equation (6-114) become

$$c(\lambda) \sinh \lambda b = \frac{2}{\pi} \int_{x'=0}^{\infty} f(x') \sin \lambda x' \, dx' \tag{6-115}$$

The substitution of $c(\lambda)$ into equation (6-113) gives

$$T(x, y) = \frac{2}{\pi} \int_{\lambda=0}^{\infty} \left[\frac{\sinh \lambda (b - y)}{\sinh \lambda b} \sin \lambda x \int_{x'=0}^{\infty} f(x') \sin \lambda x' \, dx' \right] d\lambda \tag{6-116}$$

or changing the order of integration, we have

$$T(x, y) = \frac{2}{\pi} \int_{x'=0}^{\infty} \left[f(x') \int_{\lambda=0}^{\infty} \frac{\sinh \lambda (b - y)}{\sinh \lambda b} \sin \lambda x \sin \lambda x' \, d\lambda \right] dx' \tag{6-117}$$

The integration with respect to λ has been evaluated previously [2, Section 10-11], which allows simplification of temperature solution to the form

$$T(x, y) = \frac{1}{2b} \sin \frac{\pi y}{b} \int_{x'=0}^{\infty} f(x') \left\{ \frac{1}{\cos[\pi (b - y)/b] + \cosh[\pi (x - x')/b]} - \frac{1}{\cos[\pi (b - y)/b] + \cosh[\pi (x + x')/b]} \right\} dx' \tag{6-118}$$

6-3 ONE-DIMENSIONAL HOMOGENEOUS PROBLEMS IN AN INFINITE MEDIUM FOR THE CARTESIAN COORDINATE SYSTEM

We now consider the homogeneous heat conduction problem for a 1-D infinite medium, $-\infty < x < \infty$, which is initially at a temperature $F(x)$. We are interested in the determination of the temperature $T(x,t)$ of the medium for time $t > 0$. No boundary conditions are specified for the problem since the medium extends to infinity in both directions; but the problem consists of a boundedness condition on $T(x,t)$. The mathematical formulation is given as

$$\frac{\partial^2 T(x,t)}{\partial x^2} = \frac{1}{\alpha}\frac{\partial T(x,t)}{\partial t} \quad \text{in} \quad -\infty < x < \infty \quad t > 0 \tag{6-119}$$

$$\text{IC:} \qquad T(x, t = 0) = F(x) \tag{6-120}$$

We assume a separation of variables of the form

$$T(x,t) = X(x)\Gamma(t) \tag{6-121}$$

which upon substitution into equation (6-119) and introduction of a separation constant yields

$$\frac{1}{X}\frac{d^2X}{dx^2} = \frac{1}{\alpha\Gamma}\frac{d\Gamma}{dt} = -\lambda^2 \tag{6-122}$$

The resulting ODE in the t dimension yields the desired solution

$$\Gamma(t) = C_1 e^{-\alpha\lambda^2 t} \tag{6-123}$$

while the space-variable terms produce the following ODE:

$$\frac{d^2X}{dx^2} + \lambda^2 X = 0 \quad \text{in} \quad -\infty < x < \infty \tag{6-124}$$

Two linearly independent solutions of equation (6-124) are $\cos\lambda x$ and $\sin\lambda x$, corresponding to each value of λ. As negative values of λ generate no additional solutions, we consider only $\lambda \geq 0$. The general solution of the heat conduction problem is constructed by the superposition of $X(x)\Gamma(t)$ in the form

$$T(x,t) = \int_{\lambda=0}^{\infty} [a(\lambda)\cos\lambda x + b(\lambda)\sin\lambda x]\, e^{-\alpha\lambda^2 t}\, d\lambda \tag{6-125}$$

where we have integrated over all positive values of λ. We now consider the initial condition given by equation (6-120), which yields

$$F(x) = \int_{\lambda=0}^{\infty} [a(\lambda)\cos\lambda x + b(\lambda)\sin\lambda x]\, d\lambda \tag{6-126}$$

Equation (6-126) is recognized as a *general Fourier integral* expansion of the function $F(x)$, as developed in Chapter 2 with equations (2-117)–(2-119). Provided that $F(x)$ satisfies the necessary conditions as outlined in Chapter 2, we may define the Fourier coefficients as

$$a(\lambda) = \frac{1}{\pi} \int_{x'=-\infty}^{\infty} F(x') \cos \lambda x' \, dx' \tag{6-127a}$$

and

$$b(\lambda) = \frac{1}{\pi} \int_{x'=-\infty}^{\infty} F(x') \sin \lambda x' \, dx' \tag{6-127b}$$

Equations (6-127a) and (6-127b) may be substituted into equation (6-126), the trigonometric terms combined, and the order of integration may be changed. This process yields

$$F(x) = \int_{\lambda=0}^{\infty} \left[\frac{1}{\pi} \int_{x'=-\infty}^{\infty} F(x') \cos \lambda(x - x') dx' \right] d\lambda \tag{6-128}$$

A comparison of the results in equations (6-126) and (6-128) implies that the coefficients are given together by

$$a(\lambda) \cos \lambda x + b(\lambda) \sin \lambda x = \frac{1}{\pi} \int_{x'=-\infty}^{\infty} F(x') \cos \lambda(x - x') dx' \tag{6-129}$$

Using equation (6-129), the temperature solution of equation (6-125) becomes

$$T(x, t) = \frac{1}{\pi} \int_{\lambda=0}^{\infty} \left[e^{-\alpha\lambda^2 t} \int_{x'=-\infty}^{\infty} F(x') \cos \lambda(x - x') dx' \right] d\lambda \tag{6-130}$$

In view of the integral by Dwight [1, #861.20], namely,

$$\int_{\lambda=0}^{\infty} e^{-\alpha\lambda^2 t} \cos \lambda(x - x') d\lambda = \sqrt{\frac{\pi}{4\alpha t}} \exp\left[-\frac{(x - x')^2}{4\alpha t} \right] \tag{6-131}$$

the solution given by equation (6-130) may be written in the final form

$$T(x, t) = \frac{1}{(4\pi\alpha t)^{1/2}} \int_{x'=-\infty}^{\infty} F(x') \exp\left[-\frac{(x - x')^2}{4\alpha t} \right] dx' \tag{6-132}$$

Example 6-6 One-Dimensional Infinite Medium with Region of Finite Initial Temperature

We now consider a 1-D, infinite medium, $-\infty < x < \infty$, where the region $-L < x < L$ is initially at a constant temperature T_0, and everywhere outside of this region is at zero temperature. Obtain an expression for the temperature distribution $T(x,t)$ in the infinite medium for times $t > 0$. The mathematical formulation is given as

$$\frac{\partial^2 T(x,t)}{\partial x^2} = \frac{1}{\alpha}\frac{\partial T(x,t)}{\partial t} \quad \text{in} \quad -\infty < x < \infty, \quad t > 0 \tag{6-133}$$

$$\text{IC:} \qquad T(x,t=0) = F(x) \tag{6-134a}$$

with

$$F(x) = \begin{cases} T_0 & \text{in} \quad -L < x < L \\ 0 & \text{everywhere outside this region} \end{cases} \tag{6-134b}$$

The solution is identical to the solution developed above, namely, equation (6-132), where the functional form of $F(x)$ given by equation (6-134b) may be substituted to yield the following:

$$T(x,t) = \frac{T_0}{(4\pi\alpha t)^{1/2}} \int_{x'=-L}^{L} \exp\left[-\frac{(x-x')^2}{4\alpha t}\right] dx' \tag{6-135}$$

We may define a new variable as

$$\eta = \frac{x-x'}{\sqrt{4\alpha t}} \quad \rightarrow \quad dx' = -\sqrt{4\alpha t}\, d\eta \tag{6-136}$$

Using the new variable, equation (6-135) becomes

$$T(x,t) = \frac{T_0}{2}\left[\frac{2}{\sqrt{\pi}} \int_{\eta=0}^{(L+x)/\sqrt{4\alpha t}} e^{-\eta^2} d\eta + \frac{2}{\sqrt{\pi}} \int_{\eta=0}^{(L-x)/\sqrt{4\alpha t}} e^{-\eta^2} d\eta\right] \tag{6-137}$$

which may be written in the final form

$$T(x,t) = \frac{T_0}{2}\left[\text{erf}\left(\frac{L+x}{\sqrt{4\alpha t}}\right) + \text{erf}\left(\frac{L-x}{\sqrt{4\alpha t}}\right)\right] \quad \text{in} \quad -\infty < x < \infty \tag{6-138}$$

Example 6-7 Two-Dimensional Infinite Medium at Steady-State Conditions

We now consider a 2-D, infinite rectangular strip, $-\infty < x < \infty$, $0 \le y \le b$, at steady-state conditions. The boundary at y=0 is maintained at temperature T=$f(x)$, and the boundary at y=b is maintained at zero temperature, as shown in Figure 6-6. Obtain an expression for the steady-state temperature distribution $T(x,y)$.

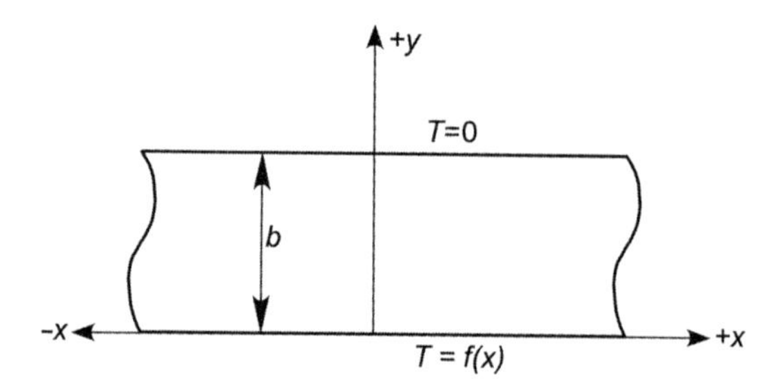

Figure 6-6 Problem description for Example 6-7.

The mathematical formulation is given as

$$\frac{\partial^2 T}{\partial x^2} + \frac{\partial^2 T}{\partial y^2} = 0 \quad \text{in} \quad -\infty < x < \infty, \quad 0 < y < b \tag{6-139}$$

$$\text{BC1:} \qquad T(y = 0) = f(x) \tag{6-140a}$$

$$\text{BC2:} \qquad T(y = b) = 0 \tag{6-140b}$$

Because the y dimension is nonhomogeneous, we will force a Fourier integral expansion in the infinite x dimension. With this in mind, we precede with a separation of the form

$$\frac{1}{Y}\frac{d^2Y}{dy^2} = \frac{-1}{X}\frac{d^2X}{dx^2} = \lambda^2 \tag{6-141}$$

The separated problems then become

$$\frac{d^2X}{dx^2} + \lambda^2 X = 0 \quad \text{in} \quad -\infty < x < \infty \tag{6-142}$$

and

$$\frac{d^2Y}{dy^2} - \lambda^2 Y = 0 \quad \text{in} \quad 0 < y < b \tag{6-143}$$

$$\text{BC2:} \qquad Y(y = b) = 0 \tag{6-144}$$

where we have introduced the separation constant λ. The solution of equation (6-142) follows the comments regarding equation (6-124), namely,

$$X(x) = C_1 \cos \lambda x + C_2 \sin \lambda x \tag{6-145}$$

where we retain both the $\cos \lambda x$ and $\sin \lambda x$ terms along with all values of λ. The solution of equation (6-143) yields

$$Y(y) = C_3 \cosh \lambda y + C_4 \sinh \lambda y \tag{6-146}$$

with BC2 yielding

$$C_4 = -C_3 \frac{\cosh \lambda b}{\sinh \lambda b} \tag{6-147}$$

Substitution of equation (6-147) into equation (6-146), along with the introduction of a hyperbolic-trigonometric substitution, yields

$$Y(y) = C_5 \sinh \lambda\,(b-y) \tag{6-148}$$

with the introduction of a new constant C_5. The complete solution is now constructed as the product of equations (6-145) and (6-148), followed by integration over all values of λ, giving

$$T(x,y) = \int_{\lambda=0}^{\infty} \sinh \lambda(b-y)[a(\lambda)\cos \lambda x + b(\lambda)\sin \lambda x]d\lambda \tag{6-149}$$

Application of the nonhomogeneous boundary condition of equation (6-140a) yields

$$f(x) = \int_{\lambda=0}^{\infty} \sinh \lambda b\,[a(\lambda)\cos \lambda x + b(\lambda)\sin \lambda x]\,d\lambda \quad \text{in} \quad -\infty < x < \infty \tag{6-150}$$

Equation (6-150) is a Fourier integral representation of $f(x)$ over the interval $-\infty < x < \infty$ in a form similar to that given by equation (6-126). The coefficients of equation (6-150) are similarly given by making use of equation (6-129), yielding

$$\sinh \lambda b\,[a(\lambda)\cos \lambda x + b(\lambda)\sin \lambda x] = \frac{1}{\pi}\int_{x'=-\infty}^{\infty} f(x')\cos \lambda(x-x')dx' \tag{6-151}$$

The substitution of these coefficients into equation (6-149) yields

$$T(x,y) = \frac{1}{\pi}\int_{\lambda=0}^{\infty}\left[\frac{\sinh \lambda(b-y)}{\sinh \lambda b}\int_{x'=-\infty}^{\infty} f(x')\cos \lambda(x-x')\,dx'\right]d\lambda \tag{6-152}$$

or changing the order of integration, we obtain

$$T(x,y) = \frac{1}{\pi}\int_{x'=-\infty}^{\infty}\left[f(x')\int_{\lambda=0}^{\infty}\frac{\sinh \lambda(b-y)}{\sinh \lambda b}\cos \lambda(x-x')\,d\lambda\right]dx' \tag{6-153}$$

The integral with respect to λ is available in the integral tables of Dwight [1, #862.41]. The temperature solution then becomes

$$T(x,y) = \frac{1}{2b}\sin\frac{y\pi}{b}\int_{x'=-\infty}^{\infty}\frac{f(x')}{\cos[\pi(b-y)/b] + \cosh[\pi(x-x')/b]}dx' \tag{6-154}$$

6-4 ONE-DIMENSIONAL HOMOGENEOUS PROBLEMS IN A SEMI-INFINITE MEDIUM FOR THE CYLINDRICAL COORDINATE SYSTEM

We now consider the solution of homogeneous, 1-D heat conduction problems for a semi-infinite region in the cylindrical coordinate system. That is, a semi-infinite region, $0 \leq r < \infty$ or $a \leq r < \infty$, that is initially at a temperature $F(r)$. Here we will make use of the Fourier–Bessel integrals that we developed in Chapter 2 for expansion of the initial conditions.

Example 6-8 One-Dimensional Problem for Semi-infinite Solid Cylinder Consider a 1-D, semi-infinite cylindrical region $0 \leq r < \infty$ that is initially at the temperature $T = F(r)$. Obtain an expression for the temperature $T(r,t)$ in the region for times $t > 0$.

The mathematical formulation of this problem is given as

$$\frac{\partial^2 T}{\partial r^2} + \frac{1}{r}\frac{\partial T}{\partial r} = \frac{1}{\alpha}\frac{\partial T}{\partial t} \quad \text{in} \quad 0 < r < \infty, \quad t > 0 \tag{6-155}$$

$$\text{BC1:} \qquad T(r \to 0) \Rightarrow \text{finite} \tag{6-156a}$$

$$\text{IC:} \qquad T(r, t = 0) = F(r) \tag{6-156b}$$

With the boundary condition BC1 considered homogeneous, we assume a separation of the form

$$T(r, t) = R(r)\Gamma(t) \tag{6-157}$$

Substitution of equation (6-157) into the PDE of equation (6-155) yields

$$\frac{1}{R}\left(\frac{d^2R}{dr^2} + \frac{1}{r}\frac{dR}{dr}\right) = \frac{1}{\alpha\Gamma}\frac{d\Gamma}{dt} = -\lambda^2 \tag{6-158}$$

The solution of the separated ODE in the t dimension yields the desired form

$$\Gamma(t) = C_1 e^{-\alpha\lambda^2 t} \tag{6-159}$$

while the r terms yield Bessel's equation of order zero, with the solution

$$R(r) = C_2 J_0(\lambda r) + C_3 Y_0(\lambda r) \tag{6-160}$$

The requirement for finiteness as $r \to 0$ stated by BC1 eliminates the $Y_0(\lambda r)$ term. Because this is not a true boundary value problem over the semi-infinite domain, we retain all positive values of λ and form the general solution by integrating over all values of λ, yielding

$$T(r, t) = \int_{\lambda=0}^{\infty} c(\lambda) J_0(\lambda r) e^{-\alpha\lambda^2 t} d\lambda \tag{6-161}$$

where we have introduced the new constant $c(\lambda) = C_1 C_2$. The application of the initial condition equation (6-156b) yields

$$F(r) = \int_{\lambda=0}^{\infty} c(\lambda) J_0(\lambda r) d\lambda \quad \text{in} \quad 0 < r < \infty \tag{6-162}$$

This is recognized as a Fourier–Bessel integral expansion of $F(r)$ as we developed in Chapter 2, equations (2-120)–(2-122). If the necessary conditions for $F(r)$ as developed in Chapter 2 are satisfied, the coefficients become

$$c(\lambda) = \lambda \int_{r'=0}^{\infty} r' J_0(\lambda r') F(r')\, dr' \tag{6-163}$$

The substitution of equation (6-163) into equation (6-161) yields the solution

$$T(r,t) = \int_{\lambda=0}^{\infty} \left[\lambda J_0(\lambda r) e^{-\alpha \lambda^2 t} \int_{r'=0}^{\infty} r' J_0(\lambda r') F(r')\, dr' \right] d\lambda \tag{6-164}$$

By changing the order of integration and making use of the following integral [Appendix IV, equation 24]

$$\int_{\lambda=0}^{\infty} \lambda J_0(\lambda r) J_0(\lambda r') e^{-\alpha \lambda^2 t}\, d\lambda = \frac{1}{2\alpha t} \exp\left(-\frac{r^2 + r'^2}{4\alpha t}\right) I_0\left(\frac{rr'}{2\alpha t}\right) \tag{6-165}$$

the solution given by equation (6-164) becomes

$$T(r,t) = \frac{1}{2\alpha t} \int_{r'=0}^{\infty} r' \exp\left(-\frac{r^2 + r'^2}{4\alpha t}\right) F(r') I_0\left(\frac{rr'}{2\alpha t}\right) dr' \tag{6-166}$$

We now consider the special case of $F(r)$ defined as

$$F(r) = \begin{cases} T_0, \text{ constant} & \text{for} \quad 0 < r < b \\ 0 & \text{for} \quad r > b \end{cases} \tag{6-167}$$

The substitution of equation (6-167) into equation (6-166) yields

$$\frac{T(r,t)}{T_0} = \frac{1}{2\alpha t} \exp\left(-\frac{r^2}{4\alpha t}\right) \int_{r'=0}^{b} r' \exp\left(-\frac{r'^2}{4\alpha t}\right) I_0\left(\frac{rr'}{2\alpha t}\right) dr' \equiv P \tag{6-168}$$

The above result is called a *P function*, which has been numerically evaluated, with the results previously tabulated [3].

Example 6-9 One-Dimensional Problem for Semi-infinite Hollow Cylinder Consider a 1-D, semi-infinite hollow cylindrical region $a \leq r < \infty$ that is initially at the temperature $T = F(r)$. For times $t > 0$, the boundary surface at $r = a$ is kept at zero temperature. Obtain an expression for the temperature $T(r,t)$ in the region for times $t > 0$.

The mathematical formulation of this problem is given as

$$\frac{\partial^2 T}{\partial r^2} + \frac{1}{r}\frac{\partial T}{\partial r} = \frac{1}{\alpha}\frac{\partial T}{\partial t} \quad \text{in} \quad a < r < \infty, \quad t > 0 \tag{6-169}$$

$$\text{BC1:} \qquad T(r = a) = 0 \tag{6-170a}$$

$$\text{IC:} \qquad T(r, t = 0) = F(r) \tag{6-170b}$$

With the boundary condition BC1 being homogeneous, we assume a separation of the form

$$T(r, t) = R(r)\Gamma(t) \tag{6-171}$$

Substitution of equation (6-171) into the PDE of equation (6-169) yields

$$\frac{1}{R}\left(\frac{d^2R}{dr^2} + \frac{1}{r}\frac{dR}{dr}\right) = \frac{1}{\alpha\Gamma}\frac{d\Gamma}{dt} = -\lambda^2 \tag{6-172}$$

As in the above example, solution of separated ODE in the t dimension yields the solution

$$\Gamma(t) = C_1 e^{-\alpha\lambda^2 t} \tag{6-173}$$

while the r terms yield Bessel's equation of order zero, with the solution

$$R(r) = C_2 J_0(\lambda r) + C_3 Y_0(\lambda r) \quad \text{in} \quad a < r < \infty \tag{6-174}$$

We explored the problem given by equation (6-174) in detail in Chapter 2 for the three relevant boundary conditions, namely, homogeneous boundaries of the first, second, or third type. The three cases are summarized in Table 2-4, with case 3 corresponding to the current type 1 boundary condition given by equation (6-170a). Eliminating one of the constants in equation (6-174) and rearranging yields the result given by case 3 of Table 2-4, namely,

$$R(r) = C_4\left[J_0(\lambda r)Y_0(\lambda a) - Y_0(\lambda r)J_0(\lambda a)\right] \tag{6-175}$$

We now form the complete solution by taking the product of the separated functions, equations (6-173) and (6-175), and integrating over all positive values of λ, which gives

$$T(r, t) = \int_{\lambda=0}^{\infty} c(\lambda)\left[J_0(\lambda r)Y_0(\lambda a) - Y_0(\lambda r)J_0(\lambda a)\right]e^{-\alpha\lambda^2 t}d\lambda \tag{6-176}$$

The initial condition equation (6-170b) is now applied to yield

$$F(r) = \int_{\lambda=0}^{\infty} c(\lambda)\left[J_0(\lambda r)Y_0(\lambda a) - Y_0(\lambda r)J_0(\lambda a)\right]d\lambda \tag{6-177}$$

which is recognized as a Fourier–Bessel integral expansion of $F(r)$. The coefficients are given by equation (2-125), which yields

$$c(\lambda) = \frac{1}{N(\lambda)}\lambda \int_{r'=a}^{\infty} r' \left[J_0(\lambda r')Y_0(\lambda a) - Y_0(\lambda r')J_0(\lambda a)\right] F(r')dr' \tag{6-178}$$

with the norm $N(\lambda)$ given by

$$N(\lambda) = J_0^2(\lambda a) + Y_0^2(\lambda a) \tag{6-179}$$

Substitution of equations (6-178) and (6-179) into equation (6-176) yields

$$T(r,t) = \int_{\lambda=0}^{\infty} \left\{ \frac{\lambda\left[J_0(\lambda r)Y_0(\lambda a) - Y_0(\lambda r)J_0(\lambda a)\right]}{J_0^2(\lambda a) + Y_0^2(\lambda a)} e^{-\alpha\lambda^2 t} \right.$$
$$\left. \times \int_{r'=a}^{\infty} r' \left[J_0(\lambda r')Y_0(\lambda a) - Y_0(\lambda r')J_0(\lambda a)\right] F(r')\, dr' \right\} d\lambda \tag{6-180}$$

Example 6-10 One-Dimensional Problem for Semi-infinite Hollow Cylinder

Consider a 1-D, semi-infinite hollow cylindrical region $a \le r < \infty$ that is initially at the temperature $T = F(r)$. For times $t > 0$, the boundary surface at $r = a$ dissipates heat by convection with convection coefficient h into a medium of zero temperature. Obtain an expression for the temperature $T(r,t)$ in the region for times $t > 0$.

The mathematical formulation of this problem is given as

$$\frac{\partial^2 T}{\partial r^2} + \frac{1}{r}\frac{\partial T}{\partial r} = \frac{1}{\alpha}\frac{\partial T}{\partial t} \quad \text{in} \quad a < r < \infty, \quad t > 0 \tag{6-181}$$

$$\text{BC1:} \quad -\left.\frac{\partial T}{\partial r}\right|_{r=a} + \left.HT\right|_{r=a} = 0 \quad \text{with} \quad H = \frac{h}{k} \tag{6-182a}$$

$$\text{IC:} \quad T(r, t=0) = F(r) \tag{6-182b}$$

With the boundary condition BC1 being homogeneous, we assume a separation of the form

$$T(r,t) = R(r)\Gamma(t) \tag{6-183}$$

Substitution of equation (6-183) into the PDE of equation (6-181) yields

$$\frac{1}{R}\left(\frac{d^2R}{dr^2} + \frac{1}{r}\frac{dR}{dr}\right) = \frac{1}{\alpha\Gamma}\frac{d\Gamma}{dt} = -\lambda^2 \tag{6-184}$$

As in Example 6-9, the solution of separated ODE in the t dimension yields

$$\Gamma(t) = C_1 e^{-\alpha\lambda^2 t} \tag{6-185}$$

while the r terms yield Bessel's equation of order zero, with the solution

$$R(r) = C_2 J_0(\lambda r) + C_3 Y_0(\lambda r) \quad \text{in} \quad a < r < \infty \tag{6-186}$$

Equation (6-186) along with boundary condition equation (6-182a) corresponds to case 1 in Table 2-4, which gives

$$R(r) = C_4 \left\{ J_0(\lambda r)[\lambda Y_1(\lambda a) + HY_0(\lambda a)] - Y_0(\lambda r)[\lambda J_1(\lambda a) + HJ_0(\lambda a)] \right\} \tag{6-187}$$

We now form the complete solution by taking the product of the separated functions, equations (6-185) and (6-187), and integrating over all positive values of λ, which gives

$$T(r,t) = \int_{\lambda=0}^{\infty} c(\lambda) R(r,\lambda) e^{-\alpha\lambda^2 t} d\lambda \tag{6-188}$$

with $R(r,\lambda)$ as defined by equation (6-187). The initial condition equation (6-182b) is now applied to yield

$$F(r) = \int_{\lambda=0}^{\infty} c(\lambda) R(r,\lambda) d\lambda \tag{6-189}$$

which is again recognized as a Fourier–Bessel integral expansion of $F(r)$. The coefficients are given by equation (2-125), which yields

$$c(\lambda) = \frac{1}{N(\lambda)} \lambda \int_{r'=a}^{\infty} r' R(r',\lambda) F(r')\, dr' \tag{6-190}$$

with the norm $N(\lambda)$ given by case 1 of Table 2-4, namely,

$$N(\lambda) = [\lambda J_1(\lambda a) + HJ_0(\lambda a)]^2 + [\lambda Y_1(\lambda a) + HY_0(\lambda a)]^2 \tag{6-191}$$

Substitution of equations (6-190) and (6-191) into equation (6-188) yields

$$T(r,t) = \int_{\lambda=0}^{\infty} \left(\frac{\lambda \left\{ J_0(\lambda r)[\lambda Y_1(\lambda a) + HY_0(\lambda a)] - Y_0(\lambda r)[\lambda J_1(\lambda a) + HJ_0(\lambda a)] \right\}}{[\lambda J_1(\lambda a) + HJ_0(\lambda a)]^2 + [\lambda Y_1(\lambda a) + HY_0(\lambda a)]^2} \right.$$
$$\left. \times\, e^{-\alpha\lambda^2 t} \int_{r'=a}^{\infty} r' \left\{ \begin{array}{l} J_0(\lambda r')[\lambda Y_1(\lambda a) + HY_0(\lambda a)] \\ -Y_0(\lambda r')[\lambda J_1(\lambda a) + HJ_0(\lambda a)] \end{array} \right\} F(r')\, dr' \right) d\lambda \tag{6-192}$$

6-5 TWO-DIMENSIONAL HOMOGENEOUS PROBLEMS IN A SEMI-INFINITE MEDIUM FOR THE CYLINDRICAL COORDINATE SYSTEM

We now consider the solution of a homogeneous, 2-D heat conduction problem for a semi-infinite domain in the z dimension in the cylindrical coordinate system. That is, a semi-infinite region, $0 \le r \le b$ and $0 \le z < \infty$, that is initially at a temperature $F(r,z)$. Here we expect a true boundary value problem in the finite r dimension and a corresponding Fourier series expansion, while we expect a Fourier integral expansion in the semi-infinite z dimension.

Example 6-11 Two-Dimensional Problem for Semi-infinite Solid Cylinder Consider a 2-D, semi-infinite cylindrical region $0 \le r \le b$ and $0 \le z < \infty$ that is initially at the temperature $T{=}F(r,z)$. For times $t > 0$, the boundaries at $r{=}b$ and $z{=}0$ are maintained at zero temperature, as illustrated in Figure 6-7. Obtain an expression for the temperature $T(r,z,t)$ in the region for times $t > 0$.

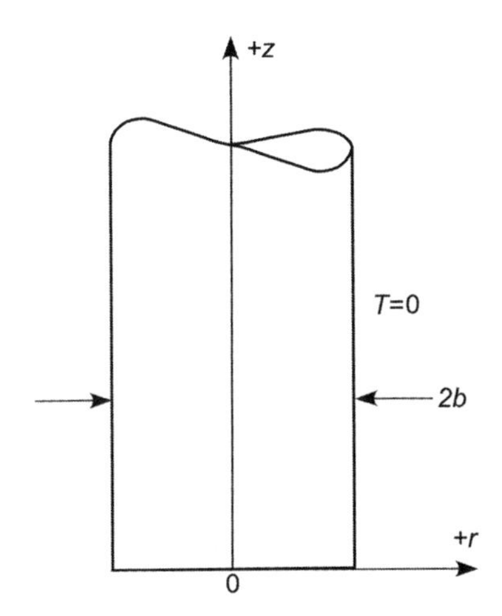

Figure 6-7 Problem description for Example 6-11.

The mathematical formulation of this problem is given as

$$\frac{\partial^2 T}{\partial r^2} + \frac{1}{r}\frac{\partial T}{\partial r} + \frac{\partial^2 T}{\partial z^2} = \frac{1}{\alpha}\frac{\partial T}{\partial t} \quad \text{in} \quad 0 \le r < b, \quad 0 < z < \infty, \quad t > 0 \tag{6-193}$$

BC1: $T(r \to 0) \Rightarrow \text{finite}$ (6-194a)

BC2: $T(r = b) = 0$ (6-194b)

BC3: $T(z = 0) = 0$ (6-194c)

IC: $T(r, z, t = 0) = F(r, z)$ (6-194d)

With the boundary conditions BC1–BC3 considered homogeneous, we assume a separation of the form

$$T(r, z, t) = R(r)Z(z)\Gamma(t) \tag{6-195}$$

which upon substitution into the PDE of equation (6-193) yields

$$\frac{1}{R}\left(\frac{d^2R}{dr^2} + \frac{1}{r}\frac{dR}{dr}\right) + \frac{1}{Z}\frac{d^2Z}{dz^2} = \frac{1}{\alpha\Gamma}\frac{d\Gamma}{dt} = -\lambda^2 \tag{6-196}$$

Solution of separated ODE in the t dimension yields the expected solution

$$\Gamma(t) = C_1 e^{-\alpha\lambda^2 t} \tag{6-197}$$

while the r terms yield Bessel's equation of order zero, with the solution

$$R(r) = C_2 J_0(\beta r) + C_3 Y_0(\beta r) \tag{6-198}$$

and the z terms yield the solution

$$Z(z) = C_4 \cos \eta y + C_5 \sin \eta y \tag{6-199}$$

with

$$\lambda^2 = \beta^2 + \eta^2 \tag{6-200}$$

The r dimension is a true boundary value problem, with the requirement for finiteness at the origin, equation (6-194a), eliminating the $Y_0(\beta r)$ term, while BC2 yields the eigenvalues β_n as follows:

$$J_0(\beta b) = 0 \quad \rightarrow \quad \beta_n \quad \text{for} \quad n = 1, 2, 3, \ldots \tag{6-201}$$

The homogeneous boundary condition BC3 dictates that $C_4 = 0$ in equation (6-199), with all positive values of η then retained given the semi-infinite domain. The general solution is then obtained forming a product solution and then summing over all eigenvalues β_n and integrating over all values of η. This gives

$$T(r, z, t) = \sum_{n=1}^{\infty} \int_{\eta=0}^{\infty} c_n(\eta) J_0(\beta_n r) \sin \eta z \, e^{-\alpha(\beta_n^2+\eta^2)t} \, d\eta \tag{6-202}$$

The application of the initial condition equation (6-194d) yields

$$F(r, z) = \sum_{n=1}^{\infty} \int_{\eta=0}^{\infty} c_n(\eta) J_0(\beta_n r) \sin \eta z \, d\eta \tag{6-203}$$

which is a Fourier series, Fourier integral expansion of the initial condition. To determine the unknown coefficients, we first use the operator

$$* \int_{r=0}^{b} r J_0(\beta_k r)\, dr$$

and utilize the property of orthogonality in the r dimension. We obtain

$$f^*(z) = \int_{\eta=0}^{\infty} c_n(\eta) \sin \eta z\, d\eta \quad \text{in} \quad 0 < z < \infty \tag{6-204}$$

where

$$f^*(z) \equiv \frac{1}{N(\beta_n)} \int_{r=0}^{b} r J_0(\beta_n r) F(r, z)\, dr \tag{6-205}$$

and where

$$N(\beta_n) \equiv \int_{r=0}^{b} r J_0^2(\beta r)\, dr = \frac{b^2 J_1^2(\beta b)}{2} \tag{6-206}$$

for the eigenvalues as defined by equation (6-201), per case 3 in Table 2-2. Equation (6-204) is now a Fourier integral representation of a prescribed function $f^*(z)$, defined on the interval $0 < z < \infty$, in terms of the function $\sin \eta z$. Such an expansion was developed in Chapter 2, equation (2-109a), and is summarized by case 3 in Table 6-1. This yields the following expression for $c_n(\eta)$:

$$c_n(\eta) = \frac{2}{\pi} \int_{z'=0}^{\infty} f^*(z') \sin \eta z'\, dz' \tag{6-207}$$

Substitution of equation (6-205) into the above yields the final expression

$$c_n(\eta) = \frac{2}{\pi} \int_{z'=0}^{\infty} \left\{ \frac{2 \sin \eta z'}{b^2 J_1^2(\beta_n b)} \int_{r'=0}^{b} r' J_0(\beta_n r') F(r', z')\, dr' \right\} dz' \tag{6-208}$$

Substitution of equation (6-208) into equation (6-202), with rearrangement of the order of integration, yields the final temperature solution

$$T(r, z, t) = \frac{4}{\pi b^2} \sum_{n=1}^{\infty} \left\{ \frac{J_0(\beta_n r)}{J_1^2(\beta_n b)} e^{-\alpha \beta_n^2 t} \int_{r'=0}^{b} \int_{z'=0}^{\infty} \left[r' J_0(\beta_n r') F(r', z') \right.\right.$$
$$\left.\left. \times \int_{\eta=0}^{\infty} e^{-\alpha \eta^2 t} \sin \eta z \sin \eta z'\, d\eta \right] dz'\, dr' \right\} \tag{6-209}$$

The last integral over η was evaluated previously, equation (6-34), to give

$$\frac{2}{\pi} \int_{\eta=0}^{\infty} e^{-\alpha \eta^2 t} \sin \eta z \sin \eta z'\, d\eta$$
$$= \frac{1}{(4\pi \alpha t)^{1/2}} \left\{ \exp\left[-\frac{(z - z')^2}{4\alpha t} \right] - \exp\left[-\frac{(z + z')^2}{4\alpha t} \right] \right\} \tag{6-210}$$

which may be substituted into equation (6-209) to yield

$$T(r,z,t) = \frac{1}{b^2(\pi\alpha t)^{1/2}} \sum_{n=1}^{\infty} \left(\frac{J_0(\beta_n r)}{J_1^2(\beta_n b)} e^{-\alpha\beta_n^2 t} \int_{r'=0}^{b} \int_{z'=0}^{\infty} \left\{ r' J_0(\beta_n r') F(r', z') \times \exp\left[-\frac{(z-z')^2}{4\alpha t} \right] - \exp\left[-\frac{(z+z')^2}{4\alpha t} \right] \right\} dz'\, dr' \right) \tag{6-211}$$

6-6 ONE-DIMENSIONAL HOMOGENEOUS PROBLEMS IN A SEMI-INFINITE MEDIUM FOR THE SPHERICAL COORDINATE SYSTEM

We finally consider the solution of a homogeneous, 1-D heat conduction problem for a semi-infinite domain in the spherical coordinate system, namely, over the domains $0 \le r < \infty$ and $b \le r < \infty$. Both domains will involve the transformation of the form $U(r,t) = rT(r,t)$ as we developed in Chapter 5, essentially converting the problems to those of the Cartesian coordinate system.

Example 6-12 One-Dimensional Problem for Semi-infinite Solid Sphere Consider a 1-D, solid spherical region $0 \le r < \infty$ that is initially at the temperature $T = F(r)$. Obtain an expression for the temperature $T(r,t)$ in the region for times $t > 0$.

The mathematical formulation of this problem is given as

$$\frac{1}{r}\frac{\partial^2(rT)}{\partial r^2} = \frac{1}{\alpha}\frac{\partial T}{\partial t} \quad \text{in} \quad 0 < r < \infty, \quad t > 0 \tag{6-212}$$

$$\text{BC1:} \qquad T(r \to 0) \Rightarrow \text{finite} \tag{6-213a}$$

$$\text{IC:} \qquad T(r, t=0) = F(r) \tag{6-213b}$$

We first make the transformation $U(r,t) = rT(r,t)$, which yields the formulation

$$\frac{\partial^2 U(r,t)}{\partial r^2} = \frac{1}{\alpha}\frac{\partial U(r,t)}{\partial t} \quad \text{in} \quad 0 < r < \infty, \quad t > 0 \tag{6-214}$$

$$\text{BC1:} \qquad U(r=0) = 0 \tag{6-215a}$$

$$\text{IC:} \qquad U(r, t=0) = rF(r) \tag{6-215b}$$

The problem given by equations (6-214) and (6-215) is identical to the problem formulation of Example 6-1, with only a slight modification of the initial condition. Accordingly, a solution of the form $U(r,t) = R(r)\Gamma(t)$ leads to the overall solution of the form

$$U(r,t) = \int_{\lambda=0}^{\infty} c(\lambda) \sin \lambda r e^{-\alpha\lambda^2 t} d\lambda \tag{6-216}$$

Application of the initial condition yields the Fourier integral expansion of the initial temperature distribution of equation (6-215b), given by

$$rF(r) = \int_{\lambda=0}^{\infty} c(\lambda) \sin \lambda r d\lambda \tag{6-217}$$

Introducing the norm from Table 6-1, the solution of the coefficients yields

$$c(\lambda) = \frac{2}{\pi} \int_{r'=0}^{\infty} r' F(r') \sin \lambda r' \, dr' \tag{6-218}$$

Substitution of equation (6-218) into equation (6-216) yields the final expression

$$U(r,t) = \frac{2}{\pi} \int_{\lambda=0}^{\infty} \left[\sin \lambda r e^{-\alpha\lambda^2 t} \int_{r'=0}^{\infty} r' F(r') \sin \lambda r' \, dr' \right] d\lambda \tag{6-219}$$

We now shift back to $T(r,t)$, yielding the overall solution

$$T(r,t) = \frac{2}{\pi} \int_{\lambda=0}^{\infty} \left[\frac{\sin \lambda r}{r} e^{-\alpha\lambda^2 t} \int_{r'=0}^{\infty} r' F(r') \sin \lambda r' \, dr' \right] d\lambda \tag{6-220}$$

The integral substitutions developed with equations (6-31)–(6-34) also enable our solution to be expressed in the alternative form

$$T(r,t) = \frac{1}{r(4\pi\alpha t)^{1/2}} \int_{r'=0}^{\infty} r' F(r') \left\{ \exp\left[-\frac{(r-r')^2}{4\alpha t} \right] - \exp\left[-\frac{(r+r')^2}{4\alpha t} \right] \right\} dr' \tag{6-221}$$

Example 6-13 One-Dimensional Problem for Semi-infinite Hollow Sphere Consider a 1-D, hollow spherical region $b \le r < \infty$ that is initially at the temperature $T=F(r)$. For times $t > 0$, the surface at $r=b$ is perfectly insulated. Obtain an expression for the temperature $T(r,t)$ in the region for times $t > 0$.

The mathematical formulation of this problem is given as

$$\frac{1}{r} \frac{\partial^2 (rT)}{\partial r^2} = \frac{1}{\alpha} \frac{\partial T}{\partial t} \quad \text{in} \quad b < r < \infty, \quad t > 0 \tag{6-222}$$

$$\text{BC1:} \quad \left. \frac{\partial T}{\partial r} \right|_{r=b} = 0 \tag{6-223a}$$

$$\text{IC:} \quad T(r, t=0) = F(r) \tag{6-223b}$$

We first make the transformation $U(r,t) = rT(r,t)$, which yields the formulation

$$\frac{\partial^2 U(r,t)}{\partial r^2} = \frac{1}{\alpha} \frac{\partial U(r,t)}{\partial t} \quad \text{in} \quad b < r < \infty, \quad t > 0 \tag{6-224}$$

$$\text{BC1:} \quad \left. \frac{\partial U}{\partial r} \right|_{r=b} - \frac{1}{b} U|_{r=b} = 0 \tag{6-225a}$$

$$\text{IC:} \quad U(r, t=0) = rF(r) \tag{6-225b}$$

where we have transformed BC1 using Table 5-2 for the case of an insulated (type 2) boundary condition. We now introduce a shift in the space coordinate given by

$$x = r - b \tag{6-226}$$

With this shift in the independent variable, the problem formulation becomes

$$\frac{\partial^2 U(x,t)}{\partial x^2} = \frac{1}{\alpha}\frac{\partial U(x,t)}{\partial t} \quad \text{in} \quad 0 < x < \infty, \quad t > 0 \tag{6-227}$$

$$\text{BC1:} \quad \left.\frac{\partial U}{\partial x}\right|_{x=0} - \frac{1}{b}U|_{x=0} = 0 \tag{6-228a}$$

$$\text{IC:} \quad U(x, t = 0) = (x + b)F(x + b) \tag{6-228b}$$

The problem defined by equations (6-227) and (6-228) was considered at the beginning of the chapter, with the solution given by equation (6-11). Our solution is readily formed with the modification of equation (6-11) by replacing H with $1/b$. The general solution for $U(x, t)$ is now given by

$$U(x,t) = \int_{\lambda=0}^{\infty} c(\lambda) e^{-\alpha\lambda^2 t}\left(\lambda \cos \lambda x + \frac{1}{b}\sin \lambda x\right) d\lambda \tag{6-229}$$

The application of the initial condition equation (6-228b) yields

$$(x + b)F(x + b) = \int_{\lambda=0}^{\infty} c(\lambda)\left(\lambda \cos \lambda x + \frac{1}{b}\sin \lambda x\right) d\lambda \tag{6-230}$$

As we developed in equation (6-13), we may express the coefficients $c(\lambda)$ by

$$c(\lambda) = \frac{1}{N(\lambda)}\int_{x'=0}^{\infty} (x' + b)F(x' + b)\left(\lambda \cos \lambda x' + \frac{1}{b}\sin \lambda x'\right) dx' \tag{6-231}$$

or

$$c(\lambda) = \frac{2}{\pi(\lambda^2 + 1/b^2)}\int_{x'=0}^{\infty} (x' + b)F(x' + b)\left(\lambda \cos \lambda x' + \frac{1}{b}\sin \lambda x'\right) dx' \tag{6-232}$$

Equations (6-229) and (6-232) may now be combined to produce the overall solution of the form

$$U(x,t) = \int_{\lambda=0}^{\infty} \left\{ \begin{array}{l} \dfrac{2\left[\lambda \cos \lambda x + (1/b)\sin \lambda x\right]}{\pi(\lambda^2 + 1/b^2)} e^{-\alpha\lambda^2 t} \\ \times \displaystyle\int_{x'=0}^{\infty} (x' + b)F(x' + b)\left[\lambda \cos \lambda x' + (1/b)\sin \lambda x'\right] dx' \end{array} \right\} d\lambda. \tag{6-233}$$

We now shift back to $T(r,t)$, yielding the overall solution

$$T(r,t)=\frac{2}{\pi r}\int_{\lambda=0}^{\infty}\left\{\begin{aligned}&\frac{\lambda\cos\lambda(r-b)+(1/b)\sin\lambda(r-b)}{\lambda^2+1/b^2}e^{-\alpha\lambda^2 t}\\&\times\int_{r'=b}^{\infty}r'F(r')\left[\lambda\cos\lambda\left(r'-b\right)+(1/b)\sin\lambda\left(r'-b\right)\right]dr'\end{aligned}\right\}d\lambda \tag{6-234}$$

REFERENCES

1. H. B. Dwight, *Tables of Integrals and Other Mathematical Data*, 4th ed., Macmillan, New York, 1961.
2. E. C. Titchmarsh, *Fourier Integrals*, 2nd ed., Clarendon, London, 1962.
3. J. I. Masters, *J. Chem. Phys.*, **23**, 1865–1874, 1955.

PROBLEMS

6-1 A semi-infinite medium, $0 \le x < \infty$, is initially at zero temperature. For times $t > 0$, the boundary surface at $x=0$ is kept at a constant temperature T_0. Obtain an expression for the temperature distribution $T(x, t)$ in the medium for times $t > 0$.

6-2 A semi-infinite medium, $0 \le x < \infty$, is initially at a uniform temperature T_0. For times $t > 0$, the boundary surface at $x=0$ dissipates heat by convection with coefficient h into a medium at fluid temperature zero. Obtain an expression for the temperature distribution $T(x, t)$ in the medium for times $t > 0$, and determine an expression for the heat flux at the surface $x=0$.

6-3 In a one-dimensional infinite medium, $-\infty < x < \infty$, initially, the region $a < x < b$ is at a constant temperature T_0, and everywhere outside this region is at zero temperature. Obtain an expression for the temperature distribution $T(x, t)$ in the medium for times $t > 0$.

6-4 Obtain an expression for the steady-state temperature distribution $T(x,y)$ in a semi-infinite strip $0 \le x \le a$, $0 \le y < \infty$, for the case when the boundary at $x=0$ is kept at a temperature $f(y)$, and the boundaries at $y=0$ and $x=a$ are kept at zero temperature.

6-5 Obtain an expression for the steady-state temperature distribution $T(x,y)$ in an infinite strip $0 \le x \le a$, $-\infty < y < \infty$, for the case when the boundary surface at $x=0$ is kept at a temperature $f(y)$ and the boundary surface at $x=a$ is kept at zero temperature.

6-6 Obtain an expression for the steady-state temperature distribution $T(x,y)$ in an infinite strip $0 \le y \le b$, $0 \le x < \infty$, for the case where the boundary at x=0 is kept at zero temperature, the boundary at y=b is insulated, and the boundary at y=0 is subjected to a surface heat flux at a rate of $f(x)$ (W/m^2).

6-7 A region $a \leq r < \infty$ in the cylindrical coordinate system is initially at a temperature $F(r)$. For times $t > 0$ the boundary at r=a is kept insulated. Obtain an expression for the temperature distribution $T(r, t)$ in the region for times $t > 0$.

6-8 A semi-infinite, solid cylinder, $0 \leq r \leq b$, $0 \leq z < \infty$, is initially at temperature $F(r, z)$. For times $t > 0$, the boundary at z=0 is kept insulated, and the boundary at r=b is dissipating heat by convection into a medium at zero temperature. Obtain an expression for the temperature distribution $T(r,z,t)$ in the solid for times $t > 0$.

6-9 A semi-infinite, hollow cylinder, $a \leq r \leq b$, $0 \leq z < \infty$, is initially at temperature $F(r, z)$. For times $t > 0$, the boundaries at z=0, r=a, and r=b are all kept at zero temperature. Obtain an expression for the temperature distribution $T(r, z, t)$ in the solid for times $t > 0$.

6-10 Obtain an expression for the steady-state temperature distribution $T(r, z)$ in a solid, semi-infinite cylinder $0 \leq r \leq b$, $0 \leq z < \infty$, when the boundary at r=b is kept at prescribed temperature $f(z)$ and the boundary at z=0 is kept at zero temperature.

6-11 Obtain an expression for the steady-state temperature, distribution $T(r, z)$ in a solid, semi-infinite cylinder $0 \leq r \leq b$, $0 \leq z < \infty$, when the boundary at z=0 is kept at temperature $f(r)$ and the boundary at r=b dissipates heat by convection into a medium at zero temperature.

6-12 Consider a region $b \leq r < \infty$ within a semi-infinite sphere (i.e., a region bounded internally by a sphere of radius r=b). Initially, the region is at a temperature $F(r)$. For times $t > 0$, the boundary surface at r=b is kept at zero temperature. Develop an expression for the temperature distribution $T(r, t)$ in the solid region for times $t > 0$.

6-13 Consider a region $b \leq r < \infty$ within a semi-infinite sphere (i.e., a region bounded internally by a sphere of radius r=b). Initially, the region is at a temperature of zero. For times $t > 0$, the boundary surface at r=b is kept at temperature of T_0. Develop an expression for the temperature distribution $T(r, t)$ in the solid region for times $t > 0$.

6-14 Consider a region $b \leq r < \infty$ within a semi-infinite sphere (i.e., a region bounded internally by a sphere of radius r=b). Initially, the region is at a temperature $F(r)$. For times $t > 0$, the boundary surface at r=b dissipates heat by convection with convection coefficient h and with a fluid at zero temperature. Develop an expression for the temperature distribution $T(r, t)$ in the solid region for times $t > 0$.

7

USE OF DUHAMEL'S THEOREM

So far we considered the solution of heat conduction problems with boundary conditions and internal energy generation that were not functions of time. However, there are many engineering problems such as heat transfer in internal combustion engine walls and space reentry problems in which the boundary condition functions are *time dependent*. In nuclear reactor fuel elements during power transients, the energy generation rate in the fuel elements varies with time. Duhamel's theorem, named after the French mathematician Jean-Marie Duhamel, who published his principal work on this topic in 1833 [1], provides a convenient approach for developing a solution to heat conduction problems with time-dependent boundary conditions and/or time-dependent energy generation, by utilizing the solution to the same problem, but with time-independent boundary conditions and/or time-independent energy generation. The method is applicable to linear problems because it is based on the superposition principle. A proof of Duhamel's theorem can be found in several references [2; 3, p. 162; 4, p. 30]. The proof given in reference 2 considers a general convection-type boundary condition from which the cases of prescribed heat flux and prescribed temperature boundary conditions are obtainable as special cases. Here we present a statement of Duhamel's theorem and then illustrate its application in the solution of heat conduction problems with time-dependent boundary condition functions and/or heat generation.

7-1 DEVELOPMENT OF DUHAMEL'S THEOREM FOR CONTINUOUS TIME-DEPENDENT BOUNDARY CONDITIONS

Consider a one-dimensional, heat conduction problem in a region R with one homogeneous boundary condition and one nonhomogenous, time-dependent boundary condition. In addition, the problem is *initially at zero temperature*, which is a necessary restriction for Duhamel's theorem as presented. The formulation for such a problem in Cartesian coordinates is given as

$$\frac{\partial^2 T(x,t)}{\partial x^2} = \frac{1}{\alpha}\frac{\partial T(x,t)}{\partial t} \quad \text{in} \quad 0 < x < L, \quad t > 0 \tag{7-1}$$

$$\text{BC1:} \quad T(x = 0, t) = f(t) \tag{7-2a}$$

$$\text{BC2:} \quad T(x = L, t) = 0 \tag{7-2b}$$

$$\text{IC:} \quad T(x, t = 0) = 0 \tag{7-2c}$$

The solution is not available to us with separation of variables because of the time-dependent, nonhomogenous boundary condition at $x = 0$. Let us consider an *auxiliary problem* where $f(t)$ is now a stepwise disturbance of unity. The formulation of this auxiliary problem, which we will denote as $\Phi(x, t)$, follows as

$$\frac{\partial^2 \Phi(x,t)}{\partial x^2} = \frac{1}{\alpha}\frac{\partial \Phi(x,t)}{\partial t} \quad \text{in} \quad 0 < x < L, \quad t > 0 \tag{7-3}$$

$$\text{BC1:} \quad \Phi(x = 0, t) = 1 \tag{7-4a}$$

$$\text{BC2:} \quad \Phi(x = L, t) = 0 \tag{7-4b}$$

$$\text{IC:} \quad \Phi(x, t = 0) = 0 \tag{7-4c}$$

This formulation corresponds to the *unit-step function* for our nonhomogeneity in the original problem, namely,

$$f(t) \to U(t) = \begin{cases} 0 & t < 0 \\ 1 & t \geq 0 \end{cases} \tag{7-5}$$

The problem of equations (7-3) and (7-4) is readily solved by separation of variables, although superposition must be used to shift the nonhomogeneity from the boundary condition to the initial condition. We note that the solution $\Phi(x, t)$ satisfies the PDE of our original problem, as well as the homogeneous boundary condition of equation (7-2b), and the initial condition of equation (7-2c). Only the time-dependent boundary condition of equation (7-2a) remains to be satisfied. We will now seek a solution of our original problem in terms of the auxiliary problem $\Phi(x, t)$ by approximating the solution at the time-dependent boundary. We illustrate this approach in Figure 7-1, which depicts the actual time-dependent boundary condition along with our approximation.

Our approximate solution is now formed in terms of $\Phi(x, t)$, giving

$$T(x,t) \cong f(0)\Phi(x,t) \quad \text{for} \quad 0 \leq t \leq \tau_1 \tag{7-6}$$

over the limited temporal domain $0 \leq t \leq \tau_1$. Because the *homogeneous* portions of the problem, namely, the PDE, BC2, and the IC, are still satisfied for

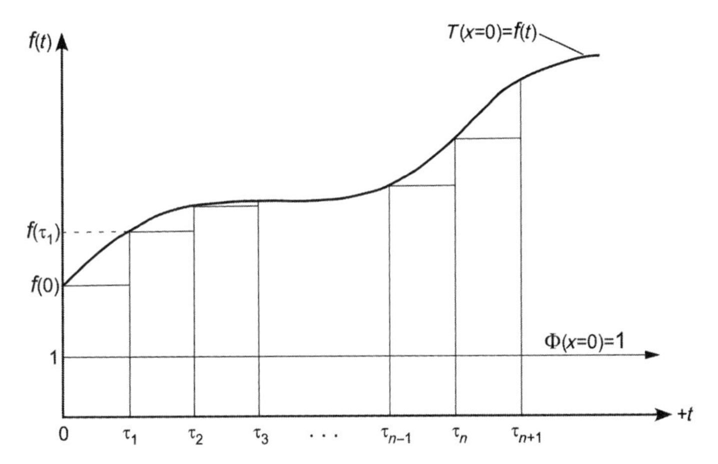

Figure 7-1 Time-dependent boundary condition at $x = 0$.

multiplication of $\Phi(x, t)$ by a constant, that is, multiplication by $f(0)$. Equation (7-6) represents an approximate solution that is exact at t=0 and, then, as time increases, progressively less accurate at the boundary x=0. This behavior is depicted in Figure 7-1 over the domain $0 \leq t \leq \tau_1$. Furthermore, our approach may be extended by adding additional increments, which we accomplish by shifting our unit-step solution. With this approach, our approximate solution becomes

$$T(x,t) \cong f(0)\Phi(x,t) + \left[f(\tau_1) - f(0)\right]\Phi(x, t-\tau_1) + \left[f(\tau_2) - f(\tau_1)\right] \times \Phi(x, t-\tau_2) + \cdots + \left[f(\tau_n) - f(\tau_{n-1})\right]\Phi(x, t-\tau_n) \tag{7-7}$$

Introducing the following notation,

$$\Delta f_m = f(\tau_m) - f(\tau_{m-1}) \tag{7-8}$$

$$\Delta \tau_m = \tau_m - \tau_{m-1} \tag{7-9}$$

our approximate solution of equation (7-7) becomes

$$T(x,t) \cong f(0)\Phi(x,t) + \sum_{m=1}^{N} \Phi(x, t-\tau_m)\left(\frac{\Delta f_m}{\Delta \tau_m}\right)\Delta \tau_m \tag{7-10}$$

We may now let our number of increments tend to infinity, $N \to \infty$, and therefore $\Delta\tau_m \to 0$, in which case the term in parenthesis becomes a derivative and the summation becomes an integral. This gives an exact solution to our original problem of the form

$$\boxed{T(x,t) = f(0)\Phi(x,t) + \int_{\tau=0}^{t} \Phi(x, t-\tau)\frac{df(\tau)}{d\tau}d\tau} \tag{7-11}$$

where equation (7-11) is *Duhamel's superposition integral*. In equation (7-11), τ is a variable of integration, noting that we have evaluated the time-dependent nonhomogeneity in terms of this variable τ. We also present an alternative form of equation (7-11) based on integration by parts, which yields

$$\boxed{T(x,t) = \int_{\tau=0}^{t} f(\tau)\frac{\partial \Phi(x, t-\tau)}{\partial t}d\tau} \tag{7-12}$$

after making use of the fact that $\Phi(x, t = 0) \equiv 0$ based on the original problem statement (i.e, our requirement of zero initial temperature), and using the relation

$$\frac{\partial \Phi(x, t-\tau)}{\partial t} = -\frac{\partial \Phi(x, t-\tau)}{\partial \tau} \tag{7-13}$$

7-2 TREATMENT OF DISCONTINUITIES

If the time-dependent boundary condition function $f(t)$ has discontinuities resulting from step changes in the applied surface temperature, heat flux, or ambient temperature for convection, then the integral appearing in Duhamel's superposition integral (7-11) needs to be broken into parts at the points of such discontinuities. Here we illustrate how to break the integral into parts at the points of discontinuities, and thereby develop an extension of Duhamel's theorem for discontinuous functions.

Consider, for example, the boundary condition function $f(t)$ that has N discontinues between t=0 and the time of interest t, as depicted in Figure 7-2.

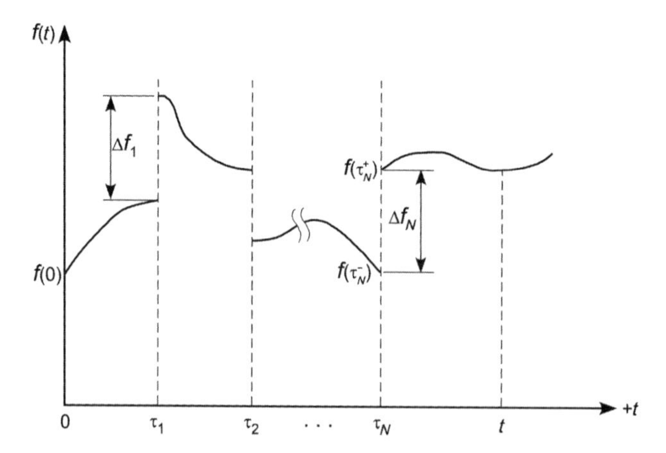

Figure 7-2 Boundary condition function $f(t)$ with discontinuities.

At each discontinuity, we will define the step change in the boundary condition, here temperature, using the notation

$$\Delta f_i = f^+(\tau_i) - f^-(\tau_i) \tag{7-14}$$

where f^- denotes the limiting value of $f(\tau_i)$ at the discontinuity as it is approached from the left, and f^+ denotes the limiting value of $f(\tau_i)$ as it is approached from the right. For example, as shown in Figure 7-2, we define $\Delta f_1 = f^+(\tau_1) - f^-(\tau_1)$, which would be a positive quantity here (i.e., positive step change), as shown. The temperature solution for time t beyond the Nth discontinuity is determined using a modified form of equation (7-11), namely,

$$\boxed{T(x,t) = \sum_{i=0}^{N} \Delta f_i \Phi(x, t-\tau_i) + \int_{\tau=0}^{t} \Phi(x, t-\tau) \frac{df(\tau)}{d\tau} d\tau} \tag{7-15}$$

where the integral is broken into parts at the discontinuities τ_1, τ_2, and so forth, as follows:

$$\begin{aligned} \int_{\tau=0}^{t} \Phi(x, t-\tau) \frac{df(\tau)}{d\tau} d\tau &= \int_{\tau=0}^{\tau_1} \Phi(x, t-\tau) \frac{df(\tau)}{d\tau} d\tau \\ &+ \int_{\tau=\tau_1}^{\tau_2} \Phi(x, t-\tau) \frac{df(\tau)}{d\tau} d\tau + \cdots + \int_{\tau=\tau_N}^{t} \Phi(x, t-\tau) \frac{df(\tau)}{d\tau} d\tau \end{aligned} \tag{7-16}$$

where the derivatives in each integral are evaluated using the functional form of $f(\tau)$ in each interval, and noting that the last integral ends at the independent variable time t. Equation (7-15) is the alternative form of Duhamel's theorem. This expression consists of the integral and summation terms and is called the *Stieltjes integral*. We note that for only a single discontinuity at the origin, $\Delta f_0 = f(0)$ and $\tau_0 = 0$, equation (7-15) reduces directly to equation (7-11). Overall in equation (7-15), the integral term is for the contribution of the continuous portion of the boundary condition function $f(t)$, and the summation term is for the contributions of the finite step changes Δf_i occurring in $f(t)$ at the discontinuities. Essentially, we have restarted Duhamel's integral solution at each discontinuity, as given by equation (7-11), by shifting our unit-step solution at each new discontinuity.

All Step Changes

We now consider a situation in which the boundary condition function $f(t)$ consists of a series of step changes Δf_i occurring at times $\tau_i = i\Delta t$ but has no continuous parts contributing to Duhamel's solution (i.e., zero slope), as illustrated in Figure 7-3.

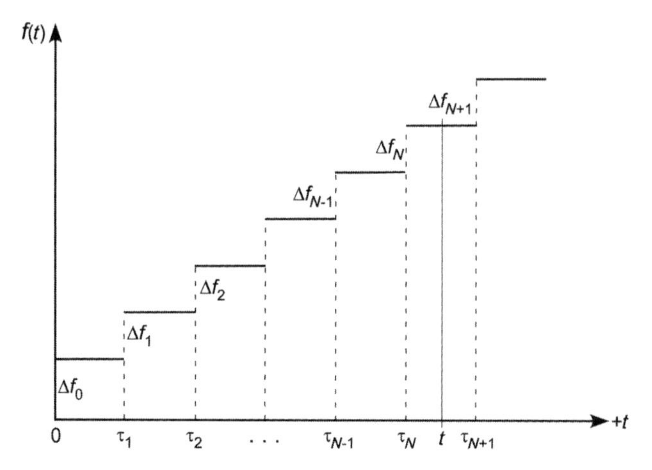

Figure 7-3 Stepwise varying boundary condition function $f(t)$.

For this specific case, the integral term of equation (7-15) drops out due to the derivatives being zero, and the solution (7-15) reduces to

$$T(x,t)=\sum_{i=0}^{N}\Phi(x,t-i\Delta t)\Delta f_i \tag{7-17}$$

where our independent variable time t is in the time interval $N\Delta t < t < (N+1)\ \Delta t$. This result can be written in a more general form as

$$T(x,t)=\sum_{i=0}^{\infty}\Phi(x,t-i\Delta t)\Delta f_i U(t-i\Delta t) \tag{7-18}$$

where

$$U(t-i\Delta t)\equiv \text{the unit-step function}$$

as defined by equation (7-5).

7-3 GENERAL STATEMENT OF DUHAMEL'S THEOREM

We present here a more general treatment of Duhamel's theorem, without the restriction of zero initial temperature. Consider the three-dimensional, nonhomogeneous heat conduction problem in a region R, with time-dependent boundary condition functions and heat generation given in the form

$$\nabla^2 T(\hat{r},t)+\frac{1}{k}g(\hat{r},t)=\frac{1}{\alpha}\frac{\partial T(\hat{r},t)}{\partial t} \qquad \text{in region } R, \qquad t>0 \tag{7-19}$$

$$k\frac{\partial T}{\partial n_i} + h_i T = f_i(\hat{r}, t) \qquad \text{on boundary } S_i, \quad t > 0 \tag{7-20a}$$

$$T(\hat{r}, t = 0) = F(\hat{r}) \qquad \text{in region } R \tag{7-20b}$$

where $\partial/\partial n_i$ is the derivative along the outward-drawn normal to the boundary surface S_i, $i = 1, 2, \ldots, N$, with N being the number of continuous boundary surfaces of the region R. Here k and h_i are the thermal conductivity and the convection coefficients at the ith surface, respectively, that are assumed to be constant. By setting $k = 0$ and $h_i = 1$, we obtain a boundary condition of the first type, and by setting $h_i = 0$, we obtain a boundary condition of the second type.

The problem given by equations (7-19) and (7-20) cannot be solved by the techniques described in the previous chapters because the nonhomogeneous terms $g(\hat{r}, t)$ and $f_i(\hat{r}, t)$ depend on time. Therefore, instead of solving this problem directly, we express its solution in terms of the solution of the simpler auxiliary problem as now defined. Let $\Phi(\hat{r}, t, \tau)$ be the solution of a modified version of the problem (7-19) and (7-20) on the assumption that the nonhomogeneous terms $g(\hat{r}, \tau)$ and $f_i(\hat{r}, \tau)$ do not depend on time; that is to say, the variable τ is merely a parameter, but not the independent variable time. Then, $\Phi(\hat{r}, t, \tau)$ is the solution of the following auxiliary problem:

$$\nabla^2\Phi(\hat{r}, t, \tau) + \frac{1}{k}g(\hat{r}, \tau) = \frac{1}{\alpha}\frac{\partial \Phi(\hat{r}, t, \tau)}{\partial t} \qquad \text{in region } R, \quad t > 0 \tag{7-21}$$

$$k\frac{\partial \Phi(\hat{r}, t, \tau)}{\partial n_i} + h_i \Phi(\hat{r}, t, \tau) = f_i(\hat{r}, \tau) \qquad \text{on boundary } S_i, \quad t > 0 \tag{7-22a}$$

$$\Phi(\hat{r}, t = 0, \tau) = F(\hat{r}) \qquad \text{in region } R \tag{7-22b}$$

where $\partial/\partial n_i$ and S_i are as defined previously, and the function $\Phi(\hat{r}, t, \tau)$ depends on the parameter τ because the functions $g(\hat{r}, \tau)$ and $f_i(\hat{r}, \tau)$ depend on τ.

The auxiliary problem of equations (7-21) and (7-22) can be solved with the techniques described in the previous chapters because $g(\hat{r}, \tau)$ and $f_i(\hat{r}, \tau)$ do not depend on time (i.e., τ is only a parameter). We now assume, therefore, that the solution $\Phi(\hat{r}, t, \tau)$ of the above auxiliary problem is available. Then, Duhamel's theorem relates the solution $\Phi(\hat{r}, t, \tau)$ of the auxiliary problem to the solution $T(\hat{r}, t)$ of the original problem by the following integral expression:

$$T(\hat{r}, t) = \frac{\partial}{\partial t}\int_{\tau=0}^{t} \Phi(\hat{r}, t - \tau, \tau)\, d\tau \tag{7-23}$$

This result can be expressed in an alternative form by performing the differentiation under the integral sign, which yields a more general expression of

Duhamel's theorem of the form

$$T(\hat{r}, t) = F(\hat{r}) + \int_{\tau=0}^{t} \frac{\partial \Phi(\hat{r}, t-\tau, \tau)}{\partial t} d\tau \tag{7-24}$$

since

$$\Phi(\hat{r}, t-\tau, \tau)\big|_{\tau=t} = \Phi(\hat{r}, 0, \tau) = F(\hat{r})$$

We now consider some special cases of equation (7-24).

1. ***Initial Temperature of Zero***. For this special case, we have $F(\hat{r}) = 0$ and equation (7-24) reduces to

$$T(\hat{r}, t) = \int_{\tau=0}^{t} \frac{\partial \Phi(\hat{r}, t-\tau, \tau)}{\partial t} d\tau \tag{7-25}$$

2. ***Initial Temperature of Zero, Problem Has Only One Nonhomogeneity***. The solid is initially at zero temperature and the problem involves only one nonhomogeneous term. Namely, if there is heat generation, all the boundary conditions for the problem are homogeneous; or, if there is no heat generation in the medium, only one of the boundary conditions is nonhomogeneous. For example, we consider a problem in which there is no heat generation, but one of the boundary conditions is nonhomogeneous, say, the one at the boundary surface S_1. The formulation is

$$\nabla^2 T(\hat{r}, t) = \frac{1}{\alpha} \frac{\partial T(\hat{r}, t)}{\partial t} \quad \text{in region } R, \quad t > 0 \tag{7-26}$$

$$k \frac{\partial T}{\partial n_i} + h_i T = \delta_{1i} f_i(t) \quad \text{on boundary } S_i, \quad t > 0 \tag{7-27a}$$

$$T(\hat{r}, t = 0) = 0 \quad \text{in region } R \tag{7-27b}$$

where $i = 1, 2, \ldots, N$, and δ_{1i} is the *Kronecker delta* defined as

$$\delta_{1i} = \begin{cases} 0 & i \neq 1 \\ 1 & i = 1 \end{cases}$$

The corresponding auxiliary problem is taken as

$$\nabla^2 \Phi(\hat{r}, t) = \frac{1}{\alpha} \frac{\partial \Phi(\hat{r}, t)}{\partial t} \quad \text{in region } R, \quad t > 0 \tag{7-28}$$

$$k \frac{\partial \Phi}{\partial n_i} + h_i \Phi = \delta_{1i} \quad \text{on boundary } S_i, \quad t > 0 \tag{7-29a}$$

$$\Phi(\hat{r}, t = 0) = 0 \quad \text{in region } R \tag{7-29b}$$

The solution $T(\hat{r}, t)$ of the problem in equations (7-26) and (7-27) is related to the solution $\Phi(\hat{r}, t)$ of the problem (7-28) and (7-29) by

$$T(\hat{r}, t) = \int_{\tau=0}^{t} f(\tau) \frac{\partial \Phi(\hat{r}, t-\tau)}{\partial t} d\tau \tag{7-30}$$

The validity of this result is apparent from the fact that if $\Phi(\hat{r}, t, \tau)$ is the solution of the problem in equations (7-28) and (7-29) for a boundary condition $\delta_{1i} f_i(\tau)$, then $\Phi(\hat{r}, t, \tau)$ is related to $\Phi(\hat{r}, t)$ by

$$\Phi(\hat{r}, t, \tau) = f(\tau)\Phi(\hat{r}, t) \tag{7-31}$$

When equation (7-31) is introduced into equation (7-25), the result of equation (7-30) is obtained, which is identical to equation (7-12) presented above for the one-dimensional case with zero initial temperature.

The physical significance of the function $\Phi(\hat{r}, t)$ governed by the auxiliary problem [(7-28) and (7-29)] is dependent on the type of boundary condition considered for the physical problem in equation (7-29a). If the boundary condition is of the *first type* [i.e., $T = \delta_{1i} f_i(t)$], then the boundary condition for the auxiliary problem is also of the first type (i.e., $\Phi = \delta_{1i}$). That is, the function $\Phi(\hat{r}, t)$ represents the response function for a solid initially at zero temperature and for times $t > 0$, with one of the boundary surfaces subjected to a unit-step change in the surface temperature. If the boundary condition for the physical problem is of the *second type* [i.e., prescribed heat flux, $k\partial T/\partial n_i = \delta_{1i} f_i(t)$], then the boundary condition for the auxiliary problem is also of the second type (i.e., $k\partial\Phi/\partial n_i = \delta_{1i}$). In this case, $\Phi(\hat{r}, t)$ represents the response function for a unit-step change in the applied surface heat flux at one of the boundary surfaces.

7-4 APPLICATIONS OF DUHAMEL'S THEOREM

We now illustrate with examples the application of Duhamel's theorem for the solution of heat conduction problems with time-dependent boundary condition functions in terms of the solution of the same problem for the time-independent, unit-step boundary condition function.

Example 7-1 One-Dimensional Rectangular Solid with Prescribed Temperature

A slab of thickness L is initially at zero temperature. For times $t > 0$, the boundary surface at $x = 0$ is kept at zero temperature, while the surface at $x = L$ is subjected to a time-varying temperature $f(t)$ defined by

$$f(t) = \begin{cases} bt & \text{for} \quad 0 < t < \tau_1 \\ 0 & \text{for} \quad t > \tau_1 \end{cases} \tag{7-32}$$

as illustrated in Figure 7-4. This is equivalent to heating the surface temperature with a programed linear rise and then rapidly (i.e., instantly) quenching back to zero.

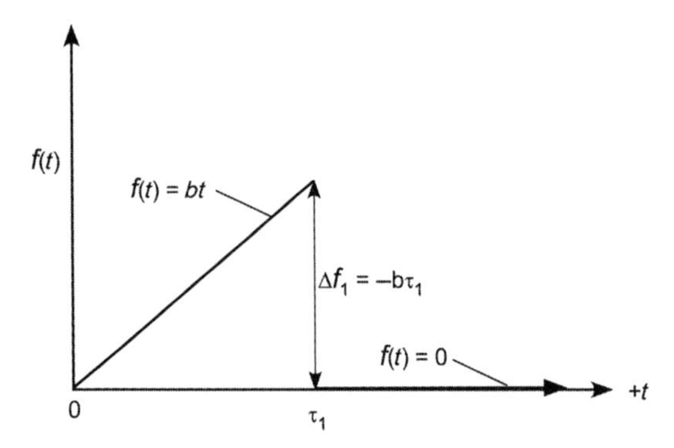

Figure 7-4 Boundary surface temperature for Example 7-1.

Using Duhamel's theorem, develop an expression for the temperature distribution $T(x,t)$ in the slab for times (i) $t < \tau_1$ and (ii) $t > \tau_1$. The mathematical formulation of this heat conduction problem is given by

$$\frac{\partial^2 T(x,t)}{\partial x^2} = \frac{1}{\alpha}\frac{\partial T(x,t)}{\partial t} \quad \text{in} \quad 0 < x < L, \quad t > 0 \tag{7-33}$$

$$\text{BC1:} \quad T(x=0,t) = 0 \tag{7-34a}$$

$$\text{BC2:} \quad T(x=L,t) = f(t) \tag{7-34b}$$

$$\text{IC:} \quad T(x,t=0) = 0 \tag{7-34c}$$

where $f(t)$ is defined by equation (7-32). The corresponding auxiliary problem becomes

$$\frac{\partial^2 \Phi(x,t)}{\partial x^2} = \frac{1}{\alpha}\frac{\partial \Phi(x,t)}{\partial t} \quad \text{in} \quad 0 < x < L, \quad t > 0 \tag{7-35}$$

$$\text{BC1:} \quad \Phi(x=0,t) = 0 \tag{7-36a}$$

$$\text{BC2:} \quad \Phi(x=L,t) = 1 \tag{7-36b}$$

$$\text{IC:} \quad \Phi(x,t=0) = 0 \tag{7-36c}$$

The solution for the auxiliary problem $\Phi(x,t)$ first requires that we split the problem because the nonhomogeneity is in the boundary condition rather than the initial condition, as required for the heat equation. This will always be the case with the unit-step boundary condition. We therefore split $\Phi(x,t)$ into two problems:

$$\Phi(x,t) = \Phi_H(x,t) + \Phi_{SS}(x) \tag{7-37}$$

where $\Phi_H(x,t)$ takes the homogeneous version of the boundary conditions, and $\Phi_{SS}(x)$ takes the nonhomogeneous boundary condition, as first presented with equation (3-129). The homogeneous problem is readily solved with separation of variables, and the steady-state ODE yields a linear temperature profile. Together, $\Phi(x, t)$ is determined as

$$\Phi(x, t) = \frac{x}{L} + \frac{2}{L}\sum_{m=1}^{\infty} \frac{(-1)^m}{\beta_m} \sin \beta_m x e^{-\alpha\beta_m^2 t} \tag{7-38}$$

where the eigenvalues are defined as $\beta_m = m\pi/L$.

The function $\Phi(x, t - \tau)$ is now obtained by replacing t with $t - \tau$ in equation (7-38),

$$\Phi(x, t-\tau) = \frac{x}{L} + \frac{2}{L}\sum_{m=1}^{\infty} \frac{(-1)^m}{\beta_m} \sin \beta_m x e^{-\alpha\beta_m^2 (t-\tau)} \tag{7-39}$$

We now apply Duhamel's theorem using equation (7-15) for the general case of discontinuities. The solutions for times $t < \tau_1$ and $t > \tau_1$ must be considered separately.

1. ***Times*** $t < \tau_1$. The boundary condition function $f(t)$ has no discontinuity in this domain; thus equation (7-15) is equivalent to equation (7-11), giving

$$T(x, t) = \cancel{f(0)}^{0}\, \Phi(x, t) + \int_{\tau=0}^{t} \Phi(x, t-\tau) \frac{df(\tau)}{d\tau} d\tau \tag{7-40}$$

where $df(\tau)/d\tau = b$ and $\Phi(x, t - \tau)$ is given by equation (7-39). Then equation (7-40) becomes

$$T(x, t) = \int_{\tau=0}^{t} \left\{ \frac{x}{L} + \frac{2}{L}\sum_{m=1}^{\infty} \frac{(-1)^m}{\beta_m} \sin \beta_m x e^{-\alpha\beta_m^2 (t-\tau)} \right\} (b)\, d\tau \tag{7-41}$$

which is simplified to the form

$$T(x, t) = b\frac{x}{L}t + \frac{2b}{L}\sum_{m=1}^{\infty} \frac{(-1)^m}{\beta_m} \sin \beta_m x e^{-\alpha\beta_m^2 t} \int_{\tau=0}^{t} e^{\alpha\beta_m^2 \tau} d\tau \tag{7-42}$$

The integration is performed to give

$$T(x, t) = b\frac{x}{L}t + \frac{2b}{L}\sum_{m=1}^{\infty} \frac{(-1)^m}{\alpha\beta_m^3} \left(1 - e^{-\alpha\beta_m^2 t}\right) \sin \beta_m x \tag{7-43}$$

over the time domain $t < \tau_1$.

2. ***Times*** $t > \tau_1$. The boundary surface function $f(t)$ has a discontinuity at time $t = \tau_1$, which is now within our domain, and equation (7-15) becomes

$$T(x,t) = \Phi(x,t)\cancel{f(0)}^0 + \Phi(x,t-\tau_1)\Delta f_1 + \int_{\tau=0}^{\tau_1} \Phi(x,t-\tau)\frac{df(\tau)}{d\tau}d\tau + \int_{\tau=\tau_1}^{t} \Phi(x,t-\tau)\cancel{\frac{df(\tau)}{d\tau}}^0 d\tau \tag{7-44}$$

where the resulting step change in $f(t)$ is a decrease in temperature, that is, $\Delta f_1 = 0 - b\tau_1 = -b\tau_1$ and where $df(\tau)/d\tau = b$ for the first integral and $df(\tau)/d\tau = 0$ for the second integral. Substituting these results into equation (7-44), we find

$$T(x,t) = -\Phi(x,t-\tau_1)(b\tau_1) + \int_{\tau=0}^{\tau_1} \Phi(x,t-\tau)(b)\,d\tau \tag{7-45}$$

where $\Phi(x,t-\tau_1)$ and $\Phi(x,t-\tau)$ are obtained from equation (7-39). With these substitutions, equation (7-45) takes the form

$$T(x,t) = -(b\tau_1)\left[\frac{x}{L} + \frac{2}{L}\sum_{m=1}^{\infty}\frac{(-1)^m}{\beta_m}\sin\beta_m x e^{-\alpha\beta_m^2(t-\tau_1)}\right] + \int_{\tau=0}^{\tau_1}\left\{\frac{x}{L} + \frac{2}{L}\sum_{m=1}^{\infty}\frac{(-1)^m}{\beta_m}\sin\beta_m x e^{-\alpha\beta_m^2(t-\tau)}\right\}(b)\,d\tau \tag{7-46}$$

Clearly, the integral is similar to the one given by equation (7-42), except the upper limit is τ_1; hence it can be performed readily. We do note, however, that since we canceled a t term in equation (7-43) following the integration, we cannot simply replace t with τ_1 in equation (7-43) to realize the second term of equation (7-46). Following proper integration and simplification, it is readily observed that as $t \to \infty$, the temperature approaches zero, as the influence of the initial temperature ramp diminishes to zero.

Example 7-2 Semi-infinite Solid with Prescribed Temperature
A semi-infinite solid, $0 \le x < \infty$, is initially at zero temperature. For times $t > 0$, the boundary surface at $x = 0$ is kept at a prescribed temperature $f(t)$. Obtain an expression for the temperature distribution $T(x,t)$ in the solid for times $t > 0$ assuming that $f(t)$ has no discontinuities. The mathematical formulation of this problem is given as

$$\frac{\partial^2 T(x,t)}{\partial x^2} = \frac{1}{\alpha}\frac{\partial T(x,t)}{\partial t} \quad \text{in} \quad 0 < x < \infty, \quad t > 0 \tag{7-47}$$

$$\text{BC1:} \quad T(x=0,t) = f(t) \tag{7-48a}$$

$$\text{BC2:} \quad T(x\to\infty,t) = 0 \tag{7-48b}$$

$$\text{IC:} \quad T(x,t=0) = 0 \tag{7-48c}$$

The unit-step auxiliary problem is taken as

$$\frac{\partial^2 \Phi(x,t)}{\partial x^2} = \frac{1}{\alpha}\frac{\partial \Phi(x,t)}{\partial t} \quad \text{in} \quad 0 < x < \infty, \quad t > 0 \tag{7-49}$$

$$\text{BC1:} \quad \Phi(x=0,t) = 1 \tag{7-50a}$$

$$\text{BC2:} \quad \Phi(x \to \infty, t) = 0 \tag{7-50b}$$

$$\text{IC:} \quad \Phi(x, t=0) = 0 \tag{7-50c}$$

Then the solution of the problem in equations (7-47) and (7-48) is given in terms of the solution of the auxiliary problem in equation (7-49), by the Duhamel's theorem (7-12), as

$$T(x,t) = \int_{\tau=0}^{t} f(\tau)\frac{\partial \Phi(x,t-\tau)}{\partial t}d\tau \tag{7-51}$$

The solution $\Phi(x,t)$ of the auxiliary problem (7-49) is obtained using superposition of the form

$$\Phi(x,t) = \Phi_H(x,t) + \Phi_{SS}(x) \tag{7-52}$$

where the steady-state solution is given by $\Phi_{SS}(x) = 1$. The steady-state solution reflects the penetration of the unity boundary condition after *infinite time* at any fixed, *finite location* of x. The formulation of the homogeneous problem becomes

$$\frac{\partial^2 \Phi_H(x,t)}{\partial x^2} = \frac{1}{\alpha}\frac{\partial \Phi_H(x,t)}{\partial t} \quad \text{in} \quad 0 < x < \infty, \quad t > 0 \tag{7-53}$$

$$\text{BC1:} \quad \Phi_H(x=0,t) = 0 \tag{7-54a}$$

$$\text{BC2:} \quad \Phi_H(x \to \infty, t) = -1 \tag{7-54b}$$

$$\text{IC:} \quad \Phi_H(x, t=0) = 0 - \Phi_{SS}(x) = -1 \tag{7-54c}$$

where the $\Phi_H(x,t)$ boundary condition at $x \to \infty$ correctly reflects the shifted initial condition. In other words, this boundary condition reflects the initial temperature at an *infinite distance* from the boundary condition at any fixed, *finite time*. The solution of $\Phi_H(x,t)$ is readily obtained from equation (6-39) by setting $T_0 = -1$ giving

$$\Phi_H(x,t) = -\text{erf}\left(\frac{x}{\sqrt{4\alpha t}}\right) \tag{7-55}$$

Combining the steady-state and homogeneous solutions now gives $\Phi(x,t)$ as

$$\Phi(x,t) = 1 - \text{erf}\left(\frac{x}{\sqrt{4\alpha t}}\right) = \text{erfc}\left(\frac{x}{\sqrt{4\alpha t}}\right) \tag{7-56}$$

where the *complementary error function* is defined, in general, as

$$\text{erfc}\,(z) = \frac{2}{\sqrt{\pi}}\int_{\xi=z}^{\infty} e^{-\xi^2}d\xi \tag{7-57}$$

In general, because the unit-step solution involves a zero initial condition and a nonhomogeneous boundary condition of unity, it is often possible to consider a solution for $\Phi(x,t)$ of the form $\Phi(x,t) = 1 - \Psi(x,t)$ where the solution $\Psi(x,t)$ satisfies an initial condition of unity and a corresponding homogeneous boundary condition. In this manner, $\Psi(x,t)$ is readily obtained from equation (6-39) by setting $T_0 = +1$ which then yields the identical solution of $\Phi(x,t)$ as given by equation (7-56). Practically speaking, these two approaches are the same in that with the first case we add 1 to a homogeneous solution with an initial condition of -1, while in the second case we subtract from 1 a homogeneous solution with an initial condition of $+1$.

Introducing equation (7-56) into equation (7-51), the solution of the original problem (7-47) becomes

$$T(x,t) = \int_{\tau=0}^{t} f(\tau)\frac{\partial}{\partial t}\left\{\operatorname{erfc}\left[\frac{x}{\sqrt{4\alpha\,(t-\tau)}}\right]\right\} d\tau \tag{7-58}$$

which now requires differentiation of the complementary error function. Using the above definition of the complementary error function, the derivative within equation (7-58) becomes

$$\frac{\partial}{\partial t}\left\{\operatorname{erfc}\left[\frac{x}{\sqrt{4\alpha\,(t-\tau)}}\right]\right\} = \frac{\partial}{\partial t}\left[\frac{2}{\sqrt{\pi}}\int_{\xi=\frac{x}{\sqrt{4\alpha(t-\tau)}}}^{\infty} e^{-\xi^2} d\xi\right] \tag{7-59}$$

Equation (7-59) results in the derivative of an integral, which is readily evaluated using the Leibniz's rule, namely,

$$\boxed{\frac{d}{dz}\int_{\alpha=u(z)}^{v(z)} f(\alpha)d\alpha = f(v)\frac{dv}{dz} - f(u)\frac{du}{dz}} \tag{7-60}$$

Applying Leibniz's rule to equation (7-59) yields

$$\frac{\partial}{\partial t}\left\{\operatorname{erfc}\left[\frac{x}{\sqrt{4\alpha(t-\tau)}}\right]\right\} = \frac{x}{\sqrt{4\pi\alpha}}\frac{1}{(t-\tau)^{3/2}}\exp\left[-\frac{x^2}{4\alpha(t-\tau)}\right] \tag{7-61}$$

Substituting the above into equation (7-58) yields the temperature as

$$T(x,t) = \frac{x}{\sqrt{4\pi\alpha}}\int_{\tau=0}^{t}\frac{f(\tau)}{(t-\tau)^{3/2}}\exp\left[-\frac{x^2}{4\alpha(t-\tau)}\right]d\tau \tag{7-62}$$

To express this result in an alternative form, a new independent variable η is defined as

$$\eta = \frac{x}{\sqrt{4\alpha(t-\tau)}} \tag{7-63}$$

where then

$$t - \tau = \frac{x^2}{4\alpha\eta^2} \quad \text{and} \quad d\tau = \frac{2}{\eta}(t - \tau)\, d\eta$$

Introducing the above expressions into equation (7-61), we obtain

$$T(x, t) = \frac{2}{\sqrt{\pi}} \int_{\eta = x/\sqrt{4\alpha t}}^{\infty} f\left(t - \frac{x^2}{4\alpha\eta^2}\right) e^{-\eta^2} d\eta \tag{7-64}$$

Periodically Varying $f(t)$. We now consider the surface temperature $f(t)$ to be periodic in time of the form

$$f(t) = T_0 \cos(\omega t - \beta) \tag{7-65}$$

The solution (7-64) becomes

$$\frac{T(x, t)}{T_0} = \frac{2}{\sqrt{\pi}} \int_{\eta = x/\sqrt{4\alpha t}}^{\infty} \cos\left[\omega\left(t - \frac{x^2}{4\alpha\eta^2}\right) - \beta\right] e^{-\eta^2} d\eta \tag{7-66a}$$

or

$$\begin{aligned} \frac{T(x, t)}{T_0} &= \frac{2}{\sqrt{\pi}} \int_{\eta=0}^{\infty} \cos\left[\omega\left(t - \frac{x^2}{4\alpha\eta^2}\right) - \beta\right] e^{-\eta^2} d\eta \\ &\quad - \frac{2}{\sqrt{\pi}} \int_{\eta=0}^{x/\sqrt{4\alpha t}} \cos\left[\omega\left(t - \frac{x^2}{4\alpha\eta^2}\right) - \beta\right] e^{-\eta^2} d\eta \end{aligned} \tag{7-66b}$$

The first definite integral can be evaluated [4, p. 65]; then our solution becomes

$$\begin{aligned} \frac{T(x, t)}{T_0} &= \exp\left[-x\left(\frac{\omega}{2\alpha}\right)^{1/2}\right] \cos\left[\omega t - x\left(\frac{\omega}{2\alpha}\right)^{1/2} - \beta\right] \\ &\quad - \frac{2}{\sqrt{\pi}} \int_{\eta=0}^{x/\sqrt{4\alpha t}} \cos\left[\omega\left(t - \frac{x^2}{4\alpha\eta^2}\right) - \beta\right] e^{-\eta^2} d\eta \end{aligned} \tag{7-67}$$

Here the second term on the right represents the transients, which diminish as $t \to \infty$, and the first term then represents the steady oscillations of temperature in the medium after the transients have passed.

In many applications, the solution for the temperature field such as given by equation (7-66) cannot be obtained in a closed form. In other cases, the heat flux at the surface, namely,

$$q_x''(x = 0) = -k \left.\frac{\partial T}{\partial x}\right|_{x=0} \tag{7-68}$$

is needed for a prescribed surface wall temperature $f(t)$. We note that at $x = 0$, the integration of $\int_{\tau=0}^{t}(f(\tau)/(t-\tau)^{3/2})d\tau$ in equation (7-62) does not exist. More specifically, taking the limit as $x \to 0$ on both sides of equation (7-62) and using the prescribed boundary condition, $T(x = 0, t) = f(t)$ for the right-hand side, one can deduce that

$$\lim_{x\to 0}\int_{\tau=0}^{t}\frac{f(\tau)}{(t-\tau)^{3/2}}\exp\left[-\frac{x^2}{4\alpha(t-\tau)}\right]d\tau \simeq \sqrt{4\pi\alpha}\frac{f(t)}{x} \tag{7-69}$$

to the leading order. Thus direct differentiation of equation (7-62) with respect to x and then setting x to zero does not lead to a usable form for the derivative at $x = 0$, as this process simply introduces more singular terms as $x \to 0$. The alternative is to differentiate equation (7-64) using Leibniz's rule, noting here that

$$d\eta = \frac{x}{4\sqrt{\alpha}}\frac{1}{(t-\tau)^{3/2}}d\tau \tag{7-70}$$

The procedure yields

$$\begin{aligned}\frac{\partial T(x,t)}{\partial x} &= \frac{2}{\sqrt{\pi}}\left(\frac{-1}{2\sqrt{\alpha t}}\right)f(t-t) + \frac{2}{\sqrt{\pi}}\int_{\eta=(x/2\sqrt{\alpha t})}^{\infty} f'\left(t-\frac{x^2}{4\alpha\eta^2}\right)\\ &\quad\times\frac{-2x}{4\alpha\eta^2}e^{-\eta^2}d\eta = -\frac{1}{\sqrt{\pi\alpha t}}f(0) + \frac{2}{\sqrt{\pi}}\int_{\tau=(x/2\sqrt{\alpha t})}^{\infty}\\ &\quad\times f'\left(t-\frac{x^2}{4\alpha\eta^2}\right)\frac{-2x}{4\alpha\dfrac{x^2}{4\alpha(t-\tau)}}e^{-\eta^2}\frac{x}{4\sqrt{\alpha}}\frac{1}{(t-\tau)^{3/2}}d\tau\\ &= -\frac{1}{\sqrt{\pi\alpha t}}f(0) - \frac{1}{\sqrt{\pi\alpha}}\int_{\tau=0}^{t} f'(\tau)\exp\left[-\frac{x^2}{4\alpha(t-\tau)}\right]\frac{1}{(t-\tau)^{1/2}}d\tau\end{aligned} \tag{7-71}$$

Evaluating equation (7-71) at $x = 0$, we now have

$$\left.\frac{\partial T}{\partial x}\right|_{x=0} = -\frac{f(0)}{\sqrt{\pi\alpha t}} - \frac{1}{\sqrt{\pi\alpha}}\int_{\tau=0}^{t}\frac{f'(\tau)}{(t-\tau)^{1/2}}d\tau \tag{7-72}$$

In the case when $f(t)$ is discontinuous with a finite jump, or contains noise from experimental measurements, the direct use of $f'(t)$ in equation (7-72) can be avoided by using an alternative form provided below. We first note that

$$\begin{aligned}\int_{\tau=0}^{t}\frac{f'(\tau)}{(t-\tau)^{1/2}}d\tau &= \int_{f(\tau)-f(t)=0}^{t}\frac{1}{(t-\tau)^{1/2}}d[f(\tau)-f(t)]\\ &= \left.\frac{f(\tau)-f(t)}{(t-\tau)^{1/2}}\right|_{\tau=0}^{t} - \frac{1}{2}\int_{\tau=0}^{t}\frac{f(\tau)-f(t)}{(t-\tau)^{3/2}}d\tau\\ &= \frac{f(t)-f(0)}{t^{1/2}} - \frac{1}{2}\int_{\tau=0}^{t}\frac{f(\tau)-f(t)}{(t-\tau)^{3/2}}d\tau\end{aligned} \tag{7-73}$$

By substituting equation (7-73) into equation (7-72), we have

$$\left.\frac{\partial T}{\partial x}\right|_{x=0} = -\frac{f(t)}{\sqrt{\pi\alpha t}} + \frac{1}{2\sqrt{\pi\alpha}}\int_{\tau=0}^{t}\frac{f(\tau)-f(t)}{(t-\tau)^{3/2}}d\tau \tag{7-74}$$

For several elementary functions $f(t)$ equation (7-74) is integrable to yield closed form expressions for the surface heat flux. Of particular interest is the function

$$f(t) = t^q \qquad \text{for} \qquad q \geq 0 \tag{7-75}$$

For such a case, the indefinite integration can be expressed using hypergeometric functions, and the resulting definite integration is given as

$$\int_{\tau=0}^{t}\frac{\tau^q - t^q}{(t-\tau)^{3/2}}d\tau = 2t^q\left[1 - \frac{\sqrt{\pi}\Gamma(q+1)}{\Gamma\left(q+\frac{1}{2}\right)}\right] \tag{7-76}$$

such that

$$\left.\frac{\partial T}{\partial x}\right|_{x=0} = -\frac{\Gamma(q+1)}{\sqrt{\alpha}\Gamma\left(q+\frac{1}{2}\right)}t^{q-1/2} \quad \text{for} \quad q \geq 0 \tag{7-77}$$

For the special case of $q = 0, \frac{1}{2}$, and 1, equation (7-77) reduces to

$$\left.\frac{\partial T}{\partial x}\right|_{x=0} = -\frac{1}{\sqrt{\pi\alpha t}} \qquad \text{for} \qquad q = 0 \tag{7-78a}$$

$$\left.\frac{\partial T}{\partial x}\right|_{x=0} = -\frac{1}{2}\sqrt{\frac{\pi}{\alpha}} \qquad \text{for} \qquad q = \frac{1}{2} \tag{7-78b}$$

$$\left.\frac{\partial T}{\partial x}\right|_{x=0} = -\frac{2\sqrt{t}}{\sqrt{\pi\alpha}} \qquad \text{for} \qquad q = 1 \tag{7-78c}$$

The expressions for $\partial T/\partial x|_{x=0}$ for several elementary functions are listed in Table 7-1, although this list is by no means exhaustive. There are other forms of $f(t)$ that can lead to closed-form expressions for the surface heat flux.

Example 7-3 One Dimensional Rectangular Solid with Prescribed Heat Flux

A slab of thickness L is initially at zero temperature. For times $t > 0$, the boundary surface at $x = L$ is kept insulated, while the surface at $x = 0$ is subjected to a time-dependent heat flux $f(t)$ of the functional form:

$$-k\left.\frac{\partial T}{\partial x}\right|_{x=0} = f(t) \equiv \begin{cases} t & \text{for} \quad 0 < t < \tau_1 \\ 0 & \text{for} \quad t > \tau_1 \end{cases} \tag{7-79}$$

TABLE 7-1 First-Order Surface Derivative for Prescribed Temperature Functions

$f(t)$ at $x=0$	$\left.\frac{\partial T}{\partial x}\right\|_{x=0}$
t^q $(q \geq 0)$	$-\frac{\Gamma(q+1)}{\sqrt{\alpha}\Gamma(q+1/2)} t^{q-1/2}$
$\sin(\omega t)^a$	$-\sqrt{\frac{2\omega}{\alpha}}\left[\cos(\omega t) C\left(\sqrt{\frac{2\omega t}{\pi}}\right)+\sin(\omega t) S\left(\sqrt{\frac{2\omega t}{\pi}}\right)\right]$
$\cos(\omega t)^a$	$\frac{-1}{\sqrt{\pi\alpha t}}-\sqrt{\frac{2\omega\pi}{\alpha}}\left[\cos(\omega t) S\left(\sqrt{\frac{2\omega t}{\pi}}\right)-\sin(\omega t) C\left(\sqrt{\frac{2\omega t}{\pi}}\right)\right]$
$\exp(-bt)^a$	$-\frac{1}{\sqrt{\pi\alpha t}}\left[1-e^{-bt}\sqrt{\pi bt}\ \mathrm{erfi}(\sqrt{bt})\right]$
$1/(1+bt)^{1/2}$	$-\frac{1}{\sqrt{\pi\alpha t}(1+bt)}$
$1/(1+bt)$	$-\frac{1}{\sqrt{\pi\alpha t}}\frac{1}{1+bt}+\frac{\sqrt{b}}{\sqrt{\pi\alpha}}\frac{\tanh^{-1}\left(\sqrt{bt}/\sqrt{1+bt}\right)}{(1+bt)^{3/2}}$
$1/(1+bt)^2$	$-\frac{1}{\sqrt{\pi\alpha t}}\frac{1-(bt/2)}{(1+bt)^2}+\frac{3}{2}\sqrt{\frac{b}{\pi\alpha}}\frac{\tanh^{-1}\left(\sqrt{bt}/\sqrt{1+bt}\right)}{(1+bt)^{5/2}}$
$\ln(1+bt)$	$-\sqrt{\frac{b}{\pi\alpha}}\frac{2}{1+bt}\tanh^{-1}\left(\frac{\sqrt{bt}}{\sqrt{1+bt}}\right)$

[a] *C(z)* and *S(z)* are Fresnel cosine and Fresnel sine functions, and erfi(t) is the imaginary error function.

Using Duhamel's theorem, we now develop an expression for the temperature distribution $T(x,t)$ in the slab for times: (i) $t < \tau_1$ and for (ii) $t > \tau_1$. The mathematical formulation of this heat conduction problem is given by

$$\frac{\partial^2 T(x,t)}{\partial x^2}=\frac{1}{\alpha}\frac{\partial T(x,t)}{\partial t} \quad \text{in} \quad 0<x<L, \quad t>0 \tag{7-80}$$

$$\text{BC1:} \quad -k\left.\frac{\partial T}{\partial x}\right|_{x=0}=f(t) \tag{7-81a}$$

$$\text{BC2:} \quad \left.\frac{\partial T}{\partial x}\right|_{x=L}=0 \tag{7-81b}$$

$$\text{IC:} \quad T(x,t=0)=0 \tag{7-81c}$$

where $f(t)$ is defined by equation (7-79). The corresponding auxiliary problem is given by

$$\frac{\partial^2 \Phi(x,t)}{\partial x^2}=\frac{1}{\alpha}\frac{\partial \Phi(x,t)}{\partial t} \quad \text{in} \quad 0<x<L, \quad t>0 \tag{7-82}$$

$$\text{BC1:} \qquad -k\left.\frac{\partial \Phi}{\partial x}\right|_{x=0} = 1 \tag{7-83a}$$

$$\text{BC2:} \qquad \left.\frac{\partial \Phi}{\partial x}\right|_{x=L} = 0 \tag{7-83b}$$

$$\text{IC:} \qquad \Phi(x, t = 0) = 0 \tag{7-83c}$$

The solution for the auxiliary problem is given by

$$\Phi(x,t) = \frac{\alpha}{Lk}t + \frac{2}{Lk}\sum_{m=1}^{\infty}\frac{\cos\beta_m x}{\beta_m^2}\left(1 - e^{-\alpha\beta_m^2 t}\right) \tag{7-84}$$

where $\beta_m = (m\pi/L)$. The solution of the problem given in equations (7-82) and (7-83) presents a difficulty with separation of variables in that the steady-state problem has no solution for a flux of unity at one surface and a perfectly insulated second surface. The problem is readily solved using the 1-D Green's function approach of the next chapter, where the first term on the right-hand side of equation (7-84) follows from the $\beta_0 = 0$ eigenvalue after the application of L'Hôpital's rule. We now apply Duhamel's theorem given by equation (7-15) for the two time domains.

1. ***Times*** $t < \tau_1$. The boundary-condition function $f(t)$ has no discontinuity beyond the initial value, and we obtain

$$T(x,t) = \Phi(x,t)\cancel{f(0)}^{0} + \int_{\tau=0}^{t}\Phi(x, t-\tau)\frac{df(\tau)}{d\tau}d\tau \qquad \text{for} \qquad t < \tau_1 \tag{7-85}$$

where $df(\tau)/d\tau = 1$, and $\Phi(x, t-\tau)$ is obtainable from equation (7-84) by replacing t with $(t-\tau)$. With these substitutions, equation (7-85) becomes

$$T(x,t) = \int_{\tau=0}^{t}\left[\frac{\alpha}{Lk}(t-\tau) + \frac{2}{Lk}\sum_{m=1}^{\infty}\frac{\cos\beta_m x}{\beta_m^2} - \frac{2}{Lk}\sum_{m=1}^{\infty}\frac{\cos\beta_m x}{\beta_m^2}e^{-\alpha\beta_m^2(t-\tau)}\right]d\tau \tag{7-86}$$

The integration is performed to give

$$T(x,t) = \frac{\alpha}{2Lk}t^2 + \frac{2}{Lk}\sum_{m=1}^{\infty}\frac{\cos\beta_m x}{\beta_m^2}t - \frac{2}{\alpha Lk}\sum_{m=1}^{\infty}\frac{\cos\beta_m x}{\beta_m^4}\left(1 - e^{-\alpha\beta_m^2 t}\right) \tag{7-87}$$

which is valid for the time domain $t < \tau_1$.

2. ***Times $t > \tau_1$***. The boundary surface function $f(t)$ has only a single discontinuity at time $t = \tau_1$, where the resulting change in $f(t)$ is a decrease of the amount $\Delta f_1 = -\tau_1$. Then, Duhamel's theorem (7-15) reduces to the form

$$T(x,t) = \Phi(x, t-\tau_1)\Delta f_1 + \int_{\tau=0}^{\tau_1} \Phi(x, t-\tau)\frac{df(\tau)}{d\tau}d\tau + \int_{\tau=\tau_1}^{t} \Phi(x, t-\tau)\cancelto{0}{\frac{df}{d\tau}}\, d\tau \tag{7-88}$$

where $df(\tau)/d\tau = 1$ for the first integral, $df(\tau)/d\tau = 0$ for the second integral, and $\Delta f_1 = -\tau_1$ as noted above. Substituting these results into equation (7-88), we obtain

$$T(x,t) = -\Phi(x, t-\tau_1)\cdot\tau_1 + \int_{\tau=0}^{\tau_1} \Phi(x, t-\tau)d\tau \tag{7-89}$$

where the functions $\Phi(x,t)$ and $\Phi(x, t-\tau_1)$ are obtainable from equation (7-84). Making these substitutions, equation (7-89) then becomes

$$T(x,t) = -\tau_1\left[\frac{\alpha}{Lk}\left(t-\tau_1\right) + \frac{2}{Lk}\sum_{m=1}^{\infty}\frac{\cos\beta_m x}{\beta_m^2} - \frac{2}{Lk}\sum_{m=1}^{\infty}\frac{\cos\beta_m x}{\beta_m^2}e^{-\alpha\beta_m^2(t-\tau_1)}\right] + \int_{\tau=0}^{\tau_1}\left[\frac{\alpha}{Lk}(t-\tau) + \frac{2}{Lk}\sum_{m=1}^{\infty}\frac{\cos\beta_m x}{\beta_m^2} - \frac{2}{Lk}\sum_{m=1}^{\infty}\frac{\cos\beta_m x}{\beta_m^2}e^{-\alpha\beta_m^2(t-\tau)}\right]d\tau \tag{7-90}$$

The integral is similar to that in equation (7-86); hence it can readily be performed.

Example 7-4 One-Dimensional Cylinder with Prescribed Temperature

A solid cylinder, $0 \le r \le b$, is initially at zero temperature. For times $t > 0$, the boundary surface at $r = b$ is kept at a prescribed temperature given by the time-dependent function $f(t)$. Obtain an expression for the temperature distribution $T(r,t)$ in the cylinder for times $t > 0$. Assume that $f(t)$ has no discontinuities.

The mathematical formulation of this problem is given by

$$\frac{\partial^2 T(r,t)}{\partial r^2} + \frac{1}{r}\frac{\partial T(r,t)}{\partial r} = \frac{1}{\alpha}\frac{\partial T(r,t)}{\partial t} \quad \text{in} \quad 0 \le r < b, \quad t > 0 \tag{7-91}$$

$$\text{BC1:} \quad T(r = b, t) = f(t) \tag{7-92a}$$

$$\text{BC2:} \quad T(r \to 0, t) \Rightarrow \text{finite} \tag{7-92b}$$

$$\text{IC:} \quad T(r, t = 0) = 0 \tag{7-92c}$$

For use of Duhamel's theorem, the auxiliary problem is taken as

$$\frac{\partial^2 \Phi(r,t)}{\partial r^2} + \frac{1}{r}\frac{\partial \Phi(r,t)}{\partial r} = \frac{1}{\alpha}\frac{\partial \Phi(r,t)}{\partial t} \quad \text{in} \quad 0 \le r < b, \quad t > 0 \tag{7-93}$$

$$\text{BC1:} \quad \Phi(r = b, t) = 1 \tag{7-94a}$$

$$\text{BC2:} \quad \Phi(r \to 0, t) \Rightarrow \text{finite} \tag{7-94b}$$

$$\text{IC:} \quad \Phi(r, t = 0) = 0 \tag{7-94c}$$

Per Duhamel's, the solution of the problem in equations (7-91) and (7-92) can be written in terms of the solution of the above auxiliary problem using equation (7-12), namely,

$$T(r,t) = \int_{\tau=0}^{t} f(\tau)\frac{\partial \Phi(r, t-\tau)}{\partial t} d\tau \tag{7-95}$$

Let us define $\Psi(r,t)$ as the solution of the problem for a solid cylinder over the same domain, $0 \le r \le b$, initially at temperature unity, and for times $t > 0$, the boundary surface at $r = b$ is kept at zero temperature. The solution for $\Psi(r,t)$ is readily obtainable from the solution of Example 4-8 by setting $F(r) = 1$, performing the integration of equation (4-150), and defining the norm per Table 2-2, case 3; we find

$$\Psi(r,t) = \frac{2}{b}\sum_{m=1}^{\infty} \frac{J_0(\beta_m r)}{\beta_m J_1(\beta_m b)} e^{-\alpha\beta_m^2 t} \tag{7-96}$$

where the eigenvalues β_m are the positive roots of

$$J_0(\beta_m b) = 0 \tag{7-97}$$

The desired solution of the auxiliary problem $\Phi(r,t)$ is now obtainable from the solution $\Psi(r,t)$ using the approach we discussed in the above example, namely, as

$$\Phi(r,t) = 1 - \Psi(r,t) = 1 - \frac{2}{b}\sum_{m=1}^{\infty} \frac{J_0(\beta_m r)}{\beta_m J_1(\beta_m b)} e^{-\alpha\beta_m^2 t} \tag{7-98}$$

Introducing equation (7-98) into equation (7-95), the solution of the original problem becomes

$$T(r,t) = \frac{2\alpha}{b}\sum_{m=1}^{\infty} e^{-\alpha\beta_m^2 t} \frac{\beta_m J_0(\beta_m r)}{J_1(\beta_m b)} \int_{\tau=0}^{t} e^{\alpha\beta_m^2 \tau} f(\tau)\, d\tau \tag{7-99}$$

where the β_m values are the roots of $J_0(\beta_m b) = 0$. The solution (7-99) does not explicitly show that $T(r,t) \to f(t)$ for $r \to b$. To obtain an alternative form of this solution, the time integration is performed by parts, yielding

$$T(r,t) = f(t)\frac{2}{b}\sum_{m=1}^{\infty}\frac{J_0(\beta_m r)}{\beta_m J_1(\beta_m b)} - \frac{2}{b}\sum_{m=1}^{\infty}\frac{J_0(\beta_m r)}{\beta_m J_1(\beta_m b)}\left[f(0)\,e^{-\alpha\beta_m^2 t} + \int_{\tau=0}^{t} e^{-\alpha\beta_m^2(t-\tau)}\frac{df(\tau)}{d\tau}d\tau\right] \tag{7-100}$$

We note, however, that the solution of $\Psi(r, t = 0)$ must be equal to the initial temperature of unity; hence $\Psi(r, t = 0) = 1$, and we therefore have

$$1 = \frac{2}{b}\sum_{m=1}^{\infty}\frac{J_0(\beta_m r)}{\beta_m J_1(\beta_m b)} \tag{7-101}$$

The substitution of equation (7-101) into the first term of equation (7-100) gives the desired closed-form expression of the form

$$T(r,t) = f(t) - \frac{2}{b}\sum_{m=1}^{\infty}\frac{J_0(\beta_m r)}{\beta_m J_1(\beta_m b)}\left[f(0)\,e^{-\alpha\beta_m^2 t} + \int_{\tau=0}^{t} e^{-\alpha\beta_m^2(t-\tau)}\frac{df(\tau)}{d\tau}d\tau\right] \tag{7-102}$$

The above solution given in this form clearly shows that $T(r = b, t) = f(t)$ in view of the eigenvalues as defined by equation (7-97).

7-5 APPLICATIONS OF DUHAMEL'S THEOREM FOR INTERNAL ENERGY GENERATION

Duhamel's theorem may also be used for treatment of time-dependent, internal energy generation provided the boundary conditions and initial temperature are homogeneous. Consider, for example, a homogeneous boundary problem in the Cartesian coordinate system, as formulated by

$$\frac{\partial^2 T(x,t)}{\partial x^2} + \frac{g(t)}{k} = \frac{1}{\alpha}\frac{\partial T(x,t)}{\partial t} \quad \text{in} \quad 0 < x < L, \quad t > 0 \tag{7-103}$$

$$\text{BC1:} \quad T(x = 0, t) \Rightarrow \text{homogeneous} \tag{7-104a}$$

$$\text{BC2:} \quad T(x = L, t) \Rightarrow \text{homogeneous} \tag{7-104b}$$

$$\text{IC:} \quad T(x, t = 0) = 0 \tag{7-104c}$$

where $g(t)$ (W/m^3) defines a time-dependent heat generation term. We now define the auxiliary problem $\Phi(x, t)$ with unity internal heat generation:

$$\frac{\partial^2 \Phi(x,t)}{\partial x^2} + \frac{1}{k} = \frac{1}{\alpha}\frac{\partial \Phi(x,t)}{\partial t} \quad \text{in} \quad 0 < x < L, \quad t > 0 \tag{7-105}$$

$$\text{BC1:} \quad \Phi(x=0,t) \Rightarrow \text{homogeneous} \tag{7-106a}$$

$$\text{BC2:} \quad \Phi(x=L,t) \Rightarrow \text{homogeneous} \tag{7-106b}$$

$$\text{IC:} \quad \Phi(x,t) = 0 \tag{7-106c}$$

where both $T(x,t)$ and $\Phi(x,t)$ have the identical homogeneous boundary conditions and zero initial condition. We now define the solution of our original problem $T(x,t)$ in terms of the unit-step generation problem $\Phi(x,t)$ by Duhamel's theorem, namely,

$$\boxed{T(x,t) = \int_{\tau=0}^{t} g(\tau) \frac{\partial \Phi(x, t-\tau)}{\partial t} d\tau} \tag{7-107}$$

where the time-dependent generation term is assumed to have no discontinuities. To illustrate the validity of this solution for heat generation, we first reformulate our governing PDEs of equations (7-103) and (7-105) into the form

$$k\left[\frac{1}{\alpha}\frac{\partial T(x,t)}{\partial t} - \frac{\partial^2 T(x,t)}{\partial x^2}\right] = g(t) \tag{7-108}$$

and

$$k\left[\frac{1}{\alpha}\frac{\partial \Phi(x,t)}{\partial t} - \frac{\partial^2 \Phi(x,t)}{\partial x^2}\right] = 1 \tag{7-109}$$

We can now make the same depiction as in Figure 7-1, but with the time-dependent heat generation $g(t)$ replacing the time-dependent boundary condition $f(t)$. Duhamel's theorem then readily follows by the same derivation.

Example 7-5 One-Dimensional Cylinder with Internal Energy Generation A solid cylinder, $0 \le r \le b$, is initially at zero temperature. For times $t > 0$, heat is generated internally throughout the solid at a rate of $g(t)$ per unit volume (W/m^3), while the boundary surface at $r = b$ is kept at zero temperature. Using Duhamel's theorem, obtain an expression for the temperature distribution $T(r,t)$ in the cylinder for times $t > 0$, assuming that $g(t)$ has no discontinuities. The mathematical formulation of this problem is given by

$$\frac{\partial^2 T(r,t)}{\partial r^2} + \frac{1}{r}\frac{\partial T(r,t)}{\partial r} + \frac{g(t)}{k} = \frac{1}{\alpha}\frac{\partial T(r,t)}{\partial t} \qquad 0 \le r < b, \quad t > 0 \tag{7-110}$$

$$\text{BC1:} \quad T(r=b,t) = 0 \tag{7-111a}$$

$$\text{BC2:} \quad T(r \to 0,t) \Rightarrow \text{finite} \tag{7-111b}$$

$$\text{IC:} \quad T(r,t=0) = 0 \tag{7-111c}$$

We now define the auxiliary problem with unit-step generation as

$$\frac{\partial^2 \Phi(r,t)}{\partial r^2} + \frac{1}{r}\frac{\partial \Phi(r,t)}{\partial r} + \frac{1}{k} = \frac{1}{\alpha}\frac{\partial \Phi(r,t)}{\partial t} \quad \text{in} \quad 0 \le r < b \quad t > 0 \tag{7-112}$$

BC1: $\Phi(r = b, t) = 0$ (7-113a)

BC2: $\Phi(r \to 0, t) \Rightarrow \text{finite}$ (7-113b)

IC: $\Phi(r, t = 0) = 0$ (7-113c)

The solution of the above auxiliary problem is readily obtainable from equations (4-233) and (4-234) by setting $g_0 = 1$, $F(r) = 0$, defining the norm using case 3 of Table 2-2, and computing the resulting integrals. With these substitutions, we find

$$\Phi(r,t) = \frac{b^2 - r^2}{4k} - \frac{2}{bk}\sum_{m=1}^{\infty} \frac{J_0(\beta_m r)}{\beta_m^3 J_1(\beta_m b)} e^{-\alpha\beta_m^2 t} \tag{7-114}$$

where the eigenvalues β_m are defined by the positive roots of

$$J_0(\beta_m b) = 0 \tag{7-115}$$

The solution of the above auxiliary problem is related to the solution of the original problem of equations (7-110) and (7-111) by Duhamel's theorem, namely, as

$$T(r,t) = \int_{\tau=0}^{t} g(\tau) \frac{\partial \Phi(r, t-\tau)}{\partial t} d\tau \tag{7-116}$$

Introducing equation (7-114) into equation (7-116), we obtain the solution as

$$T(r,t) = \frac{2\alpha}{bk}\sum_{m=1}^{\infty} e^{-\alpha\beta_m^2 t} \frac{J_0(\beta_m r)}{\beta_m J_1(\beta_m b)} \int_{\tau=0}^{t} g(\tau) e^{\alpha\beta_m^2 \tau} d\tau \tag{7-117}$$

REFERENCES

1. J. M. C. Duhamel, *J. Ec. Polytech.* **14**, 20–77, 1833.
2. R. C. Bartels and R. V. Churchill, *Bull. Am. Math. Soc.* **48**, 276–282, 1942.
3. I. N. Sneddon, *Fourier Transforms*, McGraw-Hill, New York, 1951.
4. H. S. Carslaw and J. C. Jaeger, *Conduction of Heat in Solids*, Clarendon London, 1959.

PROBLEMS

7-1 A slab, $0 \leq x \leq L$, is initially at zero temperature. For times $t > 0$, the boundary surface at $x = 0$ is subjected to a time-varying temperature $f(t) = b + ct$, while the boundary surface at $x = L$ is kept at zero temperature. Using Duhamel's theorem, develop an expression for the temperature distribution $T(x,t)$ in the slab for times $t > 0$.

7-2 A semi-infinite solid, $0 \leq x < \infty$, is initially at zero temperature. For times $t > 0$, the boundary surface at $x = 0$ is kept at temperature $T = T_0 t$, where T_0 is a constant. Using Duhamel's theorem obtain an expression for the temperature distribution $T(x,t)$ in the region for times $t > 0$.

7-3 A slab, $0 \leq x \leq L$, is initially at zero temperature. For times $t > 0$, the boundary at $x = 0$ is kept insulated, and the convection boundary condition at $x = L$ is given as $\partial T/\partial x + HT = f(t)$, where $f(t)$ is a function of time. Obtain an expression for the temperature distribution $T(x,t)$ in the slab for times $t > 0$.

7-4 A solid cylinder, $0 \leq r \leq b$, is initially at zero temperature. For times $t > 0$, the boundary condition at $r = b$ is the convection condition given as $\partial T/\partial x + HT = f(t)$, where $f(t)$ is a function of time. Obtain an expression for the temperature distribution $T(r,t)$ in the cylinder for times $t > 0$.

7-5 A solid sphere, $0 \leq r \leq b$, is initially at zero temperature, for times $t > 0$ the boundary surface $r = b$ is kept at the prescribed temperature $f(t)$, which varies with time. Obtain an expression for the temperature distribution $T(r,t)$ in the sphere for times $t > 0$.

7-6 A solid cylinder, $0 \leq r \leq b$, is initially at zero temperature. For times $t > 0$, internal energy is generated in the solid at a rate of $g(t)$ per unit volume (W/m^3), whereas the boundary surface at $r = b$ dissipates heat by convection into a medium at zero temperature. Obtain an expression for the temperature distribution $T(r,t)$ in the cylinder for times $t > 0$.

7-7 A rectangular region, $0 \leq x \leq a, 0 \leq y \leq b$, is initially at zero temperature. For times $t > 0$, the boundaries at $x = 0$ and $y = 0$ are perfectly insulated, and the boundaries at $x = a$ and $y = b$ are kept at zero temperature, while internal energy is generated in the region at a rate of $g(t)$ per unit volume (W/m^3). Obtain an expression for the temperature distribution in the region using Duhamel's theorem for time $t > 0$.

7-8 A slab of thickness L is initially at zero temperature. For times $t > 0$, the boundary surface at $x = L$ is kept at zero temperature, while the boundary surface at $x = 0$ is subjected to a time-dependent prescribed temperature

$f(t)$ defined by

$$f(t) = \begin{cases} ct & \text{for} \quad 0 < t < \tau_1 \\ 0 & \text{for} \quad t > \tau_1 \end{cases}$$

Using Duhamel's theorem, develop an expression for the temperature distribution $T(x\ t)$ for times (i) $t < \tau_1$ and (ii) $t > \tau_1$.

7-9 A semi-infinite medium, $0 < x < \infty$, is initially at zero temperature. For times $t > 0$, the boundary surface at $x = 0$ is subjected to a time-dependent prescribed temperature $f(t)$ defined by

$$f(t) = \begin{cases} ct & \text{for} \quad 0 < t < \tau_1 \\ 0 & \text{for} \quad t > \tau_1 \end{cases}$$

Using Duhamel's theorem, develop an expression for the temperature distribution $T(x,t)$ for times: (i) $t < \tau_1$, and (ii) $t > \tau_1$.

7-10 A slab of thickness L is initially at zero temperature. For times $t > 0$, the boundary surface at $x = 0$ is subjected to a time-dependent prescribed temperature $f(t)$ defined by:

$$f(t) = \begin{cases} a + bt & \text{for} \quad 0 < t < \tau_1 \\ 0 & \text{for} \quad t > \tau_1 \end{cases}$$

and the boundary at $x = L$ is kept insulated. Using Duhamel's theorem, develop an expression for the temperature distribution in the slab for times (i) $t < \tau_1$ and (ii) $t > \tau_1$.

7-11 A semi-infinite medium, $x > 0$, is initially at zero temperature. For times $t > 0$, the boundary surface at $x = 0$ is subjected to a periodically varying temperature $f(t)$ as illustrated in Figure 7-5. Develop an expression for the temperature distribution in the medium at times (i) $0 < t < \Delta t$, (ii) $\Delta t < t < 2\Delta t$, and (iii) $6\Delta t < t < 7\Delta t$.

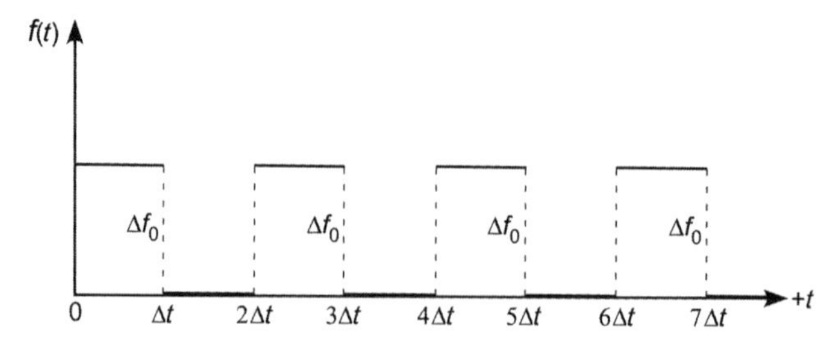

Figure 7-5 Time-dependent boundary condition for Problem 7-11.

7-12 Repeat Problem 7-5 for the case of surface temperature $f(t)$ varying with time as

$$f(t) = \begin{cases} bt & \text{for} \quad 0 < t < \tau_1 \\ 0 & \text{for} \quad t > \tau_1 \end{cases}$$

and determine the temperature distribution $T(r,t)$ in the sphere for times (i) $t < \tau_1$ and (ii) $t > \tau_1$.

7-13 A solid cylinder, $0 \leq r \leq b$, is initially at zero temperature. For times $t > 0$, the boundary condition at $r = b$ is subjected to a time-dependent prescribed temperature $f(t)$ defined by

$$f(t) = \begin{cases} T_1 + \left(T_2/\tau_1\right)t & \text{for} \quad 0 < t < \tau_1 \\ T_1 & \text{for} \quad \tau_1 < t < \tau_2 \\ T_1 - \left[T_1/\left(\tau_3 - \tau_2\right)\right]\left(t - \tau_2\right) & \text{for} \quad \tau_2 < t < \tau_3 \\ 0 & \text{for} \quad t > \tau_3 \end{cases}$$

First plot (i.e., sketch) the boundary condition $f(t)$. Using Duhamel's theorem, develop an expression for the temperature distribution $T(r,t)$ in the cylinder for times (i) $\tau_1 < t < \tau_2$ and (ii) $t > \tau_3$.

8

USE OF GREEN'S FUNCTION FOR SOLUTION OF HEAT CONDUCTION PROBLEMS

In this chapter, we first discuss the physical significance of Green's function and then present sufficiently general expressions for the solution of nonhomogeneous, transient heat conduction problems with energy generation, nonhomogeneous boundary conditions, and a given initial condition, in terms of Green's function. We present a useful method to find the Green's function for a specific problem based on the method of separation of variables. Application to one-, two-, and three-dimensional problems of finite, semi-infinite, and infinite domains is illustrated with representative examples in the rectangular, cylindrical, and spherical coordinate systems. Once the Green's function is available for a given problem, the solution for the temperature distribution is determined immediately from the analytic expressions given in this chapter. Green's function in the solution of partial differential equations of mathematical physics can be found in several references [1–11]. Overall, the Green's function approach is our most general and powerful method to solve nonhomogeneous, time-dependent conduction problems.

8-1 GREEN'S FUNCTION APPROACH FOR SOLVING NONHOMOGENEOUS TRANSIENT HEAT CONDUCTION

While the method of separation of variables is applicable to a broad class of problems, the method is not applicable to problems containing time-dependent nonhomogeneities. We consider the following 3-D, nonhomogeneous boundary value problem of heat conduction:

$$\nabla^2 T(\hat{r}, t) + \frac{1}{k} g(\hat{r}, t) = \frac{1}{\alpha} \frac{\partial T(\hat{r}, t)}{\partial t} \qquad \text{in region R}, \qquad t > 0 \qquad \text{(8-1)}$$

$$\text{BC:} \qquad k\frac{\partial T}{\partial n_i} + h_i T = h_i T_{\infty i} \equiv f_i(\hat{r}, t) \qquad \text{on surface } S_i \tag{8-2a}$$

$$\text{IC:} \qquad T(\hat{r}, t = 0) = F(\hat{r}) \tag{8-2b}$$

where $\partial/\partial n_i$ denotes differentiation along the *outward-drawn normal* to the boundary surface S_i for $i = 1, 2, \ldots, N$, with N being the number of continuous boundary surfaces of the region. For generality, it is assumed that the internal energy generation term $g(\hat{r}, t)$ and the boundary condition functions $f_i(\hat{r}, t)$ vary with both position and time. Here, k and h_i are to be treated as the thermal conductivity and the convection coefficients of the ith surface boundary, respectively, and are considered constants.

To solve the preceding heat conduction problem of equations (8-1) and (8-2), we consider the following auxiliary problem for the same region R:

$$\nabla^2 G(\hat{r}, t|\hat{r}', \tau) + \frac{1}{k}\delta(\hat{r} - \hat{r}')\delta(t - \tau) = \frac{1}{\alpha}\frac{\partial G}{\partial t} \qquad \text{in region R,} \qquad t > 0 \tag{8-3}$$

$$\text{BC:} \qquad k\frac{\partial G}{\partial n_i} + h_i G = 0 \qquad \text{on surface } S_i \tag{8-4a}$$

$$\text{IC:} \qquad G(\hat{r}, t = 0) \equiv 0 \tag{8-4b}$$

which obeys the causality requirement that Green's function G be zero for $t < \tau$ [2]. The source term in equation (8-3) is a *unit impulsive source* for the 3-D problem considered here; hence the delta function $\delta(\hat{r} - \hat{r}')$ represents a *point heat source* located at the location $\hat{r}'$, while the delta function $\delta(t - \tau)$ indicates that it is an instantaneous heat source releasing its energy spontaneously at time $t = \tau$. A comparison of equations (8-2) and (8-4) reveals that the latter is the homogeneous version of the former with regard to the boundary and initial conditions.

In the case of 2-D problems, $\delta(\hat{r} - \hat{r}')$ represents a two-dimensional delta function that characterizes a line heat source located at $\hat{r}'$, while for the 1-D problems, $\delta(r - r')$ represents a 1-D delta function which represents a plane surface heat source located at r'.

Three-Dimensional Green's Function

The physical significance of Green's function $G(\hat{r}, t|\hat{r}', \tau)$ for the 3-D problems is as follows: Green's function *represents the temperature at any location $\hat{r}$ within the domain, at any time t, due to an instantaneous point source of unit strength (i.e., 1 J), located at the point $\hat{r}'$, releasing its energy spontaneously at time $t = \tau$ into a medium of zero temperature*, as illustrated in Figure 8-1. The auxiliary problem satisfied by Green's function is valid over the same region R as the original physical problem of equations (8-1) and (8-2), but the boundary conditions (8-4b) are the homogeneous versions of the boundary conditions (8-2b), and the initial condition is zero.

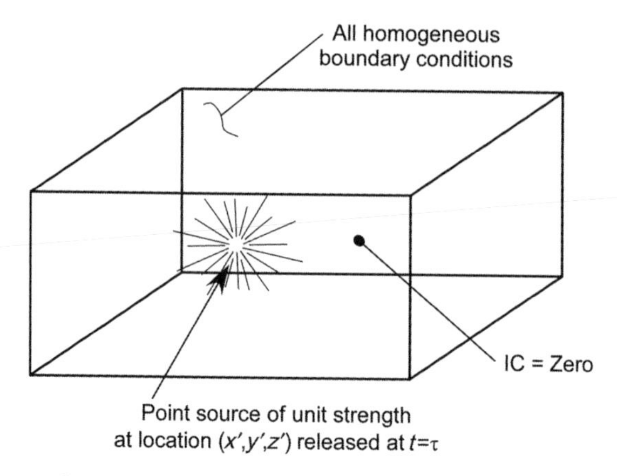

Figure 8-1 Physical problem statement for Green's function.

On the basis of this definition, the physical significance of Green's function may be interpreted as

$$G(\hat{r}, t \,|\, \hat{r}', \tau) \equiv G(\text{effect}|\,\text{impulse}) \tag{8-5}$$

The first part of the argument, $(\hat{r}, t)$, represents the *effect*, that is, the temperature in the medium at the location $\hat{r}$ and at time t, while the second part, $(\hat{r}', \tau)$, represents the *impulse*, that is, the impulsive (i.e., instantaneous) point source located at point $\hat{r}'$, releasing its heat spontaneously at time τ.

Green's function takes its name from the British mathematician George Green, who published his first related work in 1828 and continued into the 1830s. The usefulness of Green's function lies in the fact that the solution of the original problem (8-1) can be represented only in terms of Green's function. Therefore, once the Green's function is known, the temperature distribution $T(\hat{r}, t)$ in the medium may be readily computed. The mathematical proof for the developments of such expressions can be found in the literature [1, 2, 6]. Here we present only the resulting expressions to calculate the temperature solution in terms of Green's function, illustrate their use with representative examples, and describe a very simple approach for the determination of Green's functions that allows use of our solution tools developed in the previous chapters.

In the case of the 3-D, nonhomogeneous transient heat conduction problem given by equations (8-1) and (8-2), the solution for $T(\hat{r}, t)$ is expressed in terms of the 3-D Green's function $G(\hat{r}, t \,|\, \hat{r}', \tau)$ as

$$\boxed{\begin{aligned} T(\hat{r}, t) = {} & \int_{R'} G(\hat{r}, t|\hat{r}', \tau)\big|_{\tau=0} F(\hat{r}')\, dV' \\ & + \frac{\alpha}{k} \int_{\tau=0}^{t} \int_{R'} G(\hat{r}, t|\hat{r}', \tau) g(\hat{r}', \tau) dV'\, d\tau \\ & + \alpha \sum_{i=1}^{N} \left[\int_{\tau=0}^{t} \int_{S_i'} G(\hat{r}, t|\hat{r}', \tau)\big|_{r'=r_i} \frac{1}{k} f_i(\hat{r}', \tau) dA_i'\, d\tau \right] \end{aligned}} \tag{8-6}$$

where R refers to the entire volume of the region considered; S_i refers to the ith surface boundary area of the region R, $i = 1, 2, \ldots, N$, with N being the number of continuous boundary surfaces of the region; and where dV' and dA' refer to differential volume and surface elements, respectively, in the $\hat{r}'$ variables of integration. The physical significance of various terms in solution (8-6) is as follows:

1. The first term on the right-hand side of equation (8-6) is the contribution of the initial condition function $F(\hat{r})$ on the temperature distribution; that is, Green's function evaluated for $\tau = 0$ is multiplied by $F(\hat{r})$ and integrated over the region R.
2. The second term is for the contribution of the internal energy generation $g(\hat{r}, t)$ on the temperature $T(\hat{r}, t)$; that is, Green's function $G(\hat{r}, t \big| \hat{r}', \tau)$ is multiplied by the energy generation $g(\hat{r}', \tau)$, which is the generation function of variables $\hat{r}'$ and τ, and integrated over the region R and over the time from $\tau = 0$ to t.
3. The last term represents the contribution of the nonhomogeneous terms $f_i(\hat{r}, t)$ of the boundary conditions on the temperature, one term for each nonhomogeneous boundary condition. For each nonhomogeneous boundary condition, it consists of Green's function evaluated at the boundary $r' = r_i$, multiplied by $f_i(\hat{r}', \tau)$, which is the nonhomogeneous boundary function of variables $\hat{r}'$ and τ, and integrated over the boundary surface S_i and over the time from $\tau = 0$ to t.

For generality, physical problem (8-1) is formulated by considering a boundary condition of the *third type* (i.e., convective boundary) for which $f_i(\hat{r}, t) \equiv h_i T_{\infty i}(\hat{r}, t)$, where $T_{\infty i}(\hat{r}, t)$ is the spatially and temporally dependent ambient fluid temperature. Solution (8-6) is also applicable for the boundary condition of the *second type* (i.e., prescribed surface heat flux) if $f_i(\hat{r}, t)$ is interpreted as the prescribed boundary heat flux. For such a case, we first set $h_i T_{\infty i}(\hat{r}, t) = q''_{\text{boundary}} \equiv f_i(\hat{r}, t)$ in equation (8-2a), and then let $h_i = 0$ on the left-hand side, which yields

$$k \frac{\partial T}{\partial n_i} = q''_{\text{surface}} \equiv f_i(\hat{r}, t) \qquad \text{on surface } S_i \tag{8-7}$$

Likewise, Green's function solution takes the corresponding boundary condition type, but of the homogeneous form, namely,

$$\frac{\partial G}{\partial n_i} \equiv 0 \qquad \text{on surface } S_i \tag{8-8}$$

when the ith surface is a type 2 boundary condition.

In the case of boundary conditions of the *first type* (i.e., prescribed temperature) some modification is needed in the third term on the right-hand side of solution (8-6). The reason for this is that the boundary condition of the first type is obtainable from equation (8-2b) by setting $k = 0$ and $h_i = 1$, such that now

$f_i(\hat{r}, t)$ represents the prescribed surface temperature. For such a case, however, difficulty arises in setting $k = 0$ in solution (8-6) because k appears in the denominator. This difficulty can be alleviated by making the following change in the last term in solution (8-6):

$$\text{Replace} \qquad \frac{1}{k} G \bigg|_{r'=r_i} \qquad \text{by} \qquad \pm \frac{\partial G}{\partial n_i'} \bigg|_{r'=r_i} \tag{8-9}$$

where now the Green's function solution takes the corresponding homogeneous boundary condition type, namely, of the form

$$G \equiv 0 \qquad \text{on surface } S_i \tag{8-10}$$

when the ith surface is a type 1 boundary condition. The validity of this replacement is apparent if the boundary condition (8-4a) of the auxiliary problem is rearranged, prior to setting $h_i = 1$, in the form

$$\frac{1}{k} G = \pm \frac{1}{h_i} \frac{\partial G}{\partial n_i} \qquad \text{on surface } S_i \tag{8-11}$$

We have introduced the $\pm$ in equations (8-9) and (8-11) to reflect the necessary sign convention associated with convection boundary conditions, as we developed in Chapter 1; see equations (1-64).

Carslaw and Jaeger [1, pp. 291–293], provide a derivation for equation (8-6) for the special case of no internal energy generation, following the original work on applications of Green's function to heat transfer, which dates to the 1860s. Such treatment yields the first and last terms of equation (8-6), which correspond to the contributions of the initial condition and the nonhomogeneous boundary conditions, respectively, to the temperature profile. One may also develop a solution corresponding to the special case of internal energy generation in the presence of all homogeneous boundary conditions and for the case of zero initial temperature distribution. The solution of this problem takes the form of the second term only of equation (8-6). Using our principle of superposition that we developed in Chapter 3, the overall solution of our general problem of equations (8-1) and (8-2) then follows from the sum of these two cases, giving the total solution of equation (8-6). We now examine the application of the general solution (8-6) for the cases of two- and one-dimensional problems.

Two-Dimensional Green's Function

The problems defined by equations (8-1) and (8-3) are also applicable for the 2-D case, if ∇^2 is treated as a 2-D Laplacian operator and $\delta(\hat{r} - \hat{r}')$ as a 2-D delta function, that is, for example, $\delta(\hat{r} - \hat{r}') = \delta(x - x')\delta(y - y')$ in the Cartesian coordinate system, $\delta(\hat{r} - \hat{r}') = \delta(r - r')\delta(z - z')$ in the cylindrical coordinate system, and so forth.

For such a case, the physical significance of the 2-D Green's function is as follows: Green's function *represents the temperature at any location* $\hat{r}$ *within the 2-D region R, at any time t, due to an instantaneous line source of unit strength located at* $\hat{r}'$, *releasing its energy spontaneously at time* $t = \tau$ *into a medium of zero temperature*. This interpretation is similar to that for the 3-D problem considered previously, except the source is a *line heat source* of unit strength (i.e., 1 J/m). For the 2-D case, solution (8-6) reduces to

$$\begin{aligned} T(\hat{r}, t) &= \int_{A'} G(\hat{r}, t|\hat{r}', \tau)\big|_{\tau=0} F(\hat{r}')\, dA' \\ &+ \frac{\alpha}{k} \int_{\tau=0}^{t} \int_{A'} G(\hat{r}, t|\hat{r}', \tau) g(\hat{r}', \tau)\, dA'\, d\tau \\ &+ \alpha \sum_{i=1}^{N} \left[\int_{\tau=0}^{t} \int_{\substack{\text{Boundary} \\ \text{path } i}} G(\hat{r}, t|\hat{r}', \tau)\big|_{r'=r_i} \frac{1}{k} f_i(\hat{r}', \tau)\, dl_i'\, d\tau \right] \end{aligned} \tag{8-12}$$

where A is the area of the region under consideration; dl_i is the differential length along the boundary path of the ith surface boundary, $i = 1, 2, \ldots, N$, with N being the number of continuous boundary paths of region A; and where dA' refers to differential surface elements, in the $\hat{r}'$ variables of integration. The same details discussed above in relation to equation (8-6) with regard to a type 1 boundary condition also apply to equation (8-12). Therefore, for a boundary condition of the *first type*,

$$\text{Replace} \quad \left.\frac{1}{k} G\right|_{r'=r_i} \quad \text{by} \quad \pm \left.\frac{\partial G}{\partial n_i'}\right|_{r'=r_i} \tag{8-13}$$

in accordance with equation (8-9). We note that the space integrations over the initial condition function $F(\hat{r})$ and the energy generation function $g(\hat{r}, t)$ are area integrals instead of volume integrals, while the integration over the boundary condition function $f_i(\hat{r}, t)$ is now a contour integral instead of a surface integral.

One-Dimensional Green's Function

For the 1-D temperature field, the problems defined by equations (8-1) and (8-3) are also applicable if ∇^2 is now considered a 1-D Laplacian operator and $\delta(\hat{r} - \hat{r}')$ as a 1-D delta function, that is, for example, $\delta(\hat{r} - \hat{r}') = \delta(x - x')$ in the Cartesian coordinate system, $\delta(\hat{r} - \hat{r}') = \delta(r - r')$ in the cylindrical coordinate system, and so forth. For such a case, the 1-D Green's function *represents the temperature at any location x within the 1-D domain, at any time t, due to an instantaneous surface heat source of unit strength located at* x', *releasing its energy spontaneously at time* $t = \tau$ *into a medium of zero temperature*. This interpretation is similar to that for the two- and three-dimensional problems considered previously, except the source is a *surface heat source* of unit strength (i.e., 1 J/m^2). For the 1-D case, the solution (8-6) reduces to

$$\begin{aligned} T(x,t) &= \int_{L'} G(x,t|x',\tau)\big|_{\tau=0} F(x')x'^P\,dx' \\ &+ \frac{\alpha}{k}\int_{\tau=0}^{t}\int_{L'} G(x,t|x',\tau)g(x',\tau)x'^P\,dx'\,d\tau \\ &+ \alpha\sum_{i=1}^{N}\left\{\int_{\tau=0}^{t}\left[x'^P G(x,t|x',\tau)\right]\big|_{x'=x_i}\frac{1}{k}f_i(\tau)\,d\tau\right\} \end{aligned} \tag{8-14}$$

where the term x'^P is the *Sturm–Liouville weight function* given as

$$P = \begin{cases} 0 & \text{slab} \\ 1 & \text{cylinder} \\ 2 & \text{sphere} \end{cases} \tag{8-15}$$

For cylindrical and spherical problems, we will replace our spatial coordinate x with r. In equation (8-14), L refers to the domain of the 1-D region, and $G(x,t|x',\tau)\big|_{x'=x_i}$ refers to the value of Green's function G evaluated at the respective boundary point, $x' = x_i$, noting that N is now limited to 2. For a boundary condition of the *first type*,

$$\text{Replace} \quad \frac{1}{k}G\bigg|_{x'=x_i} \quad \text{by} \quad \pm\frac{\partial G}{\partial n_i'}\bigg|_{x'=x_i} \tag{8-16}$$

in accordance with equation (8-9). We note that in equation (8-14), the spatial integration over the initial condition function $F(x)$ and the energy generation function $g(x,t)$ are line integrations over the 1-D domain, while the nonhomogeneous boundary condition functions $f_i(t)$ may now only depend on time.

8-2 DETERMINATION OF GREEN'S FUNCTIONS

Once Green's function is available, the temperature distribution $T(\hat{r},t)$ in a medium is given by equations (8-6), (8-12), and (8-14) in terms of Green's function, for the three-, two-, and one-dimensional transient linear heat conduction problems, respectively. Therefore, the establishment of the proper Green's function for any given situation is an integral part of the solution methodology when utilizing the Green's function approach. Reference 1 uses the Laplace transform technique, and reference 2 describes the method of images for the determination of Green's functions. Here we present a very straightforward, yet very general approach, that utilizes the classic separation of variables technique for the determination of Green's functions, which allows us to use all the techniques developed in Chapters 2 through 5.

We consider the following, 3-D, homogeneous transient heat conduction problem:

$$\nabla^2 T(\hat{r}, t) = \frac{1}{\alpha}\frac{\partial T(\hat{r}, t)}{\partial t} \qquad \text{in region R}, \qquad t > 0 \tag{8-17}$$

$$\text{BC:} \qquad \frac{\partial T}{\partial n_i} + H_i T = 0 \qquad \text{on surface } S_i \tag{8-18a}$$

$$\text{IC:} \qquad T(\hat{r}, t = 0) = F(\hat{r}) \tag{8-18b}$$

The solution of this problem has been extensively studied in Chapters 3–5 by the method of separation of variables, and a large number of specific solutions have been already generated for a variety of situations. Suppose the solution of the homogeneous problem (8-17) and (8-18) is symbolically expressed in the form

$$T(\hat{r}, t) = \int_{R'} K(\hat{r}, \hat{r}', t) \cdot F(\hat{r}')\, dV' \tag{8-19}$$

The physical significance of equation (8-19) implies that all the terms in the solution, except the initial condition function, are lumped into a single term $K(\hat{r}, \hat{r}', t)$, which we shall call the *kernel of the integration*. The kernel $K(\hat{r}, \hat{r}', t)$, multiplied by the initial condition function $F(\hat{r}')$ and integrated over the region R, gives the solution to our problem as defined by equations (8-17) and (8-18).

Now we consider Green's function approach for the solution of the same problem, which is obtained from the general solution (8-6) as

$$T(\hat{r}, t) = \int_{R'} G(\hat{r}, t|\hat{r}', \tau)\big|_{\tau=0} \cdot F(\hat{r}')dV' \tag{8-20}$$

noting that the second and third terms are zero for the specific case of no energy generation and all homogeneous boundary conditions.

A comparison of solutions (8-19) and (8-20) implies that

$$G(\hat{r}, t|\hat{r}', \tau)\big|_{\tau=0} \equiv K(\hat{r}, \hat{r}', t) \tag{8-21}$$

We therefore conclude that the kernel $K(\hat{r}, \hat{r}', t)$, obtained by rearranging the homogeneous part of the transient heat conduction equation into the form given by equation (8-19), represents Green's function evaluated for $\tau = 0$, namely, $G(\hat{r}, t|\hat{r}', \tau)\big|_{\tau=0}$. Therefore, the solutions developed in Chapters 3–5 for the homogeneous transient heat conduction problems can readily be rearranged to the form given by equation (8-19) in order to obtain $G(\hat{r}, t|\hat{r}', \tau)\big|_{\tau=0}$. In other words, to obtain $G(\hat{r}, t|\hat{r}', \tau)\big|_{\tau=0}$, the appropriate homogeneous problem is solved and rearranged in the form of a kernel within the integral.

The general solution given by equation (8-6) requires that the full Green's function $G(\hat{r}, t|\hat{r}', \tau)$ should also be known in order to determine the contributions of the energy generation and any nonhomogeneous boundary conditions on the solution.

It has been shown by Özişik [6] that Green's function $G(\hat{r}, t|\hat{r}', \tau)$ for the transient heat conduction is obtainable from $G(\hat{r}, t|\hat{r}', \tau)\big|_{\tau=0}$ by replacing t with $t-\tau$ in the latter. That is,

$$\boxed{G(\hat{r}, t|\hat{r}', \tau) \equiv G(\hat{r}, t|\hat{r}', \tau)\big|_{\tau=0} \qquad \text{for} \qquad t \to (t-\tau)} \tag{8-22}$$

We now illustrate the determination of Green's function from the solution of homogeneous problems with specific examples. In order to alleviate the details of the solution procedure, we will often select examples from those problems that have already been solved in the previous chapters.

Example 8-1 Green's Function for a 1-D Cylinder

Consider a 1-D cylinder over the domain $0 \le r \le b$ that is initially at a temperature $F(r)$. For times $t > 0$, the boundary surface at $r = b$ is maintained at a prescribed temperature $f(t)$, while the entire volume is subjected to internal energy generation given by the function $g(r, t)$. Determine Green's function appropriate for the solution of this problem. The formulation is given as

$$\frac{1}{r}\frac{\partial}{\partial r}\left(r\frac{\partial T}{\partial r}\right) + \frac{1}{k}g(r,t) = \frac{1}{\alpha}\frac{\partial T}{\partial t} \qquad \text{in} \qquad 0 \le r < b, \qquad t > 0 \tag{8-23}$$

$$\text{BC1:} \qquad T(r \to 0) \Rightarrow \text{finite} \tag{8-24a}$$

$$\text{BC2:} \qquad T(r = b, t) = f(t) \tag{8-24b}$$

$$\text{IC:} \qquad T(r, t = 0) = F(r) \tag{8-24c}$$

To determine the desired Green's function, we consider the homogeneous version of the problem defined above, which we will denote as $\Psi(r, t)$, for the same region:

$$\frac{1}{r}\frac{\partial}{\partial r}\left(r\frac{\partial \Psi}{\partial r}\right) = \frac{1}{\alpha}\frac{\partial \Psi}{\partial t} \qquad \text{in} \qquad 0 \le r < b, \qquad t > 0 \tag{8-25}$$

$$\text{BC1:} \qquad \Psi(r \to 0) \Rightarrow \text{finite} \tag{8-26a}$$

$$\text{BC2:} \qquad \Psi(r = b, t) = 0 \tag{8-26b}$$

$$\text{IC:} \qquad \Psi(r, t = 0) = F(r) \tag{8-26c}$$

In the above formulation, we have retained the nonhomogeneous initial condition for our homogeneous $\Psi(r, t)$ problem. Without an initial condition to drive this problem, the solution of equations (8-25) and (8-26) would be the trivial solution $\Psi(r, t) = 0$. We note here, in fact, that Green's function is *independent of the initial condition*, which is readily seen by examination of equation (8-19), where the kernel excludes the initial condition.

This homogeneous problem can be readily solved by the method of separation of variables; and its solution is similar to the problem of Example 4-8. We write the solution in our general form, introducing the Fourier constants, namely,

$$\Psi(r, t) = \frac{2}{b^2} \sum_{n=1}^{\infty} e^{-\alpha\beta_n^2 t} \frac{J_0(\beta_n r)}{J_1^2(\beta_n b)} \int_{r'=0}^{b} r' J_0(\beta_n r') F(r')\, dr' \tag{8-27}$$

where the eigenvalues are the roots of the transcendental equation $J_0(\beta_n b) = 0$. We now reformulate the solution of equation (8-27) into the same general form as equation (8-19), which yields

$$\Psi(r, t) = \int_{r'=0}^{b} \left[\frac{2}{b^2} \sum_{n=1}^{\infty} e^{-\alpha\beta_n^2 t} \frac{1}{J_1^2(\beta_n b)} J_0(\beta_n r) J_0(\beta_n r') \right] F(r') r'\, dr' \tag{8-28}$$

The solution of the homogenous problem (8-25) in terms of Green's function is given, according to equation (8-14), as

$$\Psi(r, t) = \int_{r'=0}^{b} G(r, t|r', \tau)\big|_{\tau=0} F(r') r'\, dr' \tag{8-29}$$

where we have used the Sturm–Liouville weight function $p = 1$ for the cylindrical coordinate system. By comparing equations (8-28) and (8-29), we find the Green's function for $\tau = 0$,

$$G(r, t|r', \tau)\big|_{\tau=0} = \frac{2}{b^2} \sum_{n=1}^{\infty} e^{-\alpha\beta_n^2 t} \frac{1}{J_1^2(\beta_n b)} J_0(\beta_n r) J_0(\beta_n r') \tag{8-30}$$

We now replace t in equation (8-30) with $t-\tau$, as detailed in equation (8-22), to yield the desired Green's function as

$$G(r, t|r', \tau) = \frac{2}{b^2} \sum_{n=1}^{\infty} e^{-\alpha\beta_n^2 (t-\tau)} \frac{1}{J_1^2(\beta_n b)} J_0(\beta_n r) J_0(\beta_n r') \tag{8-31}$$

Example 8-2 Green's Function for a 1-D Semi-infinite Plane

Consider a 1-D semi-infinite plane over the domain $0 \le x < \infty$ that is initially at a temperature $F(x)$. For times $t > 0$, the boundary surface at $x = 0$ is maintained at a prescribed temperature $f(t)$, while the entire volume is subjected to internal energy generation given by the function $g(x,t)$. Determine Green's function appropriate for the solution of this problem. The formulation is given as

$$\frac{\partial^2 T}{\partial x^2} + \frac{1}{k} g(x, t) = \frac{1}{\alpha} \frac{\partial T}{\partial t} \quad \text{in} \quad 0 < x < \infty, \quad t > 0 \tag{8-32}$$

$$\text{BC1:} \qquad T(x=0,t) = f(t) \tag{8-33b}$$

$$\text{IC:} \qquad T(x,t=0) = F(x) \tag{8-33c}$$

To determine the desired Green's function, we consider the homogeneous version of the problem defined above for the same region:

$$\frac{\partial^2 \Psi}{\partial x^2} = \frac{1}{\alpha}\frac{\partial \Psi}{\partial t} \qquad \text{in} \qquad 0 < x < \infty, \qquad t > 0 \tag{8-34}$$

$$\text{BC1:} \qquad \Psi(x=0,t) = 0 \tag{8-35a}$$

$$\text{IC:} \qquad \Psi(x,t=0) = F(x) \tag{8-35b}$$

As in the previous example, we have retained the nonhomogeneous initial condition for our homogeneous $\Psi(x,t)$ problem. The solution of the homogeneous problem is directly obtained from equation (6-35), giving

$$\Psi(x,t) = \frac{1}{(4\pi\alpha t)^{1/2}} \int_{x'=0}^{\infty} F(x') \left\{ \exp\left[-\frac{(x-x')^2}{4\alpha t}\right] - \exp\left[-\frac{(x+x')^2}{4\alpha t}\right] \right\} dx' \tag{8-36}$$

which we rearrange to the form

$$\Psi(x,t) = \int_{x'=0}^{\infty} \frac{1}{(4\pi\alpha t)^{1/2}} \left\{ \exp\left[-\frac{(x-x')^2}{4\alpha t}\right] - \exp\left[-\frac{(x+x')^2}{4\alpha t}\right] \right\} F(x')\, dx' \tag{8-37}$$

By comparing this solution with equation (8-20), we readily conclude that $G(x,t|x',\tau)\big|_{\tau=0}$ is given by

$$G(x,t|x',\tau)\big|_{\tau=0} = \frac{1}{(4\pi\alpha t)^{1/2}} \left\{ \exp\left[-\frac{(x-x')^2}{4\alpha t}\right] - \exp\left[-\frac{(x+x')^2}{4\alpha t}\right] \right\} \tag{8-38}$$

Green's function $G(x,t|x',\tau)$ is now determined by replacing t with $t-\tau$ per equation (8-22), yielding the result:

$$G(x,t|x',\tau) = \frac{1}{[4\pi\alpha(t-\tau)]^{1/2}} \left\{ \exp\left[-\frac{(x-x')^2}{4\alpha(t-\tau)}\right] - \exp\left[-\frac{(x+x')^2}{4\alpha(t-\tau)}\right] \right\} \tag{8-39}$$

Example 8-3 Green's Function for a 1-D Plane Wall
Consider a 1-D plane wall over the domain $0 \le x \le L$ that is initially at a temperature $F(x)$. For times $t > 0$, the boundary surface at $x = 0$ is exposed to a time-dependent incident heat flux $f_1(t)$, while the surface at $x = L$ is subjected to convection heat transfer with a time-dependent fluid temperature. In addition, the entire volume is subjected to internal energy generation given by the function

$g(x,t)$. Determine Green's function appropriate for the solution of this problem. The formulation is given as:

$$\frac{\partial^2 T}{\partial x^2} = \frac{1}{\alpha}\frac{\partial T}{\partial t} \quad \text{in} \quad 0 < x < L, \qquad t > 0 \tag{8-40}$$

$$\text{BC1:} \qquad -k\left.\frac{\partial T}{\partial x}\right|_{x=0} = f_1(t) \tag{8-41a}$$

$$\text{BC2:} \qquad \left.\frac{\partial T}{\partial x}\right|_{x=L} + HT|_{x=L} = f_2(t) \tag{8-41b}$$

$$\text{IC:} \qquad T(x, t = 0) = F(x) \tag{8-41c}$$

To determine the desired Green's function, we consider the homogeneous version of the problem defined above over the same region:

$$\frac{\partial^2 \Psi}{\partial x^2} = \frac{1}{\alpha}\frac{\partial \Psi}{\partial t} \quad \text{in} \quad 0 < x < L, \qquad t > 0 \tag{8-42}$$

$$\text{BC1:} \qquad \left.\frac{\partial \Psi}{\partial x}\right|_{x=0} = 0 \tag{8-43a}$$

$$\text{BC2:} \qquad \left.\frac{\partial \Psi}{\partial x}\right|_{x=L} + H\Psi|_{x=L} = 0 \tag{8-43b}$$

$$\text{IC:} \qquad \Psi(x, t = 0) = F(x) \tag{8-43c}$$

The solution of the homogeneous problem $\Psi(x, t)$ was considered in Chapter 3 and is readily obtained from equation (3-38) after substitution for the norm, giving

$$\Psi(x, t) = \sum_{n=1}^{\infty}\left[\frac{2\left(\lambda_n^2 + H^2\right)}{L(\lambda_n^2 + H^2) + H}\cos\lambda_n x\, e^{-\alpha\lambda_n^2 t}\int_{x'=0}^{L} F(x')\cos\lambda_n x'\, dx'\right] \tag{8-44}$$

which we may rearrange to the form

$$\Psi(x, t) = \int_{x'=0}^{L}\left[\sum_{n=1}^{\infty}\frac{2\left(\lambda_n^2 + H^2\right)}{L(\lambda_n^2 + H^2) + H}\cos\lambda_n x\cos\lambda_n x' e^{-\alpha\lambda_n^2 t}\right]F(x')\, dx' \tag{8-45}$$

By comparing this solution with equation (8-20), we readily conclude that $G(x, t|x', \tau)\big|_{\tau=0}$ is given by

$$G(x, t|x', \tau)\big|_{\tau=0} = \sum_{n=1}^{\infty}\frac{2\left(\lambda_n^2 + H^2\right)}{L(\lambda_n^2 + H^2) + H}\cos\lambda_n x\cos\lambda_n x' e^{-\alpha\lambda_n^2 t} \tag{8-46}$$

Green's function $G(x, t|x', \tau)$ is now determined by replacing t with $t-\tau$ per equation (8-22), yielding the result

$$G(x, t|x', \tau) = 2\sum_{n=1}^{\infty} \frac{\lambda_n^2 + H^2}{L(\lambda_n^2 + H^2) + H} \cos \lambda_n x \cos \lambda_n x' e^{-\alpha\lambda_n^2(t-\tau)} \tag{8-47}$$

where the eigenvalues are defined by the transcendental equation

$$\lambda_n \tan \lambda_n L = H \quad \rightarrow \quad \lambda_n \quad \text{for} \quad n = 1, 2, 3, \ldots \tag{8-48}$$

8-3 REPRESENTATION OF POINT, LINE, AND SURFACE HEAT SOURCES WITH DELTA FUNCTIONS

Before continuing with our solution of the heat equation using Green's functions, it is useful here to develop some additional mathematical tools for representing various point, line, and planar heat sources. The energy source will be called an *instantaneous source* if it releases its energy spontaneously (i.e., at a single instant) or a *continuous source* if it releases its energy continuously over time. In the definition of Green's function, we also refer to a *point source*, a *line source*, and a *surface or planar source* of unit strength, in addition to the customarily used volumetric internal energy heat source that has the dimensions of W/m^3.

In order to identify such energy sources with a unified notation we introduce the symbol

$$g_B^A$$

where the superscript A refers to

$$A \equiv \begin{cases} i & \text{instantaneous} \\ c & \text{continuous} \end{cases} \tag{8-49}$$

and the subscript B denotes

$$B \equiv \begin{cases} p & \text{point} \\ L & \text{line} \\ s & \text{surface} \end{cases} \tag{8-50}$$

and where no subscript will be used for the volumetric source. Thus, based on the above notation, we write

$$g_p^{\text{i}} = \text{instantaneous point source}$$

$$g_p^{\text{c}} = \text{continuous point source}$$

$$g_L^{\text{i}} = \text{instantaneous line source}$$

$$g_s^{\text{c}} = \text{continuous surface source}$$

$$g^{\text{i}} = \text{instantaneous volumetric source}$$

$$g = \text{volumetric source}$$

and so forth, with all corresponding equivalent *volumetric source* terms having the units of W/m^3, as developed below.

In the analytic solution of temperature $T(\hat{r}, t)$ in terms of Green's functions given by equations (8-6), (8-12), and (8-14), the energy generation term $g(\hat{r}, t)$ appears under the integral sign. In order to perform the integration over a point source, surface source, instantaneous source, and so on, proper mathematical representations should be used to define such sources. Here we describe a procedure for the identification of such sources with the delta function notation and the determination of their proper dimensions.

Three-Dimensional Representations

Cartesian Coordinates Consider an instantaneous point heat source located at the position (x', y', z') and releasing its entire energy spontaneously at the instant $t = \tau$. Such a source is related to the volumetric heat source $g(x, y, z, t)$ by

$$g(x, y, z, t) \equiv g_p^{\text{i}} \delta(x - x') \delta(y - y') \delta(z - z') \delta(t - \tau) \tag{8-51}$$

where $\delta(-)$ is the *Dirac delta function*. A brief description of the properties of Dirac's delta function is given in Appendix VI, noting that the dimensions of the delta function are the inverse units of the argument. The delta function has the unique property that it is exactly zero everywhere except at an argument of zero, where it is infinite, such that the integral

$$\boxed{\int_{x=-\infty}^{\infty} \delta(x)\, dx \equiv 1} \tag{8-52}$$

We may shift the delta function to position $x = b$, where now the function $\delta(x - b)$ is zero everywhere except at $x = b$, with equation (8-52) becoming

$$\int_{x=-\infty}^{\infty} \delta(x - b)\, dx \equiv 1 \tag{8-53}$$

Finally, if the function *F(x)* is a continuous function, the following integrals hold:

$$\int_{x=-\infty}^{\infty} F(x) \delta(x)\, dx = F(0) \tag{8-54}$$

$$\int_{x=-\infty}^{\infty} F(x) \delta(x - b)\, dx = F(b) \tag{8-55}$$

and if $a \leq c \leq b$, then

$$\int_{x=a}^{b} F(x) \delta(x - c)\, dx = F(c) \tag{8-56}$$

We now consider the dimensions of equation (8-51), which yields

$$\begin{aligned} g(x,y,z,t) &\equiv g_p^{\mathrm{i}}\delta(x-x')\delta(y-y')\delta(z-z')\delta(t-\tau) \\ \mathrm{W/m^3} &= g_p^{\mathrm{i}} \quad \mathrm{m^{-1}} \quad \mathrm{m^{-1}} \quad \mathrm{m^{-1}} \quad \mathrm{s^{-1}} \end{aligned} \tag{8-57}$$

Comparison of the left-hand side and right-hand side units gives the dimensions of an instantaneous point source g_p^{i} as W · s or simply as J. The point source g_p^{i} therefore represents the total quantity of energy released by the point source.

Cylindrical Coordinates Consider an instantaneous point heat source located at position (r', ϕ', z') and releasing its entire energy spontaneously at instant $t = \tau$. Such a point source is related to the volumetric heat source $g(r, \phi, z, t)$ by

$$\begin{aligned} g(r,\phi,z,t) &\equiv g_p^{\mathrm{i}}\frac{1}{r}\delta(r-r')\delta(\phi-\phi')\delta(z-z')\delta(t-\tau) \\ \mathrm{W/m^3} &= g_p^{\mathrm{i}}\mathrm{m^{-1}} \quad \mathrm{m^{-1}} \quad - \quad \mathrm{m^{-1}} \quad \mathrm{s^{-1}} \end{aligned} \tag{8-58}$$

where again we see that the instantaneous point source g_p^{i} has units W · s or J. The term $1/r$ appearing in equation (8-58) is due to the scale factors associated with the transformation of the reciprocal of the volume element $1/dV$ from the Cartesian to a curvilinear coordinate system [12, p. 12], which in the case of the cylindrical coordinate system, gives $1/r$.

Spherical Coordinates Consider an instantaneous point heat source located at position (r', ϕ', μ') and releasing its entire energy spontaneously at instant $t = \tau$. Such a point source is related to the volumetric heat source $g(r, \phi, \mu, t)$ by

$$\begin{aligned} g(r,\phi,\mu,t) &\equiv g_p^{\mathrm{i}}\frac{1}{r^2\sqrt{1-\mu^2}}\delta(r-r')\delta(\phi-\phi')\delta(\mu-\mu')\delta(t-\tau) \\ \mathrm{W/m^3} &= g_p^{\mathrm{i}} \quad \mathrm{m^{-2}} \quad \mathrm{m^{-1}} \quad - \quad - \quad \mathrm{s^{-1}} \end{aligned} \tag{8-59}$$

where again we see that the instantaneous point source g_p^{i} has units W · s or J. The term $\left(r^2\sqrt{1-\mu^2}\right)^{-1}$ appearing in equation (8-59), as stated above, is due to the scale factors associated with the transformation of the reciprocal of the volume element $1/dV$ from the Cartesian to the spherical coordinate system. That is, $(a_r a_\phi a_\theta)^{-1} = (1 \cdot r\sin\theta \cdot r)^{-1} = (r^2\sqrt{1-\mu^2})^{-1}$, where $\mu = \cos\theta$.

In the case of a *continuous point source* g_p^{c}, the above representations have no delta function with respect to time; hence the dimension of g_p^c is W, or energy per unit time, and the delta function in time is dropped in equations (8-57)–(8-59).

One-Dimensional Representations

We now examine the representation of an instantaneous energy source and a continuous energy source in the 1-D Cartesian, cylindrical, and spherical coordinate systems.

Cartesian Coordinates Consider an instantaneous planar heat source g_s^{i} located at position x' and releasing its entire energy spontaneously at the instant $t = \tau$, as shown in Figure 8-2.

Such a source is related to the volumetric heat source $g(x, t)$ by

$$\begin{aligned} g(x, t) &\equiv g_s^{\mathrm{i}} \delta(x - x')\delta(t - \tau) \\ \mathrm{W/m^3} &= g_s^{\mathrm{i}} \quad \mathrm{m^{-1}} \qquad \mathrm{s^{-1}} \end{aligned} \tag{8-60}$$

hence g_s^{i} has the dimensions $(\mathrm{W \cdot s})/\mathrm{m}^2$ or $\mathrm{J/m^2}$. The planar source is readily converted to a *continuous source* by replacing g_s^{i} with g_s^{c}, where g_s^{c} has dimensions $\mathrm{W/m^2}$, and dropping the delta function in time.

Cylindrical Coordinates Consider an instantaneous cylindrical surface heat source g_s^{i} located at the radial position r' and releasing its entire energy spontaneously at the instant $t = \tau$, as shown in Figure 8-3.

Such a source is related to the volumetric heat source $g(r, t)$ by

$$\begin{aligned} g(r, t) &\equiv g_{\mathrm{s}}^{\mathrm{i}} \frac{1}{2\pi r} \delta(r - r')\delta(t - \tau) \\ \mathrm{W/m^3} &= g_s^{\mathrm{i}} \; \mathrm{m^{-1}} \quad \mathrm{m^{-1}} \qquad \mathrm{s^{-1}} \end{aligned} \tag{8-61}$$

We see that an instantaneous cylindrical surface heat source g_{s}^{i} has the dimensions $(\mathrm{W \cdot s})/\mathrm{m}$ or $\mathrm{J/m}$, hence energy per unit length. In equation (8-61), the additional

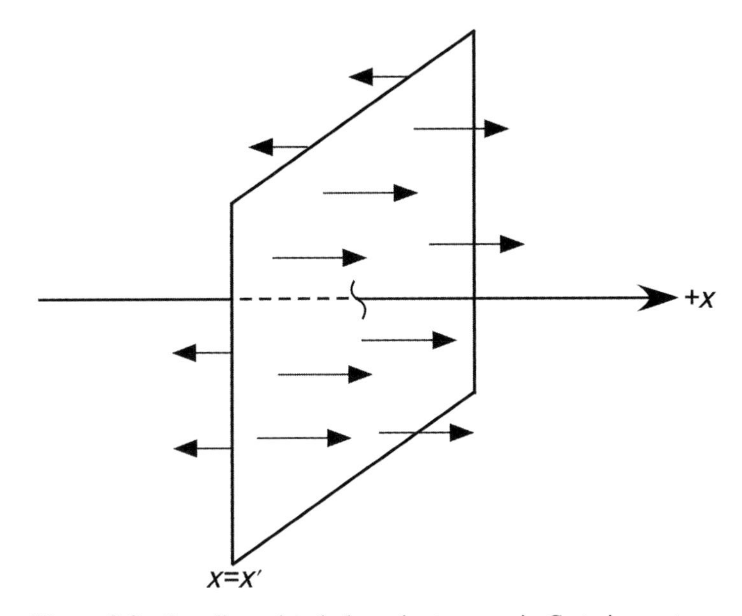

Figure 8-2 One-dimensional planar heat source in Cartesian system.

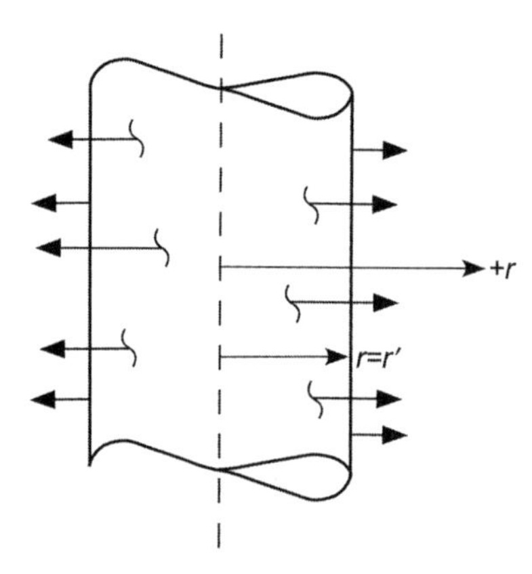

Figure 8-3 One-dimensional planar heat source in the cylindrical system.

term appearing in the denominator is associated with the scale factor of the transformation. That is, g_s^i represents the strength of the cylindrical surface source per unit length, and the quantity $g_s^{\mathrm{i}}/2\pi r$ represents the source strength per unit area. The instantaneous planar source is readily converted to a *continuous planar source* by replacing g_s^{i} with g_s^{c}, where g_s^{c} has dimensions W/m, and dropping the delta function in time.

Spherical Coordinates Consider an instantaneous spherical surface heat source g_s^{i} located at the radial position r' and releasing its entire energy spontaneously at the instant $t = \tau$. Such a source is related to the volumetric heat source $g(r, t)$ by

$$\begin{aligned} g(r,t) &\equiv g_s^{\mathrm{i}}\ \frac{1}{4\pi r^2}\delta(r-r')\delta(t-\tau) \\ \mathrm{W/m^3} &= g_s^{\mathrm{i}}\ \ \mathrm{m^{-2}} \quad\ \mathrm{m^{-1}} \qquad \mathrm{s^{-1}} \end{aligned} \tag{8-62}$$

Therefore an instantaneous spherical surface heat source g_s^{i} has the dimensions (W · s) or J; hence simply energy. In equation (8-62), the additional term appearing in the denominator is again associated with the scale factor of the transformation; hence g_s^{i} represents the strength of the spherical surface heat source, and the quantity $g_s^{\mathrm{i}}/4\pi r^2$ represents the source strength per unit area. The instantaneous planar source is readily converted to a *continuous planar source* by replacing g_s^{i} with g_s^{c}, where g_s^{c} has dimension W, and dropping the delta function in time.

One-Dimensional Distributed Heat Source Consider a 1-D, distributed heat source $g^{\mathrm{i}}(x)$ that releases its entire energy spontaneously at the instant $t = \tau$. Such a source is related to the volumetric heat source $g(x, t)$ by

$$\begin{aligned} g(x,t) &\equiv g^{\mathrm{i}}(x)\delta(t-\tau) \\ \mathrm{W/m^3} &= \ \ g_s^{\mathrm{i}} \qquad \mathrm{s^{-1}} \end{aligned} \tag{8-63}$$

hence $g^{i}(x)$ has the dimensions $(\mathrm{W\cdot s})/\mathrm{m}^3$ or $\mathrm{J/m}^3$. If the energy is released at $\tau = 0$, the distributed heat source becomes

$$g(x,t) \equiv g^{i}(x)\delta(t) \tag{8-64}$$

where $g^{i}(x)$ represents the amount of energy per unit volume ($\mathrm{J/m}^3$) released at each spatial location at instant $t = 0$.

8-4 APPLICATIONS OF GREEN'S FUNCTION IN THE RECTANGULAR COORDINATE SYSTEM

In this section we illustrate with examples the application of Green's function technique in the solution of nonhomogeneous boundary value problems of heat conduction in the Cartesian coordinate system, including time-dependent boundary conditions and energy generation terms. For convenience in the determination of Green's function, we will often consider examples for which solutions are available in Chapters 3 and 6 for their corresponding homogeneous problems.

Example 8-4 One-Dimensional Infinite Plane with Generation
Consider a 1-D infinite plane over the domain $-\infty < x < \infty$ that is initially at a temperature $F(x)$. For times $t > 0$, there is internal energy generation with the solid at the rate $g(x,t)$ per unit time, per unit volume ($\mathrm{W/m}^3$). Obtain an expression for the temperature distribution $T(x,t)$ for times $t > 0$ by the Green's function technique. The formulation is given as

$$\frac{\partial^2 T}{\partial x^2} + \frac{1}{k}g(x,t) = \frac{1}{\alpha}\frac{\partial T}{\partial t} \quad \text{in} \quad -\infty < x < \infty, \quad t>0 \tag{8-65}$$

$$\text{IC:} \quad T(x,t=0) = F(x) \tag{8-66}$$

To determine the desired Green's function, we consider the homogeneous version of the problem defined above for the same region:

$$\frac{\partial^2 \Psi}{\partial x^2} = \frac{1}{\alpha}\frac{\partial \Psi}{\partial t} \quad \text{in} \quad -\infty < x < \infty, \quad t>0 \tag{8-67}$$

$$\text{IC:} \quad \Psi(x,t=0) = F(x) \tag{8-68}$$

The solution of the homogeneous problem $\Psi(x,t)$ is directly obtained from equation (6-132), which after moving all terms into the integral, gives

$$\Psi(x,t) = \int_{x'=-\infty}^{\infty} \left\{ \frac{1}{(4\pi\alpha t)^{1/2}} \exp\left[-\frac{(x-x')^2}{4\alpha t} \right] \right\} F(x')\,dx' \tag{8-69}$$

The solution of this homogeneous problem can be written in terms of Green's function according to equation (8-20), namely,

$$\Psi(x, t) = \int_{x'=-\infty}^{\infty} G(x, t|x', \tau)\big|_{\tau=0} F(x')\, dx' \tag{8-70}$$

A comparison of equations (8-69) and (8-70) yields

$$G(x, t|x', \tau)\big|_{\tau=0} = \frac{1}{(4\pi\alpha t)^{1/2}} \exp\left[-\frac{(x-x')^2}{4\alpha t}\right] \tag{8-71}$$

The desired Green's function $G(x, t|x', \tau)$ is now determined by replacing t with $t-\tau$ per equation (8-22), yielding the result

$$G(x, t|x', \tau) = \frac{1}{[4\pi\alpha(t-\tau)]^{1/2}} \exp\left[-\frac{(x-x')^2}{4\alpha(t-\tau)}\right] \tag{8-72}$$

For the original nonhomogeneous problem, namely, a 1-D Cartesian problem, the solution is given by equation (8-14) with $p = 0$, which becomes in variable form:

$$\begin{aligned} T(x, t) = &\int_{x'=-\infty}^{\infty} G(x, t|x', \tau)\big|_{\tau=0} F(x')\, dx' \\ &+ \frac{\alpha}{k}\int_{\tau=0}^{t}\int_{x'=-\infty}^{\infty} G(x, t|x', \tau) g(x', \tau)\, dx'\, d\tau \end{aligned} \tag{8-73}$$

where the generation term and the initial temperature distribution term are both of the integration variables x' and τ. Substitution of equation (8-72) into equation (8-73) yields the overall temperature solution

$$\begin{aligned} T(x, t) = &\frac{1}{(4\pi\alpha t)^{1/2}} \int_{x'=-\infty}^{\infty} \exp\left[-\frac{(x-x')^2}{4\alpha t}\right] F(x')\, dx' \\ &+ \frac{\alpha}{k}\int_{\tau=0}^{t}\int_{x'=-\infty}^{\infty} \frac{1}{[4\pi\alpha(t-\tau)]^{1/2}} \exp\left[-\frac{(x-x')^2}{4\alpha(t-\tau)}\right] g(x', \tau)\, dx' d\tau \end{aligned} \tag{8-74}$$

We now examine some special cases of the solution given by equation (8-74).

1. There is no internal energy generation. By setting $g(x, t) = 0$, equation (8-74) reduces to

$$T(x, t) = \frac{1}{(4\pi\alpha t)^{1/2}} \int_{x'=-\infty}^{\infty} \exp\left[-\frac{(x-x')^2}{4\alpha t}\right] F(x')\, dx' \tag{8-75}$$

which is identical to our original solution for the $\Psi(x, t)$ problem as given by equation (6-132). This is correct, as the problem reduces exactly to the problem of equation (6-132) under the condition of no generation.

2. Let the medium initially be at zero temperature, and let there be an instantaneous distributed heat source $g^{\mathrm{i}}(x)$ (J/m^3) that releases its heat spontaneously at time $t = 0$. We let $F(x) = 0$ and define the heat source in consideration of equation (8-64), namely,

$$g(x,t) = g^{\mathrm{i}}(x)\delta(t) \tag{8-76}$$

Substitution of the above into equation (8-74) yields

$$T(x,t) = \frac{\alpha}{k}\int_{\tau=0}^{t}\int_{x'=-\infty}^{\infty}\frac{1}{[4\pi\alpha(t-\tau)]^{1/2}}\exp\left[-\frac{(x-x')^2}{4\alpha(t-\tau)}\right] \times g^{i}(x')\delta(\tau)\,dx'\,d\tau \tag{8-77}$$

where the τ integral is readily evaluated making use of equation (8-54), giving

$$T(x,t) = \frac{\alpha}{k}\frac{1}{(4\pi\alpha t)^{1/2}}\int_{x'=-\infty}^{\infty}\exp\left[-\frac{(x-x')^2}{4\alpha t}\right]g^{i}(x')\,dx' \tag{8-78}$$

Using the definition of thermal diffusivity ($\alpha \equiv k/\rho c$), we may rearrange equation (8-78) as follows:

$$T(x,t) = \frac{1}{(4\pi\alpha t)^{1/2}}\int_{x'=-\infty}^{\infty}\exp\left[-\frac{(x-x')^2}{4\alpha t}\right]\left[\frac{g^{i}(x')}{\rho c}\right]dx' \tag{8-79}$$

A comparison of equations (8-75) and (8-79) reveals an identical solution for the special case that

$$F(x) = \frac{g^{\mathrm{i}}(x)}{\rho c} \tag{8-80}$$

The right-hand side of equation (8-80) is the initial energy released per unit volume divided by the heat capacity per unit volume, which corresponds to a temperature rise, namely,

$$\text{Energy} = mc\Delta T \quad \rightarrow \quad \frac{\text{energy}}{\text{volume}} = \rho c\Delta T$$

Equation (8-80) implies that the heat conduction problem for an instantaneous distributed heat source $g^{\mathrm{i}}(x)$ releasing its energy at time $t = 0$ into a medium of zero temperature is equivalent to an initial value problem with the initial temperature distribution as given by equation (8-80), which is a statement of conservation of energy.

3. Let the medium initially be at zero temperature, and let there be a planar heat source $g_s^{\mathrm{c}}(t)$ (W/m^2) positioned at the location $x = a$ that releases its heat continuously starting at time $t = 0$. We let $F(x) = 0$, and define the planar heat source in consideration of equation (8-60), giving

$$g(x,t) = g_s^{\mathrm{c}}(t)\delta(x-a) \tag{8-81}$$

Substitution of the above into equation (8-74) yields

$$T(x,t) = \frac{\alpha}{k} \int_{\tau=0}^{t} \int_{x'=-\infty}^{\infty} \frac{1}{[4\pi\alpha(t-\tau)]^{1/2}} \exp\left[-\frac{(x-x')^2}{4\alpha(t-\tau)}\right] \times g_s^c(\tau)\delta(x'-a)\,dx'\,d\tau \tag{8-82}$$

where the x' integral is readily evaluated making use of equation (8-54), giving

$$T(x,t) = \frac{\alpha}{k} \int_{\tau=0}^{t} \frac{1}{[4\pi\alpha(t-\tau)]^{1/2}} \exp\left[-\frac{(x-a)^2}{4\alpha(t-\tau)}\right] g_s^c(\tau)\,d\tau \tag{8-83}$$

Example 8-5 One-Dimensional Plane Wall with Generation

Consider a 1-D plane wall over the domain $0 \le x \le L$ that is initially at a temperature $F(x)$. For times $t > 0$, the boundary surfaces at $x = 0$ and $x = L$ are maintained at temperatures given by the time-dependent functions $f_1(t)$ and $f_2(t)$, respectively. In addition, there is internal energy generation within the solid at the rate $g(x,t)$ per unit time, per unit volume (W/m^3). The problem is shown in Figure 8-4.

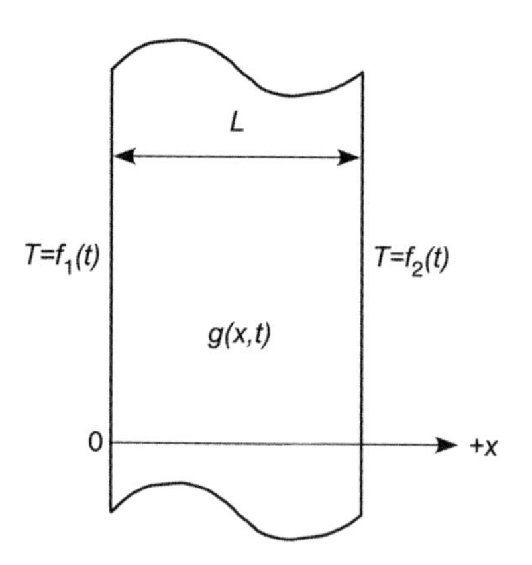

Figure 8-4 Problem description for Example 8-5.

Obtain an expression for the temperature distribution $T(x,t)$ for times $t > 0$ by the Green's function technique. The formulation is given as

$$\frac{\partial^2 T}{\partial x^2} + \frac{1}{k} g(x,t) = \frac{1}{\alpha}\frac{\partial T}{\partial t} \quad \text{in} \quad 0 < x < L, \quad t > 0 \tag{8-84}$$

BC1: $T(x=0) = f_1(t)$ (8-85a)

BC2: $T(x=L) = f_2(t)$ (8-85b)

IC: $T(x,t=0) = F(x)$ (8-85c)

To determine the desired Green's function, we consider the homogeneous version of the problem defined above over the same region:

$$\frac{\partial^2 \Psi}{\partial x^2} = \frac{1}{\alpha}\frac{\partial \Psi}{\partial t} \quad \text{in} \quad 0 < x < L, \qquad t > 0 \tag{8-86}$$

$$\text{BC1:} \qquad \Psi(x = 0) = 0 \tag{8-87a}$$

$$\text{BC2:} \qquad \Psi(x = L) = 0 \tag{8-87b}$$

$$\text{IC:} \qquad \Psi(x, t = 0) = F(x) \tag{8-87c}$$

The solution of the homogeneous problem $\Psi(x, t)$ is similar to the capstone problem of Chapter 3 and is readily obtained from equations (3-216) and (3-218) by replacing Θ_0 with $F(x)$, and noting that the norm is $L/2$. These substitutions, after rearrangement, yield the solution

$$\Psi(x, t) = \int_{x'=0}^{L} \left[\frac{2}{L}\sum_{n=1}^{\infty} \sin\lambda_n x \sin\lambda_n x' e^{-\alpha\lambda_n^2 t}\right] F(x')\,dx' \tag{8-88}$$

where the eigenvalues are given by the expression

$$\lambda_n = \frac{n\pi}{L}, \qquad n = 1, 2, 3, \ldots \tag{8-89}$$

By comparing this solution with equation (8-20), written here in the 1-D Cartesian form as

$$\Psi(x, t) = \int_{x'=0}^{L} G(x, t|x', \tau)\big|_{\tau=0} F(x')\,dx' \tag{8-90}$$

we extract $G(x, t|x', \tau)\big|_{\tau=0}$ from the kernel, which becomes

$$G(x, t|x', \tau)\big|_{\tau=0} = \frac{2}{L}\sum_{n=1}^{\infty} \sin\lambda_n x \sin\lambda_n x' e^{-\alpha\lambda_n^2 t} \tag{8-91}$$

Green's function $G(x, t|x', \tau)$ is now determined by replacing t with $t-\tau$ per equation (8-22), yielding the result

$$G(x, t|x', \tau) = \frac{2}{L}\sum_{n=1}^{\infty} \sin\lambda_n x \sin\lambda_n x' e^{-\alpha\lambda_n^2 (t-\tau)} \tag{8-92}$$

where the eigenvalues are defined by equation (8-89). Then the solution of the nonhomogeneous problem defined by (8-84) and (8-85) is given in terms of the

Green's function, according to equation (8-14) with $p = 0$, as

$$\begin{aligned} T(x,t) = & \int_{x'=0}^{L} G(x,t|x',\tau)\big|_{\tau=0} F(x')\,dx' \\ & + \frac{\alpha}{k}\int_{\tau=0}^{t}\int_{x'=0}^{L} G(x,t|x',\tau)g(x',\tau)\,dx'\,d\tau \\ & + \alpha\int_{\tau=0}^{t} \left.\frac{\partial G(x,t|x',\tau)}{\partial x'}\right|_{x'=0} f_1(\tau)\,d\tau \\ & - \alpha\int_{\tau=0}^{t} \left.\frac{\partial G(x,t|x',\tau)}{\partial x'}\right|_{x'=L} f_2(\tau)\,d\tau \end{aligned} \tag{8-93}$$

As discussed in Section 8-1, we must adjust the boundary condition terms for a type 1 boundary condition (i.e., prescribed temperature). Accordingly, we made the following two substitutions:

$$-k\left.\frac{\partial G}{\partial x'}\right|_{x'=0} = -h\,G|_{x'=0} \quad\rightarrow\quad \left.\frac{\partial G}{\partial x'}\right|_{x'=0} = \frac{1}{k}\,G|_{x'=0} \tag{8-94a}$$

$$-k\left.\frac{\partial G}{\partial x'}\right|_{x'=L} = +h\,G|_{x'=L} \quad\rightarrow\quad -\left.\frac{\partial G}{\partial x'}\right|_{x'=L} = \frac{1}{k}\,G|_{x'=L} \tag{8-94b}$$

where we have used our sign convention of matching positive conduction and convection at each boundary and have set h to unity.

Introducing the Green's function of equation (8-92) into equation (8-93) yields the temperature solution in the form

$$\begin{aligned} T(x,t) = & \frac{2}{L}\sum_{n=1}^{\infty}\sin\lambda_n x e^{-\alpha\lambda_n^2 t}\int_{x'=0}^{L}\sin\lambda_n x' F(x')\,dx' \\ & + \frac{\alpha}{k}\frac{2}{L}\sum_{n=1}^{\infty}\sin\lambda_n x e^{-\alpha\lambda_n^2 t}\int_{\tau=0}^{t}\int_{x'=0}^{L}\sin\lambda_n x' g(x',\tau)e^{\alpha\lambda_n^2\tau}\,dx'd\tau \\ & + \alpha\frac{2}{L}\sum_{n=1}^{\infty}\lambda_n\sin\lambda_n x e^{-\alpha\lambda_n^2 t}\int_{\tau=0}^{t} e^{\alpha\lambda_n^2\tau} f_1(\tau)\,d\tau \\ & - \alpha\frac{2}{L}\sum_{n=1}^{\infty}(-1)^n\lambda_n\sin\lambda_n x e^{-\alpha\lambda_n^2 t}\int_{\tau=0}^{t} e^{\alpha\lambda_n^2\tau} f_2(\tau)\,d\tau \end{aligned} \tag{8-95}$$

where the eigenvalues are given by equation (8-89). We have made two simplifications in the above equation, namely, $\cos\lambda_n x'\big|_{x'=0} = 1$ in the third term and $\cos\lambda_n x'\big|_{x'=L} = (-1)^n$ in the fourth term.

Solution (8-95) appears to vanish at the two boundaries $x = 0$ and $x = L$, instead of yielding the boundary conditions functions $f_1(t)$ and $f_2(t)$ at these locations. The reason for this is that these two terms involve series that are not uniformly convergent at the locations $x = 0$ and $x = L$. Therefore, the above solution is valid in the open interval $0 < x < L$. Such phenomena occur when the solution derives its basis from the orthogonal expansion technique with the boundary condition being utilized to develop the eigencondition. Similar results are reported on pages 102 and 103 of reference 1. This difficulty can be alleviated by integrating by parts the last two integrals in equation (8-95) and replacing the resulting series expressions by their equivalent closed form expressions. Another approach to avoid this difficulty is to remove the nonhomogeneities from the boundary condition by a splitting-up procedure as described in Section 1-7 of Özisik [12]. We now examine some special cases of our solution equation (8-95):

1. The medium is initially at zero temperature. The boundaries at $x = 0$ and $x = L$ are kept at zero temperature for times $t > 0$, and an instantaneous distributed heat source $g^{\mathrm{i}}(x)$ (J/m^3) releases its heat spontaneously at time $t = 0$. We now let $F(x) = 0$, as well as $f_1(t) = 0$ and $f_2(t) = 0$, and define the heat source in consideration of equation (8-64), namely, as

$$g(x,t) = g^{\mathrm{i}}(x)\delta(t) \tag{8-96}$$

Substitution of the above into equation (8-95) yields the solution

$$T(x,t) = \frac{\alpha}{k}\frac{2}{L}\sum_{n=1}^{\infty} \sin\lambda_n x\, e^{-\alpha\lambda_n^2 t} \int_{\tau=0}^{t}\int_{x'=0}^{L} \sin\lambda_n x' g^{i}(x')\delta(\tau)e^{\alpha\lambda_n^2\tau}\, dx'\, d\tau \tag{8-97}$$

where the τ integral is readily evaluated making use of equation (8-54), giving

$$T(x,t) = \frac{2}{L}\sum_{n=1}^{\infty} \sin\lambda_n x\, e^{-\alpha\lambda_n^2 t} \int_{x'=0}^{L} \sin\lambda_n x' \left[\frac{g^{i}(x')}{\rho c}\right] dx' \tag{8-98}$$

As discussed in Example 8-4, a comparison of equation (8-98) and the first term of equation (8-95) reveals an identical solution for the case that

$$F(x) = \frac{g^{\mathrm{i}}(x)}{\rho c} \tag{8-99}$$

which implies that the heat conduction problem for an instantaneous distributed heat source $g^{\mathrm{i}}(x)$ releasing its energy at time $t = 0$ into a medium of zero temperature is equivalent to an initial value problem with the initial temperature distribution as given by equation (8-99).

2. The medium is initially at zero temperature. The boundaries at $x = 0$ and $x = L$ are kept at zero temperature for times $t > 0$, and a planar heat source $g_s^{\mathrm{c}}(t)$ (W/m^2) is placed at the location $x = a$, with $0 < a < L$, that releases its heat continuously starting at time $t = 0$. We let $F(x) = 0$, as well as

$f_1(t) = 0$ and $f_2(t) = 0$, and define the planar heat source in consideration of equation (8-60), giving

$$g(x, t) = g_s^c(t)\delta(x - a) \tag{8-100}$$

Substitution of the above into equation (8-95) yields

$$T(x, t) = \frac{\alpha}{k}\frac{2}{L}\sum_{n=1}^{\infty} \sin\lambda_n x\, e^{-\alpha\lambda_n^2 t} \int_{\tau=0}^{t}\int_{x'=0}^{L} \sin\lambda_n x' g_s^c(\tau) \times \delta(x' - a)e^{\alpha\lambda_n^2\tau}\, dx'\, d\tau \tag{8-101}$$

where the x' integral is readily evaluated making use of equation (8-54), giving

$$T(x, t) = \frac{\alpha}{k}\frac{2}{L}\sum_{n=1}^{\infty} \sin\lambda_n x \sin\lambda_n a\, e^{-\alpha\lambda_n^2 t} \int_{\tau=0}^{t} g_s^c(\tau)e^{\alpha\lambda_n^2\tau}\, d\tau \tag{8-102}$$

Example 8-6 Two-Dimensional Cartesian Problem with Generation
Consider a 2-D rectangular region over the domain $0 \le x \le a$, $0 \le y \le b$, that is initially at zero temperature. For times $t > 0$, all of the boundary surfaces are nonhomogeneous, as shown in Figure 8-5.

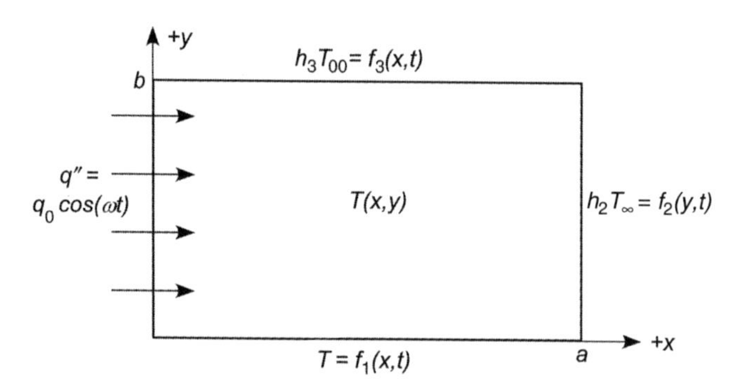

Figure 8-5 Problem description for Example 8-6.

We use Green's function to calculate the temperature distribution for $t > 0$. The mathematical formulation of this problem is given as

$$\frac{\partial^2 T}{\partial x^2} + \frac{\partial^2 T}{\partial y^2} = \frac{1}{\alpha}\frac{\partial T}{\partial t}$$

$$\text{in the region } 0 < x < a, \quad 0 < y < b \quad \text{for} \quad t > 0 \tag{8-103}$$

$$\text{BC1:}\quad -k\left.\frac{\partial T}{\partial x}\right|_{x=0} = q_0\cos\omega t \qquad \text{BC2:}\quad -k\left.\frac{\partial T}{\partial x}\right|_{x=a} = h_2\,T|_{x=a} - f_2(y,t) \tag{8-104a}$$

$$\text{BC3:}\quad T(y=0) = f_1(x,t) \qquad \text{BC4:}\quad -k\left.\frac{\partial T}{\partial y}\right|_{y=b} = h_3\,T|_{y=b} - f_3(x,t) \tag{8-104b}$$

$$\text{IC:}\quad T(x,y,t=0) = 0 \tag{8-104c}$$

To determine the appropriate Green's function, we consider the homogeneous version of this problem given as

$$\frac{\partial^2\Psi}{\partial x^2} + \frac{\partial^2\Psi}{\partial y^2} = \frac{1}{\alpha}\frac{\partial\Psi}{\partial t}$$

$$\text{in the region } 0 < x < a, \qquad 0 < y < b \qquad \text{for} \qquad t > 0 \tag{8-105}$$

$$\text{BC1:}\quad \left.\frac{\partial\Psi}{\partial x}\right|_{x=0} = 0 \qquad \text{BC2:}\quad -k\left.\frac{\partial\Psi}{\partial x}\right|_{x=a} = h_2\,\Psi|_{x=a} \tag{8-106a}$$

$$\text{BC3:}\quad \Psi(y=0) = 0 \qquad \text{BC4:}\quad -k\left.\frac{\partial\Psi}{\partial y}\right|_{y=b} = h_3\,\Psi|_{y=b} \tag{8-106b}$$

$$\text{IC:}\quad \Psi(x,y,t=0) = F(x,y) \tag{8-106c}$$

Note that we have added an initial condition to equation (8-106c) as a placeholder for determination of Green's function. As we discussed previously, without an initial condition, the solution of the homogeneous problem is zero. The homogeneous problem is similar to Example 3-8, with the adjustment of BC3 from a type 2 to a type 1. We therefore adjust the eigenfunctions and eigenvalues according to case 7 in Table 2-1, giving the solution

$$\Psi(x,y,t) = \sum_{n=1}^{\infty}\sum_{m=1}^{\infty}\frac{4(\lambda_n^2+H_2^2)(\beta_m^2+H_3^2)\cos\lambda_n x\sin\beta_m y}{\left[a(\lambda_n^2+H_2^2)+H_2\right]\left[b(\beta_m^2+H_3^2)+H_3\right]}e^{-\alpha(\beta_m^2+\lambda_n^2)t}$$

$$\times\int_{x'=0}^{a}\int_{y'=0}^{b}\cos\lambda_n x'\sin\beta_m y' F(x',y')\,dx'\,dy' \tag{8-107}$$

where the eigenvalues are defined by the transcendental equations

$$\lambda_n\tan\lambda_n a = \frac{h_2}{k} \equiv H_2 \quad\rightarrow\quad \lambda_n \qquad \text{for} \qquad n = 1,2,3,\ldots$$

$$\beta_m\tan\beta_m b = \frac{h_3}{k} \equiv H_3 \quad\rightarrow\quad \beta_m \qquad \text{for} \qquad m = 1,2,3,\ldots \tag{8-108}$$

We seek a solution to the homogeneous solution in the form of a kernel, namely,

$$\Psi(x,y,t) = \int_{x'=0}^{a}\int_{y'=0}^{b} G(x,y,t|x',y',\tau)\big|_{\tau=0}F(x',y')\,dx'dy' \tag{8-109}$$

A comparison of equations (8-107) and (8-109) gives the Green's function at $\tau = 0$,

$$G(x, y, t|x', y', \tau)\big|_{\tau=0} = \sum_{n=1}^{\infty}\sum_{m=1}^{\infty} \frac{4(\lambda_n^2 + H_2^2)(\beta_m^2 + H_3^2)}{\left[a(\lambda_n^2 + H_2^2) + H_2\right]\left[b(\beta_m^2 + H_3^2) + H_3\right]} \times \cos\lambda_n x \sin\beta_m y \cos\lambda_n x' \sin\beta_m y' e^{-\alpha(\beta_m^2+\lambda_n^2)t} \tag{8-110}$$

The desired Green's function is obtained by replacing t with $t-\tau$ in equation (8-110), which yields

$$G(x, y, t|x', y', \tau) = \sum_{n=1}^{\infty}\sum_{m=1}^{\infty} N_{nm} \cos\lambda_n x \sin\beta_m y \cos\lambda_n x' \sin\beta_m y' e^{-\alpha\eta_{nm}^2(t-\tau)} \tag{8-111}$$

where we have introduced the following terms:

$$N_{nm} = \frac{4(\lambda_n^2 + H_2^2)(\beta_m^2 + H_3^2)}{\left[a(\lambda_n^2 + H_2^2) + H_2\right]\left[b(\beta_m^2 + H_3^2) + H_3\right]} \tag{8-112}$$

and

$$\eta_{nm}^2 = \beta_m^2 + \lambda_n^2 \tag{8-113}$$

The overall temperature solution of the nonhomogeneous problem defined by equations (8-103) and (8-104) is now given in terms of the above Green's function, according to equation (8-12), which yields

$$\begin{aligned} T(x, y, t) = {} & \alpha \int_{\tau=0}^{t}\int_{y'=0}^{b} G(x, y, t|x', y', \tau)\big|_{x'=0} \frac{q_0 \cos\omega\tau}{k} dy'\, d\tau \\ & + \alpha \int_{\tau=0}^{t}\int_{y'=0}^{b} G(x, y, t|x', y', \tau)\big|_{x'=a} \frac{f_2(y', \tau)}{k} dy'\, d\tau \\ & + \alpha \int_{\tau=0}^{t}\int_{x'=0}^{a} \left.\frac{\partial G(x, y, t|x', y', \tau)}{\partial y'}\right|_{y'=0} f_1(x', \tau)\, dx'\, d\tau \\ & + \alpha \int_{\tau=0}^{t}\int_{x'=0}^{a} G(x, y, t|x', y', \tau)\big|_{y'=b} \frac{f_3(x', \tau)}{k} dx'\, d\tau \end{aligned} \tag{8-114}$$

We have omitted the terms corresponding to the initial condition and heat generation, since they are both zero, and have made the following substitution for the type 1 boundary condition at $y = 0$:

$$-k \left.\frac{\partial G}{\partial y'}\right|_{y'=0} = -h\, G|_{y'=0} \quad \rightarrow \quad \left.\frac{\partial G}{\partial y'}\right|_{y'=0} = \frac{1}{k}\, G|_{y'=0} \tag{8-115}$$

Substitution of the Green's function of equation (8-111) into equation (8-114) yields the overall temperature solution

$$T(x,y,t) = \alpha \sum_{n=1}^{\infty}\sum_{m=1}^{\infty} N_{nm} \cos\lambda_n x \sin\beta_m y \, e^{-\alpha\eta_{nm}^2 t}$$

$$\times \left[+\frac{q_0}{k}\int_{\tau=0}^{t}\int_{y'=0}^{b} \sin\beta_m y' \cos\omega\tau \, e^{\alpha\eta_{nm}^2\tau} dy'\, d\tau \right.$$

$$+ \frac{\cos\lambda_n a}{k}\int_{\tau=0}^{t}\int_{y'=0}^{b} \sin\beta_m y' f_2(y',\tau) e^{\alpha\eta_{nm}^2\tau} dy'\, d\tau$$

$$+ \beta_m \int_{\tau=0}^{t}\int_{x'=0}^{a} \cos\lambda_n x' f_1(x',\tau) e^{\alpha\eta_{nm}^2\tau}\, dx'\, d\tau$$

$$\left. + \frac{\sin\beta_m b}{k}\int_{\tau=0}^{t}\int_{x'=0}^{a} \cos\lambda_n x' f_3(x',\tau) e^{\alpha\eta_{nm}^2\tau}\, dx'\, d\tau \right] \tag{8-116}$$

where the eigenvalues are defined above by equations (8-108).

Example 8-7 Three-Dimensional Cartesian Problem with Generation
Consider a rectangular parallelepiped over the domain $0 \le x \le a$, $0 \le y \le b$, $0 \le z \le c$ that is initially at a temperature $F(x,y,z)$. For times $t > 0$, all of the boundary surfaces are maintained at zero temperature, while in addition there is internal energy generation within the solid at the rate $g(x,y,z,t)$ per unit time, per unit volume (W/m^3). Use Green's function to calculate the temperature distribution for $t > 0$.

The mathematical formulation of this problem is given as

$$\frac{\partial^2 T}{\partial x^2} + \frac{\partial^2 T}{\partial y^2} + \frac{\partial^2 T}{\partial z^2} + \frac{1}{k} g(x,y,z,t) = \frac{1}{\alpha}\frac{\partial T}{\partial t} \tag{8-117}$$

in the region $0 < x < a, \quad 0 < y < b, \quad 0 < z < c \quad$ for $\quad t > 0$

BC1: $T(x=0) = 0$ BC2: $T(x=a) = 0$ (8-118a)

BC3: $T(y=0) = 0$ BC4: $T(y=b) = 0$ (8-118b)

BC5: $T(z=0) = 0$ BC6: $T(z=c) = 0$ (8-118c)

IC: $T(x,y,z,t=0) = F(x,y,z)$ (8-118d)

To determine the appropriate Green's function, we consider the homogeneous version of this problem as

$$\frac{\partial^2 \Psi}{\partial x^2} + \frac{\partial^2 \Psi}{\partial y^2} + \frac{\partial^2 \Psi}{\partial z^2} = \frac{1}{\alpha}\frac{\partial \Psi}{\partial t} \tag{8-119}$$

in the region $0 < x < a, \quad 0 < y < b, \quad 0 < z < c \quad$ for $\quad t > 0$

$$\text{BC1:} \quad \Psi(x=0)=0 \qquad \text{BC2:} \quad \Psi(x=a)=0 \tag{8-120a}$$

$$\text{BC3:} \quad \Psi(y=0)=0 \qquad \text{BC4:} \quad \Psi(y=b)=0 \tag{8-120b}$$

$$\text{BC5:} \quad \Psi(z=0)=0 \qquad \text{BC6:} \quad \Psi(z=c)=0 \tag{8-120c}$$

$$\text{IC:} \quad \Psi(x,y,z,t=0)=F(x,y,z) \tag{8-120d}$$

The homogeneous problem is similar to Example 3-4 with the adjustment of the second boundary conditions in the y and z dimensions. Specifically, all three spatial dimensions now correspond to case 9 in Table 2-1, giving the solution

$$\Psi(x,y,z,t)=\int_{x'=0}^{a}\int_{y'=0}^{b}\int_{z'=0}^{c}\frac{8}{abc}\sum_{m=1}^{\infty}\sum_{n=1}^{\infty}\sum_{p=1}^{\infty} \times\left[e^{-\alpha(\beta_m^2+\gamma_n^2+\eta_p^2)t}\sin\beta_m x\sin\gamma_n y\sin\eta_p z\sin\beta_m x'\sin\gamma_n y'\sin\eta_p z'\right] \times F(x',y',z')\,dx'\,dy'\,dz' \tag{8-121}$$

where the eigenvalues are defined as

$$\beta_m=\frac{m\pi}{a} \qquad \gamma_n=\frac{n\pi}{b} \qquad \eta_p=\frac{p\pi}{c} \qquad \text{with } (m,n,p)=1,2,3,4,\ldots \tag{8-122}$$

We seek a solution to the homogeneous solution in the form of a kernel, namely,

$$\Psi(x,y,z,t)=\int_{x'=0}^{a}\int_{y'=0}^{b}\int_{z'=0}^{c} \times\left[G(x,y,z,t|x',y',z',\tau)\big|_{\tau=0}F(x',y',z')\right]dx'\,dy'\,dz' \tag{8-123}$$

A comparison of equations (8-121) and (8-123) gives Green's function at $\tau=0$,

$$G(x,y,z,t|x',y',z',\tau)\big|_{\tau=0}=\frac{8}{abc}\sum_{m=1}^{\infty}\sum_{n=1}^{\infty}\sum_{p=1}^{\infty} \times\left[e^{-\alpha(\beta_m^2+\gamma_n^2+\eta_p^2)t}\sin\beta_m x\sin\gamma_n y\sin\eta_p z\sin\beta_m x'\sin\gamma_n y'\sin\eta_p z'\right] \tag{8-124}$$

The desired Green's function is obtained by replacing t with $t-\tau$ in equation (8-124), which yields

$$G(x,y,z,t|x',y',z',\tau)=\frac{8}{abc}\sum_{m=1}^{\infty}\sum_{n=1}^{\infty}\sum_{p=1}^{\infty} \times\left[e^{-\alpha(\beta_m^2+\gamma_n^2+\eta_p^2)(t-\tau)}\sin\beta_m x\sin\gamma_n y\sin\eta_p z\sin\beta_m x'\sin\gamma_n y'\sin\eta_p z'\right] \tag{8-125}$$

The overall temperature solution of the nonhomogeneous problem defined by equations (8-117) and (8-118) is now given in terms of the above Green's function, according to equation (8-4), which yields

$$T(x, y, z, t) = \int_{x'=0}^{a} \int_{y'=0}^{b} \int_{z'=0}^{c} \times \left[G(x, y, z, t|x', y', z', \tau)\big|_{\tau=0} F(x', y', z') \right] dx'\, dy'\, dz' + \frac{\alpha}{k} \int_{\tau=0}^{t} \int_{x'=0}^{a} \int_{y'=0}^{b} \int_{z'=0}^{c} G(x, y, z, t|x', y', z', \tau) \times g(x', y', z', \tau)\, dx'\, dy'\, dz'\, d\tau \tag{8-126}$$

where Green's function and eigenvalues are defined as above.

8-5 APPLICATIONS OF GREEN'S FUNCTION IN THE CYLINDRICAL COORDINATE SYSTEM

In this section we illustrate with examples the application of Green's function in the solution of nonhomogeneous boundary value problems of heat conduction in the cylindrical coordinate system. For convenience in the determination of Green's function, we have often selected problems for which solutions are available in Chapter 4 for their corresponding homogeneous problem.

Example 8-8 Green's Function for a 1-D Cylinder with Generation Consider a 1-D cylinder over the domain $0 \leq r \leq b$ that is initially at a temperature $F(r)$. For times $t > 0$, the boundary surface at $r = b$ is maintained at a prescribed temperature $f(t)$, while the entire cylinder is subjected to internal energy generation at the rate $g(r, t)$ per unit time, per unit volume (W/m^3). Use Green's function to calculate temperature distribution for $t > 0$.

The formulation is given as

$$\frac{\partial^2 T}{\partial r^2} + \frac{1}{r}\frac{\partial T}{\partial r} + \frac{1}{k} g(r, t) = \frac{1}{\alpha}\frac{\partial T}{\partial t} \qquad \text{in} \qquad 0 \leq r < b, \qquad t > 0 \tag{8-127}$$

$$\text{BC1:} \qquad T(r \to 0) \Rightarrow \text{finite} \tag{8-128a}$$

$$\text{BC2:} \qquad T(r = b, t) = f(t) \tag{8-128b}$$

$$\text{IC:} \qquad T(r, t = 0) = F(r) \tag{8-128c}$$

To determine the desired Green's function, we consider the homogeneous version of the problem for the same region:

$$\frac{\partial^2 \Psi}{\partial r^2} + \frac{1}{r}\frac{\partial \Psi}{\partial r} = \frac{1}{\alpha}\frac{\partial \Psi}{\partial t} \qquad \text{in} \qquad 0 \leq r < b, \qquad t > 0 \tag{8-129}$$

$$\text{BC1:} \quad \Psi(r \to 0) \Rightarrow \text{finite} \tag{8-130a}$$

$$\text{BC2:} \quad \Psi(r = b, t) = 0 \tag{8-130b}$$

$$\text{IC:} \quad \Psi(r, t = 0) = F(r) \tag{8-130c}$$

This homogeneous problem was considered above in Example 8-1; hence we may write our solution in the form of equation (8-28), giving

$$\Psi(r, t) = \int_{r'=0}^{b} \left[\frac{2}{b^2} \sum_{n=1}^{\infty} e^{-\alpha\beta_n^2 t} \frac{1}{J_1^2(\beta_n b)} J_0(\beta_n r) J_0(\beta_n r') \right] F(r') r' \, dr' \tag{8-131}$$

where the eigenvalues are the roots of the transcendental equation $J_0(\beta_n b) = 0$. The solution of the homogenous problem (8-129) in terms of Green's function is given, according to equation (8-14), as

$$\Psi(r, t) = \int_{r'=0}^{b} G(r, t|r', \tau)\big|_{\tau=0} F(r') r' dr' \tag{8-132}$$

where we have used the Sturm–Liouville weight function $p = 1$ for the cylindrical coordinate system. By comparing equations (8-131) and (8-132), we find Green's function for $\tau = 0$, namely,

$$G(r, t|r', \tau)\big|_{\tau=0} = \frac{2}{b^2} \sum_{n=1}^{\infty} e^{-\alpha\beta_n^2 t} \frac{1}{J_1^2(\beta_n b)} J_0(\beta_n r) J_0(\beta_n r') \tag{8-133}$$

We now replace t in equation (8-133) with $t-\tau$, as detailed in equation (8-22), to yield the desired Green's function as

$$G(r, t|r', \tau) = \frac{2}{b^2} \sum_{n=1}^{\infty} e^{-\alpha\beta_n^2 (t-\tau)} \frac{1}{J_1^2(\beta_n b)} J_0(\beta_n r) J_0(\beta_n r') \tag{8-134}$$

Then the solution of the nonhomogeneous problem of equations (8-127) and (8-128) in terms of the above Green's function is given, according to equation (8-14), as

$$\begin{aligned} T(r, t) = & \int_{r'=0}^{b} G(r, t|r', \tau)\big|_{\tau=0} F(r') r' \, dr' \\ & + \frac{\alpha}{k} \int_{\tau=0}^{t} \int_{r'=0}^{b} G(r, t|r', \tau) g(r', \tau) r' \, dr' \, d\tau \\ & - \alpha \int_{\tau=0}^{t} \left[r' \frac{\partial G(r, t|r', \tau)}{\partial r'} \right] \Bigg|_{r'=b} f(\tau) \, d\tau \end{aligned} \tag{8-135}$$

where we have made a substitution for the type 1 boundary condition at $r = b$, namely,

$$-k \left. \frac{\partial G}{\partial r'} \right|_{r'=b} = +h \, G|_{r'=b} \quad \rightarrow \quad - \left. \frac{\partial G}{\partial r'} \right|_{r'=b} = \frac{1}{k} \, G|_{r'=b} \tag{8-136}$$

Introducing the above Green's function of equation (8-134) into equation (8-135), and per the derivative expression of equation (2-41b), noting that

$$\left[r'\frac{\partial G}{\partial r'}\right]_{r'=b} = -\frac{2}{b}\sum_{n=1}^{\infty} e^{-\alpha\beta_n^2(t-\tau)}\beta_n \frac{J_0(\beta_n r)}{J_1(\beta_n b)} \tag{8-137}$$

we obtain the overall temperature solution

$$\begin{aligned} T(r,t) = {} & \frac{2}{b^2}\sum_{n=1}^{\infty} e^{-\alpha\beta_n^2 t}\frac{J_0(\beta_n r)}{J_1^2(\beta_n b)}\int_{r'=0}^{b} J_0(\beta_n r')F(r')\,r'\,dr' \\ & + \frac{2\alpha}{kb^2}\sum_{n=1}^{\infty} e^{-\alpha\beta_n^2 t}\frac{J_0(\beta_n r)}{J_1^2(\beta_n b)}\int_{\tau=0}^{t}\int_{r'=0}^{b} J_0(\beta_n r')e^{\alpha\beta_n^2\tau}g(r',\tau)\,r'\,dr'\,d\tau \\ & + \frac{2\alpha}{b}\sum_{n=1}^{\infty} e^{-\alpha\beta_n^2 t}\beta_n\frac{J_0(\beta_n r)}{J_1(\beta_n b)}\int_{\tau=0}^{t} e^{\alpha\beta_n^2\tau}f(\tau)\,d\tau \end{aligned} \tag{8-138}$$

where the eigenvalues are the roots of the transcendental equation $J_0(\beta_n b) = 0$. In this solution, the first term on the right-hand side is for the effect of the initial condition function $F(r)$, and it is the same as that given by equation (8-117). The second term is for the effect of the heat generation function $g(r,t)$. The last term is for the effect of the prescribed temperature boundary-condition function $f(t)$. This solution (8-138) appears to vanish at the boundary $r = b$ instead of yielding the boundary condition function $f(t)$. As discussed above, the reason for this is that the last term in equation (8-138) involves a series that is not uniformly convergent at $r = b$. This difficulty can be alleviated by integrating the last term by parts and replacing the resulting series by its closed-form expression. An alternative approach would be to split up the original problem as discussed in Section 1–8 of Özisik [12]. We now examine three special cases of solution (8-138).

1. The cylinder has zero initial temperature, zero surface temperature, but heat is generated internally within the solid at a constant rate of g_0 (W/m^3). Making the following adjustments in equation (8-138), $F(r) = 0$, $f(t) = 0$, and $g(r,t) = g_0$, we obtain after integration, the following:

$$T(r,t) = \frac{2g_0}{kb}\sum_{n=1}^{\infty}\frac{J_0(\beta_n r)}{\beta_n^3 J_1(\beta_n b)} - \frac{2g_0}{kb}\sum_{n=1}^{\infty} e^{-\alpha\beta_n^2 t}\frac{J_0(\beta_n r)}{\beta_n^3 J_1(\beta_n b)} \tag{8-139}$$

For $t \to \infty$, the second term on the right-hand side vanishes, and the first term must therefore equal the steady-state temperature distribution in the cylinder, namely,

$$T_{ss}(r) = \frac{2g_0}{kb}\sum_{n=1}^{\infty}\frac{J_0(\beta_n r)}{\beta_n^3 J_1(\beta_n b)} \equiv \frac{g_0(b^2 - r^2)}{4k} \tag{8-140}$$

Introducing (8-140) into (8-139), the solution becomes

$$T(r,t) = \frac{g_0(b^2 - r^2)}{4k} - \frac{2g_0}{kb} \sum_{n=1}^{\infty} e^{-\alpha\beta_n^2 t} \frac{J_0(\beta_n r)}{\beta_n^3 J_1(\beta_n b)} \tag{8-141}$$

2. The cylinder has zero initial temperature, zero surface temperature, but there is a line heat source of strength $g_L^c(t)$ (W/m) situated along the centerline of the cylinder and releasing its heat continuously for times $t > 0$. For this special case we set in equation (8-138) the following:

$$F(r) = 0 \qquad f(t) = 0 \qquad \text{and} \qquad g(r,t) = g_L^c(t)\frac{1}{2\pi r}\delta(r)$$

where we have introduced the expression for a continuous line source in cylindrical coordinates located at the centerline. With these changes, equation (8-138) reduces to

$$T(r,t) = \frac{2\alpha}{kb^2} \sum_{n=1}^{\infty} e^{-\alpha\beta_n^2 t} \frac{J_0(\beta_n r)}{J_1^2(\beta_n b)} \int_{\tau=0}^{t} \int_{r'=0}^{b} J_0(\beta_n r') e^{\alpha\beta_n^2 \tau} \times g_L^c(\tau)\frac{\delta(r')}{2\pi r'} r'\, dr'\, d\tau \tag{8-142}$$

which after performing the spatial integration, recalling that $J_0(0) = 1$, yields

$$T(r,t) = \frac{\alpha}{k\pi b^2} \sum_{n=1}^{\infty} e^{-\alpha\beta_n^2 t} \frac{J_0(\beta_n r)}{J_1^2(\beta_n b)} \int_{\tau=0}^{t} e^{\alpha\beta_n^2 \tau} g_L^c(\tau) d\tau \tag{8-143}$$

3. The cylinder has zero initial temperature, zero surface temperature, but there is an instantaneous, distributed volume heat source of strength $g^{\mathrm{i}}(r)$ (J/m^3) that releases its heat spontaneously at time $t = 0$. For this case, we make the following adjustments to equation (8-138):

$$F(r) = 0 \qquad f(t) = 0 \qquad \text{and} \qquad g(r,t) = g^{\mathrm{i}}(r)\delta(t)$$

where we have used the distributed heat source of equation (8-60). Then equation (8-138) reduces to the form

$$T(r,t) = \frac{2\alpha}{kb^2} \sum_{n=1}^{\infty} e^{-\alpha\beta_n^2 t} \frac{J_0(\beta_n r)}{J_1^2(\beta_n b)} \int_{\tau=0}^{t} \int_{r'=0}^{b} J_0(\beta_n r') e^{\alpha\beta_n^2 \tau} \times g^i(r')\delta(\tau) r'\, dr'\, d\tau \tag{8-144}$$

which after performing the temporal integration yields

$$T(r,t) = \frac{2}{b^2}\sum_{n=1}^{\infty} e^{-\alpha\beta_n^2 t}\frac{J_0(\beta_n r)}{J_1^2(\beta_n b)}\int_{r'=0}^{b} J_0(\beta_n r')\frac{\alpha g^{\mathrm{i}}(r')}{k} r'\,dr' \tag{8-145}$$

A comparison of this solution with the first term in equation (8-138) reveals that the two are identical for the condition that

$$\frac{\alpha g^{\mathrm{i}}(r)}{k} = \frac{g^{\mathrm{i}}(r)}{\rho c} \equiv F(r)$$

That is to say, an instantaneous volume heat source of strength $g^{\mathrm{i}}(r)$ (J/m^3) releasing its heat spontaneously at time $t = 0$ into a medium of zero temperature is equivalent to an initial temperature distribution times the heat capacity, namely, $\rho c F(r)$, as we discussed above with equation (8-80) in the context of conservation of energy.

Example 8-9 Green's Function for a 1-D Hollow Cylinder with Generation

Consider a 1-D hollow cylinder over the domain $a \le r \le b$ that is initially at a temperature $F(r)$. For times $t > 0$, the boundary surfaces at $r = a$ and $r = b$ are maintained at zero temperature, while the entire medium is subjected to internal energy generation at the rate $g(r,t)$ per unit time, per unit volume (W/m^3). Use the Green's function to calculate the temperature distribution for $t > 0$.

The formulation is given as

$$\frac{\partial^2 T}{\partial r^2} + \frac{1}{r}\frac{\partial T}{\partial r} + \frac{1}{k}g(r,t) = \frac{1}{\alpha}\frac{\partial T}{\partial t} \quad \text{in} \quad a < r < b, \quad t > 0 \tag{8-146}$$

$$\text{BC1:} \quad T(r = a, t) = 0 \tag{8-147a}$$

$$\text{BC2:} \quad T(r = b, t) = 0 \tag{8-147b}$$

$$\text{IC:} \quad T(r, t = 0) = F(r) \tag{8-147c}$$

To determine the desired Green's function, we consider the homogeneous version of the problem for the same region:

$$\frac{\partial^2 \Psi}{\partial r^2} + \frac{1}{r}\frac{\partial \Psi}{\partial r} = \frac{1}{\alpha}\frac{\partial \Psi}{\partial t} \quad \text{in} \quad a < r < b, \quad t > 0 \tag{8-148}$$

$$\text{BC1:} \quad \Psi(r = a, t) = 0 \tag{8-149a}$$

$$\text{BC2:} \quad \Psi(r = b, t) = 0 \tag{8-149b}$$

$$\text{IC:} \quad \Psi(r, t = 0) = F(r) \tag{8-149c}$$

This homogeneous problem was considered in Example 4-10; hence we may write our solution in the form of equation (4-182), giving

$$\Psi(r,t) = \int_{r'=a}^{b} \left[\frac{\pi^2}{2} \sum_{n=1}^{\infty} \frac{\beta_n^2 J_0^2(\beta_n b) e^{-\alpha\beta_n^2 t}}{J_0^2(\beta_n a) - J_0^2(\beta_n b)} R_0(\beta_n r) R_0(\beta_n r') \right] F(r')\, r'\, dr' \tag{8-150}$$

where we have used the norm as defined by case 4 of Table 2-3, and where the eigenfunctions are given by

$$R_0(\beta_n r) = J_0(\beta_n r) Y_0(\beta_n a) - J_0(\beta_n a) Y_0(\beta_n r) \tag{8-151}$$

and the eigenvalues are given by the roots of the transcendental equation

$$J_0(\beta_n a) Y_0(\beta_n b) - J_0(\beta_n b) Y_0(\beta_n a) = 0 \tag{8-152}$$

According to equation (8-14), with $p = 1$ for cylindrical coordinates, we seek a solution to the homogeneous problem of the form

$$\Psi(r,t) = \int_{r'=a}^{b} G(r,t|r',\tau)\big|_{\tau=0} F(r')\, r'\, dr' \tag{8-153}$$

A comparison of equations (8-150) and (8-153) yields

$$G(r,t|r',\tau)\big|_{\tau=0} = \frac{\pi^2}{2} \sum_{n=1}^{\infty} e^{-\alpha\beta_n^2 t} \frac{\beta_n^2 J_0^2(\beta_n b)}{J_0^2(\beta_n a) - J_0^2(\beta_n b)} R_0(\beta_n r) R_0(\beta_n r') \tag{8-154}$$

The desired Green's function is now obtained by replacing t with $t-\tau$ in equation (8-154); we find

$$G(r,t|r',\tau) = \frac{\pi^2}{2} \sum_{n=1}^{\infty} e^{-\alpha\beta_n^2 (t-\tau)} \frac{\beta_n^2 J_0^2(\beta_n b)}{J_0^2(\beta_n a) - J_0^2(\beta_n b)} R_0(\beta_n r) R_0(\beta_n r') \tag{8-155}$$

Then the solution of the above nonhomogeneous problem (8-146) in terms of Green's function is given, according to equation (8-14), as

$$\begin{aligned} T(r,t) &= \int_{r'=a}^{b} G(r,t|r',\tau)\big|_{\tau=0} F(r') r' dr' \\ &\quad + \frac{\alpha}{k} \int_{\tau=0}^{t} \int_{r'=a}^{b} G(r,t|r',\tau) g(r',\tau) r'\, dr'\, d\tau \end{aligned} \tag{8-156}$$

with the first term accounting for the initial temperature distribution, and the second term accounting for the internal energy generation. Introducing the foregoing Green's function into equation (8-156), the solution of problem (8-146) becomes

$$T(r,t) = \frac{\pi^2}{2}\sum_{n=1}^{\infty} e^{-\alpha\beta_n^2 t}\frac{\beta_n^2 J_0^2(\beta_n b)}{J_0^2(\beta_n a) - J_0^2(\beta_n b)} R_0(\beta_n r)\int_{r'=a}^{b} R_0(\beta_n r')F(r')\,r'\,dr'$$
$$+ \frac{\pi^2\alpha}{2k}\sum_{n=1}^{\infty}\left[e^{-\alpha\beta_n^2 t}\frac{\beta_n^2 J_0^2(\beta_n b)}{J_0^2(\beta_n a) - J_0^2(\beta_n b)} R_0(\beta_n r)\right.$$
$$\left.\times\int_{\tau=0}^{t}\int_{r'=a}^{b} R_0(\beta_n r')e^{\alpha\beta_n^2\tau} g(r',\tau)\,r'\,dr'\,d\tau\right] \tag{8-157}$$

where $R_0(\beta_n r)$ are given by equation (8-151), and the eigenvalues β_n are the roots of the transcendental equation (8-152).

8-6 APPLICATIONS OF GREEN'S FUNCTION IN THE SPHERICAL COORDINATE SYSTEM

In this section we illustrate with examples the application of Green's function in the solution of nonhomogeneous boundary value problems of heat conduction in the spherical coordinate system. For convenience in the determination of Green's function, we have often selected problems for which solutions are available in Chapter 5 for their corresponding homogeneous problem.

Example 8-10 Green's Function for a 1-D Hollow Sphere with Generation Consider a 1-D hollow sphere over the domain $a \le r \le b$ that is initially at a temperature $F(r)$. For times $t > 0$, the boundary surface at $r = a$ is maintained at zero temperature, the boundary surface at $r = b$ is maintained at a temperature $f(t)$, while the entire medium is subjected to internal energy generation at the rate $g(r,t)$ per unit time, per unit volume (W/m^3). Use Green's function to calculate the temperature distribution for $t > 0$.

The formulation is given as

$$\frac{1}{r}\frac{\partial^2}{\partial r^2}(rT) + \frac{1}{k}g(r,t) = \frac{1}{\alpha}\frac{\partial T}{\partial t} \quad \text{in} \quad a < r < b \quad t > 0 \tag{8-158}$$

$$\text{BC1:} \quad T(r=a,t) = 0 \tag{8-159a}$$

$$\text{BC2:} \quad T(r=b,t) = f(t) \tag{8-159b}$$

$$\text{IC:} \quad T(r,t=0) = F(r) \tag{8-159c}$$

To determine the desired Green's function, we consider the homogeneous version of the problem for the same region:

$$\frac{1}{r}\frac{\partial^2}{\partial r^2}(r\Psi) = \frac{1}{\alpha}\frac{\partial \Psi}{\partial t} \quad \text{in} \quad a < r < b, \quad t > 0 \tag{8-160}$$

$$\text{BC1:} \quad \Psi(r = a, t) = 0 \tag{8-161a}$$

$$\text{BC2:} \quad \Psi(r = b, t) = 0 \tag{8-161b}$$

$$\text{IC:} \quad \Psi(r, t = 0) = F(r) \tag{8-161c}$$

This homogeneous problem was considered in Example 5-5; hence we may write our solution in the form of equation (5-108), which is rearranged to give

$$\Psi(r,t) = \int_{r'=a}^{b}\left[2\sum_{n=1}^{\infty}\frac{\sin\beta_n(r-a)\sin\beta_n(r'-a)}{r'r(b-a)}e^{-\alpha\beta_n^2 t}\right]F(r')r'^2\,dr' \tag{8-162}$$

where we have multiplied by the constant r'/r' to put equation (8-162) into the form of equation (8-14) for $p = 2$, consistent with the spherical coordinate system. The eigenfunctions are given by

$$\beta_n = \frac{n\pi}{b-a}, \qquad n = 1, 2, 3, \ldots \tag{8-163}$$

noting that $\beta_0 = 0$ is a trivial eigenvalue and has been dropped from the summation. According to equation (8-14), with $p = 2$ for spherical coordinates, we seek a solution to the homogeneous problem of the form

$$\Psi(r,t) = \int_{r'=a}^{b} G(r,t|r',\tau)\big|_{\tau=0}F(r')r'^2\,dr' \tag{8-164}$$

A comparison of equations (8-162) and (8-164) yields

$$G(r,t|r',\tau)\big|_{\tau=0} = 2\sum_{n=1}^{\infty}\frac{\sin\beta_n(r-a)\sin\beta_n(r'-a)}{r'r(b-a)}e^{-\alpha\beta_n^2 t} \tag{8-165}$$

The desired Green's function is now obtained by replacing t with $t-\tau$ in equation (8-165), which gives

$$G(r,t|r',\tau) = 2\sum_{n=1}^{\infty}\frac{\sin\beta_n(r-a)\sin\beta_n(r'-a)}{r'r(b-a)}e^{-\alpha\beta_n^2(t-\tau)} \tag{8-166}$$

Then the solution of the above nonhomogeneous problem defined by equations (8-158) and (8-159) in terms of this Green's function is given, according to equation (8-14), as

$$
\begin{aligned}
T(r,t) &= \int_{r'=a}^{b} G(r,t|r',\tau)\big|_{\tau=0} F(r') r'^2 dr' \\
&+ \frac{\alpha}{k}\int_{\tau=0}^{t}\int_{r'=a}^{b} G(r,t|r',\tau) g(r',\tau) r'^2\, dr'\, d\tau \\
&- \alpha \int_{\tau=0}^{t} \left[r'^2 \frac{\partial G(r,t|r',\tau)}{\partial r'} \right]\Bigg|_{r'=b} f(\tau)\, d\tau
\end{aligned}
\tag{8-167}
$$

with the first term accounting for the initial temperature distribution, the second term accounting for the internal energy generation, and the third term accounting for the nonhomogeneity at $r = b$. We have made the following substitution in the third term

$$
-k \left.\frac{\partial G}{\partial r'}\right|_{r'=b} = +h\; G|_{r'=b} \quad \rightarrow \quad -\left.\frac{\partial G}{\partial r'}\right|_{r'=b} = \frac{1}{k}\; G|_{r'=b} \tag{8-168}
$$

to reflect the type 1 boundary condition. Introducing the foregoing Green's function into equation (8-167), the solution of the problem (8-158) and (8-159) becomes

$$
\begin{aligned}
T(r,t) &= 2\sum_{n=1}^{\infty} \frac{\sin\beta_n(r-a)}{r(b-a)} e^{-\alpha\beta_n^2 t} \int_{r'=a}^{b} \sin\beta_n(r'-a) F(r')\, r'\, dr' \\
&+ \frac{2\alpha}{k}\sum_{n=1}^{\infty}\Bigg[\frac{\sin\beta_n(r-a)}{r(b-a)} e^{-\alpha\beta_n^2 t} \\
&\qquad \times \int_{\tau=0}^{t}\int_{r'=a}^{b} \sin\beta_n(r'-a) e^{\alpha\beta_n^2\tau} g(r',\tau) r'\, dr'\, d\tau \Bigg] \\
&- 2\alpha \sum_{n=1}^{\infty}\Bigg\{ \frac{\sin\beta_n(r-a)\left[b\beta_n \cos\beta_n(b-a) - \sin\beta_n(b-a) \right]}{r(b-a)} \\
&\qquad \times e^{-\alpha\beta_n^2 t} \int_{\tau=0}^{t} e^{\alpha\beta_n^2\tau} f(\tau)\, d\tau \Bigg\}
\end{aligned}
\tag{8-169}
$$

where we have performed the differentiation with respect to r' in the last term and canceled terms throughout the expression as applicable.

We now consider some special cases of our solution (8-169):

1. The hollow sphere has zero initial temperature, zero temperature at both surfaces, with a spherical surface heat source located at radius r_1, $a < r_1 < b$, with strength $g_s^c(t)$ (W) releasing its heat continuously for times $t > 0$. For this special case we set in equation (8-169) the following:

$$F(r) = 0, \qquad f(t) = 0, \qquad \text{and} \qquad g(r,t) = g_s^c(t)\frac{1}{4\pi r^2}\delta(r - r_1)$$

where we have used the expression for a continuous planar source in spherical coordinates according to equation (8-62). With these changes, equation (8-169) reduces to

$$T(r,t) = \frac{2\alpha}{k}\sum_{n=1}^{\infty}\left[\frac{\sin\beta_n(r-a)}{r(b-a)}e^{-\alpha\beta_n^2 t}\right.$$
$$\left.\times\int_{\tau=0}^{t}\int_{r'=a}^{b}\sin\beta_n(r'-a)e^{\alpha\beta_n^2\tau}g_s^c(\tau)\frac{\delta(r'-r_1)}{4\pi r'^2}r'dr'd\tau\right] \tag{8-170}$$

which after performing the spatial integration yields

$$T(r,t) = \frac{\alpha}{2\pi k r_1}\sum_{n=1}^{\infty}\frac{\sin\beta_n(r-a)\sin\beta_n(r_1-a)}{r(b-a)}e^{-\alpha\beta_n^2 t}\int_{\tau=0}^{t}g_s^c(\tau)e^{\alpha\beta_n^2\tau}\,d\tau \tag{8-171}$$

2. The hollow sphere has zero initial temperature, zero temperature at both surfaces, with an instantaneous spherical surface heat source located at radius r_1, $a < r_1 < b$, with strength g_s^i (J) releasing its heat spontaneously at time $t = 0$. For this special case we set in equation (8-169) the following:

$$F(r) = 0, \qquad f(t) = 0, \qquad \text{and} \qquad g(r,t) = g_s^i\frac{1}{4\pi r^2}\delta(r - r_1)\delta(t)$$

where we have used the expression for an instantaneous planar source in spherical coordinates according to equation (8-62). With these changes, equation (8-169) reduces to

$$T(r,t) = \frac{2\alpha}{k}\sum_{n=1}^{\infty}\left[\frac{\sin\beta_n(r-a)}{r(b-a)}e^{-\alpha\beta_n^2 t}\right.$$
$$\left.\times\int_{\tau=0}^{t}\int_{r'=a}^{b}\sin\beta_n(r'-a)e^{\alpha\beta_n^2\tau}g_s^i\frac{\delta(r'-r_1)\delta(\tau)}{4\pi r'^2}r'dr'd\tau\right] \tag{8-172}$$

which after performing the temporal and spatial integration yields

$$T(r,t) = \frac{\alpha g_s^i}{2\pi k r_1}\sum_{n=1}^{\infty}\frac{\sin\beta_n(r-a)\sin\beta_n(r_1-a)}{r(b-a)}e^{-\alpha\beta_n^2 t} \tag{8-173}$$

Example 8-11 Green's Function for a 1-D Sphere with Generation
Consider a 1-D solid sphere over the domain $0 \le r \le b$ that is initially at zero temperature. For times $t > 0$, the boundary surface at $r = b$ is maintained at zero temperature, while the entire medium is subjected to internal energy generation at the rate $g(r, t)$ per unit time, per unit volume (W/m^3). Use Green's function to calculate the temperature distribution for $t > 0$.

The formulation is given as

$$\frac{1}{r}\frac{\partial^2}{\partial r^2}(rT) + \frac{1}{k}g(r,t) = \frac{1}{\alpha}\frac{\partial T}{\partial t} \quad \text{in} \quad 0 \le r < b, \quad t > 0 \tag{8-174}$$

$$\text{BC1:} \quad T(r \to 0) \Rightarrow \text{finite} \tag{8-175a}$$

$$\text{BC2:} \quad T(r = b, t) = 0 \tag{8-175b}$$

$$\text{IC:} \quad T(r, t = 0) = 0 \tag{8-175c}$$

To determine the desired Green's function, we consider the homogeneous version of the problem for the same region:

$$\frac{1}{r}\frac{\partial^2}{\partial r^2}(r\Psi) = \frac{1}{\alpha}\frac{\partial \Psi}{\partial t} \quad \text{in} \quad 0 \le r < b, \quad t > 0 \tag{8-176}$$

$$\text{BC1:} \quad \Psi(r \to 0) \Rightarrow \text{finite} \tag{8-177a}$$

$$\text{BC2:} \quad \Psi(r = b, t) = 0 \tag{8-177b}$$

$$\text{IC:} \quad \Psi(r, t = 0) = F(r) \tag{8-177c}$$

Note that we have added an initial condition to equation (8-177c) as a placeholder for determination of Green's function, which avoids the trivial solution of Ψ. This homogeneous problem is similar to the problem considered in Example 5-3, with the change from a type 3 boundary condition to a type 1 boundary condition. Using Table 5-2 in conjunction with case 9 of Table 2-1, our solution becomes

$$\Psi(r,t) = \frac{2}{b}\sum_{n=1}^{\infty}\frac{\sin\beta_n r}{r}e^{-\alpha\beta_n^2 t}\int_{r'=0}^{b} r' F(r') \sin\beta_n r'\, dr' \tag{8-178}$$

where the eigenvalues are given

$$\beta_n = \frac{n\pi}{b}, \qquad n = 1, 2, 3, \ldots \tag{8-179}$$

The solution of the homogenous problem (8-176) in terms of Green's function is given, according to equation (8-14), as

$$\Psi(r,t) = \int_{r'=0}^{b} G(r,t|r',\tau)\big|_{\tau=0} F(r') r'^2\, dr' \tag{8-180}$$

where we have used the Sturm–Liouville weight function $p = 2$ for the spherical coordinate system. We may rearrange equation (8-178) into the form

$$\Psi(r,t) = \int_{r'=0}^{b} \left(\frac{2}{b} \sum_{n=1}^{\infty} e^{-\alpha\beta_n^2 t} \frac{\sin\beta_n r \sin\beta_n r'}{r'r} \right) F(r') r'^2 \, dr' \tag{8-181}$$

By comparing equations (8-180) and (8-181), we find the Green's function for $\tau = 0$,

$$G(r,t|r',\tau)\big|_{\tau=0} = \frac{2}{b} \sum_{n=1}^{\infty} \frac{\sin\beta_n r \sin\beta_n r'}{r'r} e^{-\alpha\beta_n^2 t} \tag{8-182}$$

We now replace t in equation (8-182) with $t-\tau$, as detailed in equation (8-22), to yield the desired Green's function as

$$G(r,t|r',\tau) = \frac{2}{b} \sum_{n=1}^{\infty} \frac{\sin\beta_n r \sin\beta_n r'}{r'r} e^{-\alpha\beta_n^2 (t-\tau)} \tag{8-183}$$

The solution of the nonhomogeneous problem of equation (8-174) in terms of the above Green's function is given, according to equation (8-14), as

$$T(r,t) = \frac{\alpha}{k} \int_{\tau=0}^{t} \int_{r'=0}^{b} G(r,t|r',\tau) g(r',\tau) r'^2 \, dr' \, d\tau \tag{8-184}$$

Introducing the above Green's function of equation (8-183) into equation (8-184), gives the overall temperature solution

$$T(r,t) = \frac{2\alpha}{kb} \sum_{n=1}^{\infty} \frac{\sin\beta_n r}{r} e^{-\alpha\beta_n^2 t} \int_{\tau=0}^{t} \int_{r'=0}^{b} \sin\beta_n r' g(r',\tau) e^{\alpha\beta_n^2 \tau} r' \, dr' \, d\tau \tag{8-185}$$

We now consider a few special cases of solution (8-185).

1. The heat source is an instantaneous, distributed volume heat source of strength $g^{\mathrm{i}}(r)$ (J/m^3) releasing its energy spontaneously at time $t = 0$. For such a source, we define the volumetric energy generation as

$$g(r,t) = g^{\mathrm{i}}(r)\delta(t) \tag{8-186}$$

as we developed in equation (8-64). Substitution into equation (8-185) gives

$$T(r,t) = \frac{2\alpha}{kb} \sum_{n=1}^{\infty} \frac{\sin\beta_n r}{r} e^{-\alpha\beta_n^2 t} \int_{\tau=0}^{t} \int_{r'=0}^{b} \sin\beta_n r' g^{\mathrm{i}}(r') \delta(\tau) e^{\alpha\beta_n^2 \tau} r' \, dr' \, d\tau \tag{8-187}$$

which after performing the τ integration yields

$$T(r,t) = \frac{2\alpha}{kb}\sum_{n=1}^{\infty}\frac{\sin\beta_n r}{r}e^{-\alpha\beta_n^2 t}\int_{r'=0}^{b}\sin\beta_n r' g^{\mathrm{i}}(r')r'dr' \tag{8-188}$$

2. The heat source is an instantaneous point heat source of strength g_p^{i} (J), which is situated at the center of the sphere, and releases its heat spontaneously at time $t = 0$. For such a source, we define the volumetric energy generation as

$$g(r,t) = \frac{g_p^{\mathrm{i}}}{4\pi r^2}\delta(r-0)\delta(t-0) \tag{8-189}$$

Substitution of the above into equation (8-185) gives

$$T(r,t) = \frac{2\alpha}{kb}\sum_{n=1}^{\infty}\frac{\sin\beta_n r}{r}e^{-\alpha\beta_n^2 t}\int_{\tau=0}^{t}\int_{r'=0}^{b}\sin\beta_n r' \frac{g_p^{\mathrm{i}}\delta(r')\delta(\tau)}{4\pi r'^2}e^{\alpha\beta_n^2\tau}r'dr'd\tau \tag{8-190}$$

which after performing the temporal and spatial integration yields

$$T(r,t) = \frac{\alpha g_p^{\mathrm{i}}}{2\pi kb}\sum_{n=1}^{\infty}\frac{\beta_n\sin\beta_n r}{r}e^{-\alpha\beta_n^2 t} \tag{8-191}$$

For the spatial integration, we made use of the following, per L'Hôpital's rule:

$$\lim_{r'\to 0}\frac{\sin\beta_n r'}{r'} = \beta_n \tag{8-192}$$

The steady-state solution of the problem is zero, which is readily observed with equation (8-191) for $t \to \infty$.

Example 8-12 Green's Function for a 2-D Sphere with Generation
Consider a 2-D solid sphere over the domain $0 \le r \le b$ and $-1 \le \mu \le 1$ that is initially at temperature $F(r,\mu)$. For times $t > 0$, the boundary surface at $r = b$ is maintained at temperature $f(\mu,t)$, while the entire medium is subjected to internal energy generation at the rate $g(r,\mu,t)$ per unit time, per unit volume (W/m^3). Use the Green's function approach to calculate the temperature distribution for $t > 0$.

The formulation is given as

$$\frac{\partial^2 T}{\partial r^2} + \frac{2}{r}\frac{\partial T}{\partial r} + \frac{1}{r^2}\frac{\partial}{\partial\mu}\left[(1-\mu^2)\frac{\partial T}{\partial\mu}\right] + \frac{1}{k}g(r,\mu,t) = \frac{1}{\alpha}\frac{\partial T}{\partial t} \tag{8-193}$$

$$\text{in} \quad 0 \le r < b, \qquad -1 \le \mu \le 1 \qquad \text{for} \qquad t > 0$$

$$\text{BC1:} \qquad T(r \to 0) \Rightarrow \text{finite} \tag{8-194a}$$

$$\text{BC2:} \qquad T(r = b, t) = f(\mu, t) \tag{8-194b}$$

$$\text{BC3:} \qquad T(\mu \to \pm 1) \Rightarrow \text{finite} \tag{8-194c}$$

$$\text{IC:} \qquad T(r, \mu, t = 0) = F(r, \mu) \tag{8-194d}$$

To determine the desired Green's function, we consider the homogeneous version of the problem for the same region:

$$\frac{\partial^2 \Psi}{\partial r^2} + \frac{2}{r}\frac{\partial \Psi}{\partial r} + \frac{1}{r^2}\frac{\partial}{\partial \mu}\left[(1-\mu^2)\frac{\partial \Psi}{\partial \mu}\right] = \frac{1}{\alpha}\frac{\partial \Psi}{\partial t} \tag{8-195}$$

$$\text{in} \qquad 0 \le r < b, \qquad -1 \le \mu \le 1 \qquad \text{for} \qquad t > 0$$

$$\text{BC1:} \qquad \Psi(r \to 0) \Rightarrow \text{finite} \tag{8-196a}$$

$$\text{BC2:} \qquad \Psi(r = b, t) = 0 \tag{8-196b}$$

$$\text{BC3:} \qquad \Psi(\mu \to \pm 1) \Rightarrow \text{finite} \tag{8-196c}$$

$$\text{IC:} \qquad \Psi(r, \mu, t = 0) = F(r, \mu) \tag{8-196d}$$

This homogeneous problem is identical to the problem considered in Example 5-7, giving the solution, after multiplication by $(r')^{1/2}$, as

$$\Psi(r, \mu, t) = \int_{r'=0}^{b} \int_{\mu'=-1}^{1} \left[\sum_{n=0}^{\infty} \sum_{p=1}^{\infty} \frac{J_{n+1/2}(\lambda_{np} r) P_n(\mu)}{N(n) N(\lambda_{np})} e^{-\alpha \lambda_{np}^2 t} \times \frac{J_{n+1/2}(\lambda_{np} r') P_n(\mu')}{(r' r)^{1/2}} \right] F(r', \mu') r'^2 dr' d\mu' \tag{8-197}$$

where we define eigenvalues from the transcendental equation

$$J_{n+1/2}(\lambda_{np} b) = 0 \quad \to \quad \lambda_{np}, \qquad p = 1, 2, 3, \ldots \qquad \text{for each } n \tag{8-198}$$

The norms of equation (8-197) are defined as

$$N(n) = \frac{2}{2n+1} \tag{8-199a}$$

$$N(\lambda_{np}) = \frac{b^2}{2} J_{n+3/2}^2(\lambda_{np} b) \tag{8-199b}$$

where we have made use of case 3 in Table 2-2 for equation (8-199b). The solution of the homogeneous problem in terms of Green's function is given, according to equation (8-12), as

$$\Psi(r, \mu, t) = \int_{r'=0}^{b} \int_{\mu'=-1}^{1} G(r, \mu, t | r', \mu', \tau)\big|_{\tau=0} F(r', \mu') r'^2 dr' d\mu' \tag{8-200}$$

where we used the relation $dA' = r'^2 dr' d\mu'$ for the 2-D spherical problem. A comparison of equations (8-197) and (8-200) yields

$$G(r, \mu, t|r', \mu', \tau)\big|_{\tau=0} = \sum_{n=0}^{\infty}\sum_{p=1}^{\infty} \frac{J_{n+1/2}(\lambda_{np} r) P_n(\mu) J_{n+1/2}(\lambda_{np} r') P_n(\mu')}{N(n) N(\lambda_{np})(r'r)^{1/2}} e^{-\alpha \lambda_{np}^2 t} \tag{8-201}$$

We now replace t in equation (8-201) with $t-\tau$, as detailed in equation (8-22), to yield the desired Green's function as

$$G(r, \mu, t|r', \mu', \tau) = \sum_{n=0}^{\infty}\sum_{p=1}^{\infty} \frac{J_{n+1/2}(\lambda_{np} r) P_n(\mu) J_{n+1/2}(\lambda_{np} r') P_n(\mu')}{N(n) N(\lambda_{np})(r'r)^{1/2}} e^{-\alpha \lambda_{np}^2 (t-\tau)} \tag{8-202}$$

The solution of the nonhomogeneous problem of equations (8-193) and (8-194) in terms of the above Green's function is given, according to equation (8-12), as

$$\begin{aligned} T(r, \mu, t) = & \int_{r'=0}^{b}\int_{\mu'=-1}^{1} G(r, \mu, t|r', \mu', \tau)|_{\tau=0} F(r', \mu') r'^2 d\mu' dr' \\ & + \frac{\alpha}{k}\int_{\tau=0}^{t}\int_{r'=0}^{b}\int_{\mu'=-1}^{1} G(r, \mu, t|r', \mu', \tau) g(r', \mu', \tau) r'^2 d\mu' dr' d\tau \\ & - \alpha \int_{\tau=0}^{t}\int_{\mu'=-1}^{1} \left[r'^2 \frac{\partial G(r, \mu, t|r', \mu', \tau)}{\partial r'} \right]\Bigg|_{r'=b} f(\mu', \tau) d\mu' d\tau \end{aligned} \tag{8-203}$$

where we have made the substitution in the third term for the type 1 boundary condition, namely,

$$-k \left.\frac{\partial G}{\partial r'}\right|_{r'=b} = +h\, G|_{r'=b} \quad \rightarrow \quad -\left.\frac{\partial G}{\partial r'}\right|_{r'=b} = \frac{1}{k}\, G|_{r'=b} \tag{8-204}$$

Introducing the above Green's function of equation (8-202) into equation (8-203), gives the overall temperature solution

$$\begin{aligned} T(r, \mu, t) = & \sum_{n=0}^{\infty}\sum_{p=1}^{\infty} \frac{J_{n+1/2}(\lambda_{np} r) P_n(\mu)}{N(n) N(\lambda_{np}) r^{1/2}} e^{-\alpha \lambda_{np}^2 t} \\ & \times \int_{r'=0}^{b}\int_{\mu'=-1}^{1} J_{n+1/2}(\lambda_{np} r') P_n(\mu') F(r', \mu') r'^{3/2} d\mu' dr' \\ & + \frac{\alpha}{k} \sum_{n=0}^{\infty}\sum_{p=1}^{\infty} \frac{J_{n+1/2}(\lambda_{np} r) P_n(\mu)}{N(n) N(\lambda_{np}) r^{1/2}} e^{-\alpha \lambda_{np}^2 t} \\ & \times \int_{\tau=0}^{t}\int_{r'=0}^{b}\int_{\mu'=-1}^{1} J_{n+1/2}(\lambda_{np} r') P_n(\mu') e^{\alpha \lambda_{np}^2 \tau} \\ & \times g(r', \mu', \tau) r'^{3/2} d\mu' dr' d\tau \end{aligned}$$

$$
\begin{aligned}
&- \alpha \sum_{n=0}^{\infty} \sum_{p=1}^{\infty} \frac{J_{n+1/2}(\lambda_{np} r) P_n(\mu)}{N(n) N(\lambda_{np}) r^{1/2}} e^{-\alpha \lambda_{np}^2 t} \\
&\times \left[b\left(n + \tfrac{1}{2}\right) J_{n+1/2}(\lambda_{np} b) - b^2 \lambda_{np} J_{n+3/2}(\lambda_{np} b) \right] \\
&\times \int_{\tau=0}^{t} \int_{\mu'=-1}^{1} P_n(\mu') e^{\alpha \lambda_{np}^2 \tau} f(\mu', \tau) d\mu' d\tau
\end{aligned} \tag{8-205}
$$

where the norms are defined by equations (8-199) and the eigenvalues are given by equation (8-198).

8-7 PRODUCTS OF GREEN'S FUNCTIONS

The multidimensional Green's functions can be obtained from the multiplication of 1-D Green's functions for all cases in the Cartesian coordinate system and for some cases in the cylindrical coordinate system; but the multiplication procedure is not possible in the spherical coordinate system. We illustrate this matter with examples in the rectangular and cylindrical coordinates.

Rectangular Coordinates

The 3-D Green's function $G(x, y, z, t|x', y', z', \tau)$ can be obtained from the product of the three 1-D Green's functions as

$$
G(x, y, z, t|x', y', z', \tau) = G_1(x, t|x', \tau) \cdot G_2(y, t|y', \tau) \cdot G_3(z, t|z', \tau) \tag{8-206}
$$

where each of the 1-D Green's functions G_1, G_2, and G_3 depends on the extent of the region (i.e., finite, semi-infinite, or infinite) and the boundary conditions associated with it (i.e., first, second, or third type). We present below a tabulation of the 3-D Green's functions in the rectangular coordinates as the product of three, 1-D Green's functions.

Region 1 $0 \le x \le a$, $0 \le y \le b$, $0 \le z \le c$:

$$
\begin{aligned}
G(x, y, z, t|x', y', z', \tau) &= \left[\sum_{m=1}^{\infty} e^{-\alpha \beta_m^2 (t-\tau)} \frac{1}{N(\beta_m)} X(\beta_m, x) X(\beta_m, x') \right] \\
&\times \left[\sum_{n=1}^{\infty} e^{-\alpha \gamma_n^2 (t-\tau)} \frac{1}{N(\gamma_n)} Y(\gamma_n, y) Y(\gamma_n, y') \right] \\
&\times \left[\sum_{p=1}^{\infty} e^{-\alpha \eta_p^2 (t-\tau)} \frac{1}{N(\eta_p)} Z(\eta_p, z) Z(\eta_p, z') \right]
\end{aligned} \tag{8-207}
$$

where the eigenfunctions, eigenconditions, and normalization integrals are obtainable from Table 2-1. In each direction, there are nine different combinations of boundary conditions; therefore the result given by equation (8-207) together with Table 2-1 represents $9 \times 9 \times 9 = 729$ different Green's functions.

Region 2 $-\infty < x < \infty,\ 0 \le y \le b,\ 0 \le z \le c$:

$$G(x, y, z, t|x', y', z', \tau) = \left\{[4\pi\alpha(t-\tau)]^{-1/2} \exp\left[-\frac{(x-x')^2}{4\alpha(t-\tau)}\right]\right\} \times \left[\sum_{n=1}^{\infty} e^{-\alpha\gamma_n^2(t-\tau)} \frac{1}{N(\gamma_n)} Y(\gamma_n, y) Y(\gamma_n, y')\right] \times \left[\sum_{p=1}^{\infty} e^{-\alpha\eta_p^2(t-\tau)} \frac{1}{N(\eta_p)} Z(\eta_p, z) Z(\eta_p, z')\right] \tag{8-208}$$

where the infinite medium Green's function, shown inside the first bracket, is obtained from equation (8-72). The result given by equation (8-208) when used together with the Table 2-1 represents $9 \times 9 = 81$ different Green's functions.

Region 3 $-\infty < x < \infty,\ -\infty < y < \infty,\ 0 \le z \le c$:

$$G(x, y, z, t|x', y', z', \tau) = \left\{[4\pi\alpha(t-\tau)]^{-1/2} \exp\left[-\frac{(x-x')^2}{4\alpha(t-\tau)}\right]\right\} \times \left\{[4\pi\alpha(t-\tau)]^{-1/2} \exp\left[-\frac{(y-y')^2}{4\alpha(t-\tau)}\right]\right\} \times \left[\sum_{p=1}^{\infty} e^{-\alpha\eta_p^2(t-\tau)} \frac{1}{N(\eta_p)} Z(\eta_p, z) Z(\eta_p, z')\right] \tag{8-209}$$

where the infinite medium Green's functions are obtained from equation (8-72). The result given by equation (8-209) when used together with Table 2-1 for the z dimension represents nine different Green's functions.

Region 4 $-\infty < x < \infty,\ -\infty < y < \infty,\ -\infty < z < \infty$:

$$G(x, y, z, t|x', y', z', \tau) = \left\{[4\pi\alpha(t-\tau)]^{-1/2} \exp\left[-\frac{(x-x')^2}{4\alpha(t-\tau)}\right]\right\} \times \left\{[4\pi\alpha(t-\tau)]^{-1/2} \exp\left[-\frac{(y-y')^2}{4\alpha(t-\tau)}\right]\right\} \times \left\{[4\pi\alpha(t-\tau)]^{-1/2} \exp\left[-\frac{(z-z')^2}{4\alpha(t-\tau)}\right]\right\} \tag{8-210}$$

Region 5 $0 \le x < \infty$: In the foregoing expressions for the 3-D Green's functions, if any one of the regions is semi-infinite, the Green's function for that region should be replaced by the semi-infinite medium Green's functions as given below, for the three unique boundary conditions at $x = 0$:

1. Boundary condition at $x = 0$ is of the first type [constructed from equation (6-35)]:

$$G(x, t|x', \tau) = [4\pi\alpha(t-\tau)]^{-1/2}\left\{\exp\left[-\frac{(x-x')^2}{4\alpha(t-\tau)}\right] - \exp\left[-\frac{(x+x')^2}{4\alpha(t-\tau)}\right]\right\} \tag{8-211}$$

2. Boundary condition at $x = 0$ is of the second type [constructed from equation (6-71)]:

$$G(x, t|x', \tau) = [4\pi\alpha(t-\tau)]^{-1/2}\left\{\exp\left[-\frac{(x-x')^2}{4\alpha(t-\tau)}\right] + \exp\left[-\frac{(x+x')^2}{4\alpha(t-\tau)}\right]\right\} \tag{8-212}$$

3. Boundary condition at $x = 0$ is of the third type [constructed from equations (6-16) and (6-17) using Table 6-1]:

$$G(x, t|x', \tau) = \int_{\beta=0}^{\infty} e^{-\alpha\beta^2(t-\tau)}\frac{1}{N(\beta)}X(\beta, x)X(\beta, x')d\beta \tag{8-213}$$

where

$$X(\beta, x) = \beta\cos\beta x + H\sin\beta x \tag{8-214}$$

$$\frac{1}{N(\beta)} = \frac{2}{\pi}\frac{1}{\beta^2 + H^2}, \qquad \text{with} \qquad H = \frac{h}{k} \tag{8-215}$$

Example 8-13 3-D Cartesian Problem Using Product Solution

Consider a rectangular parallelepiped over the domain $0 \le x \le a$, $0 \le y \le b$, $0 \le z \le c$ that is initially at a zero temperature. For times $t > 0$, all of the boundary surfaces are maintained at zero temperature, while in addition there is internal energy generation within the solid at the rate $g(x, y, z, t)$ per unit time, per unit volume (W/m^3). Develop the 3-D Green's function needed for the solution of this heat conduction equation with the Green's function approach.

Green's function for this problem is obtainable as a product of three 1-D, finite-region Green's functions subjected to the boundary condition of the first type. We use the formulation given by equation (8-207) together with case 9 of

Table 2-1, to obtain Green's function as

$$G(x, y, z, t|x', y', z', \tau) = \left[\frac{2}{a}\sum_{m=1}^{\infty} e^{-\alpha\beta_m^2(t-\tau)} \sin\beta_m x \sin\beta_m x'\right]$$
$$\times \left[\frac{2}{b}\sum_{n=1}^{\infty} e^{-\alpha\gamma_n^2(t-\tau)} \sin\gamma_n y \sin\gamma_n y'\right]$$
$$\times \left[\frac{2}{c}\sum_{p=1}^{\infty} e^{-\alpha\eta_p^2(t-\tau)} \sin\eta_p z \sin\eta_p z'\right] \tag{8-216}$$

where the eigenvalues β_m, γ_n, and η_p are the positive roots of the equations

$$\sin\beta_m a = 0 \qquad \sin\gamma_n b = 0 \qquad \sin\eta_p c = 0 \tag{8-217}$$

Cylindrical Coordinates

The multiplication of 1-D Green's functions in order to get multidimensional Green's function is possible if the problem involves only the (r,z,t) variables, that is, if the problem has azimuthal symmetry. When the problem involves (r,z,ϕ,t) variables, it is not possible to separate Green's function associated with the r and ϕ variables, due to the coupling of the order of the Bessel functions to the ϕ dimension.

We present below a tabulation of 2-D Green's functions in the (r,z,t) variables in the cylindrical coordinates developed by the multiplication of two, 1-D Green's functions.

Region 6 $0 \le r \le b$, $0 \le z \le c$:

$$G(r, z, t|r', z', \tau) = \left[\sum_{m=1}^{\infty} e^{-\alpha\beta_m^2(t-\tau)} \frac{1}{N(\beta_m)} r' R_0(\beta_m, r) R_0(\beta_m, r')\right]$$
$$\times \left[\sum_{p=1}^{\infty} e^{-\alpha\eta_p^2(t-\tau)} \frac{1}{N(\eta_p)} Z(\eta_p, z) Z(\eta_p, z')\right] \tag{8-218}$$

where the eigenfunctions, eigenvalues, and normalization integrals for the r variable are obtained from Table 2-2 by setting $\nu = 0$, and for the z variable are obtained from Table 2-1. Table 2-2 involves three different cases and Table 2-1 nine different cases; hence the result given by equation (8-218) represents $3 \times 9 = 27$ different Green's functions.

Region 7 $0 \le r < \infty, 0 \le z \le c$:

$$G(r, z, t|r', z', \tau) = \left\{ [2\alpha(t-\tau)]^{-1} r' \exp\left[-\frac{r^2 + r'^2}{4\alpha(t-\tau)} \right] I_0 \left[\frac{rr'}{2\alpha(t-\tau)} \right] \right\}$$
$$\times \left[\sum_{p=1}^{\infty} e^{-\alpha\eta_p^2(t-\tau)} \frac{1}{N(\eta_p)} Z(\eta_p, z) Z(\eta_p, z') \right] \tag{8-219}$$

where the Green's function for the r variable is constructed from the solution given by equation (6-166) and the eigenfunctions, eigenvalues, and normalization integrals associated with the z variable are obtainable from Table 2-1. Therefore, the result given by equation (8-219) represents nine different Green's functions.

Region 8 $a \le r \le b, 0 \le z \le c$:

$$G(r, z, t|r', z', \tau) = \left[\sum_{m=1}^{\infty} e^{-\alpha\beta_m^2(t-\tau)} \frac{1}{N(\beta_m)} r' R_0(\beta_m, r) R_0(\beta_m, r') \right]$$
$$\times \left[\sum_{p=1}^{\infty} e^{-\alpha\eta_p^2(t-\tau)} \frac{1}{N(\eta_p)} Z(\eta_p, z) Z(\eta_p, z') \right] \tag{8-220}$$

where the eigenfunctions, eigenvalues, and normalization integrals for the r variable are obtainable from Table 2-3 and for the z variable from Table 2-1. In Table 2-3, only the boundary conditions of the first and second type are considered because the results for the boundary condition of the third type are rather involved; hence using the available solutions, equation (8-220) represents $4 \times 9 = 36$ different Green's functions.

Region 9 $a \le r < \infty, 0 \le z \le c$:

$$G(r, z, t|r', z', \tau) = \left[\int_{\beta=0}^{\infty} e^{-\alpha\beta^2(t-\tau)} \frac{\beta}{N(\beta)} r' R_0(\beta, r) R_0(\beta, r') \right]$$
$$\times \left[\sum_{p=1}^{\infty} e^{-\alpha\eta_p^2(t-\tau)} \frac{1}{N(\eta_p)} Z(\eta_p, z) Z(\eta_p, z') \right] \tag{8-221}$$

where the Green's function for the r variable given inside the first bracket is obtained from the rearrangement of equations (2-121) and (2-122). The eigenfunctions, eigenvalues, and the normalization integral associated with the r variable are given in Table 2-4 for three different boundary conditions at $r = a$, and those associated with the z variable are given in Table 2-1 for nine different combinations of boundary conditions. Therefore, the result given by equation (8-221) represents $3 \times 9 = 27$ different Green's functions.

In the foregoing expression for the 2-D Green's function in the cylindrical coordinates, we considered only a finite region $0 \le z \le c$ for the z variable. If the z dimension is semi-infinite or infinite, the corresponding Green's function is obtained from those discussed for the rectangular coordinate system.

REFERENCES

1. H. S. Carslaw and J. C. Jaeger, *Conduction of Heat in Solids*, Clarendon, London, 1959.
2. P. M. Morse and H. Feshback, *Methods of Theoretical Physics*, McGraw-Hill, New York, 1953.
3. I. N. Sneddon, *Partial Differential Equations*, McGraw-Hill, New York, 1957.
4. J. W. Dettman, *Mathematical Methods in Physics and Engineering*, McGraw-Hill, New York, 1962.
5. R. Courant and D. Hilbert, *Methods of Mathematical Physics*, Interscience, New York, 1953.
6. M. N. Özisik, *Boundary Value Problems of Heat Conduction*, International Textbook, Scranton, PA, 1968.
7. M. D. Greenberg, *Applications of Green's Functions in Science and Engineering*, Prentice-Hall, Englewood Cliffs, NJ, 1971.
8. A. M. Aizen, I. S. Redchits, and I. M. Fedotkin, *J. Eng. Phys.* **26**, 453–458, 1974.
9. A. G. Butkovskiy, *Green's Functions and Transfer Functions Handbook*, Halstead, Wiley, New York, 1982.
10. I. Stakgold, *Green's Functions and Boundary Value Problems*, Wiley, New York, 1979.
11. J. V. Beck, K. D. Cole, A. Haji-Sheikh, and B. Litkouki, *Heat Conduction Using Green's Functions*, Hemisphere, Washington, DC, 1992.
12. M. N. Özisik, *Heat Conduction*, 2nd ed. , Wiley, New York, 1993.

PROBLEMS

8-1 A semi-infinite region $0 \le x < \infty$ is initially at temperature $F(x)$. For times $t > 0$, boundary surface at $x = 0$ is kept at zero temperature and internal energy is generated within the solid at a rate of $g(x,t)$ (W/m^3). Determine the Green's function for this problem, and using this Green's function obtain an expression for the temperature distribution $T(x,t)$ within the medium for times $t > 0$.

8-2 Repeat Problem 8-1 for the case when the boundary surface at $x = 0$ is kept perfectly insulated.

8-3 A slab, $0 \le x \le L$, is initially at temperature $F(x)$. For times $t > 0$, internal energy is generated within the solid at a rate of $g(x,t)$ (W/m^3), the boundary surface at $x = 0$ is exposed to a time-dependent heat flux

$q''(t)$, and the boundary surface at $x = L$ dissipates heat by convection into a medium at zero temperature. Using the Green's function approach, obtain an expression for the temperature distribution $T(x,t)$ in the slab for times $t > 0$.

8-4 Using the Green's function approach solve the following heat conduction problem for a 2-D rectangular region $0 \leq x \leq a$, $0 \leq y \leq b$:

$$\frac{\partial^2 T}{\partial x^2} + \frac{\partial^2 T}{\partial y^2} + \frac{1}{k} g(x, y, t) = \frac{1}{\alpha} \frac{\partial T}{\partial t}$$

$$\text{in} \quad 0 < x < a, \qquad 0 < y < b, \qquad t > 0$$

$$\text{BC1:} \quad \left.\frac{\partial T}{\partial x}\right|_{x=0} = 0$$

$$\text{BC2:} \quad \left.\frac{\partial T}{\partial y}\right|_{x=0} + H_2 \left. T\right|_{x=a} = 0$$

$$\text{BC3:} \quad T(y = 0) = 0$$

$$\text{BC4:} \quad \left.\frac{\partial T}{\partial y}\right|_{y=b} + H_4 \left. T\right|_{y=b} = 0$$

$$\text{IC:} \quad T(x, y, t = 0) = F(x, y)$$

8-5 Using the Green's function approach solve the following heat conduction problem for a 2-D, semi-infinite rectangular region:

$$\frac{\partial^2 T}{\partial x^2} + \frac{\partial^2 T}{\partial y^2} + \frac{1}{k} g(x, y, t) = \frac{1}{\alpha} \frac{\partial T}{\partial t}$$

$$\text{in} \quad 0 < x < \infty, \qquad 0 < y < b, \qquad t > 0$$

$$\text{BC1:} \quad T(x = 0) = 0$$

$$\text{BC2:} \quad \left.\frac{\partial T}{\partial x}\right|_{y=0} + H_1 \left. T\right|_{y=0} = 0$$

$$\text{BC3:} \quad T(y = b) = 0$$

$$\text{IC:} \quad T(x, y, t = 0) = F(x, y)$$

8-6 Using the Green's function approach solve the following heat conduction problem for a 1-D rectangular region:

$$\frac{\partial^2 T}{\partial x^2} + \frac{1}{k} g(x, t) = \frac{1}{\alpha} \frac{\partial T}{\partial t} \qquad \text{in} \quad 0 < x < L, \qquad t > 0$$

$$\text{BC1:} \quad T(x = 0) = 0$$

$$\text{BC2:} \qquad \left.\frac{\partial T}{\partial x}\right|_{x=L} + H\, T|_{x=L} = 0$$

$$\text{IC:} \qquad T(x, t = 0) = F(x)$$

8-7 A rectangular region $0 \leq x \leq a$, $0 \leq y \leq b$ is initially at temperature $F(x, y)$. For times $t > 0$, internal energy is generated within the solid at a rate of $g(x, y, t)$ (W/m^3), while the boundary conditions are as defined below. Obtain an expression for the temperature distribution $T(x,y,t)$ in the region for times $t > 0$.

$$\frac{\partial^2 T}{\partial x^2} + \frac{\partial^2 T}{\partial y^2} + \frac{1}{k} g(x, y, t) = \frac{1}{\alpha}\frac{\partial T}{\partial t}$$

$$\text{in} \qquad 0 < x < a, \qquad 0 < y < b, \qquad t > 0$$

$$\text{BC1:} \qquad T(x = 0) = f_1(t)$$

$$\text{BC2:} \qquad -k \left.\frac{\partial T}{\partial x}\right|_{x=a} = f_2(t)$$

$$\text{BC3:} \qquad T(y = 0) = 0$$

$$\text{BC4:} \qquad T(y = b) = f_3(t)$$

$$\text{IC:} \qquad T(x, y, t = 0) = F(x, y)$$

8-8 A 3-D infinite medium, $-\infty < x < \infty$, $-\infty < y < \infty$, $-\infty < z < \infty$, is initially at temperature $F(x,y,z)$. For times $t > 0$ internal energy is generated in the medium at a rate of $g(x,y,z,t)$ (W/m^3). Using the Green's function approach, obtain an expression for the temperature distribution $T(x, y, z, t)$ in the region for time $t > 0$. Also consider the following special cases:

a. The heat source is a continuous, point heat source of strength $g_p^{\text{c}}(t)$ (W) situated at the location (x_1, y_1, z_1) that releases its heat for times $t > 0$. Hint: $g(x, y, z, t) = g_p^{\text{c}}(t)\delta(x - x_1)\delta(y - y_1)\delta(z - z_1)$.

b. The heat source is an instantaneous, point heat source of strength g_p^{i} (J), which releases its heat spontaneously at time $t = 0$, positioned at the location (x_1, y_1, z_1).
Hint: $g(x, y, z, t) = g_p^{\text{i}}\delta(x - x_1)\delta(y - y_1)\delta(z - z_1)\delta(t)$.

8-9 Solve the following 1-D heat conduction problem for a solid cylinder $0 \leq r \leq b$ by Green's function approach:

$$\frac{\partial^2 T}{\partial r^2} + \frac{1}{r}\frac{\partial T}{\partial r} + \frac{1}{k} g(r, t) = \frac{1}{\alpha}\frac{\partial T}{\partial t} \qquad \text{in} \qquad 0 \leq r < b, \qquad t > 0$$

$$\text{BC1:} \qquad \left.\frac{\partial T}{\partial r}\right|_{r=b} + H\, T|_{r=b} = 0$$

$$\text{IC:} \qquad T(r, t = 0) = F(r)$$

8-10 Solve the following 1-D heat conduction problem for a semi-infinite solid cylinder by Green's function approach:

$$\frac{\partial^2 T}{\partial r^2} + \frac{1}{r}\frac{\partial T}{\partial r} + \frac{1}{k}g(r,t) = \frac{1}{\alpha}\frac{\partial T}{\partial t} \quad \text{in} \quad 0 \le r < \infty, \quad t > 0$$

$$\text{IC:} \quad T(r, t = 0) = F(r)$$

Also consider the following special case: Medium initially at zero temperature, the heat source is an instantaneous, line heat source of strength g_L^{i}, (J/m) situated along the z axis, which releases its heat spontaneously at time $t = 0$, that is,

$$g(r,t) = \frac{g_L^{\mathrm{i}}}{2\pi r}\delta(r-0)\delta(t-0)$$

8-11 Solve the following 1-D heat conduction problem for a semi-infinite, hollow cylinder by Green's function approach:

$$\frac{\partial^2 T}{\partial r^2} + \frac{1}{r}\frac{\partial T}{\partial r} + \frac{1}{k}g(r,t) = \frac{1}{\alpha}\frac{\partial T}{\partial t} \quad \text{in} \quad a < r < \infty, \quad t > 0$$

$$\text{BC1:} \quad T(r = a) = 0$$

$$\text{IC:} \quad T(r, t = 0) = F(r)$$

8-12 Repeat problem 8-11 for the following boundary condition:

$$\text{BC1:} \quad -\left.\frac{\partial T}{\partial r}\right|_{r=a} + H\left.T\right|_{r=a} = 0$$

8-13 Solve the following 2-D heat conduction problem for a hollow cylinder of finite length by Green's function approach:

$$\frac{\partial^2 T}{\partial r^2} + \frac{1}{r}\frac{\partial T}{\partial r} + \frac{\partial^2 T}{\partial z^2} + \frac{1}{k}g(r,z,t) = \frac{1}{\alpha}\frac{\partial T}{\partial t}$$

$$\text{in} \quad a < r < b, \quad 0 < z < c$$

$$\text{BC1:} \quad T(r = a) = 0$$

$$\text{BC2:} \quad T(r = b) = 0$$

$$\text{BC3:} \quad \left.\frac{\partial T}{\partial z}\right|_{z=0} = 0$$

$$\text{BC4:} \quad \left.\frac{\partial T}{\partial z}\right|_{z=c} + H\left.T\right|_{z=c} = 0$$

$$\text{IC:} \quad T(r, z, t = 0) = F(r, z)$$

8-14 Solve the following 2-D heat conduction problem for a long, solid cylinder by Green's function approach:

$$\frac{\partial^2 T}{\partial r^2} + \frac{1}{r}\frac{\partial T}{\partial r} + \frac{1}{r^2}\frac{\partial^2 T}{\partial \phi^2} + \frac{1}{k}g(r,\phi,t) = \frac{1}{\alpha}\frac{\partial T}{\partial t}$$

$$\text{in} \quad 0 \le r < b, \qquad 0 \le \phi \le 2\pi$$

$$\text{BC1:} \qquad \left.\frac{\partial T}{\partial r}\right|_{r=b} + H\, T|_{r=b} = 0$$

$$\text{IC:} \qquad T(r,\phi,t=0) = 0$$

8-15 Repeat Problem 8-14 for the case when the boundary surface at $r = b$ is kept at zero temperature.

8-16 Solve the following 2-D heat conduction problem for a long, cylindrical wedge by Green's function approach:

$$\frac{\partial^2 T}{\partial r^2} + \frac{1}{r}\frac{\partial T}{\partial r} + \frac{1}{r^2}\frac{\partial^2 T}{\partial \phi^2} + \frac{1}{k}g(r,\phi,t) = \frac{1}{\alpha}\frac{\partial T}{\partial t}$$

$$\text{in} \qquad 0 \le r < b, \qquad 0 < \phi < \phi_0, \qquad (\phi_0 < 2\pi)$$

$$\text{BC1:} \qquad T(r=b) = f(t)$$

$$\text{BC2:} \qquad T(\phi=0) = 0$$

$$\text{BC3:} \qquad T(\phi_0=0) = 0$$

$$\text{IC:} \qquad T(r,\phi,t=0) = 0$$

Note: $dA = r dr d\phi$ and $dl = r d\phi$ along the $r = b$ boundary.

8-17 A solid sphere $0 \le r \le b$ is initially at temperature $F(r)$. For times $t > 0$, internal energy is generated in the sphere at a rate of $g(r,t)$ (W/m^3) while the boundary surface at $r = b$ dissipates heat by convection into a medium such that $hT_\infty = f(t)$. Obtain an expression for the temperature distribution in the sphere using Green's function.

8-18 Solve the following 2-D heat conduction problem for a solid hemisphere:

$$\frac{\partial^2 T}{\partial r^2} + \frac{2}{r}\frac{\partial T}{\partial r} + \frac{1}{r^2}\frac{\partial}{\partial \mu}\left[(1-\mu^2)\frac{\partial T}{\partial \mu}\right] + \frac{1}{k}g(r,\mu,t) = \frac{1}{\alpha}\frac{\partial T}{\partial t}$$

$$\text{in} \qquad 0 \le r < b, \qquad 0 < \mu \le 1, \qquad \text{for} \qquad t > 0$$

$$\text{BC1:} \qquad T(r=b) = 0$$

$$\text{BC2:} \qquad T(\mu=0) = 0$$

$$\text{IC:} \qquad T(r,\mu,t=0) = F(r,\mu)$$

Note: $dA = r^2 dr d\mu$.

8-19 Solve the following 1-D heat conduction problem for a hollow sphere using Green's function:

$$\frac{1}{r}\frac{\partial}{\partial r^2}(rT) + \frac{1}{k}g(r,t) = \frac{1}{\alpha}\frac{\partial T}{\partial t} \qquad \text{in} \qquad a < r < b, \qquad t > 0$$

$$\text{BC1:} \qquad -\left.\frac{\partial T}{\partial r}\right|_{r=a} + H_1 \, T|_{r=a} = 0$$

$$\text{BC2:} \qquad \left.\frac{\partial T}{\partial r}\right|_{r=b} + H_2 \, T|_{r=b} = 0$$

$$\text{IC:} \qquad T(r, t = 0) = F(r)$$

Also consider the following case: The heat generation source is an instantaneous, spherical surface heat source of radius r_1 $(a < r_1 < b)$ of total strength g_s^{i} (J) that releases its heat spontaneously at $t = 0$. For this problem, consider

$$g(r,t) = g_s^{\mathrm{i}} \frac{1}{4\pi r^2} \delta(r - r_1)\delta(t)$$

8-20 Consider the following heat conduction problem for a 1-D rectangular region:

$$\frac{\partial^2 T}{\partial x^2} = \frac{1}{\alpha}\frac{\partial T}{\partial t} \qquad \text{in} \qquad 0 < x < L, \qquad t > 0$$

$$\text{BC1:} \qquad T(x = 0) = 0$$

$$\text{BC2:} \qquad T(x = L) = f(t)$$

$$\text{IC:} \qquad T(x, t = 0) = 0$$

Solve for the temperature distribution *T(x,t)* using *both* Duhamel's theorem and using the Green's function. Simplify and perform all necessary integration to show that the two solutions are *identical*.

9

USE OF THE LAPLACE TRANSFORM

The method of Laplace transform has been widely used in the solution of time-dependent heat conduction problems because the partial derivative with respect to the time variable can be removed from the differential equation of heat conduction. Although the application of the Laplace transform for the removal of the partial derivative is a relatively straightforward matter, the inversion of the transformed solution generally is rather involved unless the inversion is available in the standard Laplace transform tables. The Laplace transform technique is also readily applicable to semi-infinite domain problems, where strictly speaking, separation of variables does not apply as we discussed in previous chapters. Significantly, the Laplace transform is our final technique that is applicable for time-dependent boundary conditions, following Duhamel's theorem and Green's function.

In this chapter we present a brief description of the basic operational properties of the Laplace transformation and illustrate with numerous examples its application in the solution of one-dimensional transient heat conduction problems. The orthogonal expansion technique and the Green's function approach discussed previously often provide a much easier and straightforward method for solving such problems, but the solutions tend to converge very slowly for small times. The Laplace transformation has the advantage that it allows for the making of small-time approximations in order to obtain solutions that are strictly applicable for small times but are very rapidly convergent in the small-time domain. This aspect of the Laplace transformation approach will be emphasized later in this chapter.

The reader should consult references 1–7 for a more detailed discussion of the Laplace transform theory and references 8–13 for further applications of the Laplace transformation in the solution of heat conduction problems.

9-1 DEFINITION OF LAPLACE TRANSFORMATION

The *Laplace transform* and the *inversion formula* of a function $F(t)$ is defined by equations (9-1) and (9-2), respectively:

$$\mathcal{L}[F(t)] \equiv \overline{F}(s) = \int_{t=0}^{\infty} e^{-st} F(t)\, dt \tag{9-1}$$

and

$$\mathcal{L}^{-1}\left[\overline{F}(s)\right] \equiv F(t) = \frac{1}{2\pi i} \int_{s=\gamma-i\infty}^{\gamma+i\infty} e^{st}\overline{F}(s)ds \tag{9-2}$$

where s is the Laplace transform variable, $i \equiv \sqrt{-1}$, and γ is a positive number. The overbar denotes the transform, hence the function in the Laplace domain.

Thus, the Laplace transform of a function $F(t)$ consists of multiplying the function $F(t)$ by e^{-st} and integrating the result over t from 0 to ∞. The inversion formula consists of the complex integration as defined by equation (9-2). The inversion of Laplace transforms can be rather complex, and in general, we will rely on Laplace transform tables (see Table 9-1) for both the transform and the inversion.

Some remarks on the existence of the Laplace transform of a function $F(t)$ as defined by equation (9-1) are in order to illustrate the significance of this matter. For example, the integral (9-1) may not exist because: (1) $F(t)$ may have infinite discontinuities for some values of t; or (2) $F(t)$ may have a singularity as $t \to 0$; or (3) $F(t)$ may diverge exponentially for large t. The conditions for the existence of the Laplace transform defined by equation (9-1) may be summarized as follows:

1. Function $F(t)$ is continuous or piecewise continuous in any interval $t_1 \leq t \leq t_2$, for $t_1 > 0$.
2. $t^n \left|F(t)\right|$ is bounded as $t \to 0^+$ for some number n, such that $n < 1$.
3. Function $F(t)$ is of exponential order, namely, $e^{-\gamma t} \left|F(t)\right|$ is bounded for some positive number γ as $t \to \infty$.

For example, $F(t) = e^{t^2}$ is *not* of exponential order, that is, $e^{-\gamma t} \cdot e^{t^2}$ is unbounded at $t \to \infty$ for all positive values of γ; hence its Laplace transform *does not* exist. The Laplace transform of a function $F(t) = t^n$, when $n \leq -1$, does not exist because of condition (2), that is, $\int_0^\infty e^{-st} t^n dt$ diverges at the origin for $n \leq -1$.

Example 9-1 Laplace Transforms of Simple Functions
Determine the Laplace transform of the following functions: $F(t) = 1$, t, $e^{\pm at}$, and t^n with $n > -1$ but not necessarily an integer. According to the definition of

the Laplace transform given by equation (9-1), the Laplace transforms of these functions are given, respectively, as

$$F(t) = 1 \qquad \overline{F}(s) = \int_{t=0}^{\infty} 1 \cdot e^{-st}\,dt = -\frac{1}{s}e^{-st}\Big|_0^{\infty} = \frac{1}{s} \tag{9-3}$$

$$F(t) = t \qquad \overline{F}(s) = \int_{t=0}^{\infty} te^{-st}\,dt = \frac{1}{s^2} \tag{9-4}$$

$$F(t) = e^{\pm at} \qquad \overline{F}(s) = \int_{t=0}^{\infty} e^{\pm at}e^{-st}\,dt = \int_{t=0}^{\infty} e^{-(s\mp a)t}\,dt = \frac{1}{s \mp a} \tag{9-5}$$

$$F(t) = t^n \qquad n > -1 \qquad \overline{F}(s) = \int_{t=0}^{\infty} t^n e^{-st}\,dt \tag{9-6a}$$

where we now let $\xi = st$ and $d\xi = s\,dt$ in equation (9-6a), which yields

$$\overline{F}(s) = s^{-n-1}\int_{\xi=0}^{\infty} \xi^n e^{-\xi}\,d\xi = \frac{\Gamma(n+1)}{s^{n+1}} \tag{9-6b}$$

where the integral $\int_0^{\infty} \xi^n e^{-\xi}\,d\xi$ is the *gamma function*, denoted $\Gamma(n+1)$. The gamma function has the property $\Gamma(n+1) = n\Gamma(n)$; if n is an integer, we have $\Gamma(n+1) = n!$.

9-2 PROPERTIES OF LAPLACE TRANSFORM

Here we present some of the properties of the Laplace transform that are useful in the solution of heat conduction problems using the Laplace transformation.

Linear Property

If $\overline{F}(s)$ and $\overline{G}(s)$ are the Laplace transform of functions $F(t)$ and $G(t)$, respectively, with respect to the t variable, we may write

$$\mathcal{L}\left[c_1F(t) + c_2G(t)\right] = c_1\overline{F}(s) + c_2\overline{G}(s) \tag{9-7}$$

where c_1 and c_2 are any constants.

Example 9-2 Linear Property
By utilizing the linear property of the Laplace transform, equation (9-7), and the Laplace transform of e^{+at} given by equation (9-5), determine the Laplace transform of the functions cosh at and sinh at.

For cosh at we write

$$F_1(t) \equiv \cosh at = \frac{1}{2}(e^{at} + e^{-at}) \tag{9-8}$$

$$\overline{F}_1(s) = \mathcal{L}\left[\frac{1}{2}e^{at} + \frac{1}{2}e^{-at}\right] = \frac{1}{2}\left(\frac{1}{s-a}\right) + \frac{1}{2}\left(\frac{1}{s+a}\right) = \frac{s}{s^2 - a^2} \tag{9-9}$$

Similarly

$$F_2(t) \equiv \sinh at = \frac{1}{2}(e^{at} - e^{-at}) \tag{9-10}$$

$$\overline{F}_2(s) = \mathcal{L}\left[\frac{1}{2}e^{at} - \frac{1}{2}e^{-at}\right] = \frac{1}{2}\left(\frac{1}{s-a}\right) - \frac{1}{2}\left(\frac{1}{s+a}\right) = \frac{a}{s^2 - a^2} \tag{9-11}$$

Laplace Transform of Derivatives

The Laplace transform of the first derivative $dF(t)/dt$ of a function $F(t)$ is readily obtained by utilizing the definition of the Laplace transform, and integrating it by parts:

$$\mathcal{L}\left[\frac{dF(t)}{dt}\right] = \int_{t=0}^{\infty} \frac{dF(t)}{dt} e^{-st}\, dt = F(t)e^{-st}\Big|_0^{\infty} + s\int_{t=0}^{\infty} F(t)e^{-st}\, dt \tag{9-12a}$$

$$\boxed{\mathcal{L}\left[\frac{dF(t)}{dt}\right] \equiv s\overline{F}(s) - F(0)} \tag{9-12b}$$

where the differentiation is with respect to t, and $F(0)$ indicates the value of $F(t)$ at $t = 0^+$, namely, as we approach zero from the positive side. Thus, the Laplace transform of the first derivative of a function is equal to multiplying the transform of the function $F(t)$ by s, and subtracting from the result the value of this function at $t = 0^+$. Equation (9-12b) is important for solution of the transient heat equation with the Laplace transform method given the presence of the derivative with respect to time.

This result is now utilized to determine the Laplace transform of the second derivative of a function $F(t)$ as

$$\mathcal{L}\left[\frac{d^2F(t)}{dt^2}\right] = s\mathcal{L}\left[\frac{dF(t)}{dt}\right] - \frac{dF(t)}{dt}\bigg|_{t=0} = s\left[s\overline{F}(s) - F(0)\right] - F'(0)$$
$$= s^2\overline{F}(s) - sF(0) - F'(0) \tag{9-13}$$

Similarly, the Laplace transform of the third derivative becomes

$$\mathcal{L}\left[\frac{d^3F(t)}{dt^3}\right] = s^3\overline{F}(s) - s^2F(0) - sF'(0) - F''(0) \tag{9-14}$$

In general, the Laplace transform of the nth derivative is given as

$$\mathcal{L}\left[\frac{d^nF(t)}{dt^n}\right] = s^n\overline{F}(s) - s^{n-1}F(0) - s^{n-2}F^{(1)}(0) - s^{n-3}F^{(2)}(0)$$
$$- \cdots - F^{(n-1)}(0) \tag{9-15}$$

where

$$F^{(n)}(0) \equiv \left. \frac{d^n F(t)}{dt^n} \right|_{t=0}$$

Laplace Transform of Integrals

The Laplace transform of the integral $\int_{\tau=0}^{t} F(\tau)d\tau$ of a function $F(t)$ is determined as now described. Let

$$g(t) \equiv \int_{\tau=0}^{t} F(\tau)\, d\tau \tag{9-16a}$$

then

$$g'(t) = F(t) \tag{9-16b}$$

We take the Laplace transform of both sides of equation (9-16b) and utilize the result in equation (9-12b) to obtain

$$s\overline{g}(s) = \overline{F}(s) \tag{9-17}$$

since $g(0) = 0$ per equation (9-16a). After rearranging, we find

$$\overline{g}(s) \equiv \mathcal{L}\left[\int_{\tau=0}^{t} F(\tau)d\tau\right] = \frac{1}{s}\overline{F}(s) \tag{9-18}$$

This procedure is repeated to obtain the Laplace transform of the double integration of a function $F(t)$, namely,

$$L\left[\int_{\tau_2=0}^{t}\int_{\tau_1=0}^{\tau_2} F(\tau_1)\, d\tau_1\, d\tau_2\right] = \frac{1}{s^2}\overline{F}(s) \tag{9-19}$$

Change of Scale

Let $\overline{F}(s)$ be the Laplace transform of a function $F(t)$. Then, the Laplace transforms of functions $F(at)$ and $F[(1/a)t]$, where a is a real, positive constant, are determined as

$$\mathcal{L}[F(at)] = \int_{t=0}^{\infty} F(at)e^{-st}\, dt = \frac{1}{a}\int_{u=0}^{\infty} F(u)e^{-(s/a)u} du = \frac{1}{a}\overline{F}\left(\frac{s}{a}\right) \tag{9-20a}$$

where we set $u = at$. Similarly,

$$\mathcal{L}\left[F\left(\frac{t}{a}\right)\right] = \int_{t=0}^{\infty} F\left(\frac{t}{a}\right)e^{-st}\, dt = a\int_{u=0}^{\infty} F(u)e^{-asu} du = a\overline{F}(as) \tag{9-20b}$$

where we have now set $u = t/a$.

Example 9-3 Change of Scale
The Laplace transform of cosh t is given as $\mathcal{L}[\cosh t] \equiv s/(s^2 - 1)$. By utilizing the *change of scale* property, we may determine the Laplace transform of the functions cosh at and cosh (t/a). By utilizing equation (9-20a) we obtain

$$\mathcal{L}[\cosh at] = \frac{1}{a}\left[\frac{(s/a)}{a(s/a)^2 - 1}\right] = \frac{s}{s^2 - a^2} \tag{9-21a}$$

and by utilizing equation (9-20b) we find

$$\mathcal{L}\left[\cosh \frac{t}{a}\right] = a\left[\frac{as}{(as)^2 - 1}\right] = \frac{s^2}{s^2 - (1/a)^2} \tag{9-21b}$$

Shift Property

When the Laplace transform $\overline{F}(s)$ of a function $F(t)$ is known, the shift property enables us to write the Laplace transform of a function $e^{+at}F(t)$, where a is a constant; that is,

$$\mathcal{L}\left[e^{\pm at}F(t)\right] = \int_{t=0}^{\infty} e^{-st}e^{\pm at}F(t)\,dt = \int_{t=0}^{\infty} e^{-(s\mp a)t}F(t)\,dt = \overline{F}(s \mp a) \tag{9-22}$$

Example 9-4 Use of the Shift Property
The Laplace transform of cos bt is given as $\mathcal{L}[\cos bt] = s/(s^2 + b^2)$. By utilizing the *shift property*, we may determine the Laplace transform of the function e^{-at} cos bt. By equation (9-22) we may immediately write

$$\mathcal{L}\left[e^{-at}\cos bt\right] = \frac{s + a}{(s + a)^2 + b^2} \tag{9-23}$$

Laplace Transform of Translated Function

The *unit-step function* (or the *Heaviside unit function*) is useful in denoting the translation of a function. Figure 9-1 shows the physical significance of the unit-step functions $U(t)$ and $U(t - a)$, namely,

$$U(t) = \begin{cases} 1 & t > 0 \\ 0 & t < 0 \end{cases} \tag{9-24a}$$

$$U(t - a) = \begin{cases} 1 & t > a \\ 0 & t < a \end{cases} \tag{9-24b}$$

We now consider a function $F(t)$ defined for $t > 0$ as illustrated in Figure 9-2(a) and the translation of this function from $t = 0$ to $t = a$, as illustrated in Figure 9-2(b). The translated function $U(t - a)F(t - a)$ represents the function

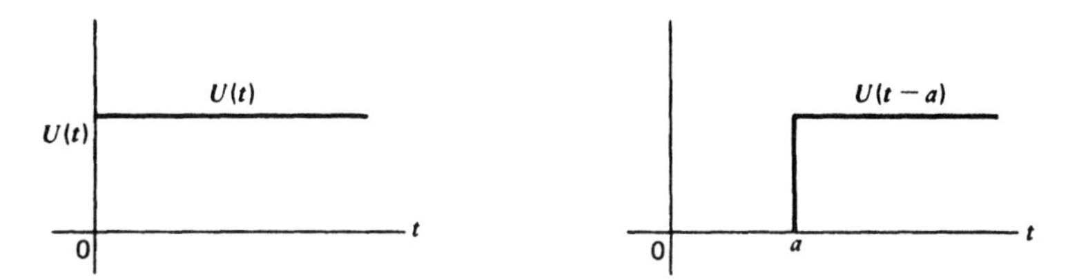

Figure 9-1 Definition of the unit-step functions $U(t)$ and $U(t-a)$.

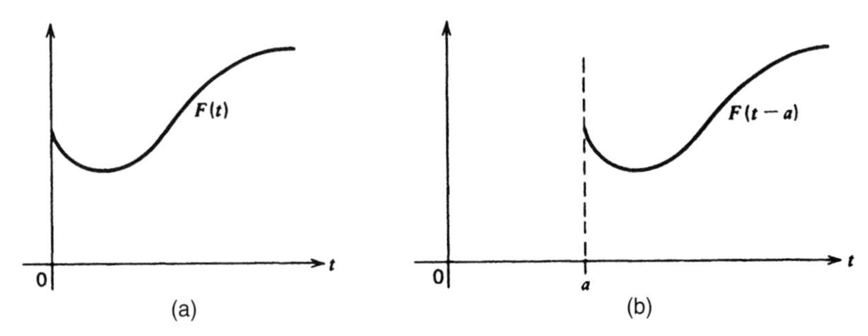

Figure 9-2 Translation of a function $F(t)$: (a) the function $F(t)$ and (b) the translation of $F(t)$ from $t=0$ to $t=a$.

$F(t)$ defined for $t>0$, translated by an amount $t=a$ in the positive t direction, namely,

$$U(t-a)F(t-a)=\begin{cases}F(t-a) & \text{for} \quad t>a\\ 0 & \text{for} \quad t<a\end{cases} \tag{9-25}$$

The Laplace transform of this translated function is determined as

$$\begin{aligned}\mathcal{L}\left[U(t-a)F(t-a)\right] &= \int_{t=0}^{\infty} e^{-st}U(t-a)F(t-a)\,dt\\ &= \int_{t=a}^{\infty} e^{-st}F(t-a)\,dt\\ &= \int_{\eta=0}^{\infty} e^{-s(a+\eta)}F(\eta)\,d\eta = e^{-as}\int_{\eta=0}^{\infty} e^{-s\eta}F(\eta)\,d\eta\\ &= e^{-as}\overline{F}(s)\end{aligned} \tag{9-26}$$

where a new variable η is defined as $\eta=t-a$. This result shows that the Laplace transform of a translated function $U(t-a)F(t-a)$ is equal to the Laplace transform $\overline{F}(s)$ of the function $F(t)$ multiplied by e^{-as}.

Similarly, the Laplace transform of a unit-step function $U(t-a)$ is given by

$$\boxed{\mathcal{L}[U(t-a)] = \frac{1}{s}e^{-as}} \tag{9-27}$$

If we let $a = 0$ in equation (9-27), we have

$$\mathcal{L}[U(t)] = \frac{1}{s}e^{-0s} = \frac{1}{s} \equiv \mathcal{L}[1] \tag{9-28}$$

Example 9-5 Transforms Involving the Unit-Step Function
Determine the Laplace transform of the following function:

$$F(t) = \begin{cases} 0 & \text{for} \quad t < 0 \\ 1 & \text{for} \quad 0 < t < 1 \\ 5 & \text{for} \quad 1 < t < 4 \\ 2 & \text{for} \quad 4 < t < 6 \\ 0 & \text{for} \quad t > 6 \end{cases} \tag{9-29}$$

which is illustrated in Figure 9-3.

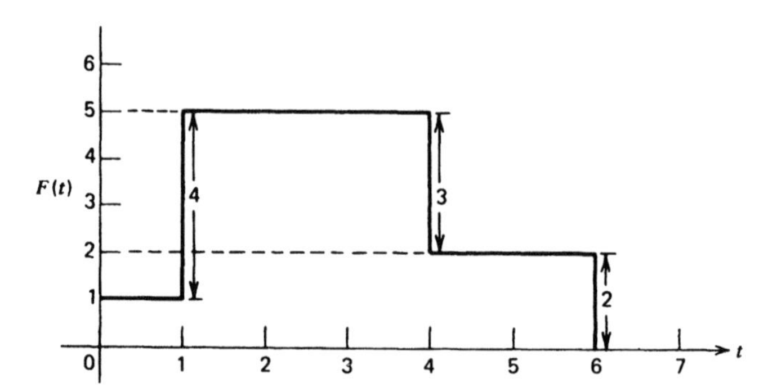

Figure 9-3 Function defined by equation (9-29).

The function given by equation (9-29) is readily represented in terms of the unit-step functions as

$$F(t) = U(t) + 4U(t-1) - 3U(t-4) - 2U(t-6) \tag{9-30}$$

and the Laplace transform of this function becomes

$$\overline{F}(s) = \frac{1}{s} + \frac{4}{s}e^{-s} - \frac{3}{s}e^{-4s} - \frac{2}{s}e^{-6s} \tag{9-31}$$

Laplace Transform of Delta Function

The delta function $\delta(x)$ is defined to be zero everywhere except at $x = 0$, as introduced in Chapter 8, such that

$$\delta(x) = 0 \qquad \text{for} \qquad x \neq 0 \tag{9-32a}$$

and

$$\int_{-\infty}^{\infty} \delta(x)\,dx \equiv 1 \tag{9-32b}$$

Useful integration formulas involving the delta function were introduced by equations (8-53)–(8-56). Making use of these expressions, the Laplace transform of the delta function $\delta(t)$ is given by

$$\mathcal{L}[\delta(t)] \equiv \bar{\delta}(s) = \int_{t=0}^{\infty} e^{-st}\delta(t)\,dt = 1 \tag{9-33}$$

while the Laplace transform of the shifted delta function is given by

$$\mathcal{L}[\delta(t-a)] = \int_{t=0}^{\infty} e^{-st}\delta(t-a)\,dt = e^{-sa} \tag{9-34}$$

Additional properties of the delta function are given in Appendix VI.

Laplace Transform of Convolution

Let $F(t)$ and $G(t)$ be two functions of t, with both functions defined for $t > 0$. The *convolution integral* or briefly the *convolution* of these two functions is denoted by the notation $F * G$ and defined by the equation

$$F * G = \int_{\tau=0}^{t} F(t-\tau)G(\tau)\,d\tau \tag{9-35a}$$

$$= \int_{\tau=0}^{t} F(\tau)G(t-\tau)\,d\tau \tag{9-35b}$$

Thus we have the commutative property $F * G \equiv G * F$. The Laplace transform of the convolution $F * G$ is given by

$$\mathcal{L}[F * G] \equiv \overline{F}(s)\overline{G}(s) \tag{9-36}$$

That is, the Laplace transform of the convolution is equal to the *product* of the respective Laplace transforms $\overline{F}(s)$ and $\overline{G}(s)$ of these two functions. The more useful form of the Laplace convolution property is given by

$$\boxed{\mathcal{L}^{-1}\left[\overline{F}(s)\overline{G}(s)\right] \equiv F(t) * G(t)} \tag{9-37}$$

which states that the inverse Laplace transform of a product of two functions in the Laplace domain is the convolution of the respective inverse Laplace transforms of the two functions in the time domain. The implication of equation (9-37) is that more complicated functions may be broken into two less complex functions for which the inverse Laplace transform readily exists, and the overall inverse Laplace is then found from the convolution.

Derivatives of Laplace Transform

We now derive an expression for the derivative of the Laplace transform of a function. Consider the Laplace transform $\overline{F}(s)$ of a function $F(t)$ given by

$$\overline{F}(s) = \int_{t=0}^{\infty} e^{-st} F(t)\, dt \tag{9-38}$$

By differentiating both sides of equation (9-38) with respect to s, we obtain

$$\frac{d\overline{F}(s)}{ds} \equiv \overline{F}'(s) = \int_{t=0}^{\infty} (-t)e^{-st} F(t)\, dt \tag{9-39a}$$

or

$$\overline{F}'(s) = \mathcal{L}[(-t)F(t)] \tag{9-39b}$$

By differentiating equation (9-39b) n times, we obtain

$$\frac{d^n\overline{F}(s)}{ds^n} \equiv \overline{F}^{(n)}(s) = \mathcal{L}\left[(-t)^n F(t)\right] \qquad n = 1, 2, 3, \ldots \tag{9-40}$$

Thus, the nth differentiation of the Laplace transform $\overline{F}(s)$ is equal to the Laplace transform of $(-t)^n F(t)$. This relation is useful in finding the inverse transforms with the aid of partial fractions and in many other applications.

Example 9-6 Derivative of the Laplace Transform
The Laplace transform of $F(t) = \sin \beta t$ is given as $\overline{F}(s) = \beta/(s^2 + \beta^2)$. Determine the Laplace transform of the function $t \sin \beta t$. By applying formula (9-39) we write

$$\mathcal{L}[(-t)F(t)] = \frac{d}{ds}\overline{F}(s) \tag{9-41}$$

$$\mathcal{L}[-t \sin \beta t] = \frac{d}{ds}\left(\frac{\beta}{s^2 + \beta^2}\right) = -\frac{2s\beta}{(s^2 + \beta^2)^2} \tag{9-42}$$

or

$$\mathcal{L}[t \sin \beta t] = \frac{2s\beta}{(s^2 + \beta^2)^2} \qquad \text{for} \qquad s > 0 \tag{9-43}$$

Integration of Laplace Transform

Consider the Laplace transform of a function $F(t)$ given by

$$\overline{F}(s) = \int_{t=0}^{\infty} e^{-st} F(t)\, dt \tag{9-44}$$

We integrate both sides of this equation with respect to s, from s to b, and obtain

$$\int_{s'=s}^{b} \overline{F}(s')\, ds' = \int_{s'=s}^{b} \int_{t=0}^{\infty} e^{-s't} F(t)\, dt\, ds' \tag{9-45}$$

which may be rearranged to the form

$$\int_{s'=s}^{b} \overline{F}(s')\, ds' = \int_{t=0}^{\infty} F(t) \left(\int_{s'=s}^{b} e^{-s't}\, ds' \right) dt = \int_{t=0}^{\infty} \frac{F(t)}{t} (e^{-st} - e^{-bt})\, dt \tag{9-46}$$

If the function $F(t)$ is such that $F(t)/t$ exists at $t \to 0$, the integral uniformly converges. Then, letting $b \to \infty$, equation (9-46) becomes

$$\int_{s'=s}^{\infty} \overline{F}(s')\, ds' = \int_{t=0}^{\infty} \left[\frac{F(t)}{t} \right] e^{-st}\, dt \equiv \mathcal{L}\left[\frac{F(t)}{t} \right] \tag{9-47}$$

Thus, the integration of the Laplace transform $\overline{F}(s)$ of a function $F(t)$ with respect to s from s to ∞ is equal to the Laplace transform of the function $F(t)/t$. This result is useful in the determination of the Laplace transform of the function $F(t)/t$ when the Laplace transform $\overline{F}(s)$ of the function $F(t)$ is known.

Example 9-7 Integration of the Laplace Transform
The Laplace transform of $\sin \beta t$ is given as $\beta / (s^2 + \beta^2)$. Determine the Laplace transform of $(1/t) \sin \beta t$. We utilize the formula (9-47)

$$\mathcal{L}\left[\frac{F(t)}{t} \right] = \int_{s'=s}^{\infty} \overline{F}(s') ds' \tag{9-48}$$

Introducing the function as given above, we obtain

$$\mathcal{L}\left[\frac{\sin \beta t}{t} \right] = \int_{s'=s}^{\infty} \frac{\beta}{s'^2 + \beta^2} ds' = \left(\tan^{-1} \frac{s'}{\beta} \right) \Bigg|_{s}^{\infty} = \frac{\pi}{2} - \tan^{-1} \left(\frac{s}{\beta} \right) \tag{9-49}$$

9-3 INVERSION OF LAPLACE TRANSFORM USING THE INVERSION TABLES

In heat conduction problems, the Laplace transformation is generally applied to the time variable. Therefore, an important step in the final analysis is the inversion of the transformed function from the Laplace variable s domain to the actual time variable t domain. To facilitate such analysis, comprehensive tables have been prepared for the inversion of the Laplace transform of a large class of functions. We present in Table 9-1 the Laplace transform of various functions that are useful in the analysis of heat conduction problems.

TABLE 9-1 Table of Laplace Transforms of Functions

Case	$\overline{F}(s)$	$F(t)$
1	$\frac{1}{s}$	1
2	$\frac{1}{s^2}$	t
3	$\frac{1}{s^n} \quad n = 1, 2, 3, \ldots$	$\frac{t^{n-1}}{(n-1)!}$
4	$\frac{1}{\sqrt{s}}$	$\frac{1}{\sqrt{\pi t}}$
5	$s^{-3/2}$	$2\sqrt{t/\pi}$
6	$s^{-(n+1/2)} \quad n = 1, 2, 3, \ldots$	$\frac{2^n}{[1 \cdot 3 \cdot 5 \cdot \ldots \cdot (2n-1)]\sqrt{\pi}} t^{n-1/2}$
7	$\frac{1}{s^n} (n > 0)$	$\frac{1}{\Gamma(n)} t^{n-1}$
8	$\frac{1}{s \mp a}$	$e^{\pm at}$
9	$\frac{1}{(s+a)^n} \quad n = 1, 2, 3, \ldots$	$\frac{t^{n-1} e^{-at}}{(n-1)!}$
10	$\frac{1}{(s+a)(s+b)} \quad (a \neq b)$	$\frac{e^{-at} - e^{-bt}}{b - a}$
11	$\frac{s}{(s+a)(s+b)} \quad (a \neq b)$	$\frac{ae^{-at} - be^{-bt}}{a - b}$
12	$\frac{1}{s^2 + a^2}$	$\frac{1}{a} \sin at$
13	$\frac{s}{s^2 + a^2}$	$\cos at$
14	$\frac{1}{s^2 - a^2}$	$\frac{1}{a} \sinh at$
15	$\frac{s}{s^2 - a^2}$	$\cosh at$
16	$\frac{1}{s(s^2 + a^2)}$	$\frac{1}{a^2}(1 - \cos at)$
17	$\frac{1}{s^2(s^2 + a^2)}$	$\frac{1}{a^3}(at - \sin at)$
18	$\frac{1}{(s^2 + a^2)^2}$	$\frac{1}{2a^3}(\sin at - at \cos at)$
19	$\frac{s}{(s^2 + a^2)^2}$	$\frac{t}{2a} \sin at$
20	$\frac{s^2}{(s^2 + a^2)^2}$	$\frac{1}{2a}(\sin at + at \cos at)$
21	$\frac{s^2 - a^2}{(s^2 + a^2)^2}$	$t \cos at$

TABLE 9-1 (*Continued*)

Case	$\overline{F}(s)$	$F(t)$
22	$\dfrac{1}{\sqrt{s}+a}$	$\dfrac{1}{\sqrt{\pi t}} - ae^{a^2 t}\operatorname{erfc}\left(a\sqrt{t}\right)$
23	$\dfrac{\sqrt{s}}{s-a^2}$	$\dfrac{1}{\sqrt{\pi t}} + ae^{a^2 t}\operatorname{erf}\left(a\sqrt{t}\right)$
24	$\dfrac{\sqrt{s}}{s+a^2}$	$\dfrac{1}{\sqrt{\pi t}} - \dfrac{2a}{\sqrt{\pi}} e^{-a^2 t}\int_0^{a\sqrt{t}} e^{\lambda^2}\,d\lambda$
25	$\dfrac{1}{\sqrt{s}(s-a^2)}$	$\dfrac{1}{a} e^{a^2 t}\operatorname{erf}\left(a\sqrt{t}\right)$
26	$\dfrac{1}{\sqrt{s}(s+a^2)}$	$\dfrac{2}{a\sqrt{\pi}} e^{-a^2 t}\int_0^{a\sqrt{t}} e^{\lambda^2}\,d\lambda$
27	$\dfrac{b^2-a^2}{(s-a^2)(b+\sqrt{s})}$	$e^{a^2 t}[b - a \operatorname{erf} a\sqrt{t}] - be^{b^2 t}\operatorname{erfc} b\sqrt{t}$
28	$\dfrac{1}{\sqrt{s}(\sqrt{s}+a)}$	$e^{a^2 t}\operatorname{erfc}\left(a\sqrt{t}\right)$
29	$\dfrac{1}{(s+a)\sqrt{s+b}}$	$\dfrac{1}{\sqrt{b-a}} e^{-at}\operatorname{erf}\left(\sqrt{b-a}\sqrt{t}\right)$
30	$\dfrac{\sqrt{s+2a}}{\sqrt{s}} - 1$	$ae^{-at}[I_1(at) + I_0(at)]$
31	$\dfrac{1}{\sqrt{s+a}\sqrt{s+b}}$	$e^{-(a+b)t/2} I_0\left(\dfrac{a-b}{2} t\right)$
32	$\dfrac{1}{\sqrt{s^2+a^2}}$	$J_0(at)$
33	$\dfrac{(\sqrt{s^2+a^2}-s)^\nu}{\sqrt{s^2+a^2}} \quad (\nu > -1)$	$a^\nu J_\nu(at)$
34	$\dfrac{s}{\left(s^2+a^2\right)^{3/2}}$	$t J_0(at)$
35	$\dfrac{1}{\left(s^2+a^2\right)^{3/2}}$	$\dfrac{tJ_1(at)}{a}$
36	$\dfrac{s}{\left(s^2-a^2\right)^{3/2}}$	$t I_0(at)$
37	$\dfrac{1}{\left(s^2-a^2\right)^{3/2}}$	$\dfrac{t I_1(at)}{a}$
38	$\dfrac{(s-\sqrt{s^2+a^2})^\nu}{\sqrt{s^2+a^2}} (\nu > -1)$	$a^\nu I_\nu(at)$
39	$\dfrac{1}{s} e^{-ks}$	$U(t-k)$ $\left[U(t) \equiv \text{unit-step function}\right]$

(continues overleaf)

TABLE 9-1 (*Continued*)

Case	$\overline{F}(s)$	$F(t)$
40	$\frac{1}{s^2}e^{-ks}$	$(t-k)\,U(t-k)$
41	1	$\delta(t)$ $[\delta(t) \equiv$ delta function$]$
42	e^{-as}	$\delta(t-a)$
43	$\frac{1}{s}e^{-k/s}$	$J_0\left(2\sqrt{kt}\right)$
44	$\frac{1}{s^{\mu}}e^{-k/s} \quad (\mu > 0)$	$\left(\frac{t}{k}\right)^{(\mu-1)/2} J_{\mu-1}\left(2\sqrt{kt}\right)$
45	$\frac{1}{s^{\mu}}e^{k/s} \quad (\mu > 0)$	$\left(\frac{t}{k}\right)^{(\mu-1)/2} I_{\mu-1}\left(2\sqrt{kt}\right)$
46	$e^{-k\sqrt{s}} \quad (k > 0)$	$\frac{k}{2\sqrt{\pi t^3}}\exp\left(-\frac{k^2}{4t}\right)$
47	$\frac{1-e^{-k\sqrt{s}}}{s} \quad (k \geq 0)$	$\operatorname{erf}\left(\frac{k}{2\sqrt{t}}\right)$
48	$\frac{1}{s}e^{-k\sqrt{s}} \quad (k \geq 0)$	$\operatorname{erfc}\left(\frac{k}{2\sqrt{t}}\right)$
49	$\frac{1}{\sqrt{s}}e^{-k\sqrt{s}} \quad (k \geq 0)$	$\frac{1}{\sqrt{\pi t}}\exp\left(-\frac{k^2}{4t}\right)$
50	$\frac{1}{s^{3/2}}e^{-k\sqrt{s}} \quad (k \geq 0)$	$2\sqrt{\frac{t}{\pi}}\exp\left(-\frac{k^2}{4t}\right) - k\operatorname{erfc}\left(\frac{k}{2\sqrt{t}}\right)$
51	$\frac{1}{s^{1+n/2}}e^{-k\sqrt{s}}(n = 0, 1, 2, \ldots, k \geq 0)$	$(4t)^{n/2} i^n \operatorname{erfc}\left(\frac{k}{2\sqrt{t}}\right)$
52	$\frac{e^{-k\sqrt{s}}}{a+\sqrt{s}} \quad (k \geq 0)$	$\frac{1}{\sqrt{\pi t}}\exp\left(-\frac{k^2}{4t}\right)$ $-ae^{ak}e^{a^2 t}\operatorname{erfc}\left(a\sqrt{t}+\frac{k}{2\sqrt{t}}\right)$
53	$\frac{e^{-k\sqrt{s}}}{\sqrt{s}(a+\sqrt{s})} \quad (k \geq 0)$	$e^{ak}e^{a^2 t}\operatorname{erfc}\left(a\sqrt{t}+\frac{k}{2\sqrt{t}}\right)$
54	$\frac{e^{-k\sqrt{s(s+a)}}}{\sqrt{s(s+a)}} \quad (k \geq 0)$	$e^{-at/2}I_0\left(\frac{1}{2}a\sqrt{t^2-k^2}\right)U(t-k)$
55	$\frac{e^{-k\sqrt{s^2+a^2}}}{\sqrt{s^2+a^2}} \quad (k \geq 0)$	$J_0\left(a\sqrt{t^2-k^2}\right)U(t-k)$
56	$\frac{e^{-k\sqrt{s^2+a^2}}}{\sqrt{s^2-a^2}} \quad (k \geq 0)$	$I_0\left(a\sqrt{t^2-k^2}\right)U(t-k)$

TABLE 9-1 (*Continued*)

Case	$\overline{F}(s)$	$F(t)$
57	$\dfrac{ae^{-k\sqrt{s}}}{s(a+\sqrt{s})} \quad (k \geq 0)$	$-e^{ak}e^{a^2t}\operatorname{erfc}\left(a\sqrt{t}+\dfrac{k}{2\sqrt{t}}\right) + \operatorname{erfc}\left(\dfrac{k}{2\sqrt{t}}\right)$
58	$\dfrac{1}{s^2}e^{-k\sqrt{s}}$	$\left(t+\dfrac{k^2}{2}\right)\operatorname{erfc}\left(\dfrac{k}{2\sqrt{t}}\right) -k\left(\dfrac{t}{\pi}\right)^{1/2}\exp\left(-\dfrac{k^2}{4t}\right)$
59	$\dfrac{1}{s}\ln s$	$-\gamma - \ln t$ $[\gamma \equiv \text{Euler's constant}]$
60	$\dfrac{-(\gamma+\ln s)}{s}$	$\ln t$
61	$\ln\dfrac{s+a}{s+b}$	$\dfrac{1}{t}(e^{-bt}-e^{-at})$
62	$\ln\dfrac{s^2+a^2}{s^2}$	$\dfrac{2}{t}(1-\cos at)$
63	$\ln\dfrac{s^2-a^2}{s^2}$	$\dfrac{2}{t}(1-\cosh at)$
64	$K_0(ks) \quad (k>0)$	$\dfrac{1}{\sqrt{t^2-k^2}}U(t-k)$
65	$K_0(k\sqrt{s}) \quad (k>0)$	$\dfrac{1}{2t}\exp\left(-\dfrac{k^2}{4t}\right)$
66	$\dfrac{1}{\sqrt{s}}K_1(k\sqrt{s}) \quad (k>0)$	$\dfrac{1}{k}\exp\left(-\dfrac{k^2}{4t}\right)$
67	$\dfrac{\sinh xs}{s\sinh as}$	$\dfrac{x}{a}+\dfrac{2}{\pi}\sum_{n=1}^{\infty}\dfrac{(-1)^n}{n}\sin\lambda_n x\cos\lambda_n t \quad \left[\lambda_n \equiv \dfrac{n\pi}{a}\right]$
68	$\dfrac{\sinh xs}{s\cosh as}$	$\dfrac{4}{\pi}\sum_{n=1}^{\infty}\dfrac{(-1)^n}{(2n-1)}\sin\lambda_n x\sin\lambda_n t$ with $\lambda_n \equiv \dfrac{(2n-1)\pi}{2a}$
69	$\dfrac{\cosh xs}{s\sinh as}$	$\dfrac{t}{a}+\dfrac{2}{\pi}\sum_{n=1}^{\infty}\dfrac{(-1)^n}{n}\cos\lambda_n x\sin\lambda_n t$ with $\lambda_n \equiv \dfrac{n\pi}{a}$

(continues overleaf)

TABLE 9-1 (*Continued*)

Case	$\overline{F}(s)$	$F(t)$
70	$\dfrac{\cosh xs}{s\cosh as}$	$1+\dfrac{4}{\pi}\sum_{n=1}^{\infty}\dfrac{(-1)^n}{(2n-1)}\cos\lambda_n x\cos\lambda_n t$ with $\lambda_n\equiv\dfrac{(2n-1)\pi}{2a}$
71	$\dfrac{\sinh xs}{s^2\sinh as}$	$\dfrac{xt}{a}+\dfrac{2a}{\pi^2}\sum_{n=1}^{\infty}\dfrac{(-1)^n}{n^2}\sin\lambda_n x\sin\lambda_n t$ with $\lambda_n\equiv\dfrac{n\pi}{a}$
72	$\dfrac{\sinh xs}{s^2\cosh as}$	$x+\dfrac{8a}{\pi^2}\sum_{n=1}^{\infty}\dfrac{(-1)^n}{(2n-1)^2}\sin\lambda_n x\cos\lambda_n t$ with $\lambda_n\equiv\dfrac{(2n-1)\pi}{2a}$
73	$\dfrac{\cosh xs}{s^2\sinh as}$	$\dfrac{t^2}{2a}+\dfrac{2a}{\pi^2}\sum_{n=1}^{\infty}\dfrac{(-1)^n}{n^2}\cos\lambda_n x\left(1-\cos\lambda_n t\right)$ with $\lambda_n\equiv\dfrac{n\pi}{a}$
74	$\dfrac{\cosh xs}{s^2\cosh as}$	$t+\dfrac{8a}{\pi^2}\sum_{n=1}^{\infty}\dfrac{(-1)^n}{(2n-1)^2}\cos\lambda_n x\sin\lambda_n t$ with $\lambda_n\equiv\dfrac{(2n-1)\pi}{2a}$
75	$\dfrac{\cosh xs}{s^3\cosh as}$	$\dfrac{t^2+x^2-a^2}{2}-\dfrac{16a^2}{\pi^3}\sum_{n=1}^{\infty}\dfrac{(-1)^n}{(2n-1)^3}\cos\lambda_n x\cos\lambda_n t$ with $\lambda_n\equiv\dfrac{(2n-1)\pi}{2a}$
76	$\dfrac{\sinh x\sqrt{s}}{\cosh a\sqrt{s}}$	$\dfrac{2}{a}\sum_{n=1}^{\infty}(-1)^n\lambda_n\sin\lambda_n x\, e^{-\lambda_n^2 t}$ with $\lambda_n\equiv\dfrac{n\pi}{a}$
77	$\dfrac{\cosh x\sqrt{s}}{\sinh a\sqrt{s}}$	$\dfrac{2}{a}\sum_{n=1}^{\infty}(-1)^{n-1}\lambda_n\cos\lambda_n x\, e^{-\lambda_n^2 t}$ with $\lambda_n\equiv\dfrac{(2n-1)\pi}{2a}$

TABLE 9-1 ***(Continued)***

Case	$\overline{F}(s)$	$F(t)$
78	$\dfrac{\sinh x\sqrt{s}}{\sqrt{s}\cosh a\sqrt{s}}$	$\dfrac{2}{a}\sum_{n=1}^{\infty}(-1)^{n-1}\sin\lambda_n x\ e^{-\lambda_n^2 t}$ with $\lambda_n \equiv \dfrac{(2n-1)\pi}{2a}$
79	$\dfrac{\cosh x\sqrt{s}}{\sqrt{s}\sinh a\sqrt{s}}$	$\dfrac{1}{a}+\dfrac{2}{a}\sum_{n=1}^{\infty}(-1)^{n}\cos\lambda_n x\ e^{-\lambda_n^2 t}$ with $\lambda_n \equiv \dfrac{n\pi}{a}$
80	$\dfrac{\sinh x\sqrt{s}}{s\sinh a\sqrt{s}}$	$\dfrac{x}{a}+\dfrac{2}{\pi}\sum_{n=1}^{\infty}\dfrac{(-1)^{n}}{n}\sin\lambda_n x\ e^{-\lambda_n^2 t}$ with $\lambda_n \equiv \dfrac{n\pi}{a}$
81	$\dfrac{\cosh x\sqrt{s}}{s\cosh a\sqrt{s}}$	$1+\dfrac{4}{\pi}\sum_{n=1}^{\infty}\dfrac{(-1)^{n}}{(2n-1)}\cos\lambda_n x\ e^{-\lambda_n^2 t}$ with $\lambda_n \equiv \dfrac{(2n-1)\pi}{2a}$
82	$\dfrac{\sinh x\sqrt{s}}{s^2\sinh a\sqrt{s}}$	$\dfrac{xt}{a}+\dfrac{2}{a}\sum_{n=1}^{\infty}\dfrac{(-1)^{n}}{\lambda_n^3}\sin\lambda_n x\left(1-e^{-\lambda_n^2 t}\right)$ with $\lambda_n \equiv \dfrac{n\pi}{a}$
83	$\dfrac{\cosh x\sqrt{s}}{s^2\cosh a\sqrt{s}}$	$\dfrac{x^2-a^2}{2}+t-\dfrac{2}{a}\sum_{n=1}^{\infty}\dfrac{(-1)^{n}}{\lambda_n^3}\cos\lambda_n x\ e^{-\lambda_n^2 t}$ with $\lambda_n \equiv \dfrac{(2n-1)\pi}{2a}$

Sources: From references 7 and 14–16.

We first present here one more useful process to assist with the inverse Laplace transform operation. If the Laplace transform $\overline{F}(s)$ of a function $F(t)$ is expressible in the form

$$\overline{F}(s) = \frac{G(s)}{H(s)} \tag{9-50}$$

where $G(s)$ and $H(s)$ are polynomials with no common factor, with $G(s)$ being a lower degree than $H(s)$, and the factors of $H(s)$ are all linear and distinct. Under such conditions, equation (9-50) can be expressed in the form

$$\overline{F}(s) = \frac{G(s)}{H(s)} = \frac{c_1}{s-a_1} + \frac{c_2}{s-a_2} + \cdots + \frac{c_n}{s-a_n} \tag{9-51}$$

Here the c_i values are independent of s. Then, by the theory of partial fractions, c_i values are determined as

$$c_i = \lim_{s \to a_i} [(s - a_i)\overline{F}(s)] \tag{9-52}$$

Clearly, if a function $\overline{F}(s)$ is expressible in partial fractions as in equation (9-51), its inversion is readily obtained by the use of the Laplace transform table, namely, using case 8.

Finally, we note that there are many occasions that the transformed function $\overline{F}(s)$ will not appear in the standard transform tables. In such cases it will be necessary to use the inversion formula (9-2) to determine the function in the time domain. Such an inversion is generally performed by the method of contour integration and the calculus of residues that require rather elaborate analysis. Therefore, the use of Laplace inversion formula (9-2) will not be considered here.

Example 9-8 Inversion Using Laplace Tables
Determine the function $F(t)$ with the Laplace transform given by

$$\overline{F}(s) = \frac{b^2}{s(s^2 + b^2)} \tag{9-53}$$

This function is not available in the Laplace transform tables in this direct form; but it can be expressible in partial fractions as

$$\overline{F}(s) = \frac{b^2}{s(s^2 + b^2)} = \frac{c_1}{s} + \frac{c_2 s + c_3}{s^2 + b^2} \tag{9-54}$$

with

$$b^2 = c_1 b^2 + c_3 s + (c_1 + c_2)s^2 \tag{9-55}$$

Equating the coefficients of like powers of s, we obtain $c_1 = 1$, $c_2 = -1$, and $c_3 = 0$. Making these substitutions, we have

$$\overline{F}(s) = \frac{1}{s} - \frac{s}{s^2 + b^2} \tag{9-56}$$

Each term on the right-hand side is readily inverted using Table 9-1, namely, cases 1 and 13, respectively. Inverting term by term, we find

$$F(t) = 1 - \cos bt \tag{9-57}$$

9-4 APPLICATION OF THE LAPLACE TRANSFORM IN THE SOLUTION OF TIME-DEPENDENT HEAT CONDUCTION PROBLEMS

In this section we illustrate with representative examples the use of the Laplace transform technique in the solution of time-dependent heat conduction problems.

In this approach, the Laplace transform is applied to remove the partial derivative with respect to the time variable, and the resulting equation is solved for the transform of temperature, after which the transform is inverted to recover the solution for the temperature distribution in the time domain. The approach is straightforward in principle, but generally the inversion is difficult unless the transform is available in the Laplace transform tables. In the following examples, typical heat conduction problems with time-dependent boundary conditions are solved by using the Laplace transform table to invert the solutions in the Laplace domain. Cartesian and spherical coordinates are considered, including both finite and semi-infinite domains.

Example 9-9 Semi-Infinite Medium in Cartesian Coordinates (Type 1 BC) A semi-infinite medium, $x \geq 0$, is initially at zero temperature. For times $t > 0$, the boundary surface at $x = 0$ is subjected to a prescribed, time-dependent temperature $f(t)$. Obtain an expression for the temperature distribution $T(x, t)$ in the medium for times $t > 0$. The mathematical formulation of this problem is given as

$$\frac{\partial^2 T(x,t)}{\partial x^2} = \frac{1}{\alpha}\frac{\partial T(x,t)}{\partial t} \quad \text{in} \quad 0 < x < \infty, \quad t > 0 \tag{9-58}$$

$$\text{BC1:} \quad T(x = 0) = f(t) \tag{9-59a}$$

$$\text{BC2:} \quad T(x \to \infty) = 0 \tag{9-59b}$$

$$\text{IC:} \quad T(x, t = 0) = 0 \tag{9-59c}$$

We recall that this problem was solved in Example 7-2 by the application of Duhamel's method. Here the Laplace transform technique is used to solve the same problem, and the standard Laplace transform table (Table 9-1) is utilized to invert the resulting solution. Taking the Laplace transform of equation (9-58) with respect to the variable t, term by term, we obtain

$$\mathcal{L}\left[\frac{\partial^2 T(x,t)}{\partial x^2}\right] = \frac{\partial^2 \overline{T}(x,s)}{\partial x^2} \tag{9-60}$$

where the x dependency is ignored with respect to the Laplace transform, and the function $\overline{T}(x, s)$ represents the temperature in the x and s dimensions, hence the transformed temperature. The right-hand side term yields

$$\mathcal{L}\left[\frac{1}{\alpha}\frac{\partial T(x,t)}{\partial t}\right] = \frac{1}{\alpha}\left[s\overline{T}(x,s) - T(0)\right] \tag{9-61}$$

where we have used equation (9-12b) for transformation of the derivative with respect to time. In view of the initial condition, $T(t = 0) = 0$. Combining the above equations, the governing equation (9-58) becomes

$$\frac{d^2\overline{T}(x,s)}{dx^2} - \frac{s}{\alpha}\overline{T}(x,s) = 0 \quad \text{in} \quad 0 < x < \infty \tag{9-62}$$

which is recognized as an ODE in the spatial variable x, in which s is considered a parameter. We must now transform the boundary conditions, yielding

$$\text{BC1:} \qquad \overline{T}(x=0,s) = \mathcal{L}\left[f(t)\right] = \overline{f}(s) \tag{9-63a}$$

$$\text{BC2:} \qquad \overline{T}(x \to \infty, s) = \mathcal{L}\left[0\right] = 0 \tag{9-63b}$$

The solution for the ODE of equation (9-62) is given by

$$\overline{T}(x,s) = C_1 e^{+\sqrt{s/\alpha}x} + C_2 e^{-\sqrt{s/\alpha}x} \tag{9-64}$$

where we have used the exponential form in keeping with our guideline for semi-infinite domain problems. Boundary condition BC2 yields C_1=0, while application of BC1 yields

$$\overline{T}(x=0) = \overline{f}(s) = C_2 \tag{9-65}$$

With these substitutions, the solution is given as

$$\overline{T}(x,s) = \overline{f}(s)e^{-\sqrt{s/\alpha}x} \tag{9-66}$$

and the solution in the time domain is now found by taking the inverse Laplace transform of equation (9-66), namely,

$$T(x,t) = \mathcal{L}^{-1}\left[\overline{T}(x,s)\right] = \mathcal{L}^{-1}\left[\overline{f}(s)e^{-\sqrt{s/\alpha}x}\right] \tag{9-67}$$

Since the functional form of $\overline{f}(s)$ is not explicitly specified, it is better to make use of the convolution property of the Laplace transform as given by equation (9-37) to invert this transform. Namely, in view of equation (9-37), we write the result in equation (9-67) as

$$T(x,t) = \mathcal{L}^{-1}\left[\overline{f}(s)e^{-\sqrt{s/\alpha}x}\right] = \mathcal{L}^{-1}\left[\overline{f}(s)\overline{g}(x,s)\right] = f(t) * g(x,t) \tag{9-68}$$

where we have defined

$$\overline{g}(x,s) = e^{-\sqrt{s/\alpha}x} \tag{9-69}$$

The transform $\overline{g}(x,s)$ is readily inverted by utilizing Table 9-1, case 46; we find

$$g(x,t) = \frac{x}{2\sqrt{\pi\alpha t^3}} e^{-x^2/4\alpha t} \tag{9-70}$$

while

$$\mathcal{L}^{-1}\left[\overline{f}(s)\right] = f(t) \tag{9-71}$$

We now utilize the definition of the convolution $f(t) * g(x,t)$ given by equation (9-35),

$$T(x,t) = \int_{\tau=0}^{t} f(\tau) g(x, t-\tau)\, d\tau \tag{9-72}$$

After replacing t by $t - \tau$ in equation (9-70), we introduce it into equation (9-72) to obtain the desired solution as

$$T(x,t) = \frac{x}{\sqrt{4\pi\alpha}} \int_{\tau=0}^{t} \frac{f(\tau)}{(t-\tau)^{3/2}} \exp\left[\frac{-x^2}{4\alpha(t-\tau)}\right] d\tau \tag{9-73}$$

This result is the same as that given by equation (7-62), which was obtained by utilizing Duhamel's theorem. The temperature $T(x,t)$ can be determined from equation (9-73) for any specified form of the function $f(t)$ by performing the integration. Sometimes it is easier to introduce the transform $\overline{f}(s)$ of the function $f(t)$ into equation (9-66) and then directly invert the result rather than performing the integration in equation (9-73). This matter is now illustrated for some special cases of the function $f(t)$:

1. $f(t) = T_0 =$ *constant:* Then, the transform of $f(t) = T_0$ is $\overline{f}(s) = T_0/s$. Introducing this result into equation (9-66), we obtain

$$\overline{T}(x,s) = \frac{T_0}{s} e^{-x\sqrt{s/\alpha}} \tag{9-74}$$

The transform (9-74) is readily inverted by utilizing Table 9-1, case 48. We obtain

$$T(x,t) = T_0 \operatorname{erfc}\left(\frac{x}{\sqrt{4\alpha t}}\right) \tag{9-75}$$

2. $f(t) = T_0 t^{1/2}$: The transform of this function is obtained from Table 9-1, case 5 as $\overline{f}(s) = T_0(\sqrt{\pi}/2)s^{-3/2}$. Introducing this result into equation (9-66) yields

$$\overline{T}(x,s) = T_0 \frac{\sqrt{\pi}}{2} s^{-3/2} e^{-x\sqrt{s/\alpha}} \tag{9-76}$$

This result is directly inverted by utilizing Table 9-1, case 50, which yields

$$T(x,t) = T_0 \left(t^{1/2} e^{-x^2/4\alpha t} - \frac{x}{2}\sqrt{\frac{\pi}{\alpha}} \operatorname{erfc}\frac{x}{\sqrt{4\alpha t}} \right) \tag{9-77}$$

Example 9-10 Semi-Infinite Medium in Cartesian Coordinates (Type 3 BC)

A semi-infinite medium, $0 \le x < \infty$, is initially at zero temperature. For times $t > 0$, the boundary surface at $x = 0$ is subjected to convection heat transfer with an environment at temperature T_∞. Obtain an expression for the temperature distribution $T(x, t)$ in the solid for times $t > 0$. We could remove the nonhomogeneity in the boundary condition by shifting the temperature by T_∞, however, to illustrate the handling of a nonhomogeneity with the Laplace technique, we will proceed. The mathematical formulation of the problem is given as

$$\frac{\partial^2 T(x,t)}{\partial x^2} = \frac{1}{\alpha}\frac{\partial T(x,t)}{\partial t} \quad \text{in} \quad 0 < x < \infty, \quad t > 0 \tag{9-78}$$

$$\text{BC1:} \quad -k\left.\frac{\partial T}{\partial x}\right|_{x=0} + h\,T|_{x=0} = hT_\infty \tag{9-79a}$$

$$\text{BC2:} \quad T(x \to \infty) = 0 \tag{9-79b}$$

$$\text{IC:} \quad T(x, t = 0) = 0 \tag{9-79c}$$

The Laplace transform of equations (9-78) and (9-79) becomes

$$\frac{d^2\overline{T}(x,s)}{dx^2} - \frac{s}{\alpha}\overline{T}(x,s) = 0 \quad \text{in} \quad 0 < x < \infty \tag{9-80}$$

$$\text{BC1:} \quad -k\left.\frac{d\overline{T}}{dx}\right|_{x=0} + h\,\overline{T}\big|_{x=0} = \frac{1}{s}hT_\infty \tag{9-81a}$$

$$\text{BC2:} \quad \overline{T}(x \to \infty, s) = 0 \tag{9-81b}$$

The solution of the ODE given by equation (9-80) is

$$\overline{T}(x,s) = C_1 e^{+\sqrt{s/\alpha}x} + C_2 e^{-\sqrt{s/\alpha}x} \tag{9-82}$$

where we have used the exponential form as in the previous problem. Boundary conditions BC1 and BC2 together yield

$$\frac{\overline{T}(x,s)}{T_\infty} = H\sqrt{\alpha}\frac{e^{-(x/\sqrt{\alpha})\sqrt{s}}}{s(H\sqrt{\alpha} + \sqrt{s})} \tag{9-83}$$

where $H \equiv h/k$. The inversion of this result is available in Table 9-1, case 57; then the solution becomes

$$\frac{T(x,t)}{T_\infty} = \operatorname{erfc}\left(\frac{x}{\sqrt{4\alpha t}}\right) - \exp\left(Hx + H^2\alpha t\right)\operatorname{erfc}\left(H\sqrt{\alpha t} + \frac{x}{\sqrt{4\alpha t}}\right) \tag{9-84}$$

Example 9-11 Rectangular Slab with Time-Dependent Boundary Condition

Consider a rectangular slab, $0 \le x \le$ L, that is initially at zero temperature. For times $t > 0$, the boundary surface at $x = 0$ is maintained at zero temperature, while the boundary at $x = L$ is subjected to a prescribed, time-dependent temperature defined by $f(t) = b + ct$. Obtain an expression for the temperature distribution $T(x, t)$ in the solid slab for times $t > 0$. The mathematical formulation of the problem is given as

$$\frac{\partial^2 T(x,t)}{\partial x^2} = \frac{1}{\alpha}\frac{\partial T(x,t)}{\partial t} \quad \text{in} \quad 0 < x < L, \quad t > 0 \tag{9-85}$$

$$\text{BC1:} \quad T(x = 0) = 0 \tag{9-86a}$$

$$\text{BC2:} \quad T(x = L) = b + ct \tag{9-86b}$$

$$\text{IC:} \quad T(x, t = 0) = 0 \tag{9-86c}$$

Taking the Laplace transform of the PDE and the boundary conditions yields

$$\frac{d^2\overline{T}(x,s)}{dx^2} - \frac{s}{\alpha}\overline{T}(x,s) = 0 \quad \text{in} \quad 0 < x < L \tag{9-87}$$

$$\text{BC1:} \quad \overline{T}(x = 0) = \mathcal{L}[0] = 0 \tag{9-88a}$$

$$\text{BC2:} \quad \overline{T}(x = L) = \mathcal{L}[b + ct] = \frac{b}{s} + \frac{c}{s^2} \tag{9-88b}$$

The solution of the ODE given by equation (9-87) is

$$\overline{T}(x,s) = C_1 \cosh\sqrt{\frac{s}{\alpha}}x + C_2 \sinh\sqrt{\frac{s}{\alpha}}x \tag{9-89}$$

where we have used the hyperbolic solutions for the finite domain slab. Boundary condition BC1 yields $C_1 = 0$, while BC2 yields

$$C_2 = \frac{b/s + c/s^2}{\sinh\sqrt{s/\alpha}L} \tag{9-90}$$

Inserting the above into equation (9-89) yields the solution in the Laplace domain, namely,

$$\overline{T}(x,s) = \frac{b/s + c/s^2}{\sinh\sqrt{s/\alpha}L}\sinh\sqrt{s/\alpha}x \tag{9-91}$$

Because equation (9-91) is not readily available in Table 9-1, we first split the problem, and then invert term by term:

$$T(x,t) = b\mathcal{L}^{-1}\left[\frac{\sinh\sqrt{s/\alpha}x}{s\sinh\sqrt{s/\alpha}L}\right] + c\mathcal{L}^{-1}\left[\frac{\sinh\sqrt{s/\alpha}x}{s^2\sinh\sqrt{s/\alpha}L}\right] \tag{9-92}$$

Equation (9-92) is now readily inverted using Table 9-1, cases 80 and 82, respectively. This yields the solution

$$T(x,t) = b\left[\frac{x}{L} + \frac{2}{\pi}\sum_{n=1}^{\infty}\frac{(-1)^n}{n}\sin\lambda_n x e^{-\alpha\lambda_n^2 t}\right]$$

$$+ c\left[\frac{xt}{L} + \frac{2}{\alpha L}\sum_{n=1}^{\infty}\frac{(-1)^n}{\lambda_n^3}\sin\lambda_n x\left(1 - e^{-\alpha\lambda_n^2 t}\right)\right] \tag{9-93}$$

where we have now defined $\lambda_n = n\pi/L$. Given the nature of the time-dependent boundary condition, we note that there is no steady-state solution to this problem.

Example 9-12 Rectangular Slab with Time-Dependent Boundary Condition

Consider a rectangular slab, $0 \le x \le L$, that is initially at temperature T_0. For times $t > 0$, the boundary surface at $x = 0$ is perfectly insulated, while the boundary at $x = L$ is subjected to a prescribed, time-dependent temperature defined by $f(t)$. Obtain an expression for the temperature distribution $T(x, t)$ in the solid slab for times $t > 0$. The mathematical formulation of the problem is given as

$$\frac{\partial^2 T(x,t)}{\partial x^2} = \frac{1}{\alpha}\frac{\partial T(x,t)}{\partial t} \quad \text{in} \quad 0 < x < L, \quad t > 0 \tag{9-94}$$

$$\text{BC1:} \quad \left.\frac{\partial T}{\partial x}\right|_{x=0} = 0 \tag{9-95a}$$

$$\text{BC2:} \quad T(x = L) = f(t) \tag{9-95b}$$

$$\text{IC:} \quad T(x, t = 0) = T_0 \tag{9-95c}$$

Taking the Laplace transform of the PDE and the boundary conditions yields

$$\frac{d^2\overline{T}(x,s)}{dx^2} - \frac{s}{\alpha}\overline{T}(x,s) = -\frac{T_0}{\alpha} \quad \text{in} \quad 0 < x < L \tag{9-96}$$

$$\text{BC1:} \quad \left.\frac{d\overline{T}}{dx}\right|_{x=0} = \mathcal{L}[0] = 0 \tag{9-97a}$$

$$\text{BC2:} \quad \overline{T}(x = L) = \mathcal{L}\left[f(t)\right] = \overline{f}(s) \tag{9-97b}$$

where we now have a nonhomogeneous term on the right-hand side of equation (9-96) resulting from the initial condition. The solution of the ODE given by equation (9-96) is of the form $\overline{T}(x,s) = \overline{T}_H(x,s) + \overline{T}_P(x,s)$, which yields

$$\overline{T}(x,s) = C_1\cosh\sqrt{\frac{s}{\alpha}}x + C_2\sinh\sqrt{\frac{s}{\alpha}}x + \frac{T_0}{s} \tag{9-98}$$

Boundary condition BC1 yields $C_2 = 0$, while BC2 yields

$$C_1 = \frac{\overline{f}(s) - T_0/s}{\cosh\sqrt{s/\alpha}L} \tag{9-99}$$

Inserting the above simplifications into equation (9-98) yields the solution in the Laplace domain, namely

$$\overline{T}(x,s) = \frac{\overline{f}(s) - T_0/s}{\cosh\sqrt{s/\alpha}L}\cosh\sqrt{\frac{s}{\alpha}}x + \frac{T_0}{s} \tag{9-100}$$

Because equation (9-100) is not readily available in Table 9-1, we first split the problem into three terms and invert term by term:

$$T(x,t) = \mathcal{L}^{-1}\left[\frac{\overline{f}(s)\cosh\sqrt{s/\alpha}x}{\cosh\sqrt{s/\alpha}L}\right] - T_0\mathcal{L}^{-1}\left[\frac{\cosh\sqrt{s/\alpha}x}{s\cosh\sqrt{s/\alpha}L}\right] + T_0\mathcal{L}^{-1}\left[\frac{1}{s}\right] \tag{9-101}$$

The second and third terms are now readily inverted by cases 81 and 1, respectively, from Table 9-1. However, since the first term is unavailable in Table 9-1, we recast this term as follows:

$$\mathcal{L}^{-1}\left[\frac{\overline{f}(s)\cosh\sqrt{s/\alpha}x}{\cosh\sqrt{s/\alpha}L}\right] = \mathcal{L}^{-1}\left[s\overline{f}(s)\frac{\cosh\sqrt{s/\alpha}x}{s\cosh\sqrt{s/\alpha}L}\right] = \mathcal{L}^{-1}\left[\overline{H}(s)\cdot\overline{G}(s)\right] \tag{9-102}$$

which is now suited to the convolution property of equation (9-37) provided the inverse Laplace transforms of $\overline{H}(s)$ and $\overline{G}(s)$ are available. The $\overline{G}(s)$ term again corresponds to case 81 from Table 9-1, while the $\overline{H}(s)$ term is rearranged to yield

$$\mathcal{L}^{-1}\left[s\overline{f}(s)\right] = \mathcal{L}^{-1}\left[s\overline{f}(s) - f(0) + f(0)\right] = \mathcal{L}^{-1}\left[s\overline{f}(s) - f(0)\right] + \mathcal{L}^{-1}\left[f(0)\right] \tag{9-103}$$

where $f(0)$ is the time-dependent boundary evaluated at $t = 0$. The two terms on the right-hand side of equation (9-103) may now be inverted as follows:

$$\mathcal{L}^{-1}\left[s\overline{f}(s) - f(0)\right] = \frac{df(t)}{dt} \tag{9-104}$$

where we have made use of the derivative property per equation (9-12b), and

$$\mathcal{L}^{-1}\left[f(0)\right] = f(t)\delta(t) \tag{9-105}$$

where we have used case 41 of Table 9-1, noting that $f(0)$ is a constant. The above simplifications may now be inserted into equation (9-101) to yield

$$T(x,t) = \left[\frac{df(t)}{dt} + f(t)\delta(t)\right] * G(x,t) - T_0G(x,t) + T_0 \tag{9-106}$$

where we define

$$G(x,t) = \mathcal{L}^{-1}\left[\frac{\cosh\sqrt{s/\alpha}x}{s\cosh\sqrt{s/\alpha}L}\right] = 1 + \frac{2}{L}\sum_{n=1}^{\infty}\frac{(-1)^n}{\lambda_n}\cos\lambda_n x e^{-\lambda_n^2 t} \qquad (9\text{-}107)$$

per case 81 of Table 9-1, where we have now defined $\lambda_n = (2n-1)\pi/2L$. Using the definition of the convolution integration, equation (9-35), our solution becomes

$$T(x,t) = \int_{\tau=0}^{t}\left[\frac{df(\tau)}{d\tau} + f(\tau)\delta(\tau)\right]G(x,t-\tau)\,d\tau - T_0 G(x,t) + T_0 \quad (9\text{-}108)$$

which, by the property of the delta function, simplifies to

$$T(x,t) = f(0)G(x,t) + \int_{\tau=0}^{t}\frac{df(\tau)}{d\tau}G(x,t-\tau)\,d\tau - T_0 G(x,t) + T_0 \quad (9\text{-}109)$$

with $G(x, t)$ defined by equation (9-107). This problem illustrates the complexities that may arise when inverting the solution from the Laplace domain, even for relatively straightforward problems. This problem is readily solved using Duhamel's theorem. To illustrate this, we recast equation (9-109) to the following:

$$T(x,t) = \left[f(0) - T_0\right]G(x,t) + \int_{\tau=0}^{t}\frac{df(\tau)}{d\tau}G(x,t-\tau)\,d\tau + T_0 \qquad (9\text{-}110)$$

Equation (9-110) is recognized as Duhamel's solution, equation (7-11), with $G(x, t)$ corresponding to the unit-step solution $\Phi(x,t)$ and $f(t)$ corresponding to the time-dependent boundary condition, noting that for Duhamel's solution it is necessary to first remove the initial condition by shifting the temperature by T_0. Solution of equations (9-94) and (9-95) for $f(t) \equiv 1$ does yield $G(x, t)$ as defined exactly by equation (9-107), hence the identical solution with Duhamel's theorem.

Example 9-13 Solid Sphere with Type 1 Boundary Condition
Consider a solid sphere of radius $r = b$ that is initially at a uniform temperature T_0. For times $t > 0$, the boundary surface at $r = b$ is kept at zero temperature. Obtain an expression for the temperature distribution $T(r, t)$ using the Laplace transform technique. The mathematical formulation of this problem is given as

$$\frac{1}{r}\frac{\partial^2}{\partial r^2}(rT) = \frac{1}{\alpha}\frac{\partial T(r,t)}{\partial t} \quad \text{in} \quad 0 \le r < b, \quad t > 0 \qquad (9\text{-}111)$$

$$\text{BC1:} \quad T(r\to 0) \Rightarrow \text{finite} \qquad (9\text{-}112a)$$

$$\text{BC2:} \quad T(r=b) = 0 \qquad (9\text{-}112b)$$

$$\text{IC:} \quad T(r,t=0) = T_0 \qquad (9\text{-}112c)$$

We first transform the problem using $U(r, t) = rT(r, t)$, which yields

$$\frac{\partial^2 U(r,t)}{\partial r^2} = \frac{1}{\alpha}\frac{\partial U(r,t)}{\partial t} \quad \text{in} \quad 0 \le r < b, \quad t > 0 \tag{9-113}$$

$$\text{BC1:} \quad U(r=0) = 0 \tag{9-114a}$$

$$\text{BC2:} \quad U(r=b) = 0 \tag{9-114b}$$

$$\text{IC:} \quad U(r, t=0) = rT_0 \tag{9-114c}$$

Taking the Laplace transform of the PDE and the boundary conditions now yields

$$\frac{d^2\overline{U}(r,s)}{dr^2} - \frac{s}{\alpha}\overline{U}(r,s) = -\frac{rT_0}{\alpha} \quad \text{in} \quad 0 < r < b \tag{9-115}$$

$$\text{BC1:} \quad \overline{U}(r=0) = \mathcal{L}[0] = 0 \tag{9-116a}$$

$$\text{BC2:} \quad \overline{U}(r=b) = \mathcal{L}[0] = 0 \tag{9-116b}$$

The solution of the ODE given by equation (9-115) requires superposition of the form $\overline{U}(r,s) = \overline{U}_H(r,s) + \overline{U}_P(r,s)$, which yields

$$\overline{U}(r,s) = C_1 \cosh\sqrt{\frac{s}{\alpha}}r + C_2 \sinh\sqrt{\frac{s}{\alpha}}r + \frac{rT_0}{s} \tag{9-117}$$

Boundary condition BC1 eliminates the cosh term, while BC2 defines the remaining constant, to yield the solution

$$\overline{U}(r,s) = \frac{rT_0}{s} - \frac{bT_0 \sinh\sqrt{s/\alpha}r}{s \sinh\sqrt{s/\alpha}b} \tag{9-118}$$

Equation (9-118) is readily inverted using cases 1 and 80, respectively, from Table 9-1 to yield

$$U(r,t) = rT_0 - bT_0\left[\frac{r}{b} + \frac{2}{\pi}\sum_{n=1}^{\infty}\frac{(-1)^n}{n}\sin\lambda_n r e^{-\alpha\lambda_n^2 t}\right] \tag{9-119}$$

where we have defined $\lambda_n = n\pi/b$. Transforming back from $U(r,t)$ to $T(r,t)$ yields

$$T(r,t) = -\frac{2bT_0}{\pi}\sum_{n=1}^{\infty}\frac{(-1)^n}{n}\frac{\sin\lambda_n r}{r}e^{-\alpha\lambda_n^2 t} \tag{9-120}$$

As $t \to \infty$, the steady-state solution of zero is obtained directly from equation (9-120). This problem is also readily solved using separation of variables, as developed in Chapter 5, to yield the identical solution upon evaluation of the resulting Fourier constants.

9-5 APPROXIMATIONS FOR SMALL TIMES

The solutions of time-dependent heat conduction problems for finite regions, such as slabs, cylinders, or spheres of finite radius, are in the form of series that converge rapidly for large values of t but converge very slowly for the small values of t, notably so near the boundaries. Therefore, such solutions are not well suited for numerical computations for very small values of time. For example, the solution of the slab problem given by

$$\frac{T(x,t)}{T_1} = \left(1 - \frac{x}{L}\right) - \frac{2}{L}\sum_{n=1}^{\infty} e^{-\alpha\lambda_n^2 t}\frac{\sin\lambda_n x}{\lambda_n} \qquad \text{where} \qquad \lambda_n = \frac{n\pi}{L} \tag{9-121}$$

converges very slowly for the values of $\alpha t/L^2$, the Fourier number, less than approximately 0.02. Therefore, for such cases, it is desirable to develop alternative forms of the solution that will converge fast for small times.

When the Laplace transform is applied to the time variable, it transforms the equation in t into an equation in s, in other words, a mapping from the t domain to the s domain. Therefore, it is instructive to examine the values of t in the time domain with the corresponding values of s in the Laplace transform domain. With this objective in mind, we now examine the Laplace transform of several functions.

Consider a function $F(t)$ that is represented as a polynomial in t in the form

$$F(t) = \sum_{k=0}^{n} a_k\frac{t^k}{k!} = a_0 + a_1\frac{t}{1!} + a_2\frac{t^2}{2!} + \cdots + a_n\frac{t^n}{n!} \tag{9-122}$$

Since the function has only a finite number of terms, we can take its Laplace transform term by term to obtain

$$\overline{F}(s) = \sum_{k=0}^{n} a_k\frac{1}{s^{k+1}} = a_0\frac{1}{s} + a_1\frac{1}{s^2} + \cdots + a_n\frac{1}{s^{n+1}} \tag{9-123}$$

according to the transform Table 9-1, case 3.

The coefficients a_0 and a_n may be determined from equations (9-122) and (9-123) as

$$a_0 = \lim_{t\to 0} F(t) = \lim_{s\to\infty} s\overline{F}(s) \tag{9-124a}$$

$$a_n = n!\lim_{t\to\infty}\frac{F(t)}{t^n} = \lim_{s\to 0} s^{n+1}\overline{F}(s) \tag{9-124b}$$

The relations given by equations (9-124) indicate that the large values of s in the Laplace transform domain correspond to small values of t in the time domain. Although the results given above are derived for a function $F(t)$, which is a polynomial, they are also applicable for other types of functions. Consider, for example, the following function and its transform:

$$F(t) = \cosh kt \qquad \text{and} \qquad \overline{F}(s) = \frac{s}{s^2 - k^2} \tag{9-125}$$

which satisfies the relation

$$\lim_{t \to 0} \cosh kt = \lim_{s \to \infty} s \frac{s}{s^2 - k^2} \tag{9-126}$$

and therefore this result is similar to that given by equation (9-124a).

An alternative approach to the above mapping discussion between the t domain and the Laplace transform domain is to consider the change of scale property of equation (9-20a), namely,

$$\mathcal{L}[F(at)] = \frac{1}{a} \overline{F}\left(\frac{s}{a}\right) \tag{9-127}$$

It is readily observed in equation (9-127) that a *large argument* in the time domain (i.e., $a \gg 1$) maps to a *small argument* in the s domain. Similarly, a large argument in the s domain (i.e., $a \ll 1$) maps to a small argument in the time domain.

These facts can be utilized to obtain an approximate solution for the function $F(t)$ valid for small times from the knowledge of its transform evaluated for large values of s as illustrated in several references [17; 4, pp. 82–85; 6; 8]. In other words, the Laplace transform of the desired function can be expanded as a large-argument, asymptotic series and then inverted term by term. For example, in the problems of slabs of finite thickness, the transform of temperature $\overline{T}(x, s)$ contains hyperbolic functions of $\sqrt{s/\alpha}$. These hyperbolic functions may be expanded in a series of negative exponentials of $\sqrt{s/\alpha}$, and the resulting expression is then inverted term by term. The solution obtained in this manner will converge fast for small times. In the problems of a solid cylinder of finite radius, for example, the transform of temperature involves Bessel functions of $\sqrt{s/\alpha}$. Then the procedure consists of using asymptotic expansions of Bessel functions in order to obtain a form involving negative exponentials of $\sqrt{s/\alpha}$, with coefficients that are series in $1/\sqrt{s/\alpha}$. The resulting expression is then inverted term by term, and the solutions obtained in this manner will converge fast for small values of time. Many examples of this procedure are given in reference 17.

Example 9-14 Rectangular Slab with Type 1 Boundary Condition

A slab, $0 \le x \le L$, is initially of zero temperature. For times $t > 0$, the boundary at $x = 0$ is kept perfectly insulted, and the boundary at $x = L$ is kept at constant temperature T_0. Obtain an expression for the temperature distribution $T(x, t)$ that is useful for small values of time.

$$\frac{\partial^2 T(x,t)}{\partial x^2} = \frac{1}{\alpha} \frac{\partial T(x,t)}{\partial t} \qquad \text{in} \qquad 0 < x < L, \qquad t > 0 \tag{9-128}$$

$$\text{BC1:} \quad \left.\frac{\partial T}{\partial x}\right|_{x=0} = 0 \tag{9-129a}$$

$$\text{BC2:} \quad T(x = L) = T_0 \tag{9-129b}$$

$$\text{IC:} \quad T(x, t = 0) = 0 \tag{9-129c}$$

Taking the Laplace transform of the PDE and the boundary conditions yields

$$\frac{d^2\overline{T}(x, s)}{dx^2} - \frac{s}{\alpha}\overline{T}(x, s) = 0 \quad \text{in} \quad 0 < x < L \tag{9-130}$$

$$\text{BC1:} \quad \left.\frac{d\overline{T}}{dx}\right|_{x=0} = \mathcal{L}[0] = 0 \tag{9-131a}$$

$$\text{BC2:} \quad \overline{T}(x = L) = \mathcal{L}\left[T_0\right] = \frac{T_0}{s} \tag{9-131b}$$

The solution of the ODE given by equation (9-130) yields

$$\overline{T}(x, s) = C_1 \cosh\sqrt{\frac{s}{\alpha}}x + C_2 \sinh\sqrt{\frac{s}{\alpha}}x \tag{9-132}$$

which after consideration of the two boundary conditions gives

$$\overline{T}(x, s) = \frac{T_0 \cosh(x\sqrt{s/\alpha})}{s \cosh(L\sqrt{s/\alpha})} \tag{9-133}$$

Equation (9-133) is readily inverted using case 81 in Table 9-1, however, the result yields a solution for $T(x, t)$ which is slowly convergent for small values of time. As depicted in Figure 9-4, the temperature solution is characterized by steep gradients near the surface for small times, which requires many terms for convergence.

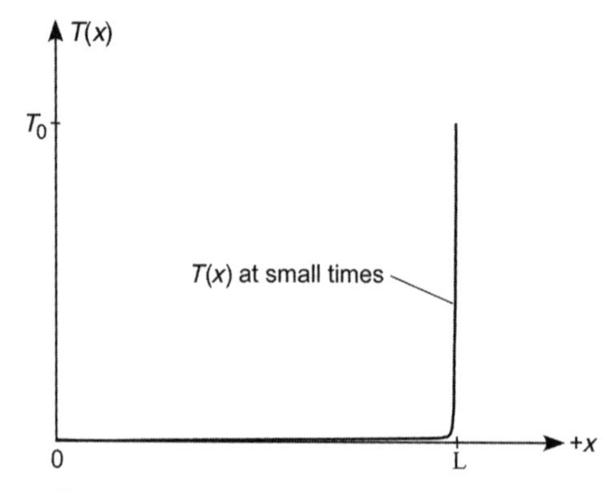

Figure 9-4 Steep gradients near the surface yield poorly convergent solutions for small times.

To obtain a solution applicable for very small times, we seek to expand this transform as an asymptotic series in negative exponentials of $\sqrt{s/\alpha}$. We first express the solution of equation (9-133) in terms of the exponentials, giving

$$\frac{\overline{T}(x,s)}{T_0} = \frac{e^{x\sqrt{s/\alpha}} + e^{-x\sqrt{s/\alpha}}}{s\left(e^{L\sqrt{s/\alpha}} + e^{-L\sqrt{s/\alpha}}\right)} \tag{9-134}$$

which after multiplying by $e^{-L\sqrt{s/\alpha}}/e^{-L\sqrt{s/\alpha}}$ yields

$$\frac{\overline{T}(x,s)}{T_0} = \frac{1}{s}\left[e^{-\sqrt{s/\alpha}(L-x)} + e^{-\sqrt{s/\alpha}(L+x)}\right]\left(\frac{1}{1+e^{-2L\sqrt{s/\alpha}}}\right) \tag{9-135}$$

We introduce here the binomial expansion

$$\boxed{\frac{1}{1+z} = 1 - z + z^2 - z^3 = \sum_{n=0}^{\infty}(-1)^n z^n} \tag{9-136}$$

where $|z| < 1$. The last term of equation (9-135) is now expanded in a binomial series, namely,

$$\left(\frac{1}{1+e^{-2L\sqrt{s/\alpha}}}\right) = \sum_{n=0}^{\infty}(-1)^n e^{-2L\sqrt{s/\alpha}\;n} \tag{9-137}$$

which upon substitution into equation (9-135) yields

$$\frac{\overline{T}(x,s)}{T_0} = \frac{1}{s}\left[e^{-\sqrt{s/\alpha}(L-x)} + e^{-\sqrt{s/\alpha}(L+x)}\right]\left[\sum_{n=0}^{\infty}(-1)^n e^{-2L\sqrt{s/\alpha}\;n}\right] \tag{9-138a}$$

or

$$\frac{\overline{T}(x,s)}{T_0} = \frac{1}{s}\sum_{n=0}^{\infty}(-1)^n e^{-[L(1+2n)-x]\sqrt{s/\alpha}} + \frac{1}{s}\sum_{n=0}^{\infty}(-1)^n e^{-[L(1+2n)+x]\sqrt{s/\alpha}} \tag{9-138b}$$

The inversion of this transform is available in Table 9-1, case 48. Inverting term by term, we obtain

$$\frac{T(x,t)}{T_0} = \sum_{n=0}^{\infty}(-1)^n \text{erfc}\left[\frac{L(1+2n)-x}{\sqrt{4\alpha t}}\right] + \sum_{n=0}^{\infty}(-1)^n \text{erfc}\left[\frac{L(1+2n)+x}{\sqrt{4\alpha t}}\right] \tag{9-139}$$

which converges rapidly for small values of t. To further explore the convergence, we examine the behavior of the complementary error function in Figure 9-5.

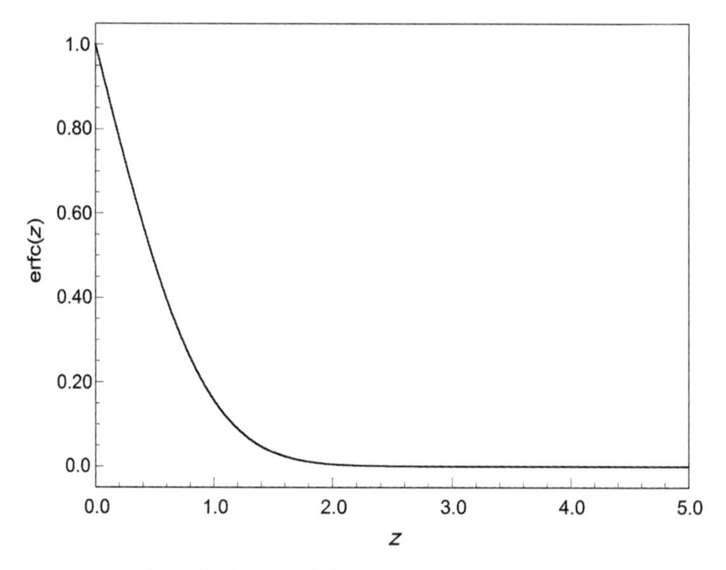

Figure 9-5 Behavior of the complementary error function.

As observed in Figure 9-5, the complementary error function rapidly converges to zero for large arguments, hence rapid convergence of our series solution of equation (9-139) for small times. The solution given by equation (9-139) is also valid for large times (i.e., small s), given that the binomial series expansion in the Laplace domain is valid for $|e^{-2L\sqrt{s/\alpha}}| < 1$, which is satisfied for $s > 0$. However, for large times, the convergence of equation (9-139) becomes slower. We note that the solution given by equation (9-139) is *not valid* for $s = 0$, which maps to infinite time (i.e., steady-state solution). It is readily observed that the erfc terms of equation (9-139) become unity as $t \to \infty$ and that there is no convergence and no valid steady-state solution.

Example 9-15 Rectangular Slab with Type 3 Boundary Condition
A slab, $0 \leq x \leq L$, is initially at uniform temperature T_0. For times $t > 0$, the boundary surface at $x = 0$ is kept insulated, and the boundary at $x = L$ dissipates heat by convection into an environment of zero temperature. Obtain an expression for the temperature distribution $T(x, t)$, which is useful for small times. The mathematical formulation of this problem is given as

$$\frac{\partial^2 T(x,t)}{\partial x^2} = \frac{1}{\alpha}\frac{\partial T(x,t)}{\partial t} \quad \text{in} \quad 0 < x < L, \quad t > 0 \tag{9-140}$$

$$\text{BC1:} \quad \left.\frac{\partial T}{\partial x}\right|_{x=0} = 0 \tag{9-141a}$$

$$\text{BC2:} \qquad \left.\frac{\partial T}{\partial x}\right|_{x=L} + H\,T|_{x=L} = 0 \tag{9-141b}$$

$$\text{IC:} \qquad T(x, t = 0) = T_0 \tag{9-141c}$$

The Laplace transform of these equations gives

$$\frac{d^2\overline{T}(x,s)}{dx^2} - \frac{s}{\alpha}\overline{T}(x,s) = -\frac{T_0}{\alpha} \qquad \text{in} \qquad 0 < x < L \tag{9-142}$$

$$\text{BC1:} \qquad \left.\frac{\partial \overline{T}}{\partial x}\right|_{x=0} = 0 \tag{9-143a}$$

$$\text{BC2:} \qquad \left.\frac{\partial \overline{T}}{\partial x}\right|_{x=L} + H\,\overline{T}\big|_{x=L} = 0 \tag{9-143b}$$

The solution of equation (9-142) follows equation (9-98), with the two boundary conditions then yielding

$$\frac{\overline{T}(x,s)}{T_0} = \frac{1}{s} - \frac{H\cosh(x\sqrt{s/\alpha})}{s\left[\sqrt{s/\alpha}\,\sinh\left(L\sqrt{s/\alpha}\right) + H\cosh\left(L\sqrt{s/\alpha}\right)\right]} \tag{9-144}$$

Since the solution is required for small times, we need to expand this transform as an asymptotic series in negative exponentials, and then invert it term by term. First, converting the hyperbolics to exponentials, we have

$$\frac{\overline{T}(x,s)}{T_0} = \frac{1}{s} - \frac{H\left(e^{x\sqrt{s/\alpha}} + e^{-x\sqrt{s/\alpha}}\right)}{s\left[\sqrt{s/\alpha}\left(e^{L\sqrt{s/\alpha}} - e^{-L\sqrt{s/\alpha}}\right) + H\left(e^{L\sqrt{s/\alpha}} + e^{-L\sqrt{s/\alpha}}\right)\right]} \tag{9-145a}$$

which we may rewrite as

$$\frac{\overline{T}(x,s)}{T_0} = \frac{1}{s} - \frac{H}{s}\,\frac{e^{-(L-x)\sqrt{s/\alpha}} + e^{-(L+x)\sqrt{s/\alpha}}}{H+\sqrt{s/\alpha}}\left[\frac{1}{1 + \dfrac{H-\sqrt{s/\alpha}}{H+\sqrt{s/\alpha}}e^{-2L\sqrt{s/\alpha}}}\right] \tag{9-145b}$$

Expanding the last term in the bracket as a binomial series, we obtain

$$\frac{\overline{T}(x,s)}{T_0} = \frac{1}{s} - \frac{H}{s}\,\frac{e^{-(L-x)\sqrt{s/\alpha}} + e^{-(L+x)\sqrt{s/\alpha}}}{H+\sqrt{s/\alpha}}\left[\sum_{n=0}^{\infty}(-1)^n\left(\frac{H-\sqrt{s/\alpha}}{H+\sqrt{s/\alpha}}\right)^n \times e^{-2L\sqrt{s/\alpha}n}\right] \tag{9-146}$$

or

$$\frac{\overline{T}(x,s)}{T_0} = \frac{1}{s} - \frac{H}{s}\frac{e^{-(L-x)\sqrt{s/\alpha}} + e^{-(L+x)\sqrt{s/\alpha}}}{H+\sqrt{s/\alpha}}$$
$$+\frac{H}{s}\frac{H-\sqrt{s/\alpha}}{\left(H+\sqrt{s/\alpha}\right)^2}\left[e^{-(3L-x)\sqrt{s/\alpha}} + e^{-(3L+x)\sqrt{s/\alpha}}\right] \tag{9-147}$$
$$-\frac{H}{s}\frac{\left(H-\sqrt{s/\alpha}\right)^2}{\left(H+\sqrt{s/\alpha}\right)^3}\left[e^{-(5L-x)\sqrt{s/\alpha}} + e^{-(5L+x)\sqrt{s/\alpha}}\right] + \cdots$$

The first few terms can readily be inverted using cases 1 and 57 from Table 9-1, which yields

$$\frac{T(x,t)}{T_0} = 1 - \left[\text{erfc}\frac{L-x}{\sqrt{4\alpha t}} - e^{H(L-x)+H^2\alpha t}\,\text{erfc}\left(H\sqrt{\alpha t} + \frac{L-x}{\sqrt{4\alpha t}}\right)\right]$$
$$-\left[\text{erfc}\frac{L+x}{\sqrt{4\alpha t}} - e^{H(L+x)+H^2\alpha t}\,\text{erfc}\left(H\sqrt{\alpha t} + \frac{L+x}{\sqrt{4\alpha t}}\right)\right] + \cdots \tag{9-148}$$

This solution converges fast for small times.

Example 9-16 Solid Sphere with Type 1 Boundary Condition

Consider a solid sphere, $0 \le r \le b$, that is initially at a uniform temperature T_0. For times $t > 0$, the boundary surface at $r = b$ is kept at zero temperature. Obtain an expression for the temperature distribution $T(r, t)$, which is useful for small times. The mathematical formulation of this problem is given as

$$\frac{1}{r}\frac{\partial^2}{\partial r^2}(rT) = \frac{1}{\alpha}\frac{\partial T(r,t)}{\partial t} \quad \text{in} \quad 0 \le r < b, \quad t > 0 \tag{9-149}$$

$$\text{BC1:} \quad T(r \to 0) \Rightarrow \text{finite} \tag{9-150a}$$

$$\text{BC2:} \quad T(r = b) = 0 \tag{9-150b}$$

$$\text{IC:} \quad T(r, t = 0) = T_0 \tag{9-150c}$$

In an identical manner as in Example 9-13, we first transform the problem using $U(r,t) = rT(r,t)$, take the Laplace transform, and then solve the ODE and transformed boundary conditions to yield

$$\overline{U}(r,s) = \frac{rT_0}{s} - \frac{bT_0 \sinh\left(r\sqrt{s/\alpha}\right)}{s \cdot \sinh\left(b\sqrt{s/\alpha}\right)} \tag{9-151}$$

Since the $U = rT$ transformation does not involve the time domain (i.e., the Laplace transform variables), we now transform back from $\overline{U}(r, s)$ to $\overline{T}(r, s)$, yielding

$$\frac{\overline{T}(r,s)}{T_0} = \frac{1}{s} - \frac{b}{r}\frac{\sinh\left(r\sqrt{s/\alpha}\right)}{s\sinh\left(b\sqrt{s/\alpha}\right)} \tag{9-152}$$

To obtain a solution that converges rapidly for small times, we expand this transform as an asymptotic series in negative exponentials and then invert term by term. The procedure is as follows:

$$\frac{\overline{T}(r,s)}{T_0} = \frac{1}{s} - \frac{b}{r}\frac{e^{r\sqrt{s/\alpha}} - e^{-r\sqrt{s/\alpha}}}{s\left(e^{b\sqrt{s/\alpha}} - e^{-b\sqrt{s/\alpha}}\right)} \tag{9-153}$$

which we rewrite as

$$\frac{\overline{T}(r,s)}{T_0} = \frac{1}{s} - \frac{b}{r}\frac{1}{s}\left[e^{-(b-r)\sqrt{s/\alpha}} - e^{-(b+r)\sqrt{s/\alpha}}\right]\left[\frac{1}{1 - e^{-2b\sqrt{s/\alpha}}}\right] \tag{9-154}$$

The last term in the bracket may be expanded as a binomial series, namely, as

$$\boxed{\frac{1}{1-z} = 1 + z + z^2 + z^3 = \sum_{n=0}^{\infty} z^n} \tag{9-155}$$

where $|z| < 1$. Using equation (9-155) yields

$$\frac{\overline{T}(r,s)}{T_0} = \frac{1}{s} - \frac{b}{r}\frac{1}{s}\left[e^{-(b-r)\sqrt{s/\alpha}} - e^{-(b+r)\sqrt{s/\alpha}}\right]\left[\sum_{n=0}^{\infty} e^{-2b\sqrt{s/\alpha}n}\right] \tag{9-156a}$$

or

$$\frac{\overline{T}(r,s)}{T_0} = \frac{1}{s} - \frac{b}{r}\sum_{n=0}^{\infty}\left\{\frac{1}{s}e^{-[b(1+2n)-r)]\sqrt{s/\alpha}} - \frac{1}{s}e^{-[b(1+2n)+r)]\sqrt{s/\alpha}}\right\} \tag{9-156b}$$

This transform is readily inverted by utilizing case 1 and case 48 from the Laplace transform Table 9-1, which yields

$$\frac{T(r,t)}{T_0} = 1 - \frac{b}{r}\sum_{n=0}^{\infty}\left[\operatorname{erfc}\frac{b(1+2n)-r}{\sqrt{4\alpha t}} - \operatorname{erfc}\frac{b(1+2n)+r}{\sqrt{4\alpha t}}\right] \tag{9-157}$$

This solution converges fast for small values of times due to the nature of the complementary error functions, as discussed in Example 9-14.

REFERENCES

1. N. W. McLachlan, *Modern Operational Calculus*, Macmillan, New York, 1948.
2. R. V. Churchill, *Operational Mathematics*, McGraw-Hill, New York, 1958.
3. M. G. Smith, *Laplace Transform Theory*, Van Nostrand, New York, 1996.
4. N. W. McLachlan, *Complex Variable Theory and Transform Calculus*, Cambridge University Press, London, 1953.
5. W. Kaplan, *Operational Methods for Linear Systems*, Addison-Wesley, Reading, MA, 1962.
6. H. S. Carslaw and J. C. Jaeger, *Operational Methods in Applied Mathematics*, Oxford University Press, London, 1948.
7. A. Erde'lyi, *Tables of Integral Transforms I*, McGraw-Hill, New York, 1954.
8. H. S. Carslaw and J. C. Jaeger, *Conduction of Heat in Solids*, 2 ed., Oxford University Press, London, 1959.
9. I. N. Sneddon, *Use of Integral Transforms*, McGraw-Hill, New York, 1972.
10. J. Irving and N. Mullineaux, *Mathematics in Physics and Engineering*, Academic, New York, 1959.
11. V. S. Arpaci, *Conduction Heat Transfer*, Addison-Wesley, Reading, MA, 1966.
12. A. V. Luikov, *Analytical Heat Diffusion Theory*, Academic, New York, 1968.
13. D. Poulikakos, *Conduction Heat Transfer*, Prentice-Hall, Englewood Cliffs, NJ 1994.
14. C. R. Wylie and L. C. Barret, *Advanced Engineering Mathematics*, McGraw-Hill, New York, 1982.
15. M. G. Greenbert, *Foundations of Applied Mathematics*, Prentice-Hall, Englewood Cliffs, NJ, 1978.
16. M. R. Spiegel, *Laplace Transforms*, Schaum's Outline Series, McGraw-Hill, New York, 1965.
17. S. Goldstein, *Proc. London Math. Soc.*, 2nd series, **34**, 51–88, 1932.

PROBLEMS

9-1 A semi-infinite medium, $0 \leq x < \infty$, is initially at uniform temperature T_0. For times $t > 0$, the boundary surface at $x = 0$ is maintained at zero temperature. Obtain an expression for the temperature distribution $T(x, t)$ in the medium for times $t > 0$ by solving this problem with the Laplace transformation.

9-2 A semi-infinite medium, $0 \leq x < \infty$, is initially at a uniform temperature T_0. For times $t > 0$, the region is subjected to a prescribed heat flux at the boundary surface $x = 0$:

$$-k\frac{\partial T}{\partial x} = f_0 = \text{constant} \qquad \text{at } x = 0$$

Obtain an expression for the temperature distribution $T(x,t)$ in the medium for times $t > 0$ by using Laplace transformation.

9-3 A semi-infinite medium, $0 \le x < \infty$, is initially at uniform temperature T_0. For times $t > 0$, the boundary surface at $x = 0$ is kept at zero temperature, while heat is generated in the medium at a constant rate of g_0 (W/m^3). Obtain an expression for the temperature distribution $T(x, t)$ in the medium for times $t > 0$ by using Laplace transformation.

9-4 A slab, $0 \le x \le L$, is initially at uniform temperature T_0. For times $t > 0$, the boundary surface at $x = 0$ is kept insulated and the boundary surface at $x = L$ is kept at zero temperature. Obtain an expression for the temperature distribution $T(x, t)$ in the slab valid for very small times.

9-5 A slab, $0 \le x \le L$, is initially at zero temperature. For times $t > 0$, heat is generated in the slab at a constant rate of g_0 (W/m^3), while the boundary surface at $x = 0$ is kept insulated and the boundary surface at $x = L$ is kept at zero temperature. Obtain an expression for the temperature distribution $T(x, t)$ in the slab (i) using the Laplace transform technique that is valid for all times and (ii) an approximation that is valid for very small times.

9-6 A slab, $0 \le x \le L$, is initially at zero temperature. For times $t > 0$, the boundary surface at $x = 0$ is kept insulated, while the boundary surface at $x = L$ is subjected to a heat flux:

$$k\frac{\partial T}{\partial x} = f_0 = \text{constant} \qquad \text{at } x = L$$

Obtain an expression for the temperature distribution $T(x, t)$ in the slab that is valid for very small times.

9-7 A solid cylinder, $0 \le r \le b$, is initially at a uniform temperature T_0. For times $t > 0$, the boundary surface at $r = b$ is kept at zero temperature. Obtain an expression for the temperature distribution $T(r, t)$ in the solid valid for very small times. Consider equation 16c of Appendix IV.

9-8 A solid cylinder, $0 \le r \le b$, is initially at a uniform temperature T_0. For times $t > 0$, the boundary surface at $r = b$ is subjected to convection boundary condition in the form

$$\frac{\partial T}{\partial r} + HT = 0 \qquad \text{at } r = b$$

Obtain an expression for the temperature distribution $T(r,t)$ in the solid valid for very small times.

9-9 A solid sphere, $0 \le r \le b$, is initially at a uniform temperature T_0. For times $t > 0$, the boundary surface at $r = b$ is kept at zero temperature. Obtain an expression for the temperature distribution $T(r, t)$ in the solid valid for very small times.

9-10 A slab, $0 \leq x \leq L$, is initially at uniform temperature T_0. For times $t > 0$, the boundary surface at $x = 0$ is maintained at constant temperature T_0, and the boundary surface at $x = L$ is maintained at a prescribed time-dependent temperature $T = T_0 \cos \omega t$, where ω is a positive constant. Obtain an expression for the temperature distribution $T(x, t)$ in the slab using the Laplace transform technique.

9-11 A slab, $0 \leq x \leq L$, is initially at uniform temperature T_0. For times $t > 0$, the boundary surface at $x = 0$ is perfectly insulated, and the boundary surface at $x = L$ is maintained at a prescribed time-dependent temperature $T = T_0 \cosh \omega t$, where ω is a positive constant. Obtain an expression for the temperature distribution $T(x, t)$ in the slab using the Laplace transform technique.

9-12 A slab, $0 \leq x \leq L$, is initially at uniform temperature of zero. For times $t > 0$, the boundary surface at $x = 0$ is perfectly insulated, and the boundary surface at $x = L$ is subjected to a prescribed time-dependent heat flux $q'' = A \cos \omega t$, where A and ω are a positive constants. Obtain an expression for the temperature distribution $T(x, t)$ in the slab using the Laplace transform technique.

10

ONE-DIMENSIONAL COMPOSITE MEDIUM

The transient temperature distribution in a composite medium consisting of several layers in contact has numerous applications in engineering. In this chapter, the mathematical formulation of one-dimensional, transient heat conduction in a composite medium consisting of M parallel layers of slabs, cylinders, or spheres is presented. The transformation of the problem with nonhomogeneous boundary conditions into the one with homogeneous boundary conditions is described. The *orthogonal expansion* technique is used to solve the homogeneous problem of composite medium of finite thickness, while the *Laplace transformation* is used to solve the homogeneous problem of composite medium of semi-infinite and infinite thickness. Finally, the Green's function approach is used for solving the nonhomogeneous problem with energy generation in the medium.

The reader should consult references 1–13 for the theory and the application of the generalized orthogonal expansion technique and Green's function approach in the solution of heat conduction problems of composite media. The use of the Laplace transform technique in the solution of composite media problems is given in references 14–17, and the application of the integral transform technique and various other approaches can be found in references 18–38.

10-1 MATHEMATICAL FORMULATION OF ONE-DIMENSIONAL TRANSIENT HEAT CONDUCTION IN A COMPOSITE MEDIUM

We consider a composite medium consisting of M parallel layers of slabs, cylinders, or spheres as illustrated in Figure 10-1. We assume the existence of contact conductance h_i at the interfaces $x = x_i$, $i = 2, 3, \ldots, M$. Initially each layer is at

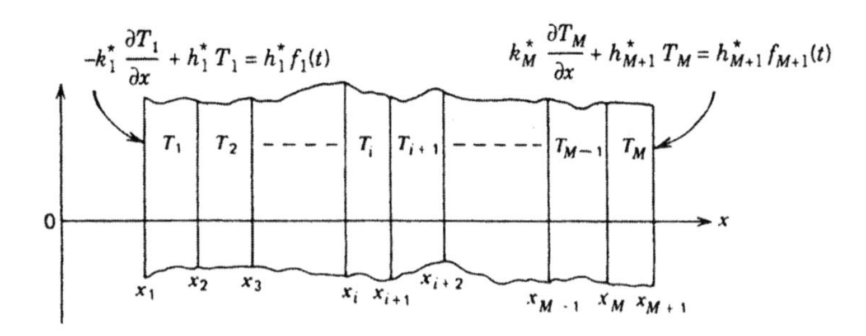

Figure 10-1 M-layer composite region.

a specified temperature $T_i(x, 0) = F_i(x)$, in $x_i < x < x_{i+1}$, $i = 1, 2, 3, \ldots, M$, for $t = 0$. For times $t > 0$, energy is generated in each layer at a rate of $g_i(x, t)$ (W/m^3), in $x_i < x < x_{i+1}$, $i = 1, 2, 3, \ldots, M$, while the energy is dissipated with convection from the two outer boundary surfaces $x = x_1$ and $x = x_{M+1}$, into ambient fluids at temperatures $f_1(t)$ and $f_{M+1}(t)$, with heat transfer coefficients h_1^* and h_{M+1}^*, respectively.

The mathematical formulation of this heat conduction problem is given as follows. The differential equations for each of the M layers are

$$\alpha_i \frac{1}{x^p} \frac{\partial}{\partial x}\left(x^p \frac{\partial T_i}{\partial x}\right) + \frac{\alpha_i}{k_i} g_i(x, t)$$
$$= \frac{\partial T_i(x, t)}{\partial t} \quad \text{in} \quad x_i < x < x_{i+1}, \quad t > 0, \quad i = 1, 2, \ldots, M \tag{10-1}$$

where

$$p = \begin{cases} 0 & \text{slab} \\ 1 & \text{cylinder} \\ 2 & \text{sphere} \end{cases}$$

The following boundary conditions are defined:

$$-k_1^* \left.\frac{\partial T_1}{\partial x}\right|_{x=x_1} + h_1^* \left.T_1\right|_{x=x_1} = h_1^* f_1(t) \qquad \text{at LHS boundary} \tag{10-2a}$$

$$\left.\begin{aligned} -k_i \left.\frac{\partial T_i}{\partial x}\right|_{x=x_{i+1}} &= h_{i+1}\left(\left.T_i\right|_{x=x_{i+1}} - \left.T_{i+1}\right|_{x=x_{i+1}}\right) \\ k_i \left.\frac{\partial T_i}{\partial x}\right|_{x=x_{i+1}} &= k_{i+1} \left.\frac{\partial T_{i+1}}{\partial x}\right|_{x=x_{i+1}} \end{aligned}\right\} \quad \begin{aligned} &\text{at the interfaces} \\ &i = 1, 2, \ldots, M-1 \end{aligned}$$

(10-2b,c)

$$k_M^* \left.\frac{\partial T_M}{\partial x}\right|_{x=x_{M+1}} + h_{M+1}^* \left.T_M\right|_{x=x_{M+1}} = h_{M+1}^* f_{M+1}(t) \qquad \text{at RHS boundary}$$

(10-2d)

and with the initial conditions

$$T_i\,(x, t = 0) = F_i\,(x) \qquad \text{in} \qquad x_i < x < x_{i+1}, \qquad i = 1, 2, \ldots, M \quad (10\text{-}3)$$

where $T_i(x,\ t)$ is the temperature of the layer i, $i = 1, 2, \ldots, M$. The problem contains M partial differential equations, $2M$ boundary conditions, and M initial conditions; hence it is mathematically well posed.

In order to distinguish the coefficients associated with the boundary conditions for the outer surfaces from those k and h for the medium and interfaces, an asterisk is used in the quantities h_1^*, h_{M+1}^*, k_1^*, and k_M^* appearing in the boundary conditions for the outer surfaces. The reason for this is that these quantities will be treated as coefficients, so that the boundary conditions of the first and second kind will be obtainable for the outer boundary surfaces by properly setting the values of these coefficients.

10-2 TRANSFORMATION OF NONHOMOGENEOUS BOUNDARY CONDITIONS INTO HOMOGENEOUS ONES

It is more convenient to solve the problems with homogeneous boundary conditions than with nonhomogeneous boundary conditions. The problem of time-dependent heat conduction for an M-layer composite medium with heat generation and nonhomogeneous outer boundary conditions can be transformed into a problem with heat generation but homogeneous boundary conditions by a procedure similar to that described by Özisik [39, Section 1-7] for the single-layer problem.

The problem defined by equations (10-1)–(10-3) has nonhomogeneous boundary conditions at the outer surfaces. In order to transform this time-dependent problem into one with homogeneous boundary conditions, we consider $T_i(x,\ t)$ constructed by the *superposition* of three simpler problems in the form

$$T_i\,(x, t) = \theta_i\,(x, t) + \phi_i\,(x)\, f_1\,(t) + \psi_i\,(x)\, f_{M+1}\,(t)$$
$$\text{in} \qquad x_i < x < x_{i+1}, \qquad i = 1, 2, \ldots, M, \qquad \text{for} \qquad t > 0 \qquad (10\text{-}4)$$

In equation (10-4), the functions $\phi_i(x)$, $\psi_i(x)$, and $\theta_i(x, t)$ are the solutions of the following three subproblems:

1. The functions $\phi_i(x)$ are the solutions of the following steady-state problem for the same region, with no heat generation, but with a single nonhomogeneous boundary condition at $x = x_1$:

$$\frac{d}{dx}\left(x^p \frac{d\phi_i}{dx}\right) = 0 \qquad \text{in} \qquad x_i < x < x_{i+1}, \qquad i = 1, 2, \ldots, M \qquad (10\text{-}5\text{a})$$

subject to the boundary conditions

$$-k_1^* \left.\frac{d\phi_1(x)}{dx}\right|_{x=x_1} + h_1^* \left.\phi_1(x)\right|_{x=x_1} = h_1^* \tag{10-5b}$$

$$\left.\begin{aligned} -k_i \left.\frac{d\phi_i}{dx}\right|_{x=x_{i+1}} &= h_{i+1}\left(\left.\phi_i\right|_{x=x_{i+1}} - \left.\phi_{i+1}\right|_{x=x_{i+1}}\right) \\ k_i \left.\frac{d\phi_i}{dx}\right|_{x=x_{i+1}} &= k_{i+1} \left.\frac{d\phi_{i+1}}{dx}\right|_{x=x_{i+1}} \end{aligned}\right\} \begin{array}{l}\text{at the interfaces}\\ i = 1, 2, \ldots,\\ \quad M-1\end{array} \tag{10-5c,d}$$

$$k_M^* \left.\frac{d\phi_M}{dx}\right|_{x=x_{M+1}} + h_{M+1}^* \left.\phi_M\right|_{x=x_{M+1}} = 0 \tag{10-5e}$$

2. The functions $\psi_i(x)$ are the solutions of the following steady-state problem for the same region, with no heat generation, but with one nonhomogeneous boundary condition at $x = x_{M+1}$:

$$\frac{d}{dx}\left(x^p \frac{d\psi_i}{dx}\right) = 0 \quad \text{in} \quad x_i < x < x_{i+1}, \quad i = 1, 2, \ldots, M \tag{10-6a}$$

subject to the boundary conditions

$$-k_1^* \left.\frac{d\psi_1}{dx}\right|_{x=x_1} + h_1^* \left.\psi_1\right|_{x=x_1} = 0 \tag{10-6b}$$

$$\left.\begin{aligned} -k_i \left.\frac{d\psi_i}{dx}\right|_{x=x_{i+1}} &= h_{i+1}\left(\left.\psi_i\right|_{x=x_{i+1}} - \left.\psi_{i+1}\right|_{x=x_{i+1}}\right) \\ k_i \left.\frac{d\psi_i}{dx}\right|_{x=x_{i+1}} &= k_{i+1} \left.\frac{d\psi_{i+1}}{dx}\right|_{x=x_{i+1}} \end{aligned}\right\} \begin{array}{l}\text{at the interfaces}\\ i = 1, 2, \ldots,\\ \quad M-1\end{array} \tag{10-6c,d}$$

$$k_M^* \left.\frac{d\psi_M}{dx}\right|_{x=x_{M+1}} + h_{M+1}^* \left.\psi_M\right|_{x=x_{M+1}} = h_{M+1}^* \tag{10-6e}$$

3. The functions $\theta_i(x, t)$ are the solutions of the following time-dependent heat conduction problem for the same region, with heat generation, but subject to homogeneous boundary conditions:

$$\alpha_i \frac{1}{x^p}\frac{\partial}{\partial x}\left(x^p \frac{\partial \theta_i}{\partial x}\right) + g_i^*(x,t) = \frac{\partial \theta_i(x,t)}{\partial t} \tag{10-7a}$$

$$\text{in} \quad x_i < x < x_{i+1}, \quad i = 1, 2, \ldots, M \quad \text{for} \quad t > 0$$

where we have defined

$$g_i^* (x, t) \equiv \frac{\alpha_i}{k_i} g_i (x, t) - \left[\phi_i (x) \frac{df_1 (t)}{dt} + \psi_i (x) \frac{df_M (t)}{dt} \right] \tag{10-7b}$$

and subject to the boundary conditions

$$- k_1^* \left. \frac{\partial \theta_1 (x)}{\partial x} \right|_{x=x_1} + h_1^* \left. \theta_1 (x) \right|_{x=x_1} = 0 \tag{10-8a}$$

$$\left. \begin{aligned} -k_i \left. \frac{\partial \theta_i}{\partial x} \right|_{x=x_{i+1}} &= h_{i+1} \left(\left. \theta_i \right|_{x=x_{i+1}} - \left. \theta_{i+1} \right|_{x=x_{i+1}} \right) \\ k_i \left. \frac{\partial \theta_i}{\partial x} \right|_{x=x_{i+1}} &= k_{i+1} \left. \frac{\partial \theta_{i+1}}{\partial x} \right|_{x=x_{i+1}} \end{aligned} \right\} \begin{aligned} &\text{at the interfaces} \\ &i = 1, 2, \ldots, \\ &\quad M-1 \end{aligned} \tag{10-8b,c}$$

$$k_M^* \left. \frac{\partial \theta_M}{\partial x} \right|_{x=x_{M+1}} + h_{M+1}^* \left. \theta_M \right|_{x=x_{M+1}} = 0 \tag{10-8d}$$

and to the initial conditions

$$\theta_i (x, t = 0) = F_i (x) - \left[\phi_i (x) f_1 (0) + \psi_i (x) f_{M+1} (0) \right] \equiv F_i^* (x) \tag{10-9}$$

in $x_i < x < x_{i+1}$, $i = 1, 2, \ldots, M$. The validity of this superposition procedure can readily be verified by introducing equation (10-4) into the original problem given by equations (10-1)–(10-3) and utilizing the above three subproblems defined by equations (10-5)–(10-8). The weighting function p is as defined above.

Example 10-1 Two-Layer, Composite Slab

A two-layer slab consists of the first layer in $0 \leq x \leq a$ and the second layer in $a \leq x \leq b$, which are in perfect thermal contact as illustrated in Figure 10-2. Let k_1 and k_2 be the thermal conductivities, and α_1 and α_2 the thermal diffusivities for the first and second layer, respectively. Initially, the first region is at temperature $F_1(x)$ and the second region at $F_2(x)$. For times $t > 0$, the boundary surface at $x = 0$ is kept at temperature $f_1(t)$, and the boundary at $x = b$ dissipates heat by convection, with a heat transfer coefficient h_3^* into an ambient environment at temperature $f_3(t)$. By applying the splitting-up procedure described above, separate this problem into (i) two steady-state problems each with one nonhomogeneous boundary condition, and (ii) one time-dependent problem with homogeneous boundary conditions and the initial condition. Figure 10-2 shows the geometry coordinates and the boundary conditions for the original problem.

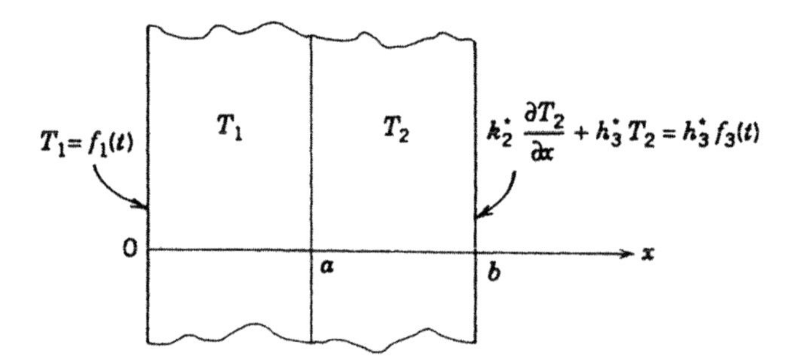

Figure 10-2 Two-layer slab with perfect thermal contact at the interface.

The mathematical formulation of this problem is given as

$$\alpha_1 \frac{\partial^2 T_1}{\partial x^2} = \frac{\partial T_1 (x, t)}{\partial t} \qquad \text{in} \qquad 0 < x < a, \qquad t > 0 \tag{10-10a}$$

$$\alpha_2 \frac{\partial^2 T_2}{\partial x^2} = \frac{\partial T_2 (x, t)}{\partial t} \qquad \text{in} \qquad a < x < b, \qquad t > 0 \tag{10-10b}$$

subject to the boundary conditions

$$T_1 (x = 0, t) = f_1 (t) \tag{10-11a}$$

$$T_1 (x = a, t) = T_2 (x = a, t) \tag{10-11b}$$

$$k_1 \left. \frac{\partial T_1}{\partial x} \right|_{x=a} = k_2 \left. \frac{\partial T_2}{\partial x} \right|_{x=a} \tag{10-11c}$$

$$k_2^* \left. \frac{\partial T_2}{\partial x} \right|_{x=b} + h_3^* \left. T_2 \right|_{x=b} = h_3^* f_3 (t) \tag{10-11d}$$

and the initial conditions

$$T_1 (x, t = 0) = F_1 (x) \qquad \text{in} \qquad 0 < x < a \tag{10-12a}$$

$$T_2 (x, t = 0) = F_2 (x) \qquad \text{in} \qquad a < x < b \tag{10-12b}$$

To transform this problem, we construct the solution of $T_i(x, t)$, $i = 1, 2$, by the superposition of the following three simpler problems in the form

$$T_i (x, t) = \theta_i (x, t) + \phi_i (x) f_1 (t) + \psi_i (x) f_3 (t) \qquad \text{in} \qquad x_i < x < x_{i+1}, \qquad i = 1, 2 \tag{10-13}$$

where $x_1 = 0$, $x_2 = a$, and $x_3 = b$. The functions $\phi_i(x)$, $\psi_i(x)$, and $\theta_i(x, t)$, for $i = 1, 2$ are the solutions of the following three simpler problems, respectively.

1. The functions $\phi_i(x)$, $i = 1, 2$ satisfy the steady-state heat conduction problem given as

$$\frac{d^2\phi_1(x)}{dx^2} = 0 \qquad \text{in} \qquad 0 < x < a \tag{10-14a}$$

$$\frac{d^2\phi_2(x)}{dx^2} = 0 \qquad \text{in} \qquad a < x < b \tag{10-14b}$$

subject to the boundary conditions

$$\phi_1(x = 0) = 1 \tag{10-15a}$$

$$\left.\begin{aligned} &\phi_1(x = a) = \phi_2(x = a) \\ &k_1 \left.\frac{d\phi_1}{dx}\right|_{x=a} = k_2 \left.\frac{d\phi_2}{dx}\right|_{x=a} \end{aligned}\right\} \tag{10-15b,c}$$

$$k_2^* \left.\frac{d\phi_2}{dx}\right|_{x=b} + h_3^* \left.\phi_2\right|_{x=b} = 0 \tag{10-15d}$$

2. The functions $\psi_i(x)$, $i = 1, 2$ satisfy the steady-state heat conduction problem given as

$$\frac{d^2\psi_1(x)}{dx^2} = 0 \qquad \text{in} \qquad 0 < x < a \tag{10-16a}$$

$$\frac{d^2\psi_2(x)}{dx^2} = 0 \qquad \text{in} \qquad a < x < b \tag{10-16b}$$

subject to the boundary conditions

$$\psi_1(x = 0) = 0 \tag{10-17a}$$

$$\left.\begin{aligned} &\psi_1(x = a) = \psi_2(x = a) \\ &k_1 \left.\frac{d\psi_1}{dx}\right|_{x=a} = k_2 \left.\frac{d\psi_2}{dx}\right|_{x=a} \end{aligned}\right\} \tag{10-17b,c}$$

$$k_2^* \left.\frac{d\psi_2}{dx}\right|_{x=b} + h_3^* \left.\psi_2\right|_{x=b} = h_3^* \tag{10-17d}$$

3. The functions $\theta_i(x, t)$, $i = 1, 2$ are the solutions of the following transient, homogeneous problem:

$$\alpha_1 \frac{\partial^2\theta_1}{\partial x^2} = \frac{\partial\theta_1(x, t)}{\partial t} + g_1^*(x, t) \qquad \text{in} \qquad 0 < x < a, \qquad t > 0 \tag{10-18a}$$

$$\alpha_2 \frac{\partial^2 \theta_2}{\partial x_2} = \frac{\partial \theta_2 (x, t)}{\partial t} + g_2^* (x, t) \qquad \text{in} \qquad a < x < b, \qquad t > 0 \tag{10-18b}$$

where, $g_i^*(x, t) = -[\phi_i(x)\, df_1/dt + \psi_i(x)\, df_3/dt]$, $i = 1, 2$, subject to the boundary conditions

$$\theta_1 (x = 0, t) = 0 \tag{10-19a}$$

$$\left.\begin{aligned} &\theta_1 (x = a, t) = \theta_2 (x = a, t) \\ &k_1 \left.\frac{\partial \theta_1}{\partial x}\right|_{x=a} = k_2 \left.\frac{\partial \theta_2}{\partial x}\right|_{x=a} \end{aligned}\right\} \tag{10-19b,c}$$

$$k_2^* \left.\frac{\partial \theta_2}{\partial x}\right|_{x=b} + h_3^* \left.\theta_2\right|_{x=b} = 0 \tag{10-19d}$$

and the initial conditions

$$\theta_1 (x, t = 0) = F_1 (x) - f_1 (0)\, \phi_1 (x) - f_3 (0)\, \psi_1 (x) \equiv F_1^* (x) \tag{10-20a}$$

over the domain $0 \le x < a$ and

$$\theta_2 (x, t = 0) = F_2 (x) - f_1 (0)\, \phi_2 (x) - f_3 (0)\, \psi_2 (x) \equiv F_2^* (x) \tag{10-20b}$$

over the domain $a < x \le b$. The validity of this superposition procedure can be verified by introducing the transformation given by equation (10-13) into the original problem given by equations (10-10)–(10-12) and utilizing the definition of the subproblems given by equations (10-14)–(10-20). Clearly, the time-dependent problem given by equations (10-18)–(10-20) has homogeneous boundary conditions.

Solving Steady-State Problems of M-Layer Slab, Cylinder, or Sphere

We now consider a steady-state problem with no energy generation, a single nonhomogeneous boundary condition of the type given by equation (10-5), but for a M-layer slab, cylinder, or sphere. The mathematical formulation is given by

$$\frac{d}{dx}\left(x^p \frac{d\phi_i}{dx}\right) = 0 \qquad \text{in} \qquad x_i < x < x_{i+1}, \qquad i = 1, 2, \ldots, M \tag{10-21}$$

subject to the boundary conditions

$$-k_1^* \left.\frac{d\phi_1 (x)}{dx}\right|_{x=x_1} + h_1^* \left.\phi_1 (x)\right|_{x=x_1} = h_1^* \tag{10-22a}$$

$$\left.\begin{aligned} -k_i \left.\frac{d\phi_i}{dx}\right|_{x=x_{i+1}} &= h_{i+1}\left(\phi_i\big|_{x=x_{i+1}} - \phi_{i+1}\big|_{x=x_{i+1}}\right) \\ k_i \left.\frac{d\phi_i}{dx}\right|_{x=x_{i+1}} &= k_{i+1} \left.\frac{d\phi_{i+1}}{dx}\right|_{x=x_{i+1}} \end{aligned}\right\} \quad \begin{aligned} &\text{at the interfaces} \\ &i = 1, 2, \ldots, M-1 \end{aligned} \tag{10-22b,c}$$

$$k_M^* \left.\frac{d\phi_M}{dx}\right|_{x=x_{M+1}} + h_{M+1}^* \, \phi_M\big|_{x=x_{M+1}} = 0 \tag{10-22d}$$

where the weighting factors are again given by

$$p = \begin{cases} 0 & \text{slab} \\ 1 & \text{cylinder} \\ 2 & \text{sphere} \end{cases} \tag{10-23}$$

The solution of the ordinary differential equation (10-21), for any layer i, is given in the respective forms:

$$\text{Slab:} \qquad \phi_i(x) = A_i + B_i x \tag{10-24a}$$

$$\text{Cylinder:} \qquad \phi_i(x) = A_i + B_i \ln x \tag{10-24b}$$

$$\text{Sphere:} \qquad \phi_i(x) = A_i + \frac{B_i}{x} \tag{10-24c}$$

The solution involves two unknown coefficients A_i and B_i for each layer i; then, for a M-layer problem, $2M$ unknown coefficients are to be determined. Substituting the solution given by equations (10-24) into the boundary conditions (10-22a–d), one obtains $2M$ equations for the determination of the $2M$ unknown coefficients A_i, B_i, for $i = 1, 2, \ldots, M$. The solution of the homogeneous transient heat conduction problems of the type given by equations (10-18)–(10-20), but for the M-layer medium, is described in the next section.

10-3 ORTHOGONAL EXPANSION TECHNIQUE FOR SOLVING M-LAYER HOMOGENEOUS PROBLEMS

We now consider the solution of the homogeneous problem of heat conduction in a composite medium consisting of M parallel layers of slabs, cylinder, or spheres in contact as illustrated in Figure 10-3. For generality, we assume contact resistance at the interfaces and convection from the outer boundaries. Let h_i be the arbitrary film coefficient (i.e., the contact conductance) at the interfaces $x = x_i, i = 2, 3, \ldots, M$, and h_1^* and h_{M+1}^* be the actual heat transfer coefficients at the outer boundaries $x = x_1$ and $x = x_{M+1}$, respectively. Each layer is homogeneous and isotropic and has thermal properties (i.e., ρ, c, k) that are constant within the layer and different from those of the adjacent layers. Initially each layer is at

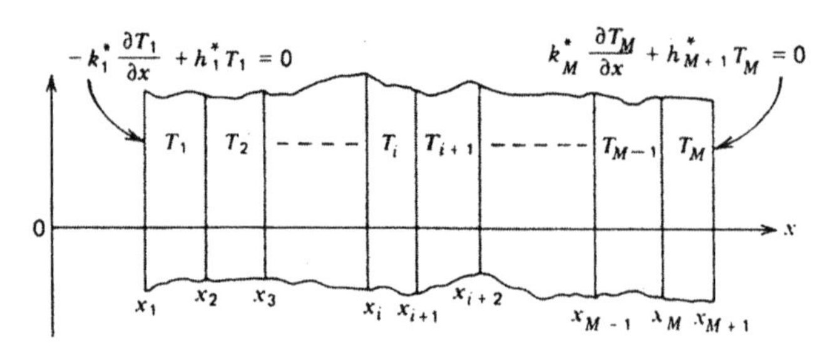

Figure 10-3 M-layer composite region.

a specified temperature $T_i(x, 0) = F_i(x)$, in $x_i < x < x_{i+1}$, $i = 1, 2, \ldots, M$. For times $t > 0$, heat is dissipated by convection from the two outer boundaries into environments at zero temperature, and there is no heat generation in the medium. We are interested in the determination of the temperature distribution $T_i(x, t)$, in the layers $i = 1, 2, \ldots, M$, for times $t > 0$. The mathematical formulation of this heat conduction problem is given by

$$\alpha_i \frac{1}{x^p} \frac{\partial}{\partial x}\left(x^p \frac{\partial T_i}{\partial x}\right) = \frac{\partial T_i\,(x,t)}{\partial t} \qquad \text{in} \qquad x_i < x < x_{i+1} \qquad \text{for} \qquad t > 0, \qquad i = 1, 2, \ldots, M \tag{10-25}$$

where again

$$p = \begin{cases} 0 & \text{slab} \\ 1 & \text{cylinder} \\ 2 & \text{sphere} \end{cases} \tag{10-26}$$

subject to the boundary conditions

$$-\,k_1^* \left.\frac{\partial T_1}{\partial x}\right|_{x=x_1} + h_1^*\, T_1\big|_{x=x_1} = 0 \tag{10-27a}$$

$$\left.\begin{aligned} -k_i \left.\frac{\partial T_i}{\partial x}\right|_{x=x_{i+1}} &= h_{i+1}\left(T_i\big|_{x=x_{i+1}} - T_{i+1}\big|_{x=x_{i+1}}\right) \\ k_i \left.\frac{\partial T_i}{\partial x}\right|_{x=x_{i+1}} &= k_{i+1} \left.\frac{\partial T_{i+1}}{\partial x}\right|_{x=x_{i+1}} \end{aligned}\right\} \quad \begin{array}{l}\text{at the interfaces} \\ i = 1, 2, \ldots, M-1\end{array} \tag{10-27b,c}$$

$$k_M^* \left.\frac{\partial T_M}{\partial x}\right|_{x=x_{M+1}} + h_{M+1}^*\, T_M\big|_{x=x_{M+1}} = 0 \tag{10-27d}$$

and the initial conditions

$$T_i\,(x, t=0) = F_i\,(x) \qquad \text{in} \qquad x_i < x < x_{i+1}, \qquad i = 1, 2, \ldots, M \tag{10-28}$$

The finite value of interface conductance h_{i+1}, in equations (10-27b) implies that the temperature is discontinuous at the interfaces. The boundary conditions

(10-27c) represents the continuity of heat flux at the interfaces; hence conservation of energy. When the interface conductance $h_{i+1} \to \infty$, the boundary condition (10-27b) reduces to

$$T_i(x = x_{i+1}) = T_{i+1}(x = x_{i+1}) \qquad \text{for} \qquad i = 1, 2, \ldots, M-1 \tag{10-29}$$

which implies the continuity of temperature, or *perfect thermal contact*, at the interfaces.

To solve the above heat conduction problem, the variables are separated in the form

$$T_i\,(x, t) = \psi_i\,(x)\,\Gamma\,(t) \tag{10-30}$$

When equation (10-30) is introduced into (10-25), we obtain

$$\alpha_i \frac{1}{x^p}\frac{1}{\psi_i}\frac{d}{dx}\left(x^p \frac{d\psi_i}{dx}\right) = \frac{1}{\Gamma}\frac{d\Gamma}{dt} \equiv -\beta^2 \tag{10-31}$$

where β is the separation constant. We recall that, in separating the variables for the case of a single-layer problem, the thermal diffusivity α was retained on the side of the equation where the time-dependent function $\Gamma(t)$ was collected. In the case of composite medium, α_i is retained on the left-hand side of equation (10-31) where the space-dependent functions $\psi_i(x)$ are collected. The reason for this is to keep the solution of the time-dependent function $\Gamma(t)$ independent of α_i, since it is discontinuous at the interfaces.

The separation given by equation (10-31) results in the following two ordinary differential equations for the determination of the functions $\Gamma(t)$ and $\psi_i\,(\beta, x)$:

$$\frac{d\Gamma\,(t)}{dt} + \beta_n^2 \Gamma\,(t) = 0 \qquad \text{for} \qquad t > 0 \tag{10-32}$$

and

$$\frac{1}{x^p}\frac{d}{dx}\left(x^p \frac{d\psi_{in}}{dx}\right) + \frac{\beta_n^2}{\alpha_i}\psi_{in} = 0 \qquad \text{in} \qquad x_i < x < x_{i+1}, \qquad i = 1, 2, \ldots, M \tag{10-33}$$

where $\psi_{in} \equiv \psi_i(\beta_n, x)$. The subscript n is included to imply that there are an infinite number of discrete values of the eigenvalues, $\beta_1 < \beta_2 < \cdots < \beta_n < \cdots$, and the corresponding eigenfunctions ψ_{in}.

The boundary conditions for equations (10-33) are obtained by introducing equation (10-31) into the boundary conditions (10-27); we find

$$-\,k_1^* \left.\frac{d\psi_{1n}}{dx}\right|_{x=x_1} + h_1^* \left.\psi_{1n}\right|_{x=x_1} = 0 \tag{10-34a}$$

$$\left.\begin{aligned} -k_i \left.\frac{d\psi_{in}}{dx}\right|_{x=x_{i+1}} &= h_{i+1}\left(\psi_{in}\big|_{x=x_{i+1}} - \psi_{i+1,n}\big|_{x=x_{i+1}}\right) \\ k_i \left.\frac{d\psi_{in}}{dx}\right|_{x=x_{i+1}} &= k_{i+1} \left.\frac{d\psi_{i+1,n}}{dx}\right|_{x=x_{i+1}} \end{aligned}\right\} \quad \begin{aligned} &\text{at the interfaces} \\ &i = 1, 2, \ldots, M-1 \end{aligned} \tag{10-34b,c}$$

$$k_M^* \left.\frac{d\psi_{Mn}}{dx}\right|_{x=x_{M+1}} + h_{M+1}^* \, \psi_{Mn}\big|_{x=x_{M+1}} = 0 \tag{10-34d}$$

The set of equations (10-33) subject to the boundary conditions (10-34) constitute an eigenvalue problem for the determination of the eigenvalues β_n and the corresponding eigenfunctions ψ_{in}.

The eigenfunctions ψ_{in} of the eigenvalue problem defined by equations (10-33) and (10-34) satisfy the following orthogonality relation [13], namely,

$$\sum_{i=1}^{M} \frac{k_i}{\alpha_i} \int_{x=x_i}^{x_{i+1}} x^p \psi_{in}(x)\, \psi_{iq}(x)\, dx = \begin{cases} 0 & \text{for} \quad n \neq q \\ N_n & \text{for} \quad n = q \end{cases} \tag{10-35}$$

where the norm N_n is defined as

$$N_n = \sum_{j=1}^{M} \frac{k_j}{\alpha_j} \int_{x=x_j}^{x_{j+1}} x^p \psi_{jn}^2(x)\, dx \tag{10-36}$$

and where ψ_{in}, ψ_{iq} are the two different eigenfunctions.

The solution for the time-variable function $\Gamma(t)$ is immediately obtained from equation (10-32) as

$$\Gamma(t) = C_1 e^{-\beta_n^2 t} \tag{10-37}$$

The general solution for the temperature distribution $T_i(x, t)$, in any region i, is constructed as

$$T_i(x, t) = \sum_{n=1}^{\infty} C_n e^{-\beta_n^2 t} \psi_{in}(x) \qquad i = 1, 2, \ldots, M \tag{10-38}$$

where the summation is over all eigenvalues β_n. This solution satisfies the differential equations (10-25) and the boundary conditions (10-27). We now constrain this solution to satisfy the initial conditions (10-28), which yields

$$F_i(x) = \sum_{n=1}^{\infty} C_n \psi_{in}(x) \qquad \text{in} \qquad x_i < x < x_{i+1}, \qquad i = 1, 2, \ldots, M \tag{10-39}$$

The Fourier coefficients C_n can be determined by utilizing the above orthogonality relation as now described. We operate on both sides of equation (10-39) by the operator

$$*\frac{k_i}{\alpha_i}\int_{x=x_i}^{x_{i+1}} x^p \psi_{iq}(x)\, dx$$

and sum the resulting expressions, for $i = 1$ to M (i.e., over all regions), to obtain

$$\sum_{i=1}^{M}\frac{k_i}{\alpha_i}\int_{x=x_i}^{x_{i+1}} x^p \psi_{iq}(x)\, F_i(x)\, dx = \sum_{n=1}^{\infty} C_n \left[\sum_{i=1}^{M}\frac{k_i}{\alpha_i}\int_{x=x_i}^{x_{i+1}} x^p \psi_{iq}(x)\, \psi_{in}(x)\, dx\right] \tag{10-40}$$

In view of the orthogonality relation (10-35), the term inside the bracket on the right-hand side of equation (10-40) vanishes for $n \neq q$ and becomes equal to N_n, the norm, for $n = q$. Then the Fourier coefficients C_n are determined as

$$C_n = \frac{1}{N_n}\sum_{i=1}^{M}\frac{k_i}{\alpha_i}\int_{x=x_i}^{x_{i+1}} x^p \psi_{in}(x)\, F_i(x)\, dx \tag{10-41}$$

Before introducing this result into equation (10-38), we change the summation index from i to j, and the dummy integration variable from x to x' in equation (10-41) to avoid confusion with the index i and the space variable x in equation (10-38). Then, the solution for the temperature distribution $T_i(x, t)$ in any region i of the composite medium is determined as

$$T_i(x, t) = \sum_{n=1}^{\infty} e^{-\beta_n^2 t}\frac{1}{N_n}\psi_{in}(x)\sum_{j=1}^{M}\frac{k_j}{\alpha_j}\int_{x'=x_j}^{x_{j+1}} x'^p \psi_{jn}(x')\, F_j(x')\, dx' \tag{10-42}$$

$$\text{in} \quad x_i < x < x_{i+1} \quad i = 1, 2, \ldots, M$$

where the norm N_n is defined as

$$N_n = \sum_{j=1}^{M}\frac{k_j}{\alpha_j}\int_{x'=x_j}^{x_{j+1}} x'^p \psi_{jn}^2(x')\, dx' \tag{10-43}$$

and

$$p = \begin{cases} 0 & \text{slab} \\ 1 & \text{cylinder} \\ 2 & \text{sphere} \end{cases} \tag{10-44}$$

An examination of this solution reveals that, for $M = 1$, equations (10-42) reduce to the solution for the single-region problem considered in the previous chapters if we set $\beta_n^2 = \alpha\gamma_n^2$, where α is the thermal diffusivity.

Green's Function for Composite Medium

The solution given by equation (10-42) can be recast to define the composite medium Green's function. That is, the solution (10-42) is rearranged as

$$T_i(x,t) = \sum_{j=1}^{M} \int_{x'=x_j}^{x_{j+1}} \frac{k_j}{\alpha_j} \left[\sum_{n=1}^{\infty} e^{-\beta_n^2 t} \frac{1}{N_n} \psi_{in}(x)\, \psi_{jn}(x') \right] x'^p F_j(x')\, dx'$$

$$\text{in} \quad x_i < x < x_{i+1}, \qquad i = 1, 2, \ldots, M \tag{10-45}$$

This result is now written more compactly, by introducing the Green's function notation, as

$$T_i(x,t) = \sum_{j=1}^{M} \int_{x'=x_j}^{x_{j+1}} G_{ij}(x,t|x',\tau)\big|_{\tau=0}\, x'^p F_j(x')\, dx' \tag{10-46}$$

where x'^p is the Sturm–Liouville weight function, and $G_{ij}(x,t|x',\tau)|_{\tau=0}$ is defined as

$$G_{ij}(x,t|x',\tau)\big|_{\tau=0} = \sum_{n=1}^{\infty} e^{-\beta_n^2 t} \frac{1}{N_n} \frac{k_j}{\alpha_j} \psi_{in}(x)\, \psi_{jn}(x') \tag{10-47}$$

$$p = \begin{cases} 0 & \text{slab} \\ 1 & \text{cylinder} \\ 2 & \text{sphere} \end{cases} \tag{10-48}$$

in the region $x_i < x < x_{i+1}$, $i = 1, 2, \ldots, M$. Thus $G_{ij}(x,t|x',\tau)|_{\tau=0}$ represents the Green's function evaluated for $\tau = 0$ associated with the solution of one-dimensional homogeneous composite medium problem defined by equations (10-25)–(10-28).

To solve the nonhomogeneous composite medium problem, such as the one with energy generation, the Green's function $G_{ij}(x,t|x',\tau)$ is needed. It is obtained from equation (10-47) by replacing t by $t - \tau$, as detailed in Chapter 8, see equation (8-22). Thus the Green's function for the problem becomes

$$G_{ij}(x,t|x',\tau) = \sum_{n=1}^{\infty} e^{-\beta_n^2 (t-\tau)} \frac{1}{N_n} \frac{k_j}{\alpha_j} \psi_{in}(x)\, \psi_{jn}(x') \tag{10-49}$$

in the region $x_i \le x \le x_{i+1}$, $i = 1, 2, \ldots, M$.

The use of Green's function in the solution of nonhomogeneous, one-dimensional composite medium problems will be demonstrated later in this chapter.

10-4 DETERMINATION OF EIGENFUNCTIONS AND EIGENVALUES

The general solution $\psi_{in}(x)$ of the eigenvalue problem given by equations (10-33) and (10-34) can be written in the form

$$\psi_{in}(x) = A_{in}\phi_{in}(x) + B_{in}\theta_{in}(x) \qquad \text{in} \qquad x_i < x < x_{i+1}, \qquad i = 1, 2, \ldots, M \tag{10-50}$$

where $\phi_{in}(x)$ and $\theta_{in}(x)$ are the two linearly independent solutions of equations (10-33) and where A_{in}, B_{in} are the coefficients. Table 10-1 lists the functions $\phi_{in}(x)$ and $\theta_{in}(x)$ for slabs, cylinders, and spheres. The heat conduction problem of an M-layer composite medium, in general, involves M solutions in the form given by equation (10-50); hence, there are $2M$ arbitrary coefficients, A_{in} and B_{in}, $i = 1, 2, \ldots, M$, to be determined. The boundary conditions (10-34) provide a system of $2M$ linear, homogeneous equations for the determination of these $2M$ coefficients; but, because the resulting system of equations is homogeneous, the coefficients can be determined only in terms of any one of them (i.e., the nonvanishing one), or within a multiple of an arbitrary constant. This arbitrariness does not cause any difficulty because the arbitrary constant will appear both in the numerator and denominator of equation (10-42) or (10-47); hence it will cancel out. Therefore, in the process of determining the coefficients A_{in} and B_{in} from the system of $2M$ homogeneous equations, any one of the nonvanishing coefficients, say, A_{in}, can be set equal to unity without loss of generality.

Finally, an additional relationship is needed for the determination of the eigenvalues β_n. This additional relationship is obtained from the requirement that the above system of $2M$ homogeneous equations has a nontrivial solution, that is, the determinant of the coefficients A_{in} and B_{in} vanishes. This condition leads to a *transcendental equation* for the determination of the eigenvalues

$$\beta_1 < \beta_2 < \beta_3 < \cdots < \beta_n < \cdots \tag{10-51}$$

Clearly, for each of these eigenvalues there are the corresponding set of values of A_{in} and B_{in}, hence the eigenfunctions $\psi_{in}(x)$. Once the eigenfunctions $\psi_{in}(x)$

TABLE 10-1 Linearly Independent Solutions $\phi_{in}(x)$ and $\theta_{in}(x)$ of Equation (10-33) for Slabs, Cylinders, and Spheres

Geometry	$\phi_{in}(x)$	$\theta_{in}(x)$
Slab	$\sin\left(\frac{\beta_n}{\sqrt{\alpha_i}}x\right)$	$\cos\left(\frac{\beta_n}{\sqrt{\alpha_i}}x\right)$
Cylinder	$J_0\left(\frac{\beta_n}{\sqrt{\alpha_i}}x\right)$	$Y_0\left(\frac{\beta_n}{\sqrt{\alpha_i}}x\right)$
Sphere	$\frac{1}{x}\sin\left(\frac{\beta_n}{\sqrt{\alpha_i}}x\right)$	$\frac{1}{x}\cos\left(\frac{\beta_n}{\sqrt{\alpha_i}x}\right)$

and the eigenvalues β_n are determined by the procedure outlined above, the temperature distribution $T_i(x, t)$ in any region i of the composite medium is determined by equations (10-42).

Example 10-2 Three-Layer Composite Slab

Consider transient heat conduction in a three-layer composite medium with perfect thermal contact at the interfaces and convection at the outer boundary surfaces. Give the eigenvalue problem and develop the equations for the determination of the coefficients A_{in}, B_{in} of equation (10-50) and the eigenvalues

$$\beta_1 < \beta_2 < \beta_3 < \cdots < \beta_n < \cdots$$

The eigenvalue problem for this transient heat conduction is similar to that given by equations (10-33) and (10-34), except $M = 3$, and the interface conductances are taken as infinite: $h_2 \to \infty$, $h_3 \to \infty$ (i.e., perfect thermal contact). Then, the eigenvalue problem becomes

$$\frac{1}{x^p}\frac{d}{dx}\left(x^p\frac{d\psi_{in}}{dx}\right) + \frac{\beta_n^2}{\alpha_i}\psi_{in} = 0 \qquad \text{in} \qquad x_i < x < x_{i+1}, \qquad i = 1, 2, 3 \tag{10-52}$$

subject to the boundary conditions

$$-k_1^* \left.\frac{d\psi_{1n}}{dx}\right|_{x=x_1} + h_1^* \left.\psi_{1n}\right|_{x=x_1} = 0 \tag{10-53a}$$

$$\left.\begin{aligned} \left.\psi_{in}\right|_{x=x_{i+1}} &= \left.\psi_{i+1,n}\right|_{x=x_{i+1}} \\ k_i \left.\frac{d\psi_{in}}{dx}\right|_{x=x_{i+1}} &= k_{i+1} \left.\frac{d\psi_{i+1,n}}{dx}\right|_{x=x_{i+1}} \end{aligned}\right\} \quad \begin{array}{c}\text{at the interfaces}\\ i = 1, 2\end{array} \tag{10-53b,c}$$

$$k_3^* \left.\frac{d\psi_{3n}}{dx}\right|_{x=x_4} + h_4^* \left.\psi_{3n}\right|_{x=x_4} = 0 \tag{10-53d}$$

where the eigenfunctions $\psi_{in}(x)$ are given by

$$\psi_{in}(x) = A_{in}\phi_{in}(x) + B_{in}\theta_{in}(x), \qquad i = 1, 2, 3 \tag{10-54}$$

and $\phi_{in}(x)$ and $\theta_{in}(x)$ are as specified in Table 10-1.

The first step in the analysis is the determination of the six coefficients A_{in}, B_{in} with $i = 1, 2, 3$. Without loss of generality, we set one of the nonvanishing coefficients, say, A_{1n} equal to unity:

$$A_{1n} = 1 \tag{10-55}$$

The eigenfunctions $\psi_{in}(x)$ given by equation (10-54) with $A_{1n} = 1$ are introduced into the boundary conditions (10-53). The resulting system of equations is expressed in the matrix form as

$$\begin{bmatrix} X_1 & Y_1 & 0 & 0 & 0 & 0 \\ \phi_{1n} & \theta_{1n} & -\phi_{2n} & -\theta_{2n} & 0 & 0 \\ k_1\phi'_{1n} & k_1\theta'_{1n} & -k_2\phi'_{2n} & -k_2\theta'_{2n} & 0 & 0 \\ 0 & 0 & \phi_{2n} & \theta_{2n} & -\phi_{3n} & -\theta_{3n} \\ 0 & 0 & k_2\phi'_{2n} & k_2\theta'_{2n} & -k_3\phi'_{3n} & -k_3\theta'_{3n} \\ 0 & 0 & 0 & 0 & X_3 & Y_3 \end{bmatrix} \begin{bmatrix} 1 \\ B_{1n} \\ A_{2n} \\ B_{2n} \\ A_{3n} \\ B_{3n} \end{bmatrix} = \begin{bmatrix} 0 \\ 0 \\ 0 \\ 0 \\ 0 \\ 0 \end{bmatrix} \tag{10-56}$$

where

$$X_1 = -k_1^*\phi'_{1n} + h_1^*\phi_{1n} \tag{10-57a}$$

$$Y_1 = -k_1^*\theta'_{1n} + h_1^*\theta_{1n} \tag{10-57b}$$

$$X_3 = k_3^*\phi'_{3n} + h_4^*\phi_{3n} \tag{10-57c}$$

$$Y_3 = k_3^*\theta'_{3n} + h_4^*\theta_{3n} \tag{10-57d}$$

and noting that the primes denote differentiation with respect to x. Only five of these equations can be used to determine the coefficients. We choose the first five of them; the resulting system of equations for the determination of these five coefficients is given in the matrix form as

$$\begin{bmatrix} Y_1 & 0 & 0 & 0 & 0 \\ \theta_{1n} & -\phi_{2n} & -\theta_{2n} & 0 & 0 \\ k_1\theta'_{1n} & -k_2\phi'_{2n} & -k_2\theta'_{2n} & 0 & 0 \\ 0 & \phi_{2n} & \theta_{2n} & -\phi_{3n} & -\theta_{3n} \\ 0 & k_2\phi'_{2n} & k_2\theta'_{2n} & -k_3\phi'_{3n} & -k_3\theta'_{3n} \end{bmatrix} \begin{bmatrix} B_{1n} \\ A_{2n} \\ B_{2n} \\ A_{3n} \\ B_{3n} \end{bmatrix} = \begin{bmatrix} -X_1 \\ -\phi_{1n} \\ -k_1\phi'_{1n} \\ 0 \\ 0 \end{bmatrix} \tag{10-58}$$

Thus, the solution of equations (10-58) gives the five coefficients B_{1n}, A_{2n}, B_{2n}, A_{3n}, and B_{3n}. The transcendental equation for the determination of the eigenvalues $\beta_1 < \beta_2 < \cdots < \beta_n < \cdots$ is obtained from the requirement that the determinant of the coefficients in the system of equations (10-56) should vanish. This condition leads to the following transcendental equation for the determination of the eigenvalues, $\beta_1 < \beta_2 < \beta_3 < \cdots \beta_n < \cdots$, namely,

$$\begin{vmatrix} X_1 & Y_1 & 0 & 0 & 0 & 0 \\ \phi_{1n} & \theta_{1n} & -\phi_{2n} & -\theta_{2n} & 0 & 0 \\ k_1\phi'_{1n} & k_1\theta'_{1n} & -k_2\phi'_{2n} & -k_2\theta'_{2n} & 0 & 0 \\ 0 & 0 & \phi_{2n} & \theta_{2n} & -\phi_{3n} & -\theta_{3n} \\ 0 & 0 & k_2\phi'_{2n} & k_2\theta'_{2n} & -k_3\phi'_{3n} & -k_3\theta'_{3n} \\ 0 & 0 & 0 & 0 & X_3 & Y_3 \end{vmatrix} = 0 \tag{10-59}$$

10-5 APPLICATIONS OF ORTHOGONAL EXPANSION TECHNIQUE

In this section, we illustrate the application of the orthogonal expansion technique described previously for the solution of transient homogeneous heat conduction problems of a two-layer cylinder and a two-layer slab.

Example 10-3 Two-Layer Cylinder
A two-layer solid cylinder as illustrated in Figure 10-4 contains an inner region $0 \le r \le a$ and an outer region $a \le r \le b$ that are in perfect thermal contact; k_1 and k_2 are the thermal conductivities, and α_1 and α_2 are the thermal diffusivities of the inner and outer regions, respectively. Initially, the inner region is at temperature $\theta_1(r, t) = F_1(r)$ and the outer region at temperature $\theta_2(r, t) = F_2(r)$. For times $t > 0$, heat is dissipated by convection from the outer surface at $r = b$ into an environment at zero temperature. Develop an expression for the temperature distribution in the cylinders for times $t > 0$.

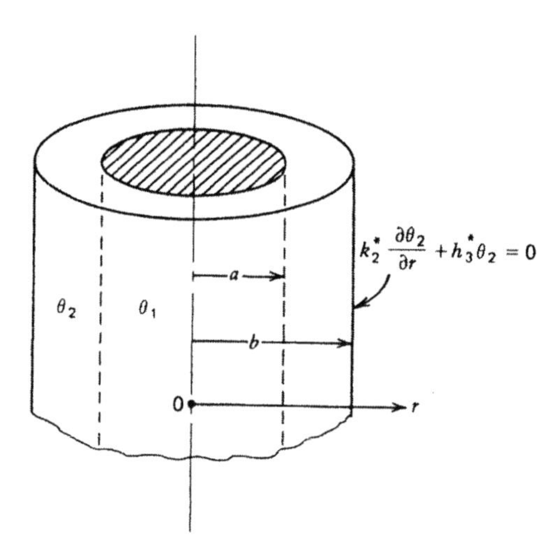

Figure 10-4 Two-layer cylinder with perfect thermal contact at the interface.

The mathematical formulation of the problem is given by

$$\frac{\alpha_1}{r}\frac{\partial}{\partial r}\left(r\frac{\partial \theta_1}{\partial r}\right) = \frac{\partial \theta_1(r,t)}{\partial t} \quad \text{in} \quad 0 \le r < a, \quad t > 0 \tag{10-60a}$$

$$\frac{\alpha_2}{r}\frac{\partial}{\partial r}\left(r\frac{\partial \theta_2}{\partial r}\right) = \frac{\partial \theta_2(r,t)}{\partial t} \quad \text{in} \quad a < r < b, \quad t > 0 \tag{10-60b}$$

The cylinder is subjected to the boundary conditions

$$\theta_1 (r \to 0, t) \Rightarrow \text{finite} \tag{10-61a}$$

$$\theta_1 (r = a, t) = \theta_2 (r = a, t) \tag{10-61b}$$

$$k_1 \left.\frac{\partial \theta_1}{\partial r}\right|_{r=a} = k_2 \left.\frac{\partial \theta_2}{\partial r}\right|_{r=a} \tag{10-61c}$$

$$k_2^* \left.\frac{\partial \theta_2}{\partial r}\right|_{r=b} + h_3^* \left.\theta_2\right|_{r=b} = 0 \tag{10-61d}$$

and to the initial conditions

$$\theta_1 (r, t = 0) = F_1 (r) \qquad \text{for} \qquad 0 < r < a \tag{10-62a}$$

$$\theta_2 (r, t = 0) = F_2 (t) \qquad \text{for} \qquad a < r < b \tag{10-62b}$$

The corresponding eigenvalue problem is taken as

$$\frac{1}{r}\frac{d}{dr}\left(r\frac{d\psi_{1n}}{dr}\right) + \frac{\beta_n^2}{\alpha_1}\psi_{1n} (r) = 0 \qquad \text{in} \qquad 0 \le r < a \tag{10-63a}$$

$$\frac{1}{r}\frac{d}{dr}\left(r\frac{d\psi_{2n}}{dr}\right) + \frac{\beta_n^2}{\alpha_2}\psi_{2n} (r) = 0 \qquad \text{in} \qquad a < r < b \tag{10-63b}$$

subject to the boundary conditions

$$\psi_{1n} (r \to 0) \Rightarrow \text{finite} \tag{10-64a}$$

$$\psi_{1n} (r = a) = \psi_{2n} (r = a) \tag{10-64b}$$

$$k_1 \left.\frac{d\psi_{1n}}{dr}\right|_{r=a} = k_2 \left.\frac{d\psi_{2n}}{dr}\right|_{r=a} \tag{10-64c}$$

$$k_2^* \left.\frac{d\psi_{2n}}{dr}\right|_{r=b} + h_3^* \left.\psi_{2n}\right|_{r=b} = 0 \tag{10-64d}$$

The general solution of the above eigenvalue problem (10-63) and (10-64), according to Table 10-1, is taken as

$$\psi_{in} (r) = A_{in} J_0 \left(\frac{\beta_n}{\sqrt{\alpha_i}} r\right) + B_{in} Y_0 \left(\frac{\beta_n}{\sqrt{\alpha_i}} r\right), \qquad i = 1, 2 \tag{10-65}$$

The boundary condition (10-64a) requires that $B_{1n} = 0$. Then the solutions $\psi_{in}(r)$ for the two regions become

$$\psi_{1n} (r) = J_0 \left(\frac{\beta_n}{\sqrt{\alpha_1}} r\right) \qquad \text{in} \qquad 0 \le r \le a \tag{10-65a}$$

$$\psi_{2n} (r) = A_{2n} J_0 \left(\frac{\beta_n}{\sqrt{\alpha_2}} r\right) + B_{2n} Y_0 \left(\frac{\beta_n}{\sqrt{\alpha_2}} r\right) \qquad \text{in} \qquad a < r < b \tag{10-65b}$$

where we have chosen $A_{1n} = 1$ for the reason stated previously. The requirement that the solutions (10-65) should satisfy the remaining three boundary conditions (10-64b–d) leads, respectively, to the following equations for the determination of these coefficients:

$$J_0\left(\frac{\beta_n a}{\sqrt{\alpha_1}}\right) = A_{2n} J_0\left(\frac{\beta_n a}{\sqrt{\alpha_2}}\right) + B_{2n} Y_0\left(\frac{\beta_n a}{\sqrt{\alpha_2}}\right) \tag{10-66a}$$

$$\frac{k_1}{k_2}\sqrt{\frac{\alpha_2}{\alpha_1}} J_1\left(\frac{\beta_n a}{\sqrt{\alpha_1}}\right) = A_{2n} J_1\left(\frac{\beta_n a}{\sqrt{\alpha_1}}\right) + B_{2n} Y_1\left(\frac{\beta_n a}{\sqrt{\alpha_2}}\right) \tag{10-66b}$$

$$-\left[A_{2n} J_1\left(\frac{\beta_n b}{\sqrt{\alpha_2}}\right) + B_{2n} Y_1\left(\frac{\beta_n b}{\sqrt{\alpha_2}}\right)\right] + \frac{h_3^*\sqrt{\alpha_2}}{k_2^*\beta_n}\left[A_{2n} J_0\left(\frac{\beta_n b}{\sqrt{\alpha_2}}\right) + B_{2n} Y_0\left(\frac{\beta_n b}{\sqrt{\alpha_2}}\right)\right] = 0 \tag{10-66c}$$

These equations are now written in the matrix form as

$$\begin{bmatrix} J_0(\gamma) & -J_0\left(\frac{a}{b}\eta\right) & -Y_0\left(\frac{a}{b}\eta\right) \\ K J_1(\gamma) & -J_1\left(\frac{a}{b}\eta\right) & -Y_1\left(\frac{a}{b}\eta\right) \\ 0 & \frac{H}{\eta}J_0(\eta) - J_1(\eta) & \frac{H}{\eta}Y_0(\eta) - Y_1(\eta) \end{bmatrix} \begin{bmatrix} 1 \\ A_{2n} \\ B_{2n} \end{bmatrix} = \begin{bmatrix} 0 \\ 0 \\ 0 \end{bmatrix} \tag{10-67}$$

where we defined

$$\gamma \equiv \frac{a\beta_n}{\sqrt{\alpha_1}}, \qquad \eta \equiv \frac{b\beta_n}{\sqrt{\alpha_2}}, \qquad H \equiv \frac{bh_3^*}{k_2^*}, \qquad K \equiv \frac{k_1}{k_2}\sqrt{\frac{\alpha_2}{\alpha_1}} \tag{10-68}$$

Any two of these equations can be used to determine the coefficients A_{2n} and B_{2n}. We choose the first two, and write the resulting equations as

$$\begin{bmatrix} J_0\left(\frac{a}{b}\eta\right) & Y_0\left(\frac{a}{b}\eta\right) \\ J_1\left(\frac{a}{b}\eta\right) & Y_1\left(\frac{a}{b}\eta\right) \end{bmatrix} \begin{bmatrix} A_{2n} \\ B_{2n} \end{bmatrix} = \begin{bmatrix} J_0(\gamma) \\ K J_1(\gamma) \end{bmatrix} \tag{10-69}$$

Then, A_{2n} and B_{2n} are obtained as

$$A_{2n} = \frac{1}{\Delta}\left[J_0(\gamma) Y_1\left(\frac{a}{b}\eta\right) - K J_1(\gamma) Y_0\left(\frac{a}{b}\eta\right)\right] \tag{10-70a}$$

$$B_{2n} = \frac{1}{\Delta}\left[K J_1(\gamma) J_0\left(\frac{a}{b}\eta\right) - J_0(\gamma) J_1\left(\frac{a}{b}\eta\right)\right] \tag{10-70b}$$

where

$$\Delta = J_0\left(\frac{a}{b}\eta\right) Y_1\left(\frac{a}{b}\eta\right) - J_1\left(\frac{a}{b}\eta\right) Y_0\left(\frac{a}{b}\eta\right) \tag{10-70c}$$

Finally, the equation for the determination of the eigenvalues is obtained from the requirement that in equation (10-67) the determinant of the coefficients should vanish. Then, the β_n values are the roots of the following transcendental equation:

$$\begin{vmatrix} J_0(\gamma) & -J_0\left(\frac{a}{b}\eta\right) & -Y_0\left(\frac{a}{b}\eta\right) \\ K J_1(\gamma) & -J_1\left(\frac{a}{b}\eta\right) & -Y_1\left(\frac{a}{b}\eta\right) \\ 0 & \frac{H}{\eta}J_0(\eta) - J_1(\eta) & \frac{H}{\eta}Y_0(\eta) - Y_1(\eta) \end{vmatrix} = 0 \tag{10-71}$$

Having established the relations for the determination of the coefficients A_{2n}, B_{2n} and the eigenvalues β_n, the eigenfunctions $\psi_{1n}(r)$ and $\psi_{2n}(r)$ are obtained according to equations (10-65). Then, the solution for the temperature $\theta_i(r, t)$, $i = 1, 2$ in any of the regions is given by equations (10-42) as

$$\theta_i(r, t) = \sum_{n=1}^{\infty} \frac{1}{N_n} e^{-\beta_n^2 t} \psi_{in}(r) \left[\frac{k_1}{\alpha_1} \int_{r'=0}^{a} r' \psi_{1n}(r') F_1(r') dr' + \frac{k_2}{\alpha_2} \int_{r'=a}^{b} r' \psi_{2n}(r') F_2(r') dr' \right], \qquad i = 1, 2 \tag{10-72}$$

where

$$N_n = \frac{k_1}{\alpha_1} \int_{r'=0}^{a} r' \psi_{1n}^2(r') dr' + \frac{k_2}{\alpha_2} \int_{r'=a}^{b} r' \psi_{2n}^2(r') dr' \tag{10-73a}$$

$$\psi_{1n}(r) = J_0\left(\frac{\beta_n}{\sqrt{\alpha_1}} r\right) \tag{10-73b}$$

$$\psi_{2n}(r) = A_{2n} J_0\left(\frac{\beta_n}{\sqrt{\alpha_2}} r\right) + B_{2n} Y_0\left(\frac{\beta_n}{\sqrt{\alpha_2}} r\right) \tag{10-73c}$$

This result is now written more compactly in terms of the Green's function as

$$\theta_i(r, t) = \int_{r'=0}^{a} G_{i1}(r, t|r', \tau)\big|_{\tau=0} F_1(r') r' dr' + \int_{r'=a}^{b} G_{i2}(r, t|r', \tau)\big|_{\tau=0} F_2(r') r' dr' \quad i = 1, 2 \tag{10-74}$$

where $G_{ij}(r, t|r', \tau)|_{\tau=0}$ is defined as

$$G_{ij}(r, t|r', \tau)\big|_{\tau=0} = \sum_{n=1}^{\infty} e^{-\beta_n^2 t} \frac{1}{N_n} \frac{k_j}{\alpha_j} \psi_{in}(r) \psi_{jn}(r') \tag{10-75}$$

Example 10-4 Two-Layer Slab
A two-layer slab consists of the first layer in $0 \leq x \leq a$ and the second layer in $a \leq x \leq b$, which are in perfect thermal contact as illustrated in Figure 10-5. Let k_1 and k_2 be the thermal conductivities, and α_1 and α_2 the thermal diffusivities for the first and second layers, respectively. Initially, the first region is at temperature $F_1(x)$ and the second region at $F_2(x)$. For times $t > 0$, the boundary surface at $x = 0$ is kept at zero temperature, and the boundary surface at $x = b$ dissipates heat by convection into a medium at zero temperature. Obtain an expression for the temperature distribution in the slab for times $t > 0$.

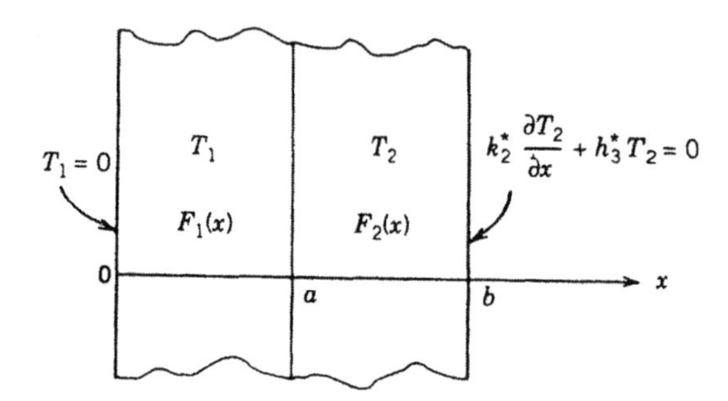

Figure 10-5 Two-layer slab in perfect thermal contact at interface.

The mathematical formulation of this problem is given as

$$\alpha_1 \frac{\partial^2 T_1}{\partial x^2} = \frac{\partial T_1(x,t)}{\partial t} \quad \text{in} \quad 0 < x < a, \quad t > 0 \tag{10-76a}$$

$$\alpha_2 \frac{\partial^2 T_2}{\partial x^2} = \frac{\partial T_2(x,t)}{\partial t} \quad \text{in} \quad a < x < b, \quad t > 0 \tag{10-76b}$$

subject to the boundary conditions

$$T_1(x = 0, t) = 0 \tag{10-77a}$$

$$T_1(x = a, t) = T_2(x = a, t) \tag{10-77b}$$

$$k_1 \left.\frac{\partial T_1}{\partial x}\right|_{x=a} = k_2 \left.\frac{\partial T_2}{\partial x}\right|_{x=a} \tag{10-77c}$$

$$k_2^* \left.\frac{\partial T_2}{\partial x}\right|_{x=b} + h_3^* \left. T_2\right|_{x=b} = 0 \tag{10-77d}$$

and the initial conditions

$$T_1(x, t = 0) = F_1(x) \quad \text{for} \quad 0 < x < a \tag{10-78a}$$

$$T_2(x, t = 0) = F_2(x) \quad \text{for} \quad a < x < b \tag{10-78b}$$

The corresponding eigenvalue problem is taken as

$$\frac{d^2\psi_{1n}}{dx^2} + \frac{\beta_n^2}{\alpha_1}\psi_{1n}(x) = 0 \qquad \text{in} \qquad 0 < x < a \tag{10-79a}$$

$$\frac{d^2\psi_{2n}}{dx^2} + \frac{\beta_n^2}{\alpha_2}\psi_{2n}(x) = 0 \qquad \text{in} \qquad a < x < b \tag{10-79b}$$

subject to the boundary conditions

$$\psi_{1n}(x = 0) = 0 \tag{10-80a}$$

$$\psi_{1n}(x = a) = \psi_{2n}(x = a) \tag{10-80b}$$

$$k_1 \left.\frac{d\psi_{1n}}{dx}\right|_{x=a} = k_2 \left.\frac{d\psi_{2n}}{dx}\right|_{x=a} \tag{10-80c}$$

$$k_2^* \left.\frac{d\psi_{2n}}{dx}\right|_{x=b} + h_3^* \left.\psi_{2n}\right|_{x=b} = 0 \tag{10-80d}$$

The general solution of the above eigenvalue problem, according to Table 10-1, is taken as

$$\psi_{in}(x) = A_{in}\sin\left(\frac{\beta_n}{\sqrt{\alpha_i}}x\right) + B_{in}\cos\left(\frac{\beta_n}{\sqrt{\alpha_i}}x\right), \qquad i = 1, 2 \tag{10-81}$$

The boundary condition (10-80a) requires that $B_{1n} = 0$. Then, the solutions ψ_{in} for the two regions are reduced to

$$\psi_{1n}(x) = \sin\left(\frac{\beta_n}{\sqrt{\alpha_1}}x\right) \qquad \text{in} \qquad 0 < x < a \tag{10-82a}$$

$$\psi_{2n}(x) = A_{2n}\sin\left(\frac{\beta_n}{\sqrt{\alpha_2}}x\right) + B_{2n}\cos\left(\frac{\beta_n}{\sqrt{\alpha_2}}x\right) \qquad \text{in} \qquad a < x < b \tag{10-82b}$$

where we have chosen $A_{1n} = 1$ for the reason stated previously. The requirement that the solutions (10-82) should satisfy the remaining boundary conditions (10-80b–d) yields the following equations for the determination of these coefficients:

$$\sin\left(\frac{\beta_n a}{\sqrt{\alpha_1}}\right) = A_{2n}\sin\left(\frac{\beta_n a}{\sqrt{\alpha_2}}\right) + B_{2n}\cos\left(\frac{\beta_n a}{\sqrt{\alpha_2}}\right) \tag{10-83a}$$

$$\frac{k_1}{k_2}\sqrt{\frac{\alpha_2}{\alpha_1}}\cos\left(\frac{\beta_n a}{\sqrt{\alpha_1}}\right) = A_{2n}\cos\left(\frac{\beta_n a}{\sqrt{\alpha_2}}\right) - B_{2n}\sin\left(\frac{\beta_n a}{\sqrt{\alpha_2}}\right) \tag{10-83b}$$

$$\left[A_{2n} \cos\left(\frac{\beta_n b}{\sqrt{\alpha_2}}\right) - B_{2n} \sin\left(\frac{\beta_n b}{\sqrt{\alpha_2}}\right)\right] + \frac{h_3^* \sqrt{\alpha_2}}{k_2^* \beta_n} \left[A_{2n} \sin\left(\frac{\beta_n b}{\sqrt{\alpha_2}}\right) + B_{2n} \cos\left(\frac{\beta_n b}{\sqrt{\alpha_2}}\right)\right] = 0 \tag{10-83c}$$

These equations are now written in the matrix form as

$$\begin{bmatrix} \sin\gamma & -\sin\left(\frac{a}{b}\eta\right) & -\cos\left(\frac{a}{b}\eta\right) \\ K\cos\gamma & -\cos\left(\frac{a}{b}\eta\right) & \sin\left(\frac{a}{b}\eta\right) \\ 0 & \frac{H}{\eta}\sin\eta + \cos\eta & \frac{H}{\eta}\cos\eta - \sin\eta \end{bmatrix} \begin{bmatrix} 1 \\ A_{2n} \\ B_{2n} \end{bmatrix} = \begin{bmatrix} 0 \\ 0 \\ 0 \end{bmatrix} \tag{10-84}$$

where we defined

$$\gamma \equiv \frac{a\beta_n}{\sqrt{\alpha_1}}, \qquad \eta \equiv \frac{b\beta_n}{\sqrt{\alpha_2}}, \qquad H \equiv \frac{bh_3^*}{k_2^*}, \qquad K \equiv \frac{k_1}{k_2}\sqrt{\frac{\alpha_2}{\alpha_1}} \tag{10-85}$$

We choose the first two of these equations to determine the coefficients A_{2n} and B_{2n}; these two equations are written as

$$\begin{bmatrix} \sin\left(\frac{a}{b}\eta\right) & \cos\left(\frac{a}{b}\eta\right) \\ \cos\left(\frac{a}{b}\eta\right) & -\sin\left(\frac{a}{b}\eta\right) \end{bmatrix} \begin{bmatrix} A_{2n} \\ B_{2n} \end{bmatrix} = \begin{bmatrix} \sin\gamma \\ K\cos\gamma \end{bmatrix} \tag{10-86}$$

Then, A_{2n} and B_{2n} are determined as

$$A_{2n} = \frac{1}{\Delta}\left[-\sin\gamma \sin\left(\frac{a}{b}\eta\right) - K\cos\gamma\cos\left(\frac{a}{b}\eta\right)\right] \tag{10-87a}$$

and

$$B_{2n} = \frac{1}{\Delta}\left[K\cos\gamma\sin\left(\frac{a}{b}\eta\right) - \sin\gamma\cos\left(\frac{a}{b}\eta\right)\right] \tag{10-87b}$$

where

$$\Delta = -\sin^2\left(\frac{a}{b}\eta\right) - \cos^2\left(\frac{a}{b}\eta\right) = -1 \tag{10-87c}$$

Finally, the equation for the determination of the eigenvalues β_n is obtained from the requirement that in matrix equation (10-84) the determinant of the coefficients should vanish. This condition yields the following transcendental equation for the determination of the eigenvalues β_n:

$$\begin{vmatrix} \sin\gamma & -\sin\left(\frac{a}{b}\eta\right) & -\cos\left(\frac{a}{b}\eta\right) \\ K\cos\gamma & -\cos\left(\frac{a}{b}\eta\right) & \sin\left(\frac{a}{b}\eta\right) \\ 0 & \frac{H}{\eta}\sin\eta+\cos\eta & \frac{H}{\eta}\cos\eta-\sin\eta \end{vmatrix} = 0 \tag{10-88}$$

The formal solution of this problem is now complete. That is, the coefficients A_{2n} and B_{2n} are given by equations (10-87), and the eigenvalues β_n by equation (10-88). Then, the eigenfunctions $\psi_{1n}(x)$ and $\psi_{2n}(x)$ defined by equations (10-82) are known, and the temperature distribution $T_i(x, t)$, $i = 1, 2$ in any one of the two regions is determined according to equation (10-42) as

$$T_i(x,t) = \sum_{n=1}^{\infty} \frac{1}{N_n} e^{-\beta_n^2 t} \psi_{in}(x) \left[\frac{k_1}{\alpha_1} \int_{x'=0}^{a} \psi_{1n}(x') F_1(x') dx' + \frac{k_2}{\alpha_2} \int_{x'=a}^{b} \psi_{2n}(x') F_2(x') dx' \right], \qquad i = 1, 2 \tag{10-89a}$$

where

$$N_n = \frac{k_1}{\alpha_1} \int_{x'=0}^{a} \psi_{1n}^2(x') dx' + \frac{k_2}{\alpha_2} \int_{x'=a}^{b} \psi_{2n}^2(x') dx \tag{10-89b}$$

$$\psi_{1n}(x) = \sin\left(\frac{\beta_n}{\sqrt{\alpha_1}} x\right) \tag{10-89c}$$

$$\psi_{2n}(x) = A_{2n} \sin\left(\frac{\beta_n}{\sqrt{\alpha_2}} x\right) + B_{2n} \cos\left(\frac{\beta_n}{\sqrt{\alpha_2}} x\right) \tag{10-89d}$$

Finally, this result can be written more compactly in terms of the Green's function as

$$T_i(x,t) = \int_{x'=0}^{a} G_{i1}(x,t|x',\tau)\big|_{\tau=0} F_1(x') dx' + \int_{x'=a}^{b} G_{i2}(x,t|x',\tau)\big|_{\tau=0} F_2(x') dx', \qquad i = 1, 2 \tag{10-90a}$$

where $G_{ij}(x, t|x', \tau)|_{\tau=0}$, the Green's function evaluated at $\tau = 0$ is given by

$$G_{ij}(x,t|x',\tau)\big|_{\tau=0} = \sum_{n=1}^{\infty} e^{-\beta_n^2 t} \frac{1}{N_n} \frac{k_j}{\alpha_j} \psi_{in}(x) \psi_{jn}(x') \tag{10-90b}$$

and where N_n, $\psi_{in}(x)$, $i = 1, 2$ are defined by equations (10-89).

10-6 GREEN'S FUNCTION APPROACH FOR SOLVING NONHOMOGENEOUS PROBLEMS

The use of Green's function is a very convenient approach for solving nonhomogeneous problems of transient heat conduction in a composite medium, if the general expression relating the solution for the temperature $T_i(x,t)$ to the Green's function is known and the appropriate Green's function is available. The general procedure is similar to that described for the case of a single-region medium, except the functional forms of the general solution and the Green's function are different.

In the following analysis, we assume that the nonhomogeneity associated with the boundary conditions is removed by a superposition procedure described previously; hence the energy generation term is the only nonhomogeneity in the problem.

We consider the following transient heat conduction problem for an M-layer composite medium with energy generation, homogeneous boundary conditions at the outer surfaces, and contact conductance at the interfaces, namely,

$$\alpha_i \frac{1}{x^p}\frac{\partial}{\partial x}\left(x^p \frac{\partial T_i}{\partial x}\right) + \frac{\alpha_i}{k_i} g_i(x,t) = \frac{\partial T_i(x,t)}{\partial t} \quad \text{in} \quad \begin{matrix} x_i < x < x_{i+1} & t>0 \\ i=1,2,\ldots,M & \end{matrix} \tag{10-91a}$$

where

$$p = \begin{cases} 0 & \text{slab} \\ 1 & \text{cylinder} \\ 2 & \text{sphere} \end{cases} \tag{10-91b}$$

and subject to the boundary conditions

$$-k_1^* \left.\frac{\partial T_1}{\partial x}\right|_{x=x_1} + h_1^* \left. T_1\right|_{x=x_1} = 0 \tag{10-92a}$$

$$\left.\begin{aligned} -k_i \left.\frac{\partial T_i}{\partial x}\right|_{x=x_{i+1}} &= h_{i+1}\left(\left.T_i\right|_{x=x_{i+1}} - \left.T_{i+1}\right|_{x=x_{i+1}}\right) \\ k_i \left.\frac{\partial T_i}{\partial x}\right|_{x=x_{i+1}} &= k_{i+1} \left.\frac{\partial T_{i+1}}{\partial x}\right|_{x=x_{i+1}} \end{aligned}\right\} \quad \begin{matrix}\text{at the interfaces} \\ i=1,2,\ldots,M-1\end{matrix} \tag{10-92b,c}$$

$$k_M^* \left.\frac{\partial T_M}{\partial x}\right|_{x=x_{M+1}} + h_{M+1}^* \left.T_M\right|_{x=x_{M+1}} = 0 \tag{10-92d}$$

and to the initial conditions

$$T_i(x,t=0) = F_i(x) \quad \text{in} \quad x_i < x < x_{i+1}, \quad i=1,2,\ldots,M \tag{10-93}$$

The appropriate eigenvalue problem for the solution of the above heat conduction problem is taken as

$$\frac{1}{x^p}\frac{d}{dx}\left(x^p \frac{d\psi_{in}}{dx}\right) + \frac{\beta_n^2}{\alpha_i}\psi_{in}(x) = 0 \quad \text{in} \quad \begin{matrix} x_i < x < x_{i+1} \\ i=1,2,\ldots,M \end{matrix} \tag{10-94}$$

subject to the boundary conditions

$$- k_1^* \left. \frac{d\psi_{1n}}{dx} \right|_{x=x_1} + h_1^* \left. \psi_{1n} \right|_{x=x_1} = 0 \tag{10-95a}$$

$$\left.\begin{aligned} -k_i \left. \frac{d\psi_{in}}{dx} \right|_{x=x_{i+1}} &= h_{i+1} \left(\left. \psi_{in} \right|_{x=x_{i+1}} - \left. \psi_{i+1,n} \right|_{x=x_{i+1}} \right) \\ k_i \left. \frac{d\psi_{in}}{dx} \right|_{x=x_{i+1}} &= k_{i+1} \left. \frac{d\psi_{i+1,n}}{dx} \right|_{x=x_{i+1}} \end{aligned}\right\} \begin{aligned} &\text{at the interfaces} \\ &i = 1, 2, \ldots, M-1 \end{aligned} \tag{10-95b,c}$$

$$k_M^* \left. \frac{d\psi_{Mn}}{dx} \right|_{x=x_{M+1}} + h_{M+1}^* \left. \psi_{Mn} \right|_{x=x_{M+1}} = 0 \tag{10-95d}$$

The solution of this multilayer transient heat conduction problem in terms of the composite medium Green's function $G_{ij}(x, t|x', \tau)$ can be obtained by proper rearrangement of the general solution given by Yener and Özisik [25] and Özisik [8]. Following this approach, we write the resulting expression in the form

$$\begin{aligned} T_i(x,t) = \sum_{j=1}^{M} \Bigg\{ & \int_{x'=x_j}^{x_{j+1}} \left. G_{ij}\left(x, t| x', \tau\right) \right|_{\tau=0} F_j\left(x'\right) x'^p dx' \\ & + \int_{\tau=0}^{t} \int_{x'=x_j}^{x_{j+1}} G_{ij}\left(x, t| x', \tau\right) \left[\frac{\alpha_j}{k_j} g_j\left(x', \tau\right) \right] x'^p dx' d\tau \Bigg\} \end{aligned} \tag{10-96}$$

in $x_i < x < x_{i+1}$, $i = 1, 2, \ldots, M$, where the composite medium Green's function $G_{ij}\left(x, t| x', \tau\right)$ is defined as

$$G_{ij}\left(x, t| x', \tau\right) = \sum_{n=1}^{\infty} e^{-\beta_n^2 (t-\tau)} \frac{1}{N_n} \frac{k_j}{\alpha_j} \psi_{in}(x) \psi_{jn}\left(x'\right) \tag{10-97a}$$

with

$$p = \begin{cases} 0 & \text{slab} \\ 1 & \text{cylinder} \\ 2 & \text{sphere} \end{cases} \tag{10-97b}$$

The norm N_n is given by

$$N_n = \sum_{j=1}^{M} \frac{k_j}{\alpha_j} \int_{x'=x_j}^{x_{j+1}} x'^p \psi_{jn}^2\left(x'\right) dx' \tag{10-98}$$

where $\psi_{in}(x)$ and $\psi_{jn}(x)$ are the eigenfunctions, the β_n values are the eigenvalues of the eigenvalue problem (10-94) and (10-95), $\left. G_{ij}\left(x, t| x', \tau\right) \right|_{\tau=0}$ is the composite medium Green's function evaluated at $\tau = 0$, and $G_{ij}\left(x, t|\, x', \tau\right)$ is the composite medium Green's function.

The function $G_{ij}(x, t|x', \tau)|_{\tau=0}$ is obtainable by rearranging the solution given by equations (10-42), of the homogeneous problem defined by equations (10-25)–(10-27), in the form

$$T_i\,(x,t) = \sum_{j=1}^{M} \int_{x'=x_j}^{x_{j+1}} \left[\sum_{n=1}^{\infty} e^{-\beta_n^2 t} \frac{1}{N_n} \frac{k_j}{\alpha_j} \psi_{in}\,(x)\, \psi_{jn}\left(x'\right) \right] F_j\left(x'\right) x'^{p} dx' \tag{10-99}$$

where x'^{p} is the Sturm–Liouville weight function with $p = 0$, 1, and 2 for slab, cylinder, and sphere, respectively. Then, the function inside the bracket in equation (10-99) is $G_{ij}\left(x, t|\, x', \tau\right)\big|_{\tau=0}$, that is,

$$G_{ij}\left(x, t|\, x', \tau\right)\Big|_{\tau=0} = \sum_{n=1}^{\infty} e^{-\beta_n^2 t} \frac{1}{N_n} \frac{k_j}{\alpha_j} \psi_{in}\,(x)\, \psi_{jn}\left(x'\right) \tag{10-100}$$

and the Green's function is obtained by replacing t by $t - \tau$ in this expression, as detailed in Chapter 8:

$$G_{ij}\left(x, t|\, x', \tau\right) = \sum_{n=1}^{\infty} e^{-\beta_n^2 (t-\tau)} \frac{1}{N_n} \frac{k_j}{\alpha_j} \psi_{in}\,(x)\, \psi_{jn}\left(x'\right) \tag{10-101}$$

We now illustrate the use of Green's function approach for developing solutions for the nonhomogeneous transient heat conduction problems of composite medium with specific examples. In order to alleviate the details of developing solutions for the corresponding homogeneous problems, we have chosen the examples from those considered in the previous sections for which solutions are already available for the homogeneous part.

Example 10-5 Two-Layer Cylinder with Generation

A two-layer solid cylinder contains an inner region $0 \leq r \leq a$ and an outer region $a \leq r \leq b$ that are in perfect thermal contact. Initially, the inner and outer regions are at temperatures $F_1(r)$ and $F_2(r)$, respectively. For times $t > 0$, heat is generated in the inner and outer regions at rates $g_1(r, t)$ and $g_2(r, t)$ (W/m^3), respectively, while heat is dissipated by convection from the outer boundary surface at $r = b$ into a medium at zero temperature. Obtain an expression for the temperature distribution in the cylinder for times $t > 0$.

The mathematical formulation of this problem is given as

$$\alpha_1 \frac{1}{r} \frac{\partial}{\partial r}\left(r \frac{\partial T_1}{\partial r} \right) + \frac{\alpha_1}{k_1} g_1\,(r,t) = \frac{\partial T_1\,(r,t)}{\partial t} \qquad \text{in} \qquad 0 < r < a, \qquad t > 0 \tag{10-102}$$

$$\alpha_2 \frac{1}{r} \frac{\partial}{\partial r}\left(r \frac{\partial T_2}{\partial r} \right) + \frac{\alpha_2}{k_2} g_2\,(r,t) = \frac{\partial T_2\,(r,t)}{\partial t} \qquad \text{in} \qquad a < r < b, \qquad t > 0 \tag{10-103}$$

subject to the boundary conditions

$$T_1\,(r \to 0) \Rightarrow \text{finite} \tag{10-104a}$$

$$T_1\,(r = a, t) = T_2\,(r = a, t) \tag{10-104b}$$

$$k_1 \left.\frac{\partial T_1}{\partial r}\right|_{r=a} = k_2 \left.\frac{\partial T_2}{\partial r}\right|_{r=a} \tag{10-104c}$$

$$k_2^* \left.\frac{\partial T_2}{\partial r}\right|_{r=b} + h_3^*\, T_2\big|_{r=b} = 0 \tag{10-104d}$$

and to the initial conditions

$$T_1\,(r, t = 0) = F_1\,(r) \qquad \text{in} \qquad 0 \le r < a \tag{10-105a}$$

$$T_2\,(r, t = 0) = F_2\,(r) \qquad \text{in} \qquad a < r < b \tag{10-105b}$$

The solution of this problem is written in terms of the Green's function, according to the general solution given by equations (10-96) and (10-97), in the form

$$\begin{aligned} T_i\,(r, t) = & \int_{r'=0}^{a} \left[G_{i1}\left(r, t|\, r', \tau\right)\right]_{\tau=0} F_1\left(r'\right) r' dr' \\ & + \int_{r'=a}^{b} \left[G_{i2}\left(r, t|\, r', \tau\right)\right]_{\tau=0} F_2\left(r'\right) r' dr' \\ & + \int_{\tau=0}^{t} \left[\int_{r'=0}^{a} G_{i1}\left(r, t|\, r', \tau\right) g_1\left(r', \tau\right) r' dr' \right. \\ & \left. + \int_{r'=a}^{b} G_{i2}\left(r, t|\, r', \tau\right) g_2\left(r', \tau\right) r' dr' \right] d\tau \qquad i = 1, 2 \end{aligned} \tag{10-106}$$

where the Green's function $G_{ij}\left(r, t|\, r', \tau\right)$ is obtainable from the solution of the homogeneous version of the problem. The homogeneous version of this problem is already considered in Example 10-3; hence the $G_{ij}\left(r, t|\, r', \tau\right)\big|_{\tau=0}$ is obtainable from equation (10-75), and $G_{ij}\left(r, t|\, r', \tau\right)$ is obtained by replacing t with $t - \tau$ in the expression $G_{ij}\left(r, t|\, r', \tau\right)\big|_{\tau=0}$. Thus Green's functions become

$$G_{ij}\left(r, t|\, r', \tau\right)\big|_{\tau=0} = \sum_{n=1}^{\infty} e^{-\beta_n^2 t} \frac{1}{N_n} \frac{k_j}{\alpha_j} \psi_{in}\,(r)\, \psi_{jn}\left(r'\right) \tag{10-107a}$$

$$G_{ij}\left(r, t|\, r', \tau\right) = \sum_{n=1}^{\infty} e^{-\beta_n^2 (t-\tau)} \frac{1}{N_n} \frac{k_j}{\alpha_j} \psi_{in}\,(r)\, \psi_{jn}\left(r'\right) \tag{10-107b}$$

where the norm N_n is obtained from equation (10-73a) as

$$N_n = \frac{k_1}{\alpha_1}\int_{r'=0}^{a} r'\psi_{1n}^2\left(r'\right)dr' + \frac{k_2}{\alpha_2}\int_{r'=a}^{b} r'\psi_{2n}^2\left(r'\right)dr' \tag{10-108}$$

The eigenfunctions $\psi_{1n}(r)$ and $\psi_{2n}(r)$ are obtained from equations (10-73b) and (10-73c), respectively:

$$\psi_{1n}(r) = J_0\left(\frac{\beta_n}{\sqrt{\alpha_1}}r\right) \tag{10-109a}$$

$$\psi_{2n}(r) = A_{2n}J_0\left(\frac{\beta_n}{\sqrt{\alpha_2}}r\right) + B_{2n}Y_o\left(\frac{\beta_n}{\sqrt{\alpha_2}}r\right) \tag{10-109b}$$

The coefficients A_{2n} and B_{2n} are given by equations (10-70a) and (10-70b), respectively. The eigenvalues β_n are the roots of the transcendental equation (10-71).

Example 10-6 Two-Layer Slab with Generation

In a two-layer slab, the first $(0 < x < a)$ and the second $(a < x < b)$ layers are in perfect thermal contact. Initially, the first layer is at temperature $F_1(x)$ and the second layer is at temperature $F_2(x)$. For times $t > 0$, the boundary at $x = 0$ is kept at zero temperature, the boundary at $x = b$ dissipates heat by convection into a medium at zero temperature, and internal energy is generated in the first layer at a rate of $g_1(x, t)$ (W/m^3). Obtain an expression for the temperature distribution in the medium for times $t > 0$.

The mathematical formulation of this problem is given as:

$$\alpha_1\frac{\partial^2 T_1}{\partial x^2} + \frac{\alpha_1}{k_1}g_1(x,t) = \frac{\partial T_1(x,t)}{\partial t} \quad \text{in} \quad 0 < x < a, \quad t > 0 \tag{10-110a}$$

$$\alpha_2\frac{\partial^2 T_2}{\partial x^2} = \frac{\partial T_2(x,t)}{\partial t} \quad \text{in} \quad a < x < b, \quad t > 0 \tag{10-110b}$$

subject to the boundary conditions

$$T_1(x = 0, t) = 0 \tag{10-111a}$$

$$T_1(x = a, t) = T_2(x = a, t) \tag{10-111b}$$

$$k_1\left.\frac{\partial T_1}{\partial x}\right|_{x=a} = k_2\left.\frac{\partial T_2}{\partial x}\right|_{x=a} \tag{10-111c}$$

$$k_2^*\left.\frac{\partial T_2}{\partial x}\right|_{x=b} + h_3^*\left.T_2\right|_{x=b} = 0 \tag{10-111d}$$

and to the initial conditions

$$T_1(x, t=0) = F_1(x) \qquad \text{for} \qquad 0 < x < a \tag{10-112a}$$

$$T_2(x, t=0) = F_2(x) \qquad \text{for} \qquad a < x < b \tag{10-112b}$$

This heat conduction problem is a special case of the general problem given by equations (10-91). Therefore, its solution is immediately obtainable in terms of the Green's functions from the general solution (10-96), for p=0, as

$$\begin{aligned} T_i(x,t) = & \int_{x'=0}^{a} \left[G_{i1}\left(x,t|x',\tau\right)\right]_{\tau=0} F_1\left(x'\right) dx' \\ & + \int_{x'=a}^{b} \left[G_{i2}\left(x,t|x',\tau\right)\right]_{\tau=0} F_2\left(x'\right) dx' \\ & + \int_{\tau=0}^{t} \int_{x'=0}^{a} G_{i1}(x,t|x',\tau) \frac{\alpha_1}{k_1} g_1(x',\tau)\, dx'\, d\tau, \quad i = 1,2 \end{aligned} \tag{10-113}$$

where Green's function is obtainable from the solution of the homogeneous version of the heat conduction problem given by equations (10-110)–(10-112). Actually, the homogeneous version of this problem is exactly the same as that considered in Example 10-1 given by equations (10-10)–(10-12). Therefore, the desired Green's function is obtainable from the result given by equation (10-90b) by replacing t with $(t - \tau)$ in this expression. We find

$$G_{ij}\left(x,t|x',\tau\right) = \sum_{n=1}^{\infty} e^{-\beta_n^2(t-\tau)} \frac{1}{N_n} \frac{k_j}{\alpha_j} \psi_{in}(x)\, \psi_{jn}\left(x'\right) \tag{10-114}$$

where the norm N_n is obtained from equation (10-89b) as

$$N_n = \frac{k_1}{\alpha_1} \int_{x'=0}^{a} \psi_{1n}^2\left(x'\right) dx' + \frac{k_2}{\alpha_2} \int_{x'=a}^{b} \psi_{2n}^2\left(x'\right) dx' \tag{10-115}$$

The eigenfunctions $\psi_{1n}(x)$ and $\psi_{2n}(x)$ are obtained from equations (10-89c) and (10-89d), respectively, as

$$\psi_{1n}(x) = \sin\left(\frac{\beta_n}{\sqrt{\alpha_1}} x\right) \tag{10-116a}$$

$$\psi_{2n}(x) = A_{2n} \sin\left(\frac{\beta_n}{\sqrt{\alpha_2}} x\right) + B_{2n} \cos\left(\frac{\beta_n}{\sqrt{\alpha_2}} x\right) \tag{10-116b}$$

The coefficients A_{2n} and B_{2n} are given by equations (10-87) and the eigenvalues β_n are the positive roots of the transcendental equation (10-88).

10-7 USE OF LAPLACE TRANSFORM FOR SOLVING SEMI-INFINITE AND INFINITE MEDIUM PROBLEMS

The Laplace transform technique is convenient for the solution of composite medium problems involving regions that are semi-infinite or infinite in extent. In this approach, the partial derivatives with respect to time are removed by the application of the Laplace transform, the resulting system of ordinary differential equations is solved, and the transforms of temperatures are inverted. However, as discussed in Chapter 9, the principal difficulty lies in the inversion of the resulting transform. In this section, we examine the solution of two-layer composite medium problems of semi-infinite and infinite extent by the Laplace transform technique and consider only those problems for which the inversion of the transforms can be performed by using the standard Laplace transform inversion tables.

Example 10-7 Two Semi-Infinite Composite Regions
Two semi-infinite regions, $x > 0$ and $x < 0$, as illustrated in Figure 10-6, are in perfect thermal contact. Initially, region 1 (i.e., $x > 0$) is at a uniform temperature T_0, and region 2 (i.e., $x < 0$) is at zero temperature. Obtain an expression for the temperature distribution in the medium for times $t > 0$.

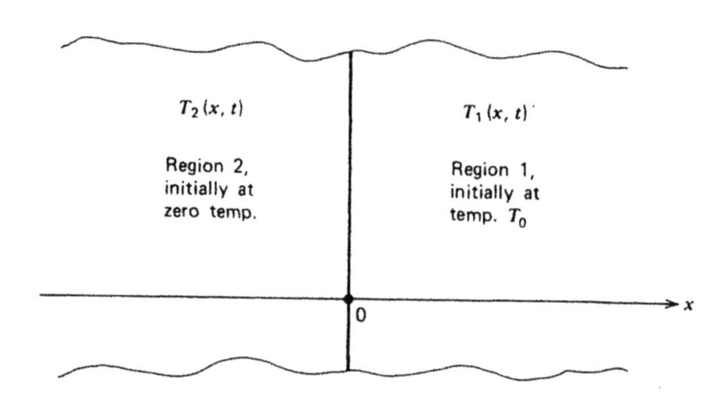

Figure 10-6 Two semi-infinite regions in perfect thermal contact.

For convenience in the analysis, we define a dimensionless temperature for both mediums as $\theta_i(x, t)$, namely,

$$\theta_i(x, t) = \frac{T_i(x, t)}{T_0} \qquad i = 1, 2 \tag{10-117}$$

The mathematical formulation of the problem, in terms of $\theta_i(x, t)$, is now given as

$$\frac{\partial^2\theta_1}{\partial x^2} = \frac{1}{\alpha_1}\frac{\partial\theta_1(x,t)}{\partial t} \quad \text{in} \quad x>0, \quad t>0 \tag{10-118a}$$

$$\frac{\partial^2\theta_2}{\partial x^2} = \frac{1}{\alpha_2}\frac{\partial\theta_2(x,t)}{\partial t} \quad \text{in} \quad x<0, \quad t>0 \tag{10-118b}$$

subject to the boundary conditions

$$\theta_1(x,t)\big|_{x=0^+} = \theta_2(x,t)\big|_{x=0^-} \tag{10-119a}$$

$$-k_1\left.\frac{\partial\theta_1}{\partial x}\right|_{x=0^+} = k_2\left.\frac{\partial\theta_2}{\partial x}\right|_{x=0^-} \tag{10-119b}$$

$$\left.\frac{\partial\theta_1}{\partial x}\right|_{x\to\infty} = \left.\frac{\partial\theta_2}{\partial x}\right|_{x\to-\infty} = 0 \tag{10-119c,d}$$

and to the initial conditions

$$\theta_1(x,t=0) = 1 \quad \text{for} \quad x>0 \tag{10-120a}$$

$$\theta_2(x,t=0) = 0 \quad \text{for} \quad x<0 \tag{10-120b}$$

The Laplace transform of the governing equations (10-118) yields

$$\frac{d^2\overline{\theta}_1(x,s)}{dx^2} = \frac{1}{\alpha_1}\left[s\overline{\theta}_1(x,s)-1\right] \quad \text{in} \quad x>0 \tag{10-121a}$$

$$\frac{d^2\overline{\theta}_2(x,s)}{dx^2} = \frac{1}{\alpha_2}s\overline{\theta}_2(x,s) \quad \text{in} \quad x<0 \tag{10-121b}$$

and the Laplace transform of the boundary conditions gives

$$\overline{\theta}_1(x,s)\big|_{x=0^+} = \overline{\theta}_2(x,s)\big|_{x=0^-} \tag{10-122a}$$

$$-k_1\left.\frac{d\overline{\theta}_1}{dx}\right|_{x=0^+} = k_2\left.\frac{d\overline{\theta}_2}{dx}\right|_{x=0^-} \tag{10-122b}$$

$$\left.\frac{d\overline{\theta}_1}{dx}\right|_{x\to\infty} = \left.\frac{d\overline{\theta}_2}{dx}\right|_{x\to-\infty} = 0 \tag{10-122c}$$

The solutions of equations (10-121) subject to the boundary conditions (10-122) are

$$\overline{\theta}_1(x,s) = \frac{1}{s} - \frac{1}{1+\beta}\frac{1}{s}e^{-(s/\alpha_1)^{1/2}x} \quad \text{for} \quad x>0 \tag{10-123a}$$

and

$$\overline{\theta}_2(x,s) = \frac{\beta}{1+\beta}\frac{1}{s}e^{-(s/\alpha_2)^{1/2}|x|} \qquad \text{for} \qquad x<0 \tag{10-123b}$$

where we define

$$\beta \equiv \frac{k_1}{k_2}\left(\frac{\alpha_2}{\alpha_1}\right)^{1/2} \tag{10-124}$$

These transforms can be inverted using the Laplace transform Table 9-1, cases 1 and 48. The resulting expressions for the temperature distribution in the medium become

$$\theta_1(x,t) \equiv \frac{T_1(x,t)}{T_0} = 1 - \frac{1}{1+\beta}\operatorname{erfc}\left(\frac{x}{2\sqrt{\alpha_1 t}}\right) \qquad \text{for} \qquad x>0 \tag{10-125a}$$

$$\theta_2(x,t) \equiv \frac{T_2(x,t)}{T_0} = \frac{\beta}{1+\beta}\operatorname{erfc}\left(\frac{|x|}{2\sqrt{\alpha_2 t}}\right) \qquad \text{for} \qquad x<0 \tag{10-125b}$$

Example 10-8 Two-Layer Composite Slab with Semi-Infinte Layer
A two-layer medium, as illustrated in Figure 10-7, is composed of region 1, $0 < x < L$, and region 2, $L < x < \infty$, which are in perfect thermal contact. Initially, region 1 is at a uniform temperature T_0 and region 2 is at zero temperature. For times $t > 0$, the boundary surface at $x = 0$ is kept insulated. Obtain an expression for the temperature distribution in the medium for times $t > 0$.

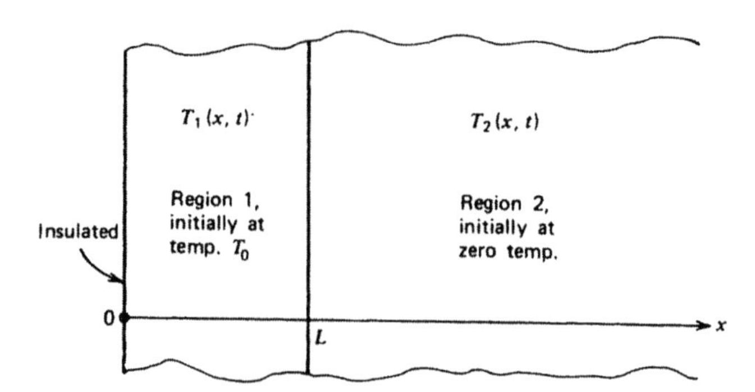

Figure 10-7 Finite and semi-infinite regions in perfect thermal contact.

We define a dimensionless temperature $\theta_i(x, t)$ as

$$\theta_i(x,t) = \frac{T_i(x,t)}{T_0} \qquad i = 1, 2 \tag{10-126}$$

With this transformation, the mathematical formulation of the problem in terms of $\theta_i(x, t)$ is given as

$$\frac{\partial^2 \theta_1}{\partial x^2} = \frac{1}{\alpha_1} \frac{\partial \theta_1 (x, t)}{\partial t} \quad \text{in} \quad 0 < x < L, \quad t > 0 \tag{10-127a}$$

$$\frac{\partial^2 \theta_2}{\partial x^2} = \frac{1}{\alpha_2} \frac{\partial \theta_2 (x, t)}{\partial t} \quad \text{in} \quad x > L, \quad t > 0 \tag{10-127b}$$

subject to the boundary conditions

$$\left. \frac{\partial \theta_1}{\partial x} \right|_{x=0} = 0 \tag{10-128a}$$

$$\theta_1 (x = L, t) = \theta_2 (x = L, t) \tag{10-128b}$$

$$k_1 \left. \frac{\partial \theta_1}{\partial x} \right|_{x=L} = k_2 \left. \frac{\partial \theta_2}{\partial x} \right|_{x=L} \tag{10-128c}$$

$$\theta_2 (x \to \infty, t) \to 0 \tag{10-128d}$$

and to the initial conditions

$$\theta_1 (x, t = 0) = 1 \quad \text{in} \quad 0 < x < L \tag{10-129a}$$

$$\theta_2 (x, t = 0) = 0 \quad \text{in} \quad L < x < \infty \tag{10-129b}$$

The Laplace transform of equations (10-127) yields

$$\frac{d^2 \bar{\theta}_1 (x, s)}{dx^2} = \frac{1}{\alpha_1} \left[s \bar{\theta}_1 (x, s) - 1 \right] \quad \text{in} \quad 0 < x < L \tag{10-130a}$$

$$\frac{d^2 \bar{\theta}_2 (x, s)}{dx^2} = \frac{1}{\alpha_2} s \bar{\theta}_2 (x, s) \quad \text{in} \quad L < x < \infty \tag{10-130b}$$

The Laplace transform of the boundary conditions gives

$$\left. \frac{d\bar{\theta}_1}{dx} \right|_{x=0} = 0 \tag{10-131a}$$

$$\bar{\theta}_1(x = L) = \bar{\theta}_2(x = L) \tag{10-131b}$$

$$k_1 \left. \frac{d\bar{\theta}_1}{dx} \right|_{x=L} = k_2 \left. \frac{d\bar{\theta}_2}{dx} \right|_{x=L} \tag{10-131c}$$

$$\bar{\theta}_2(x \to \infty) \to 0 \tag{10-131d}$$

The solution of equation (10-130a) that satisfies the boundary condition (10-131a) is taken in the form

$$\bar{\theta}_1(x,s)=\frac{1}{s}+A\cosh\left(x\sqrt{\frac{s}{\alpha_1}}\right)\qquad\text{in}\qquad 0\le x<L \tag{10-132}$$

and the solution of equation (10-130b) satisfying the boundary condition (10-131d) as

$$\bar{\theta}_2(x,s)=Be^{-x(\sqrt{s/\alpha_2})}\qquad\text{in}\qquad L<x<\infty \tag{10-133}$$

noting that we have used the hyperbolic solution for the finite-domain region and the exponential solution for the semi-infinite region, as discussed previously.

The constants A and B are determined by the application of the remaining boundary conditions (10-131b,c), which yields

$$A=-\frac{1-\gamma}{s}\frac{e^{-\sigma L}}{1-\gamma e^{-2\sigma L}} \tag{10-134a}$$

$$B=\frac{1+\gamma}{2s}e^{\sigma\mu L}\frac{1-e^{-2\sigma L}}{1-\gamma e^{-2\sigma L}} \tag{10-134b}$$

where we define

$$\sigma\equiv\sqrt{\frac{s}{\alpha_1}}\qquad \mu\equiv\sqrt{\frac{\alpha_1}{\alpha_2}} \tag{10-135a}$$

$$\gamma\equiv\frac{\beta-1}{\beta+1}\qquad \beta\equiv\frac{k_1}{k_2}\frac{1}{\mu} \tag{10-135b}$$

Introducing equations (10-134) and (10-135) into equations (10-132) and (10-133), we obtain

$$\bar{\theta}_1(x,s)=\frac{1}{s}-\frac{1-\gamma}{2s}\frac{e^{-\sigma(L-x)}+e^{-\sigma(L+x)}}{1-\gamma e^{-2\sigma L}}\qquad\text{in}\qquad 0<x<L \tag{10-136a}$$

and

$$\bar{\theta}_2(x,s)=\frac{1+\gamma}{2s}\frac{e^{-\sigma\mu(x-L)}-e^{-\sigma(2L+\mu x-\mu L)}}{1-\gamma e^{-2\sigma L}}\qquad\text{in}\qquad L<x<\infty \tag{10-136b}$$

Here we note that $|\gamma|<1$, therefore, the term $[1-\gamma\exp(-2\sigma L)]^{-1}$ can be expanded as a binomial series, and equations (10-136) become

$$\bar{\theta}_1(x,s)=\frac{1}{s}-\frac{1-\gamma}{2}\sum_{n=0}^{\infty}\gamma^n\left[\frac{e^{-\sigma[(2n+1)L-x]}}{s}+\frac{e^{-\sigma[(2n+1)L+x]}}{s}\right] \tag{10-137a}$$

in the domain $0 < x < L$, and

$$\bar{\theta}_2(x,s) = \frac{1+\gamma}{2}\sum_{n=0}^{\infty}\gamma^n\left[\frac{e^{-\sigma[2nL+\mu(x-L)]}}{s} + \frac{e^{-\sigma[(2n+2)L+\mu(x-L)]}}{s}\right] \quad \text{(10-137b)}$$

in the domain $L < x < \infty$.

The inversion of these two transforms are available in the Laplace transform Table 9-1 as cases 1 and 48. After the inversion, the temperature distribution in the medium becomes

$$\theta_1(x,t) \equiv \frac{T_1(x,t)}{T_0} = 1 - \frac{1-\gamma}{2}\sum_{n=0}^{\infty}\gamma^n\left\{\text{erfc}\left[\frac{(2n+1)L-x}{2\sqrt{\alpha_1 t}}\right] + \text{erfc}\left[\frac{(2n+1)L+x}{2\sqrt{\alpha_1 t}}\right]\right\} \quad \text{in} \quad 0 < x < L \quad \text{(10-138a)}$$

$$\theta_2(x,t) \equiv \frac{T_2(x,t)}{T_0} = \frac{1+\gamma}{2}\sum_{n=0}^{\infty}\gamma^n\left\{\text{erfc}\left[\frac{2nL+\mu(x-L)}{2\sqrt{\alpha_1 t}}\right] - \text{erfc}\left[\frac{(2n+2)L+\mu(x-L)}{2\sqrt{\alpha_1 t}}\right]\right\} \quad \text{in} \quad L < x < \infty \quad \text{(10-138b)}$$

where γ and μ are defined by equations (10-135).

REFERENCES

1. V. Vodicka, *Schweizer Arch*. **10**, 297–304, 1950.
2. V. Vodicka, *Math. Nach*. **14**, 47–55, 1955.
3. P. E. Bulavin and V. M. Kascheev, *Int. Chem. Eng*. **1**, 112–115, 1965.
4. C. W. Tittle, *J. Appl. Phys*. **36**, 1486–1488, 1965.
5. C. W. Tittle and V. L. Robinson, Analytical Solution of Conduction Problems in Composite Media, ASME Paper 65-WA-HT-52, 1965.
6. H. L. Beach, Application of the Orthogonal Expansion Technique to Conduction Heat Transfer Problems in Multilayer Cylinders, M.S. Thesis, Mech. and Aerospace Eng. Dept., North Carolina State University, Raleigh, NC 1967.
7. C. H. Moore, Heat Transfer across Surfaces in Contact: Studies of Transients in One-Dimensional Composite Systems, Ph.D. Dissertation, Mechanical Eng. Dept., Southern Methodist University, Dallas, 1967.
8. M. N. Özisik, *Boundary Value Problems of Heat Conduction*, International Textbook, Scranton, PA, 1968; Dover, New York, 1989.
9. M. H. Cobble, *J. Franklin Inst*. **290** (5), 453–465, 1970.
10. G. P. Mulholland and M. N. Cobble, *Int. J. Heat Mass Transfer* **15**, 147–160, 1972.
11. C. A. Chase, D. Gidaspow, and R. E. Peck, *Chem. Eng. Progr. Symp. Ser.* No. 92, Vol. **65**, 91–109, 1969.

12. B. S. Baker, D. Gidaspow, and D. Wasan, in *Advances in Electrochemistry and Electrochemical Engineering*, Wiley-Interscience, New York, 1971, pp. 63–156.
13. M. D. Mikhailov and M. N. Özisik, *Unified Analysis and Solutions of Heat and Mass Diffusion*, Wiley, New York, 1984.
14. J. Crank, *The Mathematics of Diffusion*, 2nd ed., Clarendon, London, 1975.
15. H. S. Carslaw and J. C. Jaeger, *Conduction of Heat in Solids*, Clarendon, London, 1959.
16. V. S. Arpaci, *Conduction Heat Transfer*, Addison-Wesley, Reading, MA., 1966.
17. A. V. Luikov, *Analytical Heat Diffusion Theory*, Academic, New York, 1968.
18. N. Y. Ölçer, *Ingenieur-Arch*. **36**, 285–293, 1968.
19. N. Y. Ölçer, *Quart. Appl. Math*. **26**, 355–371, 1968.
20. N. Y. Ölçer , *Nucl. Eng. Design* **7**, 92–112, 1968.
21. K. Senda, Family of Integral Transforms and Some Applications to Physical Problems, Technology Reports of the Osaka University, Osaka, Japan, No. 823, Vol. 18, 1968, PP. 261–286.
22. J. D. Lockwood and G. P. Mulholland, *J. Heat Transfer* **95c**, 487–491, 1973.
23. M. D. Mikhailov, *Int. J. Eng. Sci*. **11**, 235–241, 1973.
24. M. D. Mikhailov, *Int. J. Heat Mass Transfer* **16**, 2155–2164, 1973.
25. Y. Yener and M. N. Özisik, Proceedings of the 5th International Heat Transfer Conference, Tokyo, Sept. 1974.
26. J. Padovan, *AIAA J*. **12**, 1158–1160, 1974.
27. M. Ben-Amoz, *Int. J. Eng. Sci*., **12**, 633, 1974.
28. G. Horvay, R. Mani, M. A. Veluswami, and G. E. Zinsmeister, *J. Heat Transfer* **95**, 309, 1973.
29. M. Ben-Amoz, *ZAMP*, **27**, 335–345, 1976.
30. S. C. Huang and Y. P. Chang, *J. Heat Transfer* **102**, 742–748, 1980.
31. M. D. Mikhailov and M. N. Özisik, *Int. J. Heat Mass Transfer* **28**, 1039–1045, 1985.
32. J. Baker-Jarvis and R. Inguva, *J. Heat Transfer* **107**, 39–43, 1985.
33. S. C. Huang and J. P. Chang, *J. Heat Transfer* **102**, 742–748, 1980.
34. A. Haji-Sheikh and M. Mashena, *J. Heat Transfer*, **109**, 551–556, 1987.
35. M. D. Mikhailov and M. N. Özisik, *Int. J. Heat Mass Transfer* **29**, 340–342, 1988.
36. A. Haji-Sheikh and J. V. Beck, *J. Heat Transfer* **112**, 28–34, 1990.
37. M. N. Özisik, *Heat Conduction*, 2nd ed., Wiley, New York, 1993.

PROBLEMS

10-1 A two-layer solid cylinder contains the inner region, $0 \le r \le a$, and the outer region, $a \le r \le b$, which are in perfect thermal contact. Initially, the inner region is at temperature $F_1(r)$ and the outer region at temperature $F_2(r)$. For times $t > 0$, the boundary surface at $r = b$ is kept at zero temperature. Obtain an expression for the temperature distribution in the medium. Also, express the solution in terms of Green function and determine the Green function for this problem.

10-2 A two-layer slab consists of the first layer $0 \leq x \leq a$ and the second layer $a \leq x \leq b$, which are in perfect thermal contact. Initially, the first region is at temperature $F_1(x)$ and the second region is at temperature $F_2(x)$. For times $t > 0$, the outer boundaries at $x = 0$ and $x = b$ are kept at zero temperatures. Obtain an expression for the temperature distribution in the medium. Also determine the Green function for this problem.

10-3 A two-layer hollow cylinder consists of the first layer $a \leq r \leq b$ and the second layer $b \leq r \leq c$, which are in perfect thermal contact. Initially, the first region is at temperature $F_1(r)$ and the second region at temperature $F_2(r)$. For times $t > 0$, the outer boundaries at $r = a$ and $r = c$ are kept at zero temperature. Obtain an expression for the temperature distribution in the medium. Also, determine the Green function for this problem.

10-4 Repeat Problem 10-2 for the case when boundary surface at $x = 0$ is kept insulated and the boundary surface at $x = b$ dissipates heat by convection into an environment at zero temperature. Also determine the Green's function for this problem.

10-5 Repeat Problem 10-3 for the case when the boundary surface at $r = a$ is kept insulated and the boundary surface at $r = c$ dissipates heat by convection into an environment at zero temperature. Also determine the Green's function for this problem.

10-6 A two-layer solid cylinder contains the inner region $0 \leq r \leq a$ and the outer region $a \leq r \leq b$, which are in perfect thermal contact. Initially, the inner region is at temperature $F_1(r)$ and the outer region at temperature $F_2(r)$. For times $t > 0$, heat is generated in the inner region at a rate of $g_1(r, t)$ (W/m^3) while the boundary surface at $r = b$ is kept at temperature $f(t)$. By following an approach discussed in Example 10-5, transform this problem into one with homogeneous boundary condition at $r = b$.

10-7 A two-layer slab consists of the first layer $0 \leq x \leq a$ and the second layer $a \leq x \leq b$, which are in perfect thermal contact. Initially the first region is at temperature $F_1(x)$ and the second region at temperature $F_2(x)$. For times $t > 0$, heat is generated in the first region at a rate of $g_1(x, t)$, W/m^3, and in the second region at a rate of $g_2(x, t)$ (W/m^3), while the outer boundary surfaces at $x = 0$ and $x = b$ are kept at temperatures $f_1(t)$ and $f_2(t)$, respectively. Split up this problem into a steady-state problem and a time-dependent problem with heat generation, subject to homogeneous boundary conditions by following the procedure discussed in Section 10-2.

10-8 Solve Problem 10-1 with the additional condition that heat is generated in the inner region, $0 \leq r \leq a$, at a rate of $g_1(r, t)$ (W/m^3). Utilize the Green's function constructed in Problem 10-1 to solve this nonhomogeneous problem.

10-9 Solve Problem 10-2 with the additional condition that heat is generated in the first and second layers at a rate of $g_1(x, t)$ and $g_2(x, t)$ (W/m^3), respectively. Utilize the Green's function constructed in Problem 10-2 to solve this problem.

10-10 Solve Problem 10-3 with the additional condition that heat is generated in the first and second regions at a rate of $g_1(r, t)$ and $g_2(r, t)$ (W/m^3), respectively.

11

MOVING HEAT SOURCE PROBLEMS

There are numerous engineering applications, such as welding, grinding, metal cutting, firing a bullet in a gun barrel, and flame or laser hardening of metals, in which the calculation of temperature field in the solid is modeled as a problem of heat conduction involving a moving heat source. Following the pioneering works of Rosenthal [1–4] on the determination of temperature distribution in a solid resulting from arc welding, numerous papers appeared on the subject of heat transfer in solids with moving heat sources [5–29].

In machining, grinding, cutting, and sliding of surfaces, the energy generated as a result of friction heating can be modeled as a moving heat source. The determination of temperature fields around such heat sources has been studied by several investigators [14–23]. More recently lasers, because of their ability to produce high-power beams, have found applications in welding, drilling, cutting, machining of brittle materials, and surface hardening of metallic alloys. For example, in surface hardening, a high-power laser beam scans over the surface, and unique metallurgical structures may be produced by rapid cooling that occurs subsequent to the laser heating. The determination of temperature field around a moving laser beam has been studied [24–32].

The objective of this chapter is to introduce the mathematical formulation and the method of solution of heat conduction problems involving a moving heat source by considering simple, representative examples for which analytic solutions are obtainable by the method of separation of variables under quasi-stationary conditions.

11-1 MATHEMATICAL MODELING OF MOVING HEAT SOURCE PROBLEMS

A moving heat source, depending on the physical nature of the problem, can be modeled as a *point, line, surface*, or *ring* heat source that may release its energy either continuously over time or spontaneously at specified times. As discussed in Chapter 8, we use the following notation to identify various types of *continuous* heat sources:

$$g_p^{\mathrm{c}} = \text{point source} \quad (\mathrm{W})$$

$$g_L^{\mathrm{c}} = \text{line source} \quad (\mathrm{W/m})$$

$$g_s^{\mathrm{c}} = \text{surface source} \quad \left(\mathrm{W/m^2}\right)$$

where the superscript c refers to a continuous source. For an instantaneous source, we change the superscript c to i and alter the units of the source accordingly as discussed in Chapter 8, in the context of equations (8-57) and (8-58).

The spatial distribution of the strength of the heat source depends on the physical nature of the source. For example, the energy distribution in a laser beam generally is not necessarily uniform spatially. It may have a Gaussian distribution (i.e., intensity decreasing exponentially from the center of the beam with the square of the radial distance) or a doughnut shape, or a combination of these two shapes. Also, it may be a continuous source over time or activated as pulses for short periods of time.

In this section, we present the mathematical modeling of the determination of temperature fields in solids resulting from a moving point, line, and surface heat sources under the *quasi-stationary state* conditions.

Moving Point Heat Source

We consider a point heat source of constant strength g_p^{c} (W), releasing its energy continuously over time while moving along the x axis in the positive x direction with a constant velocity u, in a stationary medium that is initially at zero temperature. Figure 11-1(a) illustrates the geometry and the coordinates.

The three-dimensional heat conduction equation in the fixed x, y, z coordinate system, assuming constant properties, is taken as

$$\frac{\partial^2 T}{\partial x^2} + \frac{\partial^2 T}{\partial y^2} + \frac{\partial^2 T}{\partial z^2} + \frac{1}{k} g(x, y, z, t) = \frac{1}{\alpha} \frac{\partial T}{\partial t} \tag{11-1}$$

where $T \equiv T(x, y, z, t)$.

Let the heat source be a point heat source of constant strength g_p^{c} (W), located at $y = 0$, $z = 0$ and releasing its energy continuously as it moves along the x axis in the positive x direction with a constant velocity u. Such a point heat source is

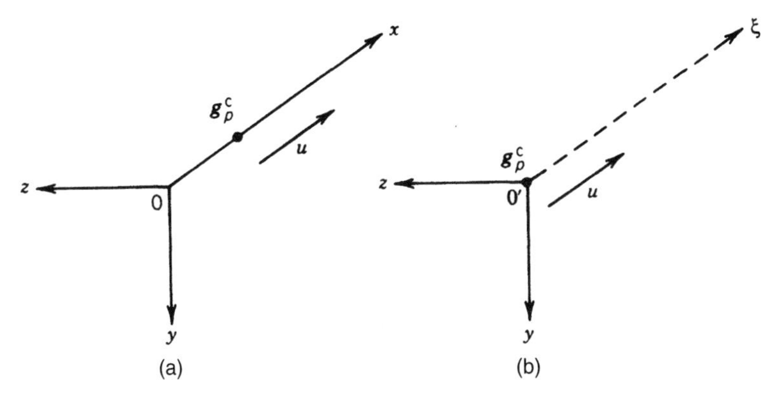

Figure 11-1 Moving point heat source: (*a*) fixed coordinates x, y, z and (*b*) moving coordinates ξ, y, z.

related to the volumetric source $g(x, y, z, t)$ by the delta function notation, see equation (8-51), as

$$g(x, y, z, t) \equiv g_p^c \delta(x - ut)\delta(y - 0)\delta(z - 0) \tag{11-2}$$

where $\delta(-)$ denotes the Dirac delta function, and where we note that each of the delta functions in equation (11-2) has the units of m^{-1}, which together with the units of g_p^c provides the correct overall units for $g(x, y, z, t)$.

Transformation of the Origin In the solution of moving heat source problems, it is convenient to let the coordinate system move with the source. This is achieved by introducing a new coordinate ξ defined by

$$\boxed{\xi = x - ut} \tag{11-3}$$

Figure 11-1(*b*) illustrates the new coordinate system ξ, y, z that moves with the source. The heat conduction equation (11-1) is transformed from the fixed x, y, z coordinate system with fixed origin 0 to the moving coordinate system ξ, y, z with moving origin 0′ by the application of the chain rule of differentiation given by

$$\frac{\partial T(\xi, y, z, t)}{\partial t} = \frac{\partial T}{\partial \xi}\frac{\partial \xi}{\partial t}^{-u} + \frac{\partial T}{\partial y}\frac{dy}{dt}^{0} + \frac{\partial T}{\partial z}\frac{dz}{dt}^{0} + \frac{\partial T}{\partial t}\frac{dt}{dt}^{1} = -u\frac{\partial T}{\partial \xi} + \frac{\partial T}{\partial t} \tag{11-4}$$

since $(\partial \xi/\partial t) = -u$, per equation (11-3). The derivatives with respect to x become

$$\frac{\partial T}{\partial x} = \frac{\partial T}{\partial \xi}\frac{\partial \xi}{\partial x}^{1} + \frac{\partial T}{\partial t}\frac{dt}{dx}^{0} + \cdots = \frac{\partial T}{\partial \xi} \tag{11-5}$$

and

$$\frac{\partial^2 T}{\partial x^2} = \frac{\partial^2 T}{\partial \xi^2} \tag{11-6}$$

while the partial derivatives with respect to y and z remain unaltered. Then, the heat conduction equation (11-1) in the coordinate system ξ, y, z moving with the source is given by

$$\frac{\partial^2 T}{\partial \xi^2} + \frac{\partial^2 T}{\partial y^2} + \frac{\partial^2 T}{\partial z^2} + \frac{1}{k} g_p^c \delta(\xi)\delta(y)\delta(z) = \frac{1}{\alpha}\left(\frac{\partial T}{\partial t} - u\frac{\partial T}{\partial \xi}\right) \tag{11-7}$$

We note that this equation is a special case of the heat conduction equation for a moving solid given by equation (1-38) in Chapter 1. In equation (11-7), the solid is moving with a velocity u in the negative ξ direction with respect to an observer located at the source; this is the reason for the negative sign in front of the velocity term in equation (11-7).

Quasi-Stationary Condition Experiments have shown that, if the solid is long enough in the direction of motion as compared to the penetration depth of the heat transfer field, the temperature distribution around the heat source soon becomes independent of time. That is, an observer stationed at the moving origin $0'$ of the ξ, y, z coordinate system fails to notice any change in the temperature distribution around him/her as the source moves on. This is identified as the quasi-stationary condition [3] and mathematically defined by setting $\partial T/\partial t = 0$. Therefore, the quasi-stationary form of equation (11-7) is obtained by setting $\partial T/\partial t = 0$, namely,

$$\frac{\partial^2 T}{\partial \xi^2} + \frac{\partial^2 T}{\partial y^2} + \frac{\partial^2 T}{\partial z^2} + \frac{1}{k} g_p^c \delta(\xi)\delta(y)\delta(z) = -\frac{u}{\alpha}\frac{\partial T}{\partial \xi} \tag{11-8}$$

Equation (11-8) can be transformed into a more convenient form by introducing a new dependent variable $\theta(\xi, y, z)$ defined as

$$\boxed{T(\xi, y, z) = \theta(\xi, y, z) e^{-(u/2\alpha)\xi}} \tag{11-9}$$

With these changes, equation (11-8) now takes the form

$$\frac{\partial^2 \theta}{\partial \xi^2} + \frac{\partial^2 \theta}{\partial y^2} + \frac{\partial^2 \theta}{\partial z^2} - \left(\frac{u}{2\alpha}\right)^2 \theta + \frac{1}{k} g_p^c \delta(\xi)\delta(y)\delta(z) e^{(u/2\alpha)\xi} = 0 \tag{11-10}$$

Here, the exponential $e^{(u/2\alpha)\xi}$ appearing in the source term can be omitted, since the term vanishes for $\xi \neq 0$ because of the delta function $\delta(\xi)$, noting that the exponential term becomes unity for $\xi = 0$.

A Moving Line Heat Source

We now consider a line heat source of constant strength g_L^c (W/m), located at the x axis and oriented parallel to the z axis as illustrated in Figure 11-2(a). The source releases its energy continuously over the time as it moves with a constant velocity u in the positive x direction. The medium is initially at zero temperature. We assume $(\partial T/\partial z) = 0$ everywhere in the medium. The two-dimensional heat conduction equation in the x, y coordinates is now taken as

$$\frac{\partial^2 T}{\partial x^2} + \frac{\partial^2 T}{\partial y^2} + \frac{1}{k} g(x, y, t) = \frac{1}{\alpha} \frac{\partial T}{\partial t} \tag{11-11}$$

where $T \equiv T(x, y, t)$. The line heat source g_L^c (W/m) is related to the equivalent volumetric source $g(x, y, t)$ (W/m^3) by the delta function notation as

$$g(x, y, t) = g_L^c \delta(y)\delta(x - ut) \tag{11-12}$$

Transformation of the Origin This heat conduction problem is now transformed from the x, y fixed coordinates to new ξ, y coordinates moving with the line heat source by the transformation

$$\xi = x - ut \tag{11-13}$$

as illustrated in Figure 11-2(b).

By following the procedure described previously, we transform the heat conduction equation (11-11) to the moving coordinate system (ξ, y) as

$$\frac{\partial^2 T}{\partial \xi^2} + \frac{\partial^2 T}{\partial y^2} + \frac{1}{k} g_L^c \delta(\xi)\delta(y) = \frac{1}{\alpha} \left(\frac{\partial T}{\partial t} - u \frac{\partial T}{\partial \xi} \right) \tag{11-14}$$

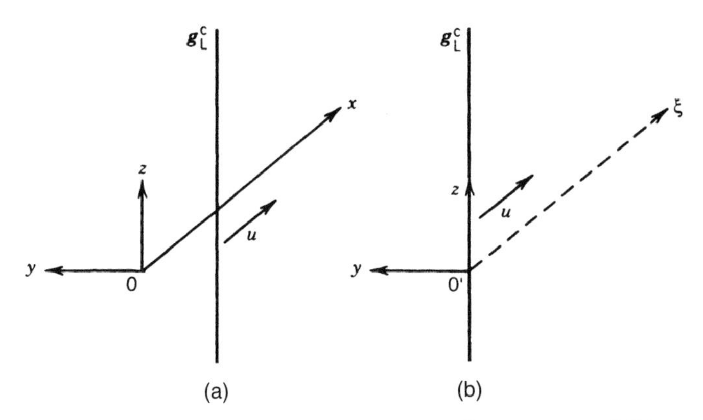

Figure 11-2 Moving line heat source: (a) fixed coordinates x, y and (b) moving coordinates ξ, y.

Quasi-Stationary Condition As discussed previously, the quasi-stationary form of equation (11-14) is obtained by setting $\partial T/\partial t = 0$. We find

$$\frac{\partial^2 T}{\partial \xi^2} + \frac{\partial^2 T}{\partial y^2} + \frac{1}{k} g_L^c \delta(\xi)\delta(y) = -\frac{u}{\alpha}\frac{\partial T}{\partial \xi} \tag{11-15}$$

This equation is transformed into a more convenient form by introducing a new dependent variable $\theta(\xi, y)$ defined as

$$T(\xi, y) = \theta(\xi, y) e^{-(u/2\alpha)\xi} \tag{11-16}$$

which upon substitution into equation (11-15), yields

$$\frac{\partial^2 \theta}{\partial \xi^2} + \frac{\partial^2 \theta}{\partial y^2} - \left(\frac{u}{2\alpha}\right)^2 \theta + \frac{1}{k} G_{\mathrm{L}} = 0 \tag{11-17}$$

where

$$G_L \equiv g_L^c \delta(\xi)\delta(y) \tag{11-18}$$

We note that the term $e^{(u/2\alpha)\xi}$ that would have appeared on the right-hand side of equation (11-18) is omitted for the reason stated previously.

A Moving Plane Surface Heat Source

We now consider a plane surface heat source of constant strength g_s^c (W/m^2), oriented perpendicular to the x axis, as illustrated in Figure 11-3. The source releases its energy continuously over the time as it moves with a constant velocity u in the positive x direction. For the one-dimensional case considered here, we assume $\partial T/\partial y = \partial T/\partial z = 0$ everywhere; hence the differential equation of heat conduction reduces to

$$\frac{\partial^2 T}{\partial x^2} + \frac{1}{k} g(x, t) = \frac{1}{\alpha}\frac{\partial T}{\partial t} \tag{11-19}$$

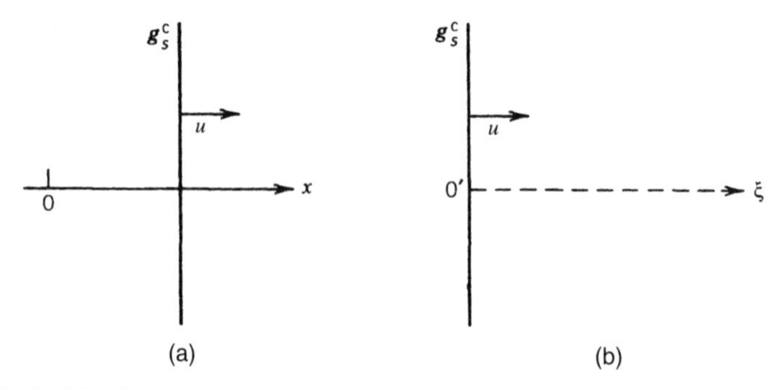

Figure 11-3 Moving plane surface heat source: (*a*) fixed coordinate x and (*b*) moving coordinate ξ.

where $T \equiv T(x, t)$. The moving continuous surface heat source g_s^c (W/m^2) is related to the equivalent volumetric source $g(x, t)$ (W/m^3) by

$$g(x, t) = g_s^c \delta(x - ut) \tag{11-20}$$

Transformation of the Origin The heat conduction equation (11-19) is transformed from the fixed x coordinate to the moving ξ coordinate by the transformation

$$\xi = x - ut \tag{11-21}$$

By following the procedure described previously, the heat conduction equation (11-19) is transformed to the moving ξ coordinate as

$$\frac{\partial^2 T}{\partial \xi^2} + \frac{1}{k} g_s^c \delta(\xi) = \frac{1}{\alpha}\left(\frac{\partial T}{\partial t} - u\frac{\partial T}{\partial \xi}\right) \tag{11-22}$$

Quasi-Stationary Condition Assuming quasi-stationary condition, equation (11-22) reduces to

$$\frac{d^2 T}{d\xi^2} + \frac{1}{k} g_s^c \delta(\xi) = -\frac{u}{\alpha}\frac{dT}{d\xi} \tag{11-23}$$

and with the application of the transformation

$$T(\xi) = \theta(\xi) e^{-(u/2\alpha)\xi} \tag{11-24}$$

equation (11-23) takes the form

$$\frac{d^2\theta}{d\xi^2} - \left(\frac{u}{2\alpha}\right)^2 \theta + \frac{1}{k} g_s^c \delta(\xi) = 0 \tag{11-25}$$

where the exponential $e^{(u/2\alpha)\xi}$, which would appear as a multiplier to the heat source, is omitted for the reasons stated previously.

11-2 ONE-DIMENSIONAL QUASI-STATIONARY PLANE HEAT SOURCE PROBLEM

In the problem of arc welding, the energy generated by the arc causes the electrode to melt; hence the problem of the temperature distribution around the arc can be modeled as a problem of moving heat source. If the electrode is long enough with respect to its diameter, the heat transfer in the first few inches of the electrode can be envisioned as being of a quasi-stationary nature. If we assume there are no surface losses from the electrode (i.e., electrode is partially insulated), the corresponding heat transfer problem can be modeled as a one-dimensional moving heat source problem governed by the heat conduction equation (11-19).

If we further assume that the quasi-stationary condition exists, the governing differential equation for this problem is taken as

$$\frac{d^2T}{d\xi^2} + \frac{1}{k} g_s^c \delta(\xi) = -\frac{u}{\alpha}\frac{dT}{d\xi}, \qquad \text{in} \qquad -\infty < \xi < \infty \tag{11-26}$$

subject to the boundary condition

$$\frac{dT}{d\xi} \to 0 \qquad \text{as} \qquad \xi \to \pm\infty \tag{11-27}$$

Applying the transformation (11-24), equation (11-26) takes the form

$$\frac{d^2\theta}{d\xi^2} - \left(\frac{u}{2\alpha}\right)^2 \theta + \frac{1}{k} g_s^c \delta(\xi) = 0 \qquad \text{in} \qquad -\infty < \xi < \infty \tag{11-28}$$

which is the same as that given by equation (11-25). The solution of this equation for $\xi \neq 0$, in other words, when the source term drops out, is taken as

$$\theta(\xi) = C_1 e^{-(u/2\alpha)\xi} + C_2 e^{(u/2\alpha)\xi} \qquad \text{for} \qquad -\infty < \xi < \infty \tag{11-29}$$

Introducing this result into equation (11-24), we obtain

$$T(\xi) = C_1 e^{-(u/\alpha)\xi} + C_2 \qquad \text{for} \qquad -\infty < \xi < \infty \tag{11-30}$$

This solution is now considered separately for the regions $\xi < 0$ and $\xi > 0$ in the forms

$$T^-(\xi) = C_1^- e^{-(u/\alpha)\xi} + C_2^- \qquad \text{for} \qquad \xi < 0 \tag{11-31a}$$

and

$$T^+(\xi) = C_1^+ e^{-(u/\alpha)\xi} + C_2^+ \qquad \text{for} \qquad \xi > 0 \tag{11-31b}$$

The unknown coefficients are determined by the application of the following boundary conditions:

$$\frac{dT^\pm}{d\xi} \to 0 \qquad \text{as} \qquad \xi \to \pm\infty \tag{11-32a,b}$$

$$T^-(\xi = 0) = T^+(\xi = 0) \qquad \text{(continuity of temperature)} \tag{11-32c}$$

$$k\left.\frac{dT^-}{d\xi}\right|_{\xi=0} - k\left.\frac{dT^+}{d\xi}\right|_{\xi=0} = g_s^c \qquad \text{(jump condition)} \tag{11-32d}$$

The last condition is obtained by integrating equation (11-26) with respect to ξ, from $\xi = -\varepsilon$ to $\xi = +\varepsilon$ and then letting $\varepsilon \to 0$. Physically, equation (11-32d) reflects conservation of energy at the infinitely thin source plane.

The application of the boundary conditions (11-32a,b), combined with the fact that $T^+ \to 0$ as $\xi \to +\infty$, gives $C_1^- = 0$ and $C_2^+ = 0$; then

$$T^-(\xi) = C_2^- \qquad \text{for} \qquad \xi < 0 \tag{11-33a}$$

and

$$T^+(\xi) = C_1^+ e^{-(u/\alpha)\xi} \qquad \text{for} \qquad \xi > 0 \tag{11-33b}$$

The requirement of continuity of temperature (11-32c) and the fact that $T^+ \to 0$ as $\xi \to +\infty$, gives $C_2^- = C_1^+ \equiv C$. The unknown constant C is determined by the application of the boundary condition (11-32d), which yields

$$C_2^- = C_1^+ \equiv C = \frac{\alpha}{uk} g_s^c \tag{11-34}$$

Then the overall solution for the problem becomes

$$T^-(\xi) = \frac{\alpha}{uk} g_s^c \qquad \text{for} \qquad \xi < 0 \tag{11-35}$$

and

$$T^+(\xi) = \frac{\alpha}{uk} g_s^c e^{-(u/\alpha)\xi} \qquad \text{for} \qquad \xi > 0 \tag{11-36}$$

Figure 11-4 shows a plot of the temperature profiles given by equations (11-35) and (11-36). Here, the rate of melting of the electrode is equivalent to the speed at which the arc moves along the electrode. The term $T^+(\xi)$ represents the temperature of a point at a distance ξ from the arc. The maximum value of temperature occurs at the moving source: $\xi = 0$. The medium remains at this maximum temperature after the source has moved further because no surface losses have been allowed in the problem.

Effects of Surface Heat Losses To illustrate the modeling of this problem for the case allowing for heat losses from the surfaces, we consider the same problem above, except we assume that heat is lost by convection from the lateral surfaces of the rod into an ambient environment at zero temperature with a heat transfer

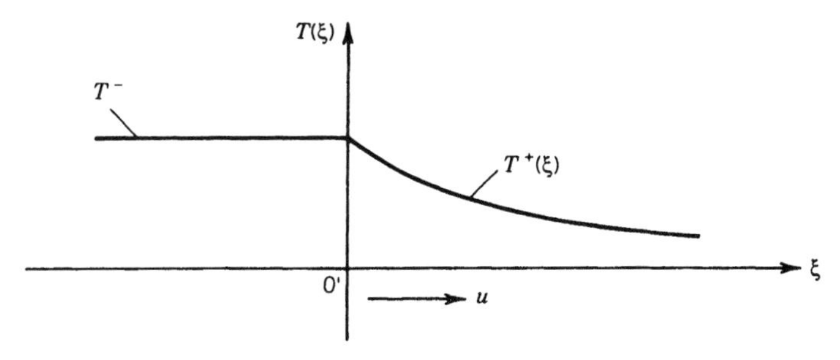

Figure 11-4 Quasi-stationary temperature distribution around a moving surface heat source.

coefficient h. If the rod has a uniform cross-section A and perimeter P, then the governing one-dimensional heat conduction equation allowing for convection losses from the lateral surfaces is given by

$$\frac{\partial^2 T}{\partial x^2} + \frac{1}{k} g(x, t) = \frac{1}{\alpha} \frac{\partial T}{\partial t} + \frac{Ph}{Ak} T \tag{11-37}$$

where the last term on the right-hand side represents convection heat losses from the lateral surfaces of the rod. The surface heat source g_s^c is related to the volumetric source $g(x, t)$ by

$$g\,(x, t) = g_s^c \delta(x - ut) \tag{11-38}$$

The equation is transformed from the fixed coordinate x to the moving coordinate ξ by the transformation

$$\xi = x - ut \tag{11-39}$$

With this substitution, the heat conduction (11-37) takes the form

$$\frac{\partial^2 T}{\partial \xi^2} + \frac{1}{k} g_s^c \delta(\xi) = \frac{1}{\alpha} \left(\frac{\partial T}{\partial t} - u \frac{\partial T}{\partial \xi} \right) + \frac{Ph}{Ak} T \tag{11-40}$$

and for the quasi-stationary condition, we have

$$\frac{d^2 T}{d\xi^2} + \frac{1}{k} g_s^c \delta(\xi) = -\frac{u}{\alpha} \frac{dT}{d\xi} + \frac{Ph}{Ak} T \tag{11-41}$$

and

$$\frac{dT}{d\xi} \to 0 \qquad \text{for} \qquad \xi \to \pm \infty \tag{11-42}$$

Applying the transformation

$$T(\xi) = \phi(\xi) e^{-(u/2\alpha)\xi} \tag{11-43}$$

equation (11-41) takes the form

$$\frac{d^2 \phi(\xi)}{d\xi^2} - m^2 \phi(\xi) + \frac{1}{k} g_s^c \delta(\xi) = 0 \qquad \text{in} \qquad -\infty < \xi < \infty \tag{11-44}$$

where

$$m = \left[\left(\frac{u}{2\alpha} \right)^2 + \frac{Ph}{Ak} \right]^{1/2} \tag{11-45}$$

The solution of equation (11-44) for $\xi \neq 0$, where the source term drops out, is given by

$$\phi(\xi) = C_1 e^{-m\xi} + C_2 e^{+m\xi} \qquad \text{for} \qquad -\infty < \xi < \infty \quad \text{and} \quad \xi \neq 0 \tag{11-46}$$

Introducing this result into equation (11-44), we obtain

$$T(\xi) = C_1 e^{-[m+(u/2\alpha)]\xi} + C_2 e^{[m-(u/2\alpha)]\xi} \qquad \text{for} \qquad -\infty < \xi < \infty \qquad (11\text{-}47)$$

It is convenient to consider this solution for the regions $\xi < 0$ and $\xi > 0$, separately, namely as

$$T^-(\xi) = C_1^- e^{-[m+(u/2\alpha)]\xi} + C_2^- e^{[m-(u/2\alpha)]\xi} \qquad \text{for} \qquad \xi < 0 \qquad (11\text{-}48a)$$

and

$$T^+(\xi) = C_1^+ e^{-[m+(u/2\alpha)]\xi} + C_2^+ e^{[m-(u/2\alpha)]\xi} \qquad \text{for} \qquad \xi > 0 \qquad (11\text{-}48b)$$

Now the boundary conditions for the determination of these four unknown coefficients are taken as

$$\frac{dT^{\pm}}{d\xi} = 0 \qquad \text{as} \qquad \xi \to \pm\infty \qquad (11\text{-}49a,b)$$

$$T^-(\xi = 0) = T^+(\xi = 0) \qquad \text{(continuity of temperature)} \qquad (11\text{-}49c)$$

$$k\left.\frac{dT^-}{d\xi}\right|_{\xi=0} - k\left.\frac{dT^+}{d\xi}\right|_{\xi=0} = g_s^c \qquad \text{jump condition} \qquad (11\text{-}49d)$$

Here, the last condition is obtained by integrating equation (11-41) from $\xi = -\varepsilon$ to $\xi = +\varepsilon$, and then letting $\varepsilon \to 0$, as discussed above. The application of the boundary conditions (11-49) to equations (11-48) gives the solution as

$$T^-(\xi) = \frac{g_s^c}{2km} e^{[m-(u/2\alpha)]\xi} \qquad \text{for} \qquad \xi < 0 \qquad (11\text{-}50a)$$

and

$$T^+(\xi) = \frac{g_s^c}{2km} e^{-[m+(u/2\alpha)]\xi} \qquad \text{for} \qquad \xi > 0 \qquad (11\text{-}50b)$$

where again

$$m = \sqrt{\left(\frac{u}{2\alpha}\right)^2 + \frac{Ph}{kA}} \qquad (11\text{-}51)$$

Clearly, equations (11-50) reduce to equations (11-35) and (11-36) for $h = 0$.

11-3 TWO-DIMENSIONAL QUASI-STATIONARY LINE HEAT SOURCE PROBLEM

We now examine a two-dimensional situation in which heat flows in the x and y directions while a line heat source of constant strength g_L^c (W/m) oriented parallel to the z axis moves along the x axis in the positive x direction with a constant velocity u. We assume $\partial T/\partial z = 0$ everywhere.

Assuming quasi-stationary conditions, the transformed energy equation is the two-dimensional version of equation (11-25); that is,

$$\frac{\partial^2\theta}{\partial\xi^2} + \frac{\partial^2\theta}{\partial y^2} - \left(\frac{u}{2\alpha}\right)^2\theta + \frac{1}{k}g_L^c\delta(\xi)\delta(y) = 0 \tag{11-52}$$

where

$$\xi = x - ut \qquad \text{and} \qquad \theta \equiv \theta(\xi, y) \tag{11-53}$$

Now $\theta(\xi, y)$ is related to the temperature $T(\xi, y)$ by

$$T(\xi, y) = \theta(\xi, y)e^{-(u/2\alpha)\xi} \tag{11-54}$$

Since the boundary conditions for $T(\xi, y)$ at infinity are given by

$$\frac{\partial T}{\partial\xi} \to 0 \qquad \text{for} \qquad \xi \to \pm\infty \tag{11-55a}$$

$$\frac{\partial T}{\partial y} \to 0 \qquad \text{for} \qquad y \to \pm\infty \tag{11-55b}$$

and equation (11-52) is symmetric with respect to the ξ and y variables, the function $\theta(\xi, y)$ therefore depends only on the distance r from the heat source. To solve this problem, we write the homogeneous portion of the differential equation (11-52) in the polar coordinates in the r variable as

$$\frac{1}{r}\frac{d}{dr}\left(r\frac{d\theta}{dr}\right) - \left(\frac{u}{2\alpha}\right)^2\theta = 0 \qquad \text{in} \qquad 0 < r < \infty \tag{11-56}$$

and treat the source term as a boundary effect at $r = 0$. To obtain a boundary condition at the origin, a circle of radius r is drawn around the line heat source, the heat released by the source is equated to the heat conducted away, and then r is allowed to go to zero. With this process, we find

$$\lim_{r\to 0}\left(-2\pi rk\frac{d\theta}{dr}\right) = g_L^c \qquad \text{as} \qquad r \to 0 \tag{11-57a}$$

and

$$\frac{d\theta}{dr} \to 0 \qquad \text{as} \qquad r \to \infty \tag{11-57b}$$

Equation (11-56) is modified Bessel's equation of order zero, and its solution satisfying the boundary condition (11-57b) is taken as

$$\theta(r) = CK_0\left(\frac{u}{2\alpha}r\right) \tag{11-58}$$

where K_0 is the modified Bessel function of the second kind of order zero, noting that we have eliminated I_0 from the solution because this function goes to positive infinity as its argument goes to infinity, see Figure 2-2.

Introducing the solution (11-58) into the boundary condition (11-57a), we find

$$-C2\pi k \lim_{r\to 0}\left[r\frac{d}{dr}K_0\left(\frac{u}{2\alpha}r\right)\right] = g_L^c \tag{11-59}$$

giving

$$(C2\pi k)(1) = g_L^c \qquad \Rightarrow \qquad C = \frac{1}{2\pi k}g_L^c \tag{11-60a}$$

since

$$K_0\left(\frac{u}{2\alpha}r\right) \to -\ln\left(\frac{u}{2\alpha}r\right) \tag{11-60b}$$

for small arguments (see Appendix IV), and

$$r\frac{d}{dr}\left[\ln\left(\frac{u}{2\alpha}r\right)\right] = (r)\left(\frac{2\alpha}{ur}\frac{u}{2\alpha}\right) = 1 \tag{11-60c}$$

Using the above relations, the solution for $\theta(r)$ becomes

$$\theta(r) = \frac{1}{2\pi k}g_L^c K_0\left(\frac{u}{2\alpha}r\right) \tag{11-61}$$

and $T(r,\xi)$ is determined according to equation (11-54) as

$$T(r,\xi) = \frac{1}{2\pi k}g_L^c K_0\left(\frac{u}{2\alpha}r\right)e^{-(u/2\alpha)\xi} \tag{11-62}$$

The two-dimensional temperature field given by equation (11-62) can have application in the arc welding of thin plates along the edges. For large values of r, equation (11-62) can be simplified by using the asymptotic value of $K_0(z)$ for large arguments (see Appendix IV):

$$K_0(z) \cong \sqrt{\frac{\pi}{2z}}e^{-z} \qquad \text{for large } z \tag{11-63}$$

11-4 TWO-DIMENSIONAL QUASI-STATIONARY RING HEAT SOURCE PROBLEM

There are many engineering applications in which the moving heat source can be modeled as a *moving ring heat source*. Consider, for example, the turning operation for a cylindrical workpiece on a lathe in order to reduce its diameter. The thermal energy released from the cutting process will cause the heating of both the tool and the workpiece. In such turning operations the relative velocity of the tool with respect to the workpiece is large in the circumferential direction. Therefore, the heat generated during the turning operation can be regarded as a *ring heat source* moving along the outer boundary in the negative z direction as illustrated in Figure 11-5. We assume azimuthal symmetry and a ring heat source

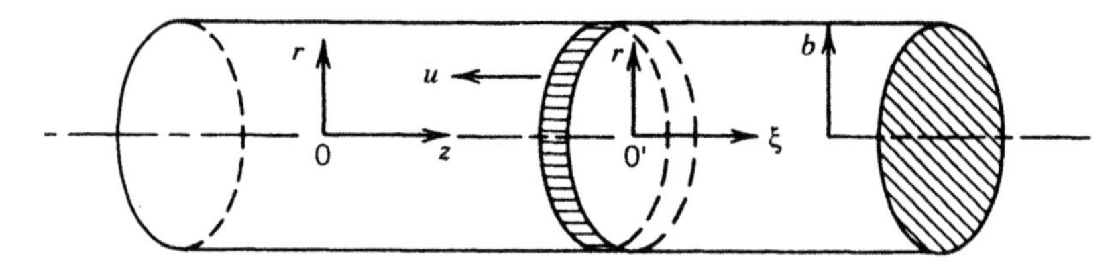

Figure 11-5 Moving ring heat source.

of constant strength Q_0(W), releasing its energy continuously as it moves with constant velocity u along the outer surface of the cylinder. We allow for convection from the outer surface of the cylinder into an ambient at zero temperature and choose the initial temperature of the solid as zero.

The mathematical formulation of this problem is given as

$$\frac{1}{r}\frac{\partial}{\partial r}\left(r\frac{\partial T}{\partial r}\right)+\frac{\partial^2 T}{\partial z^2}+\frac{1}{k}\frac{Q_0}{2\pi b}\delta(r-b)\delta(z+ut)=\frac{1}{\alpha}\frac{\partial T}{\partial t} \quad \text{in} \quad \begin{array}{c} 0<r<b \\ -\infty<z<\infty \end{array} \tag{11-64}$$

subject to the boundary conditions

$$\left.\frac{\partial T}{\partial r}\right|_{r=0}=0 \qquad \text{(symmetry)} \tag{11-65a}$$

$$k\left.\frac{\partial T}{\partial r}\right|_{r=b}+hT|_{r=b}=0 \tag{11-65b}$$

$$\frac{\partial T}{\partial z}\to 0 \qquad \text{as} \qquad z\to\pm\infty \tag{11-65c}$$

$$T(r,t=0)=0 \tag{11-65d}$$

This problem has been solved in reference 23 by using the integral transform technique. Here we describe its solution by the classical separation of variables technique. In equation (11-64), the delta function $\delta(r-b)$ denotes that the source is located at the outer surface of the cylinder, and $\delta(z+ut)$ shows its position at time t along the z axis.

The fixed coordinate system r, z is now allowed to move with the source by introducing the transformation

$$\xi = z + ut \tag{11-66}$$

In the moving ξ-coordinate system assuming a *quasi-stationary condition*, equation (11-64) reduces to

$$\frac{1}{r}\frac{\partial}{\partial r}\left(r\frac{\partial T}{\partial r}\right)+\frac{\partial^2 T}{\partial \xi^2}+\frac{1}{k}\frac{Q_0}{2\pi b}\delta(r-b)\delta(\xi)=\frac{u}{\alpha}\frac{\partial T}{\partial \xi} \quad \text{in} \quad \begin{array}{c} 0<r<b \\ -\infty<\xi<\infty \end{array} \tag{11-67}$$

subject to the following boundary conditions:

$$\left.\frac{\partial T}{\partial r}\right|_{r=0} = 0 \qquad \text{(symmetry)} \tag{11-68a}$$

$$k\left.\frac{\partial T}{\partial r}\right|_{r=b} + h\, T|_{r=b} = 0 \tag{11-68b}$$

$$\frac{\partial T}{\partial \xi} \to 0 \qquad \text{as} \qquad \xi \to \pm\infty \tag{11-68c}$$

This problem is now expressed in the *dimensionless form* as

$$\frac{1}{R}\frac{\partial}{\partial R}\left(R\frac{\partial \psi}{\partial R}\right) + \frac{\partial^2 \psi}{\partial \eta^2} + \delta(R-1)\delta(\eta) = \frac{\text{Pe}}{2}\frac{\partial \psi}{\partial \eta} \qquad \text{in} \qquad \begin{array}{c} 0 < R < 1 \\ -\infty < \eta < \infty \end{array} \tag{11-69}$$

subject to the nondimensional boundary conditions

$$\left.\frac{\partial \psi}{\partial R}\right|_{R=0} = 0 \tag{11-70a}$$

$$\left.\frac{\partial \psi}{\partial R}\right|_{R=1} + \frac{\text{Bi}}{2}\,\psi|_{R=1} = 0 \tag{11-70b}$$

$$\frac{\partial \psi}{\partial \eta} \to 0 \qquad \text{as} \qquad \eta \to \pm\infty \tag{11-70c}$$

where the various dimensionless quantities are defined as

$$\left.\begin{array}{l} \eta = \dfrac{\xi}{b}, \qquad R = \dfrac{r}{b}, \qquad \psi = \dfrac{T}{A}, \qquad A = \dfrac{Q_0}{k2\pi b} \\ \text{Pe} = \dfrac{u \cdot 2b}{\alpha} \equiv \text{Peclét number}, \qquad \text{Bi} = \dfrac{h \cdot 2b}{k} \equiv \text{Biot number} \end{array}\right\} \tag{11-71}$$

With the application of the transformation

$$\psi(R, \eta) = \theta(R, \eta)e^{-(\text{Pe}/4)\eta} \tag{11-72}$$

the differential equation (11-69) is transformed to

$$\frac{1}{R}\frac{\partial}{\partial R}\left(R\frac{\partial \theta}{\partial R}\right) + \frac{\partial^2 \theta}{\partial \eta^2} - \left(\frac{\text{Pe}}{4}\right)^2 \theta + \delta(R-1)\delta(\eta)e^{-(\text{Pe}/4)\eta} = 0 \tag{11-73}$$

where $e^{-(\text{Pe}/4)\eta}$ appearing in the source term can be omitted because the source term vanishes for $\eta \neq 0$, and $e^{-(\text{Pe}/4)\eta}$ becomes unity for $\eta = 0$. Therefore, we need to consider the solution of the homogeneous equation

$$\frac{1}{R}\frac{\partial}{\partial R}\left(R\frac{\partial \theta}{\partial R}\right) + \frac{\partial^2 \theta}{\partial \eta^2} - \left(\frac{\text{Pe}}{4}\right)^2 \theta = 0 \qquad \text{in} \qquad \begin{array}{c} 0 < R < 1 \\ -\infty < \eta < \infty \end{array} \tag{11-74}$$

subject to the boundary conditions

$$\left.\frac{\partial \theta}{\partial R}\right|_{R=0} = 0 \tag{11-75a}$$

$$\left.\frac{\partial \theta}{\partial R}\right|_{R=1} + \frac{\mathrm{Bi}}{2}\,\theta|_{R=1} = 0 \tag{11-75b}$$

$$\frac{\partial \theta}{\partial \eta} \to 0 \qquad \text{as} \qquad \eta \to \pm\,\infty \tag{11-75c}$$

in the regions $\eta < 0$ and $\eta > 0$. Let $\theta \equiv \theta^-$ be the solution for the region $\eta < 0$ and $\theta \equiv \theta^+$ be the solution for the region $\eta > 0$. The unknown coefficients associated with these solutions are determined from the requirement of continuity of temperature, namely,

$$\theta^-(\eta = 0) = \theta^+(\eta = 0) \tag{11-76}$$

and the jump condition

$$\left.\frac{\partial \theta^-}{\partial \eta}\right|_{\eta=0} - \left.\frac{\partial \theta^+}{\partial \eta}\right|_{\eta=0} = \delta(R-1) \tag{11-77}$$

This jump condition is obtained by integrating equation (11-73) from $\eta = -\varepsilon$ to $\eta = +\varepsilon$, and then letting $\varepsilon \to 0$, and physically represents conservation of energy at the source location. Once θ^+ and θ^- are determined, the dimensionless temperatures $\psi^\pm$ are determined according to the transformation given by equation (11-72).

Finally, the solution for the dimensionless temperatures $\psi^\pm(R, \eta)$ are determined as

$$\psi^\pm(R,\eta) = \sum_{n=1}^{\infty} \left[\frac{\beta_n^2}{(\mathrm{Bi}/2)^2 + \beta_n^2}\right] \frac{J_0(\beta_n R)}{J_0(\beta_n)} \frac{\exp\,(\mathrm{Pe}/4 \pm F)\,\eta}{F} \tag{11-78}$$

where the β_n values are the roots of

$$-\beta_n J_1(\beta_n) + \frac{\mathrm{Bi}}{2} J_0(\beta_n) = 0 \tag{11-79}$$

and F is defined by

$$F \equiv \sqrt{\left(\frac{\mathrm{Pe}}{4}\right)^2 + \beta_n^2} \tag{11-80}$$

with the plus and minus signs denoting the regions $\eta > 0$ and $\eta < 0$, respectively.

Here the Peclét number is a measure of the ratio of convective diffusion (i.e., due to the velocity of the moving source) to the conduction diffusion. Therefore, for the smaller Peclét number, the temperature field penetrates considerably farther "upstream" from the source than with the larger Peclét numbers.

REFERENCES

1. D. Rosenthal, *2-ene Congres National des Sciences* **201**, Brussels, 1935, pp. 1277–1292. Fédération belge des Sociétés scientifiques.
2. D. Rosenthal and R. Schmerber, *Welding J.* **17**, 2–8, 1938.
3. D. Rosenthal, *ASME Trans.* **68**, 849–866, 1946.
4. D. Rosenthal and R. H. Cameron, *Trans. ASME* **69**, 961–968, 1947.
5. E. M. Mahla, M. C. Rawland, C. A. Shook, and G. E. Doan, *Welding J.* **20**(10) (Res. Suppl.), 459, 1941.
6. W. A. Bruce, *J. Appl. Phys.* **10**, 578–585, 1939.
7. H. S. Carslaw and J. C. Jaeger, *Conduction of Heat in Solids*, Oxford, Clarendon, London, 1959.
8. J. C. Jaeger, *J. Proc. Roy. Soc. N. S. W.* **76**, 203–224, 1942.
9. R. J. Grosh, E. A. Trabant, and G. A. Hawkins, *Q. Appl. Math.* **13**, 161–167, 1955.
10. R. Weichert and K. Schönert, *Q. J. Mech. Appl. Math.* **31**(3), 363–379, 1978.
11. Y. Terauchi, H. Nadano, and M. Kohno, *Bull. JSME* **28**(245), 2789–2795, 1985.
12. N. R. DeRuisseaux and R. D. Zerkle, *Trans. ASME, J. Heat Transfer* **92**, 456–464, 1970.
13. W. Y. D. Yuen, *Math. Engg. Ind.* **1**, 1–19, 1987.
14. J. O. Outwater and M. C. Shaw, *Trans. ASME* **74**, 73, 1952.
15. E. G. Loewen and M. C. Shaw, *Trans. ASME* **76**, 217, 1954.
16. B. T. Chao and K. J. Trigger, *Trans. ASME* **80**, 311, 1958.
17. A. Cameron, A. N. Gordon, and G. T. Symm, *Proc. Roy. Soc.* **A286**, 45–61, 1965.
18. G. T. Symm, *Q. J. Mech. Appl. Math.* **20**, 381–391, 1967.
19. J. R. Barber, *Int. J. Heat Mass Transfer* **13**, 857–869, 1970.
20. I. L. Ryhming, *Acta Mech.* **32**, 261–274, 1979.
21. O. A. Tay, M. G. Stevenson, and G. de Vahl Davis, *Proc. Inst. Mech. Eng.* **188**, 627–638, 1974.
22. W. M. Mansour, M. O. N. Osman, T. S. Sansar, and A. Mazzawi, *Int. J. Prod. Rev.* **11**, 59–68, 1973.
23. R. G. Watts, ASME Paper No. 68-WA/HT-11, 1968.
24. M. F. Modest and H. Abakians, *J. Heat Transfer* **108**, 597–601, 1986.
25. J. F. Ready, *Effects of High Power Laser Radiation*, Academic, New York, 1971.
26. H. E. Cline and T. R. Anthony, *J. Appl. Phys.* **48**, 3895–3900, 1977.
27. Y. I. Nissim, A. Lietoila, R. B. Gold, and J. F. Gibbons, *J. Appl. Phys.* **51**, 274–279, 1980.
28. K. Brugger, *J. Appl. Phy.* **43**, 577–583, 1972.
29. M. F. Modest and H. Abakians, *J. Heat Transfer* **108**, 602–607, 1986.
30. M. F. Modest and H. Abakians, *J. Heat Transfer* **108**, 597–601, 1986.
31. S. Biyikli and M. F. Modest, *J. Heat Transfer* **110**, 529–532, 1988.
32. S. Roy and M. F. Modest, *J. Thermophys.* **4**, 199–203, 1990.

PROBLEMS

11-1 Consider the three-dimensional, quasi-stationary temperature field denoted $T(\xi, y, z)$ governed by the differential equation (11-8) with the boundary conditions at infinity taken as

$$\frac{\partial T}{\partial \xi} \to 0 \quad \text{for} \quad \xi \to \pm\infty, \qquad \frac{\partial T}{\partial y} \to 0 \quad \text{for} \quad y \to \pm\infty$$

$$\frac{\partial T}{\partial z} \to 0 \quad \text{for} \quad z \to \pm\infty$$

and the transformed equation (11-10) for the temperature field denoted $\theta(\xi, y, z)$. In equation (11-10), which is symmetric with respect to the variables ξ, y, and z, the function $\theta(\xi, y, z)$ depends only on the distance r from the point heat source. Then

a. Equation (11-10) can be written in the polar coordinates with respect to the r variable only; write this equation without the source term.

b. Develop the boundary condition at $r = 0$ for this equation by drawing a sphere of radius r around the point heat source, then equating the heat released by the source to the heat conducted away and letting $r \to 0$.

c. By solving this equation in the polar coordinates, develop an expression for the quasi-stationary temperature field $T(r, \xi)$ around the moving point heat source.

11-2 Develop equation (11-37) by writing an energy balance for a bar of uniform cross section with energy generation in the solid and heat dissipation from the lateral surfaces by convection with a heat transfer coefficient h into an ambient environment at zero temperature.

11-3 The temperature distribution in the gun barrel resulting from the firing of a bullet can be regarded as a problem of a point heat source moving with a constant velocity u along the axis of a solid cylinder of radius b if the base of the barrel is small enough compared to the outside radius of the barrel.

Assuming (a) constant speed and the rate of heat release by the point source, (b) no heat losses from the outer surface of the cylinder, and (c) cylinder long enough with respect to the diameter so that quasi-stationary state is established, develop the governing differential equations and the boundary conditions needed for the solution of the quasi-stationary temperature distribution in the cylinder.

11-4 Consider a boring process in order to increase the inside diameter of a hollow cylindrical workpiece. Such a problem can be modeled as a *moving ring heat source* advancing axially along the interior surface of a hollow cylinder. Assume a source of constant strength Q_0 watts, releasing its energy continuously as it moves with a constant speed u along the inner surface of the cylinder and heat loss by convection from the outer surface of the cylinder with a heat transfer coefficient h into an ambient environment at zero temperature. Initially the solid is also at zero temperature. Give the governing differential equations and the boundary conditions for the determination of the quasi-stationary temperature field in the cylinder. Note that this problem is analogous to that considered in Section 11-4, except the ring heat source is moving along the inside surface of the cylinder. Assume negligible heat loss from the inner surface of the hollow cylinder.

12

PHASE-CHANGE PROBLEMS

Transient heat transfer problems involving melting or solidification generally referred to as "phase-change" or "moving-boundary" problems are important in many engineering applications such as in the making of ice, the freezing of food, the solidification of metals in casting, thermal energy storage, processing of chemicals and plastics, crystal growth, aerodynamic ablation, casting and welding of metals and alloys, and numerous others, as well as natural phenomena such as the cooling of large masses of igneous rock. The solution of such problems is inherently difficult because the interface between the solid and liquid phases is moving as the latent heat is absorbed or released at the interface; as a result, the location of the solid–liquid interface is not known *a priori* and must follow as a part of the solution. In the solidification of pure substances, like water, the solidification takes place at a discrete temperature, and the solid and liquid phases are separated by a *clearly defined moving interface*. On the other hand, in the solidification of mixtures, alloys, and impure materials, the solidification takes place over an extended temperature range, and as a result the solid and liquid phases are separated by a moving *two-phase* region.

Early analytic works on the solution of phase-change problems include those by Lamé and Clapeyron [1] in 1831 and by Stefan [2] in 1891 in relation to ice formation. The fundamental feature of this type of problem is that the location of the boundary is both unknown and moving and that the parabolic heat conduction equation is to be solved in a region whose domain is also to be determined. Although references 1 and 2 are the early published works on this subject, the exact solution of a more general phase-change problem was discussed by F. Neumann in his lectures in the 1860s, but his lecture notes containing these solutions were not published until 1912. Since then, many phase-change problems have appeared in the literature, but the exact solutions are limited to a number of idealized situations involving semi-infinite or infinite regions, and subject to simple boundary and initial conditions [3]. Because of the nonlinearity of such problems, the superposition principle is not applicable and each case must be treated separately. When exact solutions are not available, approximate, semianalytic, and numerical methods can be used to solve the phase-change problems. We now present a brief discussion of various methods of solution of phase-change problems.

The *integral method*, which dates back to von Kármán and Pohlhausen, who used it for the approximate analysis of boundary layer equations, was applied by Goodman [5, 6] to solve a one-dimensional transient melting problem and subsequently by many other investigators [7–15] to solve various types of one-dimensional transient phase-change problems. This method provides a relatively straightforward and simple approach for approximate analysis of one-dimensional transient phase-change problems. The *variational formulation* derived by Biot [16] on the basis of an irreversible thermodynamic argument was used in the solution of one-dimensional, transient phase-change problems [17–21]. The *moving heat source method* (or the *integral equation*), originally applied by Lightfoot [22] to solve Neumann's problem, is based on the concept of representing the liberation (or absorption) of latent heat by a moving plane heat source (or sink) located at the solid–liquid interface. A general formulation of the moving heat source approach is given in reference 23, and various applications can be found in references 24–28. The *perturbation method* has been used by several investigators [29–34]; however, the analysis becomes very complicated if higher-order solutions are to be determined; also it is difficult to use this method for problems involving more than one dimension. The *embedding technique*, first introduced by Boley [35] to solve the problem of melting of a slab, has been applied to solve various phase-change problems [36–41]. The method appears to be versatile to obtain solutions for one, two, or three dimensions and to develop general starting solutions. A *variable eigenvalue approach* developed in connection with the solution of heat conduction problems involving time-dependent boundary condition parameters [42, 43] has been applied to solve one-dimensional transient phase-change problems [44]. The method is applicable to solve similar problems in the cylindrical or spherical symmetry. The *electrical network analog method* often used in early applications [45–49] has now been replaced by purely numerical methods of solution because of the availability of high-performance computer processors. A large number of purely numerical solutions of phase-change problems have been reported [50–81].

Reviews of phase-change problems up to 1965 can be found in references 82–84, while more contemporary reviews are found in references 85–89. Extensive lists of references and treatments of the fundamentals of solidification can be found in standard texts [90–94]. Experimental investigation of phase-change problems is important in order to check the validity of various analytic models, with representative experimental studies available in the literature [95–99].

In this chapter, we present a series of analytic and numerical solutions for phase-change problems in the Cartesian and cylindrical coordinate systems.

12-1 MATHEMATICAL FORMULATION OF PHASE-CHANGE PROBLEMS

To illustrate the mathematical formulation of phase-change problems, we consider first a one-dimensional solidification problem and then a one-dimensional melting problem.

Interface Condition for Phase-Change Problems

Solidification Problem A liquid having a single phase-change temperature T_m (i.e., melting temperature $\equiv$ freezing temperature) is confined to a semi-infinite region $0 < x < \infty$. Initially, the liquid is at a uniform temperature T_i that is higher than the phase-change temperature T_m. At time $t = 0$, the temperature of the boundary surface $x = 0$ is suddenly lowered to a temperature T_0, which is less than the melting/freezing temperature T_m and is maintained at that temperature for times $t > 0$. The solidification starts at the boundary surface $x = 0$, and the solid–liquid interface $x = s(t)$ moves in the *positive x direction*.

Figure 12-1(a) shows the geometry and coordinates for such a one-dimensional solidification problem. The temperatures $T_s(x,t)$ and $T_l(x,t)$ for the solid and liquid phases, respectively, are governed by the standard diffusion equations given by

$$\frac{\partial^2 T_s(x,t)}{\partial x^2} = \frac{1}{\alpha_s}\frac{\partial T_s(x,t)}{\partial t} \quad \text{in} \quad 0 < x < s(t), \quad t > 0 \tag{12-1a}$$

$$\frac{\partial^2 T_l(x,t)}{\partial x^2} = \frac{1}{\alpha_l}\frac{\partial T_l(x,t)}{\partial t} \quad \text{in} \quad s(t) < x < \infty, \quad t > 0 \tag{12-1b}$$

where we assumed constant thermophysical properties for the solid and liquid phases. Here, $s(t)$ is the location of the solid–liquid interface, which is not

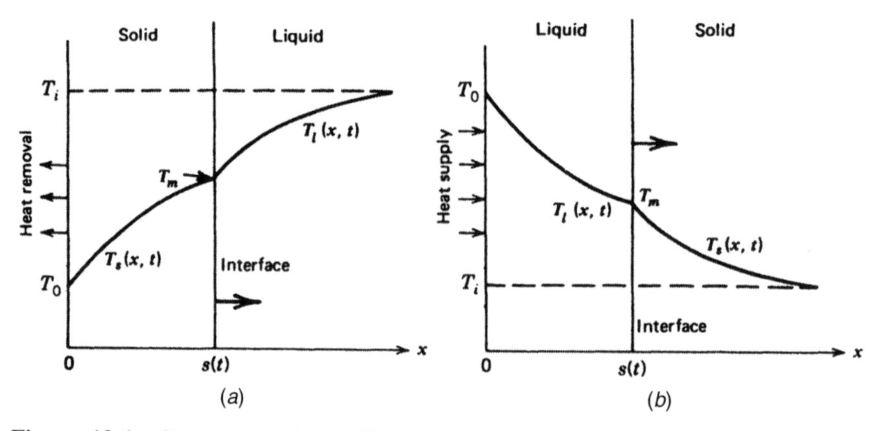

Figure 12-1 Geometry and coordinates for one-dimensional (a) solidification and (b) melting problems.

known *a priori*; hence it must be determined as a part of the solution. The subscripts s and l refer, respectively, to the solid and liquid phases. Therefore, the problem involves three unknowns: namely, $T_s(x,t)$, $T_l(x,t)$, and $s(t)$. An additional equation governing $s(t)$ is determined by considering an interface energy balance at $x = s(t)$, stated as

$$\begin{bmatrix} \text{Conduction} \\ \text{heat flux from the} \\ \text{the liquid phase} \\ \text{in the negative} \\ x \text{ direction} \end{bmatrix} + \begin{bmatrix} \text{Rate of heat} \\ \text{liberated during} \\ \text{solidification} \\ \text{per unit area of} \\ \text{interface} \end{bmatrix} = \begin{bmatrix} \text{Conduction} \\ \text{heat flux into} \\ \text{the solid phase} \\ \text{in the negative} \\ x \text{ direction} \end{bmatrix} \tag{12-2a}$$

or

$$k_l \frac{\partial T_l}{\partial x} + \rho L \frac{ds(t)}{dt} = k_s \frac{\partial T_s}{\partial x} \quad \text{at} \quad x = s(t), \quad t > 0 \tag{12-2b}$$

where L is the latent heat of fusion per unit mass (J/kg) associated with the phase change. For the time being we have neglected the density difference for the solid and liquid phases, and therefore assumed $\rho_l = \rho_s = \rho$ at the solid–liquid interface. The middle term in equation (12-2b) represents the rate at which energy is released at the moving interface due to solidification (i.e., rate of heat of fusion liberation). Equation (12-2b) may be recast as

$$\boxed{k_s \frac{\partial T_s}{\partial x} - k_l \frac{\partial T_l}{\partial x} = \rho L \frac{ds(t)}{dt} \quad \text{at} \quad x = s(t)} \tag{12-2c}$$

The continuity of temperature at the solid–liquid interface is given by

$$\boxed{T_s(x = s, t) = T_m = T_l(x = s, t)} \tag{12-3}$$

where $T_s(x = s, t)$ and $T_l(x = s, t)$ are the solid and liquid phase temperature at $x = s(t)$, respectively, and T_m is the phase-change temperature.

Summarizing, equations (12-1a), (12-1b), and (12-2c) provide three differential equations that govern the temperature distributions in the solid and liquid phases and the position $s(t)$ of the solid–liquid interface.

Equation (12-3) provides two boundary conditions. Other boundary conditions and the initial conditions are specified depending on the nature of the physical conditions at the boundary surfaces. This matter will be illustrated later in this chapter with specific examples.

Melting Problem We now consider a solid having a single phase-change temperature T_m confined to a semi-infinite region $0 < x < \infty$. Initially, the solid is at a uniform temperature T_i that is lower than the phase-change temperature T_m. At time $t = 0$, the temperature of the boundary surface $x = 0$ is suddenly raised to a temperature T_0, which is greater than the melting temperature T_m and maintained at that temperature for times $t > 0$. We assume that the coordinate system for this melting problem is arranged as illustrated in Figure 12-1(*b*), so that the solid–liquid interface moves in the *positive x direction* as in the case of the solidification problem. The governing differential equations for this problem, assuming constant properties for each phase, are given by

$$\frac{\partial^2 T_l(x,t)}{\partial x^2} = \frac{1}{\alpha_l}\frac{\partial T_l(x,t)}{\partial t} \qquad \text{in} \qquad 0 < x < s(t), \qquad t > 0 \tag{12-4a}$$

$$\frac{\partial^2 T_s(x,t)}{\partial x^2} = \frac{1}{\alpha_s}\frac{\partial T_s(x,t)}{\partial t} \qquad \text{in} \qquad s(t) < x < \infty, \qquad t > 0 \tag{12-4b}$$

and an energy balance at the solid–liquid interface $x = s(t)$ may now be written as

$$\begin{bmatrix} \text{Conduction} \\ \text{heat flux from the} \\ \text{the liquid phase} \\ \text{in the positive} \\ x \text{ direction} \end{bmatrix} - \begin{bmatrix} \text{Conduction} \\ \text{heat flux into} \\ \text{the solid phase} \\ \text{in the positive} \\ x \text{ direction} \end{bmatrix} = \begin{bmatrix} \text{Rate of heat} \\ \text{absorbed during} \\ \text{melting per} \\ \text{unit area of} \\ \text{interface} \end{bmatrix} \tag{12-5a}$$

which yields

$$\left(-k_l\frac{\partial T_l}{\partial x}\right) - \left(-k_s\frac{\partial T_s}{\partial x}\right) = \rho L\frac{ds(t)}{dt} \qquad \text{at} \qquad x = s(t), \qquad t > 0 \tag{12-5b}$$

Rearrangement of the above interface energy balance equation reveals that it is exactly the same as given by equation (12-2c); hence we again have

$$\boxed{k_s\frac{\partial T_s}{\partial x} - k_l\frac{\partial T_l}{\partial x} = \rho L\frac{ds(t)}{dt} \qquad \text{at} \qquad x = s(t)} \tag{12-5c}$$

Thus, equations (12-4) and (12-5c) provide three differential equations for the determination of the three unknowns $T_s(x,t)$, $T_l(x,t)$, and $s(t)$ for the melting problem considered here. As detailed above, equation (12-3) also holds at the interface. Appropriate boundary and initial conditions need to be specified for their solution.

We note that, in the interface energy balance equation (12-2c) or (12-5c), the term $ds(t)/dt$ represents the *velocity of the interface* in the positive x direction; hence we write

$$\boxed{\frac{ds(t)}{dt} \equiv v_x(t)} \tag{12-6}$$

Then the interface energy balance equation can be written as

$$k_s\frac{\partial T_s}{\partial x} - k_l\frac{\partial T_l}{\partial x} = \rho L v_x(t) \qquad \text{at} \qquad x = s(t) \tag{12-7}$$

Effects of Density Change The difference in the density of phases at the interface during phase change gives rise to liquid motion across the interface. Usually $\rho_s > \rho_l$, with the exception being water, bismuth, and antimony, for which $\rho_s < \rho_l$.

To illustrate the effects of density change, we consider the one-dimensional solidification problem illustrated in Figure 12-1(a). Let $\rho_s > \rho_l$ and

v_x = velocity of the interface
v_l = velocity of liquid at the interface
H_s = enthalpy of solid phase (J/kg) at interface
H_1 = enthalpy of liquid phase (J/kg) at interface

In the physical situation considered in Figure 12-1(a), the interface velocity v_x is in the positive x direction, and for $\rho_s > \rho_l$ the motion of the liquid is in the opposite direction. Then, the energy balance at the interface allowing for the contributions of diffusive and the convective energy transfer becomes

$$k_s \frac{\partial T_s}{\partial x} - k_l \frac{\partial T_l}{\partial x} = (\rho_l H_l - \rho_s H_s) v_x - \rho_l H_l v_l \qquad \text{at} \qquad x = s(t) \tag{12-8}$$

The mass conservation equation at the interface may be written as

$$(\rho_l - \rho_s) v_x = \rho_l v_l \tag{12-9a}$$

or

$$v_l = -\frac{\rho_s - \rho_l}{\rho_l} v_x \tag{12-9b}$$

Eliminating v_l from equation (12-8) by means of equation (12-9b), we obtain

$$k_s \frac{\partial T_s}{\partial x} - k_l \frac{\partial T_l}{\partial x} = \rho_s L v_x \tag{12-10}$$

since

$$H_l - H_s = L = \text{latent heat of fusion} \tag{12-11}$$

Equation (12-10) is similar to equation (12-7) except ρ is now replaced by ρ_s.

Effects of Convection Consider the solidification problem illustrated in Figure 12-1(a). If the heat transfer from the liquid phase to the solid–liquid interface is controlled by convection, and therefore diffusion in the liquid phase is neglected, the interface energy balance equation (12-2c) takes the form

$$k_s \frac{\partial T_s}{\partial x} - h(T_\infty - T_m) = \rho L \frac{ds(t)}{dt} \qquad \text{at} \qquad x = s(t) \tag{12-12}$$

where h is the heat transfer coefficient for the liquid side, T_∞ is the bulk temperature of the liquid phase, and T_m is the melting-point temperature at the interface.

In the case of the melting problem illustrated in Figure 12-1(b), if convection is dominant in the liquid phase, equation (12-12) is applicable if the minus sign before h is changed to the plus sign, giving

$$k_s \frac{\partial T_s}{\partial x} + h(T_\infty - T_m) = \rho L \frac{ds(t)}{dt} \qquad \text{at} \qquad x = s(t) \tag{12-13}$$

Nonlinearity of Interface Condition The interface boundary conditions given by equations (12-2c), (12-12), and (12-13) are nonlinear. To show the nonlinearity of these equations, we need to relate $ds(t)/dt$ to the derivative of temperatures. This is done by taking the total derivative of the interface equation (12-3), namely,

$$\left[\frac{\partial T_s}{\partial x}dx + \frac{\partial T_s}{\partial t}dt\right]_{x=s(t)} = \left[\frac{\partial T_l}{\partial x}dx + \frac{\partial T_l}{\partial t}dt\right]_{x=s(t)} = 0 \tag{12-14a}$$

or

$$\frac{\partial T_s}{\partial x}\frac{ds(t)}{dt} + \frac{\partial T_s}{\partial t} = \frac{\partial T_l}{\partial x}\frac{ds(t)}{dt} + \frac{\partial T_l}{\partial t} = 0 \qquad \text{at} \qquad x = s(t) \tag{12-14b}$$

which can be rearranged as

$$\frac{ds(t)}{dt} = -\frac{\partial T_s/\partial t}{\partial T_s/\partial x} \qquad \text{and} \qquad \frac{ds(t)}{dt} = -\frac{\partial T_l/\partial t}{\partial T_l/\partial x} \tag{12-14c}$$

Introducing these results, for example, into equation (12-2c), we obtain

$$k_s \frac{\partial T_s}{\partial x} - k_l \frac{\partial T_l}{\partial x} = -\rho L \frac{\partial T_s/\partial t}{\partial T_l/\partial x} = -\rho L \frac{\partial T_l/\partial t}{\partial T_s/\partial x} \tag{12-15}$$

The nonlinearity of the above equation is now apparent.

Generalization to Multidimension The interface energy balance equation developed above for the one-dimensional case is now generalized for the multidimensional situations. Figure 12-2 illustrates a solidification problem in a three-dimensional region. The solid and liquid phases are separated by a sharp interface defined by the equation

$$F(x, y, z, t) = 0 \tag{12-16}$$

The requirement of the continuity of temperatures at the interface becomes

$$T_s(x, y, z, t) = T_l(x, y, z, t) = T_m \qquad \text{at} \qquad F(x, y, z, t) = 0 \tag{12-17}$$

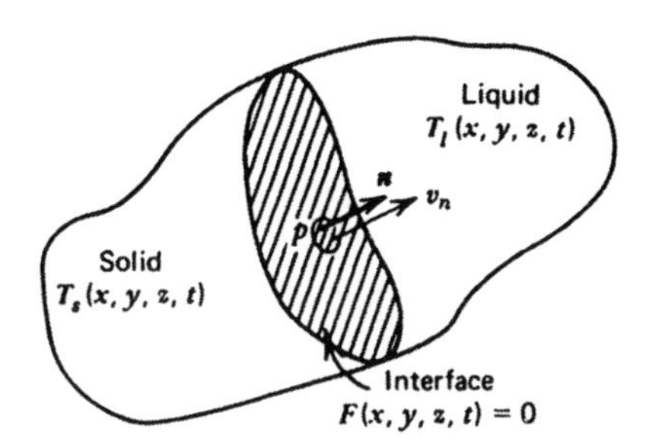

Figure 12-2 Solidification in three dimensions. Interface is moving in the direction of the surface normal n.

and the interface energy balance equation is written as

$$k_s \frac{\partial T_s}{\partial n} - k_l \frac{\partial T_l}{\partial n} = \rho L v_n \qquad \text{at} \qquad F(x, y, z, t) = 0 \tag{12-18}$$

where $\partial/\partial n$ denotes the derivative at the interface along the normal direction vector $\hat{n}$ at any location P on the interface and pointing toward the liquid region, and v_n is the velocity of this interface at the location P in the direction $\hat{n}$. Here we assumed that the densities of the solid and liquid phases are the same.

The interface energy balance equation (12-18) is not in a form suitable for developments of analytic or numerical solutions of the phase-change problems. An alternative form of this equation is given by Patel [100] as

$$\left[1 + \left(\frac{\partial s}{\partial x}\right)^2 + \left(\frac{\partial s}{\partial y}\right)^2\right]\left(k_s \frac{\partial T_s}{\partial z} - k_l \frac{\partial T_l}{\partial z}\right) = \rho L \frac{\partial s}{\partial t} \qquad \text{at} \qquad z = s(x, y, t) \tag{12-19}$$

This form of the interface energy balance equation is analogous to the form given by equation (12-2c) for the one-dimensional case; therefore, it is more suitable for numerical or analytic purposes. We now examine some special cases of equation (12-19).

For the two-dimensional problem involving (x, z, t) variables, if the location of the solid–liquid interface is specified by the relation $F(x, z, t) = z - s(x, t) = 0$, then equation (12-19) reduces to

$$\left[1 + \left(\frac{\partial s}{\partial x}\right)^2\right]\left(k_s \frac{\partial T_s}{\partial z} - k_l \frac{\partial T_l}{\partial z}\right) = \rho L \frac{\partial s}{\partial t} \qquad \text{at} \qquad z = s(x, t) \tag{12-20}$$

This equation is the same as that used in references 25, 38, and 39 for an interface boundary condition in the analysis of two-dimensional phase-change problems.

For the one-dimensional problem involving (z, t) variables, if the location of the solid–liquid interface is given by $F(z, t) = z - s(t) = 0$, equation

(12-19) reduces to

$$k_s \frac{\partial T_s}{\partial z} - k_l \frac{\partial T_l}{\partial z} = \rho L \frac{ds}{dt} \qquad \text{at} \qquad z = s(t) \tag{12-21}$$

which is identical to equation (12-2c) if z is replaced by x.

In the cylindrical coordinate system involving (r, ϕ, t) variables, if the location of the solid–liquid interface is given by $F(r, \phi, t) = r - s(\phi, t) = 0$, then the corresponding form of equation (12-19) becomes

$$\left[1 + \frac{1}{s^2}\left(\frac{\partial s}{\partial \phi}\right)^2\right]\left(k_s \frac{\partial T_s}{\partial r} - k_l \frac{\partial T_l}{\partial r}\right) = \rho L \frac{\partial s}{\partial t} \qquad \text{at} \qquad r = s(\phi, t) \tag{12-22}$$

In the cylindrical coordinate system involving (r, z, t) variables, if the location of the solid–liquid interface is given as $F(r, z, t) = z - s(r, t) = 0$, the interface equation takes the form

$$\left[1 + \left(\frac{\partial s}{\partial r}\right)^2\right]\left(k_s \frac{\partial T_s}{\partial z} - k_l \frac{\partial T_l}{\partial z}\right) = \rho L \frac{\partial s}{\partial t} \qquad \text{at} \qquad z = s(r, t) \tag{12-23}$$

Dimensionless Variables of Phase-Change Problem

The role of dimensionless variables in phase-change problems is better envisioned if the interface energy balance equation (12-2c) is expressed in the dimensionless form as

$$\frac{\partial \theta_s}{\partial \eta} - \frac{k_l}{k_s}\frac{\partial \theta_l}{\partial \eta} = \frac{1}{\text{Ste}}\frac{d\delta(\tau)}{d\tau} \tag{12-24}$$

where we define

$$\theta_i(\tau, \eta) = \frac{T_i(x, t) - T_m}{T_m - T_0}, \qquad i = s \text{ or } l; \qquad \eta = x/b \tag{12-25a}$$

and

$$\delta(\tau) = \frac{s(t)}{b}; \qquad \tau = \frac{\alpha_s t}{b^2}; \qquad \text{Ste} = \frac{C_s(T_m - T_0)}{L} \tag{12-25b}$$

Here, b is a reference or characteristic length, L is the latent heat, C_s is the specific heat, T_m is the melting temperature, T_0 is a reference temperature, $s(t)$ is the location of the solid–liquid interface, and Ste is the *Stefan number*, named after J. Stefan. The above dimensionless variables, other than the Stefan number, are similar to those frequently used in the standard heat conduction problems; the Stefan number is associated with the phase-change process.

The Stefan number signifies the importance of sensible heat relative to the latent heat. If the Stefan number is small, say, less than approximately 0.1, the heat released or absorbed by the interface during phase change is affected very little as a result of the variation of the sensible heat content of the material during the propagation of heat through the medium. For materials such as aluminum, copper, iron, lead, nickel, and tin, the Stefan number based on a temperature difference between the melting temperature and the room temperature varies from about 1 to 3. For melting or solidification processes taking place with much smaller temperature differences, the Stefan number is much smaller. For example, in phase-change problems associated with thermal energy storage, the temperature differences are small; as a result the Stefan number is generally smaller than 0.1.

12-2 EXACT SOLUTION OF PHASE-CHANGE PROBLEMS

The exact solutions of phase-change problems are limited to few idealized situations for the reasons stated previously. They are mainly for the cases of one-dimensional infinite or semi-infinite regions and simple boundary conditions, such as the prescribed temperature at the boundary surface. Exact solutions are obtainable if the problem admits a *similarity solution*, allowing the two independent variables x and t to merge into a single similarity variable $x/t^{1/2}$. Some exact solutions can be found in references 3 and 4. We present below some of the exact solutions of phase-change problems in Cartesian and cylindrical systems.

Example 12-1 Single-Phase Problem: Solidification of Supercooled Liquid Here we consider solidification of a supercooled liquid in a half-space. A supercooled liquid at a uniform temperature T_i, which is lower than the solidification (or melting) temperature T_m of the solid phase, is confined to a half-space $x > 0$. It is assumed that the solidification starts at the surface $x = 0$, and at time $t = 0$, and the solid–liquid interface moves in the positive x direction. Figure 12-3 illustrates the geometry, the coordinates, and the temperature profiles. The solid phase being at the uniform temperature T_m throughout, there is no heat transfer through it; in other words, the heat released during the solidification process is transferred into the supercooled liquid and raises its temperature. The temperature distribution is unknown only in the liquid phase; hence the problem is a *single-phase problem*. In the following analysis we determine the temperature distribution in the liquid phase and the location of the solid–liquid interface as a function of time.

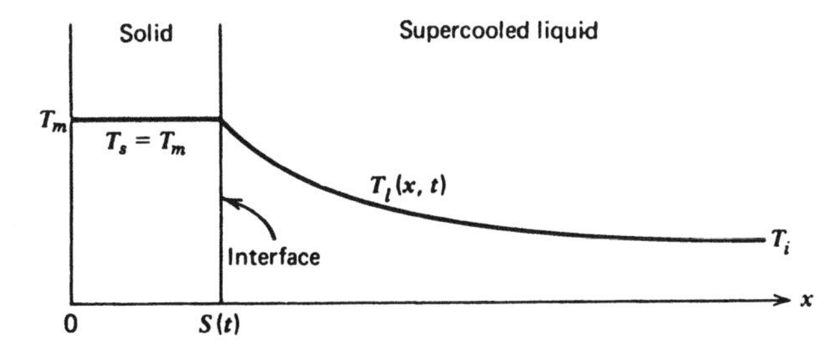

Figure 12-3 Solidification of supercooled liquid in a half-space.

Before presenting the analysis for the solution of this problem, we discuss the implications of the supercooling of a liquid. If a liquid is cooled very slowly, the bulk temperature may be lowered below the solidification temperature, and the liquid in such a state is called a *supercooled liquid*. After supercooling reaches some critical temperature, the solidification starts, and heat released during freezing raises the temperature of the supercooled liquid as the solid interface propagates. Little is known about the actual condition of the solid–liquid interface during the solidification of a supercooled liquid. During the solidification of supercooled water, the interface may grow as a dentritic surface consisting of thin, platelike crystals of ice interspersed in water rather than moving as a sharp interface [101]. Such behavior may be occasionally observed when removing bottled liquids from a freezer, in which the liquid rapidly solidifies upon perturbation of the vessel, or with the sudden solidification of a large pool of liquid wax. As a result, including the effects of irregular surface conditions in the analysis is a very complicated matter. Therefore, in the following solution only an idealized situation is considered. Namely, it is assumed that the solid–liquid interface is a sharp surface whose motion is similar to that encountered in the normal solidification process.

The mathematical formulation for the problem is given as

$$\frac{\partial^2 T_l}{\partial x^2} = \frac{1}{\alpha_l}\frac{\partial T_l(x,t)}{\partial t} \quad \text{in} \quad s(t) < x < \infty, \quad t > 0 \tag{12-26a}$$

$$T_s = T_m \quad \text{in} \quad 0 \le x < s(t), \quad t > 0 \tag{12-26b}$$

subject to the boundary conditions

$$T_l(x \to \infty, t) \to T_i \tag{12-27a}$$

$$T_l(x, t = 0) = T_i \tag{12-27b}$$

For the interface conditions, we specify

$$T_l(x = s, t) = T_m \quad \text{at} \quad x = s(t) \tag{12-28a}$$

$$-k_l \left.\frac{\partial T_l(x,t)}{\partial x}\right|_{x=s(t)} = \rho L \frac{ds(t)}{dt} \quad \text{at} \quad x = s(t) \tag{12-28b}$$

The interface equation (12-28b) states that the heat liberated at the interface as a result of solidification is exactly equal to the heat conducted into the supercooled liquid, and follows directly from equation (12-2c) for constant T_s. No equations are needed for the solid phase because it is at uniform temperature T_m. Recalling that erfc $[x/2(\alpha_l t)^{1/2}]$ is a solution of the heat conduction equation (12-26a), we choose a solution for $T_l(x,t)$ in the form

$$T_l(x,t) = A + B \operatorname{erfc}\left[\frac{x}{2(\alpha_l t)^{1/2}}\right] \tag{12-29}$$

where A and B are arbitrary constants to be determined. This solution satisfies the differential equation (12-26a). We now let $A = T_i$, which satisfies the boundary condition (12-27a) and the initial condition (12-27b), since erfc $(\infty) = 0$. If we require that the solution (12-29) should also satisfy the interface condition (12-28a), we find

$$T_m = T_i + B \text{ erfc}(\lambda) \tag{12-30a}$$

where we have defined the argument of the erfc as

$$\lambda = \frac{s(t)}{2(\alpha_l t)^{1/2}} \quad \text{or} \quad s(t) = 2\lambda(\alpha_l t)^{1/2} \tag{12-30b}$$

Since equation (12-30a) must be satisfied for *all times*, the parameter λ must be a constant. Equation (12-30a) is now solved for the coefficient B, yielding

$$B = \frac{T_m - T_i}{\text{erfc}(\lambda)} \tag{12-31}$$

and this result is introduced into equation (12-29). We obtain the solution

$$\frac{T_l(x,t) - T_i}{T_m - T_i} = \frac{\text{erfc}[x/2(\alpha_l t)^{1/2}]}{\text{erfc}(\lambda)} \tag{12-32}$$

Finally, the interface energy balance equation (12-28b) provides the additional relationship for the determination of the parameter λ. Namely, substituting $s(t)$ and $T_l(x, t)$ from equations (12-30b) and (12-32), respectively, into equation (12-28b), and making use of the following,

$$\frac{ds(t)}{dt} = \lambda\sqrt{\alpha_l}\, t^{-1/2} \tag{12-33a}$$

and the relation (see Appendix III)

$$\boxed{\frac{d \text{ erfc}(z)}{dz} = -\frac{2}{\sqrt{\pi}} e^{-z^2}} \tag{12-33b}$$

we obtain the following transcendental equation for the determination of λ:

$$\lambda e^{\lambda^2} \text{ erfc}(\lambda) = \frac{C(T_m - T_i)}{L\sqrt{\pi}} \tag{12-34}$$

Here, λ is the root of equation (12-34). Knowing the value of λ, we can determine the location of the solid–liquid interface $s(t)$ from equation (12-30b) and the temperature distribution $T_l(x, t)$ in the liquid phase from equation (12-32).

We note that the right-hand side of equation (12-34) is equal to the Stefan number divided by $\sqrt{\pi}$. In Table 12-1, we present the values of λe^{λ^2} erfc(λ) against λ. Thus, knowing the Stefan number, λ is determined from this table.

TABLE 12-1 Tabulation of Equation (12-34)

λ	$\lambda e^{\lambda^2} \text{erfc}(\lambda) = \dfrac{C(T_m - T_i)}{L\sqrt{\pi}} = \dfrac{\text{Ste}}{\sqrt{\pi}}$
0.00	0.00000E+00
0.10	8.96457E−02
0.20	1.61804E−01
0.30	2.20380E−01
0.40	2.68315E−01
0.50	3.07845E−01
0.60	3.40683E−01
0.70	3.68151E−01
0.80	3.91280E−01
0.90	4.10878E−01
1.00	4.27584E−01
1.10	4.41904E−01
1.20	4.54245E−01
1.30	4.64935E−01
1.40	4.74241E−01
1.50	4.82378E−01
1.60	4.89525E−01
1.70	4.95828E−01
1.80	5.01408E−01
1.90	5.06368E−01
2.00	5.10791E−01
2.50	5.27016E−01
3.00	5.37003E−01
3.50	5.43528E−01
4.00	5.47998E−01

Example 12-2 Single-Phase Problem: Melting at the Freezing Point
Here we consider melting in a half-space. A solid at the solidification (i.e., melting) temperature T_m is confined to a half-space $x > 0$. At time $t = 0$, the temperature of the boundary surface at $x = 0$ is raised to T_0, which is greater than T_m, and maintained at that temperature for times $t > 0$. As a result, melting starts at the surface $x = 0$ and the solid–liquid interface moves in the positive x direction. Figure 12-4 shows the coordinates and the temperature profiles. The solid phase being at a constant temperature T_m throughout, the temperature is unknown only in the liquid phase; hence the problem is a single-phase problem.

In the following analysis, the temperature distribution in the liquid phase and the location of the solid–liquid interface are determined as a function of time.

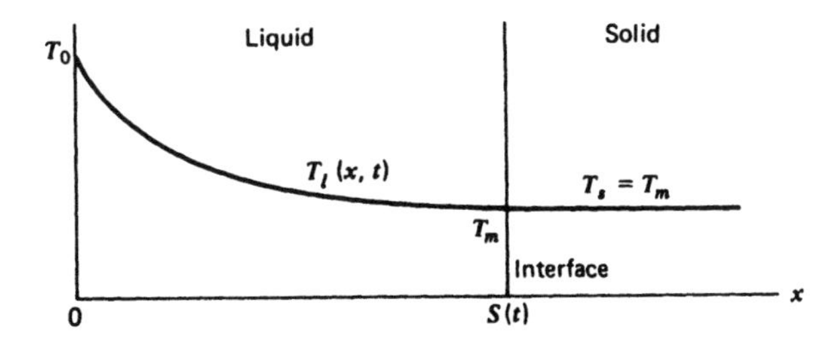

Figure 12-4 Melting in a half-space. Single-phase problem.

The mathematical formulation for the liquid phase is given as

$$\frac{\partial^2 T_l(x,t)}{\partial x^2} = \frac{1}{\alpha_l}\frac{\partial T_l(x,t)}{\partial t} \quad \text{in} \quad 0 < x < s(t), \quad t > 0 \tag{12-35a}$$

$$T_s = T_m \quad \text{in} \quad s(t) < x < \infty, \quad t > 0 \tag{12-35b}$$

subject to the boundary condition

$$T_l(x=0,t) = T_0 \tag{12-36}$$

For the interface condition, we specify

$$T_l(x=s,t) = T_m \quad \text{at} \quad x = s(t) \tag{12-37a}$$

$$-k_l \left.\frac{\partial T_l}{\partial x}\right|_{x=s(t)} = \rho L \frac{ds(t)}{dt} \quad \text{at} \quad x = s(t) \tag{12-37b}$$

Equation (12-37b) follows directly from equation (12-2c) for the case of constant T_s, and no additional equations are needed for the solid phase because it is at the melting temperature T_m throughout. If we assume a solution in the form

$$T_l(x,t) = A + B\ \text{erf}\left[\frac{x}{2(\alpha_l t)^{1/2}}\right] \tag{12-38}$$

where A and B are arbitrary constants, the differential equation (12-35a) is satisfied. If we now let $A = T_0$, the boundary condition (12-36) is satisfied since $\text{erf}(0) = 0$, giving the solution

$$T_l(x,t) = T_0 + B\ \text{erf}\left[\frac{x}{2(\alpha_l t)^{1/2}}\right] \tag{12-39}$$

If we impose the condition that this solution should also satisfy the boundary condition (12-37a) at $x = s(t)$, we obtain

$$T_m = T_0 + B \operatorname{erf}(\lambda) \tag{12-40a}$$

where we define

$$\lambda = \frac{s(t)}{2(\alpha_l t)^{1/2}} \qquad \text{or} \qquad s(t) = 2\lambda(\alpha_l t)^{1/2} \tag{12-40b}$$

Equation (12-40a) implies that λ should be a constant. Then the coefficient B is determined from equation (12-40a) as

$$B = \frac{T_m - T_0}{\operatorname{erf}(\lambda)} \tag{12-41}$$

Introducing equation (12-41) into (12-39), we obtain

$$\frac{T_l(x,t) - T_0}{T_m - T_0} = \frac{\operatorname{erf}[x/2(\alpha_l t)^{1/2}]}{\operatorname{erf}(\lambda)} \tag{12-42}$$

Finally, we utilize the interface condition (12-37b) to obtain an additional relationship for the determination of the parameter λ. When $s(t)$ and $T_l(x,t)$ from equations (12-40b) and (12-42), respectively, are introduced into equation (12-37b), and making use of the following,

$$\frac{ds(t)}{dt} = \lambda\sqrt{\alpha_l}\, t^{-1/2} \tag{12-43a}$$

and

$$\boxed{\frac{d \operatorname{erf}(z)}{dz} = \frac{2}{\sqrt{\pi}} e^{-z^2}} \tag{12-43b}$$

the following transcendental equation, similar to equation (12-34), is obtained for the determination of λ, namely,

$$\lambda e^{\lambda^2} \operatorname{erf}(\lambda) = \frac{C(T_0 - T_m)}{L\sqrt{\pi}} \tag{12-44}$$

and λ is the root of this equation. Knowing λ, $s(t)$ is then determined from equation (12-40b) and $T_l(x,t)$ from equation (12-42). As before, we note that the right-hand side of equation (12-42) is the Stefan number divided by $\sqrt{\pi}$. In Table 12-2, we present the values of λe^{λ^2} erf (λ) against λ. Thus, knowing the Stefan number, λ is determined from this table.

TABLE 12-2 Tabulation of Equation (12-44)

λ	$\lambda e^{\lambda^2}\text{erf}(\lambda) = \dfrac{C(T_0 - T_m)}{L\sqrt{\pi}} = \dfrac{\text{Ste}}{\sqrt{\pi}}$
0.00	0.00000E+00
0.10	1.13593E−02
0.20	4.63583E−02
0.30	1.07872E−01
0.40	2.01089E−01
0.50	3.34168E−01
0.60	5.19315E−01
0.70	7.74470E−01
0.80	1.12590E+00
0.90	1.61224E+00
1.00	2.29070E+00
1.10	3.24693E+00
1.20	4.61059E+00
1.30	6.58039E+00
1.40	9.46482E+00
1.50	1.37492E+01
1.60	2.02078E+01
1.70	3.00928E+01
1.80	4.54593E+01
1.90	6.97291E+01
2.00	1.08686E+02
2.50	1.29451E+03
3.00	2.43087E+04
3.50	7.31434E+05
4.00	3.55444E+07

Example 12-3 Two-Phase Problem: Solidification

We now consider solidification in a half-space such that the solution of both the solid and liquid phases is required. A liquid at a uniform temperature T_i that is greater than the melting temperature T_m of the solid phase is confined to a half-space $x > 0$. At time $t = 0$, the boundary surface at $x = 0$ is lowered to a temperature T_0, which is below T_m, and maintained at that temperature for times $t > 0$. As a result, the solidification starts at the surface $x = 0$ and the solid–liquid interface propagates in the positive x direction. Figure 12-5 illustrates the coordinates and the temperatures. This problem is a two-phase problem because the temperatures are unknown in both the solid and liquid phases. In the following analysis we determine the temperature distributions in both phases and the location of the solid–liquid interface. This problem is more general than the ones considered in the previous examples; its solution is known as *Neumann's solution*.

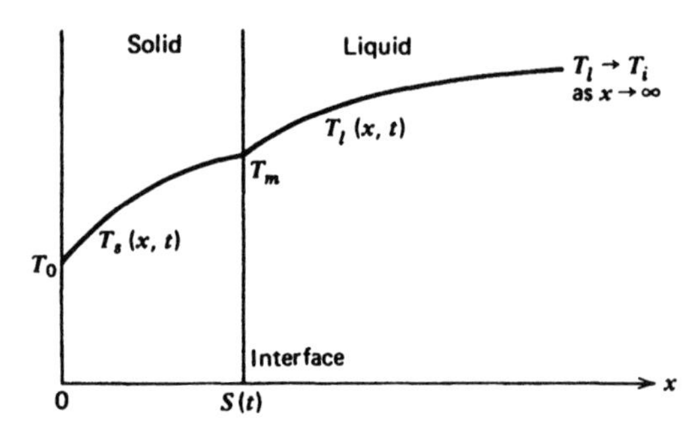

Figure 12-5 Solidification in a half-space. Two-phase problem.

The mathematical formulation of this problem for the solid phase is given as

$$\frac{\partial^2 T_s}{\partial x^2} = \frac{1}{\alpha_s} \frac{\partial T_s(x,t)}{\partial t} \quad \text{in} \quad 0 < x < s(t), \quad t > 0 \tag{12-45a}$$

$$T_s(x=0,t) = T_0 \tag{12-45b}$$

while the liquid phase is formulated as

$$\frac{\partial^2 T_l}{\partial x^2} = \frac{1}{\alpha_l} \frac{\partial T_l(x,t)}{\partial t} \quad \text{in} \quad s(t) < x < \infty, \quad t > 0 \tag{12-46a}$$

$$T_l(x \to \infty, t) \to T_i \tag{12-46b}$$

$$T_l(x, t=0) = T_i \quad \text{in} \quad x > 0 \tag{12-46c}$$

The coupling conditions at the interface $x = s(t)$ are

$$T_s(x=s(t),t) = T_l(x=s(t),t) = T_m \tag{12-47a}$$

$$k_s \left. \frac{\partial T_s}{\partial x} \right|_{x=s(t)} - k_l \left. \frac{\partial T_l}{\partial x} \right|_{x=s(t)} = \rho L \frac{ds(t)}{dt} \tag{12-47b}$$

For the solution of the solid phase, similar to Example 12-2, we choose a solution for $T_s(x,t)$ in the form

$$T_s(x,t) = T_0 + A \operatorname{erf}\left[\frac{x}{2(\alpha_s t)^{1/2}}\right] \tag{12-48}$$

which satisfies the differential equation (12-45a) and the boundary condition (12-45b). For the solution of the liquid phase, similar to Example 12-1, we choose a solution for $T_l(x,t)$ in the form

$$T_l(x,t) = T_i + B \operatorname{erfc}\left[\frac{x}{2(\alpha_l t)^{1/2}}\right] \tag{12-49}$$

which satisfies the differential equation (12-46a), the boundary condition (12-46b), and the initial condition (12-46c). The constants A and B are yet to be determined, although we still have our interface equations.

Equations (12-48) and (12-49) are first introduced into the interface condition (12-47a), yielding

$$T_0 + A \operatorname{erf}(\lambda) = T_i + B \operatorname{erfc}\left[\lambda\left(\frac{\alpha_s}{\alpha_l}\right)^{1/2}\right] = T_m \tag{12-50a}$$

where we have defined

$$\lambda = \frac{s(t)}{2(\alpha_s t)^{1/2}} \qquad \text{or} \qquad s(t) = 2\lambda(\alpha_s t)^{1/2} \tag{12-50b}$$

As before, equation (12-50a) implies that λ should be a constant. The coefficients A and B are now determined from equations (12-50) as

$$A = \frac{T_m - T_0}{\operatorname{erf}(\lambda)}, \qquad \text{and} \qquad B = \frac{T_m - T_i}{\operatorname{erfc}[\lambda(\alpha_s/\alpha_l)^{1/2}]} \tag{12-51}$$

Introducing the coefficients A and B into equations (12-48) and (12-49), we obtain the temperatures for the solid and liquid phases as

$$\frac{T_s(x,t) - T_0}{T_m - T_0} = \frac{\operatorname{erf}[x/2(\alpha_s t)^{1/2}]}{\operatorname{erf}(\lambda)} \tag{12-52a}$$

and

$$\frac{T_l(x,t) - T_i}{T_m - T_i} = \frac{\operatorname{erfc}[x/2(\alpha_l t)^{1/2}]}{\operatorname{erfc}[\lambda(\alpha_s/\alpha_l)^{1/2}]} \tag{12-52b}$$

The interface energy balance equation (12-47b) is now used to determine the relation for the evaluation of the parameter λ. That is, when $s(t)$, $T_s(x,t)$ and $T_l(x,t)$ from equations (12-50b), (12-52a), and (12-52b), respectively, are substituted into equation (12-47b), and making use of equations (12-33b) and (12-43b), we obtain the following transcendental equation for the determination of λ:

$$\frac{e^{-\lambda^2}}{\operatorname{erf}(\lambda)} + \frac{k_l}{k_s}\left(\frac{\alpha_s}{\alpha_l}\right)^{1/2} \frac{T_m - T_i}{T_m - T_0} \frac{e^{-\lambda^2(\alpha_s/\alpha_l)}}{\operatorname{erfc}[\lambda(\alpha_s/\alpha_l)^{1/2}]} = \frac{\lambda L\sqrt{\pi}}{C_s(T_m - T_0)} \tag{12-53}$$

Once λ is known from the solution of this equation, $s(t)$ is determined from equation (12-50b), $T_s(x,t)$ from equation (12-52a) and $T_l(x,t)$ from equation

(12-52b). We note that the nature of equation (12-53) does not allow parameterization by a single dimensionless number like the single-phase problems solved with the two previous examples.

Example 12-4 Solidification by a Line Heat Sink for Cylindrical System We now consider solidification by a line heat sink in an infinite medium with cylindrical symmetry. A line heat sink of strength Q (W/m) is located at $r = 0$ in a large body of liquid at a uniform temperature T_i, which is greater than the melting (i.e., solidification) temperature T_m of the medium. The heat sink is activated at time $t = 0$ to absorb heat continuously for times $t > 0$. As a result, the solidification starts at the origin $r \to 0$ and the solid–liquid interface propagates in the positive r direction. Figure 12-6 shows the coordinates and the temperature profiles. The problem has cylindrical symmetry, and since the temperatures are unknown in both regions, it is a two-phase problem. In this example, the temperature distributions in the solid and liquid phases, and the location of the solid–liquid interface as a function of time will be determined. Physically, one may consider this problem as a liquid freezing about a thin pipe containing refrigerant such that the domain of the liquid reservoir is large compared to the pipe diameter.

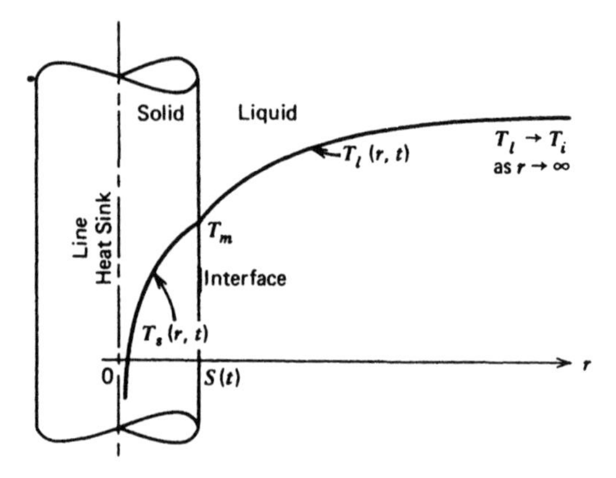

Figure 12-6 Solidification by a line heat sink in an infinite medium with cylindrical symmetry. Two-phase problem.

Paterson [102] has shown that the exact solution to the above problem is obtainable if the solution of the heat conduction equation is chosen as an *exponential integral function* in the form $E_1(r^2/4\alpha t)$. An alternative notation for $E_1(z)$ is given by $-Ei(-z)$, as discussed in references [103–104]. A tabulation of the exponential integral function $E_1(z)$ is given in Table 12-3, and a plot of function $E_1(z)$ is provided in Figure 12-7. A brief discussion of its properties is given in the note at the end of this chapter.

TABLE 12-3 Exponential Integral Function $E_1(z)^a$

z	$E_1(z)$	z	$E_1(z)$	z	$E_1(z)$	z	$E_1(z)$
0.00	∞	0.25	1.0442826	0.50	0.5597736	1.60	0.0863083
0.01	4.0379296	0.26	1.0138887	0.55	0.5033641	1.65	0.0802476
0.02	3.3547078	0.27	0.9849331				
0.03	2.9591187	0.28	0.9573083	0.60	0.4543795	1.70	0.0746546
0.04	2.6812637	0.29	0.9309182	0.65	0.4115170	1.75	0.0694887
0.05	2.4678985	0.30	0.9056767	0.70	0.3737688	1.80	0.0647131
0.06	2.2953069	0.31	0.8815057	0.75	0.3403408	1.85	0.0602950
0.07	2.1508382	0.32	0.8583352				
0.08	2.0269410	0.33	0.8361012	0.80	0.3105966	1.90	0.0562044
0.09	1.9187448	0.34	0.8147456	0.85	0.2840193	1.95	0.0524144
0.10	1.8229240	0.35	0.7942154	0.90	0.2601839	2.0	4.89005(−2)
0.11	1.7371067	0.36	0.7744622	0.95	0.2387375	2.1	4.26143
0.12	1.6595418	0.37	0.7554414				
0.13	1.5888993	0.38	0.7371121	1.00	0.2193839	2.2	3.71911
0.14	1.5241457	0.39	0.7194367	1.05	0.2018728	2.3	3.25023
0.15	1.4644617	0.40	0.7023801	1.10	0.1859909	2.4	2.84403
0.16	1.4091867	0.41	0.6859103	1.15	0.1715554	2.6	2.18502
0.17	1.3577806	0.42	0.6699973				
0.18	1.3097961	0.43	0.6546134	1.20	0.1584084	2.8	1.68553
0.19	1.2648584	0.44	0.6397328	1.25	0.1464134	3.0	1.30484
0.20	1.2226505	0.45	0.6253313	1.30	0.1354510	3.5	6.97014(−3)
0.21	1.1829020	0.46	0.6113865	1.35	0.1254168	4.0	3.77935
0.22	1.1453801	0.47	0.5978774	1.40	0.1162193	4.5	2.07340
0.23	1.1098831	0.48	0.5847843	1.45	0.1077774	5.0	1.14830
0.24	1.0762354	0.49	0.5720888	1.50	0.1000196	∞	0

[a]The figures in parentheses indicate the power of 10 by which the numbers to the left, and those below in the same column, are to be multiplied.

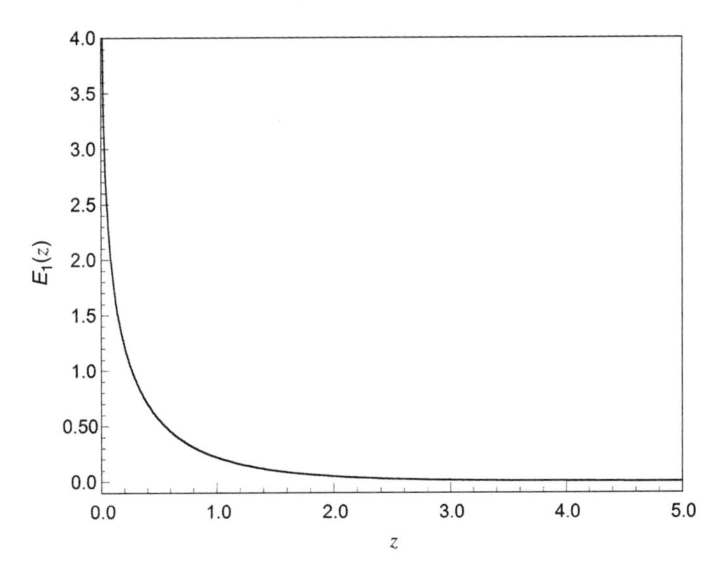

Figure 12-7 Exponential integral function $E_1(z)$.

The mathematical formulation of this problem is given for the solid phase as

$$\frac{1}{r}\frac{\partial}{\partial r}\left(r\frac{\partial T_s}{\partial r}\right)=\frac{1}{\alpha_s}\frac{\partial T_s(r,t)}{\partial t} \qquad \text{in} \qquad 0<r<s(t), \qquad t>0 \tag{12-54}$$

and for the liquid phase as

$$\frac{1}{r}\frac{\partial}{\partial r}\left(r\frac{\partial T_l}{\partial r}\right)=\frac{1}{\alpha_l}\frac{\partial T_l(r,t)}{\partial t} \qquad \text{in} \qquad s(t)<r<\infty, \qquad t>0 \tag{12-55}$$

$$T_l(r\to\infty,t)\to T_i \tag{12-56a}$$

$$T_l(r,t=0)=T_i, \qquad \text{in} \qquad r>0 \tag{12-56b}$$

For the solid–liquid interface, we specify

$$T_s(r=s,t)=T_l(r=s,t)=T_m \qquad \text{at} \qquad r=s(t) \tag{12-57a}$$

$$k_s\left.\frac{\partial T_s}{\partial r}\right|_{r=s(t)}-k_l\left.\frac{\partial T_l}{\partial r}\right|_{r=s(t)}=\rho L\frac{ds(t)}{dt} \qquad \text{at} \qquad r=s(t) \tag{12-57b}$$

We note that we have not yet specified a boundary condition in the limit as $r\to 0$. This condition will be considered below. We now choose the solutions for the solid and liquid phases, respectively, in the forms

$$T_s(r,t)=A+BE_1\left(\frac{r^2}{4\alpha_s t}\right) \qquad \text{in} \qquad 0<r<s(t) \tag{12-58a}$$

$$T_l(r,t)=T_i+CE_1\left(\frac{r^2}{4\alpha_l t}\right) \qquad \text{in} \qquad s(t)<r<\infty \tag{12-58b}$$

The derivatives of these solutions with respect to r are given as

$$\frac{\partial T_s(r,t)}{\partial r}=-\frac{2B}{r}e^{-r^2/4\alpha_s t} \tag{12-59a}$$

$$\frac{\partial T_l(r,t)}{\partial r}=-\frac{2C}{r}e^{-r^2/4\alpha_l t} \tag{12-59b}$$

where we have made use of the definition

$$\boxed{\frac{d}{dz}\left[E_1(z)\right]=\frac{d}{dz}\left(\int_{u=z}^{\infty}\frac{e^{-u}}{u}du\right)=-\frac{e^{-z}}{z}} \tag{12-60}$$

along with the chain rule. The solution (12-58a) for $T_s(r,t)$ satisfies the differential equation (12-54), while the solution (12-58b) for $T_l(r,t)$ satisfies the differential equation (12-55), the boundary condition (12-56a), and the initial condition (12-56b) since $E_1(\infty)=0$. The remaining conditions are used to determine the coefficients A, B, and C as now described. First, the energy balance around the line heat sink is written as

$$\lim_{r \to 0} \left[2\pi r k_s \frac{\partial T_s}{\partial r} \right] = Q \tag{12-61}$$

Introducing equation (12-59a) into (12-61), we find our constant B as

$$B = \frac{-Q}{4\pi k_s} \tag{12-62}$$

Equations (12-58a), (12-58b), along with equation (12-62), are introduced into the interface condition (12-57a), to yield

$$A - \frac{Q}{4\pi k_s} E_1(\lambda^2) = T_i + C E_1 \left(\frac{\lambda^2 \alpha_s}{\alpha_l} \right) = T_m \tag{12-63a}$$

where we have now defined

$$\lambda = \frac{s(t)}{2(\alpha_s t)^{1/2}} \quad \text{or} \quad s(t) = 2\lambda(\alpha_s t)^{1/2} \tag{12-63b}$$

Since equation (12-63a) should be valid for all values of time, we conclude that λ *must be a constant*, as we have done previously. The coefficients A and C are now solved from equations (12-63a); we find

$$A = T_m + \frac{Q}{4\pi k_s} E_1(\lambda^2) \tag{12-64a}$$

$$C = \frac{T_m - T_i}{E_1\left(\lambda^2 \alpha_s / \alpha_l\right)} \tag{12-64b}$$

The derivative of $s(t)$ is obtained from equation (12-63b) as

$$\frac{ds(t)}{dt} = \lambda \sqrt{\alpha_s} t^{-1/2} = \frac{2\alpha_s \lambda^2}{s} \equiv v_r \tag{12-65}$$

Introducing equations (12-64a) and (12-64b) into equations (12-58a,b), the solutions for the temperatures in the solid and liquid phases become

$$T_s(r,t) = T_m + \frac{Q}{4\pi k_s} \left[E_1(\lambda^2) - E_1\left(\frac{r^2}{4\alpha_s t} \right) \right] \quad \text{in} \quad 0 < r < s(t) \tag{12-66a}$$

and

$$T_l(r,t) = T_i + \frac{T_m - T_l}{E_1(\lambda^2 \alpha_s / \alpha_l)} E_1\left(\frac{r^2}{4\alpha_l t} \right) \quad \text{in} \quad s(t) < r < \infty \tag{12-66b}$$

Finally, when equations (12-65) and (12-66) are introduced into the interface energy balance equation (12-57b), the following transcendental equation is obtained for the determination of λ:

$$\frac{Q}{4\pi}e^{-\lambda^2} + \frac{k_l(T_m - T_i)}{E_1(\lambda^2\alpha_s/\alpha_l)}e^{-\lambda^2\alpha_s/\alpha_l} = \lambda^2\alpha_s\rho L \tag{12-67}$$

where λ is the root of this equation. Once λ is known, the location of the solid–liquid interface is determined from equation (12-63b), and the temperatures in the solid and liquid phase from equations (12-66a) and (12-66b), respectively. We note that the temperature in the solid $T_s(r \to 0) = -\infty$, in consideration of equation (12-66a), which is consistent with the physics of a line sink, noting, however, that energy is in fact conserved as $r \to 0$ by equation (12-61).

Summary Comments A scrutiny of the foregoing exact analyses reveals that in the rectangular coordinate system exact solutions are obtained for some half-space problems when the solution of the heat conduction equation is chosen as a function of $xt^{-1/2}$, namely, as $\text{erf}\left[x/2(\alpha t)^{1/2}\right]$ or $\text{erfc}\left[x/2(\alpha t)^{1/2}\right]$, with the former selected for the finite domain and the latter selected for the semi-infinite domain. In the cylindrical symmetry the corresponding solutions are in the form $E_1\left(r^2/4\alpha t\right)$, which is again a function of $rt^{-1/2}$. Paterson [102] has shown that the corresponding solution of the heat conduction equation in spherical symmetry is given in the form

$$\frac{(\alpha t)^{1/2}}{r}e^{-r^2/4\alpha t} - \frac{1}{2}\pi^{1/2}\text{erfc}\left(\frac{r}{2(\alpha t)^{1/2}}\right) \tag{12-68}$$

12-3 INTEGRAL METHOD OF SOLUTION OF PHASE-CHANGE PROBLEMS

The integral method provides a relatively simple and straightforward approach for the solution of one-dimensional transient phase-change problems and has been used for this purpose by several investigators [5–15]. The basic theory of this method is further described in Chapter 13 on the approximate solution of heat conduction problems. When it is applied to the solution of phase-change problems, the fundamental steps in the analysis remain essentially the same, except some modifications are needed in the construction of the temperature profile. In this section we illustrate the use of the integral method in the solution of phase-change problems with simple examples.

Example 12-5 Melting in a Half-Space

We first consider melting in a half-space. To give some idea on the accuracy of the integral method of solution of one-dimensional, time-dependent phase-change problems, we consider the one-phase melting problem for which the exact solution is available in Example 12-2. The problem considered is the melting of a solid confined to a half-space $x > 0$, initially at the melting temperature T_m. For times $t > 0$, the boundary surface at $x = 0$ is kept at a constant temperature T_0, which is greater than the melting temperature T_m of the solid. The melting starts at the surface $x = 0$ and the solid–liquid interface moves in the positive x direction as illustrated previously in Figure 12-4. In the following analysis we determine the location of the solid–liquid interface as a function of time.

The mathematical formulation of this problem is exactly the same as those given by equations (12-35)–(12-37). Namely, for the liquid and solid phase, the equations are given as

$$\frac{\partial^2 T_l}{\partial x^2} = \frac{1}{\alpha_l}\frac{\partial T_l(x,t)}{\partial t} \quad \text{in} \quad 0 < x < s(t), \quad t > 0 \tag{12-69a}$$

$$T_s = T_m \quad \text{in} \quad x > s(t), \quad t > 0 \tag{12-69b}$$

with the boundary condition

$$T_l(x = 0, t) = T_0 \tag{12-69c}$$

For the interface, we define

$$T_l(x = s, t) = T_m \quad \text{at} \quad x = s(t) \tag{12-70a}$$

$$-k_l \left.\frac{\partial T_l}{\partial x}\right|_{x=s(t)} = \rho L \frac{ds(t)}{dt} \quad \text{at} \quad x = s(t) \tag{12-70b}$$

We note here that the first step in the analysis with the integral method is to define a *thermal layer thickness* beyond which the temperature gradient is considered zero for practical purposes. Referring to Figure 12-4, we note that the location of the solid–liquid interface $x = s(t)$ is identical to the definition of the thermal layer, since the temperature gradient in the solid phase is zero for $x > s(t)$. Hence, we choose the region $0 \leq x \leq s(t)$ as the thermal layer appropriate for this problem and integrate the heat conduction equation (12-69a) from $x = 0$ to $x = s(t)$, to obtain

$$\left.\frac{\partial T}{\partial x}\right|_{x=s(t)} - \left.\frac{\partial T}{\partial x}\right|_{x=0} = \frac{1}{\alpha}\left[\frac{d}{dt}\left(\int_{x=0}^{s(t)} T\,dx\right) - \frac{ds(t)}{dt}\left.T\right|_{x=s(t)}\right] \tag{12-71}$$

To obtain equation (12-71), we have used the following relation:

$$\int_{x=0}^{s(t)} \frac{\partial^2 T}{\partial x^2} dx = \left.\frac{\partial T}{\partial x}\right|_0^{s(t)} \tag{12-72}$$

to generate the left-hand side and Leibniz's integral formula for the right-hand side.

For simplicity, we omitted the subscript l above and in the following analysis since we are only solving for the temperature field in the liquid phase. We note that equation (12-71) is similar to equation (13-5) to be considered in Chapter 13, and the reader is referred to the corresponding discussion for additional insight at this point. In view of the interface boundary conditions (12-70a) and (12-70b), the equation (12-72) reduces to

$$\boxed{-\frac{\rho L}{k}\frac{ds(t)}{dt} - \left.\frac{\partial T}{\partial x}\right|_{x=0} = \frac{1}{\alpha}\frac{d}{dt}\left[\theta - s(t)T_m\right]} \tag{12-73a}$$

where

$$\boxed{\theta(x) \equiv \int_{x=0}^{s(t)} T(x,t)\,dx} \tag{12-73b}$$

Equation (12-73a) is the *energy integral equation* for this problem. To solve this equation, we choose a second-degree polynomial approximation for the temperature in the form

$$T(x,t) = a + b(x-s) + c(x-s)^2 \tag{12-74}$$

where $s \equiv s(t)$. Three conditions are needed to determine these three coefficients. Equations (12-69c) and (12-70a) provide two conditions; however, the relation given by equation (12-70b) is not suitable for this purpose because if it is used, the resulting temperature profile will involve the $ds(t)/dt$ term. When such a profile is substituted into the energy integral equation, a second-order ordinary differential equation will result for $s(t)$, instead of the expected first-order equation. To alleviate this difficulty, an alternative relation is now developed [5]. The boundary condition (12-70a) is differentiated, which yields

$$dT \equiv \left[\frac{\partial T}{\partial x}dx + \frac{\partial T}{\partial t}dt\right]_{x=s(t)} = 0 \tag{12-75a}$$

or

$$\frac{\partial T}{\partial x}\frac{ds(t)}{dt} + \frac{\partial T}{\partial t} = 0 \tag{12-75b}$$

where we have again omitted the subscript l for simplicity. The term $ds(t)/dt$ is eliminated between equations (12-70b) and (12-75b) to yield

$$\left(\frac{\partial T}{\partial x}\right)^2 = \frac{\rho L}{k}\frac{\partial T}{\partial t} \qquad \text{at} \qquad x = s(t) \tag{12-76}$$

Now eliminating $\partial T/\partial t$ between equations (12-70a) and (12-76), we obtain

$$\left(\frac{\partial T}{\partial x}\right)^2 = \frac{\alpha\rho L}{k}\frac{\partial^2 T}{\partial x^2} \qquad \text{at} \qquad x = s(t) \tag{12-77}$$

This relation, together with the boundary conditions at $x = 0$ and $x = s(t)$, namely,

$$T(x = 0, t) = T_0 \tag{12-78a}$$

$$T(x = s(t), t) = T_m \tag{12-78b}$$

provide three independent relations for the determination of three unknown coefficients in equation (12-74). The resulting temperature profile becomes

$$T(x, t) = T_m + b(x - s) + c(x - s)^2 \tag{12-79a}$$

where

$$b = \frac{\alpha \rho L}{ks}[1 - (1 + \mu)^{1/2}] \tag{12-79b}$$

$$c = \frac{bs + (T_0 - T_m)}{s^2} \tag{12-79c}$$

$$\mu = \frac{2k}{\alpha \rho L}(T_0 - T_m) = \frac{2C(T_0 - T_m)}{L} \tag{12-79d}$$

Substituting the temperature profile (12-79) into the energy integral equation (12-73a), and performing the indicated operations, we obtain the following ordinary differential equation for the determination of the location of the solid–liquid interface $s(t)$, that is,

$$s\frac{ds}{dt} = 6\alpha \frac{1 - (1 + \mu)^{1/2} + \mu}{5 + (1 + \mu)^{1/2} + \mu} \tag{12-80a}$$

with

$$s(t = 0) = 0 \tag{12-80b}$$

The solution of equation (12-80a) is

$$s(t) = 2\lambda\sqrt{\alpha t} \tag{12-81a}$$

where

$$\lambda \equiv \left[3\frac{1 - (1 + \mu)^{1/2} + \mu}{5 + (1 + \mu)^{1/2} + \mu} \right]^{1/2} \tag{12-81b}$$

We note that our approximate solution (12-81a) for $s(t)$ is of the same form as the exact solution of the same problem given previously by equation (12-40b); however, the parameter λ is given by equation (12-81b) for the approximate solution, whereas it is the root of the transcendental equation (12-44), that is,

$$\lambda e^{\lambda^2} \operatorname{erf}(\lambda) = \frac{C(T_0 - T_m)}{L\sqrt{\pi}} = \frac{\mu}{2\sqrt{\pi}} \tag{12-82}$$

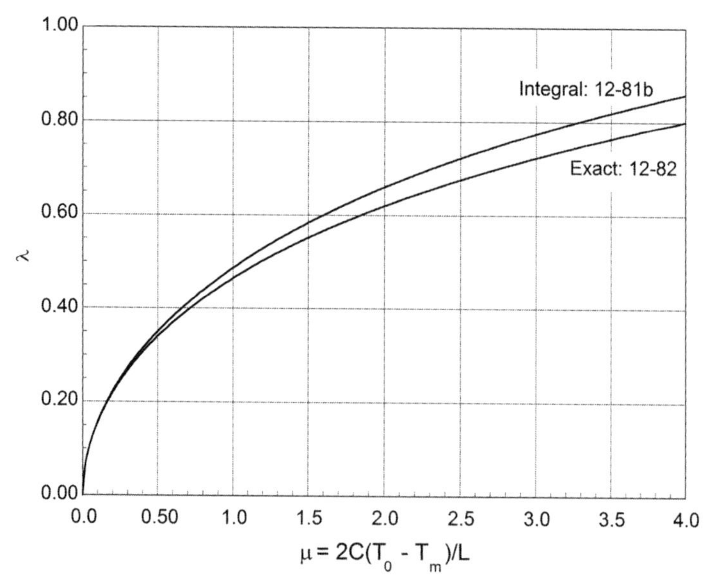

Figure 12-8 Comparison of exact and approximate solutions of the melting problem in half-space.

for the exact solution. Therefore, the accuracy of the approximate analysis can be determined by comparing the exact and approximate values of λ as a function of the quantity μ. Now, recalling the definition of the Stefan number given by equation (12-25b), we note that the parameter μ is actually twice the Stefan number. Figure 12-8 shows a comparison of the exact and approximate values of λ as a function of the parameter μ. The agreement between the exact and approximate analysis is reasonably good for the second-degree profile used here. If a cubic polynomial approximation were used, the agreement would be much closer [5].

12-4 VARIABLE TIME STEP METHOD FOR SOLVING PHASE-CHANGE PROBLEMS: A NUMERICAL SOLUTION

When analytic methods of solution are not possible or impractical, numerical techniques, such as finite differences or finite element are used for solving phase-change problems. The numerical methods of solving phase-change problems can be categorized as follows:

1. *Fixed-Grid Methods*, in which the space–time domain is subdivided into a finite number of equal grids Δx, Δt for all times. Then the moving solid–liquid interface will in general lie somewhere between two grid points at any given time. The methods of Crank [106] and Ehrlich [107]

are the examples for estimating the location of the interface by a suitable interpolation formula as a part of the solution.

2. *Variable-grid methods* in which the space–time domain is subdivided into equal intervals in one direction only and the corresponding grid size in the other direction is determined so that the moving boundary always remains at a grid point. For example, Murray and Landis [52] chose equal steps Δt in the time domain and kept the number of space intervals fixed, which in turn allowed the size of the space interval Δx to be changed (decreased or increased) as the interface moved. In an alternative approach, the space domain is subdivided into fixed equal intervals Δx, but the time step is varied such that the interface moves a distance Δx during the time interval Δt; hence it always remains at a grid point at the end of each time interval Δt. Several variations of such a variable time step approach have been reported in the literature [52, 66, 79, 80].
3. *Front-fixing method* is used in one-dimensional problems. This is essentially a coordinate transformation scheme that immobilizes the moving front and therefore alleviates the need for tracking the moving front at the expense of solving a more complicated problem by the numerical scheme [77, 106].
4. *The enthalpy method* has been used by several investigators to solve phase-change problems in situations in which the material does not have a distinct solid–liquid interface. Instead, the melting or solidification takes place over an extended range of temperatures. The solid and liquid phases are separated by a two-phase moving region. In this approach, an enthalpy function, $H(T)$, which is the total heat content of the substance, is used as a dependent variable along with the temperature. The method is also applicable for phase-change problems involving a single phase-change temperature [73–76].

In this section we present the *modified variable time step* (MVTS) method described by Gupta and Kumar [79]. We consider the solidification of a liquid initially at the melting temperature T_m^*, confined to the region $0 \leq x \leq B$. For times $t > 0$, the boundary surface at $x = 0$ is subjected to convective cooling into an ambient at a constant temperature T_∞ with a heat transfer coefficient h, while the boundary surface at $x = B$ is kept insulated or satisfies the symmetry condition. The solidification starts at the boundary surface $x = 0$, and the solid–liquid interface moves in the x direction as illustrated in Figure 12-9.

The temperature $T(x, t)$ varies only in the solid phase since the liquid region is at the melting temperature T_m^*. We are concerned with the determination of the temperature distribution $T(x, t)$ in the solid phase and location of the interface as a function of time. The mathematical formulation of this solidification problem is given as follows:

$$\text{Solid region:} \qquad \frac{\partial^2 T}{\partial x^2} = \frac{1}{\alpha}\frac{\partial T}{\partial t} \qquad \text{in} \qquad 0 < x < s(t), \qquad t > 0 \qquad (12\text{-}83a)$$

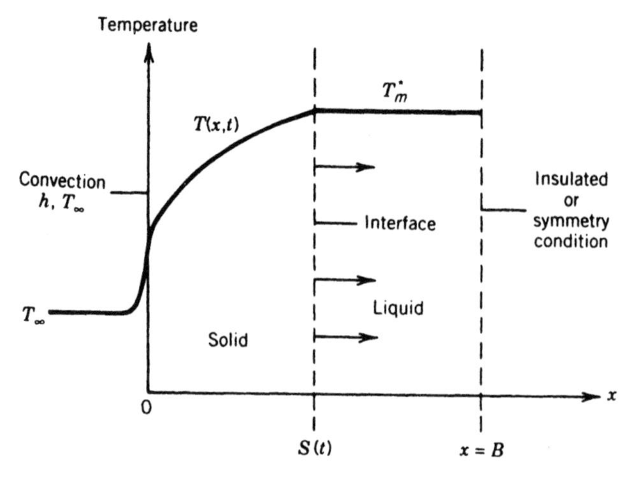

Figure 12-9 Geometry and coordinate for single-phase solidification.

$$-k\left.\frac{\partial T}{\partial x}\right|_{x=0}+hT|_{x=0}=hT_\infty \tag{12-83b}$$

Interface: $$T(x=s,t)=T_m^* \quad \text{at} \quad x=s(t) \tag{12-83c}$$

$$k\left.\frac{\partial T}{\partial x}\right|_{x=s(t)}=\rho L\frac{ds(t)}{dt} \quad \text{at} \quad x=s(t) \tag{12-83d}$$

where h is the heat transfer coefficient, $s(t)$ is the location of the solid–liquid interface, ρ is the density, L is the latent heat of fusion (i.e., solidification), k is the thermal conductivity, and α is the thermal diffusivity. We note that $T_l \equiv T_m^*$ in the region of the liquid $s(t) < x \leq B$; hence we only seek a solution $T(x, t)$ in the solid region and have therefore dropped the subscript s in the above formulation. We have assumed equal density between the solid and liquid phases.

To solve the above problem with finite differences, the "$x - t$" domain is subdivided into small intervals of constant Δx in space and variable Δt in time as illustrated in Figure 12-10. The variable time step approach requires that at each time level t_n the time step Δt_n is so chosen that the interface moves exactly a distance Δx during the time interval Δt_n; hence it always stays on the node. Therefore, we are concerned with the determination of the time step $\Delta t_n = t_{n+1} - t_n$ such that, in the time interval from t_n to t_{n+1}, the interface moves from the position $n\,\Delta x$ to the next position $(n+1)\Delta x$. We describe below first the finite-difference approximation of this solidification problem and then the determination of the time step Δt_n.

The finite-difference approximation of equations (12-83) is described below:

1. *Differential Equation (12-83a)* This differential equation can be approximated with finite differences by using either the implicit scheme or the

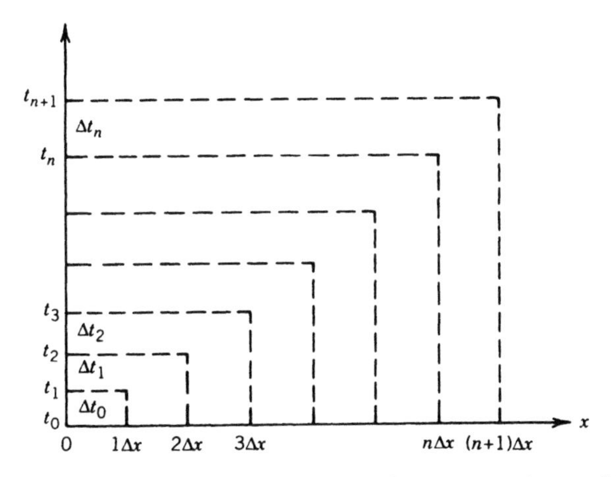

Figure 12-10 Subdivision of $x - t$ domain using constant Δx, variable Δt.

combined method. For simplicity, we prefer the implicit method and write equation (12-83a) in finite-difference form as

$$\frac{T_{i-1}^{n+1} - 2T_i^{n+1} + T_{i+1}^{n+1}}{(\Delta x)^2} = \frac{1}{\alpha}\frac{T_i^{n+1} - T_i^n}{\Delta t_n} \tag{12-84a}$$

where the following notation is adopted:

$$T(x, t_n) = T(i\Delta x, t_n) \equiv T_i^n \tag{12-84b}$$

Equation (12-84a) is rearranged as

$$[-r_n T_{i-1}^{n+1} + (1 + 2r_n)T_i^{n+1} - r_n T_{i+1}^{n+1}]^{(p)} = T_i^n \tag{12-85a}$$

where the superscript p over the bracket refers to the pth iteration, and the parameter r_n is defined as

$$r_n = \frac{\alpha\,\Delta t_n}{(\Delta x)^2}, \qquad i = 1, 2, 3, \ldots, \qquad \text{with} \qquad \Delta t_n = t_{n+1} - t_n \tag{12-85b}$$

2. *Boundary Condition at* $x = 0$ The convection boundary condition (12-83b) is rearranged as

$$\frac{\partial T}{\partial x} = HT - HT_\infty \qquad \text{where} \qquad H = \frac{h}{k} \tag{12-86a}$$

and then discretized as

$$\frac{T_1^{n+1} - T_0^{n+1}}{\Delta x} = HT_0^{n+1} - HT_\infty \tag{12-86b}$$

This result is now rearranged in the form

$$[T_1^{n+1} - (1 + H\Delta x)T_0^{n+1}]^{(p)} = -H\,\Delta x T_\infty \tag{12-86c}$$

where superscript p over the bracket denotes the pth iteration. The finite-difference equation (12-86c) is first order accurate.

3. *Interface Conditions* The condition of continuity of temperature at the interface, equation (12-83c), is written as

$$T_{n+1}^{n+1} = T_m^* = \text{melting temperature} \tag{12-87a}$$

which is valid for all times. The interface energy balance equation (12-83d) is discretized as

$$\frac{T_{n+1}^{n+1} - T_n^{n+1}}{\Delta x} = \frac{\rho L}{k}\frac{\Delta x}{\Delta t_n} \tag{12-87b}$$

which is rearranged in the form

$$[\Delta t_n]^{(p+1)} = \frac{\rho L}{k}\left[\frac{(\Delta x)^2}{T_m^* - T_n^{n+1}}\right]^{(p)} \tag{12-87c}$$

since $T_{n+1}^{n+1} = T_m^* =$ melting temperature.

Determination of Time Steps

We now describe the algorithms for the determination of time step Δt_n such that during this time step, the interface moves exactly a distance Δx.

Starting Time Step Δt_0 An explicit expression can be developed for the calculation of the first step Δt_0 as follows. Set $n = 0$ in equations (12-86c) and (12-87c), eliminate T_0^1 between the resulting two equations, and note that $T_1^1 \equiv T_m^*$. The following explicit expression is obtained for Δt_0:

$$\Delta t_0 = \frac{\rho L}{k}\frac{\Delta x(1 + H\Delta x)}{H(T_m^* - T_\infty)} \tag{12-88}$$

where $\Delta t_0 \equiv t_1 - t_0$.

Time Step Δt_1 We set $i = 1$, $n = 1$ in equation (12-85a) and note that $T_1^1 = T_2^2 = T_m^*$. Then equation (12-85a) becomes

$$[-r_1 T_0^2 + (1 + 2r_1)T_1^2]^{(p)} = (1 + r_1^{(p)})T_m^* \tag{12-89a}$$

and from the boundary condition (12-86c) for $n = 1$, we obtain

$$[-(1 + H\,\Delta x)T_0^2 + T_1^2]^{(p)} = -H\,\Delta x T_\infty. \tag{12-89b}$$

To solve equations (12-89a) and (12-89b) for T_0^2 and T_1^2, the value of $r_1^{(p)}$ is needed; however $r_1^{(p)}$ defined by equation (12-85b) depends on $\Delta t_1^{(p)}$. Therefore, iteration is needed for their solution. To start iterations, we set

$$\Delta t_1^{(0)} = \Delta t_0$$

Following the above process, $r_1^{(0)}$ is determined from equation (12-85b); using this value of $r_1^{(0)}$, equations (12-89a,b) are solved for T_0^2 and T_1^2. Knowing T_1^2, we can compute $\Delta t_1^{(1)}$ from equation (12-87c). Iterations are continued until the difference between two consecutive time steps

$$\left| \Delta t_1^{(p+1)} - \Delta t_1^{(p)} \right|$$

satisfies a specified convergence criterion.

General Time Step $\boldsymbol{\Delta t_n}$ The above results are now used in the following algorithm to calculate the time steps Δt_n at each time level t_n, $n = 2, 3, \ldots$.

1. The starting time step Δt_0 at the time level t_0 is calculated directly from the explicit expression (12-88) since all the quantities on the right-hand side of this equation are known.
2. The time steps Δt_n at the time levels t_n, $n = 2, 3, \ldots$ are calculated by iteration. A guess value Δt_n^0 is chosen as

$$\Delta t_n^{(0)} = \Delta t_{n-1}, \qquad n = 2, 3, \ldots \tag{12-90a}$$

The system of finite-difference equations (12-85) and (12-86c), together with the condition (12-87a), are solved for $i = 1, 2, 3, \ldots, n$ by setting $p = 0$, and a first estimate is obtained for the nodal temperatures

$$[T_i^{n+1}]^{(0)}, \qquad \text{for} \qquad i = 1, 2, \ldots, n \tag{12-90b}$$

We note that the system of equations is tridiagonal and hence readily solved.

3. The values of $[T_i^{n+1}]^{(0)}$ obtained from equation (12-90b) are introduced into equation (12-87c) for $p = 0$, and a first estimate for the time step $\Delta t_n^{(1)}$ is determined.
4. $\Delta t_n^{(1)}$ is used as a guess value and steps 2 and 3 are repeated to calculate a second estimate for the time step $\Delta t_n^{(2)}$.
5. Steps 2, 3, and 4 are repeated until the difference between two consecutive time steps

$$\left| \Delta t_n^{(p+1)} - \Delta t_n^{(p)} \right|$$

satisfies a specified convergence criterion.

12-5 ENTHALPY METHOD FOR SOLUTION OF PHASE-CHANGE PROBLEMS: A NUMERICAL SOLUTION

In the solution of phase-change problems considered previously, the temperature has been the sole dependent variable. That is, the energy equation has been written separately for the solid and liquid phases, and the temperatures have been coupled through the interface energy balance condition. Such a formulation gives rise to the tracking of the moving interface, and it is a difficult matter if the problem is to be solved with finite differences.

An alternative approach is the use of the enthalpy form of the energy equation along with the temperature. The advantage of the enthalpy method is that a single energy equation becomes applicable in both phases; hence there is no need to consider liquid and solid phases separately. Therefore, any numerical scheme such as the finite-difference or finite-element method can readily be adopted for the solution. In addition, the enthalpy method is capable of handling phase-change problems in which the phase change occurs over an extended temperature range rather than at a single phase-change temperature.

Figure 12-11 shows enthalpy–temperature relations for (*a*) pure crystalline substances and eutectics and (*b*) glassy substances and alloys. For pure substances the phase change takes place at a discrete temperature and hence is associated with the latent heat L. Therefore, in Figure 12-11(*a*), a jump discontinuity occurs at the melting temperature T_m^*; hence $\partial H/\partial T$ becomes infinite and the energy equation apparently is not meaningful at this point. However, it has been shown [73] that the enthalpy form of the energy equation given by

$$\nabla \cdot (k\nabla T) = \rho \frac{\partial H(T)}{\partial t} \tag{12-91}$$

is equivalent to the usual temperature form in which the heat conduction equation is written separately for the liquid and solid regions and coupled with the energy

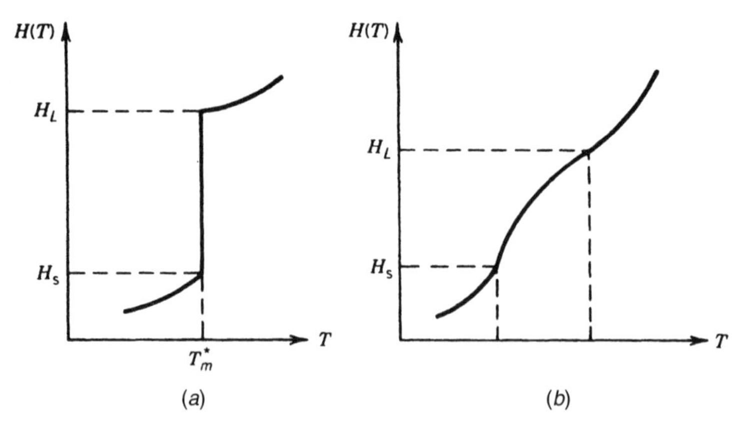

Figure 12-11 Enthalpy–temperature relationship for (*a*) pure crystalline substances and eutectics and (*b*) glassy substances and alloys.

balance equation at the solid–liquid interface. Therefore, the enthalpy method is applicable for the solution of phase-change problems involving both a distinct phase change at a discrete temperature as well as phase change taking place over an extended range of temperatures.

Figure 12-11(b) shows that for alloys and glassy substances there is no single melting-point temperature T_m^* because the phase change takes place over an extended temperature range from T_s to T_l, and a mushy zone exists between the all-solid and all-liquid regions.

To illustrate the physical significance of the enthalpy function $H(T)$ (J/kg) in relation to the case of pure substances having a single melting-point temperature T_m^*, we refer to the plot of $H(T)$ as a function of temperature as illustrated in Figure 12-11(a). When the substance is in solid form at temperature T, the substance contains a sensible heat per unit mass $C_p(T - T_m^*)$, where the melting-point temperature T_m^* is taken as the reference temperature. In the liquid form, it contains latent heat L per unit mass in addition to the sensible heat, that is, $C_p(T - T_m^*) + L$. For the specific case considered here, the enthalpy is related to temperature by

$$H = \begin{cases} C_p(T - T_m^*) & \text{for} \quad T < T_m^* \\ C_p(T - T_m^*) + L & \text{for} \quad T > T_m^* \end{cases} \tag{12-92}$$

Conversely, given the enthalpy of the substance, the corresponding temperature is determined from

$$T = \begin{cases} T_m^* + \dfrac{H}{C_p} & \text{for} \quad H < 0 \\ T_m^* & \text{for} \quad 0 \le H \le L \\ T_m^* + \dfrac{H - L}{C_p} & \text{for} \quad H > L \end{cases} \tag{12-93}$$

In the case of glassy substances and alloys, there is no discrete melting-point temperature because the phase change takes place over an extended range of temperatures, as illustrated in Figure 12-11(b). Such a relationship between $H(T)$ and T is obtained from either experimental data or standard physical tables. In general, enthalpy is a nonlinear function of temperature. Therefore an enthalpy versus temperature variation needs to be available. Assuming linear release of latent heat over the mushy region, the variation of $H(T)$ with temperature can be taken as

$$H = \begin{cases} C_p T & \text{for} \quad T < T_s & \text{solid region} \\ C_p T + \dfrac{T - T_s}{T_l - T_s} L & \text{for} \quad T_s \le T \le T_l & \text{mushy region} \\ C_p T + L & \text{for} \quad T > T_l & \text{liquid region} \end{cases} \tag{12-94}$$

where L is the latent heat, and T_s and T_l are the pure solid- and pure liquid-phase temperatures, respectively.

To solve the phase-change problem with the enthalpy method, an explicit or an implicit finite-difference scheme can be used. The implicit scheme is generally preferred because of its ability to accommodate a wide range of time steps without the restriction of the stability criteria. We present below the *implicit enthalpy* method for solving the one-dimensional, two-phase solidification problem for a substance having a single phase-change temperature T_m^*.

Implicit Enthalpy Method for Solidification at a Single Phase-Change Temperature

We consider one-dimensional solidification of a liquid having a single melting-point temperature T_m^* and confined to the region $0 \le x \le L$. Initially, the liquid is at a uniform temperature T_0 that is greater than the melting temperature T_m^* of the liquid. For times $t > 0$, the boundary surface at $x = 0$ is kept at a temperature T_b that is lower than the melting temperature T_m^* of the substance. The boundary condition at $x = L$ satisfies the symmetry requirement. For simplicity, the properties are assumed to be constant.

The enthalpy formulation of this phase-change problem is given by

$$\rho \frac{\partial H}{\partial t} = k \frac{\partial^2 T}{\partial x^2}, \qquad \text{in} \qquad 0 < x < L, \qquad t > 0 \tag{12-95a}$$

$$T(x = 0, t) = T_b \tag{12-95b}$$

$$\left. \frac{\partial T}{\partial x} \right|_{x=L} = 0 \tag{12-95c}$$

$$T(t = 0) = T_0 \quad \text{or} \quad H(t = 0) = H_0 \qquad \text{for} \qquad 0 \le x \le L \tag{12-95d}$$

To approximate this problem with finite differences, the region $0 \le x \le L$ is subdivided into M equal parts each of width $\Delta x = L/M$.

The finite-difference approximation of the differential equation (12-95a) using the implicit scheme is given by

$$\rho \frac{H_i^{n+1} - H_i^n}{\Delta t} = k \frac{T_{i-1}^{n+1} - 2T_i^{n+1} + T_{i+1}^{n+1}}{(\Delta x)^2} \tag{12-96}$$

where the subscript $i = 1, 2, \ldots, M - 1$ denotes the spatial discretization and the superscript $n = 1, 2, \ldots$ denotes the time discretization. The solution of equation (12-96) for the enthalpy H_i^{n+1} gives

$$H_i^{n+1} = H_i^n + \frac{k\,\Delta t}{\rho(\Delta x)^2}[F^*(H_{i-1}^{n+1}) - 2F^*(H_i^{n+1}) + F^*(H_{i+1}^{n+1})] \tag{12-97}$$

for $i = 1, 2, 3, \ldots, M - 1$, where the notation

$$T = F^*(H) \tag{12-98}$$

denotes that the temperature T is related to the enthalpy H. The system of equations (12-97) can be written more compactly in the vector form as

$$\hat{H}^{n+1} = \hat{H}^n + \Delta t \hat{F}(\hat{H}^{n+1}) \tag{12-99a}$$

where $\hat{H}$ is a vector whose components are the nodal enthalpies H_i and $\hat{F}$ is a function with ith component given by

$$F_i(\hat{H}) = \frac{k}{\rho(\Delta x)^2}[F^*(H_{i-1}) - 2F^*(H_i) + F^*(H_{i+1})] \tag{12-99b}$$

For a substance having a single phase-change temperature T_m^*, the temperature is related to the enthalpy by

$$T = \begin{cases} \dfrac{H}{C_p} & H < C_p T_m^* \\ T_m^* & C_p T_m^* \le H \le (C_p T_m^* + L) \\ \dfrac{H - L}{C_p} & H > (C_p T_m^* + L) \end{cases} \tag{12-100}$$

Equivalently, equation (12-100) can be written as

$$H(T) = \begin{cases} C_p T & T < T_m^* \\ C_p T + L & T > T_m^* \end{cases} \tag{12-101}$$

The difference between these equations, and that given by equations (12-93), is that in the latter temperature T_m^* is used as the reference temperature. The finite-difference equations (12-99), together with the appropriate boundary and initial conditions for the problem and the "temperature–enthalpy" relations given by equations (12-100), constitute a set of equations for the determination of nodal enthalpies H_i^{n+1} at the time level $n + 1$, from the knowledge of the enthalpies H_i^n in the previous time level. These equations being nonlinear, an iterative scheme is needed to solve for $\hat{H}^{n+1}$. Furthermore, if it is required that the solid–liquid interface move one and only one spatial step Δx during each consecutive time step Δt, iteration becomes necessary to establish the magnitude of each time step accordingly. Voller and Cross [74] used enthalpy formulation for a one-dimensional solidification problem with a single phase-change temperature which led to very accurate solutions. We present below the equations needed to perform such iterations.

Equation (12-99) is written in the form

$$\hat{G}(\hat{H}^{n+1}) \equiv \hat{H}^n + \Delta t \hat{F}(\hat{H}^{n+1}) - \hat{H}^{n+1} = \hat{0} \tag{12-102}$$

To calculate $\hat{H}^{n+1}$, the modification of Newton's method is applied:

$$\hat{H}^{n+1,k+1} = \hat{H}^{n+1,k} - \omega \frac{\hat{G}(\hat{H}^{n+1,k})}{\hat{G}'(\hat{H}^{n+1,k})} \tag{12-103a}$$

where ω is the relaxation parameter, the superscript k denotes the number of iterations, and n is the number of discretization steps on time. The derivative $\hat{G}'$ with respect to $\hat{H}^{n+1,k}$ is determined as

$$\hat{G}'(\hat{H}^{n+1,k}) = \Delta t \frac{\partial \hat{F}(\hat{H}^{n+1,k})}{\partial \hat{H}^{n+1,k}} - \hat{I} \equiv \hat{J} - \hat{I} \tag{12-103b}$$

where $\hat{I}$ is the identity matrix, and $\hat{J}$ is the Jacobian matrix whose components are given by

$$\hat{J}_{i,l} = \Delta t \left. \frac{\partial F_i}{\partial H_l} \right|_{\hat{H}=\hat{H}^{n+1,k}} \tag{12-104}$$

where $F_i(\hat{H})$ is as defined by equation (12-99b).

Then the equation for the determination of the ith component of enthalpy $H_i^{n+1,k+1}$ becomes

$$H_i^{n+1,k+1} = H_i^{n+1,k} + \omega \frac{H_i^{n,k} - H_i^{n+1,k} + \Delta t F_i(\hat{H}^{n+1,k})}{1 - J_{ii}} \tag{12-105a}$$

where

$$J_{ii} = \Delta t \left. \frac{\partial F_i}{\partial H_i} \right|_{\hat{H}=\hat{H}^{n+1,k}} \tag{12-105b}$$

The Algorithm

To start the iterations on $\hat{H}$, an initial estimate on the components of enthalpy is chosen as

$$H_i^{n+1,0} = H_i^n + \Delta t F_i(\hat{H}^n) \tag{12-106}$$

where $F_i(\hat{H}^n)$ is as defined by equation (12-99b). Iterations are carried out by using equations (12-105) until a specified convergence criterion is achieved.

Also we need to perform iterations on the size of the time step Δt_k such that the interface will move one and only one grid point over the duration of this time step. This requirement can be satisfied by noting that at each time step one and only one nodal enthalpy takes the value $CT_m^* + \frac{1}{2}L$.

Suppose the calculations are carried out up to the time level n, and that the nodal enthalpies H_i^n are determined for all nodal points i at time t, namely, the time level n. Let Δt_i denote the time step during which the interface moves by one spatial step Δx, from the node i to the node $i + 1$. Then we have

$$\Delta t_i = t_{i+1} - t_i \tag{12-107}$$

With the above conditions, the iterations on the size of the time step Δt are performed in the following manner:

1. The initial guess for the size of the time step Δt_k^0 is taken as

$$\Delta t_i^0 = \Delta t_{i-1} \tag{12-108}$$

2. The enthalpy distribution $\hat{H}^{t+\Delta t_i^m}$, where the superscript m on Δt refers to the mth iteration on the time step, is determined from the solution of equations (12-105) and (12-106). Here, the mth time step Δt_i^m is computed using an iterative scheme given by

$$\Delta t_i^{m+1} = \Delta t_i^m + \omega^* \Delta t_i^m \left(\frac{H_{i+1}^{t+\Delta t_i^m}}{CT_{\mathrm{m}}^* + (L/2)} - 1 \right) \tag{12-109}$$

where ω^* is the relaxation parameter associated with the time step iterations.

3. When the value of $H_{i+1}^{t+\Delta t_i^m}$ converges to $CT_{\mathrm{m}}{}^* + (L/2)$, the corresponding enthalpy values at all nodes are considered to be the solution for the time $t + \Delta t_i^m$.
4. Once the enthalpy values are available at the nodes, the corresponding values of node temperatures T_i are determined from the temperature–enthalpy relation given by equations (12-100).

Implicit Enthalpy Method for Solidification over an Extended Temperature Range

If the phase change takes place over an extended temperature range, there is a mushy zone between the solidus (i.e., solid) and liquidus (i.e., liquid) regions. In such a case, the enthalpy $H(T)$ is a smooth continuous function or piecewise continuous function. Assuming a linear variation of latent heat over the mushy region, the variation of $H(T)$ with temperature can be taken as that given by equations (12-94):

$$H = \begin{cases} C_p T & \text{for} \quad T < T_s \quad \text{solid region} \\ C_p T + \dfrac{T - T_s}{T_1 - T_s} L & \text{for} \quad T_s \le T \le T_l \quad \text{mushy region} \\ C_p T + L & \text{for} \quad T > T_l \quad \text{liquid region} \end{cases} \tag{12-110}$$

and the corresponding relations for temperature as a function of enthalpy becomes

$$T = \begin{cases} \dfrac{H}{C_p} & \text{for} \quad H < C_p T_s \\ \dfrac{H(T_l - T_s) + L T_s}{C_p(T_l - T_s) + L} & \text{for} \quad C_p T_s \le H \le (C_p T_l + L) \\ \dfrac{H - L}{C_p} & \text{for} \quad H > (C_p T_l + L) \end{cases} \tag{12-111}$$

Then the algorithm described previously is applicable if equation (12-99b) is used together with equations (12-111). Readers should consult reference 74 for a comparison of *explicit enthalpy* and *implicit enthalpy* methods of solution for phase change at a single temperature and constant properties with the *exact analytic solution* of a one-dimensional solidification problem.

REFERENCES

1. G. Lamé and B. P. Clapeyron, *Ann, Chem. Phys*. **47**, 250–256, 1831.
2. J. Stefan, *Ann. Phys. Chemie (Wiedemannsche Annalen)* **42**, 269–286, 1891.
3. H. S. Carslaw and J. C. Jaeger, *Conduction of Heat in Solids*, 2nd ed., Clarendon, London, 1959.
4. J. Crank, *Free and Moving Boundary Problems*, Oxford University Press, New York, 1984.
5. T. R. Goodman, *Trans. Am. Soc. Mech. Eng*. **80**, 335–342, 1958.
6. T. R. Goodman, *J. Heat Transfer* **83c**, 83–86, 1961.
7. T. R. Goodman and J. Shea, *J. Appl. Mech*. **32**, 16–24, 1960.
8. G. Poots, *Int. J. Heat Mass Transfer* **5**, 339–348, 1962.
9. G. Poots, *Int. J. Heat Mass Transfer* **5**, 525, 1962.
10. R. H. Tien, *Trans. Metall. Soc. AIME* **233**, 1887–1891, 1965.
11. R. H. Tien and G. E. Geiger, *J. Heat Transfer* **89c**, 230–234, 1967.
12. R. H. Tien and G. E. Geiger, *J. Heat Transfer* **90c**, 27–31, 1968.
13. S. H. Cho and J. E. Sunderland, *J. Heat Transfer* **91c**, 421–426, 1969.
14. J. C. Muehlbauer, J. D. Hatcher, D. W. Lyons, and J. E. Sunderland, *J. Heat Transfer* **95c**, 324–331, 1973.

15. K. Mody and M. N. Özisik, *Lett. Heat Mass Transfer* **2**, 487–493, 1975.
16. M. A. Biot, *J. Aeronaut. Sci*. **24**, 857–873, 1957.
17. T. J. Lardner, *AIAA J*. **1**, 196–206, 1963.
18. W. Zyskowski, The Transient Temperature Distribution in One-Dimensional Heat-Conduction Problems with Nonlinear Boundary Conditions, ASME Paper No. 68-HT-6, 1968.
19. M. A. Biot and H. Daughaday, *J. Aerospace Sci*. **29**, 227–229, 1962.
20. M. A. Biot, *Variational Principles in Heat Transfer*, Oxford University Press, London, 1970.
21. A. Prasad and H. C. Agrawal, *AIAA J*. **12**, 250–252, 1974.
22. N. M. H. Lightfoot, *Proc. London Math. Soc*. **31**, 97–116, 1929.
23. M. N. Özisik, *J. Heat Transfer*, **100C**, 370–371, 1978.
24. Y. K. Chuang and J. Szekely, *Int. J. Heat Mass Transfer* **14**, 1285–1295, 1971.
25. K. A. Rathjen and L. M. Jiji, *J. Heat Transfer* **93c**, 101–109, 1971.
26. Y. K. Chuang and J. Szekely, *Int. J. Heat Mass Transfer* **15**, 1171–1175, 1972.
27. H. Budhia and F. Kreith, *Int. J. Heat Mass Transfer* **16**, 195–211, 1973.
28. L. T. Rubenstein, *The Stefan Problem*, Trans. Math. Monographs Vol. 27, Am. Math. Soc., Providence, RI, 1971, pp. 94–181.
29. K. A. Rathjen and L. M. Jiji, Transient Heat Transfer in Fins Undergoing Phase Transformation, *Fourth International Heat Transfer Conference, Paris-Versailles*, **2**, 1970.
30. R. I. Pedroso and G. A. Domoto, *J. Heat Transfer* **95**, 42, 1973.
31. R. I. Pedroso and G. A. Domoto, *Int. J. Heat Mass Transfer* **16**, 1037, 1973.
32. L. M. Jiji and S. Weimbaum, A Nonlinear Singular Perturbation Theory for Non-Similar Melting or Freezing Problems, Conduction Cu-3, 5th International Heat Transfer Conference, Tokyo, 1974.
33. D. S. Riley, F. T. Smith, and G. Poots, *Int. J. Heat Mass Transfer* **17**, 1507, 1974.
34. C. L. Hwang and Y. P. Shih, *Int. J. Heat Mass Transfer* **18**, 689–695, 1975.
35. B. A. Boley, *J. Math. Phys*. **40**, 300–313, 1961.
36. B. A. Boley, *Int. J. Eng. Sci*. **6**, 89–111, 1968.
37. J. M. Lederman and B. A. Boley, *Int. J. Heat Mass Transfer* **13**, 413–427, 1970.
38. D. L. Sikarshie and B. A. Boley, *Int. J. Solids Structures* **1**, 207–234, 1965.
39. B. A. Boley and H. P. Yagoda, *Q. Appl. Math*. **27**, 223–246, 1969.
40. Y. F. Lee and B. A. Boley, *Int. J. Eng. Sci*. **11**, 1277, 1973.
41. A. N. Güzelsu and A. S. Çakmak, *Int. J. Solids and Structures* **6**, 1087, 1970.
42. M. N. Özisik and R. L. Murray, *J. Heat Transfer* **96c**, 48–51, 1974.
43. Y. Yener and M. N. Özisik, On the Solution of Unsteady Heat Conduction in Multi-Region Finite Media with Time Dependent Heat Transfer Coefficient, Proceedings of 5th International Heat Transfer Conference, Tokyo, September 3–7, 1974.
44. M. N. Özisik and S. Güçeri, *Can. J. Chem. Eng*. **55**, 145–148, 1977.
45. F. Kreith and F. E. Romie, *Proc. Phys. Soc. Lond*. **68(B)**, 283, 1955.
46. D. R. Otis, Solving the Melting Problem Using the Electric Analog to Heat Conduction, in *Heat Transfer and Fluid Mechanics Institute*, Stanford Univ., Stanford, CA, 1956.

47. G. Liebmann, *ASME Trans*. **78**, 1267, 1956.
48. D. C. Baxter, The Fusion Times of Slabs and Cylinders, ASME Paper No. 61-WA-179, 1961.
49. C. F. Bonilla and A. L. Strupczewski, *Nucl. Struc. Eng*. **2**, 40–47, 1965.
50. J. Douglas and T. M. Gallie, *Duke Math. J*. **22**, 557, 1955.
51. J. Crank, *Quart. J. Mech. Appl. Math*. **10**, 220, 1957.
52. W. D. Murray and F. Landis, *Trans. Am. Soc. Mech. Eng*. **81**, 106, 1959.
53. G. S. Springer and D. R. Olson, Method of Solution of Axisymmetric Solidification and Melting Problems, ASME Paper No. 63-WA-246, 1962.
54. G. S. Springer and D. R. Olson, Axisymmetric Solidification and Melting of Materials, ASME Paper No. 63-WA-185, 1963.
55. D. N. de G. Allen and R. T. Severn, *Q. J. Mech. Appl. Math*. **15**, 53, 1962.
56. W. D. Seider and S. W. Churchill, The Effect of Insulation on Freezing Motion, 7th National Heat Transfer Conference, Cleveland, 1964.
57. C. Bonacina, G. Comini, A. Fasano, and M. Primicerio, *Int. J. Heat Mass Transfer* **16**, 1825–1832, 1973.
58. G. M. Dusinberre, A Note on Latent Heat in Digital Computer Calculations, ASME Paper No. 58-HT-7.
59. N. R. Eyres, D. R. Hartree, J. Ingham, R. Jackson, R. J. Jarjant, and J. B. Wagstaff, *Phil. Trans. Roy. Soc*. **A240**, 1–57, 1948.
60. D. C. Baxter, *J. Heat Transfer* **84c**, 317–326, 1962.
61. L. C. Tao, *AIChE J*. **13**, 165, 1967.
62. L. C. Tao, *AIChE J*. **14**, 720, 1968.
63. G. S. Springer, *Int. J. Heat Mass Transfer* **12**, 521, 1969.
64. S. H. Cho and J. E. Sunderland, *Int. J. Heat Mass Transfer* **13**, 123, 1970.
65. A. Lazaridis, *Int. J. Heat Mass Transfer* **13**, 1459–1477, 1970.
66. J. Crank and R. S. Gupta, *J. Int. Maths. Appl*. **10**, 296–304, 1972.
67. G. G. Sackett, *SIAM J. Num. Anal*. **8**, 80–96, 1971.
68. G. H. Meyer, *Num. Math*. **16**, 248–267, 1970.
69. J. Crank and R. D. Phahle, *Bull. Inst. Math. Appl*. **9**, 12–14, 1973.
70. J. Szekely and M. J. Themelis, *Rate Phenomena in Process Metallurgy*, Wiley-Interscience, New York, 1971, Chapter 10.
71. R. D. Atthey, *J. Inst. Math. Appl*. **13**, 353–366, 1974.
72. G. H. Meyer, *SIAM J. Num. Anal*. **10**, 522–538, 1973.
73. N. Shamsunder and E. M. Sparrow, *J. Heat Transfer* **97c**, 333–340, 1975.
74. V. Voller and M. Cross, *Int. J. Heat Mass Transfer* **24**, 545–556, 1981.
75. V. Voller and M. Cross, *Int. J. Heat Mass Transfer* **26**, 147–150, 1983.
76. K. H. Tacke, *Int. J. Num. Meth. Eng*. **21**, 543–554, 1985.
77. R. M. Furzeland, *J. Inst. Math. Appl*. **26**, 411–429, 1980.
78. R. S. Gupta and D. Kumar, *Comp. Mech. Appl. Mech. Eng*. **23**, 101–109, 1980.
79. R. S. Gupta and D. Kumar, *Int. J. Heat Mass Transfer* **24**, 251–259, 1981.
80. Q. Pham, *Int. J. Heat Mass Transfer* **28**, 2079–2084, 1985.
81. D. Poirier and M. Salcudean, *J. Heat Transfer* **110**, 562–570, 1988.

82. B. A. Boley, in *Proceedings of the 3rd Symposium on Naval Structural Mechanics*, Pergamon Press, New York, 1963.
83. S. B. Bankoff, *Advances in Chemical Engineering*, Vol. 5, Academic, New York, 1964.
84. J. C. Muehlbauer and J. E. Sunderland, *Appl. Mech. Rev*. **18**, 951–959, 1965.
85. R. Viskanta, *J. Heat Transfer—Trans. ASME* **110**, 1205–1219, 1988.
86. R. Viskanta, *JSME Int. J. Series B—Fluids Thermal Eng.* **33**, 409–423, 1990.
87. H. Hu and S. A. Argyropoulos, *Modelling Simulation Mat. Sci. Eng.* **4**, 371–396, 1996.
88. A. E. Delgado and D. W. Sun, *J. Food Eng.* **47**, 157–174, 2001.
89. B. Zalba, J. M. Marin, L. F. Cabeza, and H. Mehling, *Appl. Thermal Eng.* **23**, 251–282, 2003.
90. W. Kurz and D. J. Fisher, *Fundamentals of Solidification*, Trans. Tech. Publications, Aerdermannsdorf, Switzerland, 1989.
91. M. C. Flemings, *Solidification Processing*, McGraw-Hill, New York, 1974.
92. A. Ohno, *Solidification*, Springer, Berlin, 1988.
93. I. Minkoff, *Solidification and Cast Structure*, Wiley, New York, 1986.
94. M. Rappaz, *Int. Math. Rev*. **34**, 93–123, 1989.
95. L. T. Thomas and J. W. Westwater, *Chem. Eng. Prog. Symp*. **59m**, 155–164, 1963.
96. D. V. Boger and J. E. Westwater, *J. Heat Transfer* **89c**, 81–89, 1967.
97. L. M. Jiji, K. A. Rathjen, and T. Drezewiecki, *Int. J. Heat Mass Transfer* **13**, 215–218, 1970.
98. J. A. Bailey and J. R. Davila, *Appl. Sci. Res*. **25**, 245–261, 1971.
99. J. Szekely and A. S. Jassal, *Met. Trans. B* **9B**, 389–398, 1978.
100. P. D. Patel, *AIAA J*. **6**, 2454, 1968.
101. R. R. Gilpin, *Int. J. Heat Mass Transfer* **20**, 693–699, 1977.
102. S. Paterson, *Proc. Glasgow Math. Assoc* **1**, 42–47, 1952–53.
103. N. N. Levedev, *Special Functions and Their Applications*, Prentice-Hall, Englewood Cliffs, NJ, 1965.
104. M. Abramowitz and I. A. Stegun, *Handbook of Mathematical Functions*, NBS Applied Mathematics Series 55, U.S. Government Printing Office, Washington, DC, 1964, p. 239.
105. S. M. Selby, *Standard Mathematical Tables*, Chemical Rubber Company, Cleveland, Ohio, 1971, p. 515.
106. J. Crank, *J. Mech. Appl. Math*. **10**, 220–231, 1957.
107. L. W. Ehrlick, *J. Assn. Comp. Math*. **5**, 161–176, 1958.

PROBLEMS

12-1 Verify that the interface energy balance equation (12-2c) is also applicable for the melting problem illustrated in Figure 12-1(b).

12-2 In the melting problem illustrated in Figure 12-1(b), if the heat transfer on the liquid side is by convection and on the solid phase is by pure conduction, derive the interface energy balance equation. Take the bulk temperature of the liquid side as T_∞ and the heat transfer coefficient as h.

12-3 Solve exactly the phase-change problem considered in Example 12-2 for the case of solidification in a half-space $x > 0$. That is, a liquid at the melting temperature T_m^* is confined to a half-space $x > 0$. At time $t = 0$ the boundary at $x = 0$ is lowered to a temperature T_0 below T_m^* and maintained at that temperature for times $t > 0$. Determine the temperature distribution in the solid phase and the location of the solid–liquid interface as a function of time.

12-4 Solve exactly the problem considered in Example 12-3 for the case of melting. That is, a solid in $x > 0$ is initially at a uniform temperature T_i lower than the melting temperature T_m^*. For times $t > 0$ the boundary surface at $x = 0$ is kept at a constant temperature T_0, which is higher than the melting temperature T_m^*. Determine the temperature distribution in the liquid and solid phases, and the location of the solid–liquid interface as a function of time.

12-5 Solve exactly the problem considered in Example 12-4 for the case of melting. That is, a line heat source of strength Q (W/m) is situated at $r = 0$ in an infinite medium that is at a uniform temperature T_i lower than the melting temperature T_m. The melting will start at $r \to 0$, and the solid–liquid interface will move in the positive r direction. Determine the temperature distribution in the solid and liquid phases, and the location of the solid–liquid interface as a function of time.

12-6 Using the integral method of solution, solve the solidification Problem 12-3 and obtain an expression for the location of the solid–liquid interface. Compare this result with that obtained in Example 12-5 for the case of melting.

12-7 A solid confined in a half-space $x > 0$ is initially at the melting temperature T_m^*. For times $t > 0$ the boundary surface at $x = 0$ is subjected to a heat flux in the form

$$-k \left.\frac{\partial T}{\partial x}\right|_{x=0} = H \equiv \text{constant}$$

Using the integral method of solution and a second-degree polynomial approximation for the temperature, obtain an expression for the location of the solid–liquid interface as a function of time.

12-8 Consider one-dimensional solidification of a liquid having a single melting-point temperature T_m^*, confined to the region $0 \le x \le L$. Initially, the liquid is at a uniform temperature T_0 that is higher than the melting temperature T_m^* of the liquid. For times $t > 0$, the boundary surface at $x = 0$ is kept at a temperature $T = T_b$ that is lower than the melting temperature T_m^* of the substance. The boundary at $x = L$ satisfies the symmetry requirement. The properties are assumed to be constant. Develop the finite-difference formulation of this problem by using the *explicit enthalpy method*; that is, use the explicit finite-difference scheme to discretize the differential equations.

12-9 Repeat Problem 12-8 for the case of a material having phase change over an extended temperature range.

NOTE

1. The exponential integral function $E_1(z)$ is defined here for a positive, real argument as

$$E_1(z) = \int_{u=z}^{\infty} \frac{e^{-u}}{u}\, du = \int_{u=1}^{\infty} \frac{e^{-zu}}{u}\, du \qquad \text{for} \qquad z > 0 \tag{1}$$

The function $E_1(z)$ decreases monotonically from the value $E_1(0) = \infty$ to $E_1(\infty) = 0$ as z is varied from $z = 0$ to $z \to \infty$ as shown in Table 12-1 and plotted in Figure 12-7. The derivative of $E_1(z)$ with respect to z is given as

$$\frac{d}{dz}\left[E_1(z)\right] = \frac{d}{dz}\left[\int_{u=z}^{\infty} \frac{e^{-u}}{u} du\right] = -\frac{e^{-z}}{z} \tag{2}$$

The notation $E_1(z)$ has been used in reference 104 (p. 228), and its polynomial approximations are given for $0 \le x \le 1$ and $1 \le x < \infty$ [104, p. 231]. A tabulation of $E_1(z)$ function is given in references 104 [p. 239] and 105 [p. 515].

13

APPROXIMATE ANALYTIC METHODS

Analytic solutions, whether exact or approximate, are always useful in engineering analysis because they provide a better insight into the physical significance of various parameters affecting the problem. When exact analytic solutions are impossible or too difficult to obtain or the resulting analytic solutions are too complicated for computational purposes, approximate analytic solutions provide a powerful alternative approach to handle such problems.

There are numerous approximate analytic methods for solving the partial differential equations governing the engineering problems. In this chapter we present the *integral method*, the *Galerkin method*, and the method of *partial integration* and illustrate their applications with representative examples. The accuracy of an approximate solution cannot be assessed unless the results are compared with the exact solution. Therefore, in order to give some idea of the accuracy of the approximate analysis, simple problems for which exact solutions are available are first solved with the approximate methods and the results are compared with the exact solutions. The applications to the solution of more complicated, nonlinear problems are then considered.

13-1 INTEGRAL METHOD: BASIC CONCEPTS

The use of the integral method for the solution of partial differential equations dates back to von Kármán and Pohlhausen, who applied the method for the approximate analysis of boundary layer momentum and energy equations of fluid mechanics [1]. Landahl [2] used it in the field of biophysics to solve the diffusion equation in connection with the spread of a concentrate. Merk [3] applied this approach to solve a two-dimensional steady-state melting problem, and Goodman [4, 5] used it for the solution of a one-dimensional transient melting problem. Since then, this method has been applied in the solution of various types of one-dimensional transient heat conduction problems [6–16], melting and solidification problems [16–25], and heat and momentum transfer problems involving melting of ice in seawater and the melting and extrusion of polymers [26–29].

The method is simple, straightforward, and easily applicable to both linear and nonlinear one-dimensional transient boundary value problems of heat conduction for certain boundary conditions [30–31]. The results are approximate, but several solutions obtained with this method when compared with the exact solutions have confirmed that the accuracy is generally acceptable for many engineering applications. In this section we first present the basic concepts involved in the application of this method by solving a simple transient heat conduction problem for a semi-infinite medium. The method is then applied to the solution of various one-dimensional, time-dependent heat conduction problems. The application to the solution of melting, solidification, and ablation problems was considered previously in Chapter 12 on moving-boundary problems.

When the differential equation of heat conduction is solved exactly in a given region subject to specified boundary and initial conditions, the resulting solution is satisfied exactly at each point over the considered domain; however, with the integral method the solution is satisfied only *on the average* over the region. We now summarize the basic steps in the analysis with the integral method when it is applied to the solution of a one-dimensional, transient heat conduction problem in a semi-infinite medium subject to some prescribed boundary and uniform initial conditions, but no heat generation.

1. The differential equation of heat conduction is integrated over a phenomenological distance $\delta(t)$, called the *thermal layer* in order to remove from the differential equation the derivative with respect to the space variable. The thermal layer is defined as the distance beyond which, for practical purposes, there is no heat flow; hence the initial temperature distribution remains unaffected beyond $\delta(t)$. The resulting equation is called the *energy integral equation* (it is also called the *heat balance integral*).
2. A suitable profile is chosen for the temperature distribution over the thermal layer. A polynomial profile is generally preferred for this purpose, noting that experience has shown that there is no significant improvement in the accuracy of the solution to choose a polynomial greater than the fourth degree. The coefficients in the polynomial are determined in terms of the thermal layer thickness $\delta(t)$ by utilizing the actual (or if necessary, derived) boundary conditions.
3. When the temperature profile thus constructed is introduced into the energy integral equation and the indicated operations are performed, an ordinary differential equation is obtained for the thermal layer thickness $\delta(t)$ with time as the independent variable. The solution of this differential equation

subject to the appropriate initial condition [i.e., in this case $\delta(t) = 0$ for $t = 0$] gives $\delta(t)$ as a function of time.

4. Once $\delta(t)$ is available from step 3, the temperature distribution $T(x, t)$ is known as a function of time and position in the medium, and the heat flux at the surface is readily determined. Experience has shown that the method is more accurate for the determination of heat flux than the temperature profile.

13-2 INTEGRAL METHOD: APPLICATION TO LINEAR TRANSIENT HEAT CONDUCTION IN A SEMI-INFINITE MEDIUM

To illustrate the mathematical details of the basic steps discussed above for the application of the integral method, in the following example we consider a problem of transient heat conduction in a semi-infinite medium with no energy generation.

Example 13-1 Integral Method for Semi-Infinite Medium
A semi-infinite medium $x \geq 0$ is initially at a uniform temperature T_i. For times $t > 0$, the boundary surface is kept at constant temperature T_0 as illustrated in Figure 13-1. Develop expressions for the temperature distribution and the surface heat flux with the integral method by using a cubic polynomial approximation for the temperature profile.

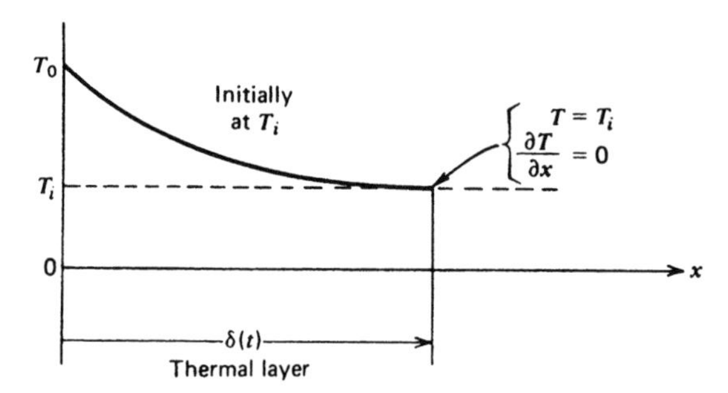

Figure 13-1 Definition of thermal penetration layer for heat conduction in a semi-infinite medium.

The mathematical formulation of this problem is given as

$$\frac{\partial^2 T(x,t)}{\partial x^2} = \frac{1}{\alpha}\frac{\partial T(x,t)}{\partial t} \quad \text{in} \quad x > 0, \quad t > 0 \tag{13-1}$$

$$\text{BC:} \quad T(x = 0, t) = T_0 \tag{13-2a}$$

$$\text{IC:} \quad T(x, t = 0) = T_i \quad \text{in} \quad x \geq 0 \tag{13-2b}$$

We now solve this problem with the integral method by following the basic steps discussed above. We first integrate equation (13-1) with respect to the space variable from $x = 0$ to $x = \delta(t)$, noting first that

$$\int_{x=0}^{\delta(t)} \frac{\partial^2 T}{\partial x^2} dx = \left. \frac{\partial T}{\partial x} \right|_0^{\delta(t)} \tag{13-3}$$

we then have

$$\left. \frac{\partial T}{\partial x} \right|_{x=\delta(t)} - \left. \frac{\partial T}{\partial x} \right|_{x=0} = \frac{1}{\alpha} \int_{x=0}^{\delta(t)} \frac{\partial T}{\partial t} dx \tag{13-4}$$

When the integral on the right-hand side is performed using Leibniz's integral formula, we obtain

$$\left. \frac{\partial T}{\partial x} \right|_{x=\delta} - \left. \frac{\partial T}{\partial x} \right|_{x=0} = \frac{1}{\alpha} \left[\frac{d}{dt} \left(\int_{x=0}^{\delta(t)} T\, dx \right) - \frac{d\delta}{dt} T \Big|_{x=\delta} \right] \tag{13-5}$$

By the definition of thermal layer, as illustrated in Figure 13-1, we have

$$\left. \frac{\partial T}{\partial x} \right|_{x=\delta} = 0 \qquad \text{and} \qquad T|_{x=\delta} = T_i \tag{13-6}$$

For convenience in the analysis, we now define

$$\theta(t) \equiv \int_{x=0}^{\delta(t)} T(x,t)\, dx \tag{13-7}$$

Introducing equations (13-6) and (13-7) into (13-5), we obtain

$$\boxed{-\alpha \left. \frac{\partial T}{\partial x} \right|_{x=0} = \frac{d}{dt} \left(\theta - T_i \delta \right)} \tag{13-8}$$

which is called the *energy integral equation* for the problem considered here. To better explain the physics of the energy integral equation, we rearrange equation (13-8) by making use of the definition of the thermal diffusivity, to yield

$$-k \left. \frac{\partial T}{\partial x} \right|_{x=0} = \rho C \frac{d}{dt} \left(\theta - T_i \delta \right) \tag{13-9}$$

which we now consider in terms of Figure 13-2.

The shaded area in Figure 13-2 represents the thermal energy per unit area that has been added to the system since $t = 0$, namely, $\rho C(\theta - T_i\delta)$. Equation (13-9) is then interpreted as the rate at which energy is entering the system per unit area via conduction (i.e., left-hand side) being equal to the rate of storage of energy per unit area (i.e., right-hand side). However, unlike the actual heat equation, the energy balance is now over the entire domain *on average* (i.e., global), rather than at any single location.

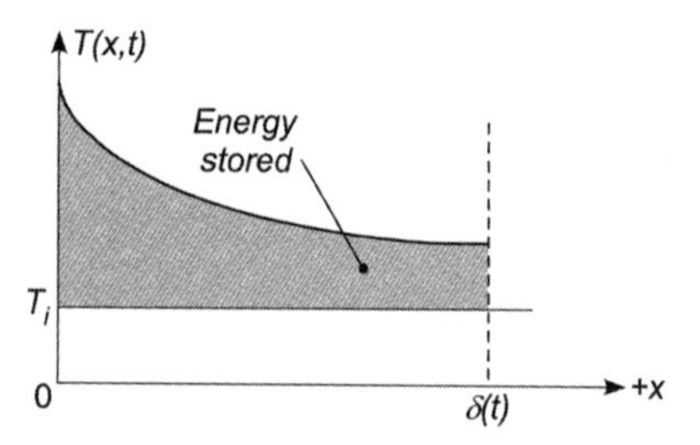

Figure 13-2 Conservation of energy associated with the energy integral equation.

We now choose a cubic polynomial representation for *T(x, t)* in the form

$$T(x,t) = a + bx + cx^2 + dx^3 \qquad \text{in} \qquad 0 \le x \le \delta(t) \tag{13-10}$$

where the coefficients are in general functions of time. Four conditions are needed to determine these four coefficients in terms of $\delta(t)$. Three of these conditions are obtained from the boundary conditions at $x = 0$, and at the edge of the thermal layer $x = \delta(t)$ as follows:

$$\begin{aligned} &(1) \qquad T|_{x=0} = T_0 \\ &(2) \qquad T|_{x=\delta} = T_i \\ &(3) \qquad \left.\frac{\partial T}{\partial x}\right|_{x=\delta} = 0 \end{aligned} \tag{13-11}$$

The fourth condition may be derived (i.e., *derived boundary condition*) by evaluating the differential equation (13-1) at $x = 0$, and by making use of the fact that $T = T_0 =$ constant at $x = 0$; hence the derivative of temperature with respect to the time vanishes at $x = 0$, and we obtain

$$(4) \qquad \left.\frac{\partial^2 T}{\partial x^2}\right|_{x=0} = 0 \tag{13-12}$$

Clearly, a fourth condition could also be derived by evaluating the differential equation (13-1) at $x = \delta(t)$ and utilizing the fact that $T = T_i =$ constant, by definition, at $x = \delta$. This matter will be discussed later in the analysis. The application of the four boundary condition equations (13-11) and (13-12) yields

$$T(x=0) = T_0 \quad \rightarrow \quad a = T_0 \tag{13-13a}$$

$$\left.\frac{\partial^2 T}{\partial x^2}\right|_{x=0} = 0 \quad \rightarrow \quad c = 0 \tag{13-13b}$$

$$\left.\begin{array}{l}\left.\dfrac{\partial T}{\partial x}\right|_{x=\delta}=0\\[2ex] T(x=\delta)=T_i\end{array}\right\} \rightarrow \begin{array}{l} d=\dfrac{T_0-T_i}{2\delta^3}\\[2ex] b=\dfrac{-3\left(T_0-T_i\right)}{2\delta}\end{array} \tag{13-13c}$$

and the final temperature profile becomes

$$\frac{T(x,t)-T_i}{T_0-T_i}=1-\frac{3}{2}\left(\frac{x}{\delta}\right)+\frac{1}{2}\left(\frac{x}{\delta}\right)^3 \tag{13-14}$$

We are now ready to plug the temperature profile (13-14) into the energy integral equation (13-8), although we first evaluate the following two terms:

$$\theta=\int_{x=0}^{\delta(t)} T(x,t)\,dx=\frac{3}{8}\left(T_0-T_i\right)\delta(t)+T_i\delta(t) \tag{13-15a}$$

and

$$\left.\frac{\partial T}{\partial x}\right|_{x=0}=-\frac{3}{2}\left(T_0-T_i\right)\frac{1}{\delta(t)} \tag{13-15b}$$

Substituting the above into the energy integral equation (13-8), we obtain the following ordinary differential equation for $\delta(t)$:

$$4\alpha=\delta(t)\frac{d\delta}{dt} \qquad \text{for} \qquad t>0 \tag{13-16a}$$

subject to the initial condition

$$\delta(t=0)=0 \tag{13-16b}$$

The solution of equations (13-16) is readily achieved by separating and integrating the ODE, which gives

$$\delta(t)=\sqrt{8\alpha t} \tag{13-17}$$

Knowing $\delta(t)$, we now determine the temperature distribution $T(x, t)$ according to equation (13-14). Finally, we may determine the heat flux $q''(x,t)$ at the surface $x=0$ as given by application of Fourier's law, namely,

$$q''(x=0,t)=-k\left.\frac{\partial T}{\partial x}\right|_{x=0}=\frac{3k}{2}\frac{T_0-T_i}{\delta(t)} \tag{13-18a}$$

where $\delta(t)$ is defined above by equation (13-17).

Other Profiles

In the foregoing example, we considered a cubic polynomial representation for $T(x, t)$ that involved four unknown coefficients and required four conditions for their determination. Three of these conditions given by equation (13-11) are the *natural conditions* for the problem, and the fourth condition equation (13-12) is a *derived condition* obtained by evaluating the differential equation at $x = 0$. It is also possible to derive an alternative fourth condition by evaluating the differential equation at $x = \delta$, yielding

$$\left.\frac{\partial^2 T}{\partial x^2}\right|_{x=\delta} = 0 \tag{13-19}$$

Therefore, it is also possible to use the above three natural conditions (13-11) together with the alternative derived condition (13-19) to obtain an *alternative cubic temperature profile* in the form

$$\frac{T(x,t) - T_i}{T_0 - T_i} = \left(1 - \frac{x}{\delta}\right)^3 \tag{13-20a}$$

where now

$$\delta = \sqrt{24\alpha t} \tag{13-20b}$$

If a fourth-degree polynomial representation is used for $T(x, t)$, the resulting five coefficients are determined by the application of the five conditions given by equations (13-11), (13-12), and (13-19), and the following temperature profile is obtained

$$\frac{T(x,t) - T_i}{T_0 - T_i} = 1 - 2\left(\frac{x}{\delta}\right) + 2\left(\frac{x}{\delta}\right)^3 - \left(\frac{x}{\delta}\right)^4 \tag{13-21a}$$

where we now define

$$\delta = \sqrt{\tfrac{40}{3}\alpha t} \tag{13-21b}$$

Comparison with Exact Solution In the foregoing analysis we developed two different cubic temperature profiles given by equations (13-14) and (13-20) and a fourth-degree profile given by equation (13-21). One can also develop another approximate solution by utilizing a second-degree polynomial representation. The question regarding which one of these approximate solutions is more accurate cannot be answered until each of these solutions is compared with the exact solution of the problem, which is given by

$$\frac{T(x,t) - T_i}{T_0 - T_i} = 1 - \operatorname{erf}\left(\frac{x}{\sqrt{4\alpha t}}\right) \tag{13-22}$$

Figure 13-3 shows a comparison of these approximate temperature distributions with the exact solution. The agreement is better for small values of the parameter $x/\sqrt{4\alpha t}$. Overall, the fourth-degree polynomial approximation agrees better with the exact solution. The cubic polynomial representation utilizing the

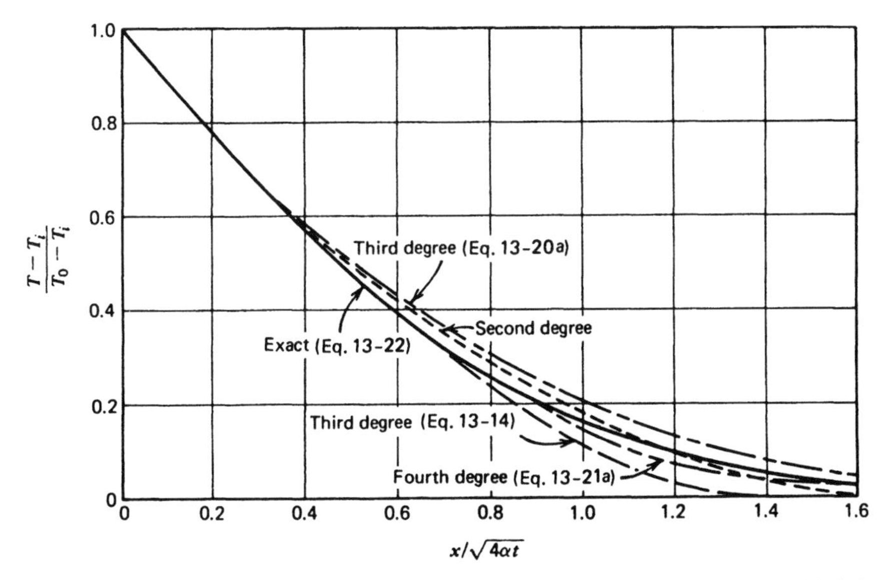

Figure 13-3 Comparison of exact and approximate analytical solutions for a semi-infinite region.

condition at $x = 0$ seems to agree with the exact solution better than the one utilizing the condition at $x = \delta$.

The heat flux at the boundary surface $x = 0$ is a quantity of practical interest, and for all three of the temperature profiles considered above, it may be expressed in the form

$$q''(x = 0, t) = -k \left. \frac{\partial T}{\partial x} \right|_{x=0} = B \frac{k(T_0 - T_i)}{\sqrt{\alpha t}} \tag{13-23}$$

Table 13-1 gives the values of the constant B as calculated from the above exact and approximate solutions. The fourth-degree polynomial approximation represents the heat flux with an error of approximately 3%, which is acceptable for most engineering applications.

Cylindrical and Spherical Symmetry The use of polynomial representation for temperature, although giving reasonably good results in the rectangular coordinate system, will yield significant error in the problems of cylindrical and spherical symmetry [11]. This is to be expected since the volume into which the heat diffuses does not remain the same for equal increments of r in the cylindrical and spherical coordinate systems. This situation may be remedied by modifying the temperature profiles as

Cylindrical symmetry: $$T(r, t) = (\text{polynomial in } r)(\ln r) \tag{13-24a}$$

Spherical symmetry: $$T(r, t) = \frac{\text{polynomial in } r}{r} \tag{13-24b}$$

TABLE 13-1 Error Involved in the Surface Heat Flux

Heat Flux Calculated per Temperature Profile	B as Defined by Equation (13-23)	Percent Error Involved
Exact [equation (13-22)]	$\frac{1}{\sqrt{\pi}} \cong 0.565$	0
Cubic approximation [equation (13-14)]	$\frac{3}{2\sqrt{8}} \cong 0.530$	−6.2
Cubic approximation [equation (13-20)]	$\frac{3}{\sqrt{24}} \cong 0.612$	+8.3
Fourth-degree approximation [equation (13-21)]	$\frac{2}{\sqrt{\frac{40}{3}}} \cong 0.548$	−3.0

Since the problems with spherical symmetry can be transformed into a problem in the rectangular coordinate system, as discussed in Chapter 5, one needs to be concerned with such a modification only for the cylindrical symmetry.

Problems with Energy Generation

The integral method is also applicable for the solution of one-dimensional transient heat conduction problems with a uniform energy generation that may be constant or time dependent over the region. The following example illustrates the application to a problem with energy generation in the medium.

Example 13-2 Semi-Infinite Problem with Heat Generation
A semi-infinite region, $x > 0$, is initially at a constant temperature T_i. For times $t > 0$, heat is generated within the solid at a rate of $g(t)$ (W/m^3), while the boundary at $x = 0$ is kept at a constant temperature T_0. Obtain an expression for the temperature distribution *T(x, t)* in the medium using the integral method.

The mathematical formulation of this problem is given as

$$\alpha \frac{\partial^2 T}{\partial x^2} + \frac{\alpha}{k} g(t) = \frac{\partial T}{\partial t} \quad \text{in} \quad x > 0, \quad t > 0 \tag{13-25a}$$

$$\text{BC:} \quad T(x = 0, t) = T_0 \tag{13-25b}$$

$$\text{IC:} \quad T(x, t = 0) = T_i \quad \text{for} \quad x \geq 0 \tag{13-25c}$$

This problem is illustrated in Figure 13-4. We note here that since heat generation is present throughout the entire domain, the temperature is rising at a steady rate beyond the thermal penetration distance $\delta(t)$. We define this rise in temperature above the initial temperature as ΔT, for $x > \delta(t)$, due to the internal energy.

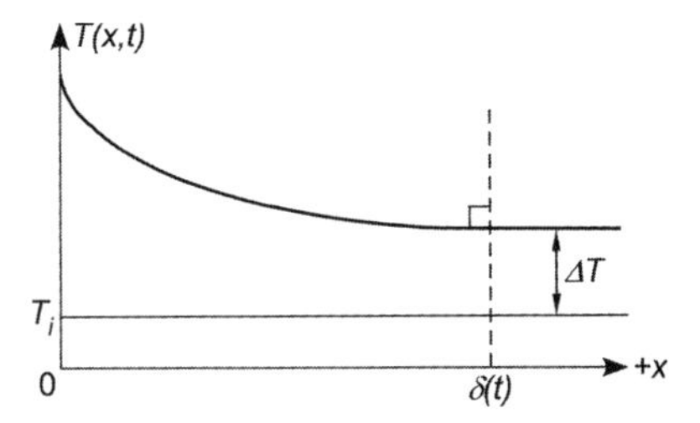

Figure 13-4 Thermal penetration distance in semi-infinite medium with internal energy generation.

In view of Figure 13-4, we now define two additional boundary conditions, namely,

$$T(x = \delta, t) = T|_{x=\delta} \equiv f(t) \tag{13-25d}$$

$$\left.\frac{\partial T}{\partial x}\right|_{x=\delta} = 0 \tag{13-25e}$$

We are now ready to integrate the heat equation (13-25a) from $x = 0$ to $x = \delta(t)$, noting that

$$\int_{x=0}^{\delta(t)} \frac{\alpha}{k} g(t)\, dx = \frac{\alpha}{k} g(t) x \Big|_{0}^{\delta(t)} = \frac{\alpha}{k} g(t)\delta(t) \tag{13-26}$$

since $g(t)$ does not depend on x and is therefore removed from the integral. Treating the remaining terms as we did previously, we obtain the *energy integral equation*

$$-\alpha \left.\frac{\partial T}{\partial x}\right|_{x=0} + \frac{\alpha}{k} g(t)\delta(t) = \frac{d\theta}{dt} - \frac{d\delta}{dt} T|_{x=\delta} \tag{13-27}$$

where we utilized the boundary condition (13-25e), $dT/dx = 0$ at $x = \delta$, and have again defined

$$\theta(t) \equiv \int_{x=0}^{\delta(t)} T(x, t)\, dx \tag{13-28}$$

We note that equation (13-27) is similar to equation (13-5) except for the generation term and the fact that the term $T|_{x=\delta}$ is now a function of time. We now develop an expression for the term $T|_{x=\delta}$ by considering conservation of energy in the domain $x > \delta(t)$. We first define

$$G(t) \equiv \int_{t'=0}^{t} g(t')\, dt' \tag{13-29}$$

where $G(t)$ represents the total amount of energy generated per unit volume (J/m^3) from $t = 0$ to any time t. In the region $x > \delta(t)$, conservation of energy becomes

$$\rho C \Delta T = \rho C \left(T|_{x=\delta} - T_i \right) \equiv G(t) \tag{13-30}$$

which yields the expression

$$T|_{x=\delta} = T_i + \frac{G(t)}{\rho C} \tag{13-31}$$

We also note that the term $g(t)\delta(t)$ on the left-hand side of equation (13-27) can be written as

$$g(t)\delta(t) = \delta(t)\frac{dG(t)}{dt} \tag{13-32a}$$

since

$$\frac{dG(t)}{dt} = g(t) - \cancel{g(0)}^0 \tag{13-32b}$$

Equations (13-31) and (13-32a) are now introduced into equation (13-27), giving

$$-\alpha \left.\frac{\partial T}{\partial x}\right|_{x=0} = \frac{d\theta}{dt} - \frac{\alpha}{k}\delta(t)\frac{dG}{dt} - \left(T_i + \frac{\alpha}{k}G\right)\frac{d\delta}{dt} \tag{13-33}$$

where we have made the substitution $\alpha/k = 1/\rho C$. We note that

$$\delta\frac{dG}{dt} - G\frac{d\delta}{dt} = \frac{d}{dt}(\delta G) \tag{13-34}$$

which upon substitution into equation (13-33) yields

$$\boxed{-\alpha \left.\frac{\partial T}{\partial x}\right|_{x=0} = \frac{d}{dt}\left(\theta - \frac{\alpha}{k}G\delta - T_i\delta\right)} \tag{13-35}$$

which is the *energy integral equation* for the considered problem. If $g(t) = 0$, equation (13-35) reduces directly to equation (13-8). Physically, this again represents conservation of energy over the entire domain, with the additional energy generated internally now accounted for on the right-hand side. For a solution, we assume a cubic polynomial representation for $T(x, t)$ in the form

$$T(x,t) = a_1 + a_2x + a_3x^2 + a_4x^3 \tag{13-36}$$

which now requires four boundary conditions to evaluate these four coefficients. Three are readily defined as

$$\begin{aligned} &1) \quad T(x=0) = T_0 \\ &2) \quad \left.\frac{\partial T}{\partial x}\right|_{x=\delta} = 0 \\ &3) \quad T|_{x=\delta} = T_i + \frac{\alpha}{k}G(t) \end{aligned} \tag{13-37}$$

For our derived boundary condition, we consider the heat equation at $x = \delta$, namely,

$$\left[\frac{\partial^2 T}{\partial x^2} + \frac{1}{k} g(t) = \frac{1}{\alpha}\frac{\partial T}{\partial t}\right]_{x=\delta} \tag{13-38a}$$

and note that per conservation of energy

$$\left[\frac{g(t)}{\rho C} = \frac{\partial T}{\partial t}\right]_{x=\delta} \tag{13-38b}$$

which upon substitution into the above gives our fourth boundary condition as

$$4) \quad \left.\frac{\partial^2 T}{\partial x^2}\right|_{x=\delta} = 0 \tag{13-38c}$$

Applying equations (13-37) and (13-38c), the temperature profile is determined as

$$T(x,t) = T_0 + \left[1 - \left(1 - \frac{x}{\delta}\right)^3\right] F \qquad \text{in} \qquad 0 \le x \le \delta \tag{13-39a}$$

where we define $F(t)$, a known function of time, as

$$F(t) \equiv (T_i - T_0) + \frac{\alpha}{k} G \tag{13-39b}$$

Introducing equation (13-39a) into equation (13-35) and performing the indicated operations, we obtain the following differential equation for $\delta(t)$:

$$12\alpha F^2 = (F\delta)\frac{d(F\delta)}{dt} \qquad \text{for} \qquad t > 0 \tag{13-40}$$

The solution of equation (13-40), subject to the initial condition $\delta(t = 0) = 0$, gives

$$\delta(t) = \left[\frac{24\alpha \int_{t'=0}^{t} F^2(t')\,dt'}{F^2(t)}\right]^{1/2} \tag{13-41}$$

Equations (13-39) together with equation (13-41) give the temperature distribution in the medium as a function of time and position. For the special case of no heat generation, equations (13-39) and (13-40), respectively, reduce to

$$\frac{T(x,t) - T_i}{T_0 - T_i} = \left(1 - \frac{x}{\delta}\right)^3 \tag{13-42a}$$

and

$$\delta(t) = \sqrt{24\alpha t} \tag{13-42b}$$

which are exactly the same as equations (13-20) in the previous example.

13-3 INTEGRAL METHOD: APPLICATION TO NONLINEAR TRANSIENT HEAT CONDUCTION

Another advantage of the integral method is that it can handle the nonlinear problems quite readily. In the following two examples we illustrate the application of the integral method to the solution of nonlinear heat conduction problems. In the first example the nonlinearity is due to the boundary condition, in the second it is due to the differential equation.

Example 13-3 Semi-Infinite Region with Nonlinear Boundary Condition A semi-infinite medium is initially at uniform temperature T_i. For times $t > 0$, the boundary surface at $x = 0$ is subjected to a heat flux that is a prescribed function of time and of the surface temperature. We seek an expression for the surface temperature $T_s(t)$ for times $t > 0$ using the integral method.

The mathematical formulation of this problem is given as

$$\frac{\partial^2 T}{\partial x^2} = \frac{1}{\alpha}\frac{\partial T}{\partial t} \quad \text{in} \quad x > 0, \quad \text{for} \quad t > 0 \tag{13-43a}$$

$$\text{BC:} \quad -\left.\frac{\partial T}{\partial x}\right|_{x=0} = f(T_s, t) \tag{13-43b}$$

$$\text{IC:} \quad T(x, t = 0) = T_i \quad \text{in} \quad x \geq 0 \tag{13-43c}$$

Here the boundary condition function $f(T_s, t)$ contains the thermal conductivity 1/k and is a function of time t and of the boundary surface temperature $T_s(x = 0) \equiv T_s(t)$.

The integration of the differential equation (13-43a) over the thermal layer $\delta(t)$ gives

$$\left.\frac{\partial T}{\partial x}\right|_{x=\delta} - \left.\frac{\partial T}{\partial x}\right|_{x=0} = \frac{1}{\alpha}\left[\frac{d}{dt}\left(\int_{x=0}^{\delta} T\,dx\right) - \frac{d\delta}{dt}\,T|_{x=\delta}\right] \tag{13-44}$$

In view of the conditions

$$\begin{aligned} &(1) \quad \left.\frac{\partial T}{\partial x}\right|_{x=\delta} = 0 \\ &(2) \quad T|_{x=\delta} = T_i \\ &(3) \quad -\left.\frac{\partial T}{\partial x}\right|_{x=0} = f(T_s, t) \end{aligned} \tag{13-45}$$

equation (13-44) becomes

$$\alpha f(T_s, t) = \frac{d}{dt}(\theta - T_i\delta) \tag{13-46a}$$

where

$$\theta(t) \equiv \int_{x=0}^{\delta} T\,dx \tag{13-46b}$$

Equation (13-46) becomes the energy integral equation for the considered problem.

To solve this equation, we choose a cubic polynomial representation for $T(x,t)$ given as

$$T(x,t) = a_1 + a_2 x + a_3 x^2 + a_4 x^3 \tag{13-47}$$

These four coefficients are determined by utilizing the three conditions (13-45), together with the derived condition

$$\left.\frac{\partial^2 T}{\partial x^2}\right|_{x=\delta} = 0 \tag{13-48}$$

The resulting temperature profile becomes

$$T(x,t) - T_i = \frac{\delta f(T_s,t)}{3}\left(1 - \frac{x}{\delta}\right)^3 \quad \text{in} \quad 0 \le x \le \delta \tag{13-49}$$

and for $x = 0$, this relation gives

$$T_s(t) - T_i = \frac{\delta f(T_s,t)}{3} \tag{13-50}$$

From equations (13-49) and (13-50), we write

$$\frac{T(x,t) - T_i}{T_s(t) - T_i} = \left(1 - \frac{x}{\delta}\right)^3 \tag{13-51}$$

Introducing equation (13-51) into equation (13-46a), performing the indicated operations, and eliminating $\delta(t)$ from the resulting expression by means of equation (13-50), we obtain the following first-order ordinary differential equation for the determination of the surface temperature T_s, namely,

$$\frac{4}{3}\alpha f(T_s,t) = \frac{d}{dt}\left[\frac{(T_s - T_i)^2}{f(T_s,t)}\right] \quad \text{for} \quad t > 0 \tag{13-52a}$$

with initial condition

$$T_s(t = 0) = T_i \tag{13-52b}$$

Equation (13-52a) can be integrated numerically if the boundary condition function $f(T_s,t)$ depends on both the surface temperature and the time. For the special case of $f(T_s,t)$ being a function of surface temperature only, namely,

$$f(T_s,t) = f(T_s) \tag{13-53}$$

equation (13-52a) is written as

$$\frac{4}{3}\alpha f(T_s) = \frac{d}{dT_s}\left[\frac{(T_s - T_i)^2}{f(T_s)}\right]\frac{dT_s}{dt} \tag{13-54a}$$

or

$$\frac{4}{3}\alpha = \frac{2(T_s - T_i)f(T_s) - f'(T_s)(T_s - T_i)^2}{f^3(T_s)}\frac{dT_s}{dt} \quad \text{for} \quad t > 0 \tag{13-54b}$$

with initial condition

$$T_s(t = 0) = T_i \tag{13-54c}$$

The integration of equation (13-54) establishes the relation between the surface temperature $T_s(t)$ and the time t as

$$\frac{4}{3}\alpha t = \int_{T_s=T_i}^{T_s} \frac{2(T_s - T_i)f(T_s) - f'(T_s)(T_s - T_i)^2}{f^3(T_s)}\, dT_s \tag{13-55}$$

where $f'(T_s)$ denotes differentiation with respect to T_s.

Example 13-4 Semi-Infinite Medium with Temperature-Dependent Properties

A semi-infinite medium, $x > 0$, is initially at zero temperature. For times $t > 0$, the boundary surface is subjected to a prescribed heat flux that varies with time. The thermal properties $k(T)$, $C(T)$, and $\rho(T)$ are all assumed to depend on temperature. Obtain an expression for the temperature distribution in the medium using the integral method.

The mathematical formulation of this problem is given as

$$\frac{\partial}{\partial x}\left(k\frac{\partial T}{\partial x}\right) = \rho C\frac{\partial T}{\partial t} \quad \text{in} \quad x > 0, \quad t > 0 \tag{13-56a}$$

$$\text{BC:} \quad -k\left.\frac{\partial T}{\partial x}\right|_{x=0} = f(t) \tag{13-56b}$$

$$\text{IC:} \quad T(t = 0) = 0 \quad \text{for} \quad x \geq 0 \tag{13-56c}$$

where $k \equiv k(T)$, $C \equiv C(T)$, and $\rho \equiv \rho(T)$. By now applying the transformation

$$U = \int_{T'=0}^{T} \rho C\, dT' \tag{13-57}$$

the system (13-56) is transformed into

$$\frac{\partial}{\partial x}\left(\alpha\frac{\partial U}{\partial x}\right) = \frac{\partial U}{\partial t} \quad \text{in} \quad x > 0, \quad t > 0 \tag{13-58a}$$

$$\text{BC:} \qquad -\left.\frac{\partial U}{\partial x}\right|_{x=0} = \frac{1}{\alpha_s} f(t) \tag{13-58b}$$

$$\text{IC:} \qquad U(t=0) = 0 \qquad \text{for} \qquad x \geq 0 \tag{13-58c}$$

where $\alpha \equiv \alpha(U)$, and α_s refers to the value of α at the boundary surface $x = 0$. Equation (13-58a) is now integrated over the thermal layer $\delta(t)$, yielding

$$\left[\alpha \frac{\partial U}{\partial x}\right]_{x=0}^{\delta} = \frac{d}{dt}\left(\int_{x=0}^{\delta} U\,dx - U\Big|_{\delta}\,\delta\right) \tag{13-59}$$

In view of the conditions

$$(1) \qquad \left.\frac{\partial U}{\partial x}\right|_{x=\delta} = 0$$

$$(2) \qquad U|_{x=\delta} = 0 \tag{13-60}$$

$$(3) \qquad \left[\alpha \frac{\partial U}{\partial x}\right]_{x=0} = -f(t)$$

equation (13-59) becomes

$$\frac{d\theta}{dt} = f(t) \tag{13-61a}$$

where

$$\theta(t) \equiv \int_{x=0}^{\delta} U\,dx \tag{13-61b}$$

Equation (13-61a) is the energy integral equation for the considered problem. To solve this equation, we choose a cubic polynomial representation for $U(x, t)$ as

$$U(x,t) = a_1 + a_2 x + a_3 x^2 + a_4 x^3 \tag{13-62}$$

The four coefficients are determined by utilizing the following four conditions:

$$U\Big|_{x=\delta} = 0, \qquad \left.\frac{\partial U}{\partial x}\right|_{x=\delta} = 0, \qquad \left.\frac{\partial U}{\partial x}\right|_{x=0} = -\frac{f(t)}{\alpha_s}, \qquad \text{and} \qquad \left.\frac{\partial^2 T}{\partial x^2}\right|_{x=\delta} = 0 \tag{13-63}$$

Applying to equation (13-62), the corresponding profile becomes

$$U(x,t) = \frac{\delta(t) f(t)}{3\alpha_s}\left[1 - \frac{x}{\delta(t)}\right]^3 \qquad \text{in} \qquad 0 \leq x \leq \delta \tag{13-64}$$

By substituting equation (13-64) into equations (13-61) and performing the indicated operations, we obtain the following differential equation for the determination of the thermal layer thickness $\delta(t)$:

$$\frac{d}{dt}\left[\frac{\delta^2 f(t)}{12\alpha_s}\right] = f(t) \qquad \text{for} \qquad t > 0 \tag{13-65a}$$

with

$$\delta(t = 0) = 0 \tag{13-65b}$$

The solution of equation (13-65) is

$$\delta(t) = \left[\frac{12\alpha_s}{f(t)} \int_{t'=0}^{t} f(t')\, dt' \right]^{1/2} \tag{13-66}$$

This equation cannot yet be used to calculate the thermal layer thickness $\delta(t)$ directly because it involves α_s, the thermal diffusivity evaluated at the surface temperature, U_s, which is still unknown. To circumvent this difficulty an additional relationship is needed between α_s and U_s; such a relationship is obtained as now described. For $x = 0$, equation (13-64) gives

$$U_s = \frac{\delta f(t)}{3\alpha_s} \tag{13-67}$$

Eliminating $\delta(t)$ between equations (13-66) and (13-67), we obtain

$$U_s\sqrt{\alpha_s} = \left[\frac{4}{3} f(t) \int_{t'=0}^{t} f(t')\, dt' \right]^{1/2} \tag{13-68}$$

The computational procedure is now as follows:

1. The α_s is known as a function of T_s and therefore of U_s. Then the left-hand side of equation (13-68) can be regarded to depend on α_s only.
2. Equation (13-68) is then used to compute α_s as a function of time.
3. Knowing α_s at each time, equation (13-66) is used to calculate $\delta(t)$.
4. Knowing $\delta(t)$, equation (13-64) is used to calculate U.
5. Knowing U, equation (13-57) is used to determine the actual temperature distribution $T(x, t)$.

13-4 INTEGRAL METHOD: APPLICATION TO A FINITE REGION

In the previous examples, we considered the application of the integral method for the solution of transient heat conduction in a semi-infinite medium in which the thickness of the thermal layer $\delta(t)$ would increase without bound. However, for the problem of a slab, $0 \le x \le L$, for example, with the surface $x = L$ insulated, the analysis is exactly the same as that described for a semi-infinite region so long as the thickness of the thermal layer remains less than the thickness of the plate; however, as soon as the thickness of the thermal layer becomes equal to that of the slab, that is, $\delta(t) = L$, the thermal layer has no physical significance. A different analysis is then required for times beyond the moment for which $\delta(t)$ first reaches L. This matter is illustrated with the following example.

Example 13-5 Integral Method for Finite Region
A slab, $0 \le x \le L$, is initially at a uniform temperature T_i. For times $t > 0$, the boundary surface at $x = 0$ is kept at a constant temperature T_0 while the boundary surface at $x = L$ is kept insulated. Obtain an expression for the temperature distribution in the slab for times $t > 0$ by using the integral method.

The mathematical formulation of this problem is given as

$$\frac{\partial^2 T}{\partial x^2} = \frac{1}{\alpha}\frac{\partial T}{\partial t} \quad \text{in} \quad 0 < x < L, \quad t > 0 \tag{13-69a}$$

$$\text{BC1:} \quad T(x = 0, t) = T_0 \tag{13-69b}$$

$$\text{BC2:} \quad \left.\frac{\partial T}{\partial x}\right|_{x=L} = 0 \tag{13-69c}$$

$$\text{IC:} \quad T(x, t = 0) = T_i \quad \text{in} \quad 0 \le x \le L \tag{13-69d}$$

For the reasons discussed above, the analysis is now performed in *two distinct stages*: the *first stage*, during which the thermal layer thickness does not exceed the slab thickness [i.e., $\delta(t) \le L$], and the *second stage*, during which the time scale of the problem exceeds the time at which $\delta(t) = L$.

The First Stage. For the case $\delta(t) \le L$, we integrate equation (13-69a) over the thermal layer thickness and obtain

$$-\alpha \left.\frac{\partial T}{\partial x}\right|_{x=0} = \frac{d}{dt}(\theta - T_i\delta) \tag{13-70a}$$

where

$$\theta = \int_{x=0}^{\delta} T(x, t)\, dx \tag{13-70b}$$

The energy integral equation thus obtained is exactly the same as that given by equation (13-8) for the semi-infinite region. We choose a cubic profile for the temperature as given by equation (13-10), apply the conditions given by equations (13-11) to determine the coefficients, and utilize equation (13-70) to obtain the thermal layer thickness as discussed for the semi-infinite region. The resulting temperature profile becomes

$$\frac{T(x, t) - T_i}{T_0 - T_i} = 1 - \frac{3}{2}\left(\frac{x}{\delta}\right) + \frac{1}{2}\left(\frac{x}{\delta}\right)^3 \tag{13-71a}$$

where

$$\delta = \sqrt{8\alpha t} \tag{13-71b}$$

This solution is valid for $0 \le x \le \delta(t)$, as long as $\delta(t) \le L$. The time t_L when $\delta(t) = L$ is obtained from equation (13-71b) by setting $\delta(t) = L$, that is,

$$\boxed{t_L = \frac{L^2}{8\alpha}} \tag{13-72}$$

Clearly, the solution (13-71) is not applicable for times $t > t_L$.

The Second Stage. For times $t > t_L$, the concept of thermal layer has no physical significance. The analysis for the second stage may be performed in the following manner. We integrate the differential equation (13-69a) from $x = 0$ to $x = L$, which yields

$$-\alpha \left. \frac{\partial T}{\partial x} \right|_{x=0} = \frac{d}{dt}(\theta - T_i L) \tag{13-73a}$$

where we now define

$$\theta \equiv \int_{x=0}^{L} T(x,t)\,dx \tag{13-73b}$$

A comparison of equation (13-73a) with equation (13-70a) reveals that in the latter, the plate thickness L has replaced $\delta(t)$; hence there is no thermal layer. The temperature $T(x, t)$ is again expressed by a polynomial. Suppose, as before, we choose a cubic polynomial representation in the form

$$T(x,t) = a + bx + cx^2 + dx^3 \qquad \text{in} \qquad 0 < x < L, \qquad t > t_L \tag{13-74}$$

where the coefficients are generally functions of time. In this case, we have no thermal layer to be determined from the solution of the differential equation (13-73). Therefore, we choose only three conditions:

$$\begin{aligned} &1) \qquad T|_{x=0} = T_0 \\ &2) \qquad \left. \frac{\partial T}{\partial x} \right|_{x=L} = 0 \\ &3) \qquad \left. \frac{\partial^2 T}{\partial x^2} \right|_{x=0} = 0 \end{aligned} \tag{13-75}$$

These conditions are applied to the cubic profile given by equation (13-74), and all the coefficients are expressed in terms of one of them, say $b \equiv b(t)$. We find the following profile:

$$T(x,t) = T_0 + bL\left[\frac{x}{L} - \frac{1}{3}\left(\frac{x}{L}\right)^3\right] \qquad \text{in} \qquad 0 \le x \le L \tag{13-76}$$

which is expressed in the dimensionless form as

$$\frac{T(x,t) - T_i}{T_0 - T_i} = 1 + \eta(t)\left[\frac{x}{L} - \frac{1}{3}\left(\frac{x}{L}\right)^3\right] \qquad \text{in} \qquad 0 \le x \le L \tag{13-77a}$$

where

$$\eta(t) \equiv \frac{Lb(t)}{T_0 - T_i} \tag{13-77b}$$

When the profile (13-77a) is introduced into equation (13-73a) and the indicated operations are performed, the following ordinary differential equation is obtained for the determination of $\eta(t)$:

$$\frac{d\eta(t)}{dt} + \frac{12\alpha}{5L^2}\eta(t) = 0 \qquad \text{for} \qquad t \geq t_L \tag{13-78}$$

The initial condition needed to solve this differential equation is determined from the requirement that the temperature defined by equation (13-77a) at $x = L$ and at the time $t = t_L = L^2/8\alpha$ should be equal to the initial condition $T = T_i$. With this initial condition, we find

$$\eta(t = t_L) = -\frac{3}{2} \qquad \text{for} \qquad t = t_L = \frac{L^2}{8\alpha} \tag{13-79}$$

The solution of the differential equation (13-78) subject to the initial condition (13-79) gives

$$\eta(t) = -\frac{3}{2}\exp\left[-\left(\frac{12\alpha}{5L^2}t - \frac{3}{10}\right)\right] \tag{13-80}$$

Thus equation (13-77a) with $\eta(t)$ as given by equation (13-80) represents the temperature distribution in the slab for times $t > t_L \equiv L^2/8\alpha$. Figure 13-5 shows a comparison of the exact and approximate solutions.

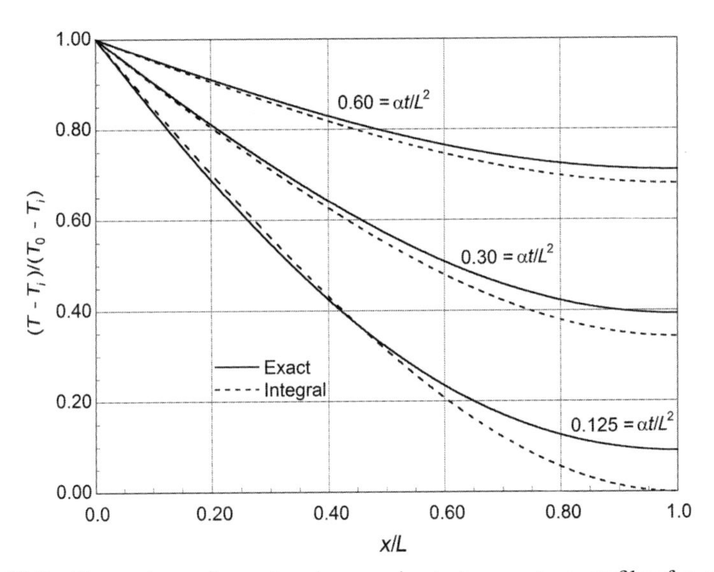

Figure 13-5 Comparison of exact and approximate temperature profiles for a slab of thickness L.

13-5 APPROXIMATE ANALYTIC METHODS OF RESIDUALS

When the exact solution T cannot be obtained for an ordinary or partial differential equation, or computationally it is desirable to seek an approximate solution, an approximate *trial solution* $\tilde{T}$ containing a finite number of undetermined coefficients $c_1, c_2, \ldots, c_n$ can be constructed by the superposition of some *basis functions* such as polynomials, trigonometric functions, and the like. The *trial solution* is so selected that it satisfies exactly the essential boundary conditions for the problem; however, the trial solution does not satisfy the differential equation and leads to a *residual* R because it is not the exact solution. Here we define the residual as the difference between the exact satisfaction of the differential equation and the approximate satisfaction of the differential equation by the trial solution. Therefore, for the true solution, the residuals vanish identically. Accordingly, the problem of constructing an approximate trial solution becomes one of determining the unknown coefficients $c_1, c_2, \ldots, c_n$ so that the residual stays "close" to zero throughout the domain of the solution. Depending on the number of terms taken for the trial solution, the type of base functions used, and the way the unknown coefficients are determined, several different approximate solutions are possible for a given problem.

Different schemes have been proposed for the determination of the unknown coefficients $c_1, c_2, \ldots, c_n$ associated with the construction of the trial family of approximate solution $\tilde{T}$. To illustrate the basic approaches followed in the various approximate methods of solution, we examine the following simple problem considered in reference 55:

$$\frac{dT(t)}{dt} + T(t) = 0 \qquad \text{for} \qquad t > 0 \tag{13-81a}$$

$$\text{IC:} \qquad T(t = 0) = 1 \tag{13-81b}$$

The exact solution of this problem is given by

$$T(t) = e^{-t} \tag{13-82}$$

We wish to obtain an approximate solution for this problem in the specific interval $0 < t < 1$. To construct a trial solution $\tilde{T}(t)$, we choose the *basis functions* to be polynomials in t (e.g., t and t^2) and create a trial solution of the form

$$\tilde{T}(t) = c_0 + c_1 t + c_2 t^2 \tag{13-83}$$

The above trial solution contains only two undetermined coefficients c_1 and c_2, as the coefficient c_0 is used only to satisfy the nonhomogeneous boundary/initial condition equation (13-81b) and is not included as our basis functions; hence $c_0 = 1$, and we have

$$\tilde{T}(t) = 1 + c_1 t + c_2 t^2 \tag{13-84}$$

The trial solution (13-84) now satisfies the condition (13-81b) for all values of c_1 and c_2. We then introduce our trial solution equation (13-84) into the differential equation (13-81a), which yields

$$\left(c_1 + 2c_2 t\right) + \left(1 + c_1 t + c_2 t^2\right) = 0 \tag{13-85}$$

When the left-hand side is exactly equal to zero, an exact solution is realized. However, when the left-hand side does not equal zero, we define the *residual* $R(c_1, c_2, t)$, namely,

$$\left(c_1 + 2c_2 t\right) + \left(1 + c_1 t + c_2 t^2\right) = R(c_1, c_2, t) \tag{13-86a}$$

or rearranging, as

$$R(c_1, c_2, t) \equiv 1 + (1 + t)c_1 + (2t + t^2)c_2 \tag{13-86b}$$

This residual vanishes only with the exact solution for the problem. Now, the problem of finding an approximate solution for the problem (13-81) in the interval $0 < t < 1$ becomes one of adjusting the values of c_1 and c_2 so that the residual $R(c_1, c_2, t)$ stays "close" to zero throughout the interval $0 < t < 1$. Various schemes have been proposed for the determination of these unknown coefficients; when c_1 and c_2 are known, the trial solution $\tilde{T}(t)$ given by equation (13-84) becomes the approximate solution for the problem.

We briefly describe below some of the popular schemes for the determination of the unknown coefficients.

Collocation Method If the trial solution contains n undetermined coefficients, n different locations are selected where the residual $R(c_1, c_2, t)$ is forced to vanish, thus providing n simultaneous algebraic equations for the determination of the n coefficients $c_1, c_2, \ldots, c_n$. The basic assumption is that the residual does not deviate much from zero between the collocation locations (where it identically equals zero). For the specific example considered above, we select the collocation locations $t = \frac{1}{3}$ and $t = \frac{2}{3}$, which are evenly spaced throughout the domain. Introducing these values into the residual equation (13-86b), we obtain

$$R(c_1, c_2, \tfrac{1}{3}) = 1 + \tfrac{4}{3}c_1 + \tfrac{7}{9}c_2 = 0 \tag{13-87a}$$

and

$$R(c_1, c_2, \tfrac{2}{3}) = 1 + \tfrac{5}{3}c_1 + \tfrac{16}{9}c_2 = 0 \tag{13-87b}$$

Thus we have two algebraic equations for the determination of the two unknown coefficients c_1 and c_2. Simultaneous solution gives

$$c_1 = -\tfrac{27}{29} \cong -0.9310 \qquad \text{and} \qquad c_2 = \tfrac{9}{29} \cong 0.3103$$

Introducing these coefficients into equation (13-84), we obtain the approximate solution for the problem (13-81) in the interval $0 < t < 1$, based on the collocation method as

$$\tilde{T}(t) = 1 - 0.9310t + 0.3103t^2 \tag{13-88}$$

Checking the values in the center of the domain, namely at $t = \frac{1}{2}$, we find that $\tilde{T}(t = \frac{1}{2}) = 0.6121$ and that the exact solution, per equation (13-82), yields $T(t = \frac{1}{2}) = 0.6025$, an agreement of about 1%. We can also check the agreement at $t = \frac{1}{3}$, which yields $\tilde{T}(t = \frac{1}{3}) = 0.7241$ and $T(t = \frac{1}{3}) = 0.7165$, also for an agreement of about 1%. We note here that even though $t = \frac{1}{3}$ was used as a location for the implementation of the collocation scheme, that does *not ensure that the solution is exact at that location*. In fact, only the differential equation is solved exactly at $t = \frac{1}{3}$. Examination of the ODE, equation (13-81a), reveals that the sum of the derivative of T and the value of T equal zero. With this in mind, we see that value of the trial solution is less than the exact solution at $t = \frac{1}{3}$ and that the slope of the trial solution is therefore reduced accordingly.

Least-Squares Method Referring to the simple example considered above, the coefficients c_1 and c_2 are determined from the requirement that the *integral of the square of the residual* $R(c_1, c_2, t)$ given by equation (13-86b) is minimized over the interval $0 < t < 1$. Using a traditional approach to minimization of a function, we set

$$\frac{1}{2}\frac{\partial}{\partial c_1}\int_{t=0}^{1} R^2(c_1, c_2, t)\, dt = \int_{t=0}^{1} R\frac{\partial R}{\partial c_1}\, dt = 0 \tag{13-89a}$$

and

$$\frac{1}{2}\frac{\partial}{\partial c_2}\int_{t=0}^{1} R^2(c_1, c_2, t)\, dt = \int_{t=0}^{1} R\frac{\partial R}{\partial c_2}\, dt = 0 \tag{13-89b}$$

where we have introduced the factor of one-half for convenience. Such an approach may be shown to yield a minimum. Applying the above to equation (13-86b) gives

$$\int_{t=0}^{1} R\frac{\partial R}{\partial c_1}\, dt = \frac{3}{2} + \frac{7}{3}c_1 + \frac{9}{4}c_2 = 0 \tag{13-90a}$$

and

$$\int_{t=0}^{1} R\frac{\partial R}{\partial c_2}\,dt = \frac{4}{3} + \frac{9}{4}c_1 + \frac{38}{15}c_2 = 0 \tag{13-90b}$$

Again we have two algebraic equations for the determination of the two unknown constants c_1 and c_2. A simultaneous solution gives

$$c_1 = -0.9427 \qquad \text{and} \qquad c_2 = 0.3110$$

Introducing these values into equation (13-84), we obtain the approximate solution for the problem (13-81b) in the interval $0 < t < 1$ as

$$\tilde{T}(t) = 1 - 0.9427t + 0.3110t^2 \tag{13-91}$$

Again checking the values in the center of the domain, namely, at $t = \frac{1}{2}$, we find that $\tilde{T}(t = \frac{1}{2}) = 0.6064$, an agreement of about 1/100%. Checking the agreement at $t = \frac{1}{3}$, we find $\tilde{T}(t = \frac{1}{3}) = 0.7203$, for an agreement of about 0.5%. Since the least-squares method provides a minimization of the residual over an integral, we have no knowledge about the satisfaction of the ODE at any particular location as we did with the previous approach.

Rayleigh–Ritz Method This method requires the variational formulation of the differential equation so that the boundary conditions for the problem are incorporated into the variational form. Once the variational form is available, the trial solution given by equation (13-84) is introduced into the variational expression $J(c_1, c_2)$, and this result is minimized as

$$\frac{\partial J(c_1, c_2)}{\partial c_1} = 0 \tag{13-92a}$$

$$\frac{\partial J(c_1, c_2)}{\partial c_2} = 0 \tag{13-92b}$$

Thus, we have two algebraic equations for the determination of the two unknown coefficients c_1 and c_2. Generally, the difficulty with this approach is the determination of the corresponding variational expression. If the variational form cannot be found, the scheme is not applicable. The principles of variational calculus are discussed in several texts [32–36], and the application to the solution of heat conduction problems can be found in several other references [37–56]. Next, we consider the Galerkin method [57], which leads to the same approximate solution as the Rayleigh-Ritz method without requiring the variational form of the problem.

Galerkin Method The method requires that the weighted averages of the residual $R(c_1, c_2, t)$ should vanish over the interval considered. The weighting functions

$w_1(t)$ and $w_2(t)$ are taken as the *same basis functions* used to construct the trial solution $\tilde{T}(t)$. For the specific problem considered here, t and t^2 are the weight functions to be used for the integration of the residual $R(c_1, c_2, t)$ over the interval $0 < t < 1$. Thus, the Galerkin method becomes

$$\int_{t=0}^{1} t R(c_1, c_2, t)\, dt = 0 \tag{13-93a}$$

$$\int_{t=0}^{1} t^2 R(c_1, c_2, t)\, dt = 0 \tag{13-93b}$$

Application of the above expressions for our residual equation (13-86b) yields

$$\int_{t=0}^{1} t \left[1 + (1+t)c_1 + (2t + t^2)c_2\right] dt = \tfrac{1}{2} + \tfrac{5}{6}c_1 + \tfrac{11}{12}c_2 = 0 \tag{13-94a}$$

$$\int_{t=0}^{1} t^2 \left[1 + (1+t)c_1 + (2t + t^2)c_2\right] dt = \tfrac{1}{3} + \tfrac{7}{12}c_1 + \tfrac{9}{20}c_2 = 0 \tag{13-94b}$$

Equations (13-94) provide two algebraic equations for the determination of the unknown coefficients c_1 and c_2. A simultaneous solution gives

$$c_1 = -0.9143 \qquad \text{and} \qquad c_2 = -0.2857$$

Introducing these coefficients into equation (13-84), the approximate solution becomes

$$\tilde{T}(t) = 1 - 0.9143t + 0.2857t^2 \tag{13-95}$$

We also note here that the weight functions $w_1(t)$ and $w_2(t)$ can be interpreted as

$$w_i(t) = \frac{\partial \tilde{T}(t)}{\partial c_i} \qquad i = 1, 2 \tag{13-96}$$

Again checking the values in the center of the domain, namely at $t = \frac{1}{2}$, we find that $\tilde{T}(t = \frac{1}{2}) = 0.6143$, an agreement of about 1.3%. Checking the agreement at $t = \frac{1}{3}$, we find $\tilde{T}(t = \frac{1}{3}) = 0.7180$, for an agreement of about 0.2%. Since the Galerkin method provides a minimization of the weighted residual over an integral, we also have no knowledge about the satisfaction of the ODE at any particular location, similar to the previous approach.

In Figures 13-6 and 13-7 we present a plot of the exact solution, along with our trial solutions for the collocation method, equation (13-88), the least-squares method, equation (13-91), and the Galerkin method, equation (13-95), along with the exact solution given by equation (13-82). We see that all of the trial solutions cross the exact solution at some point in the domain, hence exact agreement at some location, although that location is not known a priori.

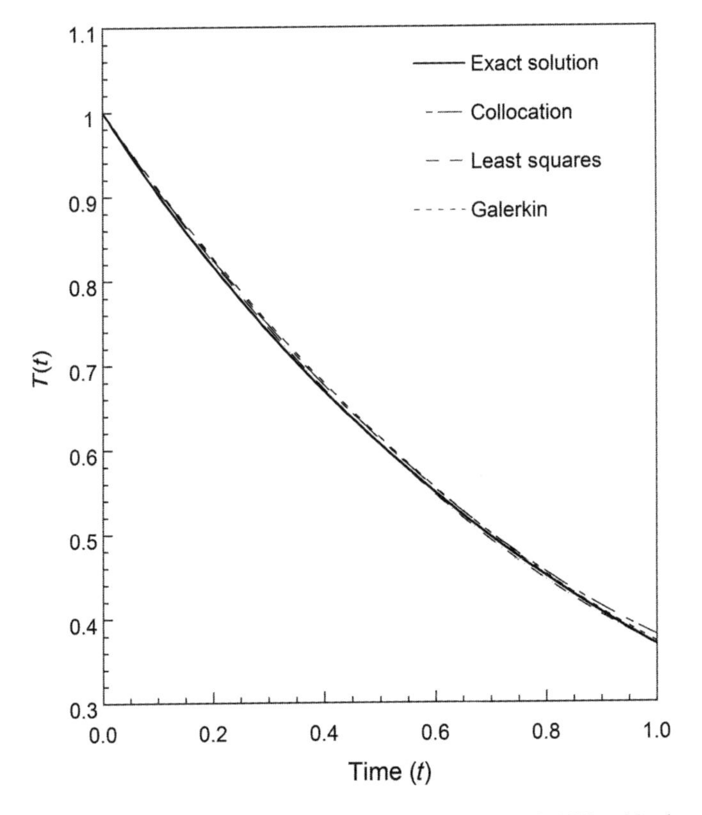

Figure 13-6 Comparison of the exact solution [equation (13-82)] with the collocation [equation (13-88)], the least-squares [equation (13-91)], and the Galerkin [equation (13-95)] solutions over the full domain of the problem.

As noted above, the Galerkin method does not require the variational form of the problem and yields the same result as the Rayleigh–Ritz method, therefore the problem setup is easier and more direct. We now present further applications of the Galerkin method and discuss the construction of the basis functions.

13-6 THE GALERKIN METHOD

In the previous section, we illustrated with a simple example the basic concepts in the application of some of the popular approximate analytic methods for the solution of differential equations. Here we focus attention on the Galerkin method, especially on its application to more general problems, and the methods of determining the trial functions.

The method is perfectly universal; it can be applied to elliptic, hyperbolic and parabolic equations, nonlinear problems, as well as complicated boundary

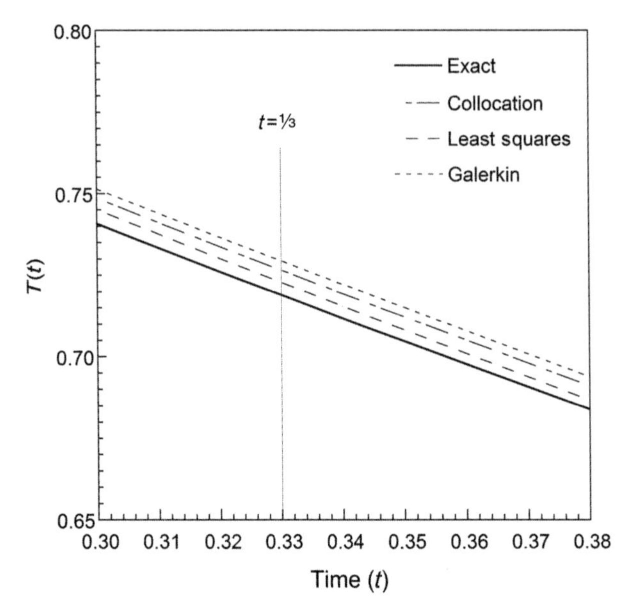

Figure 13-7 Closeup of Figure 13-6 showing the four solutions in the vincinity of $t = \frac{1}{3}$.

conditions. The reader should consult references [16, 37, 53–55] for a discussion of the theory and application of the Galerkin method and references [58–69] for its application in the solution of various types of boundary value problems.

Application to Steady-State Heat Conduction

We consider a steady-state heat conduction problem given in the form

$$\nabla^2 T(\hat{r}) + AT(\hat{r}) + \frac{1}{k} g(\hat{r}) = 0 \qquad \text{in} \qquad R \tag{13-97a}$$

$$k\frac{\partial T}{\partial n} + hT = f(\hat{r}_s) \qquad \text{on boundary } S \tag{13-97b}$$

where $\partial/\partial n$ denotes the derivative along the outward-drawn normal to the boundary surface S. Clearly, the problem defined by equations (13-97) covers a wide range of steady-state heat conduction problems as special cases.

Let $\phi_j(\hat{r})$, $j = 1, 2, 3, \ldots$, be a set of *basis functions*. We construct the n-term trial solution $\tilde{T}_n(\hat{r})$ in the form

$$\tilde{T}_n(\hat{r}) = \psi_0(\hat{r}) + \sum_{j=1}^{n} c_j \phi_j(\hat{r}) \tag{13-98}$$

where the function $\psi_0(\hat{r})$ is included to satisfy the nonhomogeneous part of the boundary condition (13-97b), and the basis functions $\phi_j(\hat{r})$ satisfy the homogeneous part. When all the boundary conditions are homogeneous, the function $\psi_0(\hat{r})$ is not needed. The subscript n in the trial solution $\tilde{T}_n(\hat{r})$ denotes that it is an n-term trial solution.

When the trial solution (13-98) is substituted into the differential equation (13-97a), a residual $R(c_1, c_2, \ldots, c_n; \hat{r})$ is left because $\tilde{T}_n(\hat{r})$ is not an exact solution. We obtain

$$\nabla^2 \tilde{T}_n(\hat{r}) + A\tilde{T}_n(\hat{r}) + \frac{1}{k} g(\hat{r}) \equiv R(c_1, c_2, \ldots, c_n; \hat{r}) \neq 0 \tag{13-99}$$

Then the Galerkin method for the determination of the n unknown coefficients $c_1, c_2, \ldots, c_n$ is given by

$$\int_R \phi_j(\hat{r}) \left[\nabla^2 \tilde{T}_n(\hat{r}) + A\tilde{T}_n(\hat{r}) + \frac{1}{k} g(\hat{r}) \right] dV = 0, \qquad j = 1, 2, \ldots, n \tag{13-100a}$$

which is written more compactly in the form

$$\int_R \phi_j(\hat{r}) R(c_1, c_2, \ldots, c_n; \hat{r})\, dV = 0, \qquad j = 1, 2, \ldots, n \tag{13-100b}$$

Equations (13-100) provide n algebraic equations for the determination of n unknown coefficients $c_1, c_2, \ldots, c_n$. If the problem can be solved by the separation of variables, and the basis functions $\phi_j(\hat{r})$ are taken to be the eigenfunctions for the problem, then the solution generated by the Galerkin method becomes the exact solution for the problem as the number of terms n approaches infinity. However, in general, the eigenfunctions for the problem are not available; hence the question arises regarding what kind of functions should be chosen as the basis functions to construct the trial solution.

The functions $\phi_j(\hat{r})$ should satisfy the homogeneous part of the boundary conditions and should be linearly independent over the given region R. In addition, the functions $\phi_j(\hat{r})$, if possible, should belong to a class of functions that are complete in the considered region. They should be continuous in the region and should have continuous first and second derivatives. They may be polynomials, trigonometric, circular, or spherical functions, but they should satisfy the homogeneous part of the boundary conditions for the problem. We now present some guidelines for selection of the basis functions for select classes of boundary conditions.

Construction of Basis Functions When Boundary Conditions Are All of the First Kind

In regions having simple geometries, such as a slab, cylinder, sphere, or rectangle, the functions $\phi_j(\hat{r})$ can be taken as the eigenfunctions obtained by the separation of variables, as available from Chapters 3 to 5. Thus, the functions $\phi_j(\hat{r})$ can be used as the basis functions to construct the trial solution $\tilde{T}_n(\hat{r})$ for the problem. However, there are many situations in which the boundaries of the region are irregular; as a result, it becomes very difficult to find basis functions $\phi_j(\hat{r})$ that will satisfy the homogeneous boundary conditions. Here we present a methodology to construct the basis functions $\phi_j(\hat{r})$ for such situations in two-dimensional problems.

Let a function $\omega(x, y)$ be a continuous function, and have continuous derivatives with respect to x and y within the region, and in addition satisfy the homogeneous boundary condition of the first kind at the boundaries of the region:

$$\omega(x, y) > 0 \quad \text{in} \quad R \tag{13-101a}$$

$$\omega(x, y) = 0 \quad \text{on boundary } S \tag{13-101b}$$

Once the functions $\omega(x, y)$ are available, the basis functions $\phi_j(x, y)$ can be constructed by the products of $\omega(x, y)$ with various powers of x and y in the following form:

$$\begin{aligned} \phi_1 &= \omega(x, y) \\ \phi_2 &= x \cdot \omega(x, y) \\ \phi_3 &= y \cdot \omega(x, y) \\ \phi_4 &= x^2 \cdot \omega(x, y) \\ \phi_5 &= xy \cdot \omega(x, y) \\ \phi_6 &= y^2 \cdot \omega(x, y), \ldots \end{aligned} \tag{13-102}$$

The functions $\phi_j(x, y)$, $j = 1, 2, \ldots, n$ as constructed in this manner satisfy the homogeneous part of the boundary conditions for the problem, have continuous derivatives in x and y, and it is proved in reference 53 (p. 276) that they constitute a *complete* system of functions. Then, the problem becomes one of determining the auxiliary functions $\omega(x, y)$. These functions can be determined by utilizing the equations for the contour of the boundary as now described.

Region Having a Single Continuous Contour If the region has a single continuous contour, such as a circle, the equation of the boundary can always be expressed in the form

$$F(x, y) = 0 \quad \text{on boundary } S \tag{13-103}$$

Clearly, the function $F(x, y)$ is continuous, has partial derivatives with respect to x and y, and vanishes at the boundary of the region R. Then, the function $\omega(x, y)$ can be chosen as

$$\omega(x, y) = \pm F(x, y) \tag{13-104}$$

such that $\omega(x, y) > 0$ per equation (13-101a).

For example, for a circular region of radius b with the center at the origin, the equation for the circle is given as

$$x^2 + y^2 = b^2 \tag{13-105a}$$

which may be written in the form

$$b^2 - x^2 - y^2 = 0 \tag{13-105b}$$

The above expression for the contour readily satisfies the equation

$$F(x, y) \equiv b^2 - x^2 - y^2 = 0 \tag{13-106}$$

and the function $\omega(x, y)$ is then taken as

$$\omega(x, y) = b^2 - x^2 - y^2 \tag{13-107}$$

Region Having a Contour as Convex Polynomial Consider a region in the form of a convex polynomial, and let the equations for the sides be given in the form

$$\begin{aligned} F_1 &\equiv a_1 x + b_1 y + d_1 = 0 \\ F_2 &\equiv a_2 x + b_2 y + d_2 = 0 \\ &\vdots \\ F_n &\equiv a_n x + b_n y + d_n = 0 \end{aligned} \tag{13-108}$$

Then, the function $\omega(x, y)$ chosen in the form

$$\omega(x, y) = \pm F_1(x, y) F_2(x, y) \cdots F_n(x, y) \tag{13-109}$$

appropriately vanishes at every point on the boundary and satisfies the homogeneous part of the boundary conditions of the first kind for the region.

Region Having a Contour as Nonconvex Polynomial The construction of the function $\omega(x, y)$ for this case is more involved because the function $\omega(x, y)$ has to be assigned piecewise in different parts of the region. Further discussion of this matter can be found in reference 53 (p. 278).

Example 13-6 Construction of Basis Functions for Different Geometries Construct the functions $\omega(x, y)$ as discussed above for the four different geometries shown in Figure 13-8.

The equations of the respective contours for each of the four geometries shown in Figure 13-8(*a*)–(*d*) are given, respectively, as

$$a - x = 0 \qquad a + x = 0 \qquad b - y = 0 \qquad b + y = 0 \tag{13-110a}$$

$$y - \alpha x = 0 \qquad y + \beta x = 0 \qquad L - x = 0 \tag{13-110b}$$

$$x = 0 \qquad y = 0 \qquad 1 - \frac{x}{a} - \frac{y}{b} = 0 \tag{13-110c}$$

$$R_1^2 - x^2 - y^2 = 0 \qquad R_2^2 - (x - L)^2 - y^2 = 0 \tag{13-110d}$$

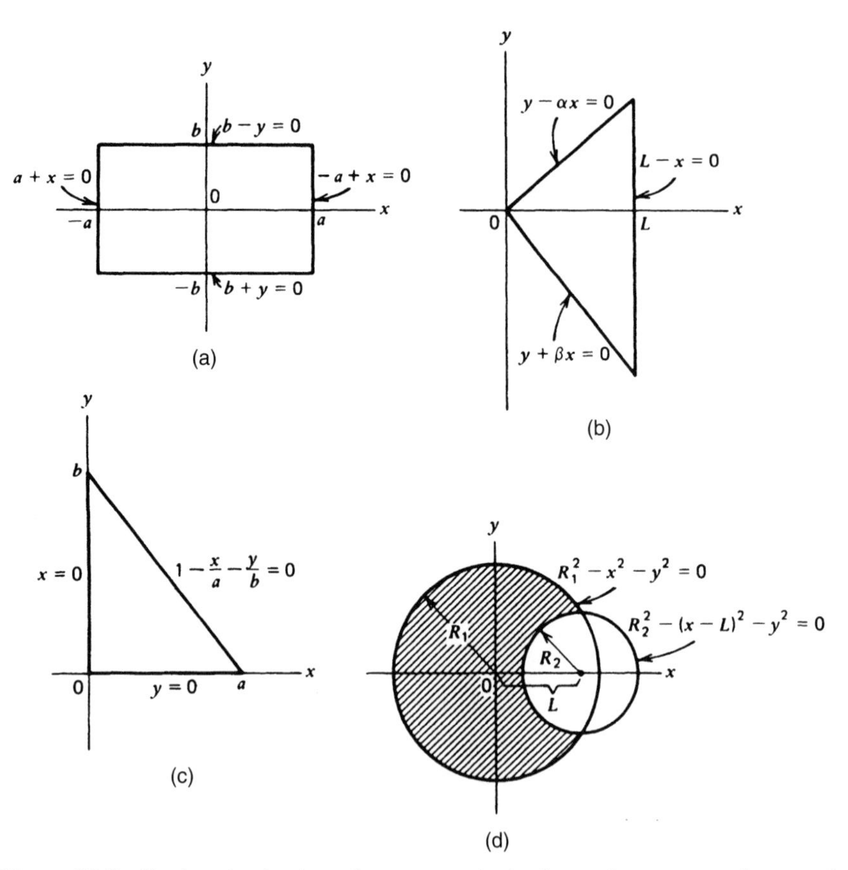

Figure 13-8 Regions having boundary contour in the form of a convex polygon and a region bounded by two circles: (*a*) rectangular region, (*b*) triangular region, (*c*) triangular region, and (*d*) a crescent-shaped region between two intersecting circles.

Then the corresponding functions $\omega(x, y)$ for each of these geometries shown in Figure 13-8(*a*)–(*d*) are given, respectively, as

$$\omega(x, y) = (a^2 - x^2)(b^2 - y^2) \tag{13-111a}$$

$$\omega(x, y) = (y - \alpha x)(y - \beta x)(L - x) \tag{13-111b}$$

$$\omega(x, y) = xy\left(1 - \frac{x}{a} - \frac{y}{b}\right) \tag{13-111c}$$

$$\omega(x, y) = \left(R_1^2 - x^2 - y^2\right)\left(R_2^2 - (x - L)^2 - y^2\right) \tag{13-111d}$$

Construction of Basis Functions for Boundary Conditions of the Third Kind

We consider one-dimensional steady-state heat conduction in a slab of thickness L, subjected to convection into a medium at zero temperature. The boundary conditions at $x = 0$ and $x = L$ are given as

$$-\left.\frac{dT}{dx}\right|_{x=0} + H_1 \, T|_{x=0} = 0 \tag{13-112a}$$

$$\left.\frac{dT}{dx}\right|_{x=L} + H_2 \, T|_{x=L} = 0 \tag{13-112b}$$

If such a heat conduction problem is to be solved by the Galerkin method, the first two trial functions $\phi_1(x)$ and $\phi_2(x)$ must be chosen as

$$\phi_1(x) = x^2 \left(x - L - \frac{L}{2 + H_2 L} \right) \tag{13-113a}$$

$$\phi_2(x) = (L - x)^2 \left(x + \frac{L}{2 + H_1 L} \right) \tag{13-113b}$$

and the remaining $\phi_3, \phi_4, \ldots$ can be taken as

$$\phi_j(x) = x^j (L - x)^2, \qquad j = 3, 4, 5, \ldots \tag{13-114}$$

Then, the trial solution is constructed as

$$\tilde{T}_n(x) = \phi_1 + \phi_2 + \sum_{j=3}^{n} c_j \phi_j \qquad \text{in} \qquad 0 \le x \le L \tag{13-115}$$

which now satisfies the boundary conditions (13-113) for all values of c_j, $j \ge 3$.

For other combinations of the boundary conditions of the first, second, and third kinds, functions $\phi_j(x)$ are constructed with similar considerations.

Integration Formula

In performing computations associated with the application of the Galerkin method, the following integration formula is useful [53, p. 269]:

$$\boxed{\int_{x=0}^{L} x^k (L - x)^m \, dx = \frac{k!m!}{(k + m + 1)!} L^{k+m+1}} \tag{13-116}$$

We now present three examples of the Galerkin method for a range of differential equations.

Example 13-7 One-Dimensional Cartesian Problem

Consider the following one-dimensional, steady-state heat conduction problem:

$$\frac{d^2T}{dx^2} + AT + Bx = 0 \qquad \text{in} \qquad 0 < x < 1 \tag{13-117a}$$

$$\text{BC1:} \qquad T(x=0)=0 \tag{13-117b}$$

$$\text{BC2:} \qquad T(x=1)=0 \tag{13-117c}$$

where A and B are constants. We solve this problem by the Galerkin method using one- and two-term trial solutions, and compare the approximate results with the exact solution of the problem for the case $A = B = 1$.

The application of the Galerkin method gives

$$\int_{x=0}^{1} \phi_j(x)\left(\frac{d^2\tilde{T}}{dx^2} + A\tilde{T} + Bx\right) dx = 0, \qquad j = 1, 2, \ldots \tag{13-118}$$

where $\phi_j(x)$ are a set of basis functions and $\tilde{T}(x)$ is the trial solution. For these two boundary conditions, we have $\omega(x, y) = x(1-x)$, and our basis functions are chosen as

$$\phi_1 = x(1-x) \qquad \text{and} \qquad \phi_2 = x^2(1-x) \tag{13-119}$$

which satisfy the homogeneous boundary conditions (13-117b,c) for the problem.

One-Term Trial Solution We choose the trial solution as

$$\tilde{T}_1(x) = c_1\phi_1(x) \tag{13-120}$$

where the basis function $\phi_1(x)$, as defined by equation (13-119), is taken as

$$\phi_1(x) = x(1-x) = x - x^2 \tag{13-121}$$

Then we have

$$\tilde{T}_1(x) = c_1 x - c_1 x^2 \tag{13-122a}$$

$$\frac{d^2\tilde{T}_1}{dx^2} = -2c_1 \tag{13-122b}$$

Introducing equations (13-122a,b) into equation (13-118), the one-term Galerkin method of solution gives

$$\int_{x=0}^{1} (x - x^2)\left[-2c_1 + A(c_1 x - c_1 x^2) + Bx\right] dx = 0 \tag{13-123}$$

Performing this integration and solving for c_1, we obtain

$$c_1 = \frac{B}{4[1-(A/10)]} \tag{13-124}$$

and the one-term trial solution becomes

$$\tilde{T}_1(x) = \frac{B}{4[1-(A/10)]}x(1-x) \tag{13-125}$$

For the special case $A = B = 1$, this result reduces to

$$\tilde{T}_1(x) = \frac{1}{3.6}x(1-x) \tag{13-126}$$

Two-Term Trial Solution The trial solution is taken as

$$\tilde{T}_2(x) = c_1\phi_1(x) + c_2\phi_2(x) \tag{13-127}$$

where the basis functions $\phi_1(x)$ and $\phi_2(x)$ are chosen as

$$\phi_1(x) = x(1-x) \qquad \text{and} \qquad \phi_2(x) = x^2(1-x) \tag{13-128}$$

Then we have

$$\tilde{T}_2(x) = c_1(x - x^2) + c_2(x^2 - x^3) \tag{13-129a}$$

and

$$\frac{d^2\tilde{T}_2(x)}{dx^2} = -2c_1 + 2c_2 - 6c_2 x \tag{13-129b}$$

Introducing equations (13-129a,b) into equations (13-118) we obtain

$$\int_{x=0}^{1} (x - x^2)\left[\frac{d^2\tilde{T}_2}{dx^2} + A\tilde{T}_2(x) + Bx\right] dx = 0 \qquad \text{for} \qquad j = 1 \tag{13-130a}$$

$$\int_{x=0}^{1} (x^2 - x^3)\left[\frac{d^2\tilde{T}_2}{dx^2} + A\tilde{T}_2(x) + Bx\right] dx = 0 \qquad \text{for} \qquad j = 2 \tag{13-130b}$$

where

$$\begin{aligned}\frac{d^2\tilde{T}_2}{dx^2} + A\tilde{T}_2(x) + Bx = {} & (-2c_1 + 2c_2 - 6c_2 x) \\ & + A(c_1 x - c_1 x^2 + c_2 x^2 - c_2 x^3) + Bx\end{aligned} \tag{13-130c}$$

When the integrations are performed, equations (13-130a,b) provide two algebraic equations for the determination of the two unknown coefficients c_1 and c_2, namely,

$$\left(1 - \frac{A}{10}\right)c_1 + \frac{1}{2}\left(1 - \frac{A}{10}\right)c_2 = \frac{B}{4} \tag{13-131a}$$

$$\left(1 - \frac{A}{10}\right)c_1 + \frac{6}{15}\left(2 - \frac{A}{7}\right)c_2 = \frac{3}{10}B \tag{13-131b}$$

For the special case of $A = B = 1$, these coefficients are

$$c_1 = \frac{71}{369} \qquad \text{and} \qquad c_2 = \frac{7}{41}$$

and the two-term trial solution becomes

$$\tilde{T}_2(x) = x(1-x)\left(\frac{71}{369} + \frac{7}{41}x\right) \tag{13-132}$$

The Exact Solution The exact solution of this problem is given by

$$T(x) = \frac{B}{A}\left(\frac{\sin A^{1/2}x}{\sin A^{1/2}} - x\right) \tag{13-133}$$

which for the case $A = B = 1$ becomes

$$T(x) = \frac{\sin x}{\sin 1} - x \tag{13-134}$$

We present in Table 13-2 a comparison of the one-term and two-term approximate solutions, along with the exact result. Clearly, the accuracy is significantly improved using a two-term solution.

TABLE 13-2 Comparison of Approximate and Exact Solutions of Example 13-7 for the Case with $A = B = 1$

x	T Exact	T_1 Approx.	% Error	T_2 Approx.	% Error
0.25	0.04400	0.0521	+18.4	0.04408	+0.18
0.50	0.06974	0.0694	−0.48	0.06944	−0.43
0.75	0.06005	0.0521	−13.2	0.06009	+0.06
0.85	0.04282	0.0354	−17.3	0.04302	+0.46

Example 13-8 One-Dimensional Cylindrical Problem
Consider the following steady-state heat conduction problem for a solid cylinder:

$$\frac{1}{r}\frac{d}{dt}\left(r\frac{dT}{dr}\right) + \left(1 - \frac{1}{r^2}\right)T = 0 \quad \text{in} \quad 1 < r < 2 \tag{13-135a}$$

$$\text{BC1:} \quad T(r=1) = 4 \tag{13-135b}$$

$$\text{BC2:} \quad T(r=2) = 8 \tag{13-135c}$$

We now solve this problem by the Galerkin method using one-term trial solution and compare this approximate result with the exact solution of the problem. The application of the Galerkin method gives

$$\int_{r=1}^{2} \phi_1(r)\left[\frac{1}{r}\frac{d}{dr}\left(r\frac{d\tilde{T}}{dr}\right) + \left(1 - \frac{1}{r^2}\right)\tilde{T}\right] dr = 0 \tag{13-136}$$

The one-term trial solution is taken in the form

$$\tilde{T}_1(r) = \psi_0(r) + c_1\phi_1(r) \tag{13-137}$$

where the function $\psi_0(r)$ satisfies the nonhomogeneous part of the boundary conditions, equations (13-135b,c). We assume a simple polynomial, $\psi_0(r) = A + Br$, which upon application of equations (13-135b,c) yields

$$\psi_0(r) = 4r \tag{13-138}$$

The first basis function $\phi_1(r)$ must satisfy the homogeneous form of the boundary conditions. We consider the inner contour r=1, giving $F_1(r) = 1 - r$, and the outer contour as r= 2, giving $F_2(r) = 2 - r$. This gives $\omega(r) = \pm F_1(r)F_2(r)$, where we select the sign that yields a positive value in the domain; hence we take

$$\omega(r) = (r - 1)(2 - r) \tag{13-139a}$$

and our basis function becomes

$$\phi_1(r) = (r - 1)(2 - r) \tag{13-139b}$$

Substituting the trial solution (13-139b) into equation (13-136), performing the integration, and solving the result for c_1, we obtain $c_1 = 3.245$. Then, the one-term approximate solution becomes

$$\tilde{T}_1(r) = 3.245(r - 1)(2 - r) + 4r \tag{13-140}$$

The exact solution of this problem is

$$T(r) = 14.43J_1(r) + 3.008Y_1(r) \tag{13-141}$$

where $J_1(r)$ and $Y_1(r)$ are the Bessel functions. A comparison of the approximate and exact solutions at the locations $r = 1.2$, 1.5, and 1.8 shows that the agreement is within 0.03%. Therefore, in this example even the one-term approximation gives a very good result.

Example 13-9 Two-Dimensional Problem with Generation
We now solve the steady-state heat conduction problem in a rectangular region ($-a \leq x \leq a$; $-b \leq y \leq b$) with heat generation at a constant rate of g (W/m^3), and with the boundaries kept at zero temperature, using the Galerkin method, and compare the result with the exact solution.

The mathematical formulation of the problem is

$$\frac{\partial^2 T}{\partial x^2} + \frac{\partial^2 T}{\partial y^2} + \frac{1}{k}g = 0 \qquad \text{in} \qquad -a < x < a, \qquad -b < y < b \tag{13-142a}$$

$$\text{BC1:} \quad T(x=-a, y) = 0 \tag{13-142b}$$

$$\text{BC2:} \quad T(x=a, y) = 0 \tag{13-142c}$$

$$\text{BC3:} \quad T(x, y=-b) = 0 \tag{13-142d}$$

$$\text{BC4:} \quad T(x, y=b) = 0 \tag{13-142e}$$

The solution of this problem by the Galerkin method is written as

$$\int_{x=-a}^{a} \int_{y=-b}^{b} \phi_j(x, y) \left(\frac{\partial^2 \tilde{T}}{\partial x^2} + \frac{\partial^2 \tilde{T}}{\partial y^2} + \frac{1}{k} g \right) dx\, dy = 0 \tag{13-143}$$

To construct the trial solution, we consider the contours along the four boundaries, which yield $F_1(x, y) = a - x$, $F_2(x, y) = a + x$, $F_3(x, y) = b - y$, and $F_4(x, y) = b + y$. We then construct

$$\omega(x, y) = (a - x)(a + x)(b - y)(b + y) \tag{13-144}$$

and note that $\omega(x, y) > 0$ within the domain of the problem. We consider a one-term trial solution taken as

$$\tilde{T}_1(x, y) = c_1 \phi_1(x, y) \tag{13-145}$$

where the function ϕ_1 is obtained directly from equation (13-144) as

$$\phi_1(x, y) = (a^2 - x^2)(b^2 - y^2) \tag{13-146}$$

Introducing this trial solution into equation (13-143) and performing the integrations, we obtain

$$c_1 = \frac{5}{8} \frac{g/k}{a^2 + b^2} \tag{13-147}$$

Using the above, the one-term approximate solution becomes

$$\tilde{T}_1(x, y) = \frac{5}{8} \frac{g/k}{a^2 + b^2} (a^2 - x^2)(b^2 - y^2) \tag{13-148}$$

The exact solution of this problem is

$$T(x, y) = \frac{g}{k} \left(\frac{a^2 - x^2}{2} \right) - 2a^2 \frac{g}{k} \sum_{n=0}^{\infty} \frac{(-1)^n}{\beta_n^3} \frac{\cosh\left[\beta_n y/a\right] \cos\left[\beta_n x/a\right]}{\cosh\left[\beta_n b/a\right]} \tag{13-149}$$

where $\beta_n = [(2n+1)\pi]/2$ defines the eigenvalues. To compare these two results, we consider the center temperature (i.e., $x = 0$, $y = 0$) for the case $a = b$, which yields

$$\text{Approximate:} \qquad \tilde{T}_1(0,0) = \frac{5}{16}\frac{ga^2}{k} = 0.3125\frac{ga^2}{k} \tag{13-150a}$$

$$\text{Exact:} \qquad T(0,0) = \frac{ga^2}{k}\left[\frac{1}{2} - 2\sum_{n=0}^{\infty}\frac{(-1)^n}{\beta_n^3\cosh\beta_n}\right] = 0.293\frac{ga^2}{k} \tag{13-150b}$$

The error involved with a one-term solution is about 6.7%. For a two-term trial solution, the temperature distribution may be taken in the form

$$\tilde{T}_2(x,y) = (c_1 + c_2x^2)(a^2 - x^2)(b^2 - y^2) \tag{13-151}$$

and the calculations are performed in a similar manner to determine the coefficients c_1 and c_2.

13-7 PARTIAL INTEGRATION

In the previous section, the Galerkin method has been applied to the solution of two-dimensional, steady-state heat conduction problems by using a trial solution $\tilde{T}(x,y)$ in the x and y variables; as a result, the problem has been reduced to the solution of a set of algebraic equations for the determination of the unknown coefficients c_1, $c_2, \ldots, c_n$. A more accurate approximation is obtainable if a one-dimensional trial function is used either in the x variable $\tilde{T}(x)$ or the y variable $\tilde{T}(y)$, and the problem is reduced to the solution of an ordinary differential equation for the determination of a function $Y(y)$ or $X(x)$. One advantage of such an approach is that, in situations when the functional form of the temperature profile cannot be chosen *a priori* in one direction, it is left to be determined according to the character of the problem for the solution of the resulting ordinary differential equation.

The partial integration approach is also applicable for the approximate solution of transient heat conduction problems. Here we illustrate the application of the partial integration technique with some representative examples.

Example 13-10 Two-Dimensional Problem with Generation
We now solve the steady-state heat conduction problem considered in Example 13-9 with the Galerkin method using partial integration with respect to the y variable and solving the resulting ordinary differential equation in the x variable.

The Galerkin method when applied to the differential equation (13-142a) by partial integration with respect to the y variable gives

$$\int_{y=-b}^{b}\phi_j(y)\left(\frac{\partial^2\tilde{T}}{\partial x^2} + \frac{\partial^2\tilde{T}}{\partial y^2} + \frac{g}{k}\right)dy = 0 \qquad \text{in} \qquad -a < x < a \tag{13-152}$$

We consider only a one-term trial solution $\tilde{T}_1(x, y)$ chosen as

$$\tilde{T}_1(x, y) = \phi_1(y)X(x) \tag{13-153a}$$

where, based on equation (13-146), we now use

$$\phi_1(y) = b^2 - y^2 \tag{13-153b}$$

This trial solution satisfies the boundary conditions at $y = \pm b$, while the function $X(x)$ is yet to be determined. Introducing the trial solution (13-153) into equation (13-152) and performing the indicated operations, we obtain

$$\frac{d^2X(x)}{dx^2} - \frac{5}{2b^2}X(x) = -\frac{5g}{4b^2k} \quad \text{in} \quad -a < x < a \tag{13-154}$$

subject to the boundary conditions

$$\text{BC1:} \quad X(x = -a, y) = 0 \tag{13-155a}$$

$$\text{BC2:} \quad X(x = a, y) = 0 \tag{13-155b}$$

The solution of the problem given by equations (13-154) and (13-155) is

$$X(x) = \frac{g}{2k}\left\{1 - \frac{\cosh\left[\sqrt{2.5}x/b\right]}{\cosh\left[\sqrt{2.5}a/b\right]}\right\} \tag{13-156}$$

From equation (13-153a), the one-term trial solution becomes

$$\tilde{T}_1(x, y) = \frac{g}{2k}(b^2 - y^2)\left\{1 - \frac{\cosh\left[\sqrt{2.5}x/b\right]}{\cosh\left[\sqrt{2.5}a/b\right]}\right\} \tag{13-157}$$

and the temperature at the center (i.e., $x = y = 0$), for the case of $a = b$, becomes

$$\tilde{T}_1(0, 0) = 0.3026\frac{ga^2}{k} \tag{13-158}$$

This result involves an error of approximately 3.3%, whereas the one-term approximation obtained in the previous example by the application of the Galerkin method for both x and y variables involves an error of approximately 6.7%. Thus the solution by partial integration improves the accuracy at this location, as well as in general.

Example 13-11 Two-Dimensional Cylindrical Problem
We consider the following steady-state heat conduction problem for a segment (i.e., wedge) of a cylinder, $0 \le r \le 1$, $0 \le \theta \le \theta_0$, in which heat is generated

at a constant rate of g (W/m^3), and all the boundary surfaces are kept at zero temperature. The formulation of this problem is

$$\frac{1}{r}\frac{\partial}{\partial r}\left(r\frac{\partial T}{\partial r}\right)+\frac{1}{r^2}\frac{\partial^2 T}{\partial \theta^2}+\frac{g}{k}=0 \quad \text{in} \quad \begin{array}{l} 0 \le r < 1 \\ 0 < \theta < \theta_0 \end{array} \tag{13-159a}$$

$$\text{BC1:} \quad T(r \to 0, \theta) \Rightarrow \text{finite} \tag{13-159b}$$

$$\text{BC2:} \quad T(r = 1, \theta) = 0 \tag{13-159c}$$

$$\text{BC3:} \quad T(r, \theta = 0) = 0 \tag{13-159d}$$

$$\text{BC4:} \quad T(r, \theta = \theta_0) = 0 \tag{13-159e}$$

We solve this problem using the Galerkin method by partial integration with respect to the θ variable and then compare the approximate result with the exact solution. The Galerkin method is now applied to the differential equation (13-159a) by partial integration with respect to the variable θ, which gives

$$\int_{\theta=0}^{\theta_0}\phi_1(\theta)\left[\frac{1}{r}\frac{\partial}{\partial r}\left(r\frac{\partial \tilde{T}}{\partial r}\right)+\frac{1}{r^2}\frac{\partial^2 \tilde{T}}{\partial \theta^2}+\frac{g}{k}\right]d\theta = 0 \quad \text{in} \quad 0 \le r < 1 \tag{13-160}$$

We consider a one-term trial solution taken as

$$\tilde{T}(r,\theta) = F(r)\phi_1(\theta) \tag{13-161a}$$

where

$$\phi_1(\theta) = \sin\left(\frac{\pi\theta}{\theta_0}\right) \tag{13-161b}$$

The trial solution thus chosen satisfies the boundary conditions at $\theta = 0$ and $\theta = \theta_0$, however, the function $F(r)$ has yet to be determined. Introducing the trial solution (13-161) into (13-160) and performing the integration, we obtain

$$\frac{1}{r}\frac{d}{dr}\left(r\frac{dF}{dr}\right)-\frac{\beta^2}{r^2}F(r) = -\frac{4g}{k\pi} \quad \text{in} \quad 0 \le r < 1 \tag{13-162a}$$

subject to

$$F(r = 1) = 0 \tag{13-162b}$$

where we define

$$\beta \equiv \frac{\pi}{\theta_o} \tag{13-162c}$$

Equation (13-162a) requires the sum of a homogeneous solution (Cauchy's equation) and particular solution. The particular solution of equation (13-162a) is

$$F_p = \frac{4g}{\pi k}\frac{r^2}{\beta^2 - 4} \tag{13-163}$$

and the complete solution for $F(r)$ is then constructed as

$$F(r) = c_1 r^{\beta} + c_2 r^{-\beta} + \frac{4g}{\pi k}\frac{r^2}{\beta^2 - 4} \tag{13-164}$$

In consideration of equation (13-159b), $c_2 = 0$, from the requirement that the solution should remain finite at the origin. Now c_1 is determined by the application of the boundary condition at $r = 1$ to give

$$c_1 = -\frac{4g}{\pi k}\frac{1}{\beta^2 - 4} \tag{13-165}$$

Then the solution for $F(r)$ is obtained as

$$F(r) = \frac{4g}{\pi k}\frac{r^2 - r^{\beta}}{\beta^2 - 4} \tag{13-166}$$

and the one-term trial solution $\tilde{T}(r, \theta)$ becomes

$$\tilde{T}(r, \theta) = \frac{4g}{\pi k}\frac{r^2 - r^{(\pi/\theta_0)}}{(\pi/\theta_0)^2 - 4}\sin\left(\frac{\pi\theta}{\theta_0}\right) \tag{13-167}$$

The exact solution of the problem (13-159) is given as

$$T(r, \theta) = \frac{4g}{\pi k}\sum_{\substack{n=1,3\\ \text{odd}}}^{\infty}\frac{1}{n}\frac{r^2 - r^{(n\pi/\theta_0)}}{(n\pi/\theta_0)^2 - 4}\sin\left(\frac{n\pi\theta}{\theta_0}\right) \tag{13-168}$$

A comparison of the above two solutions reveals that the one-term approximate solution represents the first term only ($n=1$) of the exact solution.

Example 13-12 Two-Dimensional Cartesian Problem

We solve the steady-state heat conduction problem with constant rate of heat generation (W/m^3) for a region bounded by $x = 0$, $x = a$, $y = 0$, and $y = f(x)$, for the boundary conditions as shown in Figure 13-9, using the Galerkin method by partial integration with respect to the y variable.

The mathematical formulation of this problem is given as

$$\frac{\partial^2 T}{\partial x^2} + \frac{\partial^2 T}{\partial y^2} + \frac{g}{k} = 0 \quad \text{in} \quad 0 < x < a, \quad 0 < y < f(x) \tag{13-169a}$$

$$\text{BC1:} \quad T(x = 0, y) = 0 \tag{13-169b}$$

$$\text{BC2:} \quad T(x = a, y) = 0 \tag{13-169c}$$

$$\text{BC3:} \quad T(x, y = f(x)) = 0 \tag{13-169d}$$

$$\text{BC4:} \quad \left.\frac{\partial T(x, y)}{\partial y}\right|_{y=0} = 0 \tag{13-169e}$$

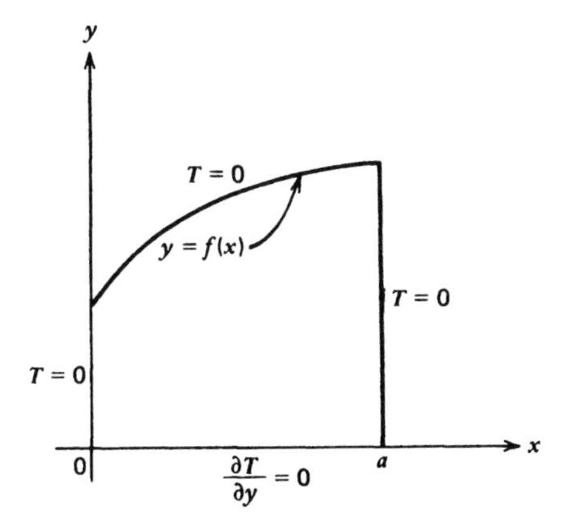

Figure 13-9 Region and boundary conditions for Example 13-12.

The Galerkin method is now applied to the differential equation (13-169a) by partial integration with respect to the y variable. We obtain

$$\int_{y=0}^{f(x)} \phi_j(y) \left(\frac{\partial^2 \tilde{T}}{\partial x^2} + \frac{\partial^2 \tilde{T}}{\partial y^2} + \frac{g}{k} \right) dy = 0 \tag{13-170}$$

We consider one-term trial solution taken as

$$\tilde{T}_1(x, y) = X(x) \cdot \phi_1(y) \tag{13-171a}$$

where we define

$$\phi_1(y) = [y^2 - f^2(x)] \tag{13-171b}$$

Clearly, this trial solution satisfies the boundary conditions at $y = 0$ and $y = f(x)$; however, the function $X(x)$ is yet to be determined. Introducing the trial solution (13-171) into equation (13-170), and performing the indicated operations, we obtain the following ordinary differential equation for the determination of the function $X(x)$, namely,

$$\frac{2}{5} f^2 \frac{d^2 X}{dx^2} + 2ff' \frac{dX}{dx} + (ff'' + f'^2 - 1)X = -\frac{g}{2k} \quad \text{in} \quad 0 < x < a \tag{13-172a}$$

subject to the boundary conditions

$$\text{BC1:} \quad X(x = 0) = 0 \tag{13-172b}$$

$$\text{BC2:} \quad X(x = a) = 0 \tag{13-172c}$$

Once the function $f(x)$ defining the form of the boundary arc is specified, this equation can be solved and the function $X(x)$ can be determined. For example, if we define $y = f(x) = b$, the problem corresponds to a rectangular region, and the ODE of equation (13-172a) reduces to

$$\frac{d^2X}{dx^2} - \frac{5}{2b^2}X = -\frac{5g}{4b^2k} \quad \text{in} \quad 0 < x < a \tag{13-173a}$$

$$\text{BC1:} \quad X(x = 0) = 0 \tag{13-173b}$$

$$\text{BC2:} \quad X(x = a) = 0 \tag{13-173c}$$

which is the same as that given by equations (13-154); and the one-term approximate solution becomes

$$\tilde{T}_1(x, y) = (y^2 - b^2)X(x) \tag{13-174}$$

where $X(x)$ is as given by equation (13-156).

13-8 APPLICATION TO TRANSIENT PROBLEMS

In this last section, we now illustrate the application of the Galerkin method to the solution of time-dependent problems with the following two examples.

Example 13-13 Time-Dependent Cartesian Problem
A slab in the region $0 \le x \le 1$ is initially at a temperature $T(x) = T_0(1 - x^2)$. For times $t > 0$, the boundary at $x = 0$ is kept insulated, and the boundary at $x = 1$ is kept at zero temperature. Using the Galerkin method combined with partial integration, we obtain an approximate solution for the temperature distribution $\tilde{T}(x, t)$ in the slab and compare it with the exact solution $T(x, t)$.

The mathematical formulation of this problem is given as

$$\frac{\partial^2 T}{\partial x^2} = \frac{1}{\alpha}\frac{\partial T(x, t)}{\partial t} \quad \text{in} \quad 0 < x < 1, \quad t > 0 \tag{13-175a}$$

$$\text{BC1:} \quad \left.\frac{\partial T(x, t)}{\partial x}\right|_{x=0} = 0 \tag{13-175b}$$

$$\text{BC2:} \quad T(x = 1, t) = 0 \tag{13-175c}$$

$$\text{IC:} \quad T(x, t = 0) = T_0(1 - x^2) \quad \text{in} \quad 0 \le x \le 1 \tag{13-175d}$$

We apply the Galerkin method to equation (13-175a) with partial integration with respect to the x variable and obtain

$$\int_{x=0}^{1} \phi_j(x)\left(\frac{\partial^2 \tilde{T}}{\partial x^2} - \frac{1}{\alpha}\frac{\partial \tilde{T}}{\partial t}\right) dx = 0 \tag{13-176}$$

We choose a one-term trial solution $\tilde{T}_1(x,t)$ as

$$\tilde{T}_1(x,t) = T_0 f(t)\phi_1(x) \tag{13-177a}$$

where

$$\phi_1(x) = 1 - x^2 \tag{13-177b}$$

$$f(t=0) = 1 \tag{13-177c}$$

with the function $f(t)$ yet to be determined. Clearly, the trial solution chosen as above satisfies the initial condition, and the two boundary conditions for the problem. Substituting the trial solution (13-177) into equation (13-176), and performing the indicated operations, we obtain the differential equation for $f(t)$ as

$$\frac{df(t)}{dt} + \frac{5\alpha}{2} f(t) = 0 \qquad \text{for} \qquad t > 0 \tag{13-178a}$$

with

$$f(t=0) = 1 \tag{13-178b}$$

The solution for $f(t)$ is

$$f(t) = e^{-(5\alpha/2)t} \tag{13-179}$$

and the one-term approximate solution $\tilde{T}_1(x,t)$ becomes

$$\tilde{T}(x,t) = T_0(1 - x^2)e^{-(5\alpha/2)t} \tag{13-180}$$

The exact solution of the problem (13-175) is obtained as

$$T(x,t) = 4T_0 \sum_{n=0}^{\infty} (-1)^n \frac{1}{\beta_n^3} e^{-\alpha\beta_n^2 t} \cos \beta_n x \tag{13-181a}$$

where we define

$$\beta_n = \frac{(2n+1)\pi}{2} \tag{13-181b}$$

We list in Table 13-3 a comparison of this approximate solution with the exact solution. Even the one-term approximate solution is in reasonably good agreement with the exact solution. Improved approximations can be obtained by choosing a higher-order trial solution of the form

$$\tilde{T}_n(x,t) = T_0 \sum_{j=1}^{n} f_j(t)\phi_j(x) \tag{13-182}$$

where the functions $\phi_j(x)$ satisfy the boundary conditions for the problem, and the function $f_j(t)$, with $f_j(0) = 1$, is determined from the resulting ordinary differential equations obtained after the application of the Galerkin method with partial integration with respect to the x variable.

TABLE 13-3 Agreement between Approximate and Exact Solutions of Example 13-13

x	$\alpha t = 0.01$	$\alpha t = 0.1$	$\alpha t = 1$
0.2	+1%	−1%	+4.4%
0.6	+2%	+5.5%	+3.1%

Example 13-14 Time-Dependent Cylindrical Problem

In our final problem, we consider the transient heat conduction problem for a solid cylinder, $0 \le r \le 1$, with heat generation within the medium, as given in the *dimensionless* form as

$$\frac{1}{r}\frac{\partial}{\partial r}\left(r\frac{\partial T}{\partial r}\right) + G(r) = \frac{\partial T(r,t)}{\partial t} \quad \text{in} \quad 0 < r < 1, \quad t > 0 \quad (13\text{-}183\text{a})$$

$$\text{BC1:} \quad T(r \to 0, t) \Rightarrow \text{finite} \quad (13\text{-}183\text{b})$$

$$\text{BC2:} \quad T(r = 1, t) = 0 \quad (13\text{-}183\text{c})$$

$$\text{IC:} \quad T(r, t = 0) = 0 \quad \text{in} \quad 0 \le r \le 1 \quad (13\text{-}183\text{d})$$

We now solve this problem by the combined application of the Laplace transform and the Galerkin method. The Laplace transform of this problem with respect to the time variable is

$$\mathcal{L}(\overline{T}) \equiv \frac{d}{dr}\left(r\frac{d\overline{T}}{dr}\right) - sr\overline{T} + \frac{1}{s}rG(r) = 0 \quad \text{in} \quad 0 < r < 1 \quad (13\text{-}184\text{a})$$

$$\text{BC1:} \quad \overline{T}(r \to 0, s) \Rightarrow \text{finite} \quad (13\text{-}184\text{b})$$

$$\text{BC2:} \quad \overline{T}(r = 1, s) = 0 \quad (13\text{-}184\text{c})$$

where $\overline{T}(r, s)$ is the Laplace transform of $T(r, t)$, and s is the Laplace transform variable, as described in Chapter 9. The application of the Galerkin method to equation (13-184a) is written as

$$\int_{r=0}^{1} \phi_j(r)\mathcal{L}[\tilde{\overline{T}}(r,s)]\, dr = 0 \quad (13\text{-}185)$$

where $\tilde{\overline{T}}(r, s)$ is the trial solution for $\overline{T}(r, s)$, and $\phi_j(r)$ are the functions that satisfy the boundary conditions for the problem, and from which the trial solution is constructed. In this example, we show that if the proper function is chosen for $\phi_j(r)$, and sufficient number of $\phi_j(r)$ terms are included to construct the trial solution, it is possible to obtain the exact solution for the problem. We now choose $\phi_j(r)$ as

$$\phi_j(r) = J_0(\beta_j r) \tag{13-186a}$$

where the β_j values are the roots of

$$J_o(\beta_j) = 0 \tag{13-186b}$$

Then, each of the functions $\phi_j(r)$ satisfies the boundary conditions (13-184b) and (13-184c) for the problem. We construct the trial solution $\tilde{\bar{T}}(r, s)$ in terms of the $\phi_j(r)$ functions as

$$\tilde{\bar{T}}(r, s) = \sum_j c_j \phi_j(r) = \sum_j c_j J_0(\beta_j r) \tag{13-187}$$

where the summation is taken over the permissible values of β_j as defined by the roots of equation (13-186b). Introducing equations (13-187) and (13-186a) into equation (13-185), we obtain

$$\int_{r=0}^{1} \mathcal{L}\left[\sum_j c_j J_0(\beta_j r)\right] J_0(\beta_i r)\, dr = 0, \qquad i = 1, 2, \ldots \tag{13-188a}$$

or

$$\begin{aligned}\sum_j c_j \int_{r=0}^{1} \left\{\frac{d}{dr}\left[r\frac{dJ_0(\beta_j r)}{dr}\right] - sr J_0(\beta_j r)\right\} J_0(\beta_i r)\, dr \\ +\frac{1}{s}\int_{r=0}^{1} rG(r) J_0(\beta_i r)\, dr = 0\end{aligned} \tag{13-188b}$$

or

$$\begin{aligned}-\sum_j c_j(\beta_j^2 + s)\int_{r=0}^{1} r J_0(\beta_j r) J_0(\beta_i r)\, dr \\ +\frac{1}{s}\int_{r=0}^{1} rG(r) J_0(\beta_i r)\, dr = 0\end{aligned} \tag{13-188c}$$

The first integral is evaluated as

$$\int_{r=0}^{1} r J_0(\beta_j r) J_0(\beta_i r)\, dr = \begin{cases} 0 & i \neq j \\ \frac{1}{2} J_1^2(\beta_i) & i = j \end{cases} \tag{13-189}$$

Introducing (13-189) into (13-188c), the summation drops out, and we obtain

$$c_i = \frac{2}{s(s + \beta_i^2) J_1^2(\beta_i)} \int_{r=0}^{1} rG(r) J_0(\beta_i r)\, dr \tag{13-190}$$

We introduce equation (13-190) into (13-187) after changing i to j and r to r' to obtain

$$\tilde{\bar{T}}(r,s) = 2\sum_j \frac{1}{s(s+\beta_j^2)} \frac{J_0(\beta_j r)}{J_1^2(\beta_j)} \int_{r'=0}^{1} r' G(r') J_0(\beta_j r')\, dr' \qquad \text{(13-191a)}$$

or

$$\tilde{\bar{T}}(r,s) = 2\sum_j \frac{1}{\beta_j^2}\left(\frac{1}{s} - \frac{1}{s+\beta_j^2}\right) \frac{J_0(\beta_j r)}{J_1^2(\beta_j)} \int_{r'=0}^{1} r' G(r') J_0(\beta_j r')\, dr' \qquad \text{(13-191b)}$$

The Laplace transform can be inverted by means of the Laplace transform Table 9-1, cases 1 and 8; we obtain

$$\tilde{T}(r,t) = 2\sum_j \frac{1}{\beta_j^2}(1 - e^{-\beta_j^2 t}) \frac{J_0(\beta_j r)}{J_1^2(\beta_j)} \int_{r'=0}^{1} r' G(r') J_0(\beta_j r')\, dr' \qquad \text{(13-192)}$$

where the summation is over all eigenvalues β_j's, which are the positive roots of

$$J_0(\beta_j) = 0 \qquad \text{(13-193)}$$

We note here that the solution obtained in this manner is in fact the exact solution of this problem.

REFERENCES

1. H. Schlichting, *Boundary Layer Theory* 6th ed., McGraw-Hill, New York, 1968, Chapter 13.
2. H. D. Landahl, *Bull. Math. Biophys.* **15**, 49–61, 1953.
3. H. J. Merk, *Appl. Sci. Res.* **4**, Section A, 435–452, 1954.
4. T. R. Goodman, *Trans. ASME* **80**, 335–342, 1958.
5. T. R. Goodman, *J. Heat Transfer* **83c**, 83–86, 1961.
6. T. R. Goodman, in *Advances in Heat Transfer*, Vol. 1, T. F. Irvine and J. P. Hartnett (eds), Academic, New York, 1964, pp. 52–120.
7. W. C. Reynolds and T. A. Dolton, The Use of Integral Methods in Transient Heat Transfer Analysis, Department of Mechanical Engineering, Report No. 36, Stanford University, Stanford, Calif., Sept. 1, 1958.
8. K. T. Yang, *Trans. ASME* **80**, 146–147, 1958.
9. K. T. Yang, *International Developments in Heat Transfer*, Vol. 1, ASME, New York, 1963, pp. 18–27.
10. P. J. Schneider, *J. Aerospace Sci.* **27**, 546–549, 1960.
11. T. J. Lardner and F. B. Pohle, *J. Appl. Mech.* 310–312, June 1961.
12. R. Thorsen and F. Landis, *Int. J. Heat Mass Transfer* **8**, 189–192, 1965.

13. B. Persson and L. Persson, Calculation of the Transient Temperature Distribution in Convectively Heated Bodies With Integral Method, ASME Paper No. 64-HT-19, 1964.
14. F. A. Castello, An Evaluation of Several Methods of Approximating Solutions to the Heat Conduction Equation, ASME Paper No. 63-HT-44, 1963.
15. H. H. Bengston and F. Kreith, *J. Heat Transfer* **92c**, 182–184, 1970.
16. M. N. Özisik, *Boundary Value Problems of Heat Conduction*, International Textbook, Scranton, PA, 1968.
17. M. Altman, *Chem. Eng. Prog. Symp. Series*, **57**, 16–23, 1961.
18. G. Poots, *Int. J. Heat Mass Transfer* **5**, 339–348, 1962.
19. G. Poots, *Int. J. Heat Mass Transfer* **5**, 525, 1962.
20. R. H. Tien and G. E. Geiger, *J. Heat Transfer* **89c**, 230–234, 1967.
21. R. H. Tien and G. E. Geiger, *J. Heat Transfer* **90c**, 27–31, 1968.
22. J. C. Muehlbauer, J. D. Hatcher, D. W. Lyons, and J. E. Sunderland, *J. Heat Transfer* **95c**, 324–331, 1973.
23. S. Cho and J. E. Sunderland, Melting or Freezing of Finite Slabs, ASME Paper 68-WA/HT-37, Dec. 1968.
24. S. H. Cho and J. E. Sunderland, *J. Heat Transfer* **91c**, 421–426, 1969.
25. K. Mody and M. H. Özisik, *Lett. Heat Mass Transfer* **2**, 487–493, 1975.
26. O. M. Griffin, *J. Heat Transfer* **95c**, 317–322, 1973.
27. O. M. Griffin, *Poly. Eng. Sci.* **12**, 140–149, 1972.
28. O. M. Griffin, *Proceedings of the 5th International Heat Transfer Conference*, Vol. 1, Tokyo, 1974, pp. 211–215.
29. O. M. Griffin, *Int. J. Heat Mass Transfer* **20**, 675–683, 1977.
30. W. F. Ames, *Nonlinear Partial Differential Equations in Engineering*, Academic, New York, 1965.
31. W. F. Ames, *Nonlinear Ordinary Differential Equations in Transport Process*, Academic, New York, 1968.
32. R. Weinstock, *Calculus of Variations*, McGraw-Hill, New York, 1952.
33. I. M. Gelfand and S. V. Fomin, *Calculus of Variations*, Prentice-Hall, Englewood Cliffs, NJ, 1963.
34. P. M. Morse and H. Feshbach, *Methods of Theoretical Physics*, McGraw-Hill, New York, 1953.
35. A. R. Forsythe, *Calculus of Variations*, Dover, New York, 1960.
36. S. G. Mikhlin, *Variational Methods in Mathematical Physics*, Macmillan, New York, 1964.
37. R. S. Schechter, *The Variational Method in Engineering*, McGraw-Hill, New York, 1967.
38. M. A. Biot, *J. Aeronaut. Sci.* **24**, 857–873, 1957.
39. M. A. Biot, *J. Aeronaut. Sci.* **26**, 367–381, 1959.
40. P. D. Richardson, *J. Heat Transfer* **86c**, 298–299, 1964.
41. V. S. Arpaci and C. M. Vest, Variational Formulation of Transformed Diffusion Problems, ASME Paper No. 67-HT-77.
42. T. J. Lardner, *AIAA J.* **1**, 196–206, 1963.

43. M. E. Gurtin, *Quart. Appl. Math.* **22**, 252–256, 1964.
44. B. A. Finlayson and L. E. Scriven, *Int. J. Heat Mass Transfer* **10**, 799–821, 1967.
45. D. F. Hays and H. N. Curd, *Int. J. Heat Mass Transfer* **11**, 285–295, 1968.
46. P. Rafalski and W. Zyszkowski, *AIAA J.* **6**, 1606, 1968.
47. W. Zyszkowski, *J. Heat Transfer* **91c**, 77–82, 1969.
48. D. Djukic and B. Vujanovic, *Z. Angew. Math. Mech.* **51**, 611–616, 1971.
49. B. Vujanovic and A. M. Straus, *AIAA J.* **9**, 327–330, 1971.
50. B. Krajewski, *Int. J. Heat Mass Transfer* **18**, 495–502, 1975.
51. W. Ritz, *J. Reine Angewandte Math.* **135**, 1–61, 1908.
52. N. Kryloff, *Memorial Sci. Math. Paris* **49**, 1931.
53. L. V. Kantorovich and V. I. Krylov, *Approximate Methods of Higher Analysis*, Wiley, New York, 1964.
54. L. Collatz, *The Numerical Treatment of Differential Equations* (English trans.), Springer, Heidelberg, 1960.
55. S. H. Crandall, *Engineering Analysis*, McGraw-Hill, New York, 1956.
56. V. S. Arpaci, *Conduction Heat Transfer*, Addison-Wesley, Reading, MA, 1966.
57. G. B. Galerkin, *Vestnik Inzhenerov Teckhnikov*, 879, 1915.
58. N. W. McLachlan, *Phil. Mag.* **36**, 600, 1945.
59. D. Dicker and M. B. Friedman, *J. Appl. Mech.* **30**, 493–499, 1963.
60. F. Erdogan, *J. Heat Transfer* **85c**, 203–208, 1963.
61. L. J. Snyder, T. W. Spiggs, and W. E. Stewart, *AICHE J.* **10**, 535–540, 1964.
62. B. A. Finlayson and L. E. Scriven, *Chem. Eng. Sci.* **20**, 395–404, 1965.
63. P. A. Laura and A. J. Faulstich, *Int. J. Heat Mass Transfer* **11**, 297–303, 1968.
64. A. H. Eraslan, *J. Heat Transfer* **91c**, 212–220, 1969.
65. A. H. Eraslan, *AIAA J.* **10**, 1759–1766, 1966.
66. M. Mashena and A. Haji-Sheikh, *Int. J. Heat Mass Transfer* **29**, 317–329, 1986.
67. A. Haji-Sheikh and M. Mashena, *J. Heat Transfer* **109**, 551–556, 1987.
68. A. Haji-Sheikh and R. Lakshminarayanan, *J. Heat Transfer* **109**, 557–562, 1987.
69. S. Nomura and A. Haji-Sheikh, *J. Heat Transfer* **110**, 110–112, 1988.

PROBLEMS

13-1 A semi-infinite region $x > 0$ is initially at zero temperature. For times $t > 0$, the convection boundary condition at the surface $x = 0$ is given as $-k(\partial T/\partial x) + hT = f_1$, where f_1 = constant. Obtain an expression for the temperature distribution $T(x, t)$ in the medium using the integral method with a cubic polynomial representation for temperature.

13-2 A semi-infinite medium $x > 0$ is initially at a uniform temperature T_i. For times $t > 0$, the boundary surface at $x = 0$ is subjected to a prescribed heat flux, that is, $-k(\partial T/\partial x) = f(t)$ at $x = 0$, where $f(t)$ varies with time. Obtain an expression for the temperature distribution $T(x, t)$ in the medium using the integral method and a cubic polynomial representation for $T(x, t)$.

13-3 A region exterior to a cylindrical hole of radius $r = b$ (i.e., $r > b$) is initially at zero temperature. For times $t > 0$ the boundary surface at $r = b$ is kept at a constant temperature T_0. Obtain an expression for the temperature distribution in the medium using the integral method with a second-degree polynomial representation modified by $\ln r$ for $T(x, t)$.

13-4 A semi-infinite medium $x > 0$ is initially at zero temperature. For times $t > 0$ heat is generated in the medium at a constant rate of g (W/m^3), while heat is removed from the boundary surface at $x = 0$ as $k(\partial T/\partial x) = f =$ constant. Obtain an expression for the temperature distribution $T(x, t)$ in the medium for times $t > 0$, using the integral method and a cubic polynomial representation for $T(x, t)$.

13-5 Consider a heat conduction problem for a semi-infinite medium $x > 0$ with the fourth-power radiative heat transfer at the boundary surface $x = 0$ defined as

$$\frac{\partial^2 T}{\partial x^2} = \frac{1}{\alpha}\frac{\partial T}{\partial t} \qquad \text{in} \quad x > 0, \quad t > 0$$

$$k\frac{\partial T}{\partial x} = \sigma\varepsilon(T_s^4 - T_\infty^4) \qquad \text{at} \quad x = 0, \quad t > 0$$

$$T = T_i \qquad \text{for} \quad t = 0, \quad x \geq 0$$

where T_s is the surface temperature. Apply the formal solution given by equation (13-55) for the solution of this problem. For the case of $T_\infty = 0$, by performing the resulting integration analytically obtain an expression for the surface temperature T_s as a function of time.

13-6 Consider the following steady-state heat conduction problem for a rectangular region $0 < x < a$, $0 < y < b$:

$$\frac{\partial^2 T}{\partial x^2} + \frac{\partial^2 T}{\partial y^2} = 0 \qquad \text{in} \quad 0 < x < a, \quad 0 < y < b$$

$$T = 0 \qquad \text{at} \quad y = 0, \quad y = b$$

$$\frac{\partial T}{\partial x} = 0 \qquad \text{at} \quad x = 0$$

$$T = T_0 \sin\left(\frac{3\pi y}{b}\right) \qquad \text{at} \quad x = a$$

Solve this problem by the Galerkin method using partial integration with respect to the y variable for a trial function chosen in the form $\tilde{T}_1(x, y) = f(x)\sin(3\pi y/b)$ and compare this result with the exact solution.

13-7 Solve the following steady-state heat conduction problem:

$$\frac{\partial^2 T}{\partial x^2} + \frac{\partial^2 T}{\partial y^2} + \frac{1}{k} g = 0 \qquad \text{in the region shown in Figure 13-8 } (b)$$

$$T = 0 \qquad \text{on the boundaries}$$

using the Galerkin method and a one-term trial solution chosen in the form

$$\tilde{T}_1(x, y) = c_1(y - \alpha x)(y + \beta x)(L - x)$$

13-8 Solve the following steady-state heat conduction problem:

$$\frac{\partial^2 T}{\partial x^2} + \frac{\partial^2 T}{\partial y^2} + \frac{1}{k} g = 0 \qquad \text{in the region shown in Figure 13-8 } (c) \text{ for } a = b = 1$$

$$T = 0 \qquad \text{on the boundaries}$$

using the Galerkin method and a one-term trial solution.

14

INTEGRAL TRANSFORM TECHNIQUE

The solution of partial differential equations of heat conduction by the classical method of separation of variables is not always convenient when the equation and the boundary conditions involve nonhomogeneities. It is for this reason that we considered the Green's function approach for the solution of linear, nonhomogeneous boundary value problems of heat conduction. The *integral transform technique* also provides a systematic, efficient, and straightforward approach for the solution of both homogeneous and nonhomogeneous, steady-state, and time-dependent boundary value problems of heat conduction. In this method, the second partial derivatives with respect to the space variables are generally removed from the partial differential equation of heat conduction by the application of the integral transformation. For example, in time-dependent problems, the partial derivatives with respect to the space variables are removed and the partial differential equation is reduced to a first-order ordinary differential equation in the time variable for the transform of the temperature. The ordinary differential equation is solved subject to the transformed initial conditions, and the result is inverted successively to obtain the solution for the temperature. The inversion process is straightforward because the inversion formulas are available at the onset of the problem. The procedure is also applicable to the solution of steady-state heat conduction problems involving more than one space variable. In such cases the partial differential equation of heat conduction is reduced to an ordinary differential equation in one of the space variables. The resulting ordinary differential equation for the transformed temperature is solved, and the solution is inverted to obtain the temperature distribution.

The integral transform technique derives its basis from the classical method of separation of variables. That is, the integral transform pairs needed for the solution of a given problem are developed by considering the representation of an arbitrary function in terms of the eigenfunctions of the corresponding eigenvalue problem. Therefore, the eigenfunctions, eigenvalues, and the normalization integrals developed in Chapters 2–5 for the solution of homogeneous problems will be utilized for the construction of the integral transform pairs.

The fundamental theory of the integral transform technique is given in several texts [1–3] and a summary of various transform pairs and transform tables are presented in various references [4–8]. The literature on the use of the integral transform technique for the solution of heat conduction problems is ever growing. The reader should consult references 9–23 for the general solution of three-dimensional problems of finite regions. Its applications for the solution of specific heat conduction problems in the rectangular [24, 25], cylindrical [26–29], and spherical [30] coordinate systems are also given. Some useful convolution properties of integral transforms are discussed in references 31–35.

In this chapter, a general method of analysis of three-dimensional, time-dependent heat conduction problems of finite region by the integral transform technique is presented first. Its applications for the solution of problems of finite, semi-infinite, and infinite regions in the rectangular, cylindrical, and spherical coordinate systems are then presented systematically. Additional applications can be found in the references 37–40.

14-1 USE OF INTEGRAL TRANSFORM IN THE SOLUTION OF HEAT CONDUCTION PROBLEMS

In this section, we present the use of the integral transform technique in the solution of three-dimensional, time-dependent, nonhomogeneous boundary value problems of heat conduction with constant coefficients in finite regions. We consider the following heat conduction problem:

$$\nabla^2 T(\hat{r},t) + \frac{1}{k} g(\hat{r},t) = \frac{1}{\alpha}\frac{\partial T(\hat{r},t)}{\partial t} \qquad \text{in region } R, \qquad t>0 \qquad (14\text{-}1\text{a})$$

$$k_i \frac{\partial T(\hat{r}_i,t)}{\partial n_i} + h_i T(\hat{r}_i,t) = f_i(\hat{r}_i,t) \qquad \text{on boundary } S_i \qquad (14\text{-}1\text{b})$$

$$T(\hat{r},t=0) = F(\hat{r}) \qquad \text{in region } R \qquad (14\text{-}1\text{c})$$

where $i = 1, 2, \ldots, N$, and N is the number of continuous boundary surfaces of the region R ($N = 1$ for a semi-infinite medium, $N = 2$ for a slab, $N = 4$ for a 2-D rectangular region, etc.); $\partial/\partial n_i$ denotes the normal derivative at the boundary surface S_i in the *outward* direction; h_i and k_i are the boundary condition coefficients at the boundary surface S_i; k is the thermal conductivity; α is the thermal diffusivity; $f_i(\hat{r}_i,t)$ is a specified boundary condition function; $F(\hat{r})$ is a specified initial condition function; and $g(\hat{r},t)$ is the heat generation term. We use the notation $\hat{r}_i$ to represent our spatial coordinate vector at a specified value of one coordinate, say r_i; hence this vector is a function of only two independent variables for the general 3-D problem.

The basic steps in the solution of this problem with the integral transform technique can be summarized as follows:

1. Appropriate integral transform pair is developed.
2. By the application of integral transformation, the partial derivatives with respect to the space variables are removed from the heat conduction equation; thus reducing it to an ordinary differential equation for the transform of temperature.
3. The resulting ordinary differential equation is solved subject to the transformed initial condition. When the transform of the temperature is inverted by the inversion formula, the desired solution is obtained. The procedure is now described in detail.

Development of Integral Transform Pair The integral transform pair needed for the solution of the above heat conduction problem can be developed by considering the following eigenvalue problem:

$$\nabla^2\psi(\hat{r}) + \lambda^2\psi(\hat{r}) = 0 \qquad \text{in region } R \tag{14-2a}$$

$$k_i\frac{\partial\psi(\hat{r}_i)}{\partial n_i} + h_i\psi(\hat{r}_i) = 0 \qquad \text{on boundary } S_i \tag{14-2b}$$

where $i = 1, 2, \ldots, N$, and $k_i, h_i, \partial/\partial n_i$ are as defined previously. We note that this eigenvalue problem is obtainable by the separation of the homogeneous version of the heat conduction problem (14-1). The eigenfunctions $\psi(\lambda_m, \hat{r})$ of this eigenvalue problem satisfy the following orthogonality condition (see note 1 at the end of this chapter for a proof of this orthogonality relation):

$$\int_R \psi(\lambda_m, \hat{r})\psi(\lambda_n, \hat{r})\,dv = \begin{cases} 0 & \text{for} \quad m \neq n \\ N(\lambda_m) & \text{for} \quad m = n \end{cases} \tag{14-3a}$$

where the *normalization integral* $N(\lambda_m)$ is defined as

$$N(\lambda_m) = \int_R [\psi(\lambda_m, \hat{r})]^2\,dV \tag{14-3b}$$

We now consider the representation of a function $T(\hat{r}, t)$, defined in the finite region R, in terms of the eigenfunctions $\psi(\lambda_m, \hat{r})$ in the form

$$T(\hat{r}, t) = \sum_{m=1}^{\infty} C_m(t)\psi(\lambda_m, \hat{r}) \qquad \text{in } R \tag{14-4}$$

where the summation is taken over the entire discrete spectrum of eigenvalues λ_m. To determine the unknown coefficients we operate on both sides of equation (14-4) by the operator

$$* \int_R \psi(\lambda_n, \hat{r})\, dV$$

that is, multiply by $\psi(\lambda_n, \hat{r})$ and integrate over the region R, and then utilize the orthogonality relation (14-3) to obtain

$$C_m(t) = \frac{1}{N(\lambda_m)} \int_R \psi(\lambda_m, \hat{r}) T(\hat{r}, t)\, dV \tag{14-5}$$

This expression is introduced into equation (14-4), and the resulting representation is split up into two parts to define the *integral transform pair* in the space variable $\hat{r}$ for the function $T(\hat{r}, t)$ as

$$\text{Inversion formula:} \qquad T(\hat{r}, t) = \sum_m \frac{\psi(\lambda_m, \hat{r})}{N(\lambda_m)} \overline{T}(\lambda_m, t) \tag{14-6a}$$

$$\text{Integral transform:} \qquad \overline{T}(\lambda_m, t) = \int_{R'} \psi(\lambda_m, \hat{r}') T(\hat{r}', t)\, dV' \tag{14-6b}$$

where $\overline{T}(\lambda_m, t)$ is called the integral transform of the function $T(\hat{r}, t)$ with respect to the space variable $\hat{r}$. It is to be noted that in the above formal representation, the summation is actually a triple, a double, or a single summation; and the integral is a volume, a surface, or a line integral for the three-, two-, or one-dimensional regions, respectively. In the Cartesian coordinate system, the eigenfunctions $\psi(\lambda_m, \hat{r})$ and the normalization integral $N(\lambda_m)$ are composed of the *products* of one-dimensional eigenfunctions and normalization integrals, respectively.

Integral Transform of Heat Conduction Problem Having established the appropriate integral transform pair as given above, the next step in the analysis is the removal of the partial derivatives with respect to the space variables from the differential equation (14-1a) by the application of the integral transform (14-6b). That is, both sides of equation (14-1a) are multiplied by $\psi(\hat{r})$ and integrated over the region R:

$$\int_R \psi_m(\hat{r}) \nabla^2 T(\hat{r}, t)\, dV + \frac{1}{k} \int_R \psi_m(\hat{r}) g(\hat{r}, t)\, dV = \frac{1}{\alpha} \frac{\partial}{\partial t} \int_R \psi_m(\hat{r}) T(\hat{r}, t)\, dV \tag{14-7}$$

where $\psi(\hat{r}) \equiv \psi(\lambda_m, \hat{r})$. By utilizing the definition of the integral transform (14-6b), this expression is written as

$$\int_R \psi_m(\hat{r}) \nabla^2 T(\hat{r}, t)\, dV + \frac{1}{k} \overline{g}(\lambda_m, t) = \frac{1}{\alpha} \frac{d\overline{T}(\lambda_m, t)}{dt} \tag{14-8}$$

where $\overline{g}(\lambda_m, t)$ and $\overline{T}(\lambda_m, t)$ are the integral transforms of the function $g(\hat{r}, t)$ and $T(\hat{r}, t)$, respectively, defined as

$$\overline{g}(\lambda_m, t) = \int_R \psi_m(\hat{r}) g(\hat{r}, t)\, dV \tag{14-9a}$$

$$\overline{T}(\lambda_m, t) = \int_R \psi_m(\hat{r}) T(\hat{r}, t)\, dV \tag{14-9b}$$

The integral on the left-hand side of equation (14-8) can be evaluated by making use of Green's theorem, expressed as

$$\int_R \psi_m(\hat{r}) \nabla^2 T(\hat{r}, t)\, dV = \int_R T \nabla^2 \psi_m(\hat{r})\, dV + \sum_{i=1}^{N} \int_{S_i'} \left(\psi_m \frac{\partial T}{\partial n_i} - T \frac{\partial \psi_m}{\partial n_i} \right) ds_i' \tag{14-10}$$

where $i = 1, 2, \ldots, N$, and N is the number of continuous boundary surfaces of the region R. Various terms on the right-hand side of equation (14-10) are evaluated as now described. The integral $\int_R T \nabla^2 \psi_m\, dV$ is evaluated by writing equation (14-2a) for the eigenfunction $\psi_m(\hat{r})$, multiplying both sides by $T(\hat{r}, t)$, integrating over the region R, and utilizing the definition of the integral transform. We find

$$\int_R T \nabla^2 \psi_m(\hat{r})\, dV = -\lambda_m^2 \int_R T \psi_m(\hat{r})\, dV = -\lambda_m^2 \overline{T}(\lambda_m, t) \tag{14-11a}$$

The surface integral in equation (14-10) is evaluated by making use of the boundary conditions (14-1b) and (14-2b). That is, equation (14-1b) is multiplied by $\psi_m(\hat{r})$, equation (14-2b) is multiplied by $T(\hat{r}, t)$, and the results are subtracted; we then obtain

$$\psi_m \frac{\partial T}{\partial n_i} - T \frac{\partial \psi_m}{\partial n_i} = \frac{\psi_m(\hat{r}_i)}{k_i} f_i(\hat{r}_i, t) \tag{14-11b}$$

Equations (14-11) are introduced into equation (14-10), giving

$$\int_R \psi_m(\hat{r}) \nabla^2 T(\hat{r}, t)\, dV = -\lambda_m^2 \overline{T}(\lambda_m, t) + \sum_{i=1}^{N} \int_{S_i'} \frac{\psi_m(\hat{r'}_i)}{k_i} f_i(\hat{r'}_i, t)\, ds_i' \tag{14-12}$$

Substituting equation (14-12) into (14-8), we obtain

$$\frac{d\overline{T}(\lambda_m, t)}{dt} + \alpha \lambda_m^2 \overline{T}(\lambda_m, t) = A(\lambda_m, t) \qquad \text{for} \qquad t > 0 \tag{14-13a}$$

where

$$A(\lambda_m, t) \equiv \frac{\alpha}{k} \overline{g}(\lambda_m, t) + \alpha \sum_{i=1}^{s} \int_{S_i'} \frac{\psi_m(\hat{r'}_i)}{k_i} f_i\left(\hat{r'}_i, t\right) ds_i' \tag{14-13b}$$

Thus by the application of the integral transform technique, we removed from the heat conduction equation (14-1a) all the partial derivatives with respect to the space variables and reduced it to a first-order ordinary differential equation (14-13a) for the transform $\overline{T}(\lambda_m, t)$ of the temperature. In the process of integral transformation, we utilized the boundary conditions (14-1b); therefore the boundary conditions for the problem are incorporated in this result. The integral transform of the initial condition (14-1c) becomes

$$\overline{T}(\lambda_m, t = 0) = \int_R \psi_m(\hat{r}) F(\hat{r})\, dV \equiv \overline{F}(\lambda_m) \tag{14-13c}$$

Solution for Transform and Inversion The solution of equation (14-13a) subject to the transformed initial condition (14-13c) gives the transform $\overline{T}(\lambda_m, t)$ of temperature as

$$T(\lambda_m, t) = e^{-\alpha\lambda_m^2 t} \left[\overline{F}(\lambda_m) + \int_{t'=0}^{t} e^{\alpha\lambda_m^2 t'} A(\lambda_m, t')\, dt' \right] \tag{14-14}$$

Introducing this integral transform into the inversion formula (14-6a), we obtain the solution of the boundary value problem of heat conduction, equations (14-1), in the form

$$\boxed{T(\hat{r}, t) = \sum_{m=1}^{\infty} \frac{1}{N(\lambda_m)} e^{-\alpha\lambda_m^2 t} \psi_m(\hat{r}) \left[\overline{F}(\lambda_m) + \int_{t'=0}^{t} e^{\alpha\lambda_m^2 t'} A(\lambda_m, t')\, dt' \right]} \tag{14-15a}$$

where we define

$$A(\lambda_m, t') \equiv \frac{\alpha}{k} \overline{g}(\lambda_m, t') + \alpha \sum_{i=1}^{N} \int_{S_i'} \frac{\psi_m(\hat{r}_i')}{k_i} f_i\left(\hat{r}_i', t'\right) ds_i' \tag{14-15b}$$

$$\overline{F}(\lambda_m) \equiv \int_{R'} \psi_m(\hat{r}') F(\hat{r}')\, dV' \tag{14-15c}$$

$$\overline{g}(\lambda_m, t') \equiv \int_{R'} \psi_m(\hat{r}') g(\hat{r}', t')\, dV' \tag{14-15d}$$

$$N(\lambda_m) \equiv \int_{R'} [\psi_m(\hat{r}')]^2\, dV' \tag{14-15e}$$

and the summation is taken over all eigenvalues. This solution is derived formally for a boundary condition of the third type for all boundaries. If some of the boundaries are of the second type (i.e., prescribed flux) and some of the third type, the general form of the solution remains unchanged. However, some modification is needed in the term $A(\lambda_m, t')$ if the problem involves boundary conditions of the first type (i.e., prescribed temperature). Suppose the boundary condition for the surface $i = 1$ is of the first type; this implies that the boundary condition coefficient k_1 should be set equal to zero. This situation causes difficulty in the interpretation of the term $A(\lambda_m, t')$ given by equation (14-15b) because k_1 appears in the denominator. This difficulty can be alleviated by making the following change in equation (14-15b):

When $k_1 = 0$ (i.e., for a type 1 boundary), replace

$$\frac{\psi_m(\hat{r}'_1)}{k_1} \qquad \text{by} \qquad -\frac{1}{h_1}\frac{\partial \psi_m(\hat{r}'_1)}{\partial n'_1} \tag{14-15f}$$

The validity of this replacement becomes apparent if we rearrange the boundary condition (14-2b) of the eigenvalue problem for $i = 1$, in the form

$$\frac{\psi_m(\hat{r}_1)}{k_1} = -\frac{1}{h_1}\frac{\partial \psi_m(\hat{r}_1)}{\partial n_1}$$

on boundary S_1. Finally, when all boundary conditions are of the second type, the interpretation of the general solution (14-15) requires special consideration. The reason for this, $\lambda_0 = 0$ is also an eigenvalue corresponding to the eigenfunction $\psi_0 = \text{constant} \neq 0$, for this particular case. This matter will be illustrated later in this section.

One-Dimensional Finite Region

We now consider the one-dimensional version of the heat conduction problem (14-1) in the space variable x_1 (i.e., x, y, z, r, etc.) for a finite region R_1. Let $\psi(\beta_m, x_1)$ be the eigenfunctions, $N(\beta_m)$ be the normalization integral, β_m the eigenvalues, and $w(x_1)$ the Sturm–Liouville weighting function of the one-dimensional version of the eigenvalue problem (14-2). As was discussed in Chapter 2, the eigenfunctions $\psi(\beta_m, x_1)$ are orthogonal with respect to the weighting function $w(x_1)$:

$$\int_{R_1} w(x_1)\psi(\beta_m, x_1)\psi(\beta_n, x_1)\, dx_1 = \begin{cases} 0 & \text{for} \quad \beta_m \neq \beta_n \\ N(\beta_m) & \text{for} \quad \beta_m = \beta_n \end{cases} \tag{14-16a}$$

where

$$N(\beta_m) \equiv \int_{R_1} w(x_1)[\psi(\beta_m, x_1)]^2\, dx_1 \tag{14-16b}$$

Suppose we wish to represent a function $T(x_1, t)$, defined in the finite interval R_1, in terms of the eigenfunctions $\psi(\beta_m, x_1)$. Such a representation is immediately obtained by utilizing the above orthogonality condition, and the result is written as

$$T(x_1, t) = \sum_{m=1}^{\infty} \frac{\psi(\beta_m, x_1)}{N(\beta_m)} \int_{R'_1} w(x'_1)\psi(\beta_m, x'_1)T(x'_1, t)\, dx'_1 \qquad \text{in region } R_1 \tag{14-16c}$$

The desired integral transform pair is constructed by splitting up this representation into two parts as

$$\text{Inversion formula:} \qquad T(x_1, t) = \sum_{m=1}^{\infty} \frac{\psi(\beta_m, x_1)}{N(\beta_m)} \overline{T}(\beta_m, t) \tag{14-17a}$$

$$\text{Integral transform:} \qquad \overline{T}(\beta_m, t) = \int_{R_1'} w(x_1')\psi(\beta_m, x_1')T(x_1', t)\, dx_1' \tag{14-17b}$$

The solution of the one-dimensional version of the heat conduction problem (14-1) is obtained from the general solution (14-15) as

$$T(x_1, t) = \sum_{m=1}^{\infty} \frac{\psi(\beta_m, x_1)}{N(\beta_m)} e^{-\alpha\beta_m^2 t} \left[\overline{F}(\beta_m) + \int_{t'=0}^{t} e^{\alpha\beta_m^2 t'} A(\beta_m, t')\, dt' \right] \tag{14-18a}$$

where we define

$$A(\beta_m, t') = \frac{\alpha}{k}\overline{g}(\beta_m, t') + \alpha \left\{ \left[w(x_1') \frac{\psi(\beta_m, x_1')}{k} \right]_{x_1'=s_1} f_1(t') + \left[w(x_1') \frac{\psi(\beta_m, x_1')}{k} \right]_{x_1'=s_2} f_2(t') \right\} \tag{14-18b}$$

$$\overline{F}(\beta_m) = \int_{R_1'} w(x_1')\psi(\beta_m, x_1')F(x_1')\, dx_1' \tag{14-18c}$$

$$\overline{g}(\beta_m, t') = \int_{R_1'} w(x_1')\psi(\beta_m, x_1')g(x_1', t')\, dx_1' \tag{14-18d}$$

$$N(\beta_m) = \int_{R_1'} w(x_1')[\psi(\beta_m, x_1')]^2\, dx_1' \tag{14-18e}$$

If the boundary condition is of the first kind at any of the boundaries, the following adjustment should be made in equation (14-18b). When k = 0 replace

$$\frac{\psi(\beta_m, x_1')}{k} \qquad \text{by} \qquad -\frac{1}{h_i} \frac{\partial\psi(\beta_m, x_1')}{\partial n_1'} \tag{14-19}$$

where $i = 1$ or 2, and $\partial/\partial n_1'$ denotes derivative along the outward-drawn normal to the boundary surface noting h_i is then unity.

Boundary Condition of the Second Kind for All Boundaries

When the boundary conditions for a finite region are all of the second kind, that is, all h_i values are zero in the heat conduction problem (14-1), then the eigenvalue problem (14-2) takes the form

$$\nabla^2\psi(\hat{r}) + \lambda^2\psi(\hat{r}) = 0 \qquad \text{in region } R \tag{14-20a}$$

$$\frac{\partial\psi(\hat{r}_i)}{\partial n_i} = 0 \qquad \text{on all boundaries } S_i \tag{14-20b}$$

For this particular case, $\lambda_0 = 0$ is also an eigenvalue corresponding to the eigenfunction $\psi_0 = \text{constant} \neq 0$. The validity of this statement can be verified by integrating equation (14-20a) over the region R, applying Green's theorem to change the volume integral to the surface integral, and then utilizing the boundary conditions (14-20b). We obtain

$$\lambda^2 \int_R \psi(\hat{r})\,dV = -\int_R \nabla^2\psi(\hat{r})\,dV = -\sum_{i=1}^{N}\int_{s_i}\frac{\partial\psi}{\partial n_i}ds_i = 0 \tag{14-21}$$

Clearly, $\lambda_0 = 0$ is also an eigenvalue corresponding to the eigenfunction $\psi_0 = \text{constant} \neq 0$. We can set $\psi_0 = 1$, because ψ_0 will cancel out when it is introduced into the solution (14-15) for the eigenvalue $\lambda_0 = 0$. Then, the general solution (14-15) for the case of all boundary conditions is of the second kind [i.e., $\partial T/\partial n_i = (1/k_i) f_i(\hat{r}_i, t)$] on all sides S_i, takes the form

$$\begin{aligned} T(\hat{r},t) &= \frac{1}{N_0}\left[\overline{F}(\lambda_0) + \int_{t'=0}^{t} A(\lambda_0,t)\,dt'\right] \\ &+ \sum_{m=1}^{\infty}\frac{1}{N_m}e^{-\alpha\lambda_m^2}\psi(\lambda_m,\hat{r})\left[\overline{F}(\lambda_m) + \int_{t'=0}^{t} e^{\alpha\lambda_m^2 t'}A(\lambda_m,t')\,dt'\right] \end{aligned} \tag{14-22}$$

where we have

$$N_0 = \int_{R'} dV' \tag{14-23a}$$

$$\overline{F}(\lambda_0) = \int_{R'} F(\hat{r}')\,dV' \tag{14-23b}$$

$$A(\lambda_0,t') = \frac{\alpha}{k}\int_{R'} g(\hat{r}',t')\,dV' + \alpha\sum_{i=1}^{N}\int_{S_i'}\frac{f_i(\hat{r}_i',t')}{k_i}ds_i' \tag{14-23c}$$

and the functions $A(\lambda_m, t')$, $\overline{F}(\lambda_m)$, and N_m are as defined by equations (14-15b), (14-15c), and (14-15e), respectively.

The average temperature over the region R is defined as

$$T_{\text{avg}}(t) = \frac{\int_R T(\hat{r},t)\,dV}{\int_R dV} \tag{14-24a}$$

and the solution (14-22) is introduced into this expression. If we take into account the following relations obtained from equation (14-21), namely,

$$\int_R \psi(\lambda_m, \hat{r})\, dV = 0 \qquad \text{for} \qquad \lambda_m \neq 0 \tag{14-24b}$$

$$\int_R \psi(\lambda_0)\, dV = \int_R dV \qquad \text{since} \qquad \psi_0 = 1 \qquad \text{for} \qquad \lambda_0 = 0 \tag{14-24c}$$

then equation (14-24a) gives

$$T_{\text{avg}}(t) = \frac{\overline{F}(\lambda_0) + \int_{t'=0}^{t} A(\lambda_0, t')\, dt'}{\int_R dV} \tag{14-25a}$$

This result implies that the first term in the solution (14-22) resulting from the eigenvalue $\lambda_0 = 0$ is the average value of $T(\hat{r}, t)$ over the finite region R. For the special case of no heat generation and all insulated boundaries (i.e., $\partial T/\partial n_i = 0$), the quantity $A(\lambda_0, t')$ vanishes and equation (14-25a) reduces to

$$T_{\text{avg}}(t) = \frac{\int_R F(\hat{r})\, dV}{\int_R dV} \tag{14-25b}$$

Clearly, the expressions given by equations (14-25) are the generalization of the special case considered in Example 4-8 for problems with insulated boundaries and correctly reflect conservation of energy.

14-2 APPLICATIONS IN THE RECTANGULAR COORDINATE SYSTEM

The general analysis developed in the previous section is now applied for the solution of time-dependent heat conduction problems in the rectangular coordinate system. The one-dimensional cases are considered first for the finite, semi-infinite, and infinite regions. The multidimensional problems involving any combinations of the finite, semi-infinite, and infinite regions for the x, y, and z directions are then handled by the successive application of the one-dimensional integral transform.

One-Dimensional Problems of Finite Region

We consider the following heat conduction problem for a slab, $0 \leq x \leq L$:

$$\frac{\partial^2 T}{\partial x^2} + \frac{1}{k} g(x, t) = \frac{1}{\alpha} \frac{\partial T(x, t)}{\partial t} \qquad \text{in} \qquad 0 < x < L, \qquad t > 0 \tag{14-26a}$$

$$\text{BC1:} \qquad -k_1 \left.\frac{\partial T}{\partial x}\right|_{x=0} + h_1 \left. T\right|_{x=0} = f_1(t) \tag{14-26b}$$

$$\text{BC2:} \qquad k_2 \left.\frac{\partial T}{\partial x}\right|_{x=L} + h_2 \left. T\right|_{x=L} = f_2(t) \tag{14-26c}$$

$$\text{IC:} \qquad T(x, t = 0) = F(x) \qquad \text{in} \qquad 0 \le x \le L \tag{14-26d}$$

The eigenvalue problem associated with the solution of this problem is exactly the same as that given by case 1 of Table 2-1. Clearly, this eigenvalue problem is the one-dimensional version of the general eigenvalue problem (14-2). To construct the desired integral transform pair for the solution of the above heat conduction problem, we need the representation of an arbitrary function, defined in the interval $0 \le x \le L$, in terms of the eigenfunctions of the eigenvalue problems of Table 2-1. Such a representation has already been developed in Chapter 3: see, for example, equations (3-124) and (3-126). Then the integral transform pair for the function $T(x, t)$ with respect to the x variable is readily obtained by splitting up the representation into two parts as

$$\text{Inversion formula} \qquad T(x, t) = \sum_{m=1}^{\infty} \frac{X(\beta_m, x)}{N(\beta_m)} \overline{T}(\beta_m, t) \tag{14-27a}$$

$$\text{Integral transform} \qquad \overline{T}(\beta_m, t) = \int_{x'=0}^{L} X(\beta_m, x') T(x', t)\, dx' \tag{14-27b}$$

where

$$N(\beta_m) \equiv \int_{x=0}^{L} [X(\beta_m, x)]^2\, dx \tag{14-27c}$$

The functions $X(\beta_m, x)$, $N(\beta_m)$, and the eigenvalues β_m are obtainable from Table 2-1 for the nine different combinations of boundary conditions at $x = 0$ and $x = L$.

To solve the heat conduction problem (14-26), we take the integral transform of equation (14-26a) by the application of the transform (14-27b). That is, we multiply both sides of equation (14-26a) by $X(\beta_m, x)$, and integrate over the region $0 \le x \le L$. The resulting expression contains the term

$$\int_{x=0}^{L} X(\beta_m, x) \frac{\partial^2 T}{\partial x^2}\, dx$$

which is evaluated as discussed in Section 14-1 by making use of Green's theorem (or integrating it by parts twice), utilizing the eigenvalue problem discussed above, and the boundary conditions (14-26b) and (14-26c) of the above heat conduction problem. Then, the integral transform of equation (14-26a) leads to the following ordinary differential equation for the transform $\overline{T}(\beta_m, t)$ of temperature

$$\frac{d\overline{T}(\beta_m, t)}{dt} + \alpha\beta_m^2\overline{T}(\beta_m, t) = A(\beta_m, t) \qquad \text{for} \qquad t > 0 \tag{14-28a}$$

$$\text{IC}: \qquad \overline{T}(\beta_m, t = 0) = \overline{F}(\beta_m) \tag{14-28b}$$

where $\overline{F}(\beta_m)$ is the integral transform of the initial condition function $F(x)$ and $A(\beta_m, t)$ is defined below. The solution of equations (14-28) gives the transform of temperature $\overline{T}(\beta_m, t)$, and when this result is inverted by the inversion formula (14-27a), the solution of the heat conduction problem (14-26) becomes

$$T(x, t) = \sum_{m=1}^{\infty} \frac{X(\beta_m, x)}{N(\beta_m)} e^{-\alpha\beta_m^2 t} \left[\overline{F}(\beta_m) + \int_{t'=0}^{t} e^{\alpha\beta_m^2 t'} A(\beta_m, t')\, dt'\right] \tag{14-29a}$$

where we define

$$A(\beta_m, t') = \frac{\alpha}{k}\overline{g}(\beta_m, t') + \alpha\left[\left.\frac{X(\beta_m, x)}{k_1}\right|_{x=0} f_1(t') + \left.\frac{X(\beta_m, x)}{k_2}\right|_{x=L} f_2(t')\right] \tag{14-29b}$$

$$\overline{F}(\beta_m) = \int_{x'=0}^{L} X(\beta_m, x')F(x')\, dx' \tag{14-29c}$$

$$\overline{g}(\beta_m, t') = \int_{x'=0}^{L} X(\beta_m, x')g(x', t')\, dx' \tag{14-29d}$$

$$N(\beta_m) = \int_{x'=0}^{L} [X(\beta_m, x')]^2\, dx' \tag{14-29e}$$

If the boundary conditions at $x = 0$ or $x = L$ or both are of the first type, the following changes should be made in the term $A(\beta_m, t')$ defined by equation (14-29b):

$$\text{When } k_1 = 0, \text{ replace} \qquad \left.\frac{X(\beta_m, x)}{k_1}\right|_{x=0} \qquad \text{by} \qquad \left.\frac{dX(\beta_m, x)}{dx}\right|_{x=0} \tag{14-29f}$$

$$\text{When } k_2 = 0, \text{ replace} \qquad \left.\frac{X(\beta_m, x)}{k_2}\right|_{x=0} \qquad \text{by} \qquad -\left.\frac{dX(\beta_m, x)}{dx}\right|_{x=L} \tag{14-29g}$$

We also note that the solution given by equations (14-29) is also immediately obtainable from the general solution (14-18) by setting $\psi(\beta_m, x_1) = X(\beta_m, x)$, $\psi(\beta_m, x_1') = X(\beta_m, x_1')$, and $w(x_1') = 1$.

The eigenfunctions $X(\beta_m, x)$, the normalization integral $N(\beta_m)$, and the eigenvalues β_m appearing in the solution (12-39) are obtainable from Table 2-1 in Chapter 2 for the nine different combinations of the boundary conditions.

Alternative Solution In some cases it is desirable to split up the solution, $T(x, t)$ as

$$T(x,t) = \sum_{j=0}^{2} T_{0j}(x,t) + T_H(x,t) - \sum_{j=0}^{2} T_j(x,t) \tag{14-30}$$

where the functions $T_{0j}(x, t)$ are the solutions of the following quasi-steady-state problem:

$$\frac{\partial^2 T_{0j}(x,t)}{\partial x^2} + \delta_{0j}\frac{g(x,t)}{k} = 0 \quad \text{in} \quad 0 < x < L \tag{14-31a}$$

$$\text{BC1:} \quad -k_1 \left.\frac{\partial T_{0j}}{\partial x}\right|_{x=0} + h_1 \left. T_{0j}\right|_{x=0} = \delta_{1j} f_1(t) \tag{14-31b}$$

$$\text{BC2:} \quad k_2 \left.\frac{\partial T_{0j}}{\partial x}\right|_{x=L} + h_2 \left. T_{0j}\right|_{x=L} = \delta_{2j} f_2(t) \tag{14-31c}$$

$$\delta_{ij} = \begin{cases} 1 & \text{for} \quad i = j \\ 0 & \text{for} \quad i \neq j \end{cases} \quad \text{and} \quad i, j = 0, 1, 2$$

The function $T_H(x, t)$ is the solution of the following homogeneous problem:

$$\frac{\partial^2 T_H(x,t)}{\partial x^2} = \frac{1}{\alpha}\frac{\partial T_H(x,t)}{\partial t} \quad \text{in} \quad 0 < x < L, \quad t > 0 \tag{14-32a}$$

$$\text{BC1:} \quad -k_1 \left.\frac{\partial T_H}{\partial x}\right|_{x=0} + h_1 \left. T_H\right|_{x=0} = 0 \tag{14-32b}$$

$$\text{BC2:} \quad k_2 \left.\frac{\partial T_H}{\partial x}\right|_{x=L} + h_2 \left. T_H\right|_{x=L} = 0 \tag{14-32c}$$

$$\text{IC:} \quad T_H(x, t=0) = F(x) - \sum_{j=0}^{2} T_{0j}(x, t=0) \quad \text{in} \quad 0 \le x \le L \tag{14-32d}$$

The functions $T_j(x, t)$ are related to the function $\theta_j(x, \tau, t)$ by the following relation:

$$T_j(x,t) = \int_{\tau=0}^{t} \left.\frac{\partial \theta_j(x, \tau', t-\tau)}{\partial \tau'}\right|_{\tau'=\tau} d\tau \tag{14-33}$$

where $\theta_j(x, \tau, t)$ is the solution of the following homogeneous problem:

$$\frac{\partial^2 \theta_j}{\partial x^2} = \frac{1}{\alpha}\frac{\partial \theta_j(x,\tau,t)}{\partial t} \quad \text{in} \quad 0 < x < L, \quad t > 0 \tag{14-34a}$$

$$\text{BC1:} \qquad -k_1 \left. \frac{\partial \theta_j}{\partial x} \right|_{x=0} + h_1 \left. \theta_j \right|_{x=0} = 0 \tag{14-34b}$$

$$\text{BC2:} \qquad k_2 \left. \frac{\partial \theta_j}{\partial x} \right|_{x=L} + h_2 \left. \theta_j \right|_{x=L} = 0 \tag{14-34c}$$

$$\text{IC:} \qquad \theta_j(x, \tau, t = 0) = T_{0j}(x, \tau) \qquad \text{in} \qquad 0 \le x \le L \tag{14-34d}$$

where $j = 0, 1, 2$. In the following examples we consider some special cases of the one-dimensional problem (14-26).

Example 14-1 One-Dimensional Slab with Generation
Obtain the solution of the following heat conduction problem for a slab by utilizing the general solutions given previously:

$$\frac{\partial^2 T}{\partial x^2} + \frac{1}{k} g\,(x, t) = \frac{1}{\alpha} \frac{\partial T\,(x, t)}{\partial t} \qquad \text{in} \qquad 0 < x < L, \qquad t > 0 \tag{14-35a}$$

$$\text{BC1:} \qquad \left. \frac{\partial T}{\partial x} \right|_{x=0} = 0 \tag{14-35b}$$

$$\text{BC2:} \qquad T(x = L) = 0 \tag{14-35c}$$

$$\text{IC:} \qquad T(t = 0) = F(x) \qquad \text{in} \qquad 0 \le x \le L \tag{14-35d}$$

The solution of this problem is immediately obtainable from the solution (14-29) as

$$T\,(x, t) = \sum_{m=0}^{\infty} \frac{X(\beta_m, x)}{N(\beta_m)} e^{-\alpha \beta_m^2 t} \left[\overline{F}(\beta_m) + \frac{\alpha}{k} \int_{t'=0}^{t} e^{\alpha \beta_m^2 t'} \overline{g}(\beta_m, t')\, dt' \right] \tag{14-36}$$

where the integral transforms $\overline{F}(\beta_m)$ and $\overline{g}(\beta_m, t')$ are defined by equations (14-29c) and (14-29d), respectively. The eigenfunctions $X(\beta_m, x)$, the normalization integral $N(\beta_m)$, and the expression defining the eigenvalues β_m, are obtained from Table 2-1, case 6, as

$$X(\beta_m, x) = \cos \beta_m x, \qquad \frac{1}{N(\beta_m)} = \frac{2}{L} \qquad \text{and} \qquad \cos \beta_m L = 0 \tag{14-37}$$

Introducing equations (14-37) into (14-36), we find

$$T\,(x, t) = \frac{2}{L} \sum_{m=1}^{\infty} e^{-\alpha \beta_m^2 t} \cos \beta_m x \left\{ \int_{x'=0}^{L} F(x') \cos \beta_m x' dx' + \frac{\alpha}{k} \int_{t'=0}^{t} \int_{x'=0}^{L} g(x', t') \cos \beta_m x' e^{\alpha \beta_m^2 t'}\, dx'\, dt' \right\} \tag{14-38}$$

where the β_m values are the positive roots of $\cos \beta_m L = 0$; hence they are given by $\beta_m = (2m-1)\pi/2L$, for $m = 1, 2, 3 \ldots$.

Example 14-2 One-Dimensional Slab with Time-Dependent Boundary Condition

Obtain the solution of the following heat conduction problem for a slab

$$\frac{\partial^2 T}{\partial x^2} = \frac{1}{\alpha}\frac{\partial T(x,t)}{\partial t} \quad \text{in} \quad 0 < x < L, \quad t > 0 \tag{14-39a}$$

$$\text{BC1:} \quad \left.\frac{\partial T}{\partial x}\right|_{x=0} = 0 \tag{14-39b}$$

$$\text{BC2:} \quad T(x = L) = f_2(t) \tag{14-39c}$$

$$\text{IC:} \quad T(t = 0) = 0 \quad \text{in} \quad 0 \le x \le L \tag{14-39d}$$

We now consider the case when the surface temperature is given by $f_2(t) = \gamma t$, where γ is a constant. We solve this problem using both the solution (14-29) and its alternative form (14-30).

Approach 1 The solution of this problem is immediately obtainable from the solution (14-29) as

$$T(x,t) = -\alpha \sum_{m=1}^{\infty} \frac{X(\beta_m, x)}{N(\beta_m)} e^{-\alpha\beta_m^2 t} \int_{t'=0}^{t} e^{\alpha\beta_m^2 t'} \left[\frac{dX(\beta_m, x)}{dx}\right]_{x=L} f_2(t')\, dt' \tag{14-40}$$

where use is made of equation (14-29g) since the boundary condition at $x = L$ is of the first kind. The eigenfunctions $X(\beta_m, x)$, the normalization integral $N(\beta_m)$, and the eigenvalues β_m are the same as those given by equation (14-37). Then the solution (14-40) becomes

$$T(x,t) = \frac{2\alpha}{L} \sum_{m=1}^{\infty} (-1)^{m-1} e^{-\alpha\beta_m^2 t} \beta_m \cos \beta_m x \int_{t'=0}^{t} e^{\alpha\beta_m^2 t'} f_2(t')\, dt' \tag{14-41a}$$

since $\beta_m = (2m-1)\pi/2L$ and $dX/dx|_{x=L} = -\beta_m \sin \beta_m L = -\beta_m (-1)^{m-1}$. For $f_2(t) = \gamma t$, this result reduces to

$$T(x,t) = \frac{2\alpha\gamma}{L} \sum_{m=1}^{\infty} (-1)^{m-1} e^{-\alpha\beta_m^2 t} \beta_m \cos \beta_m x \int_{t'=0}^{t} t' e^{\alpha\beta_m^2 t'}\, dt' \tag{14-41b}$$

The integral term is evaluated as

$$\int_{t'=0}^{t} t' e^{\alpha\beta_m^2 t'}\, dt' = e^{\alpha\beta_m^2 t}\left(\frac{t}{\alpha\beta_m^2} - \frac{1}{\alpha^2\beta_m^4}\right) + \frac{1}{\alpha^2\beta_m^4} \tag{14-42}$$

Then, the solution (14-41b) takes the form

$$T(x,t) = \gamma t \frac{2}{L} \sum_{m=1}^{\infty} (-1)^{m-1} \frac{\cos \beta_m x}{\beta_m} - \frac{\gamma}{\alpha} \frac{2}{L} \sum_{m=1}^{\infty} (-1)^{m-1} \frac{\cos \beta_m x}{\beta_m^3} + \frac{2\gamma}{\alpha L} \sum_{m=1}^{\infty} (-1)^{m-1} e^{-\alpha \beta_m^2 t} \frac{\cos \beta_m x}{\beta_m^3} \tag{14-43}$$

Closed-form expressions for the two series are given as (see note 2 at end of this chapter for the derivation of these closed-form expressions)

$$\frac{2}{L} \sum_{m=1}^{\infty} (-1)^{m-1} \frac{\cos \beta_m x}{\beta_m} = 1 \tag{14-44a}$$

and

$$\frac{2}{L} \sum_{m=1}^{\infty} (-1)^{m-1} \frac{\cos \beta_m x}{\beta_m^3} = -\frac{1}{2}(x^2 - L^2) \tag{14-44b}$$

Introducing these results into equation (14-43), the solution becomes

$$T(x,t) = \gamma t + \frac{\gamma}{2\alpha}(x^2 - L^2) + \frac{2\gamma}{\alpha L} \sum_{m=1}^{\infty} (-1)^{m-1} e^{-\alpha \beta_m^2 t} \frac{\cos \beta_m x}{\beta_m^3} \tag{14-45}$$

Approach 2 We now solve the problem (14-39) by utilizing the alternative form of the solution given by equations (14-30). Then we have

$$T(x,t) = T_{02}(x,t) + T_H(x,t) - T_2(x,t) \tag{14-46}$$

where the function $T_{02}(x,t)$ satisfies the following quasi-steady-state problem:

$$\frac{\partial^2 T_{02}}{\partial x^2} = 0 \quad \text{in} \quad 0 < x < L \tag{14-47a}$$

$$\text{BC1:} \quad \left. \frac{\partial T_{02}}{\partial x} \right|_{x=0} = 0 \tag{14-47b}$$

$$\text{BC2:} \quad T_{02}(x = L) = f_2(t) = \gamma t \tag{14-47c}$$

the solution of which is

$$T_{02}(x,t) = \gamma t \tag{14-48}$$

The function $T_H(x,t)$ satisfies the following homogeneous problem:

$$\frac{\partial^2 T_H}{\partial x^2} = \frac{1}{\alpha} \frac{\partial T_H}{\partial t} \quad \text{in} \quad 0 < x < L, \quad t > 0 \tag{14-49a}$$

$$\text{BC1:} \qquad \left.\frac{\partial T_H}{\partial x}\right|_{x=0} = 0 \tag{14-49b}$$

$$\text{BC2:} \qquad T_H(x = L) = 0 \tag{14-49c}$$

$$\text{IC:} \qquad T_H(t = 0) = -T_{02}(x, t = 0) = 0 \qquad \text{in} \qquad 0 \leq x \leq L \tag{14-49d}$$

which has a trivial solution; hence

$$T_H(x, t) = 0 \tag{14-50}$$

Finally $T_2(x, t)$ is related to the function $\theta_2(x, t)$ by

$$T_2(x, t) = \int_{\tau=0}^{t} \left.\frac{\partial \theta_2(x, \tau', t - \tau)}{\partial \tau'}\right|_{\tau'=\tau} d\tau \tag{14-51}$$

where $\theta_2(x, \tau, t)$ is the solution of the following homogeneous problem:

$$\frac{\partial^2 \theta_2}{\partial x^2} = \frac{1}{\alpha}\frac{\partial \theta_2}{\partial t} \qquad \text{in} \qquad 0 < x < L \qquad t > 0 \tag{14-52a}$$

$$\text{BC1:} \qquad \left.\frac{\partial \theta_2}{\partial x}\right|_{x=0} = 0 \tag{14-52b}$$

$$\text{BC2:} \qquad \theta_2(x = L) = 0 \tag{14-52c}$$

$$\text{IC:} \qquad \theta_2(t = 0) = T_{02}(x, \tau) = \gamma\tau \qquad \text{in} \qquad 0 \leq x \leq L \tag{14-52d}$$

When equations (14-52) are solved and $\theta_2(x, \tau, t)$ is introduced into equation (14-51), we obtain

$$T_2(x, t) = \frac{2\gamma}{\alpha L} \sum_{m=1}^{\infty} (-1)^{m-1} (1 - e^{-\alpha\beta_m^2 t}) \frac{\cos \beta_m x}{\beta_m^3} \tag{14-53}$$

Introducing equations (14-48), (14-50), and (14-53) into equation (14-46), we find

$$T(x, t) = \gamma t - \frac{\gamma}{\alpha}\frac{2}{L} \sum_{m=1}^{\infty} (-1)^{m-1} \frac{\cos \beta_m x}{\beta_m^3} + \frac{2\gamma}{\alpha L} \sum_{m=1}^{\infty} (-1)^{m-1} e^{-\alpha\beta_m^2 t} \frac{\cos \beta_m x}{\beta_m^3} \tag{14-54}$$

and when the closed-form expression (14-44b) is introduced, the solution (14-54) becomes

$$T(x, t) = \gamma t - \frac{\gamma}{2\alpha}(x^2 - L^2) + \frac{2\gamma}{\alpha L} \sum_{m=1}^{\infty} (-1)^{m-1} e^{-\alpha\beta_m^2 t} \frac{\cos \beta_m x}{\beta_m^3} \tag{14-55}$$

which is identical to equation (14-45).

One-Dimensional Problems of Semi-Infinite and Infinite Regions

The integral transform technique developed for the solution of heat conduction problems of finite regions is now extended for the solution of problems of semi-infinite regions. Only one of the space variables, the x variable, needs to be considered because the same results are applicable for the solution of problems involving y or z variables.

***Region* $0 \le x \le \infty$.** To illustrate the basic concepts, we consider the solution of the following one-dimensional, time-dependent heat conduction problem for a semi-infinite region:

$$\frac{\partial^2 T}{\partial x^2} + \frac{1}{k} g(x,t) = \frac{1}{\alpha} \frac{\partial T(x,t)}{\partial t} \quad \text{in} \quad 0 < x < \infty \quad t > 0 \tag{14-56a}$$

$$\text{BC:} \quad -k_1 \left. \frac{\partial T}{\partial x} \right|_{x=0} + h_1 \left. T \right|_{x=0} = f_1(t) \tag{14-56b}$$

$$\text{IC:} \quad T(x, t=0) = F(x) \quad \text{in} \quad 0 \le x \le \infty \tag{14-56c}$$

Basic steps in the solution of this problem can be summarized as follows:

1. Develop the appropriate integral transform pair. The integral transform pair is developed by considering the eigenvalue problem appropriate for the problem (14-56) and then representing the function $T(x, t)$, defined in the interval $0 \le x \le \infty$, in terms of the eigenfunctions of this eigenvalue problem, and then by splitting up the representation into two parts as the *inversion formula* and the *integral transform*.
2. Remove the partial derivative $\partial^2 T/\partial x^2$ from the differential equation (14-56a) by the application of the integral transform and utilizing the eigenvalue problem and the boundary conditions for the heat conduction problem.
3. Solve the resulting ordinary differential equation for the transform of temperature subject to the transformed initial condition. Invert the transform of temperature by the inversion formula to obtain the desired solution.

Step 1 is immediately obtainable from the results available in Chapter 6. That is, the eigenvalue problem is given by Table 6-1, and the representation of a function in the region $0 \le x \le \infty$ is given by equations (6-12) and (6-13). Then, the integral-transform pair with respect to the x variable of the function $T(x, t)$ is immediately obtained according to these equations as

$$\text{Inversion formula:} \quad T(x,t) = \int_{\beta=0}^{\infty} \frac{X(\beta, x)}{N(\beta)} \overline{T}(\beta, t)\, d\beta \tag{14-57a}$$

$$\text{Integral transform:} \quad \overline{T}(\beta, t) = \int_{x'=0}^{\infty} X(\beta, x') T(x', t)\, dx' \tag{14-57b}$$

where the functions $X(\beta, x)$ and $N(\beta)$ are listed in Table 6-1 for three different boundary conditions at $x = 0$. We note that the eigenvalues β for a semi-infinite medium are *continuous*, and as a result, the inversion formula is an integral over β from zero to infinity instead of a summation over the discrete eigenvalues as for the finite region.

Step 2 involves taking the integral transform of equation (14-56a) by the application of the transform (14-57b); that is, we multiply both sides of equation (14-56a) by $X(\beta, x)$ and integrate with respect to x from $x = 0$ to ∞ to obtain

$$\int_{x=0}^{\infty} X(\beta, x)\frac{\partial^2 T}{\partial x^2}\,dx + \frac{1}{k}\overline{g}(\beta, t) = \frac{1}{\alpha}\frac{d\overline{T}(\beta, t)}{dt} \tag{14-57c}$$

The integral on the left is performed by integrating it by parts twice and utilizing the eigenvalue problem (2-48) and the boundary condition (14-56b). The reader is referred to note 3 at end of this chapter for the details of this portion of the analysis. Then the resulting equation, and the transform of the initial condition (14-56c), respectively, become

$$\frac{d\overline{T}(\beta, t)}{dt} + \alpha\beta^2\overline{T}(\beta, t) = \frac{\alpha}{k}\overline{g}(\beta, t) + \alpha\left.\frac{X(\beta, x)}{k_1}\right|_{x=0} f_1(t) \qquad \text{for} \qquad t > 0 \tag{14-58a}$$

$$\text{IC:} \qquad \overline{T}(\beta, t = 0) = \overline{F}(\beta) \tag{14-58b}$$

Step 3 is now the solution of equations (14-58) for $\overline{T}(\beta, t)$, and the result is inverted by the inversion formula (14-57b), to obtain the solution as

$$T(x, t) = \int_{\beta=0}^{\infty} \frac{X(\beta, x)}{N(\beta)} e^{-\alpha\beta_m^2 t}\left[\overline{F}(\beta) + \int_{t'=0}^{t} e^{\alpha\beta^2 t'} A(\beta, t')\,dt'\right] \tag{14-59a}$$

where

$$A(\beta, t') = \frac{\alpha}{k}\overline{g}(\beta, t') + \alpha\left.\frac{X(\beta, x')}{k_1}\right|_{x'=0} f_1(t') \tag{14-59b}$$

$$\overline{F}(\beta) = \int_{x'=0}^{\infty} X(\beta, x')F(x')\,dx' \tag{14-59c}$$

$$\overline{g}(\beta, t') = \int_{x'=0}^{\infty} X(\beta, x')g(x', t')\,dx' \tag{14-59d}$$

If the boundary condition at $x = 0$ is of the first type (i.e., $k_1 = 0$), the following change should be made in the term $A(\beta, t')$:

$$\text{Replace} \qquad \left.\frac{X(\beta, x')}{k_1}\right|_{x'=0} \qquad \text{by} \qquad \left.\frac{dX(\beta, x')}{dx'}\right|_{x'=0} \tag{14-59e}$$

The functions $X(\beta, x)$ and $N(\beta)$ are obtainable from Table 6-1 for three different boundary conditions at $x = 0$.

Region $-\infty < x < \infty$. We now consider the following heat conduction problem for an infinite medium:

$$\frac{\partial^2 T}{\partial x^2} + \frac{1}{k} g(x,t) = \frac{1}{\alpha} \frac{\partial T(x,t)}{\partial t} \quad \text{in} \quad -\infty < x < \infty, \quad t > 0 \tag{14-60a}$$

$$\text{IC:} \quad T(x, t = 0) = F(x) \quad \text{in the region} \tag{14-60b}$$

The eigenvalue problem appropriate for the solution of this problem is given by equation (6-124), and the representation of a function $F^*(x)$, defined in the interval $-\infty < x < \infty$, in terms of the eigenfunctions of this eigenvalue problem, is given by equations (6-126) and (6-127) as

$$F^*(x) = \frac{1}{\pi} \int_{\beta=0}^{\infty} \int_{x'=-\infty}^{\infty} F^*(x') \cos \beta(x' - x)\, dx'\, d\beta \tag{14-61}$$

This representation is expressed in the alternative form (see note 4 at the end of this chapter for the derivation) as

$$F^*(x) = \frac{1}{2\pi} \int_{\beta=-\infty}^{\infty} e^{-i\beta x} \left[\int_{x'=-\infty}^{\infty} e^{i\beta x'} F^*(x')\, dx' \right] d\beta \tag{14-62}$$

where we define $i = \sqrt{-1}$.

This expression is now utilized to define the integral transform pair for the temperature $T(x, t)$ with respect to the x variable as

$$\text{Inversion formula} \quad T(x,t) = \frac{1}{2\pi} \int_{\beta=-\infty}^{\infty} e^{-i\beta x} \overline{T}(\beta, t)\, d\beta \tag{14-63a}$$

$$\text{Integral transform} \quad \overline{T}(\beta, t) = \int_{x'=-\infty}^{\infty} e^{i\beta x'} T(x', t)\, dx' \tag{14-63b}$$

Taking the integral transform of the heat conduction problem (14-60) according to the transform (14-63b), we obtain

$$\frac{d\overline{T}(\beta,t)}{dt} + \alpha\beta^2 \overline{T}(\beta,t) = \frac{\alpha}{k} \overline{g}(\beta,t) \quad \text{for} \quad t > 0 \tag{14-64a}$$

$$\text{IC:} \quad \overline{T}(\beta, t = 0) = \overline{F}(\beta) \tag{14-64b}$$

When this equation is solved for $\overline{T}(\beta, t)$ and the result is inverted by the inversion formula (14-63a), we obtain the solution of the heat conduction problem (14-60) as

$$T(x,t) = \frac{1}{2\pi} \int_{\beta=-\infty}^{\infty} e^{-\alpha\beta^2 t - i\beta x} \left[\overline{F}(\beta) + \frac{\alpha}{k} \int_{t'=0}^{t} e^{\alpha\beta^2 t'} \overline{g}(\beta, t')\, dt' \right] d\beta \tag{14-65a}$$

where we have

$$\overline{F}(\beta) = \int_{x'=-\infty}^{\infty} e^{i\beta x'} F(x')\, dx' \tag{14-65b}$$

$$\overline{g}(\beta, t') = \int_{x'=-\infty}^{\infty} e^{i\beta x'} g(x', t')\, dx' \tag{14-65c}$$

The order of integration is changed, and the result is rearranged as

$$T(x,t) = \frac{1}{2\pi} \int_{x'=-\infty}^{\infty} F(x') \int_{\beta=-\infty}^{\infty} e^{-\alpha\beta^2 t - i\beta(x-x')}\, d\beta\, dx' + \frac{1}{2\pi}\frac{\alpha}{k} \int_{t'=0}^{t} \int_{x'=-\infty}^{\infty} g(x',t') \int_{\beta=-\infty}^{\infty} e^{-\alpha\beta^2(t-t') - i\beta(x-x')}\, d\beta\, dx'\, dt' \tag{14-66}$$

We now make use of the following integral:

$$\frac{1}{2\pi} \int_{\beta=-\infty}^{\infty} e^{-\alpha\beta^2 t - i\beta x}\, d\beta = \frac{1}{(4\pi\alpha t)^{1/2}} e^{-x^2/4\alpha t} \tag{14-67}$$

and the solution (14-66) becomes

$$T(x,t) = \frac{1}{(4\pi\alpha t)^{1/2}} \int_{x'=-\infty}^{\infty} \exp\left[-\frac{(x-x')^2}{4\alpha t}\right] F(x')\, dx' + \frac{\alpha}{k} \int_{t'=0}^{t} \frac{1}{[4\pi\alpha(t-t')]^{1/2}} \int_{x'=-\infty}^{\infty} \exp\left[-\frac{(x-x')^2}{4\alpha(t-t')}\right] g(x',t')\, dx'\, dt' \tag{14-68}$$

Example 14-3 Semi-Infinite Slab with Generation

We seek the solution of the following heat conduction problem for a semi-infinite region:

$$\frac{\partial^2 T}{\partial x^2} + \frac{1}{k} g(x,t) = \frac{1}{\alpha} \frac{\partial T(x,t)}{\partial t} \quad \text{in} \quad 0 < x < \infty, \quad t > 0 \tag{14-69a}$$

$$\text{BC:} \quad \left.\frac{\partial T}{\partial x}\right|_{x=0} = 0 \tag{14-69b}$$

$$\text{IC:} \quad T(t=0) = F(x) \quad \text{in} \quad 0 \le x < \infty \tag{14-69c}$$

The solution of this problem is immediately obtainable from equations (14-59) as

$$T(x,t) = \int_{\beta=0}^{\infty} \frac{X(\beta,x)}{N(\beta)} e^{-\alpha\beta^2 t} \left[\overline{F}(\beta) + \frac{\alpha}{k} \int_{t'=0}^{t} e^{\alpha\beta^2 t'} \overline{g}(\beta,t')\, dt' \right] d\beta \tag{14-70a}$$

where we define

$$\overline{F}(\beta) = \int_{x'=0}^{\infty} X(\beta, x') F(x')\, dx' \tag{14-70b}$$

$$\overline{g}(\beta, t') = \int_{x'=0}^{\infty} X(\beta, x') g(x', t')\, dx' \tag{14-70c}$$

The functions $X(\beta, x)$ and $N(\beta)$ are determined from case 2, Table 6-1 as

$$X(\beta, x) = \cos \beta x \qquad \text{and} \qquad \frac{1}{N(\beta)} = \frac{2}{\pi} \tag{14-71}$$

Introducing equations (14-71) into (14-70), the solution becomes

$$\begin{aligned} T(x, t) &= \frac{2}{\pi} \int_{\beta=0}^{\infty} e^{-\alpha\beta^2 t} \cos \beta x \int_{x'=0}^{\infty} F(x') \cos \beta x'\, dx'\, d\beta \\ &+ \frac{2\alpha}{\pi k} \int_{\beta=0}^{\infty} e^{-\alpha\beta^2 t} \cos \beta x \int_{t'=0}^{t} \int_{x'=0}^{\infty} g(x', t') \cos \beta x' e^{\alpha\beta^2 t'}\, dx'\, dt'\, d\beta \end{aligned} \tag{14-72}$$

In this expression, the orders of integration can be changed and the integrations with respect to β can be performed by making use of the following relation, which we obtained by adding equations (2-57b) and (2-57c), namely,

$$\begin{aligned} &\frac{2}{\pi} \int_{\beta=0}^{\infty} e^{-\alpha\beta^2 t} \cos \beta x \cos \beta x'\, d\beta \\ &= \frac{1}{(4\pi\alpha t)^{1/2}} \left\{ \exp\left[-\frac{(x - x')^2}{4\alpha t} \right] + \exp\left[-\frac{(x + x')^2}{4\alpha t} \right] \right\} \end{aligned} \tag{14-73}$$

Then, the solution (14-72) takes the form

$$\begin{aligned} T(x, t) &= \frac{1}{(4\pi\alpha t)^{1/2}} \int_{x'=0}^{\infty} F(x') \left\{ \exp\left[-\frac{(x - x')^2}{4\alpha t} \right] + \exp\left[-\frac{(x + x')^2}{4\alpha t} \right] \right\} dx' \\ &+ \frac{\alpha}{k} \int_{t'=0}^{t} \frac{1}{[4\pi\alpha(t - t')]^{1/2}} \int_{x'=0}^{\infty} g(x', t') \\ &\cdot \left\{ \exp\left[-\frac{(x - x')^2}{4\alpha(t - t')} \right] + \exp\left[-\frac{(x + x')^2}{4\alpha(t - t')} \right] \right\} dx'\, dt' \end{aligned} \tag{14-74}$$

Several special cases are obtainable from this solution depending on the functional forms of the heat generation term and the initial condition function.

Multidimensional Problems

The solution of multidimensional, time-dependent heat conduction problems by the integral transform technique is readily handled by the successive application of one-dimensional integral transforms to remove from the equation one of the partial derivatives with respect to the space variable in each step. In the rectangular coordinate system, the order of the integral transformation with respect to the space variables is immaterial. This matter is now illustrated with examples.

Example 14-4 Two-Dimensional Semi-Infinite Strip with Generation Obtain the solution $T(x, y, t)$ of the following heat conduction problem for a semi-infinite rectangular strip, $0 \le x < \infty$, $0 \le y \le b$:

$$\frac{\partial^2 T}{\partial x^2} + \frac{\partial^2 T}{\partial y^2} + \frac{g(x,y,t)}{k} = \frac{1}{\alpha}\frac{\partial T}{\partial t} \quad \text{in} \quad 0 < x < \infty \quad 0 < y < b \quad t > 0 \tag{14-75a}$$

$$\text{BC:} \quad T = 0 \quad \text{at} \quad \text{all boundaries} \tag{14-75b}$$

$$\text{IC:} \quad T(t = 0) = 0 \quad \text{in the region} \tag{14-75c}$$

The integral transform pair for $T(x, y, t)$ with respect to the x variable is defined, see equations (14-57), as

$$\text{Inversion formula} \qquad T(x,y,t) = \int_{\beta=0}^{\infty} \frac{X(\beta,x)}{N(\beta)} \overline{T}(\beta,y,t)\, d\beta \tag{14-76a}$$

$$\text{Integral transform} \qquad \overline{T}(\beta,y,t) = \int_{x'=0}^{\infty} X(\beta,x') T(x',y,t)\, dx' \tag{14-76b}$$

and the integral transform pair for $\overline{T}(\beta, y, t)$ with respect to the y variable is defined, see equation (14-27), as

$$\text{Inversion formula} \qquad \overline{T}(\beta,y,t) = \sum_{n=1}^{\infty} \frac{Y(\gamma_n,y)}{N(\gamma_n)} \tilde{\overline{T}}(\beta,\gamma_n,t) \tag{14-76c}$$

$$\text{Integral transform} \qquad \tilde{\overline{T}}(\beta,\gamma_n,t) = \int_{y'=0}^{b} Y(\gamma_n,y') \overline{T}(\beta,y',t)\, dy' \tag{14-76d}$$

where the overbar denotes the transform with respect to the x variable and the tilde with respect to the y variable.

We now take the integral transform of the problem (14-75) first with respect to the x variable using the transform (14-76b) and then with respect to the y variable using the transform (14-76d) to obtain

$$\frac{d\tilde{\overline{T}}}{dt} + \alpha(\beta^2 + \gamma_n^2)\tilde{\overline{T}}(\beta,\gamma_n,t) = \frac{\alpha}{k}\tilde{\overline{g}}(\beta,\gamma_n,t) \qquad \text{for} \qquad t > 0 \tag{14-77a}$$

$$\text{IC:} \qquad \tilde{\overline{T}}(\beta,\gamma_n,t=0) = 0 \tag{14-77b}$$

Equation (14-77) is solved for $\tilde{\bar{T}}$, and successively inverted by the inversion formulas (14-76c) and (14-76a), to find the solution of the problem (14-75) as

$$T(x, y, t) = \int_{\beta=0}^{\infty} \sum_{n=1}^{\infty} \frac{X(\beta, x)Y(\gamma_n, y)}{N(\beta)N(\gamma_n)} e^{-\alpha(\beta^2+\gamma_n^2)t} \cdot \frac{\alpha}{k} \int_{t'=0}^{t} e^{\alpha(\beta^2+\gamma_n^2)t'} \tilde{\bar{g}}(\beta, \gamma_n, t')\, dt'\, d\beta \tag{14-78a}$$

where the double transform $\tilde{\bar{g}}(\beta, \gamma_n, t)$ is defined as

$$\tilde{\bar{g}}(\beta, \gamma_n, t') = \int_{y'=0}^{b} \int_{x'=0}^{\infty} X(\beta, x')Y(\gamma_n, y')g(x', y', t')\, dx' dy' \tag{14-78b}$$

The functions $X(\beta, x)$ and $N(\beta)$ are obtained from case 3, Table 6-1, as

$$X(\beta, x) = \sin \beta x \qquad \frac{1}{N(\beta)} = \frac{2}{\pi} \tag{14-79a}$$

and the functions $Y(\gamma_n, y)$ and $N(\gamma_n)$ are obtained from case 9, Table 2-1, as

$$Y(\gamma_n, y) = \sin \gamma_n y, \qquad \frac{1}{N(\gamma_n)} = \frac{2}{b} \tag{14-79b}$$

and the γ_n values are the positive roots of $\sin \gamma_n b = 0$. Introducing the results (14-79) into equation (14-78), the solution becomes

$$T(x, y, t) = \frac{4\alpha}{\pi bk} \int_{\beta=0}^{\infty} \sum_{n=1}^{\infty} e^{-\alpha(\beta^2+\gamma_n^2)t} \sin \beta x \sin \gamma_n y \cdot \int_{t'=0}^{t} e^{\alpha(\beta^2+\gamma_n^2)t'} \int_{y'=0}^{b} \int_{x'=0}^{\infty} g(x', y', t') \sin \beta x' \sin \gamma_n y'\, dx' dy'\, dt'\, d\beta \tag{14-80}$$

In this solution, the integration with respect to β can be performed by making use of the following result, see equation (2-57d), namely,

$$\frac{2}{\pi} \int_{\beta=0}^{\infty} e^{-\alpha\beta^2(t-t')} \sin \beta x \sin \beta x'\, d\beta = \frac{1}{[4\pi\alpha(t-t')]^{1/2}} \left\{ \exp\left[-\frac{(x-x')^2}{4\alpha(t-t')} \right] - \exp\left[-\frac{(x+x')^2}{4\alpha(t-t')} \right] \right\} \tag{14-81}$$

Then, the solution (14-80) takes the form

$$T(x, y, t) = \frac{2\alpha}{bk} \sum_{n=1}^{\infty} e^{-\alpha\gamma_n^2 t} \sin \gamma_n y \int_{t'=0}^{t} \frac{e^{\alpha\gamma_n^2 t'}}{[4\pi\alpha(t - t')]^{1/2}}$$

$$\cdot \int_{y'=0}^{b} \int_{x'=0}^{\infty} g(x', y', t') \sin \gamma_n y' \tag{14-82}$$

$$\cdot \left\{ \exp\left[-\frac{(x - x')^2}{4\alpha(t - t')} \right] - \exp\left[-\frac{(x + x')^2}{4\alpha(t - t')} \right] \right\} dx' dy' dt'$$

Example 14-5 Three-Dimensional Finite Domain with Generation
Obtain the solution $T(x, y, z, t)$ of the following heat conduction problem for a rectangular parallelepiped of domain $0 \le x \le a, 0 \le y \le b, 0 \le z \le c$, given by

$$\frac{\partial^2 T}{\partial x^2} + \frac{\partial^2 T}{\partial y^2} + \frac{\partial^2 T}{\partial z^2} + \frac{g(x, y, z, t)}{k} = \frac{1}{\alpha} \frac{\partial T}{\partial t} \quad \text{in} \quad \begin{matrix} 0 < x < a, & 0 < y < b \\ 0 < z < c, & t > 0 \end{matrix} \tag{14-83a}$$

$$\text{BC:} \quad T = 0 \quad \text{at} \quad \text{all boundaries} \tag{14-83b}$$

$$\text{IC:} \quad T(t = 0) = 0 \quad \text{in the region} \tag{14-83c}$$

This problem can be solved by the successive application of the one-dimensional integral transform to the x, y, and z variables, solving the resulting ordinary differential equation, and then inverting the transform of temperature successively. It is also possible to write the solution immediately from the general solution (14-15) by setting the following:

$$\psi_m(\hat{r}) \to X(\beta_m, x)Y(\gamma_n, y)Z(\eta_p, z)$$

$$\lambda_m^2 \to (\beta_m^2 + \gamma_n^2 + \eta_p^2)$$

$$\sum_m \to \sum_{m=1}^{\infty} \sum_{n=1}^{\infty} \sum_{p=1}^{\infty} \quad \text{and} \quad \int_{R'} dV' \to \int_{x'=0}^{a} \int_{y'=0}^{b} \int_{z'=0}^{c} dx'\, dy'\, dz'$$

We then obtain

$$T(x, y, z, t) = \sum_{m=1}^{\infty} \sum_{n=1}^{\infty} \sum_{p=1}^{\infty} \frac{X(\beta_m, x)Y(\gamma_n, y)Z(\eta_p, z)}{N(\beta_m)N(\gamma_n)N(\eta_p)} e^{-\alpha(\beta_m^2 + \gamma_n^2 + \eta_p^2)t}$$

$$\cdot \frac{\alpha}{k} \int_{t'=0}^{t} e^{\alpha(\beta_m^2 + \gamma_n^2 + \eta_p^2)t'} \overline{\overline{\overline{g}}}(\beta_m, \gamma_n, \eta_p, t')\, dt' \tag{14-84a}$$

where the triple transform is defined as

$$\bar{\bar{\bar{g}}}(\beta_m, \gamma_n, \eta_p, t') = \int_{x'=0}^{a} \int_{y'=0}^{b} \int_{z'=0}^{c} X(\beta_m, x') Y(\gamma_n, y') Z(\eta_p, z') \cdot g(x', y', z', t')\, dx'\, dy'\, dz' \tag{14-84b}$$

The eigenfunctions, the normalization integrals, and the eigenvalues are obtained from case 9, Table 2-1, as

$$X(\beta_m, x) = \sin \beta_m x, \qquad N(\beta_m) = \frac{a}{2}, \qquad \sin \beta_m a = 0 \tag{14-85a}$$

$$Y(\gamma_n, y) = \sin \gamma_n y, \qquad N(\gamma_n) = \frac{b}{2}, \qquad \sin \gamma_n b = 0 \tag{14-85b}$$

$$Z(\eta_p, z) = \sin \eta_p z, \qquad N(\eta_p) = \frac{c}{2}, \qquad \sin \eta_p c = 0 \tag{14-85c}$$

14-3 APPLICATIONS IN THE CYLINDRICAL COORDINATE SYSTEM

To solve the heat conduction problems in the cylindrical coordinate system with the integral transform technique, appropriate integral transform pairs are needed in the r, ϕ, and z variables. The integral transform pairs for the z variable depends on whether the range of z is finite, semi-infinite, or infinite as well as the boundary conditions associated with it. Since the transform pairs for the z variable are exactly the same as those discussed previously for the rectangular coordinate system, this matter is not considered here any further. Therefore, in this section we develop the integral transform pairs for the r and ϕ variables and illustrate their application to the solution of heat conduction problems involving (r, t), (r, ϕ, t), (r, z, t), and (r, ϕ, z, t) variables.

Problems in (r, t) Variables

The one-dimensional, time-dependent heat conduction problems in the r variable may be confined to any one of the regions $0 \le r \le b, a \le r \le b, 0 \le r < \infty$, and $a \le r < \infty$. The integral transform pair for each of these cases is different. Therefore, we develop the appropriate transform pairs and illustrate the methods of solution for each of these cases.

Problems of Region $0 \le r \le b$ We consider the following heat conduction problem for a solid cylinder of radius $r = b$:

$$\frac{\partial^2 T}{\partial r^2} + \frac{1}{r}\frac{\partial T}{\partial r} + \frac{g(r,t)}{k} = \frac{1}{\alpha}\frac{\partial T}{\partial t} \quad \text{in} \quad 0 \le r < b, \qquad t > 0 \tag{14-86a}$$

$$\text{BC:} \qquad k_2 \left. \frac{\partial T}{\partial r} \right|_{r=b} + h_2 \, T|_{r=b} = f_2(t) \tag{14-86b}$$

$$\text{IC:} \qquad T(t=0) = F(r) \qquad \text{in} \qquad 0 \le r \le b \tag{14-86c}$$

The appropriate eigenvalue problem is given by Table 2-2 for the case $\nu = 0$ since the problem considered here possesses azimuthal symmetry. The integral transform pair with respect to the r variable for the function $T(r, t)$ is determined according to equations (4-149) and (4-150) by setting $\nu = 0$ and then letting $J_0(\beta_m, r) \to R_0(\beta_m, r)$. We obtain

$$\text{Inversion formula} \qquad T(r,t) = \sum_{m=1}^{\infty} \frac{R_0(\beta_m, r)}{N(\beta_m)} \overline{T}(\beta_m, t) \tag{14-87a}$$

$$\text{Integral transform} \qquad \overline{T}(\beta_m, t) = \int_{r'=0}^{b} r' R_0(\beta_m, r) T(r', t) \, dr' \tag{14-87b}$$

where the functions $R_0(\beta_m, r)$, $N(\beta_m)$, and the eigenvalues β_m are obtainable from Table 2-2 for three different boundary conditions after setting $\nu = 0$.

To solve problem (14-86), we take integral transform of equation (14-86a) according to the transform (14-87b). That is, we operate on both sides of equation (14-86a) by the operator $* \int_{r=0}^{b} r R_0(\beta_m, r) \, dr$, noting the weighting function of r, and obtain

$$\int_{r=0}^{b} r R_0(\beta_m, r) \left(\frac{\partial^2 T}{\partial r^2} + \frac{1}{r} \frac{\partial T}{\partial r} \right) dr + \frac{1}{k} \overline{g}(\beta_m, t) = \frac{1}{\alpha} \frac{d\overline{T}(\beta, t)}{dt} \tag{14-88}$$

The integral on the left is evaluated either by integrating it by parts twice or by using Green's theorem and then utilizing the boundary conditions of case 1 in Table 2-2 for $\nu = 0$ and (14-86b); we find

$$\int_{r=0}^{b} r R_0(\beta_m, r) \left(\frac{\partial^2 T}{\partial r^2} + \frac{1}{r} \frac{\partial T}{\partial r} \right) dr = -\beta_m^2 \overline{T}(\beta_m, r) + b \left. \frac{R_0(\beta_m, r)}{k_2} \right|_{r=b} f_2(t) \tag{14-89}$$

Introducing this expression into equation (14-88), and taking the integral transform of the initial condition (14-86c), we obtain

$$\frac{d\overline{T}(\beta_m, t)}{dt} + \alpha \beta_m^2 \overline{T}(\beta_m, t) = \frac{\alpha}{k} \overline{g}(\beta_m, t) + \alpha b \left. \frac{R_0(\beta_m, r)}{k_2} \right|_{r=b} f_2(t) \tag{14-90a}$$

$$\text{IC:} \qquad \overline{T}(\beta_m, t=0) = \overline{F}(\beta) \tag{14-90b}$$

Equation (14-90) is solved for $\overline{T}(\beta_m, t)$ and inverted by the inversion formula (14-87a), to yield the solution of the problem (14-86) as

$$T(r,t) = \sum_{m=1}^{\infty} \frac{R_0(\beta_m, r)}{N(\beta_m)} e^{-\alpha\beta_m^2 t} \left[\overline{F}(\beta_m) + \int_{t'=0}^{t} e^{\alpha\beta_m^2 t'} A(\beta_m, t')\, dt' \right] \tag{14-91a}$$

where we define

$$A(\beta_m, t') = \frac{\alpha}{k} \overline{g}(\beta_m, t') + \alpha b \left. \frac{R_0(\beta_m, r)}{k_2} \right|_{r=b} f_2(t') \tag{14-91b}$$

$$\overline{F}(\beta_m) = \int_{r'=0}^{b} r' R_0(\beta_m, r') F(r')\, dr' \tag{14-91c}$$

$$\overline{g}(\beta_m, t') = \int_{r'=0}^{b} r' R_0(\beta_m, r') g(r', t')\, dr' \tag{14-91d}$$

$$N(\beta_m) = \int_{r'=0}^{b} r' [R_0(\beta_m, r')]^2\, dr' \tag{14-91e}$$

Here, $R_0(\beta_m, r)$, $N(\beta_m)$, and β_m are obtained from Table 2-2 by setting $\nu = 0$. For a boundary condition of the first type at $r = b$, the following change should be made in equation (14-91b):

$$\text{Replace} \quad \left. \frac{R_0(\beta_m, r)}{k_2} \right|_{r=b} \quad \text{by} \quad -\left. \frac{dR_0(\beta_m, r)}{dr} \right|_{r=b} \tag{14-91f}$$

Problems of Region $a \le r \le b$ We now consider the heat conduction problem for a hollow cylinder $a \le r \le b$ given as

$$\frac{\partial^2 T}{\partial r^2} + \frac{1}{r}\frac{\partial T}{\partial r} + \frac{g(r,t)}{k} = \frac{1}{\alpha}\frac{\partial T}{\partial t} \quad \text{in} \quad a < r < b, \quad t > 0 \tag{14-92a}$$

$$\text{BC1:} \quad -k_1 \left. \frac{\partial T}{\partial r} \right|_{r=a} + h_1 \left. T \right|_{r=a} = f_1(t) \tag{14-92b}$$

$$\text{BC2:} \quad k_2 \left. \frac{\partial T}{\partial r} \right|_{r=b} + h_2 \left. T \right|_{r=b} = f_2(t) \tag{14-92c}$$

$$\text{IC:} \quad T(t=0) = F(r) \quad \text{in} \quad a \le r \le b \tag{14-92d}$$

The eigenvalue problem is given by Table 2-3 for $\nu = 0$, and the integral transform pair is obtained according to equations (4-183) and (4-184) following substitution for the correct eigenfunctions per Table 2-3. We find

Inversion formula:
$$T(r,t)=\sum_{m=1}^{\infty}\frac{R_0(\beta_m,r)}{N(\beta_m)}\overline{T}(\beta_m,t) \tag{14-93a}$$

Integral transform:
$$\overline{T}(\beta_m,t)=\int_{r'=a}^{b} r'R_0(\beta_m,r')T(r',t)\,dr' \tag{14-93b}$$

where the functions $R_0(\beta_m,r)$, $N(\beta_m)$, and the eigenvalues β_m are obtainable directly from Table 2-3 by setting $v=0$ for any combination of boundary conditions of the first and second types.

We now take the integral transform of the system (14-92) by the application of the transform (14-93b), utilize the eigenvalue problem of Table 2-3 for $v=0$ as described previously, solve for the transform of temperature, and invert the result by the inversion formula (14-93a) to obtain the solution for the temperature as

$$T(r,t)=\sum_{m=1}^{\infty}\frac{R_0(\beta_m,r)}{N(\beta_m)}e^{-\alpha\beta_m^2 t}\left[\overline{F}(\beta_m)+\int_{t'=0}^{t}e^{\alpha\beta_m^2 t'}A(\beta_m,t')\,dt'\right] \tag{14-94a}$$

where we define

$$A(\beta_m,t')=\frac{\alpha}{k}\overline{g}(\beta_m,t')+\alpha\left[a\left.\frac{R_0(\beta_m,r)}{k_1}\right|_{r=a}f_1(t')+b\left.\frac{R_0(\beta_m,r)}{k_2}\right|_{r=b}f_2(t')\right] \tag{14-94b}$$

$$\overline{F}(\beta_m)=\int_{r'=a}^{b} r'R_0(\beta_m,r')F(r')\,dr' \tag{14-94c}$$

$$\overline{g}(\beta_m,t')=\int_{r'=a}^{b} r'R_0(\beta_m,r')g(r',t')\,dr' \tag{14-94d}$$

$$N(\beta_m)=\int_{r'=a}^{b} r'[R_0(\beta_m,r')]^2\,dr' \tag{14-94e}$$

Here, $R_0(\beta_m,r)$, $N(\beta_m)$, and β_m are obtainable from Table 2-3 by setting $v=0$. For a boundary condition of the first type, the following changes should be made in equation (14-94b):

When $k_1=0$, replace $\left.\frac{R_0(\beta_m,r)}{k_1}\right|_{r=a}$ by $\left.\frac{dR_0(\beta_m,r)}{dr}\right|_{r=a}$ (14-94f)

When $k_2=0$, replace $\left.\frac{R_0(\beta_m,r)}{k_2}\right|_{r=b}$ by $-\left.\frac{dR_0(\beta_m,r)}{dr}\right|_{r=b}$ (14-94g)

Problems of Region 0 ≤ r ≤ ∞ We consider the following heat conduction problem for an infinite region $0 \le r < \infty$:

$$\frac{\partial^2 T}{\partial r^2} + \frac{1}{r}\frac{\partial T}{\partial r} + \frac{g(r,t)}{k} = \frac{1}{\alpha}\frac{\partial T}{\partial t} \quad \text{in} \quad 0 \le r < \infty, \quad t > 0 \tag{14-95a}$$

$$\text{IC:} \quad T(t=0) = F(r) \quad \text{in} \quad 0 \le r < \infty \tag{14-95b}$$

The appropriate eigenvalue problem is given by equation (6-158) for $v = 0$, and the integral transform pair is constructed according to the representation given by equations (2-120) and (2-121) for $v = 0$; we obtain

$$\text{Inversion formula} \qquad T(r,t) = \int_{\beta=0}^{\infty} \beta J_0(\beta r)\overline{T}(\beta,t)\,d\beta \tag{14-96a}$$

$$\text{Integral transform} \qquad \overline{T}(\beta,t) = \int_{r'=0}^{\infty} r' J_0(\beta r')T(r',t)\,dr' \tag{14-96b}$$

We take the integral transform of the system (14-95) by the application of the transform (14-96b), utilize the eigenvalue problem (6-158), see Example 6-8, for $v = 0$ as discussed previously, solve for the transform of the temperature, and invert the result by the inversion formula (14-96a). We obtain

$$T(r,t) = \int_{\beta=0}^{\infty} \beta J_0(\beta r) e^{-\alpha\beta^2 t}\left[\overline{F}(\beta) + \frac{\alpha}{k}\int_{t'=0}^{t} e^{\alpha\beta^2 t'}\overline{g}(\beta,t')\,dt'\right]d\beta \tag{14-97a}$$

where we have

$$\overline{F}(\beta) = \int_{r'=0}^{\infty} r' J_0(\beta r')F(r')\,dr' \tag{14-97b}$$

$$\overline{g}(\beta,t') = \int_{r'=0}^{\infty} r' J_0(\beta r')g(r',t')\,dr' \tag{14-97c}$$

Introducing equations (14-97b,c) into equation (14-97a), and changing the order of integrations, we find

$$\begin{aligned} T(r,t) &= \int_{r'=0}^{\infty} r' F(r')\left[\int_{\beta=0}^{\infty} e^{-\alpha\beta^2 t}\beta J_0(\beta r)J_0(\beta r')\,d\beta\right]dr' \\ &+ \frac{\alpha}{k}\int_{r'=0}^{\infty}\int_{t'=0}^{t} r' g(r',t')\left[\int_{\beta=0}^{\infty} e^{-\alpha\beta^2(t-t')}\beta J_0(\beta r)J_0(\beta r')\,d\beta\right]dt'\,dr' \end{aligned} \tag{14-98}$$

We now consider the following integral [36, p. 395], namely,

$$\int_{\beta=0}^{\infty} e^{-\alpha\beta^2 t}\beta J_v(\beta r)J_v(\beta r')\,d\beta = \frac{1}{2\alpha t}\exp\left(-\frac{r^2+r'^2}{4\alpha t}\right) I_v\left(\frac{rr'}{2\alpha t}\right) \tag{14-99}$$

By setting $v = 0$ in equation (14-99) and introducing the resulting expression into equation (14-98), we obtain

$$T(r,t) = \frac{1}{2\alpha t}\int_{r'=0}^{\infty} r' \exp\left(-\frac{r^2 + r'^2}{4\alpha t}\right) F(r') I_0\left(\frac{rr'}{2\alpha t}\right) dr'$$

$$+ \frac{1}{2k}\int_{r'=0}^{\infty}\int_{t'=0}^{t} \frac{r'}{t-t'} \exp\left[-\frac{r^2+r'^2}{4\alpha(t-t')}\right] g(r',t') I_0\left[\frac{rr'}{2\alpha(t-t')}\right] dt'\, dr' \tag{14-100}$$

Problems of Region $a \le r < \infty$ We finally consider the following heat conduction problem for a semi-infinite region $a \le r < \infty$:

$$\frac{\partial^2 T}{\partial r^2} + \frac{1}{r}\frac{\partial T}{\partial r} + \frac{g(r,t)}{k} = \frac{1}{\alpha}\frac{\partial T}{\partial t} \quad \text{in} \quad a < r < \infty, \quad t > 0 \tag{14-101a}$$

$$\text{BC:} \quad -k_1 \left.\frac{\partial T}{\partial r}\right|_{r=a} + h_1 \left.T\right|_{r=a} = f_1(t) \tag{14-101b}$$

$$\text{IC:} \quad T(t=0) = F(r) \quad \text{in} \quad a \le r < \infty \tag{14-101c}$$

The eigenvalue problem is given by Table 2-4, and the desired integral transform pair is obtained according to the equations (6-177) and (6-178) as

$$\text{Inversion formula} \qquad T(r,t) = \int_{\beta=0}^{\infty} \frac{\beta}{N(\beta)} R_0(\beta,r)\overline{T}(\beta,t)\, d\beta \tag{14-102a}$$

$$\text{Integral transform} \qquad \overline{T}(\beta,t) = \int_{r'=a}^{\infty} r' R_0(\beta,r') T(r',t)\, dr' \tag{14-102b}$$

where the functions $R_0(\beta,r)$ and $N(\beta)$ are available from Table 2-4 for the three different boundary conditions at $r = a$. The problem is solved by taking the integral transform of the system (14-101) according to the transform (14-102b), utilizing the eigenvalue problem as discussed above, solving for the transform of the temperature, and inverting the transform by the inversion formula (14-102a). We obtain

$$T(r,t) = \int_{\beta=0}^{\infty} \frac{\beta}{N(\beta)} R_0(\beta,r) e^{-\alpha\beta^2 t}\left[\overline{F}(\beta) + \int_{t'=0}^{\infty} e^{\alpha\beta^2 t'} A(\beta,t')\, dt'\right] d\beta \tag{14-103a}$$

where we define

$$A(\beta,t') = \frac{\alpha}{k}\overline{g}(\beta,t') + \alpha a \left.\frac{R_0(\beta,r)}{k_1}\right|_{r=a} f_1(t') \tag{14-103b}$$

$$\overline{F}(\beta) = \int_{r'=a}^{\infty} r' R_0(\beta,r') F(r')\, dr' \tag{14-103c}$$

$$\overline{g}(\beta,t') = \int_{r'=a}^{\infty} r' R_0(\beta,r') g(r',t')\, dr' \tag{14-103d}$$

Problems in (r, ϕ, t) Variables

When the partial derivatives with respect to the r and ϕ variables are to be removed from the heat conduction equation, the order of integral transformation is important. *It should be applied first with respect to the ϕ variable and then to the r variable*. Therefore, we need the integral transform pair that will remove from the differential equation the partial derivative with respect to the ϕ variable, that is, $\partial^2 T/\partial\phi^2$. The ranges of the ϕ variable in the cylindrical coordinate system include $0 \le \phi \le 2\pi$ as in the case of problems of *full cylinder,* and $0 \le \phi \le \phi_0$, for $\phi_0 < 2\pi$, as in the case of problems of a *portion of the cylinder*. The integral transform pairs for each of these two situations are different. Therefore, we first develop the integral-transform pairs with respect to the ϕ variable for these two cases and then present its application in the solution of heat conduction problems.

Transform Pair for $0 \le \phi \le 2\pi$ In this case since the region in the ϕ variable is a full circle, no boundary conditions are specified in ϕ except the requirement that the function should be cyclic with a period of 2π. The appropriate eigenvalue problem in ϕ is given by equations (4-132) and (4-133), and the representation of a function in the interval $0 \le \phi \le 2\pi$ in terms of the eigenfunctions of this eigenvalue problem is given by equation (4-35). Therefore, the integral transform pair with respect to the ϕ variable for the function $T(r, \phi, t)$ is obtained by combining the representation of equations (4-35) – (4-38) using a trigonometric substitution, giving

Inversion formula
$$T(r, \phi, t) = \frac{1}{\pi} \sum_{v} \overline{T}(r, v, t) \tag{14-104a}$$

Integral transform
$$\overline{T}(r, v, t) = \int_{\phi'=0}^{2\pi} \cos v(\phi - \phi') T(r, \phi', t)\, d\phi' \tag{14-104b}$$

where $v = 0, 1, 2, 3 \ldots$, and we replace π by 2π for the case of $v = 0$.

Transform Pair for $0 \le \phi \le \phi_0$ ($\phi_0 < 2\pi$) The region being a portion of a circle, boundary conditions are needed at $\phi = 0$ and $\phi = \phi_0$. Here, we consider boundary conditions of the *first* and *second* type only. For example, for boundary conditions of the first type at both boundaries, $\phi = 0$ and $\phi = \phi_0$, the eigenvalue problem for the ϕ variable is given by

$$\frac{d^2\Phi(\phi)}{d\phi^2} + v^2\Phi(\phi) = 0 \quad \text{in} \quad 0 < \phi < \phi_0 \ (< 2\pi) \tag{14-105a}$$

$$\text{BC1:} \quad \Phi(\phi = 0) = 0 \tag{14-105b}$$

$$\text{BC2:} \quad \Phi\left(\phi = \phi_0\right) = 0 \tag{14-105c}$$

which is exactly of the same form as that given by Table 2-1 for the one-dimensional finite region $0 \le x \le L$ in the rectangular coordinate system.

Therefore, the integral transform pair in the ϕ variable for the function $T(r, \phi, t)$, defined in the interval $0 \leq \phi \leq \phi_0$, is taken as

Inversion formula $$T(r, \phi, t) = \sum_{\nu} \frac{\Phi(\nu, \phi)}{N(\nu)} \overline{T}(r, \nu, t) \tag{14-106a}$$

Integral transform $$\overline{T}(r, \nu, t) = \int_{\phi'=0}^{\phi_0} \Phi(\nu, \phi') T(r, \phi', t) \, d\phi' \tag{14-106b}$$

where

$$N(\nu) = \int_{\phi=0}^{\phi_0} [\Phi(\nu, \phi)]^2 \, d\phi \tag{14-106c}$$

For any combination of boundary conditions of the first and second type, the functions $\Phi(\nu, \phi)$, $N(\nu)$ and the eigenvalues ν are obtainable from Table 2-1, by appropriate change in the notation.

Having established the integral transform pairs needed for the removal of the differential operator $\partial^2 T / \partial \phi^2$ from the heat conduction equation, we now proceed to the solution of heat conduction problems involving (r, ϕ, t) variables.

Problems of Region $0 \leq r \leq b$, $0 \leq \phi \leq 2\pi$ We consider the following time-dependent heat conduction problem for a solid cylinder of radius $r = b$ in which temperature varies both r and ϕ variables:

$$\frac{\partial^2 T}{\partial r^2} + \frac{1}{r}\frac{\partial T}{\partial r} + \frac{1}{r^2}\frac{\partial^2 T}{\partial \phi^2} + \frac{g(r, \phi, t)}{k} = \frac{1}{\alpha}\frac{\partial T(r, \phi, t)}{\partial t}$$

$$\text{in} \quad 0 \leq r < b, \quad 0 \leq \phi \leq 2\pi, \quad t > 0 \tag{14-107a}$$

$$\text{BC:} \quad k_2 \left.\frac{\partial T}{\partial r}\right|_{r=b} + h_2 \left. T \right|_{r=b} = f_2(\phi, t) \tag{14-107b}$$

$$\text{IC:} \quad T(r, \phi, t = 0) = F(r, \phi) \quad \text{in the region} \tag{14-107c}$$

This problem is now solved by successive application of the integral transforms with respect to the ϕ and r variables. The integral transform pair in the ϕ variable for the function $T(r, \phi, t)$ is given by equation (14-104); hence we have

Inversion formula $$T(r, \phi, t) = \frac{1}{\pi} \sum_{\nu} \overline{T}(r, \nu, t) \tag{14-108a}$$

Integral transform $$\overline{T}(r, \nu, t) = \int_{\phi'=0}^{2\pi} \cos \nu (\phi - \phi') \, T(r, \phi', t) \, d\phi' \tag{14-108b}$$

where $\nu = 0, 1, 2, 3 \ldots$, and where we replace π by 2π for $\nu = 0$. The integral transform of the system (14-107) by the application of the transform (14-108b) yields (see note 5 at the end of this chapter for the details)

$$\frac{\partial^2 \overline{T}}{\partial r^2} + \frac{1}{r}\frac{\partial \overline{T}}{\partial r} - \frac{\nu^2}{r^2}\overline{T} + \frac{\overline{g}(r, \nu, t)}{k} = \frac{1}{\alpha}\frac{\partial \overline{T}(r, \nu, t)}{\partial t} \quad \text{in} \quad 0 \le r < b, \quad t > 0 \tag{14-109a}$$

$$\text{BC:} \quad k_2 \left.\frac{\partial \overline{T}}{\partial r}\right|_{r=b} + h_2 \left.\overline{T}\right|_{r=b} = \overline{f}_2(\nu, t) \tag{14-109b}$$

$$\text{IC:} \quad \overline{T}(r, \nu, t = 0) = \overline{F}(r, \nu) \quad \text{in} \quad 0 \le r \le b \tag{14-109c}$$

where the overbar denotes the integral transform with respect to the ϕ variable.

The integral transform pair in the r variable for the function $\overline{T}(r, \nu, t)$ is obtainable according to equations (4-149) and (4-150) by setting $\nu = 0$ and then letting $J_0(\beta_m, r) \to R_0(\beta_m, r)$. We find

$$\text{Inversion formula} \quad \overline{T}(r, \nu, t) = \sum_{m=1}^{\infty} \frac{R_\nu(\beta_m, r)}{N(\beta_m)} \tilde{\overline{T}}(\beta_m, \nu, t) \tag{14-110a}$$

$$\text{Integral transform} \quad \tilde{\overline{T}}\left(\beta_m, \nu, t\right) = \int_{r'=0}^{b} r' R_\nu\left(\beta_m, r'\right) \overline{T}\left(r', \nu, t\right) dr' \tag{14-110b}$$

Here, the tilde denotes the integral transform with respect to the r variable. $R_\nu(\beta_m, r)$ and β_m are the eigenfunctions and eigenvalues associated with the eigenvalue problem given by Table 2-2. The functions $R_\nu(\beta_m, r)$, $N(\beta_m)$ and the eigenvalues β_m are now obtainable from Table 2-2 for the three different boundary conditions at $r = b$.

The integral transform of the system (14-109) by the application of the transform (14-110b) yields (see note 6 at the end of this chapter for the details)

$$\frac{d\tilde{\overline{T}}}{dt} + \alpha\beta_m^2 \tilde{\overline{T}}\left(\beta_m, \nu, t\right) = \frac{\alpha}{k}\tilde{\overline{g}}\left(\beta_m, \nu, t\right) + \alpha b \left.\frac{R_\nu\left(\beta_m, r\right)}{k_2}\right|_{r=b} \overline{f}_2(\nu, t) \tag{14-111a}$$

$$\text{IC:} \quad \tilde{\overline{T}}\left(\beta_m, \nu, t = 0\right) = \tilde{\overline{F}}\left(\beta_m, \nu\right) \tag{14-111b}$$

Equations (14-111) are solved for $\tilde{\overline{T}}(\beta_m, \nu, t)$, and the resulting double transform is successively inverted by the inversion formulas (14-110a) and (14-108a). Then, the solution of the problem (14-107) becomes

$$T(r, \phi, t) = \frac{1}{\pi} \sum_{\nu} \sum_{m=1}^{\infty} \frac{R_\nu\left(\beta_m, r\right)}{N\left(\beta_m\right)} e^{-\alpha\beta_m^2 t} \cdot \left[\tilde{\overline{F}}\left(\beta_m, \nu\right) + \int_{t'=0}^{t} e^{\alpha\beta_m^2 t'} A\left(\beta_m, \nu, t'\right) dt'\right] \tag{14-112a}$$

where $\nu = 0, 1, 2, 3\ldots$, and where we replace π by 2π for $\nu = 0$, and we define

$$A\left(\beta_m, \nu, t'\right) = \frac{\alpha}{k}\tilde{\bar{g}}\left(\beta_m, \nu, t'\right) + \alpha b \left.\frac{R_\nu\left(\beta_m, r\right)}{k_2}\right|_{r=b} \bar{f}\left(\nu, t'\right) \tag{14-112b}$$

$$\bar{f}(\nu, t) = \int_{\phi'=0}^{2\pi} f_2\left(\phi', t\right)\cos\nu\left(\phi - \phi'\right)d\phi' \tag{14-112c}$$

$$\tilde{\bar{F}}\left(\beta_m, \nu\right) = \int_{r'=0}^{b}\int_{\phi'=0}^{2\pi} r' R_\nu\left(\beta_m, r'\right)\cos\nu\left(\phi - \phi'\right)F\left(r', \phi'\right)d\phi'\,dr' \tag{14-112d}$$

$$\tilde{\bar{g}}\left(\beta_m, \nu, t'\right) = \int_{r'=0}^{b}\int_{\phi'=0}^{2\pi} r' R_\nu\left(\beta_m, r'\right)\cos\nu\left(\phi - \phi'\right)g\left(r', \phi', t'\right)d\phi'\,dr' \tag{14-112e}$$

and the functions $R_\nu(\beta_m, r)$, $N(\beta_m)$ and the eigenvalues β_m are obtainable from Table 2-2.

For a boundary condition of the first type at $r = b$, the following changes should be made in equation (14-112b). When $k_2 = 0$,

$$\text{replace} \quad \left.\frac{R_\nu\left(\beta_m, r\right)}{k_2}\right|_{r=b} \quad \text{by} \quad -\left.\frac{dR_\nu\left(\beta_m, r\right)}{dr}\right|_{r=b} \tag{14-112f}$$

Problems of Region $a \le r \le b$, $0 \le \phi \le 2\pi$ The extension of the above analysis for solid cylinder to the solution of time-dependent heat conduction problem of a hollow cylinder $a \le r \le b$ in which temperature varies with both r and ϕ variables is a straightforward matter. Clearly, the heat conduction problem (14-107) will involve an additional boundary condition at $r = a$. The definition of the integral transform pair (14-108) remains the same, but that of (14-110) is modified by changing the lower limit of the integration to $r = a$; then the functions $R_\nu(\beta_m, r)$, $N(\beta_m)$ and the eigenvalues β_m are to be obtained from Table 2-3. As a result, the solution (14-112) will include an additional term in the definition of $A(\beta_m, \nu, t')$ for the effects of the boundary condition at $r = a$, and the lower limit of the integrations with respect to r' will be $r' = a$.

Problems of Region $0 \le r \le b$, $0 \le \phi \le \phi_0$ $(< 2\pi)$ We now consider the solution by the integral transform technique of the following time-dependent heat conduction problem for a portion of a solid cylinder of radius $r = b$, in the region $0 \le \phi \le \phi_0$ $(< 2\pi)$, namely,

$$\frac{\partial^2 T}{\partial r^2} + \frac{1}{r}\frac{\partial T}{\partial r} + \frac{1}{r^2}\frac{\partial^2 T}{\partial \phi^2} + \frac{g(r, \phi, t)}{k} = \frac{1}{\alpha}\frac{\partial T(r, \phi, t)}{\partial t}$$

$$\text{in} \quad 0 \le r < b, \quad 0 < \phi < \phi_0, \quad t > 0 \tag{14-113a}$$

BC1 : $T(\phi = 0) = 0$ (14-113b)

BC2: $T(\phi = \phi_0) = 0$ (14-113c)

BC3: $k_4 \left.\frac{\partial T}{\partial r}\right|_{r=b} + h_4 \left. T \right|_{r=b} = f_4(\phi, t)$ (14-113d)

IC: $T(r, \phi, t = 0) = F(r, \phi)$ in the region (14-113e)

The integral transform pair in the ϕ variable for the function $T(r, \phi, t)$ is obtained from equations (14-106) as

Inversion formula $$T(r, \phi, t) = \sum_{\nu} \frac{\Phi(\nu, \phi)}{N(\nu)} \overline{T}(r, \nu, t) \tag{14-114a}$$

Integral transform $$\overline{T}(r, \nu, t) = \int_{\phi'=0}^{\phi_0} \Phi\left(\nu, \phi'\right) T\left(r, \phi', t\right) d\phi' \tag{14-114b}$$

where the functions $\Phi(\nu, \phi)$, $N(\nu)$ and the eigenvalues ν are obtainable from Table 2-1 by appropriate change of the notation (i.e., $L \to \phi_0$, $\beta_m \to \nu$, $x \to \phi$). We note that the eigenvalues ν for this case are not integers but are determined according to the eigenvalue equations given in Table 2-1.

The integral transform of the system (14-113) by the application of the transform (14-114b) yields

$$\frac{\partial^2 \overline{T}}{\partial r^2} + \frac{1}{r}\frac{\partial \overline{T}}{\partial r} - \frac{\nu^2}{r^2}\overline{T} + \frac{1}{k}\overline{g}(r, \nu, t) = \frac{1}{\alpha}\frac{\partial \overline{T}(r, \nu, t)}{\partial t}$$

in $0 \le r < b, \quad t > 0$ (14-115a)

BC: $k_4 \left.\frac{\partial \overline{T}}{\partial r}\right|_{r=b} + h_4 \left. \overline{T} \right|_{r=b} = \overline{f}_4(\nu, t)$ (14-115b)

IC: $\overline{T}(r, \nu, t = 0) = \overline{F}(r, \nu)$ in $0 \le r \le b$ (14-115c)

where the overbar denotes the integral transform of the function with respect to the ϕ variable.

The integral transform pair in the r variable for the function $\overline{T}(r, \nu, t)$ is immediately obtained from the transform pair (14-110) as

Inversion formula $$\overline{T}(r, \nu, t) = \sum_{m=1}^{\infty} \frac{R_\nu\left(\beta_m, r\right)}{N\left(\beta_m\right)} \tilde{\overline{T}}(\beta_m, \nu, t) \tag{14-116a}$$

Integral transform $$\tilde{\overline{T}}(\beta_m, \nu, t) = \int_{r'=0}^{b} r' R_\nu\left(\beta_m, r'\right) \overline{T}\left(r', \nu, t\right) dr' \tag{14-116b}$$

where the tilde denotes the integral transform with respect to the r variable. The functions $R_\nu(\beta_m, r)$, $N(\beta_m)$ and the eigenvalues β_m are obtainable from Table 2-2.

The integral transform of the system (14-115) by the application of the transform (14-116b) yields (i.e., the procedure is similar to that described in note 6 at the end of this chapter)

$$\frac{d\tilde{\bar{T}}}{dt} + \alpha\beta_m^2 \tilde{\bar{T}}(\beta_m, \nu, t) = A(\beta_m, \nu, t) \qquad \text{for } t > 0 \tag{14-117a}$$

$$\text{IC}: \qquad \tilde{\bar{T}}\left(\beta_m, \nu, t = 0\right) = \tilde{\bar{F}}\left(\beta_m, \nu\right) \tag{14-117b}$$

where

$$A(\beta_m, \nu, t) \equiv \frac{\alpha}{k}\tilde{\bar{g}}(\beta_m, \nu, t) + \alpha b \left.\frac{R_\nu\left(\beta_m, r\right)}{k_4}\right|_{r=b} \bar{f}_4(\nu, t) \tag{14-118}$$

Equation (14-117) is solved for $\tilde{\bar{T}}(\beta_m, \nu, t)$, and the resulting double transform of the temperature is successively inverted by the inversion formulas (14-116a) and (14-114a). Then, the solution of the problem (14-113) becomes

$$T(r, \phi, t) = \sum_\nu \sum_{m=1}^{\infty} \frac{\Phi\left(\nu, \phi\right) R_\nu\left(\beta_m, r\right)}{N\left(\nu\right) N\left(\beta_m\right)} e^{-\alpha\beta_m^2 t} \cdot \left[\tilde{\bar{F}}\left(\beta_m, \nu\right) + \int_{t'=0}^{t} e^{\alpha\beta_m^2 t'} A\left(\beta_m, \nu, t'\right) dt'\right] \tag{14-119}$$

where $A(\beta_m, \nu, t')$ is defined by equations (14-118), and $\tilde{\bar{F}}, \tilde{\bar{g}}$ are the double transforms:

$$\tilde{\bar{H}}\left(\beta_m, \nu\right) = \int_{r'=0}^{b} \int_{\phi'=0}^{\phi_0} r' R\left(\beta_m, r'\right) \Phi\left(\nu, \phi'\right) H\left(r', \phi'\right) d\phi'\, dr' \tag{14-120}$$

with $H \equiv F$ or $H \equiv g$, for the initial condition and generation, respectively.

The overbar denotes the integral transform with respect to the ϕ variable and the tilde the integral transform with respect to the r variable as defined by equations (14-114b) and (14-116b), respectively.

For a boundary condition of the first type at any of these boundaries, the usual replacements should be made in the definition of $A(\beta_m, \nu, t)$ given by equation (14-118a).

Problems of Region $a \le r \le b$, $0 \le \phi \le \phi_0$ ($\phi_0 < 2\pi$) The extension of the above solution to the problem of time-dependent heat conduction in a hollow cylinder $a \le r \le b$, confined to a region $0 \le \phi \le \phi_0$ ($< 2\pi$), is a straightforward matter. The heat conduction problem (14-113) will include an additional boundary condition at $r = a$. The definition of the integral transform pair (14-114) remains the same, but that given by equations (14-116) is modified by changing the lower limit of the integration to $r = a$; then, the function $R_v(\beta_m, r)$, $N(\beta_m)$ and the eigenvalues β_m are obtained from Table 2-3.

Problems of Region 0 ≤ r < ∞, 0 ≤ φ ≤ 2π We now consider the solution of the following time-dependent heat conduction problem for an infinite medium in which temperature varies with both r and ϕ variables:

$$\frac{\partial^2 T}{\partial r^2} + \frac{1}{r}\frac{\partial T}{\partial r} + \frac{1}{r^2}\frac{\partial^2 T}{\partial \phi^2} + \frac{g(r, \phi, t)}{k} = \frac{1}{\alpha}\frac{\partial T(r, \phi, t)}{\partial t}$$

$$\text{in} \quad 0 \le r < \infty, \quad 0 \le \phi \le 2\pi, \quad t > 0 \qquad (14\text{-}121a)$$

$$\text{IC:} \quad T(r, \phi, t = 0) = F(r, \phi) \quad \text{in the region} \qquad (14\text{-}121b)$$

The integral transform pair in the ϕ variable for the function $T(r, \phi, t)$ is given by equation (14-104); hence we have

$$\text{Inversion formula} \quad \overline{T}(r, \phi, t) = \frac{1}{\pi}\sum_{v}\overline{T}(r, v, t) \qquad (14\text{-}122a)$$

$$\text{Integral transform} \quad \overline{T}(r, v, t) = \int_{\phi'=0}^{2\pi} \cos v\left(\phi - \phi'\right) T\left(r, \phi', t\right) d\phi' \qquad (14\text{-}122b)$$

where $v = 0, 1, 2, 3 \ldots$, and where we replace π by 2π for $v = 0$. The integral transform of the system (14-121) by the application of the transform (14-122b) yields

$$\frac{\partial^2 \overline{T}}{\partial r^2} + \frac{1}{r}\frac{\partial \overline{T}}{\partial r} - \frac{v^2}{r^2}\overline{T} + \frac{\overline{g}(r, v, t)}{k} = \frac{1}{\alpha}\frac{\partial \overline{T}(r, v, t)}{\partial t}$$

$$\text{in} \quad 0 \le r < \infty, \quad t > 0 \qquad (14\text{-}123a)$$

$$\text{IC:} \quad \overline{T}(r, v, t = 0) = \overline{F}(r, v) \quad \text{in} \quad 0 \le r < \infty \qquad (14\text{-}123b)$$

where the overbar denotes the integral transform with respect to the ϕ variable.

The integral transform pair in the r variable for the function $\overline{T}(r, v, t)$ is constructed according to the representation given by equations (2-120) and (2-121). We find

$$\text{Inversion formula} \quad \overline{T}(r, v, t) = \int_{\beta=0}^{\infty} \beta J_v(\beta r)\,\tilde{\overline{T}}(\beta, v, t)\, d\beta \qquad (14\text{-}124a)$$

$$\text{Integral transform} \quad \tilde{\overline{T}}(\beta, v, t) = \int_{r'=0}^{\infty} r' J_v\left(\beta r'\right) \overline{T}\left(r', v, t\right) dr' \qquad (14\text{-}124b)$$

where the tilde denotes the integral transform with respect to the r variable and the eigenvalue problem associated with this transform pair is given by equations (6-158) and (6-160).

The integral transform of the system (14-123) by the application of the transform (14-124b) gives

$$\frac{d\tilde{\bar{T}}}{dt} + \alpha\beta^2\tilde{\bar{T}}(\beta, \nu, t) = \frac{\alpha}{k}\tilde{\bar{g}}(\beta, \nu, t) \qquad \text{for} \qquad t > 0 \tag{14-125a}$$

$$\text{IC:} \qquad \tilde{\bar{T}}(\beta, \nu, t = 0) = \tilde{\bar{F}}(\beta, \nu) \tag{14-125b}$$

Equation (14-125) is solved for $\tilde{\bar{T}}(\beta, \nu, t)$, and the resulting double transform is successively inverted by the inversion formulas (14-124a) and (14-122a). Then the solution of the problem (14-121) becomes

$$\begin{aligned} T(r, \phi, t) = \frac{1}{\pi}\sum_{\nu}\int_{\beta=0}^{\infty} \beta J_\nu(\beta r)\, e^{-\alpha\beta^2 t} \\ \cdot\left[\tilde{\bar{F}}(\beta, \nu) + \frac{\alpha}{k}\int_{t'=0}^{t} e^{\alpha\beta^2 t'}\tilde{\bar{g}}(\beta, \nu, t')\, dt'\right] d\beta \end{aligned} \tag{14-126a}$$

where $\nu = 0, 1, 2, 3\ldots$, and we replace π by 2π for $\nu = 0$ and define

$$\tilde{\bar{F}}(\beta, \nu) = \int_{r'=0}^{\infty}\int_{\phi'=0}^{2\pi} r' J_\nu(\beta r') \cos\nu(\phi - \phi')\, F(r', \phi')\, d\phi'\, dr' \tag{14-126b}$$

$$\tilde{\bar{g}}(\beta, \nu, t') = \int_{r'=0}^{\infty}\int_{\phi'=0}^{2\pi} r' J_\nu(\beta r') \cos\nu(\phi - \phi')\, g(r', \phi', t')\, d\phi'\, dr' \tag{14-126c}$$

Introducing equations (14-126b,c) into equation (14-126a) and changing the order of integrations, we obtain

$$\begin{aligned} T(r, \phi, t) = &\frac{1}{\pi}\sum_{\nu}\int_{r'=0}^{\infty}\int_{\phi'=0}^{2\pi} r' \cos\nu(\phi - \phi')\, F(r', \phi') \\ &\cdot\left[\int_{\beta=0}^{\infty} e^{-\alpha\beta^2 t}\beta J_\nu(\beta r)\, J_\nu(\beta r')\, d\beta\right] d\phi'\, dr' \\ &+ \frac{1}{\pi}\frac{\alpha}{k}\sum_{\nu}\int_{r'=0}^{\infty}\int_{\phi'=0}^{2\pi}\int_{t'=0}^{t} r' \cos\nu(\phi - \phi')\, g(r', \phi', t') \\ &\cdot\left[\int_{\beta=0}^{\infty} e^{-\alpha\beta^2(t-t')}\beta J_\nu(\beta r)\, J_\nu(\beta r')\, d\beta\right] dt' d\phi'\, dr' \end{aligned} \tag{14-127}$$

The terms inside the brackets can be evaluated by utilizing the expression (14-99). Then the solution (14-127) becomes

$$T(r,\phi,t) = \frac{1}{2\pi\alpha t}\sum_{\nu}\int_{r'=0}^{\infty}\int_{\phi'=0}^{2\pi} r'\cos\nu(\phi-\phi')\,F(r',\phi') \cdot \exp\left(-\frac{r^2+r'^2}{4\alpha t}\right) I_\nu\left(\frac{rr'}{2\alpha t}\right) d\phi'\,dr'$$
$$+\frac{1}{2\pi k}\sum_{\nu}\int_{r'=0}^{\infty}\int_{\phi'=0}^{2\pi}\int_{t'=0}^{t}\frac{r'}{t-t'}\cos\nu(\phi-\phi')\,g(r',\phi',t') \cdot \exp\left[-\frac{r^2+r'^2}{4\alpha(t-t')}\right] I_\nu\left(\frac{rr'}{2\alpha(t-t')}\right) dt'\,d\phi'\,dr' \tag{14-128}$$

where $\nu = 0, 1, 2, 3\ldots$, and where we replace π by 2π for $\nu = 0$. Several special cases are obtainable from this solution.

Problems in (r, z, t) Variables

The solution of time-dependent heat conduction problems in the (r, z) variables with the integral transform technique is now a straightforward matter. The integral transform pairs with respect to the r variable are the same as those developed in this section for the problems having azimuthal symmetry, and those with respect to the z variable are the same as those for the rectangular coordinate system. Also, the order of integral transformation with respect to the r and z variables is immaterial. We illustrate this matter with the following example.

Example 14-6 Two-Dimensional Solid Cylinder with Generation
Consider the solution of the following heat conduction problem for a solid cylinder of radius $r = b$ and height $z = L$:

$$\frac{\partial^2 T}{\partial r^2}+\frac{1}{r}\frac{\partial T}{\partial r}+\frac{\partial^2 T}{\partial z^2}+\frac{g(r,z,t)}{k}=\frac{1}{\alpha}\frac{\partial T(r,z,t)}{\partial t}$$
$$\text{in}\quad 0\le r<b,\quad 0<z<L,\quad t>0 \tag{14-129a}$$

$$\text{BC:}\quad T(r,z,t)=0\quad\text{on all boundaries},\quad t>0 \tag{14-129b}$$

$$\text{IC:}\quad T(r,z,t=0)=F(r,z)\quad\text{in the region} \tag{14-129c}$$

The integral transform pair for the removal of partial derivatives with respect to the r variable in the region $0 \le r \le b$ is the same as that given by equations (14-87). Therefore, the transform pair with respect to the r variable for the function $T(r, z, t)$ is given as

$$\text{Inversion formula} \quad T(r, z, t) = \sum_{m=1}^{\infty} \frac{R_0(\beta_m, r)}{N(\beta_m)} \overline{T}(\beta_m, z, t) \tag{14-130a}$$

$$\text{Integral transform} \quad \overline{T}(\beta_m, z, t) = \int_{r'=0}^{b} r' R_0(\beta_m, r') T(r', z, t)\, dr' \tag{14-130b}$$

where $R_0(\beta_m, r)$, $N(\beta_m)$, and β_m are obtainable from Table 2-2 by setting $\nu = 0$, and the overbar denotes transform with respect to the r variable. The integral transform of the system (14-129) by the application of the transform (14-130b) yields

$$-\beta_m^2 \overline{T}(\beta_m, z, t) + \frac{\partial^2 \overline{T}}{\partial z^2} + \frac{\overline{g}(\beta_m, z, t)}{k} = \frac{1}{\alpha} \frac{\partial \overline{T}(\beta_m, z, t)}{\partial t}$$

$$\text{in} \quad 0 < z < L, \quad t > 0 \tag{14-131a}$$

$$\text{BC1:} \quad \overline{T}(\beta_m, z = 0, t) = 0 \tag{14-131b}$$

$$\text{BC2:} \quad \overline{T}(\beta_m, z = L, t) = 0 \tag{14-131b}$$

$$\text{IC:} \quad \overline{T}(\beta_m, z, t = 0) = \overline{F}(\beta_m, z) \quad \text{in} \quad 0 < z < L \tag{14-131c}$$

The integral transform pair with respect to the z variable in the region $0 \le z \le L$ is obtained from equations (14-27) as

$$\text{Inversion formula} \quad \overline{T}(\beta_m, z, t) = \sum_{p=1}^{\infty} \frac{Z(\eta_p, z)}{N(\eta_p)} \tilde{\overline{T}}(\beta_m, \eta_p, t) \tag{14-132a}$$

$$\text{Integral transform} \quad \tilde{\overline{T}}(\beta_m, \eta_p, t) = \int_{z'=0}^{L} Z(\eta_p, z') \overline{T}(\beta_m, z', t)\, dz' \tag{14-132b}$$

where $Z(\eta_p, z)$, $N(\eta_p)$, and η_p are obtainable from Table 2-1, and the tilde denotes the integral transform with respect to the z variable. The integral transform of the system (14-131), by the application of the transform (14-132b), is

$$\frac{d\tilde{\overline{T}}}{dt} + \alpha(\beta_m^2 + \eta_p^2) \tilde{\overline{T}}(\beta_m, \eta_p, t) = \frac{\alpha}{k} \tilde{\overline{g}}(\beta_m, \eta_p, t) \quad t > 0 \tag{14-133a}$$

$$\text{IC:} \quad \tilde{\overline{T}}(\beta_m, \eta_p, t = 0) = \tilde{\overline{F}}(\beta_m, \eta_p) \tag{14-133b}$$

Equations (14-133) are solved for $\overline{T}(\beta_m, \eta_p, t)$, and the resulting double transform is successively inverted by the inversion formulas (14-132a) and (14-130a).

Then the solution of the problem (14-129) becomes

$$T(r, z, t) = \sum_{m=1}^{\infty} \sum_{p=1}^{\infty} \frac{R_0(\beta_m, r) Z(\eta_p, z)}{N(\beta_m) N(\eta_p)} e^{-\alpha(\beta_m^2 + \eta_p^2)t} \cdot \left[\tilde{\bar{F}}(\beta_m, \eta_p) + \frac{\alpha}{k} \int_{t'=0}^{t} e^{\alpha(\beta_m^2 + \eta_p^2)t'} \tilde{\bar{g}}(\beta_m, \eta_p, t') \, dt' \right] \quad \text{(14-134a)}$$

where the double transforms are defined as

$$\tilde{\bar{H}} = \int_{z'=0}^{L} \int_{r'=0}^{b} r' R_0(\beta_m, r') Z(\eta_p, z') H(r', z') \, dr' \, dz' \quad \text{(14-134b)}$$

with $H \equiv F$ or $H \equiv g$ for the initial condition and generation, respectfully.

From Table 2-2, case 3, for $v = 0$ we have

$$R_0(\beta_m, r) = J_0(\beta_m r), \qquad \frac{1}{N(\beta_m)} = \frac{2}{b^2 J_1^2(\beta_m b)}$$

and the β_m values are the roots of $J_0(\beta_m b) = 0$. From Table 2-1, case 9, we have

$$Z(\eta_p, z) = \sin \eta_p z \qquad \frac{1}{N(\eta_p)} = \frac{2}{L}$$

and the η_p values are the roots of $\sin \eta_p L = 0$.

Problems in (r, ϕ, z, t) Variables

The solution of heat conduction problems in (r, ϕ, z, t) variables is readily handled with the integral transform technique. The basic steps in the analysis are summarized.

1. The partial derivative with respect to the z variable is removed by the application of transform in the z variable. The appropriate transform pairs are the same as those given for the rectangular coordinate system.
2. The partial derivative with respect to the ϕ variable is removed by the application of transform in the ϕ variable. If the range of ϕ is $0 \le \phi \le 2\pi$, the transform pair is given by equations (14-104). If the range of ϕ is $0 \le \phi \le \phi_0$, $(\phi_0 \le 2\pi)$, the transform pair is given by equations (14-106) for boundary conditions of the first and the second type.
3. The partial derivatives with respect to the r variable are removed by the application of transform in the r variable. The transform pair to be used depends on the range of the r variable, that is, $0 \le r \le b$, $a \le r \le b$, or for $0 \le r < \infty$. For example, the transform pair is as given by equations (14-110) for $0 \le r \le b$, or given by equations (14-124) for $0 \le r < \infty$.

4. The resulting ordinary differential equation with respect to the time variable is solved subject to the triple transformed initial condition. The triple transform of temperature obtained in this manner is successively inverted with respect to the r, ϕ, and z variables to obtain the solution for $T(r, \phi, z, t)$.

14-4 APPLICATIONS IN THE SPHERICAL COORDINATE SYSTEM

To solve heat conduction problems in the spherical coordinate system with the integral transform technique, appropriate integral transform pairs are needed in the r, μ, and ϕ variables. In this section we develop such integral transform pairs and illustrate their application to the solution of heat conduction problems involving (r, t), (r, μ, t), and (r, μ, ϕ, t) variables.

Problems in (r, t) Variables

The time-dependent heat conduction problems involving only the r variable can be transformed into a one-dimensional, time-dependent heat conduction problem in the rectangular coordinate system by defining a new variable $U(r, t) = rT(r, t)$ as discussed in Section 5-3. The resulting heat conduction problem in the rectangular coordinate system is readily solved with the integral transform technique as described previously. Therefore, the solution of the problems in (r, t) variables is not considered here any further.

Problems in (r, μ, t) Variables

The differential equation of heat conduction in the (r, μ, t) variables is taken in the form

$$\frac{\partial^2 T}{\partial r^2} + \frac{2}{r}\frac{\partial T}{\partial r} + \frac{1}{r^2}\frac{\partial}{\partial \mu}\left[\left(1-\mu^2\right)\frac{\partial T}{\partial \mu}\right] + \frac{g\,(r, \mu, t)}{k} = \frac{1}{\alpha}\frac{\partial T\,(r, \mu, t)}{\partial t} \tag{14-135}$$

By defining a new variable $V(r, \mu, t)$, see Chapter 5, as

$$V\,(r, \mu, t) = r^{1/2}T\,(r, \mu, t) \tag{14-136}$$

equation (14-135) is transformed into

$$\frac{\partial^2 V}{\partial r^2} + \frac{1}{r}\frac{\partial V}{\partial r} - \frac{1}{4}\frac{V}{r^2} + \frac{1}{r^2}\frac{\partial}{\partial \mu}\left[\left(1-\mu^2\right)\frac{\partial V}{\partial \mu}\right] + \frac{r^{1/2}g}{k} = \frac{1}{\alpha}\frac{\partial V}{\partial t} \tag{14-137}$$

where $g \equiv g(r, \mu, t)$ and $V \equiv V(r, \mu, t)$.

The partial derivatives with respect to the space variables can be removed from this equation by the successive application of integral transforms with respect to the μ and r variables. *The order of transformation is important in this case; it is applied first to the μ variable and then to the r variable*. Then we need to develop the integral transform pairs only with respect to the μ variable for the following cases: The range of μ variable is $-1 \leq \mu \leq 1$ as in the case of the *full sphere* and is in the range $0 \leq \mu \leq 1$ as in the case of the *hemisphere*.

Transform Pair for $-1 \leq \mu \leq 1$ This case corresponds to the full sphere. Therefore, no boundary conditions are specified in the μ variable except the requirement that the function should remain finite at $\mu = \pm 1$. The integral transform pair in the μ variable for the function $V(r, \mu, t)$ is constructed by considering the representation of this function in a form similar to that given by equations (5-159) and (5-161) and then splitting up the representation into two parts. We find

$$\text{Inversion formula} \qquad V(r, \mu, t) = \sum_{n=0}^{\infty} \frac{2n+1}{2} P_n(\mu) \overline{V}(r, n, t) \tag{14-138a}$$

$$\text{Integral transform} \qquad \overline{V}(r, n, t) = \int_{\mu'=-1}^{1} P_n(\mu') V(r, \mu', t) \, d\mu' \tag{14-138b}$$

where $P_n(\mu)$ is the Legendre polynomial and $n = 0, 1, 2, 3 \ldots$.

Transform Pair for $0 \leq \mu \leq 1$ This case corresponds to the hemisphere. The integral transform pair is determined by splitting up the expansion given by equations (2-99) and (2-100), giving

$$\text{Inversion formula} \qquad V(r, \mu, t) = \sum_{n} (2n+1) P_n(\mu) \overline{V}(r, n, t) \tag{14-139a}$$

$$\text{Integral transform} \qquad \overline{V}(r, n, t) = \int_{\mu'=0}^{1} P_n(\mu') V(r, \mu', t) \, d\mu' \tag{14-139b}$$

where the values of n are $n = 1, 3, 5, \ldots$ (i.e., odd integers) for boundary condition of the first type at $\mu = 0$ and $n = 0, 2, 4, \ldots$ (i.e., even integers) for boundary condition of the second type at $\mu = 0$.

Example 14-7 Solid Sphere with Generation

We consider the solution of the following time-dependent heat conduction problem for a solid sphere of radius $r = b$:

$$\frac{\partial^2 T}{\partial r^2} + \frac{2}{r}\frac{\partial T}{\partial r} + \frac{1}{r^2}\frac{\partial}{\partial \mu}\left[(1-\mu)^2 \frac{\partial T}{\partial \mu}\right] + \frac{g(r, \mu, t)}{k} = \frac{1}{\alpha}\frac{\partial T(r, \mu, t)}{\partial t}$$

$$\text{in} \quad 0 \leq r < b, \quad -1 \leq \mu \leq 1, \quad t > 0 \tag{14-140}$$

$$\text{BC:} \quad T(r = b, \mu, t) = 0 \tag{14-141a}$$

$$\text{IC:} \quad T(r, \mu, t = 0) = F(r, \mu) \quad \text{in the region} \tag{14-141b}$$

Here we have considered a homogeneous boundary condition of the first type for simplicity in the analysis; however, the analysis for a boundary condition of the third type is performed in a similar manner.

A new dependent variable $V(r, \mu, t)$ is defined as

$$V(r, \mu, t) = r^{1/2} T(r, \mu, t) \tag{14-142}$$

Then, the problem in equations (14-140) and (14-141) is transformed into

$$\frac{\partial^2 V}{\partial r^2} + \frac{1}{r}\frac{\partial V}{\partial r} - \frac{1}{4}\frac{V}{r^2} + \frac{1}{r^2}\frac{\partial}{\partial \mu}\left[(1-\mu^2)\frac{\partial V}{\partial \mu}\right] + \frac{r^{1/2} g(r, \mu, t)}{k} = \frac{1}{\alpha}\frac{\partial V}{\partial t}$$

$$\text{in} \quad 0 \le r < b, \quad -1 \le \mu \le 1, \quad t > 0 \tag{14-143a}$$

$$\text{BC:} \quad V(r = b, \mu, t) = 0 \tag{14-143b}$$

$$\text{IC:} \quad V(r, \mu, t = 0) = r^{1/2} F(r, \mu) \quad \text{in the region} \tag{14-143c}$$

The integral transform pair with respect to the μ variable for $-1 \le \mu \le 1$ is obtained from equations (14-138) as

$$\text{Inversion formula} \qquad V(r, \mu, t) = \sum_{n=0}^{\infty} \frac{2n+1}{2} P_n(\mu)\overline{V}(r, n, t) \tag{14-144a}$$

$$\text{Integral transform} \qquad \overline{V}(r, n, t) = \int_{\mu'=-1}^{1} P_n(\mu') V(r, \mu', t) d\mu' \tag{14-144b}$$

where $n = 0, 1, 2, 3 \ldots$.

We take the integral transform of the system (14-143) by the application of the transform (14-144b) to obtain (see note 7 at the end of this chapter for details)

$$\frac{\partial^2 \overline{V}}{\partial r^2} + \frac{1}{r}\frac{\partial \overline{V}}{\partial r} - \frac{(n+\frac{1}{2})^2}{r^2}\overline{V} + \frac{\overline{g}^*(r, n, t)}{k} = \frac{1}{\alpha}\frac{\partial \overline{V}(r, n, t)}{\partial t}$$

$$\text{in} \quad 0 \le r < b \quad t > 0 \tag{14-145a}$$

$$\text{BC:} \quad \overline{V}(r = b, n, t) = 0 \tag{14-145b}$$

$$\text{IC:} \quad \overline{V}(r, n, t = 0) = \overline{F}^*(r, n) \quad \text{in} \quad 0 \le r \le b \tag{14-145c}$$

where we define

$$g^*(r, \mu, t) = r^{1/2} g(r, \mu, t) \qquad F^*(r, \mu) = r^{1/2} F(r, \mu) \tag{14-145d}$$

and the overbar denotes the integral transform with respect to the μ variable according to the transform (14-144b).

Equation (14-145a) is similar in form to equation (14-109a) in the cylindrical coordinate system. Therefore, the integral transform pair needed to remove the partial derivatives with respect to the r variable is immediately obtainable from equation (14-110) or (14-116) by setting $v \equiv n + \frac{1}{2}$. We find

$$\text{Inversion formula} \qquad \overline{V}(r, n, t) = \sum_{p=1}^{\infty} \frac{R_{n+1/2}(\lambda_{np}, r)}{N(\lambda_{np})} \tilde{\overline{V}}(\lambda_{np}, n, t) \tag{14-146a}$$

$$\text{Integral transform} \qquad \tilde{\overline{V}}(\lambda_{np}, n, t) = \int_{r'=0}^{b} r' R_{n+1/2}(\lambda_{np}, r') \overline{V}(r', n, t)\, dr' \tag{14-146b}$$

where the tilde denotes the integral transform with respect to the r variable. The functions $R_{n+1/2}(\lambda_{np}, r)$, $N(\lambda_{np})$, and the eigenvalues λ_{np} for the boundary condition of the first type considered in the problem (14-145) are obtainable from Table 2-2, case 3, as

$$R_{n+1/2}(\lambda_{np}, r) = J_{n+1/2}(\lambda_{np} r), \qquad \frac{1}{N(\lambda_{np})} = \frac{2}{b^2 [J_{n+1/2}(\lambda_{np} b)]^2} \tag{14-146c}$$

and the λ_{np} values are the positive roots for each integer value n, of

$$J_{n+1/2}(\lambda_{np} b) = 0 \tag{14-146d}$$

Taking the integral transform of the system (14-145) by the application of the transform (14-146b), we obtain

$$\frac{d\tilde{\overline{V}}}{dt} + \alpha \lambda_{np}^2 \tilde{\overline{V}}(\lambda_{np}, n, t) = \frac{\alpha}{k} \tilde{\overline{g}}^* \qquad \text{for} \qquad t > 0 \tag{14-147a}$$

$$\text{IC:} \qquad \tilde{\overline{V}}(\lambda_{np}, n, t = 0) = \tilde{\overline{F}}^* \tag{14-147b}$$

Equation (14-147) is solved for $\tilde{\overline{V}}(\lambda_{np}, n, t)$, the resulting double transform is inverted successively by the inversion formulas (14-146a) and (14-144a) to obtain $V(r, \mu, t)$. When $V(r, \mu, t)$ is transformed by the expression (14-142d) into $T(r, \mu, t)$, the solution of the problem (14-141) is obtained as

$$T(r, \mu, t) = \sum_{n=0}^{\infty} \sum_{p=1}^{\infty} \frac{(2n+1) J_{n+1/2}(\lambda_{np} r) P_n(\mu)}{r^{1/2} b^2 [J_{n+1/2}(\lambda_{np} b)]^2} e^{-\alpha \lambda_{np}^2 t} \cdot \left[\tilde{\overline{F}}^* + \frac{\alpha}{k} \int_{t'=0}^{t} e^{\alpha \lambda_{np}^2 t'} \tilde{\overline{g}}^*\, dt' \right] \tag{14-148a}$$

where we define

$$\tilde{\bar{F}}^* = \int_{r'=0}^{b} \int_{\mu'=-1}^{1} r'^{3/2} J_{n+1/2}(\lambda_{np} r') P_n(\mu') F(r', \mu') \, d\mu' \, dr' \tag{14-148b}$$

$$\tilde{\bar{g}}^* = \int_{r'=0}^{b} \int_{\mu'=-1}^{1} r'^{3/2} J_{n+1/2}(\lambda_{np} r') P_n(\mu') g(r', \mu', t') \, d\mu' \, dr' \tag{14-148c}$$

for $n = 0, 1, 2 \ldots$, and where the λ_{np} values are the roots of equation (14-146d). For the case of no heat generation, this solution reduces to that given by equation (4-88).

Example 14-8 Hemisphere with Heat Generation
We consider the following time-dependent heat conduction equation for a hemisphere of radius $r = b$:

$$\frac{\partial^2 T}{\partial r^2} + \frac{2}{r}\frac{\partial T}{\partial r} + \frac{1}{r^2}\frac{\partial}{\partial \mu}\left[(1-\mu^2)\frac{\partial T}{\partial \mu}\right] + \frac{g(r, \mu, t)}{k} = \frac{1}{\alpha}\frac{\partial T(r, \mu, t)}{\partial t}$$

$$\text{in} \quad 0 \le r < b, \quad 0 \le \mu \le 1, \quad t > 0 \tag{14-149a}$$

$$\text{BC1:} \quad T(r = \text{b}, \mu, t) = 0 \tag{14-149b}$$

$$\text{BC2:} \quad T(r, \mu = 0, t) = 0 \tag{14-149c}$$

$$\text{IC:} \quad T(r, \mu, t = 0) = F(r, \mu) \quad \text{in the region} \tag{14-149d}$$

The basic steps for the solution of this problem are exactly the same as those described above for the solution of problem (14-140). The only difference is that, the range of μ being $0 \le \mu \le 1$, the integral transform pair with respect to the μ variable is determined according to equations (14-139) as

$$\text{Inversion formula} \quad V(r, \mu, t) = \sum_n (2n+1) P_n(\mu) \overline{V}(r, n, t) \tag{14-150a}$$

$$\text{Integral transform} \quad \overline{V}(r, n, t) = \int_{\mu'=0}^{1} P_n(\mu') V(r, \mu', t) d\mu' \tag{14-150b}$$

where the values of n are $n = 1, 3, 5, \ldots$ (i.e., odd integers) since the boundary condition at $\mu = 0$ is of the first type.

The integral transform pair with respect to the r variable is taken the same as that given by equations (14-146). The system (14-149) is transformed from the $T(r, \mu, t)$ variable into the $V(r, \mu, t)$ variable. Then, by the application of the integral transform with respect to the μ and r variables, an ordinary differential equation is obtained for the double transform $\tilde{\overline{V}}(\lambda_{np}, n, t)$. The resulting ordinary differential equation is solved, and the double transform is successively inverted by the inversion formulas (14-146a) and (14-150a) to obtain $V(r, \mu, t)$. When $V(r, \mu, t)$ is transformed by the transformation (14-142), the solution $T(r, \mu, t)$ of the problem (14-149) is obtained as

$$T(r, \mu, t) = \sum_{n=1,3,5...}^{\infty} \sum_{p=1}^{\infty} \frac{2(2n+1) J_{n+1/2}(\lambda_{np} r) P_n(\mu)}{r^{1/2} b^2 [J_{n+1/2}(\lambda_{np} b)]^2} e^{-\alpha \lambda_{np}^2 t} \cdot \left[\tilde{\bar{F}}^* + \frac{\alpha}{k} \int_{t'=0}^{t} e^{\alpha \lambda_{np}^2 t'} \tilde{\bar{g}}^* \, dt' \right] \tag{14-151a}$$

where

$$\tilde{\bar{F}}^* = \int_{r'=0}^{b} \int_{\mu'=0}^{1} r'^{3/2} J_{n+1/2}(\lambda_{np} r') P_n(\mu') F(r', \mu') \, d\mu' \, dr' \tag{14-151b}$$

$$\tilde{\bar{g}}^* = \int_{r'=0}^{b} \int_{\mu'=0}^{1} r'^{3/2} J_{n+1/2}(\lambda_{np} r') P_n(\mu') g(r', \mu', t') \, d\mu' \, dr' \tag{14-151c}$$

and the λ_{np} values are the roots for each value of n, of

$$J_{n+1/2}(\lambda_{np} b) = 0 \tag{14-151d}$$

We note that for the case of no heat generation, the solution (14-151) reduces to that given by equations (4-98).

Problems in (r, μ, ϕ, t) Variables

The differential equation of heat conduction in the (r, μ, ϕ, t) variables is taken in the form

$$\frac{\partial^2 T}{\partial r^2} + \frac{2}{r} \frac{\partial T}{\partial r} + \frac{1}{r^2} \frac{\partial}{\partial \mu} \left[(1 - \mu^2) \frac{\partial T}{\partial \mu} \right] + \frac{1}{r^2 (1 - \mu^2)} \frac{\partial^2 T}{\partial \phi^2} + \frac{g(r, \mu, \phi, t)}{k} = \frac{1}{\alpha} \frac{\partial T(r, \mu, \phi, t)}{\partial t} \tag{14-152}$$

We consider the problem of *full sphere*; hence choose the ranges of μ and ϕ variables as $0 \le \phi \le 2\pi$ and $-1 \le \mu \le 1$. The range of the r variable may be finite or infinite. Now a new variable $V(r, \mu, \phi, t)$ is again defined as

$$V(r, \mu, \phi, t) = r^{1/2} T(r, \mu, \phi, t) \tag{14-153}$$

Then equation (14-152) is transformed into

$$\frac{\partial^2 V}{\partial r^2}+\frac{1}{r}\frac{\partial V}{\partial r}-\frac{1}{4}\frac{V}{r^2}+\frac{1}{r^2}\frac{\partial}{\partial \mu}\left[(1-\mu^2)\frac{\partial V}{\partial \mu}\right]$$
$$+\frac{1}{r^2(1-\mu^2)}\frac{\partial^2 V}{\partial \phi^2}+\frac{r^{1/2}g}{k}=\frac{1}{\alpha}\frac{\partial V}{\partial t} \qquad (14\text{-}154)$$

where $0 \le \phi \le 2\pi$, $-1 \le \mu \le 1$, and the range of r is finite or infinite.

The partial derivatives with respect to the space variables can be removed from this equation by the application of integral transform with respect to the ϕ, μ, and r variables. For this particular case the order of transformation is important. *That is, the transformation should be applied first with respect to the ϕ variable, then to the μ variable, and finally to the r variable*. The procedure is as follows.

Removal of the Derivative in the ϕ Variable The range of the ϕ variable being in $0 \le \phi \le 2\pi$, the transform pair with respect to the ϕ variable is obtained from equations (14-104) as

Inversion formula $$V(r,\mu,\phi,t)=\frac{1}{\pi}\sum_{m=1}^{\infty}\overline{V}(r,\mu,m,t) \qquad (14\text{-}155a)$$

Integral transform $$\overline{V}(r,\mu,m,t)=\int_{\phi'=0}^{2\pi}\cos m(\phi-\phi')V(r,\mu,\phi',t)d\phi' \qquad (14\text{-}155b)$$

where $m = 0, 1, 2, 3 \ldots$, and where we replace π by 2π for $m = 0$. The eigenvalue problem associated with this transform pair is the same as that given by equations (14-122).

The integral transform of equation (14-154) by the application of the transform (14-155b) is

$$\frac{\partial^2 \overline{V}}{\partial r^2}+\frac{1}{r}\frac{\partial \overline{V}}{\partial r}-\frac{1}{4}\frac{\overline{V}}{r^2}+\frac{1}{r^2}\left\{\frac{\partial}{\partial \mu}\left[(1-\mu^2)\frac{\partial \overline{V}}{\partial \mu}\right]-\frac{m^2}{1-\mu^2}\overline{V}\right\}+\frac{\overline{g}^*}{k}=\frac{1}{\alpha}\frac{\partial \overline{V}}{\partial t} \qquad (14\text{-}156)$$

where $g^* \equiv r^{1/2}g(r,\mu,\phi,t)$, $\overline{V} \equiv \overline{V}(r,\mu,m,t)$, and the overbar denotes the integral transform with respect to the ϕ variable. Thus, by the application of the transform (14-155b), we removed from the differential equation the partial derivative with respect to the ϕ variable, that is, $\partial^2 V/\partial \phi^2$.

Removal of the Derivative in the μ Variable In equation (14-156), the differential operator with respect to the μ variable is in the form

$$\frac{\partial}{\partial \mu}\left[(1-\mu^2)\frac{\partial \overline{V}}{\partial \mu}\right]-\frac{m^2}{1-\mu^2}\overline{V}$$

and the range of μ is $-1 \leq \mu \leq 1$. The integral transform pair to remove this differential operator can be constructed by considering the representation of a function, defined in the interval $-1 \leq \mu \leq 1$, in terms of the eigenfunctions of Legendre's associated differential equation, see equation (2-80), namely,

$$\frac{d}{d\mu}\left[(1-\mu^2)\frac{dM}{d\mu}\right]+\left[n(n+1)-\frac{m^2}{1-\mu^2}\right]M=0 \tag{14-157}$$

and then splitting up the representation into two parts. The resulting integral transform pair with respect to the μ variable for the function $\overline{V}(r,\mu,m,t)$ is given as

Inversion formula $$\overline{V}(r,\mu,m,t)=\sum_{m=1}^{\infty}\frac{P_n^m(\mu)}{N(m,n)}\tilde{\overline{V}}(r,n,m,t) \tag{14-158a}$$

Integral transform $$\tilde{\overline{V}}(r,n,m,t)=\int_{\mu'=-1}^{1}P_n^m(\mu')\overline{V}(r,\mu',m,t)\,d\mu' \tag{14-158b}$$

where

$$\frac{1}{N(m,n)}=\frac{2n+1}{2}\frac{(n-m)!}{(n+m)!} \quad \text{for} \quad n \geq m \tag{14-158c}$$

and where n and m are integers, $P_n^m(\mu)$ is the associated Legendre polynomical of degree n, order m, of the first kind; and the tilde denotes the integral transform with respect to the μ variable.

Taking the integral transform of equation (14-156) by the application of the transform (14-158b) and utilizing equation (14-157) we obtain (see note 8 at the end of this chapter for details)

$$\frac{\partial^2\tilde{\overline{V}}}{\partial r^2}+\frac{1}{r}\frac{\partial\tilde{\overline{V}}}{\partial r}-\frac{(n+\frac{1}{2})^2}{r^2}\tilde{\overline{V}}+\frac{\tilde{\overline{g}}^*}{k}=\frac{1}{\alpha}\frac{\partial\tilde{\overline{V}}(r,n,m,t)}{\partial t} \tag{14-159}$$

where the tilde denotes the transform with respect to the μ variable.

Removal of the Derivative in the r Variable The differential operator with respect to the r variable can readily be removed from equation (14-159) by the application of an appropriate transform in the r variable developed previously for the solution of problems in the cylindrical coordinate system. The form of the transform pair depends on the range of r, whether it is finite or infinite. We now illustrate the application with an example given below.

Example 14-9 Three-Dimensional Sphere with Generation
Solve the following time-dependent, three-dimensional heat conduction problem for a solid sphere of radius $r = b$:

$$\frac{\partial^2 T}{\partial r^2}+\frac{2}{r}\frac{\partial T}{\partial r}+\frac{1}{r^2}\frac{\partial}{\partial\mu}\left[(1-\mu^2)\frac{\partial T}{\partial\mu}\right]+\frac{1}{r^2(1-\mu^2)}\frac{\partial^2 T}{\partial\phi^2}+\frac{g}{k}=\frac{1}{\alpha}\frac{\partial T}{\partial t}$$

$$\text{in} \quad 0 \leq r < b, \quad -1 \leq \mu \leq 1, \quad 0 \leq \phi \leq 2\pi, \quad t > 0 \tag{14-160a}$$

$$\text{BC:} \qquad T(r=b)=0 \tag{14-160b}$$

$$\text{IC:} \qquad T(t=0)=F(r,\mu,\phi) \text{ in the region} \tag{14-160c}$$

where we have $g \equiv g(r,\mu,\phi,t)$ and $T \equiv T(r,\mu,\phi,t)$. By defining a new dependent variable as

$$V(r,\mu,\phi,t)=r^{1/2}T(r,\mu,\phi,t) \tag{14-161}$$

the system (14-160) is transformed into

$$\frac{\partial^2 V}{\partial r^2}+\frac{1}{r}\frac{\partial V}{\partial r}-\frac{1}{4}\frac{V}{r^2}+\frac{1}{r^2}\frac{\partial}{\partial \mu}\left[(1-\mu^2)\frac{\partial V}{\partial \mu}\right]$$
$$+\frac{1}{r^2(1-\mu^2)}\frac{\partial^2 V}{\partial \phi^2}+\frac{r^{1/2}g}{k}=\frac{1}{\alpha}\frac{\partial V}{\partial t}$$
$$\text{in} \quad 0\le r<b, \qquad -1\le\mu\le 1, \qquad 0\le\phi\le 2\pi, \qquad t>0 \tag{14-162a}$$

$$\text{BC:} \qquad V(r=b)=0 \tag{14-162b}$$

$$\text{IC:} \qquad V(t=0)=r^{1/2}F(r,\mu,\phi)\text{in the region} \tag{14-162c}$$

The integral transform of this system with respect to the ϕ variable by the application of the transform (14-155b) yields

$$\frac{\partial^2 \overline{V}}{\partial r^2}+\frac{1}{r}\frac{\partial \overline{V}}{\partial r}-\frac{1}{4}\frac{\overline{V}}{r^2}+\frac{1}{r^2}\left\{\frac{\partial}{\partial \mu}\left[(1-\mu^2)\frac{\partial \overline{V}}{\partial \mu}\right]-\frac{m^2}{1-\mu^2}\overline{V}\right\}$$
$$+\frac{\overline{g}^*}{k}=\frac{1}{\alpha}\frac{\partial \overline{V}}{\partial t} \tag{14-163a}$$
$$\text{in} \quad 0\le r<b, \qquad -1\le\mu\le 1, \qquad t>0$$

$$\text{BC:} \qquad \overline{V}(r=b)=0 \tag{14-163b}$$

$$\text{IC:} \qquad \overline{V}(t=0)=\overline{F}^*(r,\mu,m) \quad \text{in} \quad 0\le r\le b \quad -1\le\mu\le 1 \tag{14-163c}$$

where we have $g^* \equiv r^{1/2}g(r,\mu,\phi,t)$, $F^* \equiv r^{1/2}F(r,\mu,\phi)$, $\overline{V} \equiv \overline{V}(r,\mu,m,t)$, and the overbar denotes the transform with respect to the ϕ variable. Now, the integral transform of the system (14-163) with respect to the μ variable, by the application of the transform (14-158b), gives

$$\frac{\partial^2 \tilde{\overline{V}}}{\partial r^2}+\frac{1}{r}\frac{\partial \tilde{\overline{V}}}{\partial r}-\frac{(n+\frac{1}{2})^2}{r^2}\tilde{\overline{V}}+\frac{\tilde{\overline{g}}^*}{k}=\frac{1}{\alpha}\frac{\partial \tilde{\overline{V}}}{\partial t} \tag{14-164a}$$
$$\text{in} \quad 0\le r<b, \qquad t>0$$

$$\text{BC:} \qquad \tilde{\overline{V}}(r=b)=0 \tag{14-164b}$$

$$\text{IC:} \qquad \tilde{\overline{V}}(t=0)=\tilde{\overline{F}}^*(r,n,m) \qquad 0\le r\le b \tag{14-164c}$$

where $\tilde{\bar{V}} \equiv V(r, n, m, t)$, and the tilde denotes transform with respect to the μ variable. This system is now exactly of the same form as that given by equations (14-145). To remove the differential operator with respect to the r variable the appropriate integral transform pair is exactly the same as that given by equations (14-146). Therefore, taking the integral transform of the system (14-164) by the application of the transform (14-146b), we find

$$\frac{d\overset{x}{\tilde{\bar{V}}}}{dt} + \alpha\lambda_{np}^2 \overset{x}{\tilde{\bar{V}}}(\lambda_{np}, n, m, t) = \frac{\alpha}{k}\overset{x}{\tilde{\bar{g}}}^* \qquad \text{for} \qquad t > 0 \tag{14-165a}$$

$$\text{IC:} \qquad \overset{x}{\tilde{\bar{V}}}(t = 0) = \overset{x}{\tilde{\bar{F}}}(\lambda_{np}, n, m) \tag{14-165b}$$

where the superscript x denotes the integral transform with respect to the r variable. Equation (14-165) is solved for

$$\overset{x}{\tilde{\bar{V}}}(\lambda_{np}, n, m, t)$$

The resulting triple transform is successively inverted by the inversion formulas (14-146a), (14-158a), and (14-155a) to obtain $V(r, \mu, \phi, t)$. When the function $V(r, \mu, \phi, t)$ is transformed by the expression (14-161), we obtain the solution $T(r, \mu, \phi, t)$ of the problem (14-160) as

$$T(r, \mu, \phi, t) = \frac{1}{\pi}\sum_{n=0}^{\infty}\sum_{m=0}^{n}\sum_{p=1}^{\infty} \frac{J_{n+1/2}(\lambda_{np}r)P_n^m(\mu)}{r^{1/2}N(m, n)N(\lambda_{np})} e^{-\alpha\lambda_{np}^2 t} \cdot \left(\overset{x}{\tilde{\bar{F}}} + \frac{\alpha}{k}\int_{t'=0}^{t} e^{-\alpha\lambda_{np}^2 t'}\, \overset{x}{\tilde{\bar{g}}}^*\, dt'\right) \tag{14-166a}$$

where we replace π by 2π for the case of $m = 0$. The various quantities in equation (14-166a) are now defined as

$$\overset{x}{\tilde{\bar{F}}} = \int_{r'=0}^{b}\int_{\mu'=-1}^{1}\int_{\phi'=0}^{2\pi} r'^{3/2} J_{n+1/2}(\lambda_{np}r')P_n^m(\mu') \cdot \cos m(\phi - \phi')F(r', \mu', \phi')\, d\phi' d\mu'\, dr' \tag{14-166b}$$

$$\overset{x}{\tilde{\bar{g}}}^* = \int_{r'=0}^{b}\int_{\mu'=-1}^{1}\int_{\phi'=0}^{2\pi} r'^{3/2} J_{n+1/2}(\lambda_{np}r')P_n^m(\mu') \cdot \cos m(\phi - \phi')g(r', \mu', \phi', t')d\phi' d\mu'\, dr' \tag{14-166c}$$

with the norms

$$\frac{1}{N(m,n)} = \frac{2n+1}{2}\frac{(n-m)!}{(n+m)!} \tag{14-166d}$$

$$\frac{1}{N(\lambda_{np})} = \frac{2}{b^2[J_{n+1/2}(\lambda_{np}b)]^2} \tag{14-166e}$$

and the λ_{np} values are the positive roots for each integer n, of

$$J_{n+1/2}(\lambda_{np}b) = 0 \tag{14-166f}$$

For the case of no heat generation, this solution reduces to that given by equations (4-118).

14-5 APPLICATIONS IN THE SOLUTION OF STEADY-STATE PROBLEMS

The integral transform technique is also very effective in the solution of multi dimensional, steady-state heat conduction problems because, by the successive application of the integral transform, the partial differential equation is reduced to an ordinary differential equation in one of the space variables. The resulting ordinary differential equation is solved for the transform of the temperature, which is then inverted successively to obtain the desired solution. This procedure is now illustrated with examples.

Example 14-10 Two-Dimensional Rectangular Region
Solve the following steady-state heat conduction problem for a rectangular region $0 \le x \le a$, $0 \le y \le b$:

$$\frac{\partial^2 T}{\partial x^2} + \frac{\partial^2 T}{\partial y^2} = 0 \quad \text{in} \quad 0 < x < a, \quad 0 < y < b \tag{14-167a}$$

$$\text{BC1:} \quad T(x=0) = 0 \tag{14-167b}$$

$$\text{BC2:} \quad T(x=a) = 0 \tag{14-167c}$$

$$\text{BC3:} \quad T(y=0) = f(x) \tag{14-167d}$$

$$\text{BC4:} \quad T(y=b) = 0 \tag{14-167e}$$

In this example, we prefer to take the integral transform with respect to the x variable because in the resulting ordinary differential equation for the transform of T, the boundary condition at $y = 0$ becomes a constant; hence its integration is readily performed. The integral transform pair with respect to the x variable, for $0 \le x \le a$, of function $T(x, y)$ is defined as

$$\text{Inversion formula} \qquad T(x, y) = \sum_{m=1}^{\infty} \frac{X(\beta_m, x)}{N(\beta_m)} \overline{T}(\beta_m, y) \tag{14-168a}$$

$$\text{Integral transform} \qquad \overline{T}(\beta_m, y) = \int_{x'=0}^{a} X(\beta_m, x') T(x', y)\, dx' \tag{14-168b}$$

where $X(\beta_m, x)$, $N(\beta_m)$, and β_m are obtained from Table 2-1, case 9, as

$$X(\beta_m, x) = \sin \beta_m x, \qquad \frac{1}{N(\beta_m)} = \frac{2}{a} \qquad \text{and} \qquad \sin \beta_m a = 0 \tag{14-168c}$$

The integral transform of the system (14-167), by the application of the transform (14-168b), yields

$$\frac{d^2 \overline{T}}{dy^2} - \beta_m^2 \overline{T}(\beta_m, y) = 0 \qquad \text{in} \qquad 0 < y < b \tag{14-169a}$$

$$\text{BC1:} \qquad \overline{T}(y = 0) = \overline{f}(\beta_m) \tag{14-169b}$$

$$\text{BC2:} \qquad \overline{T}(y = b) = 0 \tag{14-169c}$$

The solution of equations (14-169) is

$$\overline{T}(\beta_m, y) = \overline{f}(\beta_m) \frac{\sinh \beta_m (b - y)}{\sinh \beta_m b} \tag{14-170}$$

The inversion of this result with the inversion formula (14-168a) gives

$$T(x, y) = \frac{2}{a} \sum_{m=1}^{\infty} \sin \beta_m x \frac{\sinh \beta_m (b - y)}{\sinh \beta_m b} \int_{x'=0}^{a} \sin \beta_m x' f(x')\, dx' \tag{14-171a}$$

where we have

$$\beta_m = m\pi / a, \qquad m = 1, 2, 3 \ldots \tag{14-171b}$$

Example 14-11 Two-Dimensional Solid Cylinder
Solve the following steady-state heat conduction problem for a long solid cylinder:

$$\frac{\partial^2 T}{\partial r^2} + \frac{1}{r} \frac{\partial T}{\partial r} + \frac{1}{r^2} \frac{\partial^2 T}{\partial \phi^2} = 0 \qquad \text{in} \qquad 0 \le r < b, \qquad 0 \le \phi \le 2\pi \tag{14-172a}$$

$$\text{BC:} \qquad k_2 \left. \frac{\partial T}{\partial r} \right|_{r=b} + h_2 \left. T \right|_{r=b} = f_2(\phi) \tag{14-172b}$$

The integral transform pair with respect to the ϕ variable over the range $0 \leq \phi \leq 2\pi$ is obtained from equations (14-104) as

$$\text{Inversion formula} \qquad T(r,\phi) = \frac{1}{\pi}\sum_{\nu} \overline{T}(r,\nu) \tag{14-173a}$$

$$\text{Integral transform} \qquad \overline{T}(r,\nu) = \int_{\phi'=0}^{2\pi} \cos\nu(\phi-\phi')T(r,\phi')\,d\phi' \tag{14-173b}$$

where $\nu = 0, 1, 2, 3\ldots$, and where we replace π by 2π for $\nu = 0$. The integral transform of the system (14-172), by the application of the transform (14-173b) yields

$$\frac{d^2\overline{T}}{dr^2} + \frac{1}{r}\frac{d\overline{T}}{dr} - \frac{\nu^2}{r^2}\overline{T}(r,\nu) = 0 \qquad \text{in} \qquad 0 \leq r < b \tag{14-174a}$$

$$\text{BC:} \qquad k_2 \left.\frac{d\overline{T}}{dr}\right|_{r=b} + h_2\,\overline{T}\big|_{r=b} = \overline{f}_2(\nu) \tag{14-174b}$$

The solution of equations (14-174) is

$$\overline{T}(r,\nu) = b\left(\frac{r}{b}\right)^{\nu}\frac{\overline{f}_2(\nu)}{k_2\nu + h_2 b} \tag{14-175}$$

The inversion of this result by the inversion formula (14-173a) gives the temperature distribution as

$$T(r,\phi) = \frac{1}{\pi}\sum_{\nu} b\left(\frac{r}{b}\right)^{\nu}\frac{\overline{f}_2(\nu)}{k_2\nu + h_2 b} \tag{14-176a}$$

where

$$\overline{f}_2(\nu) = \int_{\phi'=0}^{2\pi} \cos\nu(\phi-\phi')f_2(\phi')d\phi' \tag{14-176b}$$

for $\nu = 0, 1, 2, 3\ldots$, and where we replace π by 2π for $\nu = 0$.

Example 14-12 Two-Dimensional Solid Hemisphere

Solve the following steady-state heat conduction problem in spherical coordinates for a solid hemisphere of radius $r = b$:

$$\frac{\partial}{\partial r}\left(r^2\frac{\partial T}{\partial r}\right) + \frac{\partial}{\partial \mu}\left[\left(1-\mu^2\right)\frac{\partial T}{\partial \mu}\right] = 0 \qquad 0 \leq r < b, \qquad 0 < \mu \leq 1 \tag{14-177a}$$

$$\text{BC1:} \qquad \left.\frac{\partial T}{\partial \mu}\right|_{\mu=0} = 0 \tag{14-177b}$$

$$\text{BC2:} \qquad T(r=b,\mu) = f(\mu) \tag{14-177c}$$

This problem is the same as that considered in system (5-21). The integral transform pair with respect to the μ variable for $0 \le \mu \le 1$ and the boundary condition of the second type at $\mu = 0$ is obtained from equations (14-150) as

$$\text{Inversion formula} \qquad T(r,\mu) = \sum_n (2n+1) P_n(\mu) \overline{T}(r,n) \tag{14-178a}$$

$$\text{Integral transform} \qquad \overline{T}(r,n) = \int_{\mu'=0}^{1} P_n(\mu') T(r,\mu')\, d\mu' \tag{14-178b}$$

where $n = 0, 2, 4, 6 \ldots$ (i.e., even integers). The integral transform of the system (14-177) by the application of the transform (14-178b) yields

$$\frac{d}{dr}\left(r^2 \frac{d\overline{T}}{dr}\right) - n(n+1)\overline{T}(r,n) = 0 \qquad 0 \le r < b \tag{14-179a}$$

$$\text{BC}: \qquad \overline{T}(r=b,n) = \overline{f}(n) \tag{14-179b}$$

The solution of equations (14-179) is

$$\overline{T}(r,n) = \left(\frac{r}{b}\right)^n \overline{f}(n) \tag{14-180}$$

The inversion of this result by the inversion formula (14-178a) gives the solution for the temperature as

$$T(r,\mu) = \sum_{n=0,2,4\ldots}^{\infty} (2n+1) P_n(\mu) \left(\frac{r}{b}\right)^n \int_{\mu'=0}^{1} P_n(\mu') f(\mu')\, d\mu' \tag{14-181}$$

REFERENCES

1. M. D. Mikhailov and M. N. Özisik, *Unified Analysis and Solutions of Heat and Mass Diffusion*, Wiley, New York, 1984.
2. I. N. Sneddon, *The Use of Integral Transforms*, McGraw-Hill, New York, 1972.
3. E. C. Titchmarsh, *Fourier Integrals*, 2nd ed., Clarendon, London, 1948.
4. A. Erdelyi, W. Magnus, F. Oberhettinger, and F. G. Tricomi, *Tables of Integrals Transforms*, McGraw-Hill, New York, 1954.
5. C. J. Tranter, *Integral Transforms in Mathematical Physics*, Wiley, New York, 1962.
6. V. A. Ditkin and A. P. Produnikov, *Integral Transforms and Operational Calculus*, Pergamon, New York, 1965.
7. M. N. Özisik, Integral Transform in the Solution of Heat-Conduction Equation in the Rectangular coordinate System, ASME Paper 67-WA/HT-46, 1967.
8. M. N. Özisik , *Boundary Value Problems of Heat Conduction*, International Textbook, Scranton, PA, 1968.

9. A. C. Eringen, *Q. J. Math. Oxford* **5**(2), 120–129, 1954.
10. R. V. Churchill, *Mich. Math. J.* **3**, 85, 1955–1956.
11. A. McD. Mercer, *Q. J. Math. Oxford* **14**, 9–15, 1963.
12. N. Y. Ölçer, *Österr. Ing.-Arch* **18**, 104–113, 1964.
13. N. Y. Ölçer, *Int. J. Heat Mass Transfer* **7**, 307–314, 1964.
14. N. Y. Ölçer , *Int. J. Heat Mass Transfer* **8**, 529–556, 1965.
15. N. Y. Ölçer, *J. Math Phys.* **46**, 99–106, 1967.
16. M. D. Mikhailov, *Int. J. Eng. Sci.* **10**, 577–591, 1972.
17. M. D. Mikhailov, *Int. J. Heat Mass Transfer* **16**, 2155–2164, 1973.
18. M. D. Mikhailov, *Int. J. Heat Mass Transfer* **17**, 1475–1478, 1974.
19. M. D. Mikhailov, *Int. J. Eng. Sci.* **11**, 235–241, 1973.
20. M. N. Özisik and R. L. Murray, *J. Heat Transfer* **96c**, 48–51, 1974.
21. Y. Yener and M. N. Özisik, On the Solution of Unsteady Heat Conduction in Multi-Region Media with Time Dependent Heat Transfer Coefficient, *Proc. 5th. Int. Heat Trans. Conference*, Tokyo, Sept. 1974 Cu 2.5, pp. 188–192.
22. M. D. Mikhailov, *Int. J. Heat Mass Transfer* **18**, 344–345, 1975.
23. M. D. Mikhailov, *Int. J. Heat Mass Transfer* **20**, 1409–1415, 1977.
24. N. Y. Ölçer, *Int. J. Heat Mass Transfer* **12**, 393–411, 1969.
25. K. Kobayashi, N. Ohtani, and J. Jung, *Nucl. Sci. Eng.* **55**, 320–328, 1974.
26. I. N. Sneddon, *Phil. Mag.* **37**, 17, 1946.
27. G. Cinelli, *Int. J. Engl. Sci.* **3**, 539–559, 1965.
28. N. Y. Ölçer, *Brit. J. Appl. Phys.* **18**, 89–105, 1967.
29. N. Y. Ölçer, *Proc. Comb. Phil. Soc.* **64**, 193–202, 1968.
30. N. Y. Ölçer, *J. Heat Transfer* **91c**, 45–50, 1969.
31. A. W. Jacobson, *Quart. Appl. Math.* **7**, 293–302, 1949.
32. C. J. Tranter, *Quart. J. Math. Oxford* **1**, 1–8, 1950.
33. R. V. Churchill and C. L. Dolph, *Proc. Am. Math. Soc.* **5**, 93, 1954.
34. R. V. Churchill, *J. Math. Phys.* **33**, 165–178, 1954–1955.
35. R. Courant and D. Hilbert, *Methods of Mathematical Physics*, Vol. 1, Interscience, New York, 1953, p. 106.
36. G. N. Watson, *A Treatise on the Theory of Bessel Functions*, Cambridge University Press, London, 1966.
37. M. D. Mikhailov, M. N. Özisik, and N. L. Vulchanov, *Int. J. Heat Mass Transfer* **26**, 1131–1141, 1983.
38. M. D. Mikhailov and M. N. Özisik, *Int. J. Heat Mass Transfer* **28**, 1039–1045, 1985.
39. M. D. Mikhailov and M. N. Özisik, *J. Franklin Ins.* **321**, 299–307, 1986.
40. R. M. Cotta and M. N. Özisik, *Can. J. Chem. Eng.*, **64**, 734–742, 1986.

PROBLEMS

14-1 Solve the one-dimensional, time-dependent heat conduction problem for a slab $0 \leq x \leq L$, which is initially at zero temperature, and for times

$t > 0$ the boundaries at $x = 0$ and $x = L$ are kept at temperatures zero and $f_2(t)$, respectively. Consider the case when the surface temperature is given by $f_2(t) = \gamma t$, where γ is a constant.

14-2 Solve the one-dimensional, time-dependent heat conduction problem for a slab $0 \le x \le L$, which is initially at zero temperature, and for times $t > 0$ heat is generated in the medium at a rate of $g(x, t)$, W/m^3, while the boundary surface at $x = 0$ is kept insulated and the boundary surface at $x = L$ is kept at zero temperature. Consider the case when the heat source is an instantaneous plane heat source of total strength g_s^i W · s/m^2, situated at $x = b$, and release its heat spontaneously at time $t = 0$, that is, $g(x, t) = g_s^i \delta(x - b)\delta(t)$.

14-3 A semi-infinite medium, $0 \le x < \infty$, is initially at zero temperature. For times $t > 0$ the boundary at $x = 0$ is kept at zero temperature, while heat is generated in the medium at a rate of $g(x, t)$, W/m^3. Obtain an expression for the temperature distribution $T(x, t)$ in the medium. Consider the cases where (1) the heat source is a continuous plane surface heat source of strength $g_s^c(t)$W/m^2, which is situated at $x = b$, that is, $g(x, t) = g_s^c(t)\delta(x - b)$, and (2) the heat source is a constant heat source, that is, $g(x, t) = g_0 =$ constant, W/m^3.

14-4 An infinite medium $-\infty < x < \infty$ is initially at zero temperature. A plane surface heat source of strength $g_s^c(t)$, W/m^2, situated at $x = 0$, releases heat continuously for times $t > 0$. Obtain an expression for the temperature distribution $T(x, t)$ in the medium for times $t > 0$ [i.e., $g(x, t) = g_s(t)\delta(x)$].

14-5 A rectangular region $0 \le x \le a, 0 \le y \le b$ is initially at zero temperature. For times $t > 0$ heat is generated in the medium at a rate of $g(x, y, t)$ W/m^3, while the boundaries are kept at zero temperature. Obtain an expression for the temperature distribution $T(x, y, t)$ in the region. Also consider the special case when the heat source is an instantaneous line heat source g_L^i of strength W · s/m, situated at (x_1, y_1) within the region and releases its heat spontaneously at time $t = 0$, that is, $g(x, y, t) = g_L^i \delta(x - x_1)\delta(y - y_1)\delta(t)$.

14-6 A three-dimensional infinite medium $-\infty < x < \infty$, $-\infty < y < \infty$, $-\infty < z < \infty$ is initially at zero temperature. For times $t > 0$, heat is generated in the medium at a rate of $g(x, y, z, t)$, W/m^3. Obtain an expression for the temperature distribution $T(x, y, z, t)$ in the medium. Also consider the special case when the heat source is an instantaneous point heat source of strength g_s^i W · s, situated at $x = 0$, $y = 0$, $z = 0$ and releasing its heat spontaneously at time $t = 0$, that is, $g(x, y, z, t) = g_p^i \delta(x)\delta(y)\delta(z)\delta(t)$.

14-7 A solid cylinder $0 \le r \le b$, is initially at zero temperature. For times $t > 0$, heat is generated within the region at a rate of $g(r, t)$, W/m^3, while the boundary at $r = b$ is kept at zero temperature. Obtain an expression for the temperature distribution $T(r, t)$ in the cylinder. Consider the special cases where (1) the heat is generated at a constant rate g_0, W/m^3, in the region, and (2) the heat source is a line heat source of strength $g_L(t)$, W/m, situated along the axis of the cylinder, that is,

$$g(r, t) = \frac{1}{2\pi r} g_L(t)\delta(r)$$

14-8 A long solid cylinder, $0 \leq r \leq b$, is initially at temperature $F(r)$. For times $t > 0$ the boundary at $r = b$ is kept insulated. Obtain an expression for the temperature distribution $T(r, t)$ in the cylinder.

14-9 A long hollow cylinder, $a \leq r \leq b$, is initially at temperature $F(r)$. For times $t > 0$ the boundaries at $r = a$ and $r = b$ are kept insulated. Obtain an expression for the temperature distribution $T(r, t)$ in the region.

14-10 A long hollow cylinder, $a \leq r \leq b$, is initially at zero temperature. For times $t > 0$ heat is generated in the medium at a rate of $g(r, t)$, W/m^3, while the boundaries at $r = a$ and $r = b$ are kept at zero temperature. Obtain an expression for the temperature distribution $T(r, t)$ in the cylinder. Consider the special cases where (1) the heat generation rate is constant, that is, g_0 = constant, and (2) the heat source is an instantaneous cylindrical heat source of radius $r = r_1$ (i.e., $a < r_1 < b$) of strength g_s^i, W · s/m, per linear length of the cylinder, which is situated inside the cylinder coaxially and releases its heat spontaneously at time $t = 0$, that is,

$$g(r, t) = \frac{1}{2\pi r} g_s^i \delta(r - r_1)\delta(t)$$

14-11 An infinite region, $0 \leq r < \infty$, is initially at zero temperature. For times $t > 0$ heat is generated in the medium at a rate of $g(r, t)$, W/m^3. Obtain an expression for the temperature distribution $T(r, t)$ in the medium for times $t > 0$. Consider the special cases where (1) the heat source is of constant strength, that is, $g(r, t) = g_0$ = constant, and (2) the heat source is an instantaneous line heat source of strength g_L^i, Ws/m, situated along the z axis in the medium and releases its heat spontaneously at time $t = 0$, that is, $g(r, t) = (\frac{1}{2\pi r}) g_L^i \delta(r)\delta(t)$.

14-12 The region $a \leq r < \infty$ is initially at zero temperature. For times $t > 0$, heat is generated in the medium at a rate of $g(r, t)$, W/m^3, while the boundary surface at $r = a$ is kept at zero temperature. Obtain an expression for the temperature distribution $T(r, t)$ in the medium for times $t > 0$. Consider the special case of constant heat generation in the medium.

14-13 The cylindrical region $0 \leq r \leq b$, $0 \leq \phi \leq 2\pi$ is initially at temperature $F(r, \phi)$. For times $t > 0$ the boundary surface at $r = b$ is kept insulated.

Obtain an expression for the temperature distribution $T(r, \phi, t)$ in the region for times $t > 0$.

14-14 The cylindrical region $0 \leq r \leq b$, $0 \leq \phi \leq 2\pi$ is initially at zero temperature. For times $t > 0$, heat is generated in the medium at a rate of $g(r, \phi, t)$, W/m^3, while the boundary at $r = b$ is kept at zero temperature. Obtain an expression for the temperature distribution $T(r, \phi, t)$ in the region for times $t > 0$.

14-15 The cylindrical region $a \leq r \leq b$, $0 \leq \phi \leq 2\pi$ is initially at temperature $F(r, \phi)$. For times $t > 0$, the boundaries at $r = a$ and $r = b$ are kept at zero temperatures. Obtain an expression for the temperature distribution $T(r, \phi, t)$ in the region for times $t > 0$.

14-16 The cylindrical region consisting of a portion of a cylinder, $0 \leq r \leq b$, $0 \leq \phi \leq \phi_0$ (where $\phi_0 < 2\pi$) is initially at zero temperature. For times $t > 0$ heat is generated in the medium at a rate of $g(r, \phi, t)$, W/m^3, while all boundary surfaces are kept at zero temperature. Obtain an expression for the temperature distribution $T(r, \phi, t)$ in the region for times $t > 0$. Also consider the special case of $g(r, \phi, t) = g_0 =$ constant.

14-17 The cylindrical region consisting of a portion of a cylinder, $a \leq r \leq b$, $0 \leq \phi \leq \phi_0$ (where $\phi_0 < 2\pi$) is initially at temperature $F(r, \phi)$. For times $t > 0$ all boundary surfaces are kept at zero temperature. Obtain an expression for the temperature distribution $T(r, \phi, t)$ in the region for times $t > 0$. Also consider the special case of uniform initial temperature distribution, that is, $F(r, \phi) = T_0 =$ constant.

14-18 The cylindrical region $a \leq r \leq b$, $0 \leq \phi \leq \phi_0$ (where $\phi_0 < 2\pi$) is initially at zero temperature. For times $t > 0$ heat is generated in the medium at a rate of $g(r, \phi, t)$, W/m^3, while the boundaries are kept at zero temperature. Obtain an expression for the temperature distribution $T(r, \phi, t)$ in the medium for times $t > 0$.

14-19 A cylindrical region $0 \leq r \leq b$, $0 \leq z \leq L$ is initially at zero temperature. For times $t > 0$ heat is generated in the medium at a rate of $g(r, z, t)$ W/m^3, while the boundary surface at $z = 0$ is kept insulated and all the remaining boundaries are kept at zero temperature. Obtain an expression for the temperature distribution $T(r, z, t)$ in the region.

14-20 A cylindrical region $0 \leq r \leq b$, $0 \leq z < \infty$ is initially at temperature $F(\mathbf{r}, z)$. For times $t > 0$ all the boundary surfaces are kept at zero temperature. Obtain an expression for the temperature distribution $T(r, z, t)$ in the region for times $t > 0$.

14-21 A hemispherical region $0 \leq r \leq b$, $0 \leq \mu \leq 1$ is initially at temperature $F(r, \mu)$. For times $t > 0$ the boundary surface at $\mu = 0$ is kept insulated and the boundary surface at $r = b$ is kept at zero temperature. Obtain an expression for the temperature distribution $T(r, \mu, t)$ in the hemisphere for times $t > 0$.

14-22 A hemispherical region $0 \le r \le b,\ 0 \le \mu \le 1$ is initially at zero temperature. For times $t > 0$ heat is generated in the medium at a rate of $g(r, \mu, t)$, W/m^3, while the boundary surface at $\mu = 0$ is kept insulated and the boundary at $r = b$ is kept at zero temperature. Obtain an expression for the temperature distribution $T(r, \mu, t)$ in the region for times $t > 0$.

14-23 A hollow hemispherical region $a \le r \le b,\ 0 \le \mu \le 1$ is initially at zero temperature. For times $t > 0$ heat is generated in the medium at a rate of $g(r, \mu, t)$, W/m^3, while the boundaries are kept at zero temperature. Obtain an expression for the temperature distribution $T(r, \mu, t)$ in the region.

14-24 A solid sphere of radius $r = b$ is initially at temperature $F(r, \mu, \phi)$. For times $t > 0$ the boundary surface at $r = b$ is kept insulated. Obtain an expression for the temperature distribution $T(r, \mu, \phi, t)$ in the sphere for times $t > 0$.

14-25 Solve for the steady-state temperature distribution $T(x, y)$ in a rectangular strip $0 \le y \le b,\ 0 \le x < \infty$ subject to the boundary conditions $T = f(y)$ at $x = 0$ and $T = 0$ at $y = 0$ and $y = b$.

14-26 Solve for the steady-state temperature distribution $T(r, \mu, \phi)$ in a solid sphere of radius $r = b$ subject to the boundary condition $T = f(\mu, \phi)$ at the boundary surface $r = b$.

NOTES

1. To prove the orthogonality relation given by equations (14-3), equation (14-2a) is written for two different eigenfunctions $\psi_m(\hat{r})$ and $\psi_n(\hat{r})$, corresponding to two different eigenvalues λ_m and λ_n as

$$\nabla^2\psi_m(\hat{r}) + \lambda_m^2\psi_m(\hat{r}) = 0 \quad \text{in} \quad R \tag{1a}$$

$$\nabla^2\psi_n(\hat{r}) + \lambda_n^2\psi_n(\hat{r}) = 0 \quad \text{in} \quad R \tag{1b}$$

The first equation is multiplied by $\psi_n(\hat{r})$, the second by $\psi_m(\hat{r})$, and the results are subtracted and integrated over the region R:

$$(\lambda_m^2 - \lambda_n^2)\int_R \psi_m(\hat{r})\psi_n(\hat{r})\,dV = \int_R [\psi_m\nabla^2\psi_n - \psi_n\nabla^2\psi_m]\,dV \tag{2}$$

The volume integral on the right is changed to surface integral by Green's theorem. We find

$$\begin{aligned}(\lambda_m^2 - \lambda_n^2)\int_R \psi_m(\hat{r})\psi_n(\hat{r})\,dV &= \int_s \left(\psi_m\frac{\partial\psi_n}{\partial n} - \psi_n\frac{\partial\psi_m}{\partial n}\right) ds \\ &= \sum_{i=1}^{N}\int_{s_i}\left(\psi_m\frac{\partial\psi_n}{\partial n_i} - \psi_n\frac{\partial\psi_m}{\partial n_i}\right) ds_i\end{aligned} \tag{3}$$

The boundary condition (14-2b) is written for two different eigenfunctions $\psi_m(\hat{r})$ and $\psi_n(\hat{r})$:

$$k_i \frac{\partial \psi_m}{\partial n_i} + h_i \psi_m = 0 \tag{4a}$$

$$k_i \frac{\partial \psi_n}{\partial n_i} + h_i \psi_n = 0 \tag{4b}$$

The first is multiplied by ψ_n, the second by ψ_n, and the results are subtracted

$$\psi_m \frac{\partial \psi_n}{\partial n_i} - \psi_n \frac{\partial \psi_m}{\partial n_i} = 0 \tag{5}$$

when this result is introduced into equation (3), we obtain

$$(\lambda_m^2 - \lambda_n^2) \int_R \psi_m(\hat{r}) \psi_n(\hat{r})\, dV = 0 \tag{6}$$

Thus we have

$$\int_R \psi_m(\hat{r}) \psi_n(\hat{r})\, dV = 0 \qquad \text{for} \qquad m \neq n \tag{7}$$

2. The closed-form expressions given by equations (14-44) can be derived as now described. Consider the problem

$$\frac{\partial^2 \theta}{\partial x^2} = \frac{1}{\alpha} \frac{\partial \theta}{\partial t} \qquad \text{in} \qquad 0 < x < L, \qquad t > 0 \tag{1a}$$

$$\text{BC:} \qquad \left. \frac{\partial \theta}{\partial x} \right|_{x=0} = 0 \qquad t > 0 \tag{1b}$$

$$\text{BC:} \qquad \theta(x = L) = 0 \qquad t > 0 \tag{1c}$$

$$\text{IC:} \qquad \theta(t = 0) = 1 \qquad \text{in} \qquad 0 \le x \le L \tag{1d}$$

The solution of this problem is given as

$$\begin{aligned} \theta(x, t) &= \frac{2}{L} \sum_{m=1}^{\infty} e^{-\alpha \beta_m^2 t} \cos \beta_m x \int_{x'=0}^{L} 1 \cos \beta_m x'\, dx' \\ &= \frac{2}{L} \sum_{m=1}^{\infty} e^{-\alpha \beta_m^2 t} \frac{\cos \beta_m x}{\beta_m} (-1)^{m-1} \end{aligned} \tag{2}$$

where $\beta_m = (2m - 1)\pi/2L$. For $t = 0$ equation (2) should be equal to the initial condition (1d); hence

$$\frac{2}{L} \sum_{m=1}^{\infty} (-1)^{m-1} \frac{\cos \beta_m x}{\beta_m} = 1 \tag{3}$$

which is the result given by equation (14-44a). We now consider the problem given by equations (1) for an initial condition $(x^2 - L^2)$. The solution becomes

$$\theta(x, t) = \frac{2}{L} \sum_{m=1}^{\infty} e^{-\alpha \beta_m^2 t} \cos \beta_m x \int_{x'=0}^{L} (x'^2 - L^2) \cos \beta_m x'\, dx' \tag{4}$$

After performing the integration we obtain

$$\theta(x, t) = -\frac{4}{L} \sum_{m=1}^{\infty} e^{-\alpha \beta_m^2 t} (-1)^{m-1} \frac{\cos \beta_m x}{\beta_m^3} \tag{5}$$

For $t = 0$, we have $\theta(x, t) = x^2 - L^2$; then

$$x^2 - L^2 = -\frac{4}{L} \sum_{m=1}^{\infty} (-1)^{m-1} \frac{\cos \beta_m x}{\beta_m^3} \tag{6}$$

which is the result given by equation (14-44b).

3. The integral transform of equation (14-56a) according to the definition of the integral transform (14-57b) is

$$\int_{x=0}^{\infty} X(\beta, x) \frac{\partial^2 T}{\partial x^2}\, dx + \frac{1}{k} \overline{g}(\beta, t) = \frac{1}{\alpha} \frac{d\overline{T}(\beta, t)}{dt} \tag{1}$$

The first term on the left is evaluated by integrating it by parts twice:

$$\begin{aligned} \int_{x=0}^{\infty} X(\beta, x) \frac{\partial^2 T}{\partial x^2}\, dx &= \left[X \frac{\partial T}{\partial x} \right]_0^{\infty} - \int_{x=0}^{\infty} \frac{dX}{dx} \frac{\partial T}{\partial x} dx \\ &= \left[X \frac{\partial T}{\partial x} - T \frac{dX}{dx} \right]_0^{\infty} + \int_{x=0}^{\infty} T \frac{d^2 X}{dx^2}\, dx \end{aligned} \tag{2}$$

The term inside the bracket vanishes at the upper limit; it is evaluated at the lower limit by utilizing the boundary conditions (14-56b) and (Table 6-1); we obtain

$$\left[X \frac{\partial T}{\partial x} - T \frac{dX}{dx} \right]_0^{\infty} = \left. \frac{X(\beta, x)}{k_1} \right|_{x=0} f_1(t) \tag{3}$$

The second term on the right in equation (2) is evaluated by multiplying equation (6-6) by $T(x, t)$ and utilizing the definition of the integral transform (14-57b). We obtain

$$\int_{x=0}^{\infty} T \frac{\partial^2 X}{dx^2}\, dx = -\beta^2 \int_{x=0}^{\infty} T X\, dx = -\beta^2 \overline{T}(\beta, t) \tag{4}$$

Introducing equations (2)–(4) into (1), we find

$$\frac{d\overline{T}(\beta, t)}{dt} + \alpha \beta^2 \overline{T}(\beta, t) = \frac{\alpha}{k} \overline{g}(\beta, t) + \left. \frac{X(\beta, x)}{k_1} \right|_{x=0} f_1(t) \tag{5}$$

which is the result given by equation (14-58a).

4. We consider the representation

$$F^*(x) = \frac{1}{\pi} \int_{\beta=0}^{\infty} \int_{x'=-\infty}^{\infty} F^*(x') \cos \beta (x - x')\, dx'\, d\beta \tag{1}$$

The integral of the cosine term with respect to β is expressed as a complex integral in the form

$$\int_{\beta=0}^{L} \cos\beta(x'-x)\,d\beta = \frac{1}{2}\int_{\beta=-L}^{L} \cos\beta(x'-x)\,d\beta = \frac{1}{2}\int_{\beta=-L}^{L} e^{i\beta(x'-x)}\,d\beta \tag{2a}$$

since

$$\int_{\beta=-L}^{L} \sin\beta(x'-x)\,d\beta = 0 \tag{2b}$$

Then, in the limit $L \to \infty$, equation (1) can be written as

$$F^*(x) = \frac{1}{2\pi}\int_{\beta=-\infty}^{\infty} e^{-i\beta x}\int_{x'=-\infty}^{\infty} e^{i\beta x'} F^*(x')\,dx'\,d\beta \tag{3}$$

which is the result given by equation (14-62).

5. The integral transform of equation (14-107a) with respect to the ϕ variable, by the application of the transform (14-108b), is

$$\frac{\partial^2 \overline{T}}{\partial r^2} + \frac{1}{r}\frac{\partial \overline{T}}{\partial r} + \frac{1}{r^2}\int_{\phi=0}^{2\pi} \cos\nu(\phi-\phi')\frac{\partial^2 T}{\partial \phi^2}d\phi + \frac{\overline{g}}{k} = \frac{1}{\alpha}\frac{\partial \overline{T}}{\partial t} \tag{1}$$

The integral term is evaluated as follows. Let $\Phi(\phi) \equiv \cos\nu(\phi-\phi')$. Then, the integral term becomes

$$\begin{aligned}\int_{\phi=0}^{2\pi} \Phi(\phi)\frac{\partial^2 T}{\partial \phi^2}d\phi &= \left[\Phi\frac{\partial T}{\partial \phi}\right]_0^{2\pi} - \int_{\phi=0}^{2\pi} \frac{d\Phi}{d\phi}\frac{\partial T}{\partial \phi}d\phi \\ &= \left[\Phi\frac{\partial T}{\partial \phi} - T\frac{d\Phi}{d\phi}\right]_0^{2\pi} + \int_{\phi=0}^{2\pi} T\frac{d^2\Phi}{d\phi^2}d\phi \\ &= \int_{\phi=0}^{2\pi} T\frac{d^2\Phi}{d\phi^2}d\phi \end{aligned} \tag{2}$$

since the terms in the bracket vanish because the functions are cyclic with a period of 2π. To evaluate this integral, we consider the eigenvalue problem given as

$$\frac{d^2\Phi(\phi)}{d\phi^2} + \nu^2\Phi(\phi) = 0 \qquad \text{in} \qquad 0 \le \phi \le 2\pi \tag{3}$$

The function $\Phi(\phi)$ is cyclic with a period of 2π. We multiply this equation by T, integrate with respect to ϕ from 0 to 2π , and utilize the definition of the integral transform (14-108b) to obtain

$$\int_{\phi=0}^{2\pi} T\frac{d^2\Phi(\phi)}{d\phi^2}d\phi = -\nu^2\int_{\phi=0}^{2\pi} \Phi(\phi)T\,d\phi = -\nu^2\overline{T} \tag{4}$$

Introducing equations (1) and (4) into (1), we find

$$\frac{\partial^2 \overline{T}}{\partial r^2} + \frac{1}{r}\frac{\partial \overline{T}}{\partial r} - \frac{\nu^2}{r^2}\overline{T} + \frac{\overline{g}}{k} = \frac{1}{\alpha}\frac{\partial \overline{T}}{\partial t} \tag{5}$$

which is the result given by equation (14-109a).

6. The integral transform of equation (14-109a) by the application of the transform (14-110b) is

$$\int_{r=0}^{b} rR_\nu(\beta_m, r)\left[\frac{\partial^2 \overline{T}}{\partial r^2} + \frac{1}{r}\frac{\partial \overline{T}}{\partial r} - \frac{\nu^2}{r^2}\overline{T}\right] dr + \frac{\tilde{\overline{g}}}{k} = \frac{1}{\alpha}\frac{d\tilde{\overline{T}}}{dt}. \tag{1}$$

The integral term is evaluated by integrating it by parts twice, utilizing the eigenvalue problem for the R_ν function and the boundary conditions as

$$\begin{aligned} I &\equiv \int_{r=0}^{b} rR_\nu\left[\frac{\partial^2 \overline{T}}{\partial r^2} + \frac{1}{r}\frac{\partial \overline{T}}{\partial r} - \frac{\nu^2}{r^2}\overline{T}\right] dr \\ &= \left[r\left(R_\nu\frac{\partial \overline{T}}{\partial r} - \overline{T}\frac{dR_\nu}{dr}\right)\right]_0^b + \int_{r=0}^{b} r\left(\frac{d^2R_\nu}{dr^2} + \frac{1}{r}\frac{dR_\nu}{dr} - \frac{\nu^2}{r^2}R_\nu\right)\overline{T}\, dr \\ &= b\left(R_\nu\frac{\partial \overline{T}}{\partial r} - \overline{T}\frac{dR_\nu}{\partial r}\right)\Bigg|_{r=b} + \int_{r=0}^{b} r\left(\frac{d^2R_\nu}{dr^2} + \frac{1}{r}\frac{dR_\nu}{dr} - \frac{\nu^2}{r^2}R_\nu\right)\overline{T}\, dr \end{aligned} \tag{2}$$

since the term inside the bracket vanishes at the lower limit. From the eigenvalue problem [i.e., equation (2-36b)] we have

$$\frac{d^2R_\nu}{dr^2} + \frac{1}{r}\frac{dR_\nu}{dr} + \left(\beta_m^2 - \frac{\nu^2}{r^2}\right)R_\nu = 0 \qquad \text{in} \qquad 0 \le r < b \tag{3}$$

Multiplying equation (3) by $r\overline{T}$, integrating it with respect to r from $r = 0$ to $r = b$, and utilizing the definition of the integral transform (14-110b) we obtain

$$\begin{aligned} &\int_{r=0}^{b} r\left(\frac{d^2R_\nu}{dr^2} + \frac{1}{r}\frac{dR_\nu}{dr} - \frac{\nu^2}{r^2}R_\nu\right)\overline{T}\, dr \\ &\qquad = -\beta_m^2\int_{r=0}^{b} rR_\nu\overline{T}\, dr = -\beta_m^2\overline{T} \end{aligned} \tag{4}$$

From the boundary conditions (14-109b) and (2-37b) we have

$$k_2\frac{\partial \overline{T}}{\partial r} + h_2\overline{T} = \overline{f}_2(\nu, t) \qquad \text{at} \qquad r = b \tag{5a}$$

$$k_2\frac{dR_\nu}{dr} + h_2R_\nu = 0 \qquad \text{at} \qquad r = b \tag{5b}$$

Multiplying (5a) by R_ν and (5b) by $\overline{T}$ and subtracting the results, we obtain

$$\left[R_\nu\frac{\partial \overline{T}}{\partial r} - \overline{T}\frac{dR_\nu}{dr}\right]_{r=b} = b\left.\frac{R_\nu}{k_2}\right|_{r=b}\overline{f}_2(\nu, t) \tag{6}$$

Introducing equations (6), (4), and (2) into equation (1), we find

$$-\beta_m^2\tilde{\overline{T}}(\beta_m, \nu, t) + b\left.\frac{R_\nu(\beta_m, r)}{k_2}\right|_{r=b}\overline{f}_2(\nu, t) + \frac{\tilde{\overline{g}}(\beta_m, \nu, t)}{k} = \frac{1}{\alpha}\frac{d\tilde{\overline{T}}}{dt}$$

or

$$\frac{d\tilde{\overline{T}}(\beta_m, \nu, t)}{dt} + \alpha\beta_m^2 d\tilde{\overline{T}}(\beta_m, \nu, t) = \frac{\alpha}{k}\tilde{\overline{g}}(\beta_m, \nu, t) + \alpha b \left. \frac{R_\nu(\beta_m, r)}{k_2} \right|_{r=b} \overline{f}_2(\nu, t) \tag{7}$$

which is the result given by equation (14-111a).

7. When taking the integral transform of the differential equation (14-143a) by the application of the transform (14-144b), we need to consider only the removal of the following differential operator:

$$\nabla^2 V \equiv -\frac{V}{4} + \frac{\partial}{\partial\mu}\left[(1-\mu^2)\frac{\partial V}{\partial\mu}\right] \quad \text{in} \quad -1 \leq \mu \leq 1 \tag{1}$$

since the integral transform of the remaining terms is straightforward. The integral transform of this operator under the transform (14-144b) is

$$\overline{\nabla^2 V} = -\frac{1}{4}\overline{V} + \int_{\mu=-1}^{1} P_n(\mu)\frac{\partial}{\partial\mu}\left[(1-\mu^2)\frac{\partial V}{\partial\mu}\right] d\mu \tag{2}$$

The integration is performed by integrating it by parts twice:

$$\overline{\nabla^2 V} = -\frac{1}{4}\overline{V} + \left[(1-\mu^2)\left(P_n\frac{\partial V}{\partial\mu} - V\frac{dP_n}{d\mu}\right)\right]_{-1}^{1} + \int_{\mu=-1}^{1} V\frac{d}{d\mu}\left[(1-\mu^2)\frac{dP_n}{d\mu}\right] d\mu \tag{3}$$

The terms inside the bracket vanish at both limits. The integral term is evaluated by noting that the $P_n(\mu)$ function satisfies the Legendre equation:

$$\frac{d}{d\mu}\left[(1-\mu^2)\frac{dP_n}{d\mu}\right] + n(n+1)P_n(\mu) = 0 \tag{4}$$

Multiplying this equation by V and integrating from $\mu = -1$ to 1, we find

$$\int_{\mu=-1}^{1} V\frac{d}{d\mu}\left[(1-\mu^2)\frac{dP_n}{d\mu}\right] d\mu = -n(n+1)\int_{\mu=-1}^{1} P_n(\mu)V d\mu = -n(n+1)\overline{V} \tag{5}$$

Introducing equation (5) into (3), we obtain

$$\overline{\nabla^2 V} = -\tfrac{1}{4}\overline{V} - n(n+1)\overline{V} = -(n+\tfrac{1}{2})^2\overline{V} \tag{6}$$

which is the term that appears in equation (14-145a).

8. When taking the integral transform of the equation (14-156) by the application of the transform (14-158b) we need to consider only the removal of the following differential operator:

$$\nabla^2 \overline{V} \equiv -\frac{1}{4}\overline{V} + \left\{ \frac{\partial}{\partial \mu} \left[(1-\mu^2) \frac{\partial \overline{V}}{\partial \mu} - \frac{m^2 \overline{V}}{1-\mu^2} \right] \right\} \quad \text{in} \quad -1 \le \mu \le 1 \tag{1}$$

The integral transform of this operator under the transform (14-158b) is

$$\widetilde{\nabla^2 \overline{V}} = -\frac{1}{4}\tilde{\overline{V}} + \int_{\mu=-1}^{1} P_n^m \left\{ \frac{\partial}{\partial \mu} \left[(1-\mu^2) \frac{\partial \overline{V}}{\partial \mu} \right] \right\} d\mu - \int_{\mu=-1}^{1} \frac{m^2}{1-\mu^2} P_n^m \overline{V}\, d\mu \tag{2}$$

The first integral is performed by integrating by parts twice:

$$\widetilde{\nabla^2 \overline{V}} \equiv -\frac{1}{4}\tilde{\overline{V}} + \left[(1-\mu^2) \left(P_n^m \frac{\partial \overline{V}}{\partial \mu} - \overline{V} \frac{dP_n^m}{d\mu} \right) \right]_{-1}^{1} + \int_{\mu=-1}^{1} \overline{V} \left\{ \frac{\partial}{\partial \mu} \left[(1-\mu^2) \frac{dP_n^m}{d\mu} \right] - \frac{m^2}{1-\mu^2} \overline{V} \right\} d\mu \tag{3}$$

The second term in the bracket vanishes at both limits. The integral term is evaluated by noting that the $P_n^m(\mu)$ function satisfies Legendre's associated differential equation (14-157):

$$\frac{d}{d\mu} \left[(1-\mu^2) \frac{dP_n^m}{d\mu} \right] + \left[n(n+1) - \frac{m^2}{1-\mu^2} \right] P_n^m = 0 \tag{4}$$

Multiplying this equation by $\overline{V}$ and integrating from $\mu = -1$ to 1, we find

$$\int_{\mu=-1}^{1} \overline{V} \left\{ \frac{d}{d\mu} \left[(1-\mu^2) \frac{dP_n^m}{d\mu} \right] - \frac{m^2}{1-\mu^2} P_n^m \right\} d\mu = -n(n+1) \int_{\mu=-1}^{1} P_n^m \overline{V} d\mu = -n(n+1) \tilde{\overline{V}} \tag{5}$$

Introducing equation (5) into (3), we obtain

$$\widetilde{\nabla^2 \overline{V}} \equiv -\tfrac{1}{4}\tilde{\overline{V}} - n(n+1)\tilde{\overline{V}} = -(n+\tfrac{1}{2})^2 \tilde{\overline{V}} \tag{6}$$

which is the term that appears in equation (14-159).

15

HEAT CONDUCTION IN ANISOTROPIC SOLIDS

In the previous chapters we have considered heat conduction in solids that are said to be *isotropic*; that is, the thermal conductivity does not depend on direction. However, there are many natural and synthetic materials in which the thermal conductivity varies with direction; they are called *anisotropic* materials. For example, crystals, wood, sedimentary rocks, metals that have undergone heavy cold pressing, laminated sheets, cables, heat shielding materials for space vehicles, fiber-reinforced composite structures, and many others are anisotropic materials. In wood, the thermal conductivity is different along the grain, across the grain, and circumferentially. In laminated sheets the thermal conductivity is not the same along and across the laminations. Therefore, heat conduction in anisotropic materials has numerous important applications in various branches of science and engineering.

Most of the earlier work has been limited to the problems of one-dimensional heat flow in crystal physics [1, 2]. The differential equation of heat conduction for anisotropic solids involves cross derivatives of the space variables; therefore, the general analysis of heat conduction in anisotropic solids is complicated. When the cross derivatives are absent from the heat conduction equation, as in the case of *orthotropic* solids, the analysis of heat transfer is significantly simplified and has been considered in several references [3–12]. Several works have appeared in the literature on the solution of heat conduction in anisotropic media [13–30]. Experimental work on heat diffusion in anisotropic solids is very limited; the available work [2, 5, 17] deals with either the one-dimensional situation or the orthotropic materials. More recently, considerable attention has been given to carbon nanotube structures, superconductors, and graphene [31–37], which can all exhibit appreciable anisotropy.

In this chapter, we present the differential equation of heat conduction and the boundary conditions for anisotropic solids, discuss the thermal conductivity coefficients for crystal structures, and illustrate the solution of the steady-state and time-dependent heat conduction problems in anisotropic solids with representative examples.

15-1 HEAT FLUX FOR ANISOTROPIC SOLIDS

The heat flux in isotropic solids, as discussed in Chapter 1, obeys Fourier's law

$$\hat{q}'' = -k\nabla T \tag{15-1}$$

where the thermal conductivity is independent of direction, and the heat flux vector $\hat{q}''$ is normal to the isothermal surface passing through the spatial position considered.

In the case of anisotropic solids, the component of the heat flux, say, q_1'', along the $0x_1$ direction, depends in general on a linear combination of the temperature gradients along the $0x_1$, $0x_2$, and $0x_3$ directions. With this consideration, the general expressions for the three components of the heat flux q_1'', q_2'', and q_3'' along the $0x_1$, $0x_2$, and $0x_3$ directions in the rectangular coordinate system are given, respectively, as [38]

$$-q_1'' = k_{11}\frac{\partial T}{\partial x_1} + k_{12}\frac{\partial T}{\partial x_2} + k_{13}\frac{\partial T}{\partial x_3} \tag{15-2a}$$

$$-q_2'' = k_{21}\frac{\partial T}{\partial x_1} + k_{22}\frac{\partial T}{\partial x_2} + k_{23}\frac{\partial T}{\partial x_3} \tag{15-2b}$$

$$-q_3'' = k_{31}\frac{\partial T}{\partial x_1} + k_{32}\frac{\partial T}{\partial x_2} + k_{33}\frac{\partial T}{\partial x_3} \tag{15-2c}$$

which can be written more compactly in the form

$$q_i'' = -\sum_{j=1}^{3} k_{ij}\frac{\partial T}{\partial x_j} \qquad \text{for} \qquad i = 1, 2, 3 \tag{15-3}$$

Therefore, for an anisotropic solid the heat flux vector $\hat{q}''$ is not necessarily normal to the isothermal surface passing through the point considered. The thermal conductivity of an anisotropic solid involves nine components, k_{ij}, called the *conductivity coefficients*, that are considered to be the components of a second-order tensor $\bar{\bar{k}}$:

$$\bar{\bar{k}} \equiv \begin{vmatrix} k_{11} & k_{12} & k_{13} \\ k_{21} & k_{22} & k_{23} \\ k_{31} & k_{32} & k_{33} \end{vmatrix} \tag{15-4a}$$

From Onsager's [38] principles of thermodynamics of irreversible processes, it is shown that when the fluxes (i.e., q_i'') and the forces (i.e., $\partial T/\partial x_i$) are related to each other linearly as given by equations (15-2), the phenomenological coefficients obey the reciprocity relation. A discussion of the application of Onsager's reciprocity relation for the thermal conductivity coefficients associated with heat conduction in anisotropic solids is given by Casimir [39]. Therefore, the conductivity coefficients k_{ij} can be considered to obey the reciprocity relation

$$k_{ij} = k_{ji} \qquad i, j = 1, 2, 3 \tag{15-4b}$$

Furthermore, as discussed in reference 40, according to irreversible thermodynamics, the coefficients k_{11}, k_{22}, and k_{33} are positive, that is,

$$k_{ii} > 0 \tag{15-4c}$$

and the magnitude of the coefficients k_{ij}, for $i \neq j$, is limited by the requirement [38]

$$k_{ii}k_{jj} > k_{ij}^2 \qquad \text{for} \qquad i \neq j \tag{15-4d}$$

The expression for the heat flux components, given by equation (15-3) for the rectangular coordinate system, can readily be generalized for the orthogonal curvilinear coordinate system (u_1, u_2, u_3) as

$$q_i'' = -\sum_{j=1}^{3} \frac{1}{a_j} k_{ij} \frac{\partial T}{\partial u_j} \qquad i = 1, 2, 3 \tag{15-5}$$

where a_j are the scale factors discussed in Chapter 1.

For the (x_1, x_2, x_3) *rectangular coordinate* system, equation (15-5) reduces to equations (15-2).

For the (r, ϕ, z) *cylindrical coordinate* system, we set $u_1 = r$, $u_2 = \phi$, $u_3 = z$, and $a_1 = 1$, $a_2 = r$, $a_3 = 1$; then equation (15-5) gives

$$-q_r'' = k_{11}\frac{\partial T}{\partial r} + k_{12}\frac{1}{r}\frac{\partial T}{\partial \phi} + k_{13}\frac{\partial T}{\partial z} \tag{15-6a}$$

$$-q_\phi'' = k_{21}\frac{\partial T}{\partial r} + k_{22}\frac{1}{r}\frac{\partial T}{\partial \phi} + k_{23}\frac{\partial T}{\partial z} \tag{15-6b}$$

$$-q_z'' = k_{31}\frac{\partial T}{\partial r} + k_{32}\frac{1}{r}\frac{\partial T}{\partial \phi} + k_{33}\frac{\partial T}{\partial z} \tag{15-6c}$$

For the (r, ϕ, θ) *spherical coordinate system,* we set $u_1 = r$, $u_2 = \phi$, $u_3 = \theta$, and $a_1 = 1$, $a_2 = r\sin\theta$, $a_3 = r$, and obtain

$$-q_r'' = k_{11}\frac{\partial T}{\partial r} + k_{12}\frac{1}{r\sin\theta}\frac{\partial T}{\partial \phi} + k_{13}\frac{1}{r}\frac{\partial T}{\partial \theta} \tag{15-7a}$$

$$-q''_\phi = k_{21}\frac{\partial T}{\partial r} + k_{22}\frac{1}{r\sin\theta}\frac{\partial T}{\partial \phi} + k_{23}\frac{1}{r}\frac{\partial T}{\partial \theta} \tag{15-7b}$$

$$-q''_\theta = k_{31}\frac{\partial T}{\partial r} + k_{32}\frac{1}{r\sin\theta}\frac{\partial T}{\partial \phi} + k_{33}\frac{1}{r}\frac{\partial T}{\partial \theta} \tag{15-7c}$$

15-2 HEAT CONDUCTION EQUATION FOR ANISOTROPIC SOLIDS

The differential equation of heat conduction for an anisotropic solid in the orthogonal curvilinear coordinate system (u_1, u_2, u_3) is given as

$$-\frac{1}{a_1a_2a_3}\left[\frac{\partial}{\partial u_1}\left(a_2a_3q''_1\right) + \frac{\partial}{\partial u_2}\left(a_1a_3q''_2\right) + \frac{\partial}{\partial u_3}\left(a_1a_2q''_3\right)\right] + g = \rho C\frac{\partial T}{\partial t} \tag{15-8}$$

where q''_1, q''_2, and q''_3 are the three components of the heat flux vector defined by equation (15-5), g is the heat generation term, and the other quantities are as defined previously.

We now present the explicit forms of the heat conduction equation (15-8) for the rectangular, cylindrical, and spherical coordinates for the case of constant conductivity coefficients.

Rectangular Coordinate System

For the (x, y, z) rectangular coordinate system, we set $u_1 = x$, $u_2 = y$, $u_3 = z$, and $a_1 = a_2 = a_3 = 1$; then equation (15-8), with q''_i, given by equation (15-5), yields

$$k_{11}\frac{\partial^2 T}{\partial x^2} + k_{22}\frac{\partial^2 T}{\partial y^2} + k_{33}\frac{\partial^2 T}{\partial z_2} + \left(k_{12} + k_{21}\right)\frac{\partial^2 T}{\partial x\,\partial y} + \left(k_{13} + k_{31}\right)\frac{\partial^2 T}{\partial x\,\partial z}$$
$$+ \left(k_{23} + k_{32}\right)\frac{\partial^2 T}{\partial y\,\partial z} + g\,(x, y, z, t) = \rho C\frac{\partial T(x, y, z, t)}{\partial t} \tag{15-9}$$

where $k_{12} = k_{21}$, $k_{13} = k_{31}$, and $k_{23} = k_{32}$ by the reciprocity relation.

Cylindrical Coordinate System

For the (r, ϕ, z) cylindrical coordinate system, we set $u_1 = r$, $u_2 = \phi$, $u_3 = z$, and $a_1 = 1$, $a_2 = r$, $a_3 = 1$. Then, from equations (15-8) and (15-5) we obtain

$$k_{11}\frac{1}{r}\frac{\partial}{\partial r}\left(r\frac{\partial T}{\partial r}\right) + k_{22}\frac{1}{r^2}\frac{\partial^2 T}{\partial \phi^2} + k_{33}\frac{\partial^2 T}{\partial z^2} + \left(k_{12} + k_{21}\right)\frac{1}{r}\frac{\partial^2 T}{\partial \phi\,\partial z}$$
$$+ \left(k_{13} + k_{31}\right)\frac{\partial^2 T}{\partial r\,\partial z} + \frac{k_{13}}{r}\frac{\partial T}{\partial z} + \left(k_{23} + k_{32}\right)\frac{1}{r}\frac{\partial^2 T}{\partial \phi\,\partial z} + g\,(r, \phi, z, t)$$
$$= \rho C\frac{\partial T(r, \phi, z, t)}{\partial t} \tag{15-10}$$

where $k_{ij} = k_{ji}$, $i \neq j$.

Spherical Coordinate System

For the (r, ϕ, θ) spherical coordinate system, we set $u_1 = r, u_2 = \phi, u_3 = \theta$, and $a_1 = 1, a_2 = r \sin\theta, a_3 = r$; then from equations (15-8) and (15-5) we obtain

$$
\begin{aligned}
&k_{11}\frac{1}{r^2}\frac{\partial}{\partial r}\left(r^2\frac{\partial T}{\partial r}\right) + k_{22}\frac{1}{r^2\sin^2\theta}\frac{\partial^2 T}{\partial \phi^2} + k_{33}\frac{1}{r^2\sin\theta}\frac{\partial}{\partial \theta}\left(\sin\theta\frac{\partial T}{\partial \theta}\right) \\
&+\frac{k_{12}+k_{21}}{r\sin\theta}\frac{\partial^2 T}{\partial r\,\partial \phi} + k_{12}\frac{1}{r^2\sin\theta}\frac{\partial T}{\partial \phi} + \frac{k_{13}+k_{31}}{r}\frac{\partial^2 T}{\partial r\,\partial \theta} + k_{13}\frac{1}{r^2}\frac{\partial T}{\partial \theta} \\
&+\left(k_{23}+k_{32}\right)\frac{1}{r^2\sin\theta}\frac{\partial^2 T}{\partial \theta\,\partial \phi} + k_{31}\frac{\cos\theta}{r\sin\theta}\frac{\partial T}{\partial r} + g\,(r, \phi, \theta, t) \\
&= \rho C\frac{\partial T\,(r, \phi, \theta, t)}{\partial t}
\end{aligned}
\tag{15-11}
$$

where $k_{ij} = k_{ji}$, $i \neq j$.

15-3 BOUNDARY CONDITIONS

The boundary conditions for the heat conduction equation for an anisotropic medium may be of the first, second, or third type. We consider a boundary surface S_i normal to the coordinate axis u_i. The boundary condition of the third type can be written as

$$
\mp\delta_i k_{\text{ref}}\cdot\left(\frac{\partial T}{\partial n^*}\right) + h_i T = f_i \qquad \text{on boundary } S_i \tag{15-12}
$$

where

$$
\frac{\partial T}{\partial n^*} \equiv \sum_{j=1}^{3}\frac{1}{a_j}\frac{k_{ij}}{k_{\text{ref}}}\frac{\partial T}{\partial u_j} \tag{15-13}
$$

and where

a_j = scale factor

k_{ref} = reference conductivity that may be chosen as k_{11}, k_{22}, or k_{33}

δ_i = zero or unity; that is, by setting $\delta_i = 0$, the boundary condition of the first type is obtained

Here $\partial/\partial n^*$ is the derivative as defined by equation (15-13). The choice of *plus* or *minus* sign in equation (15-12) depends on whether the outward-drawn normal to the boundary surface S_i is pointing in the *positive* or *negative* u_i direction, respectively. We illustrate the boundary conditions for the anisotropic medium with specific examples given below.

Example 15-1 Convection Boundary Conditions in Anisotropic Slab
Here we write the boundary conditions of the third type for an anisotropic slab at the boundary surfaces $x = 0$ and $x = L$. For the (x, y, z) rectangular coordinate system, we write

$$-\left(k_{11}\frac{\partial T}{\partial x} + k_{12}\frac{\partial T}{\partial y} + k_{13}\frac{\partial T}{\partial z}\right) + h_1 T = f_1 \quad \text{at} \quad x = 0 \tag{15-14a}$$

$$+\left(k_{11}\frac{\partial T}{\partial x} + k_{12}\frac{\partial T}{\partial y} + k_{13}\frac{\partial T}{\partial z}\right) + h_2 T = f_2 \quad \text{at} \quad x = L \tag{15-14b}$$

Equations (15-14) can be written more compactly in the form given by equation (15-12) by setting $k_{\text{ref}} \equiv k_{11}$:

$$-k_{11}\left.\frac{\partial T}{\partial n^*}\right|_{x=0} + h_1 \left.T\right|_{x=0} = f_1 \tag{15-14c}$$

$$k_{11}\left.\frac{\partial T}{\partial n^*}\right|_{x=L} + h_2 \left.T\right|_{x=L} = f_2 \tag{15-14d}$$

where

$$\frac{\partial}{\partial n^*} \equiv \frac{\partial}{\partial x} + \varepsilon_{12}\frac{\partial}{\partial y} + \varepsilon_{13}\frac{\partial}{\partial z} \tag{15-14e}$$

$$\varepsilon_{ij} \equiv k_{ij}/k_{11} \tag{15-14f}$$

Example 15-2 Convection Boundary Condition in Anisotropic Cylinder
We now write the boundary conditions of the third type for an anisotropic hollow cylinder at the boundary surfaces $r = a$ and $r = b$. For the (r, ϕ, z) cylindrical coordinate system, we take the scale factors as $a_1 = 1, a_2 = r$, and $a_3 = 1$ and write

$$-\left.\left(k_{11}\frac{\partial T}{\partial r} + k_{12}\frac{1}{r}\frac{\partial T}{\partial \phi} + k_{13}\frac{\partial T}{\partial z}\right)\right|_{r=a} + h_1 \left.T\right|_{r=a} = f_1 \tag{15-15a}$$

$$+\left.\left(k_{11}\frac{\partial T}{\partial r} + k_{12}\frac{1}{r}\frac{\partial T}{\partial \phi} + k_{13}\frac{\partial T}{\partial z}\right)\right|_{r=b} + h_2 \left.T\right|_{r=b} = f_2 \tag{15-15b}$$

Equations (15-15) can be written more compactly in the form given by equation (15-12) by setting $k_{\text{ref}} \equiv k_{11}$:

$$-k_{11}\left.\frac{\partial T}{\partial n^*}\right|_{r=a} + h_1 \left.T\right|_{r=a} = f_1 \tag{15-15c}$$

$$k_{11} \left. \frac{\partial T}{\partial n^*} \right|_{r=b} + h_2 \, T|_{r=b} = f_2 \tag{15-15d}$$

where

$$\frac{\partial}{\partial n^*} \equiv \frac{\partial}{\partial r} + \varepsilon_{12} \frac{1}{r} \frac{\partial}{\partial \phi} + \varepsilon_{13} \frac{\partial}{\partial z} \tag{15-15e}$$

$$\varepsilon_{ij} = \frac{k_{ij}}{k_{11}} \tag{15-15f}$$

15-4 THERMAL RESISTIVITY COEFFICIENTS

In the previous sections, we expressed each component of the heat flux vector as a linear sum of temperature gradients along the $0x_1, 0x_2$, and $0x_3$ axes as given by equation (15-2). Sometimes it is desirable to express the temperature gradient in a given direction as linear combination of the heat flux components in the $0x_1, 0x_2$, and $0x_3$ directions. To obtain such a relationship in the (x_1, x_2, x_3) rectangular coordinate system, we write equations (15-2) in matrix notation as

$$-\left[k_{ij}\right]\left[\frac{\partial T}{\partial x_i}\right] = \left[q_i''\right] \tag{15-16a}$$

or

$$-\left[\frac{\partial T}{\partial x_i}\right] = \left[k_{ij}\right]^{-1}\left[q_i''\right] \tag{15-16b}$$

Let r_{ij} be the elements of the inverse matrix $[k_{ij}]^{-1}$; then equation (15-16b) is written explicitly as

$$-\frac{\partial T}{\partial x_1} = r_{11}q_1'' + r_{12}q_2'' + r_{13}q_3'' \tag{15-17a}$$

$$-\frac{\partial T}{\partial x_2} = r_{21}q_1'' + r_{22}q_2'' + r_{23}q_3'' \tag{15-17b}$$

$$-\frac{\partial T}{\partial x_3} = r_{31}q_1'' + r_{32}q_2'' + r_{33}q_3'' \tag{15-17c}$$

where the coefficients r_{ij} are called the *thermal resistivity coefficients*. The coefficients r_{ij} can be determined in terms of k_{ij} by the matrix inversion procedure. Since $k_{ij} = k_{ji}$, it can be shown that r_{ij}'s are given by

$$r_{ij} = (-1)^{i+j} \frac{a_{ij}}{\Delta} \tag{15-18a}$$

where Δ is the symmetrical thermal conductivity tensor given by

$$\Delta \equiv \begin{vmatrix} k_{11} & k_{12} & k_{13} \\ k_{21} & k_{22} & k_{23} \\ k_{31} & k_{32} & k_{33} \end{vmatrix} \tag{15-18b}$$

and where a_{ij} is the cofactor obtained from Δ by omitting the ith row and the jth column. As in the case of thermal conductivity coefficients, k_{ij}, the thermal resistivity coefficients, r_{ij}, obey the reciprocity relation, $r_{ij} = r_{ji}$.

To illustrate the application of equation (15-18), we write below the thermal resistivity coefficient r_{12} and r_{11} in terms of the thermal conductivity coefficients as

$$r_{12} = (-1)^3 \frac{\begin{vmatrix} k_{21} & k_{23} \\ k_{31} & k_{33} \end{vmatrix}}{\Delta} = \frac{k_{23}k_{31} - k_{21}k_{33}}{\Delta} \tag{15-19a}$$

and

$$r_{11} = (-1)^2 \frac{\begin{vmatrix} k_{22} & k_{23} \\ k_{32} & k_{33} \end{vmatrix}}{\Delta} = \frac{k_{22}k_{33} - k_{23}k_{32}}{\Delta} \tag{15-19b}$$

15-5 DETERMINATION OF PRINCIPAL CONDUCTIVITIES AND PRINCIPAL AXES

We consider the heat conduction equation for an anisotropic solid in the x_1, x_2, x_3 rectangular coordinate system written as

$$k_{11}\frac{\partial^2 T}{\partial x_1^2} + k_{22}\frac{\partial^2 T}{\partial x_2^2} + k_{33}\frac{\partial^2 T}{\partial x_3^2} + \left(k_{12} + k_{21}\right)\frac{\partial^2 T}{\partial x_1\,\partial x_2} + \left(k_{13} + k_{31}\right)\frac{\partial^2 T}{\partial x_1\,\partial x_3}$$
$$+ \left(k_{23} + k_{32}\right)\frac{\partial^2 T}{\partial x_2\,\partial x_3} + g = \sigma C\frac{\partial T}{\partial t} \tag{15-20}$$

where $k_{ij} = k_{ji}$. When the conductivity matrix given by equation (15-4a) is symmetric, it is possible to find a new system of rectangular coordinates ξ_1, ξ_2, and ξ_3 that can transform it to a diagonal form as

$$\begin{vmatrix} k_1 & 0 & 0 \\ 0 & k_2 & 0 \\ 0 & 0 & k_3 \end{vmatrix} \tag{15-21}$$

where k_1, k_2, and k_3 are called the *principal conductivities* along the principal coordinate axes ξ_1, ξ_2, and ξ_3, respectively. Then the heat conduction equation (15-20), in terms of the principal coordinates, becomes

$$\frac{\partial^2 T}{\partial \xi_1^2} + \frac{\partial^2 T}{\partial \xi_2^2} + \frac{\partial^2 T}{\partial \xi_3^2} + g = \rho C \frac{\partial T}{\partial t} \tag{15-22}$$

The principal conductivities k_1, k_2, and k_3 are determined in the following manner. Let the thermal conductivity matrix be denoted by

$$\bar{\bar{k}} \equiv \begin{vmatrix} k_{11} & k_{12} & k_{13} \\ k_{21} & k_{22} & k_{23} \\ k_{31} & k_{32} & k_{33} \end{vmatrix} \tag{15-23}$$

Then the principal conductivities k_1, k_2, and k_3 are the eigenvalues of the following equation:

$$\begin{vmatrix} k_{11}-\lambda & k_{12} & k_{13} \\ k_{21} & k_{22}-\lambda & k_{23} \\ k_{31} & k_{32} & k_{33}-\lambda \end{vmatrix} = 0 \tag{15-24}$$

This is a cubic equation in λ and has three roots. Each of these roots is a real number because the conductivity coefficients k_{ij} are real numbers (see reference 41 for proof) and each corresponds to a principal conductivity, that is, $\lambda_1 = k_1, \lambda_2 = k_2$, and $\lambda_3 = k_3$ along the principal axes ξ_1, ξ_2, and ξ_3, respectively. The principal axes ξ_1, ξ_2, and ξ_3 are determined in the following manner:

Let l_1, l_2, l_3 be the direction cosines of the principal axis $0\xi_1$ with respect to the axes $0x_1, 0x_2, 0x_3$, and $\lambda_1 = k_1$ be the principal conductivity along the direction $0\xi_1$. Then l_1, l_2, l_3 satisfy the relation

$$\begin{vmatrix} k_{11}-\lambda_1 & k_{12} & k_{13} \\ k_{21} & k_{22}-\lambda_1 & k_{23} \\ k_{31} & k_{32} & k_{33}-\lambda_1 \end{vmatrix} \cdot \begin{vmatrix} l_1 \\ l_2 \\ l_3 \end{vmatrix} = 0 \tag{15-25a}$$

which provides three homogeneous equations for the three unknowns l_1, l_2, l_3; only two of these equations are linearly independent. An additional relation is obtained from the requirement that the direction cosines satisfy

$$l_1^2 + l_2^2 + l_3^2 = 1 \tag{15-25b}$$

Thus the three direction cosines of the principal axis $0\xi_1$ are determined from equations (15-25). The procedure is repeated with $\lambda_2 = k_2$ for the determination of m_1, m_2, m_3 of the principal axis $0\xi_2$ and with $\lambda_3 = k_3$ for n_1, n_2, n_3 of the principal axes $0\xi_3$.

15-6 CONDUCTIVITY MATRIX FOR CRYSTAL SYSTEMS

With symmetry considerations, crystals can be grouped into seven distinct systems identified as triclinic, monoclinic, orthorhombic hexagonal, tetragonal, trigonal, and cubic systems. Readers should consult references 1 and 2 for an in-depth discussion of this matter. Here we are concerned with the thermal conductivity tensors associated with such systems and summarize the results as follows:

1. ***Triclinic*** In this system there are no limitations imposed on the conductivity coefficients by symmetry considerations; hence all nine components of k_{ij} can be nonzero, and we have

$$\bar{\bar{k}} \equiv \begin{vmatrix} k_{11} & k_{12} & k_{13} \\ k_{21} & k_{22} & k_{23} \\ k_{31} & k_{32} & k_{33} \end{vmatrix} \tag{15-26}$$

2. ***Monoclinic*** Some of the components become zero with symmetry considerations; hence we have

$$\bar{\bar{k}} \equiv \begin{vmatrix} k_{11} & k_{12} & 0 \\ k_{21} & k_{22} & 0 \\ 0 & 0 & k_{23} \end{vmatrix} \tag{15-27}$$

3. ***Orthorhombic*** The conductivity coefficients are given by

$$\bar{\bar{k}} \equiv \begin{vmatrix} k_{11} & 0 & 0 \\ 0 & k_{22} & 0 \\ 0 & 0 & k_{33} \end{vmatrix} \tag{15-28}$$

4. ***Cubic*** In this system we have $k_{11} = k_{22} = k_{33}$; hence we write

$$\bar{\bar{k}} \equiv \begin{vmatrix} k_{11} & 0 & 0 \\ 0 & k_{11} & 0 \\ 0 & 0 & k_{11} \end{vmatrix} \tag{15-29}$$

5. ***Hexagonal, Tetragonal, and Trigonal*** For this system, we have

$$\bar{\bar{k}} \equiv \begin{vmatrix} k_{11} & k_{12} & 0 \\ -k_{12} & k_{11} & 0 \\ 0 & 0 & k_{33} \end{vmatrix} \tag{15-30}$$

It was previously stated that, whenever the heat flux law of the form given by equations (15-2) holds, the classical thermodynamic considerations lead to the reciprocity relationship given by equation (15-4b). In the case of crystals,

TABLE 15-1 Values of Principal Conductivities for Some Crystals at 30°C, in W/(m·K)

Crystal[a]	System	$k_1 k_2$	k_3
Quartz	Trigonal	6.5	11.3
Calcite	Trigonal	4.2	5.0
Bismuth	Trigonal	9.2	6.7
Graphite	Hexagonal	355	89

[a]For crystals listed here $k_1 = k_2$. From International Critical Tables (1929), Vol. 5, p. 231.

the results on the conductivity coefficients given above have been derived from the considerations of macroscopic symmetry. Since no general proof is available to show that the coefficients are symmetric, it has been necessary to rely on experiments. If the relation given by equation (15-4b) should apply, then it implies that $k_{12} = 0$ in equation (15-30) and $k_{21} = k_{12}$ in equation (15-27). Experimentally, principal conductivities are always found to be positive. Table 15-1 lists the values of principal conductivities for some crystals.

15-7 TRANSFORMATION OF HEAT CONDUCTION EQUATION FOR ORTHOTROPIC MEDIUM

The heat conduction equation for an orthotropic medium can be transformed to a standard heat conduction equation for an isotropic solid as described below. We now consider the heat conduction equation for an orthotropic medium in the rectangular coordinate system given by

$$k_1 \frac{\partial^2 T}{\partial x^2} + k_2 \frac{\partial^2 T}{\partial y^2} + k_3 \frac{\partial^2 T}{\partial z^2} + g = \rho C \frac{\partial T}{\partial t} \tag{15-31}$$

New independent variables X, Y, and Z are defined as

$$X = x\left(\frac{k}{k_1}\right)^{1/2}, \qquad Y = y\left(\frac{k}{k_2}\right)^{1/2}, \qquad \text{and} \qquad Z = z\left(\frac{k}{k_3}\right)^{1/2} \tag{15-32}$$

where k is a reference conductivity. Equation (15-31) becomes

$$k\left(\frac{\partial^2 T}{\partial X^2} + \frac{\partial^2 T}{\partial Y^2} + \frac{\partial^2 T}{\partial Z^2}\right) + g = \rho C \frac{\partial T}{\partial t} \tag{15-33}$$

which looks like the standard heat conduction equation for an isotropic solid. However, the choice of the reference thermal conductivity is not arbitrary. The reason for this is that a volume element in the original space "$dx\,dy\,dz$" transforms, under the transformation (15-32), into

$$\frac{(k_1 k_2 k_3)^{1/2}}{k^{3/2}} dX\, dY\, dZ \tag{15-34}$$

If the quantities ρC and the generation term g defined on the basis of unit volume should have the same physical significance, we should then have

$$\frac{(k_1 k_2 k_3)^{1/2}}{k^{3/2}} = 1 \quad \text{or} \quad k = (k_1 k_2 k_3)^{1/3} \tag{15-35a,b}$$

Then the heat conduction equation (15-33) takes the form

$$(k_1 k_2 k_3)^{1/3} \left(\frac{\partial^2 T}{\partial X^2} + \frac{\partial^2 T}{\partial Y^2} + \frac{\partial^2 T}{\partial Z^2} \right) + g = \rho C \frac{\partial T}{\partial t} \tag{15-36}$$

where X, Y, and Z are as defined by equation (15-32). This implies an isotropic medium of thermal conductivity $(k_1 k_2 k_3)^{1/3}$.

Several other ways of arriving at the result given by equation (15-35) are discussed in reference 8. Similar transformations are applicable to transform the equation into the standard form for the cylindrical and spherical coordinate systems.

Under the transformation discussed above, the solution of the resulting heat conduction equation is a straightforward matter, but the transformation of the solution to the original physical space requires additional commutations according to the transformation used. That is, the corresponding isotropic heat conduction problem of thermal conductivity $(k_1 k_2 k_3)^{1/3}$ is readily solved. The region is then distorted according to the transformation of equation (15-32).

15-8 SOME SPECIAL CASES

We now examine some special situations that may give some insight into the physical significance of heat flow in an anisotropic medium.

Temperature Depending Only on x_1 and x_2

For such a case we have $(\partial T/\partial x_3) = 0$; then equations (15-2) for the heat flux components reduce to

$$-q_1'' = k_{11} \frac{\partial T}{\partial x_1} + k_{12} \frac{\partial T}{\partial x_2} \tag{15-37a}$$

$$-q_2'' = k_{21} \frac{\partial T}{\partial x_1} + k_{22} \frac{\partial T}{\partial x_2} \tag{15-37b}$$

$$-q_3'' = k_{31} \frac{\partial T}{\partial x_1} + k_{32} \frac{\partial T}{\partial x_2} \tag{15-37c}$$

This result implies that there is still a heat flux component q_3'' in the x_3 direction even though there is no temperature gradient in that direction. The heat conduction equation (15-9) simplifies to

$$k_{11}\frac{\partial^2 T}{\partial x_1^2} + k_{22}\frac{\partial^2 T}{\partial x_2^2} + \left(k_{12} + k_{21}\right)\frac{\partial^2 T}{\partial x_1\,\partial x_2} = \rho C\frac{\partial T}{\partial t} \tag{15-38}$$

where we assumed no energy generation in the medium.

Temperature Depending Only on x_1

For such a case, we have $\partial T/\partial x_2 = 0$, $\partial T/\partial x_3 = 0$; then equation (15-21) for the heat flux components reduces to

$$-q_1'' = k_{11}\frac{\partial T}{\partial x_1}, \qquad -q_2'' = k_{21}\frac{\partial T}{\partial x_1}, \qquad -q_3'' = k_{31}\frac{\partial T}{\partial x_1} \tag{15-39a,b,c}$$

This result implies that there is heat flow in the x_2 and x_3 directions even though temperature gradients are assumed to be zero in those directions. The heat conduction equation (15-9) reduces to

$$k_{11}\frac{\partial^2 T}{\partial x_1^2} = \rho C\frac{\partial T}{\partial t} \tag{15-40}$$

where we have assumed no energy generation. This equation is similar to the one-dimensional heat conduction equation for an isotropic medium. A physical situation simulating one-dimensional heat flow through an isotropic solid can be realized as follows.

Consider a large, thin plate of crystal placed between two highly conducting materials maintained at constant uniform temperatures T_1 and T_2 as illustrated in Figure 15-1. Since the crystal is thin and large, the isothermal surfaces are parallel to the large faces of the crystal except in the region near the edges. If the plate thickness is small compared to the lateral dimensions, the edge effects become negligible. We note that the temperature gradient vector ∇T is along the $0x_1$ axis, but the heat flux vector $\hat{q}''$ is not parallel to ∇T. The total heat flux flowing normal to the plate is q_1'' since the heat flux components q_2'' and q_3'' do not carry heat in that direction. Then the quantities that can readily be measured with experiments are $\partial T/\partial x_1$ and q_1''; hence under steady-state conditions, equation (15-39a), that is,

$$-q_1'' = k_{11}\frac{\partial T}{\partial x_1} \tag{15-41}$$

can be used to determine the conductivity coefficient k_{11}.

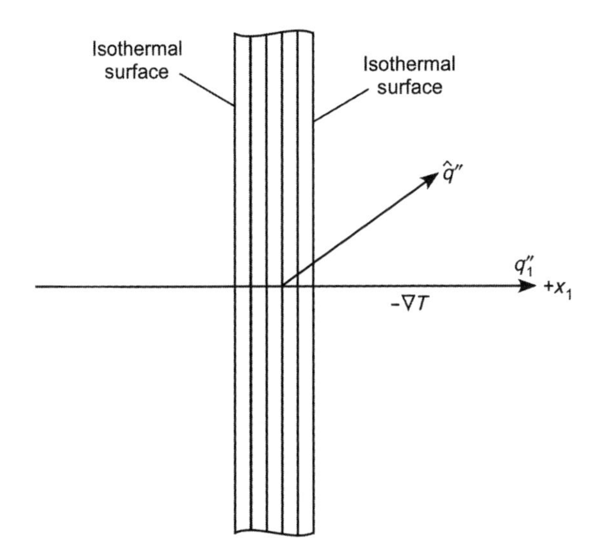

Figure 15-1 Heat flow across a large thin crystal plate.

The variation of k_{11} with the orientation of the $0x_1$ axis with reference to the principal axis is given by the following relation [1–4]:

$$k_{11} = l_1^2 k_1 + l_2^2 k_2 + l_3^2 k_3 \tag{15-42}$$

where k_1, k_2, and k_3 are the principal conductivities, and l_1, l_2, and l_3 are the direction cosines of the $0x_1$ axis relative to the principal axes $0\xi_1, 0\xi_2$, and $0\xi_3$, respectively.

Heat Flow in the x_1 Direction

The physical situation simulating such a condition can be realized by considering a long, thin crystal rod with two ends kept at different constant temperatures and the lateral surfaces insulated as illustrated in Figure 15-2.

The heat flow is along the $0x_1$ direction only since the lateral surface of the rod is insulated. Then we have

$$q_2'' = q_3'' = 0 \tag{15-43}$$

which implies that the heat flux vector $\hat{q}''$ is along the $0x_1$ axis. When the results given by equations (15-43) are introduced into equations (15-17), the three components of the temperature gradient vector become

$$-\frac{\partial T}{\partial x_1} = r_{11} q_1'', \qquad -\frac{\partial T}{\partial x_2} = r_{21} q_1'', \qquad -\frac{\partial T}{\partial x_3} = r_{31} q_1'' \tag{15-44a,b,c}$$

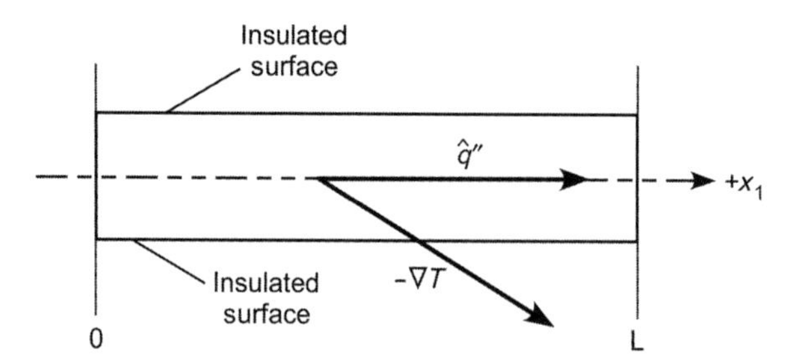

Figure 15-2 Heat flow along a thin, long rod.

Here, the temperature gradient $\partial T/\partial x_1$ and the heat flux q_1'' in the x_1 direction along the rod are the measurable quantities. Then equation (15-44a) can be used to determine the resistivity coefficient r_{11}. The variation of r_{11} with the orientation of the $0x_1$ axis with reference to the principal axes is given by the relation [1–4]

$$r_{11} = l_1^2 r_1 + l_2^2 r_2 + l_3^2 r_3 \tag{15-44d}$$

where r_1, r_2, and r_3 are the principal resistivities and l_1, l_2, and l_3 are the direction cosines of the $0x_1$ axis relative to the principal axes $0\xi_1$, $0\xi_2$, and $0\xi_3$, respectively.

15-9 HEAT CONDUCTION IN AN ORTHOTROPIC MEDIUM

In the case of noncrystalline anisotropic solids, such as wood, the thermal conductivities k_1, k_2, and k_3 are in the mutually perpendicular directions. Then the three components of the heat flux (q_1'', q_2'', q_3'') are given in the (u_1, u_2, u_3) orthogonal curvilinear coordinate system as

$$q_1'' = -\frac{k_1}{a_1}\frac{\partial T}{\partial u_1}, \qquad q_2'' = -\frac{k_2}{a_2}\frac{\partial T}{\partial u_2}, \qquad q_3'' = -\frac{k_3}{a_3}\frac{\partial T}{\partial u_3} \tag{15-45a}$$

where a_1, a_2, a_3 are the scale factors, as discussed in Chapter 1.

Introducing equation (15-45a) into the energy equation (15-8), the heat conduction equation for an orthotropic solid becomes

$$\frac{1}{a_1 a_2 a_3}\left[\frac{\partial}{\partial u_1}\left(\frac{a_2 a_3}{a_1}k_1\frac{\partial T}{\partial u_1}\right) + \frac{\partial}{\partial u_2}\left(\frac{a_1 a_3}{a_2}k_2\frac{\partial T}{\partial u_2}\right) + \frac{\partial}{\partial u_3}\left(\frac{a_1 a_2}{a_3}k_3\frac{\partial T}{\partial u_3}\right)\right] + g = \rho C\frac{\partial T}{\partial t} \tag{15-45b}$$

Assuming k_1, k_2, k_3 are constant, equation (15-45b) for the rectangular, cylindrical, and spherical coordinates takes the following forms, respectively:

Rectangular Coordinate System (x, y, z)

$$k_1 \frac{\partial^2 T}{\partial x^2} + k_2 \frac{\partial^2 T}{\partial y^2} + k_3 \frac{\partial^2 T}{\partial z^2} + g = \rho C \frac{\partial T}{\partial t} \tag{15-46a}$$

Cylindrical coordinate system (r, ϕ, z)

$$k_1 \frac{1}{r} \frac{\partial}{\partial r} \left(r \frac{\partial T}{\partial r} \right) + k_2 \frac{1}{r^2} \frac{\partial^2 T}{\partial \phi^2} + k_3 \frac{\partial^2 T}{\partial z^2} + g = \rho C \frac{\partial T}{\partial t} \tag{15-46b}$$

Spherical coordinate system (r, ϕ, θ)

$$\begin{aligned} & k_1 \frac{1}{r^2} \frac{\partial}{\partial r} \left(r^2 \frac{\partial T}{\partial r} \right) + k_2 \frac{1}{r^2 \sin^2 \theta} \frac{\partial^2 T}{\partial \phi^2} \\ & + k_3 \frac{1}{r^2 \sin \theta} \frac{\partial}{\partial \theta} \left(\sin \theta \frac{\partial T}{\partial \theta} \right) + g = \rho C \frac{\partial T}{\partial t} \end{aligned} \tag{15-46c}$$

We illustrate below with examples the solution of heat conduction in orthotropic medium for both the steady-state and time-dependent situations.

Example 15-3 Point Heat Source in Cartesian System
Consider a point source of strength Q watts, located at the origin of the rectangular coordinate system, releasing its heat continuously over time at a constant rate in an orthotropic medium. In the regions away from the source the region is at a temperature T_∞. We seek to develop an expression for the steady-state temperature distribution in the solid.

We consider the transformed equation (15-33). For the steady-state problem in the region outside the origin where there is no energy generation, we have

$$\frac{\partial^2 T}{\partial X^2} + \frac{\partial^2 T}{\partial Y^2} + \frac{\partial^2 T}{\partial Z^2} = 0$$

$$\text{in} \quad 0 < X < \infty, \quad 0 < Y < \infty, \quad 0 < Z < \infty \tag{15-47}$$

where we define

$$X = \left(\frac{k}{k_1} \right)^{1/2} x, \qquad Y = \left(\frac{k}{k_2} \right)^{1/2} y, \qquad Z = \left(\frac{k}{k_3} \right)^{1/2} z \tag{15-48a}$$

with

$$k = \left(k_1 k_2 k_3\right)^{1/3} \tag{15-48b}$$

The boundary condition at the origin is obtained by drawing a small sphere of radius R around the point source and equating the rate of energy released by the source to the heat conducted into the medium:

$$\left(4\pi R^2\right)\left(-k\frac{\partial T}{\partial R}\right) = Q \qquad \text{as} \qquad R \to 0 \tag{15-49}$$

where $R = (X^2 + Y^2 + Z^2)^{1/2}$. The boundary condition at infinity is given as

$$T(R \to \infty) \to T_\infty \tag{15-50}$$

Equation (15-47) is Laplace's equation, and its solution satisfying the boundary condition (15-50) is written as

$$T\,(R) = \frac{C}{R} + T_\infty \tag{15-51}$$

where the unknown constant C is determined by the application of the boundary condition (15-49) as

$$\left(4\pi R^2\right)\left(k\frac{C}{R^2}\right) = Q \tag{15-52}$$

or

$$C = \frac{1}{k}\frac{Q}{4\pi} \tag{15-53}$$

After equation (15-53) is introduced into (15-51), the solution becomes

$$T\,(R) - T_\infty = \frac{Q}{4\pi}\frac{1}{kR} \tag{15-54a}$$

or

$$T\,(x, y, z) - T_\infty = \frac{Q}{4\pi}\left(k_1 k_2 k_3\right)^{-1/2}\left(\frac{x^2}{k_1} + \frac{y^2}{k_2} + \frac{z^2}{k_3}\right)^{-1/2} \tag{15-54b}$$

Clearly, $T(R) - T_\infty$ decreases with increasing distance R from the origin.

Figure 15-3 shows that the heat flux vector $\hat{q}''$ is along the R coordinate lines and the maximum temperature gradient ∇T is normal to the ellipsoidal isothermal surfaces. We note that the vectors $\hat{q}''$ and ∇T are not necessarily parallel to each other.

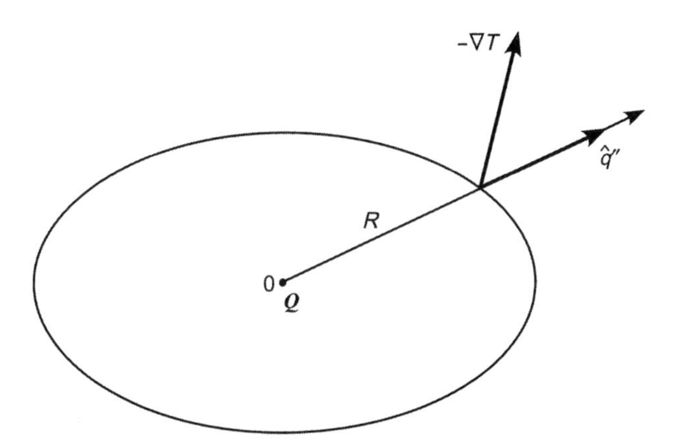

Figure 15-3 Ellipsoidal isothermal surfaces around a point source, Q.

Example 15-4 Steady-State Orthotropic Cartesian Problem
Consider the steady-state heat conduction problem for an orthotropic rectangular region $0 \leq x \leq a, 0 \leq y \leq b$ in which heat is generated at a constant rate of g_0 (W/m^3). Boundaries at $x = 0$ and $y = 0$ are kept insulated, and those at $x = a$ and $y = b$ are dissipating heat by convection into an environment at zero temperature. The orthotropic thermal conductivities in the $0x$ and $0y$ directions are, respectively, k_1 and k_2. We now seek an expression for the steady-state temperature distribution in the region. The mathematical formulation of this problem is given as

$$\frac{\partial^2 T}{\partial x^2} + \frac{1}{\varepsilon^2}\frac{\partial^2 T}{\partial y^2} = -\frac{g_0}{k_1} \quad \text{in} \quad 0 < x < a, \quad 0 < y < b \tag{15-55}$$

$$\text{BC1:} \quad \left.\frac{\partial T}{\partial x}\right|_{x=0} = 0 \tag{15-56}$$

$$\text{BC2:} \quad \left.\frac{\partial T}{\partial x}\right|_{x=a} + H_1 \, T|_{x=a} = 0 \tag{15-57}$$

$$\text{BC3:} \quad \left.\frac{\partial T}{\partial y}\right|_{y=0} = 0 \tag{15-58}$$

$$\text{BC4:} \quad \left.\frac{\partial T}{\partial y}\right|_{y=b} + H_2 \, T|_{y=b} = 0 \tag{15-59}$$

where we define

$$\varepsilon^2 \equiv \frac{k_1}{k_2}, \qquad H_1 = \frac{h_1}{k_1}, \qquad H_2 = \frac{h_2}{k_2}$$

We now define the integral transform pair with respect to the x variable as

$$\text{Transform:} \qquad \overline{T}\left(\beta_{m,} y\right) = \int_{x'=0}^{a} X\left(\beta_m, x'\right) T\left(x', y\right) dx' \tag{15-60a}$$

$$\text{Inversion:} \qquad T\left(x, y\right) = \sum_{m=1}^{\infty} \frac{1}{N\left(\beta_m\right)} X\left(\beta_m, x\right) \overline{T}\left(\beta_m, y\right) \tag{15-60b}$$

where $X(\beta_m, x)$, $N(\beta_m)$, and β_m are obtained from Table 2-1, case 4, as

$$X\left(\beta_m, x\right) = \cos\beta_m x, \qquad \frac{1}{N\left(\beta_m\right)} = 2\frac{\beta_m^2 + H_1^2}{a\left(\beta_m^2 + H_1^2\right) + H_1} \tag{15-60c}$$

and the eigenvalues β_m values are the roots of

$$\beta_m \tan\beta_m a = H_1 \tag{15-60d}$$

Taking the integral transform of system (15-55)–(15-59) by the application of the transform (15-60a), we obtain

$$\frac{d^2\overline{T}}{dy^2} - \beta_m^2\varepsilon^2\overline{T}\left(\beta_m, y\right) = -\frac{\varepsilon^2}{k_1}\overline{g}_0 \qquad \text{in} \qquad 0 < y < b \tag{15-61a}$$

$$\text{BC1:} \qquad \left.\frac{d\overline{T}}{dy}\right|_{y=0} = 0 \tag{15-61b}$$

$$\text{BC2:} \qquad \left.\frac{d\overline{T}}{dy}\right|_{y=b} + H_2 \left.\overline{T}\right|_{y=b} = 0 \tag{15-61c}$$

The solution of the system (15-61) is

$$\overline{T}\left(\beta_m, y\right) = \frac{1}{k_1\beta_m^2}\overline{g} - \frac{1}{k_1\beta_m^2}\overline{g}\frac{\cosh\beta_m\varepsilon y}{\left(\beta_m\varepsilon/H_2\right)\sinh\beta_m\varepsilon b + \cosh\beta_m\varepsilon b} \tag{15-62a}$$

where

$$\overline{g} = \int_{x=0}^{a} g_0\cos\beta_m x\, dx = \frac{\sin\beta_m a}{\beta_m} g_0 \tag{15-62b}$$

The inversion of (15-62) by the inversion formula (15-60b) yields

$$\begin{aligned} T\left(x, y\right) = {} & \frac{g_0}{k_1}\sum_{m=1}^{\infty}\frac{1}{N\left(\beta_m\right)}\frac{\cos\beta_m x\sin\beta_m a}{\beta_m^3} \\ & - \frac{g_0}{k_1}\sum_{m=1}^{\infty}\frac{1}{\beta_m^3 N\left(\beta_m\right)}\frac{\cos\beta_m x\sin\beta_m a\cosh\beta_m\varepsilon y}{\left(\beta_m\varepsilon/H_2\right)\sinh\beta_m\varepsilon b + \cosh\varepsilon b} \end{aligned} \tag{15-63}$$

A closed-form expression for the first summation on the right is determined as (see note 1 at end of this chapter)

$$\sum_{m=1}^{\infty} \frac{1}{N(\beta_m)} \frac{\cos\beta_m x \sin\beta_m a}{\beta_m^3} = \frac{a}{H_1} + \frac{1}{2}\left(a^2 - x^2\right) \tag{15-64}$$

Then the solution (15-63) takes the form

$$T(x, y) = \frac{g_0 a}{k_1 H_1} + \frac{g_0}{2k_1}\left(a^2 - x^2\right) - \frac{2g_0}{k_1}\sum_{m=1}^{\infty} \frac{1}{\beta_m^3} \frac{\beta_m^2 + H_1^2}{a\left(\beta_m^2 + H_1^2\right) + H_1}$$
$$\times \frac{\cos\beta_m x \sin\beta_m a \cosh\beta_m \varepsilon y}{\left(\beta_m \varepsilon / H_2\right)\sinh\beta_m \varepsilon b + \cosh\beta_m \varepsilon b} \tag{15-65a}$$

where the β_m values are the positive roots of

$$\beta_m \tan\beta_m a = H_1 \tag{15-65b}$$

Example 15-5 Time-Dependent, Orthotropic Problem

Consider the time-dependent heat conduction problem for an orthotropic rectangular region $0 \le x \le a, 0 \le y \le b$. Initially the region is at a uniform temperature T_0. For times $t > 0$, the boundaries at $x = 0$ and $y = 0$ are kept insulated and those at $x = a$ and $y = b$ are dissipating heat by convection into an environment at zero temperature, while heat is generated in the region at a constant rate of g_0 (W/m^3). The orthotropic thermal conductivities in the $0x$ and $0y$ directions are, respectively, k_1 and k_2. We now seek an expression for the time-dependent temperature distribution $T(x, y, t)$ in the region for times $t > 0$. The mathematical formulation of this problem is given as

$$\frac{\partial^2 T}{\partial x^2} + \frac{1}{\varepsilon^2}\frac{\partial^2 T}{\partial y^2} + \frac{g_0}{k_1} = \frac{1}{\alpha_1}\frac{\partial T}{\partial t}$$

$$\text{in} \quad 0 < x < a, \quad 0 < y < b, \quad t > 0 \tag{15-66a}$$

$$\text{BC1:} \quad \left.\frac{\partial T}{\partial x}\right|_{x=0} = 0 \tag{15-66b}$$

$$\text{BC2:} \quad \left.\frac{\partial T}{\partial x}\right|_{x=a} + H_1 \left.T\right|_{x=a} = 0 \tag{15-66c}$$

$$\text{BC3:} \quad \left.\frac{\partial T}{\partial y}\right|_{y=0} = 0 \tag{15-66d}$$

$$\text{BC4:} \quad \left.\frac{\partial T}{\partial y}\right|_{y=b} + H_2 \left.T\right|_{y=b} = 0 \tag{15-66e}$$

$$\text{IC:} \quad T(t = 0) = T_0 \quad \text{in the region} \tag{15-66f}$$

where we define

$$\varepsilon^2 \equiv \frac{k_1}{k_2}, \qquad H_1 = \frac{h_1}{k_1}, \qquad H_2 = \frac{h_2}{k_2}, \qquad \alpha_1 = \frac{k_1}{\rho C}$$

It is convenient to split up this problem into two simpler problems, namely, as

$$T(x, y, t) = T_{ss}(x, y) + T_H(x, y, t) \tag{15-67}$$

where the steady-state temperature $T_{ss}(x, y)$ is the solution of the following problem:

$$\frac{\partial^2 T_{ss}}{\partial x^2} + \frac{1}{\varepsilon^2}\frac{\partial^2 T_{ss}}{\partial y^2} + \frac{g_0}{k_1} = 0 \quad \text{in} \quad 0 < x < a, \quad 0 < y < b \tag{15-68a}$$

$$\text{BC1:} \quad \left.\frac{\partial T_{ss}}{\partial x}\right|_{x=0} = 0 \tag{15-68b}$$

$$\text{BC2:} \quad \left.\frac{\partial T_{ss}}{\partial x}\right|_{x=a} + H_1 \left. T_{ss}\right|_{x=a} = 0 \tag{15-68c}$$

$$\text{BC3:} \quad \left.\frac{\partial T_{ss}}{\partial y}\right|_{y=0} = 0 \tag{15-68d}$$

$$\text{BC4:} \quad \left.\frac{\partial T_{ss}}{\partial y}\right|_{y=b} + H_2 \left. T_{ss}\right|_{y=b} = 0 \tag{15-68e}$$

and the transient temperature $T_H(x, y, t)$ is the solution of the following homogeneous problem:

$$\frac{\partial^2 T_H}{\partial x^2} + \frac{1}{\varepsilon^2}\frac{\partial^2 T_H}{\partial y^2} = \frac{1}{\alpha_1}\frac{\partial T_H}{\partial t} \quad \text{in} \quad 0 < x < a, \quad 0 < y < b, \quad t > 0 \tag{15-69a}$$

$$\text{BC1:} \quad \left.\frac{\partial T_H}{\partial x}\right|_{x=0} = 0 \tag{15-69b}$$

$$\text{BC2:} \quad \left.\frac{\partial T_H}{\partial x}\right|_{x=a} + H_1 \left. T_H\right|_{x=a} = 0 \tag{15-69c}$$

$$\text{BC3:} \quad \left.\frac{\partial T_H}{\partial y}\right|_{y=0} = 0 \tag{15-69d}$$

$$\text{BC4:} \quad \left.\frac{\partial T_H}{\partial y}\right|_{y=b} + H_2 \left. T_H\right|_{y=b} = 0 \tag{15-69e}$$

$$\text{IC:} \quad T_H(t = 0) = T_0 - T_{ss}(x, y) \equiv F(x, y) \quad \text{in the region} \tag{15-69f}$$

The steady-state problem (15-68) is exactly the same as that considered in Example 15-4; therefore its solution is immediately obtainable from equation (15-65). The homogeneous problem defined by equations (15-69) can readily be solved by the integral transform technique as now described. We define the integral transform pair with respect to the x variable as

$$\text{Transform:} \qquad \overline{T}\left(\beta_m, y, t\right) = \int_{x'=0}^{a} X\left(\beta_m, x'\right) T\left(x', y, t\right) dx' \tag{15-70a}$$

$$\text{Inversion:} \qquad T\left(x, y, t\right) = \sum_{m=1}^{\infty} \frac{1}{N\left(\beta_m\right)} X\left(\beta_m, x\right) \overline{T}\left(\beta_m, y, t\right) \tag{15-70b}$$

where

$$X\left(\beta_m, x\right) = \cos \beta_m x, \qquad \frac{1}{N\left(\beta_m\right)} = 2\frac{\beta_m^2 + H_1^2}{a\left(\beta_m^2 + H_1^2\right) + H_1} \tag{15-70c}$$

and the β_m values are the positive roots of

$$\beta_m \tan \beta_m a = H_1 \tag{15-70d}$$

The integral transform pair with respect to the y variable is defined as

$$\text{Transform:} \qquad \tilde{\overline{T}}\left(\beta_m, \gamma_n, t\right) = \int_{y'=0}^{b} Y\left(\gamma_n, y'\right) \overline{T}\left(\beta_m, y', t\right) dy' \tag{15-71a}$$

$$\text{Inversion:} \qquad \overline{T}\left(\beta_m, y, t\right) = \sum_{n=1}^{\infty} \frac{Y\left(\gamma_n, y\right)}{N\left(\gamma_n\right)} \tilde{\overline{T}}\left(\beta_m, \gamma_n, t\right) \tag{15-71b}$$

where

$$Y\left(\gamma_n, y\right) = \cos \gamma_n y, \qquad \frac{1}{N\left(\gamma_n\right)} = 2\frac{\gamma_n^2 + H_2^2}{b\left(\gamma_n^2 + H_2^2\right) + H_2} \tag{15-71c}$$

and the γ_n values are the positive roots of

$$\gamma_n \tan \gamma_n b = H_2 \tag{15-71d}$$

The integral transform of the system (15-69) with respect to the x variable by the application of the transform (15-70a) is

$$-\beta_m^2 \overline{T}_H\left(\beta_m, y, t\right) + \frac{1}{\varepsilon^2}\frac{\partial^2 \overline{T}_H}{\partial y^2} = \frac{1}{\alpha_1}\frac{\partial \overline{T}_H}{\partial t} \quad \text{in} \quad 0 < y < b, \quad t > 0 \tag{15-72a}$$

$$\text{BC1:} \qquad \left.\frac{\partial \overline{T}_H}{\partial y}\right|_{y=0} = 0 \tag{15-72b}$$

$$\text{BC2:} \qquad \left.\frac{\partial \overline{T}_H}{\partial y}\right|_{y=b} + H_2\, \overline{T}_H\big|_{y=b} = 0 \tag{15-72c}$$

$$\text{IC:} \qquad \overline{T}_H(t=0) = \overline{F}\left(\beta_m, y\right) \qquad \text{in} \qquad 0 \le y \le b \tag{15-72d}$$

The integral transform of the system (15-72) with respect to the y variable by the application of the transform (15-71a) gives

$$-\beta_m^2 \tilde{\overline{T}}_H\left(\beta_m, \gamma_n, t\right) - \frac{1}{\varepsilon^2}\gamma_n^2 \tilde{\overline{T}}_H = \frac{1}{\alpha_1}\frac{d\tilde{\overline{T}}_H}{dt} \tag{15-73a}$$

or

$$\frac{d\tilde{\overline{T}}_H}{dt} + \alpha_1 \lambda_m^2 \tilde{\overline{T}}_H = 0 \qquad \text{for} \qquad t > 0 \tag{15-73b}$$

$$\text{IC:} \qquad \tilde{\overline{T}}_H\left(\beta_m, \gamma_n, t=0\right) = \tilde{\overline{F}}\left(\beta_m, \gamma_n\right) \tag{15-73c}$$

where we define

$$\lambda_{mn}^2 = \beta_m^2 + \frac{1}{\varepsilon^2}\gamma_n^2 \tag{15-73d}$$

The solution of equations (15-73) is

$$\tilde{\overline{T}}_H\left(\beta_m, \gamma_n, t\right) = e^{-\alpha_1 \lambda_{mn}^2 t}\, \tilde{\overline{F}}\left(\beta_m, \gamma_n\right) \tag{15-74}$$

The inversion of equation (15-74) successively by the inversion formulas (15-71b) and (15-70b) gives the solution for $T_H(x, y, t)$ as

$$T_H\left(x, y, t\right) = \sum_{m=1}^{\infty}\sum_{n=1}^{\infty} \frac{e^{-\alpha_1 \lambda_{mn}^2 t}}{N\left(\beta_m\right) N\left(\gamma_n\right)} \cos \beta_m x \cos \gamma_n y$$

$$\times \int_{x'=0}^{a}\int_{y'=0}^{b} \cos \beta_m x' \cos \gamma_n y' F\left(x', y'\right) dx'\, dy' \tag{15-75}$$

where $N(\beta_m)$ and $N(\gamma_n)$ are defined by equations (15-70c) and (15-71c), respectively; β_m and γ_n are the roots of the transcendental equations (15-70d) and (15-71d), respectively; and λ_{mn}^2 is defined by equation (15-73d). The function $F(x', y')$ being specified according to equations (15-69f) and (15-65), the integral with respect to the space variables in equation (15-65) can be evaluated analytically or numerically.

15-10 MULTIDIMENSIONAL HEAT CONDUCTION IN AN ANISOTROPIC MEDIUM

The multidimensional heat conduction equation for the case of general anisotropy involves cross derivatives, whereas the boundary conditions may contain various partial derivatives with respect to the space variables. As a result, the analytic solution of the multidimensional heat conduction problem for the general anisotropic case is difficult to obtain, especially for finite regions. However, the solutions can be obtained for special situations involving semi-infinite or infinite regions as illustrated in the following examples.

Example 15-6 Two-Dimensional Anisotropic Cartesian Problem
We consider a two-dimensional, time-dependent heat conduction problem for an anisotropic region $0 \le x \le \infty$, $-\infty < y < \infty$ in the rectangular coordinate system. The medium is initially at temperature $F(x, y)$, and for times $t > 0$, the boundary surface at $x = 0$ is kept at zero temperature. We solve for an expression for the temperature distribution $T(x, y, t)$ in the region for times $t > 0$.

Since no temperature variation is considered in the z direction, we have $\partial T/\partial z = 0$. Then, the heat conduction equation (15-9) reduces to

$$\frac{\partial^2 T}{\partial x^2} + \varepsilon_{22}\frac{\partial^2 T}{\partial y^2} + 2\varepsilon_{12}\frac{\partial^2 T}{\partial x \partial y} = \frac{1}{\alpha_{11}}\frac{\partial T}{\partial t} \tag{15-76a}$$

$$\text{in} \quad 0 < x < \infty, \quad -\infty < y < \infty, \quad t > 0$$

with the boundary and initial conditions

$$\text{BC1:} \quad T(x = 0) = 0 \tag{15-76b}$$

$$\text{IC:} \quad T(t = 0) = F(x, y) \quad \text{in} \quad 0 \le x < \infty, \quad -\infty < y < \infty \tag{15-76c}$$

where we have defined

$$\varepsilon_{ij} = \frac{k_{ij}}{k_{11}}, \quad k_{ij} = k_{ji}, \quad \text{and} \quad \alpha_{11} = \frac{k_{11}}{\rho C} \tag{15-76d}$$

We note that the differential equation involves one cross derivative, and the region in the y direction is infinite in extent. Therefore, the integral transform with respect to the y variable can be applied to remove from the equations the first and second partial derivatives with respect to the y variable. The integral transform pair with respect to the y-variable is defined as [see equation (14-63)]

$$\text{Inversion:} \quad T(x, y, t) = \frac{1}{2\pi}\int_{\gamma=-\infty}^{\infty} e^{-i\gamma y}\overline{T}(x, \gamma, t)\, d\gamma \tag{15-77a}$$

$$\text{Transform:} \quad \overline{T}(x, \gamma, t) = \int_{y'=-\infty}^{\infty} e^{i\gamma y'} T\left(x, y', t\right) dy' \tag{15-77b}$$

where the overbar now denotes the integral transform with respect to the y variable.

The integral transform of the system (15-76) by the application of the transform (15-77b) (see note 2 at the end of this chapter for the transform of the second and the first derivatives with respect to the y variable) yields

$$\frac{\partial^2 \overline{T}}{\partial x^2} - \gamma^2 \varepsilon_{22} \overline{T} - 2i\gamma\varepsilon_{12} \frac{\partial \overline{T}}{\partial x} = \frac{1}{\alpha_{11}} \frac{\partial \overline{T}}{\partial t} \quad \text{in} \quad 0 < x < \infty, \quad t > 0 \tag{15-78a}$$

$$\text{BC:} \quad \overline{T}(x = 0) = 0 \tag{15-78b}$$

$$\text{IC:} \quad \overline{T}(t = 0) = \overline{F}(x, \gamma) \quad \text{in} \quad 0 \le x < \infty \tag{15-78c}$$

where $\overline{T} \equiv \overline{T}(x, \gamma, t)$. The partial derivative $\partial \overline{T}/\partial x$ can be removed from this equation by defining a new variable $\overline{w}(x, \gamma, t)$ as

$$\overline{T}(x, \gamma, t) = \overline{w}(x, \gamma, t)\, e^{i\gamma\varepsilon_{12}x} \tag{15-79}$$

The system (15-78) is now transformed to

$$\frac{\partial^2 \overline{w}}{\partial x^2} - \gamma^2 \left(\varepsilon_{22} - \varepsilon_{12}^2\right) \overline{w} = \frac{1}{\alpha_{11}} \frac{\partial \overline{w}}{\partial t} \quad \text{in} \quad 0 < x < \infty, \quad t > 0 \tag{15-80a}$$

$$\text{BC:} \quad \overline{w}(x = 0) = 0 \tag{15-80b}$$

$$\text{IC:} \quad \overline{w}(t = 0) = e^{-i\gamma\varepsilon_{12}x}\, \overline{F}(x, \gamma) \quad \text{in} \quad 0 \le x < \infty \tag{15-80c}$$

To remove the partial derivative with respect to the x variable from this system, the integral transform pair with respect to the x variable for the region $0 < x < \infty$ is defined as [see equations (14-57) and Table 6-1, case 3]

$$\text{Inversion:} \quad \overline{w}(x, \gamma, t) = \frac{2}{\pi} \int_{\beta=0}^{\infty} \sin \beta x\, \tilde{\overline{w}}(\beta, \gamma, t)\, d\beta \tag{15-81a}$$

$$\text{Transform:} \quad \tilde{\overline{w}}(\beta, \gamma, t) = \int_{x'=0}^{\infty} \sin \beta x'\, \overline{w}\left(x', \gamma, t\right) dx' \tag{15-81b}$$

where the *tilde* denotes the transform with respect to the x variable. The integral transform of the system (15-80) by the application of the transform (15-81b) gives

$$\frac{d\tilde{\overline{w}}}{dt} + \alpha_{11}\gamma^2 \tilde{\overline{w}}(\beta, \gamma, t) = 0 \quad \text{for} \quad t > 0 \tag{15-82a}$$

with

$$\text{IC:} \quad \tilde{\overline{w}}(t = 0) = \tilde{\overline{H}}(\beta, \gamma) \tag{15-82b}$$

and where the following hold:

$$\lambda^2 \equiv \beta^2 + \gamma^2 \left(\varepsilon_{22} - \varepsilon_{12}^2\right) \tag{15-83a}$$

$$\varepsilon_{22} - \varepsilon_{12}^2 > 0 \qquad \text{according to equation (15-4d)} \tag{15-83b}$$

$$\tilde{\overline{H}}(\beta, \gamma) = \int_{x'=0}^{\infty} e^{-i\gamma\varepsilon_{12}x'} \overline{F}\left(x', \gamma\right) \sin \beta x' \, dx' \tag{15-83c}$$

$$\overline{F}\left(x', \gamma\right) = \int_{y'=-\infty}^{\infty} e^{i\gamma y'} F\left(x', y'\right) dy' \tag{15-83d}$$

The solution of equation (15-82) is

$$\tilde{\overline{w}}(\beta, \gamma, t) = \tilde{\overline{H}}(\beta, \gamma) \, e^{-\alpha_{11}\lambda^2 t} \tag{15-84}$$

The inversion of equation (15-84) by the inversion formula (15-81a), and then the application of equation (15-79), yields

$$\overline{T}(x, \gamma, t) = \frac{2}{\pi} \int_{\beta=0}^{\infty} \sin \beta x \, \tilde{\overline{H}}(\beta, \gamma) \, e^{-\alpha_{11}\lambda^2 t + i\gamma\varepsilon_{12}x} d\beta \tag{15-85}$$

This result is inverted by the inversion formula (15-77a), the explicit form of $\tilde{\overline{H}}(\beta, \gamma)$ is introduced, and the order of integrations is rearranged:

$$\begin{aligned} T(x, y, t) = &\int_{x'=0}^{\infty} \int_{y'=-\infty}^{\infty} F\left(x', y'\right) \\ &\times \left\{ \frac{1}{2\pi} \int_{\gamma=-\infty}^{\infty} e^{-\alpha_{11}\left(\varepsilon_{22}-\varepsilon_{12}^2\right)\gamma^2 t - i\gamma[(y-y')-\varepsilon_{12}(x-x')]} d\gamma \right\} \\ &\times \left\{ \frac{2}{\pi} \int_{\beta=0}^{\infty} e^{-\alpha_{11}\beta^2 t} \sin \beta x \, \sin \beta x' d\beta \right\} dy' \, dx' \end{aligned} \tag{15-86}$$

In this result, the integrals with respect to the variables γ and β can be evaluated by making use of the integrals given by equations (14-67) and (14-81), respectively, that is

$$\frac{1}{2\pi} \int_{\gamma=-\infty}^{\infty} e^{-\gamma^2 \alpha t - i\gamma z} d\gamma = \frac{1}{(4\pi\alpha t)^{1/2}} e^{-z^2/4\alpha t} \tag{15-87a}$$

and

$$\begin{aligned} &\frac{2}{\pi} \int_{\beta=0}^{\infty} e^{-\beta^2 \alpha t} \sin \beta x \sin \beta x' \, d\beta \\ &= \frac{1}{(4\pi\alpha t)^{1/2}} \left\{ \exp\left[-\frac{\left(x - x'\right)^2}{4\alpha t} \right] - \exp\left[-\frac{\left(x + x'\right)^2}{4\alpha t} \right] \right\} \end{aligned} \tag{15-87b}$$

Then, the solution (15-86) takes the form

$$
\begin{aligned}
T(x, y, t) = & \frac{1}{\left[4\pi\alpha_{11}\left(\varepsilon_{22}-\varepsilon_{12}^2\right)t\right]^{1/2}\left(4\pi\alpha_{11}t\right)^{1/2}} \int_{x'=0}^{\infty}\int_{y'=-\infty}^{\infty} F\left(x', y'\right) \\
& \times \exp\left\{-\frac{\left[\left(y-y'\right)^2-\varepsilon_{12}\left(x-x'\right)^2\right]^2}{4\pi\alpha_{11}\left(\varepsilon_{22}-\varepsilon_{12}^2\right)t}\right\} \\
& \times \left\{\exp\left[-\frac{\left(x-x'\right)^2}{4\alpha_{11}t}\right]-\exp\left[-\frac{\left(x+x'\right)^2}{4\alpha_{11}t}\right]\right\} dy'\,dx' \qquad (15\text{-}88)
\end{aligned}
$$

Example 15-7 Anisotropic Medium with Generation
An anisotropic medium $0 \leq x < \infty, -\infty < y < \infty$ is initially at zero temperature. For times $t > 0$, heat is generated in the medium at a rate of $g(x, y, t)$ (W/m^3) while the boundary surface at $x = 0$ is kept at zero temperature. We solve for an expression for the temperature distribution $T(x, y, t)$ in the region for times $t > 0$.

The mathematical formulation of the heat conduction problem is given as

$$
\frac{\partial^2 T}{\partial x^2} + \varepsilon_{22}\frac{\partial^2 T}{\partial y^2} + 2\varepsilon_{12}\frac{\partial^2 T}{\partial x\,\partial y} + \frac{1}{k_{11}} g(x, y, t) = \frac{1}{\alpha_{11}}\frac{\partial T}{\partial t}
$$

$$
\text{in} \quad 0 < x < \infty, \qquad -\infty < y < \infty, \qquad t > 0 \qquad (15\text{-}89a)
$$

$$
\text{BC:} \quad T(x = 0) = 0 \qquad (15\text{-}89b)
$$

$$
\text{IC:} \quad T(t = 0) = 0 \quad \text{in} \quad 0 \leq x < \infty, \qquad -\infty < y < \infty \qquad (15\text{-}89c)
$$

where we have defined

$$
\varepsilon_{ij} = \frac{k_{ij}}{k_{11}}, \qquad k_{ij} = k_{ji}, \qquad \text{and} \qquad \alpha_{11} = \frac{k_{11}}{\rho C} \qquad (15\text{-}89d)
$$

This problem is similar to that considered in Example 15-6, except for the heat generation and the zero initial condition. Therefore, the integral transform pairs defined in the previous example are applicable for the solution of this problem. The integral transform of the system (15-89) by the application of the transform (15-77b) gives

$$
\frac{\partial^2 \overline{T}}{\partial x^2} - \gamma^2\varepsilon_{22}\overline{T} - 2i\gamma\varepsilon_{12}\frac{\partial \overline{T}}{\partial x} + \frac{1}{k_{11}}\overline{g}(x, \gamma, t) = \frac{1}{\alpha_{11}}\frac{\partial \overline{T}}{\partial t}
$$

$$
\text{in} \quad 0 < x < \infty, \qquad t > 0 \qquad (15\text{-}90a)
$$

$$\text{BC:} \qquad \overline{T}(x=0)=0 \tag{15-90b}$$

$$\text{IC:} \qquad \overline{T}(t=0)=0 \qquad \text{in} \qquad 0 \le x < \infty \tag{15-90c}$$

where $\overline{T}=\overline{T}(x,\gamma,t)$. The partial derivative $\partial\overline{T}/\partial x$ can be removed from equation (15-90a) by the application of the transform (15-79). Then the system (15-90) is transformed to

$$\frac{\partial^2\overline{w}}{\partial x^2}-\gamma^2\left(\varepsilon_{22}-\varepsilon_{12}^2\right)\overline{w}+\frac{1}{k_{11}}e^{-i\gamma\varepsilon_{12}x}\overline{g}\left(x,\gamma,t\right)=\frac{1}{\alpha_{11}}\frac{\partial\overline{w}}{\partial t}$$

$$\text{in} \qquad 0<x<\infty, \qquad t>0 \tag{15-91a}$$

$$\text{BC:} \qquad \overline{w}(x=0)=0 \tag{15-91b}$$

$$\text{IC:} \qquad \overline{w}(t=0)=0 \qquad \text{in} \qquad 0 \le x < \infty \tag{15-91c}$$

The partial derivative with respect to the x variable is removed from equation (15-91a) by the application of the transform (15-81b). Then, the system (15-91) is reduced to the following ordinary differential equation:

$$\frac{d\tilde{\overline{w}}}{\partial t}+\alpha_{11}\lambda^2\tilde{\overline{w}}\left(\beta,\gamma,t\right)=\frac{\alpha_{11}}{k_{11}}\tilde{\overline{G}}\left(\beta,\gamma,t\right) \qquad \text{for} \qquad t>0 \tag{15-92a}$$

$$\text{IC:} \qquad \tilde{\overline{w}}\left(\beta,\gamma,t=0\right)=0 \tag{15-92b}$$

where we define

$$\lambda^2\equiv\beta^2+\gamma^2\left(\varepsilon_{22}-\varepsilon_{12}^2\right) \tag{15-93a}$$

$$\varepsilon_{22}-\varepsilon_{12}^2>0 \tag{15-93b}$$

$$\tilde{\overline{G}}\left(\beta,\gamma,t\right)=\int_{x'=0}^{\infty}e^{-i\gamma\varepsilon_{12}x'}\overline{g}\left(x',\gamma,t\right)\sin\beta x'\,dx' \tag{15-93c}$$

$$\overline{g}\left(x',\gamma,t\right)=\int_{y'=-\infty}^{\infty}e^{i\gamma y'}g\left(x',y',t\right)dy' \tag{15-93d}$$

The solution of equations (15-92) is

$$\tilde{\overline{w}}\left(\beta,\gamma,t\right)=e^{-\alpha_{11}\lambda^2 t}\int_{t'=0}^{t}\frac{\alpha_{11}}{k_{11}}\tilde{\overline{G}}\left(\beta,\gamma,t'\right)e^{\alpha_{11}\lambda^2 t'}\,dt' \tag{15-94}$$

The inversion of this result by the inversion formula (15-81a), and then the application of equation (15-79), yields

$$T\left(x,\gamma,t\right)=\frac{2}{\pi}\int_{\beta=0}^{\infty}\int_{t'=0}^{t}\frac{\alpha_{11}}{k_{11}}\sin\beta x\,\tilde{\overline{G}}(\beta,\gamma,t)\,e^{-\alpha_{11}\lambda^2(t-t')+i\gamma\varepsilon_{12}x}\,dt'\,d\beta \tag{15-95}$$

This result is inverted by the inversion formula (15-77a), the explicit form of $\tilde{\overline{G}}(\beta, \gamma, t)$ defined by equation (15-93) is introduced, and the order of the integrations is rearranged to yield

$$T(x, y, t) = \int_{x'=0}^{\infty}\int_{y'=-\infty}^{\infty}\int_{t'=0}^{t} g(x', y', t') \times \left\{\frac{1}{2\pi}\int_{\gamma=-\infty}^{\infty} e^{-\alpha_{11}\left(\varepsilon_{22}-\varepsilon_{12}^2\right)\gamma^2(t-t')-i\gamma[(y-y')-\varepsilon_{12}(x-x')]}d\gamma\right\} \times \left\{\frac{2}{\pi}\int_{\beta=0}^{\infty} e^{-\alpha_{11}\beta^2(t-t')}\sin\beta x\sin\beta x'\,d\beta\right\} dt'\,dy'\,dx' \tag{15-96}$$

The integrals with respect to the variables γ and β can be evaluated by making use of the integrals (15-87a) and (15-87b); then the solution (15-96) takes the form

$$T(x, y, t) = \frac{1}{\left[4\pi\alpha_{11}\left(\varepsilon_{22}-\varepsilon_{12}^2\right)(t-t')\right]^{1/2}\left[4\pi\alpha_{11}(t-t')\right]^{1/2}} \times \int_{x'=0}^{\infty}\int_{y'=-\infty}^{\infty}\int_{t'=0}^{t} g(x', y', t')\exp\left\{-\frac{\left[(y-y')-\varepsilon_{12}(x-x')\right]^2}{4\pi\alpha_{11}\left(\alpha_{22}-\varepsilon_{12}^2\right)(t-t')}\right\} \times \left\{\exp\left[-\frac{(x-x')^2}{\alpha_{11}(t-t')}\right]-\exp\left[-\frac{(x+x')^2}{4\alpha_{11}(t-t')}\right]\right\} dt'\,dy'\,dx' \tag{15-97}$$

Example 15-8 Anisotropic Cylinder
An anisotropic cylindrical region $0 \le r \le b$, $-\infty < z < \infty$ is initially at temperature $F(r, z)$. For times $t > 0$, the boundary surface at $r = b$ is kept at zero temperature. We seek an expression for the temperature distribution $T(r, z, t)$ in the cylinder for times $t > 0$. Since there is no azimuthal variation of temperature, we have $\partial T/\partial\phi = 0$. Then the heat conduction equation (15-10) becomes

$$\frac{1}{r}\frac{\partial}{\partial r}\left(r\frac{\partial T}{\partial r}\right) + \varepsilon_{33}\frac{\partial^2 T}{\partial z^2} + 2\varepsilon_{13}\frac{\partial^2 T}{\partial r\,\partial z} + \varepsilon_{13}\frac{1}{r}\frac{\partial T}{\partial z} = \frac{1}{\alpha_{11}}\frac{\partial T}{\partial t} \tag{15-98a}$$

$$\text{in} \quad 0 \le r < b, \quad -\infty < z < \infty, \quad t > 0$$

with the following boundary and initial conditions:

$$\text{BC:} \quad T(r \to 0) \Rightarrow \text{finite} \tag{15-98b}$$

$$\text{BC:} \quad T(r = b) = 0 \tag{15-98c}$$

$$\text{IC:} \quad T(t = 0) = F(r, z) \quad \text{in} \quad 0 \le r < b, \quad -\infty < z < \infty \tag{15-98d}$$

where we have defined

$$\varepsilon_{ij} = \frac{k_{ij}}{k_{11}}, \qquad k_{ij} = k_{ji}, \qquad \alpha_{11} = \frac{k_{11}}{\rho C} \tag{15-98e}$$

This problem is now solved by the application of integral transform technique as now described. The integral transform pair with respect to the z variable over the domain $-\infty < z < \infty$ is defined as

$$\text{Inversion:} \qquad T(r, z, t) = \frac{1}{2\pi} \int_{\gamma=-\infty}^{\infty} e^{-i\gamma z} \overline{T}(r, \gamma, t)\, d\gamma \tag{15-99a}$$

$$\text{Transform:} \qquad \overline{T}(r, \gamma, t) = \int_{z'=-\infty}^{\infty} e^{i\gamma z'} T\left(r, z', t\right) dz' \tag{15-99b}$$

The integral transform of the system (15-98) by the application of the transform (15-99b) yields

$$\frac{\partial^2 \overline{T}}{\partial r^2} + \frac{1}{r}\frac{\partial \overline{T}}{\partial r} - \gamma^2 \varepsilon_{33} \overline{T} - 2i\gamma\varepsilon_{13}\frac{\partial \overline{T}}{\partial r} - i\gamma\frac{\varepsilon_{13}}{r}\overline{T} = \frac{1}{\alpha_{11}}\frac{\partial \overline{T}}{\partial t} \tag{15-100a}$$

or

$$\frac{\partial^2 \overline{T}}{\partial r^2} + \left(\frac{1}{r} - 2i\gamma\varepsilon_{13}\right)\frac{\partial \overline{T}}{\partial r} - \left(\frac{i\gamma\varepsilon_{13}}{r} + \gamma^2\varepsilon_{33}\right)\overline{T} = \frac{1}{\alpha_{11}}\frac{\partial \overline{T}}{\partial t}$$

$$\text{in} \quad 0 \le r < b, \qquad t > 0 \tag{15-100b}$$

$$\text{BC:} \qquad \overline{T}(r = b) = 0 \tag{15-100b}$$

$$\text{IC:} \qquad \overline{T}(t = 0) = \overline{F}(r, \gamma) \qquad \text{in} \qquad 0 \le r < b \tag{15-100c}$$

where $\overline{T} \equiv \overline{T}(r, \gamma, t)$. A new variable $\overline{w}(r, \gamma, t)$ is defined as

$$\overline{T}(r, \gamma, t) = \overline{w}(r, \gamma, t)\, e^{i\gamma\varepsilon_{13} r} \tag{15-101}$$

Then the system (15-100) is transformed to

$$\frac{\partial^2 \overline{w}}{\partial r^2} + \frac{1}{r}\frac{\partial \overline{w}}{\partial r} - \left(\varepsilon_{33} - \varepsilon_{13}^2\right)\gamma^2 \overline{w} = \frac{1}{\alpha_{11}}\frac{\partial \overline{w}}{\partial t} \qquad \text{in} \qquad 0 \le r < b, \qquad t > 0 \tag{15-102a}$$

$$\text{BC:} \qquad \overline{w}(r = b) = 0 \tag{15-102b}$$

$$\text{IC:} \qquad \overline{w}(t = 0) = e^{-i\gamma\varepsilon_{13} r}\overline{F}(r, \gamma) \qquad \text{in} \qquad 0 \le r \le b \tag{15-102c}$$

To remove the partial derivative with respect to the r variable, the integral transform pair is defined as [see equations (14-87) and Table 2-2, case 3]

$$\text{Inversion:} \qquad \overline{w}(r,\gamma,t) = \sum_{m=1}^{\infty} \frac{1}{N(\beta_m)} J_0(\beta_m r)\,\tilde{\overline{w}}(\beta_m,\gamma,t) \tag{15-103a}$$

$$\text{Transform:} \qquad \tilde{\overline{w}}(\beta_m,\gamma,t) = \int_{r'=0}^{b} r' J_0(\beta_m r')\,\overline{w}(r',\gamma,t)\,dr' \tag{15-103b}$$

where

$$\frac{1}{N(\beta_m)} = \frac{2}{b^2}\frac{1}{J_0'^2(\beta_m b)} = \frac{2}{b^2 J_1^2(\beta_m b)} \tag{15-103c}$$

and the β_m values are the roots of

$$J_0(\beta_m b) = 0 \tag{15-103d}$$

The integral transform of the system (15-102) by the application of transform (15-103b) is

$$\frac{d\tilde{\overline{w}}}{dt} + \alpha_{11}\lambda^2\tilde{\overline{w}}(\beta_m,\gamma,t) = 0 \qquad \text{for} \qquad t>0 \tag{15-104a}$$

$$\text{IC:} \qquad \tilde{\overline{w}}(\beta_m,\gamma,t=0) = \tilde{\overline{H}}(\beta_m,\gamma) \tag{15-104b}$$

where we define the following:

$$\lambda^2 \equiv \beta_m^2 + \gamma^2\left(\varepsilon_{33} - \varepsilon_{13}^2\right) \tag{15-105a}$$

$$\varepsilon_{33} - \varepsilon_{13}^2 > 0 \tag{15-105b}$$

$$\tilde{\overline{H}}(\beta_m,\gamma) = \int_{r'=0}^{b} r' J_0(\beta_m, r')\,e^{-i\gamma\varepsilon_{13}r'}\,\overline{F}(r',\gamma)\,dr' \tag{15-105c}$$

$$\overline{F}(r',\gamma) = \int_{z'=-\infty}^{\infty} e^{i\gamma z'} F(r',z')\,dz' \tag{15-105d}$$

The solution of equation (15-104) is

$$\tilde{\overline{w}}(\beta_m,\gamma,t) = e^{-\alpha_{11}\lambda^2 t}\,\tilde{\overline{H}}(\beta_m,\gamma) \tag{15-106}$$

The inversion of (15-106) by the inversion formula (15-103a) gives

$$\overline{w}(r,\gamma,t) = \sum_{m=1}^{\infty} \frac{1}{N(\beta_m)} J_0(\beta_m r)\,e^{-\alpha_{11}\lambda^2 t}\,\tilde{\overline{H}}(\beta_m,\gamma) \tag{15-107}$$

This result is introduced into equation (15-101) to obtain

$$\overline{T}(r,\gamma,t)=\sum_{m=1}^{\infty}\frac{1}{N(\beta_m)}J_0(\beta_m r)\,e^{-\alpha_{11}\lambda^2 t+i\gamma\varepsilon_{13}r}\,\widetilde{\overline{H}}(\beta_m,\gamma) \tag{15-108}$$

The inversion of equation (15-108) by the inversion formula (15-99a) gives

$$T(r,z,t)=\frac{1}{2\pi}\sum_{m=1}^{\infty}\int_{\gamma=-\infty}^{\infty}\frac{1}{N(\beta_m)}J_0(\beta_m r)\,\widetilde{\overline{H}}(\beta_m,\gamma)\,e^{-\alpha_{11}\lambda^2 t-i\gamma(z-\varepsilon_{13}r)}d\gamma \tag{15-109}$$

where

$$\lambda^2\equiv\beta_m^2+\gamma^2\left(\varepsilon_{33}-\varepsilon_{13}^2\right)$$

The term $\widetilde{\overline{H}}(\beta_m,\gamma)$ defined by equations (15-105c) is introduced into equation (15-109), and the order of integrations is rearranged, giving

$$\begin{aligned}T(r,z,t)=&\sum_{m=1}^{\infty}e^{-\alpha_{11}\beta_m^2 t}\frac{J_0(\beta_m r)}{N(\beta_m)}\int_{z'=-\infty}^{\infty}\int_{r'=0}^{b}\Bigg[r'J_0(\beta_m r')F(r',z')\\&\times\left\{\frac{1}{2\pi}\int_{\gamma=-\infty}^{\infty}e^{-\gamma^2\alpha_{11}(\varepsilon_{33}-\varepsilon_{13})t-i\gamma[(z-z')+\varepsilon_{13}(r'-r)]}d\gamma\right\}\Bigg]dr'\,dz'\end{aligned} \tag{15-110}$$

The integral with respect to γ can be evaluated according to equation (15-87a), and the solution becomes

$$\begin{aligned}T(r,z,t)=&\frac{1}{\sqrt{4\pi\alpha_{11}\left(\varepsilon_{33}-\varepsilon_{13}^2\right)t}}\sum_{m=1}^{\infty}e^{-\alpha_{11}\beta_m^2 t}\frac{J_0(\beta_m r)}{N(\beta_m)}\\&\times\int_{z'=-\infty}^{\infty}\int_{r'=0}^{b}r'J_0(\beta_m r')F(r',z')\\&\times\exp\left\{-\frac{\left[(z-z')+\varepsilon_{13}(r'-r)\right]^2}{4\alpha_{11}\left(\varepsilon_{33}-\varepsilon_{13}^2\right)t}\right\}dr'\,dz'\end{aligned} \tag{15-111}$$

where $N(\beta_m)$ is given by equation (15-103c), and the β_m values are the roots of equation (15-103d).

REFERENCES

1. W. A. Wooster, *A Textbook in Crystal Physics*, Cambridge University Press, London, 1938.
2. J. F. Nye, *Physical Properties of Crystals*, Clarendon London, 1957.

3. H. S. Carslaw and J. C. Jeager, *Conduction of Heat in Solids*, Clarendon, London, 1959.
4. M. N. Özisik, *Boundary Value Problems of Heat Conduction*, International Textbook, Scranton, PA, 1968; Dover, New York, 1989.
5. W. H. Giedt and D. R. Hornbaker, *ARS J*. **32**, 1902–1909, 1962.
6. K. J. Touryan, *AIAA J*. **2**, 124–126, 1964.
7. B. Venkatraman, S. A. Patel, and F. V. Pohle, *J. Aerospace Sci*. **29**, 628–629, 1962.
8. B. T. Chao, *Appl. Sci. Res*. **A12**, 134–138, 1963.
9. H. F. Cooper, Joulean Heating of an Infinite Rectangular Rod with Orthotropic Thermal Properties, ASME Paper No. 66-WA/HT-14, 1966.
10. H. F. Cooper, Transient and Steady State Temperature Distribution in Foil Wound Solenoids and Other Electric Apparatus of Rectangular Cross-Section, 1965 *IEEE International Convention Record*, Part 10, March 1965, pp. 67–75.
11. N. Vutz and S. W. Angrist, Thermal Contact Resistance of Anisotropic Materials, ASME Paper 69-HT-47, 1969.
12. R. C. Pfahl, *Int. J. Heat Mass Transfer* **18**, 191–204, 1975.
13. B. F. Blackwell, *An Introduction to Heat Conduction in an Anisotropic Medium*, SC-RR-69-542, Sandia Lab., Albuquerque, N.M, Oct. 1969.
14. Y. P. Chang, C. S. Kang, and D. J. Chen, *Int. J. Heat Mass Transfer* **16**, 1905–1918, 1973.
15. J. Padovan, *J. Heat Transfer* **96c**, 428–431, 1974.
16. J. Padovan, *AIAA J*. **10**, 60–64, 1972.
17. K. Katayama, Transient Heat Conduction in Anisotropic Solids, *Proceedings of the 5th International Heat Transfer Conference*, Tokyo, Sept. 1974, Cu 1.4, pp. 137–141.
18. M. H. Cobble, *Int. J. Heat Mass Transfer* **17**, 379–380, 1974.
19. G. P. Mulholland and B. P. Gupta, *J. Heat Transfer* **99c**, 135–137, 1977.
20. Y. P. Chang and C. H. Tsou, *J. Heat Transfer* **99c**, 132–134, 1977.
21. Y. P. Chang and C. H. Tsou, *J. Heat Transfer* **99c**, 41–47, 1977.
22. Y. P. Chang, *Int. J. Heat Mass Transfer* **20**, 1019–1025, 1977.
23. M. N. Özisik and S. M. Shouman, *J. Franklin Inst.* **309**, 457–472, 1980.
24. J. Padovan, *J. Heat Transfer* **96c**, 313–318, 1974.
25. G. P. Mulholland, Diffusion Through Laminated Orthotropic Cylinders, *Proceedings of the 5th International Heat Transfer Conference*, Tokyo, 1974, Cu 4.3, pp. 250–254.
26. M. H. Sadd and I. Miskioglu, *J. Heat Transfer* 100, 553–555, 1978.
27. K. C. Poon, R. C. H. Tsou, and Y. P. Chang, *J. Heat Transfer* **101**, 340–345, 1979.
28. M. D. Mikhailov and M. N. Özisik, *Lett. Heat Mass Transfer* **8**, 329–335, 1981.
29. S. C. Huang and Y. P. Chang, *J. Heat Transfer* **106**, 646–648, 1984.
30. W. S. Wang and T. W. Chou, *J. Composite Mat.* **19**, 424–442, 1985.
31. S. J. Hagen, Z. Z. Wang, and N. P. Ong, *Phys. Rev. B* **40**, 9389–9392, 1989.
32. K. E. Goodson and Y. S. Ju, *Ann. Rev. of Mat. Sci.* **29**, 261–293, 1999.
33. S. Berber, Y. K. Kwon, and D. Tomanek, *Phys. Rev. Lett.* **84**, 4613–4616, 2000.

34. J. Hone, M. C. Llaguno, N. M. Nemes, A. T. Johnson, J. E. Fischer, D. A. Walters, M. J. Casavant, J. Schmidt, R. E. Smalley, *Appl. Phys. Lett.* **77**, 666–668, 2000.
35. J. Hone, M. C. Llaguno, M. J. Biercuk, A. T. Johnson, B. Battlog, Z. Benes, J. E. Fischer, *Appl. Phys. A — Mat. Sci. Proc*. **74**, 339–343, 2002.
36. J. E. Fischer, W. Zhou, J. Vavro, M.C. Llaguno, C. Guthy, R. Haggenmueller, M. J. Casavant, D. E. Walters, R. E. Smalley, et al., *J. Appl. Phys*. **93**, 2157–2163, 2003.
37. A. A. Balandin, S. Ghosh, W. Z. Bao, I. Calizo, D. Teweldebrhan, F. Miao, C. N. Lou, et al., *Nano Lett*. **8**, 902–907, 2008.
38. L. Onsagar, *Phys Rev*. **37**, 405–426, 1931; **38**, 2265–2279, 1931.
39. H. B. G. Casimir, *Rev. Mod. Phys*. **17**, 343–350, 1945.
40. I. Prigogine, *Thermodynamics of Irreversible Processes*, Wiley-Interscience, New York, 1961.
41. L. P. Eisenhart, *Coordinate Geometry*, Dover, New York, 1962.
42. A. R. Amir-Moe'z and A. L. Fass, *Elements of Linear Spaces*, Pergamon, New York, 1962.
43. S. H. Maron and C. F. Prutton, *Principles of Physical Chemistry*, Macmillan, New York, 1958.

PROBLEMS

15-1 Write the expressions for the three components of the heat flux, q_i'', $i = 1, 2, 3$, for an anisotropic medium in the following orthogonal coordinate systems: (1) prolate spheroid; and (2) oblate spheroid.

15-2 Write the time-dependent heat conduction equation for an anisotropic medium with constant conductivity coefficients for the following cases:

a. In the cylindrical coordinate system when temperature is a function of r, ϕ variables.

b. In the spherical coordinate system when temperature is a function of r, ϕ variables.

15-3 Write the boundary conditions of the third kind for an anisotropic solid at the following boundary surfaces.

a. At the boundary surfaces $z = 0$, $z = L$, and $r = b$ of a solid cylinder of radius b, height L.

b. At the surface $r = b$ of a solid sphere.

15-4 Write the thermal resistivity coefficients r_{11}, r_{13}, and r_{23} in terms of the thermal conductivity coefficients k_{ij}.

15-5 Consider two-dimensional steady-state heat conduction in an orthotropic rectangular solid in the region $0 \leq x \leq a, 0 \leq y \leq b$ with thermal conductivities k_1 and k_2 in the x and y directions, respectively. The boundaries at $x = 0$, $x = a$, and $y = b$ are kept at zero temperature, while the boundary at $y = 0$ is maintained at a temperature $T = f(x)$. Develop an expression for the steady-state temperature distribution $T(x, y)$ in the solid.

15-6 Consider steady-state heat conduction in an orthotropic solid cylinder $0 \leq r \leq b, 0 \leq z \leq L$ in which heat is generated at a uniform rate of g_0 W/m^3 while the boundaries are kept at zero temperature. The thermal conductivity coefficients in the r and z directions are k_1 and k_2, respectively. Obtain an expression for the steady-state temperature distribution $T(r, z)$ in the cylinder.

15-7 Consider an orthotropic region $0 \leq x < \infty, 0 \leq y \leq \infty$, which is initially at temperature $F(x, y)$, and for times $t > 0$ the boundaries at $x = 0$ and $y = 0$ are kept at zero temperature. The thermal conductivity coefficients for the x and y directions are k_1 and k_2, respectively. Obtain an expression for the temperature distribution $T(x, y, t)$ in the medium for times $t > 0$.

15-8 An orthotropic solid cylinder $0 \leq r \leq b, 0 \leq z \leq L$ is initially at temperature $F(r, z)$. For times $t > 0$ the boundaries are kept at zero temperature. The thermal conductivity coefficients for the r and z directions and k_1 and k_2, respectively. Obtain an expression for the temperature distribution $T(r, z, t)$ in the solid for times $t > 0$.

15-9 Consider two-dimensional-steady state heat conduction in an orthotropic solid cylinder of radius $r = b$ and height $z = L$ with thermal conductivities k_1 and k_2 in the r and z directions, respectively. The boundary surfaces at $r = b$ and $z = L$ are kept at zero temperatures, while the boundary surface at $z = 0$ is kept at temperature $T = f(r)$. Develop an expression for the steady-state temperature $T(r, z)$.

15-10 Consider time-dependent, two-dimensional heat conduction problem for an anisotropic medium $0 \leq x < \infty, -\infty < y < \infty$, which is initially at temperature $F(x, y)$, and for times $t > 0$ the boundary surface at $x = 0$ is kept insulated. Obtain an expression for the temperature distribution $T(x, y, t)$ in the medium for times $t > 0$.

15-11 Consider time-dependent, two-dimensional heat conduction problem for an anisotropic region $0 \leq x < \infty, -\infty < y < \infty$ that is initially at zero temperature. For times $t > 0$, heat is generated in the medium at a rate of $g(x, y, t)$ W/m^3, while the boundary at $x = 0$ is kept insulated. Obtain an expression for the temperature distribution $T(x, y, t)$ in the medium for times $t > 0$.

15-12 Consider time-dependent, two-dimensional heat conduction in an anisotropic hollow cylinder $a \leq r \leq b, -\infty < z < \infty$, which is initially at temperature $F(r, z)$. For times $t > 0$, the boundaries at $r = a$ and $r = b$ are kept at zero temperature. Obtain an expression for the temperature distribution $T(r, z, t)$ in the medium for times $t > 0$.

15-13 Transform the heat conduction equation

$$k_{11}\frac{\partial^2 T}{\partial x^2} + k_{22}\frac{\partial^2 T}{\partial y^2} + g = \pi C\frac{\partial T}{\partial t}$$

into a one similar to that for the isotropic medium.

NOTES

1. The closed-form expression given by equation (15-74) is determined as now described. We consider the following heat conduction problem:

$$\frac{d^2T}{dx^2} + \frac{g_0}{k_1} = 0 \quad \text{in} \quad 0 < x < a \tag{1a}$$

$$\text{BC1:} \quad \left.\frac{dT}{dx}\right|_{x=0} = 0 \tag{1b}$$

$$\text{BC2:} \quad \left.\frac{dT}{dx}\right|_{x=a} + H_1 \left. T\right|_{x=a} = 0 \tag{1c}$$

This problem is solved both by direct integration and using the integral transform technique as given below.

a. When it is solved by direct integration, we obtain

$$T = \frac{g_0 a}{k_1 H_1} + \frac{g_0}{2k_1}\left(a^2 - x^2\right) \tag{2}$$

b. To solve the system, equation (1), by the integral transform technique, we take its transform by the application of transform (15-70a) and obtain

$$\overline{T} = \frac{1}{k_1 \beta^2}\overline{g} \tag{3a}$$

where

$$\overline{g} = \int_{x=0}^{a} g_0 \cos\beta_m x\, dx = \frac{\sin\beta_m a}{\beta_m} g_0 \tag{3b}$$

Introducing the transform (3) into the inversion formula (15-60b), we obtain the solution as

$$T = \frac{g_0}{k_1}\sum_{n=0}^{\infty}\frac{1}{N\left(\beta_m\right)}\frac{\cos\beta_m x \sin\beta_m a}{\beta_m^3} \tag{4}$$

Since equations (2) and (4) are the solution of the same problem, by equating them, we obtain

$$\sum_{m=1}^{\infty}\frac{1}{N\left(\beta_m\right)}\frac{\cos\beta_m x \sin\beta_m a}{\beta_m^3} = \frac{a}{H_1} + \frac{1}{2}\left(a^2 - x^2\right) \tag{5}$$

which is the result given by equation (15-64).

2. The integral transform of $\partial^2 T/\partial y^2$ by the application of the transform (15-77b) is determined as

$$\int_{y=-\infty}^{\infty} e^{i\gamma y}\frac{\partial^2 T}{\partial y^2}dy = \left[\frac{\partial T}{\partial y}e^{i\gamma y} - i\gamma T e^{i\gamma y}\right]_{y=-\infty}^{\infty} - \gamma^2\int_{y=-\infty}^{\infty} e^{i\gamma y}T\,dy$$

$$= -\gamma^2\int_{y=-\infty}^{\infty} e^{i\gamma y}T\,dy = -\gamma^2\overline{T} \tag{1}$$

To obtain this result, we integrated by parts twice, assumed that T and $\partial T/\partial y$ both vanish as $y \to \pm\infty$, and utilized the definition of the transform (15-77b). The integral transform of $\partial^2 T/\partial x\partial y$ is determined as

$$\int_{y=-\infty}^{\infty} e^{i\gamma y}\frac{\partial^2 T}{\partial x\partial y}dy = \left[\frac{\partial T}{\partial x}e^{i\gamma y}\right]_{-\infty}^{\infty} - i\gamma\frac{\partial}{\partial x}\int_{y=-\infty}^{\infty} e^{i\gamma y}T\,dy$$

$$= -i\gamma\frac{\partial}{\partial x}\int_{y=-\infty}^{\infty} e^{i\gamma y}T\,dy = -i\gamma\frac{\partial\overline{T}}{\partial x} \tag{2}$$

where we assumed that $\partial T/\partial x$ vanish at $y \to \pm\infty$.

16

INTRODUCTION TO MICROSCALE HEAT CONDUCTION

As we described in Chapter 1, conduction is a specific mode of heat transfer in which the energy exchange takes place in solids or quiescent fluids from the region of high temperature to the region of low temperature due to the presence of a *temperature gradient* within the system. Accordingly, conduction represents the *diffusion* of heat within the medium, with energy cascading at the microscopic level from regions of higher energy to regions of lower energy. For a known temperature distribution $T(\hat{r}, t)$, we defined the flow of heat by Fourier's law, namely, $q''(\hat{r}, t) = -k\, \nabla T(\hat{r}, t)$, which has served as our cornerstone for the formulation of the heat equation and related boundary conditions. Our treatment up to this point invokes the continuum assumption, as developed in Section 1-3, in which our volumes and length scales of interest contain a large quantity of individual atoms, enabling thermodynamic properties to be defined on a statistically valid basis, and allowing heat to diffuse through the material based on the resulting well-defined temperature gradient. However, natural questions to ask are at what point does the continuum assumption break down, and at what length scale is Fourier's law no longer a valid expression to quantify the flow of heat?

In the last two decades, engineering and science have moved to the nanoscale [1–3], bringing temporal and spatial length scales in the nano-, pico-, and femtoscales, which in the limit correspond to physical length scales at the dimensions of a single atom, and physical time scales on the order of the very time scales of electronic transitions, internal relaxation, and intermolecular energy transfer. Structures such as nanowires [4, 5], thin films and layered structures [6, 7], and quantum structures such as quantum dots [8, 9] all challenge the validity of the continuum model and therefore require modifications to our heat transfer laws or the introduction of entirely new theories. These topics and theories in the context of energy transfer bring us to the regime known as *microscale heat transfer*. Several excellent references are available on this topic, including the classic 1998 book *Microscale Energy Transport* [10] and several more contemporary monographs [11–14].

In this chapter, our goal is to provide a concise introduction to the topic of microscale heat transfer, including a brief introduction to the relevant length scales, a discussion of the physics of energy carriers, development of guidelines for the limitations of Fourier's law, and an introduction to the various microscale heat transfer regimes.

16-1 MICROSTRUCTURE AND RELEVANT LENGTH SCALES

In the context of conduction heat transfer, we may consider a wide range of length scales, as depicted in Figure 16-1. The largest length scale, which we define as the *characteristic length scale* L_c of the problem, corresponds to the actual physical dimension of interest in the heat transfer problem and is often equal to the domain of the problem (e.g., the diameter of the cylinder as depicted in Fig. 16-1). For a one-dimensional Cartesian problem (i.e., planar wall), the characteristic length could be the thickness of the wall. In general, a characteristic length may be approximated by the ratio of the volume-to-surface area, namely,

$$L_c \cong \frac{\text{volume}}{\text{surface area}} \tag{16-1}$$

which yields $L_c = D/6$ for a sphere of diameter D, and $L_c = L/6$ for a cubic volume with sides of length L. There is no upper limit on the characteristic

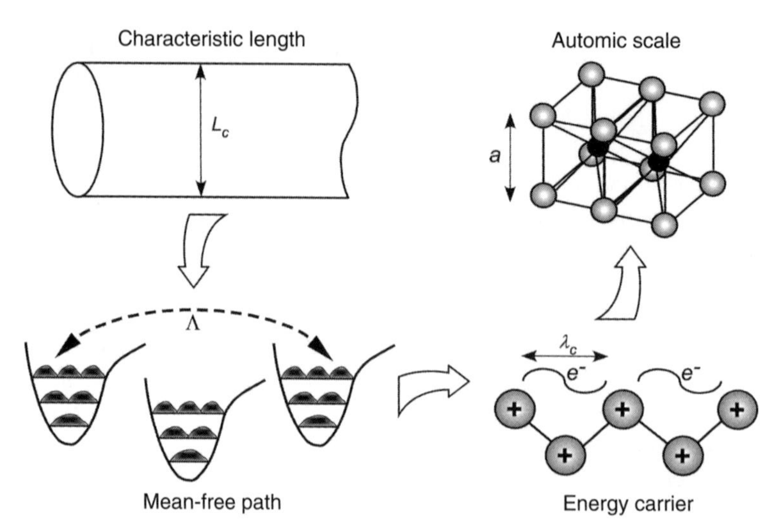

Figure 16-1 Length scales from the largest to smallest dimensions relevant to conduction heat transfer.

length scale, considering, for example, the one-dimensional semi-infinite medium problems.

The next two smaller length scales depicted in Figure 16-1 represent the mean free path of the energy carrier Λ and the energy carrier length scale λ_c, respectively. These will be discussed in detail in the following section. Finally, the smallest dimension in Figure 16-1 represents the atomic length scale, which is the smallest scale of interest to us within the more variable dimensions of the material's microstructure. There are many excellent references regarding the microstructure of materials that describe the atomic structure, physical and chemical properties, and related nomenclature and theory [15–17]. Here we only present a brief overview to define the relevant length scales of interest and to provide representative physical dimensions for a range of materials, limiting our discussion to solids.

Crystalline Microstructure

All solid materials, at the most fundamental level, are composed of atoms bound together. Solids are considered *crystals*, or crystalline solids, when their constituent atoms are assembled in a well-ordered, periodic array. Materials are considered a *single crystal* when the periodic structure extends largely uninterrupted throughout the entire domain of the crystal. An example of single crystals is found in gemstones such as diamonds. A *polycrystalline material* is one in which the local periodic structure is limited to relatively short distances (e.g., tens of microns), and the overall solid is comprised of many such crystalline regions coming together with random orientations. Many common materials, including most metals, are polycrystalline. Materials with no long-range crystalline order are called *amorphous solids*, or glassy materials, and the ordering in atomic bonding structure is confined to very small collections of atoms

Within all solids, the basic structural unit at the atomic level is called the *unit cell*, which corresponds to the actual arrangement and bonding of the constituent atoms. The unit cells are three-dimensional structures that are then repeated to create the solid; hence the unit cell represents the building blocks for creation of long-range order within crystalline solids. The unit cells are physically characterized by their arrangement (i.e., geometry) and size and chemically by the nature of the chemical bonds between the atoms. Of greatest interest to us is the characteristic size of the unit cell, which is defined in terms of the *lattice parameters* or lattice constants, which themselves correspond to the lengths of the edges of the unit cell. For simple cubic structures, a single lattice constant a defines the unit cell dimension, while for more complicated geometries such as orthorhombic or hexagonal unit cells, up to three lattice constants are necessary. Common cubic structures include the body-centered cubic (bcc) and the close-packed structures such as the hexagonal close-packed (hcp) and the face-centered cubic (fcc) structures. Figure 16-2 depicts a simple bcc crystal along with the zincblende structure, an example of the close-packed fcc structure. In the zincblende structure, the two different elements form sp^3 tetrahedral bonds,

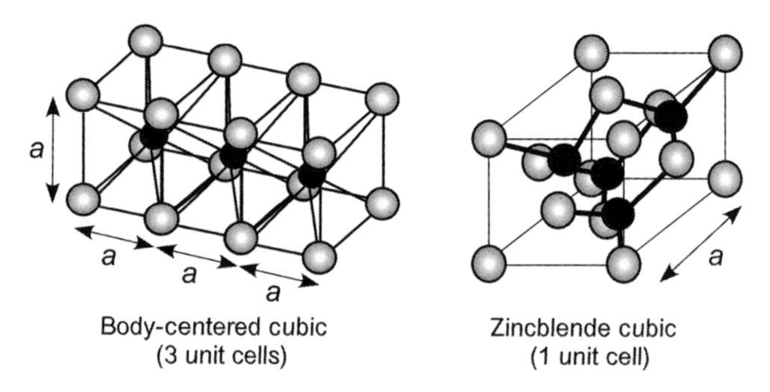

Figure 16-2 Unit cell structures.

and we note that when the atoms are all identical, this is called the diamond structure.

The actual dimensions of the unit cells (i.e., the lattice constants) are dependent on the specific atom or atoms within the unit cell and the nature of the chemical bonds. The chemical bonds play a key role in the resulting electrical, optical, and thermal properties of the material and may be categorized as *covalent*, *ionic*, and *metallic*. In covalent bonds, one or more electrons are shared between the two bonding atoms, which may be similar or dissimilar; hence the electrons may be shared equally or unequally. In ionic bonds, the electrons are not shared, but rather there is a net charge transfer. Common examples include ionic solids such as sodium chloride (Na^+Cl^-) in which one electron is transferred from the sodium atom to the chlorine atom. The third bond we consider is the metallic bond, common to many metals, which may be thought of as a limiting case of covalent bonding in which electrons are shared among all atoms in an *electron gas*. The atoms have a slight positive charge and create a period potential, through which the electrons freely move, giving an overall collective nature to the bonding. The electron gas is available to readily convey a current, making metals in general strong conductors of electricity. As we will discuss below, the electrons are also available to transport thermal energy, making many metals strong conductors of heat as well. A convenient unit of energy to characterize bonding energies is the *electron volt* (eV), which is defined as 1.6×10^{-19} J. The three bonds discussed above range in strength from about 1 to 10 eV, making them what we consider *strong bonds*, as opposed to weaker *van der Waals* forces, which are on the order of 0.01 eV.

We summarize the lattice constants for a broad range of materials in Table 16-1, classifying the various compounds in terms of their crystal structure. Overall, Table 16-1 provides a summary of our first length scale of interest, the atomic scale characterized by the lattice constant, namely,

$$\boxed{2\,\text{Å} \lesssim a \lesssim 7\,\text{Å}} \tag{16-2}$$

where we define the unit angstrom as $1\,\text{Å} = 1 \times 10^{-10}$ m. We see in Table 16-1 that the lattice constant is about 5 Å on average, or about 0.5 nm.

TABLE 16-1 Microstructure of Select Materials

Material	Lattice Constant, a (Å)	Crystal Structure
Fe	2.87	bcc
Cr	2.88	bcc
Mo	3.15	bcc
Cs	6.05	bcc
BeCu	2.70	bcc
CuZn (β-brass)	2.94	bcc
CsCl	4.12	bcc
Cu	3.61	fcc
Pd	3.89	fcc
Au	4.08	fcc
Ag	4.09	fcc
Pb	4.95	fcc
Zn	2.66	hcp
Mg	3.21	hcp
Zr	3.32	hcp
MgO	4.21	Cubic (rocksalt)
AgCl	5.55	Cubic (rocksalt)
NaCl	5.64	Cubic (rocksalt)
BaS	6.39	Cubic (rocksalt)
KBr	6.60	Cubic (rocksalt)
C (diamond)	3.57	Cubic (diamond)
Si	5.43	Cubic (diamond)
ZnSe	5.67	Cubic (zincblende)
CdS	5.82	Cubic (zincblende)
HgTe	6.43	Cubic (zincblende)

Sources: From reference 10, 12, and 13.

Crystalline Order

As discussed briefly above, the crystal structure is described by the periodic nature of the unit cell, which we may consider as single crystals with very long range order, polycrystalline materials with short-range order, noting distinct grain boundaries, and amorphous or glassy materials with no overall organized structure or periodicity. Single crystals have lower free energy than polycrystals, and, as a result, polycrystalline materials may be reorganized to obtain long-range order (i.e., approaching single crystals) by heating in a process called *annealing*. Since the single crystalline structure is preferred thermodynamically, the conversion is a favorable process, although the annealing process promotes the rate of conversion.

The above discussion of crystalline materials and the overall crystal structure is an idealization with regard to single crystals, as all crystals are characterized by disruptions to the periodic structure in the form of imperfections. This disruption includes grain dislocations, which are a nonalignment of two regions of otherwise ideal crystal structure. For polycrystalline materials in general, these dislocations become grain boundaries, which separate the long-term crystal order. Point defects are characterized by missing atoms, extra atoms, or substitutional defects in which a different atom than expected is present at the indicated location. Overall, all defects can cause considerable effects on the resulting thermal transport, as discussed in the following sections of this chapter, as they disrupt the transfer of energy through the material microstructure.

Amorphous solids are considered to have no ordered crystal structure or periodicity, which is characteristic of common glasses such as fused silica and soda–lime glass. Amorphous solids can also include thin metal–oxide films. In general, amorphous solids can be formed by rapid cooling, such that there is no time for the local crystal nuclei to grow in concert with any long-range order. As noted above, the strong disorder of amorphous solids can greatly influence the thermal transport properties.

16-2 PHYSICS OF ENERGY CARRIERS

We now consider the physics of how energy is actually stored and transported in various materials. We first present the two energy carriers associated with heat conduction, namely, the *electron* and the *phonon*, and then discuss the transport and storage of energy characteristic of each carrier. Importantly, we discuss the associated length scales with each carrier, which will be fundamental parameters in our establishing guidelines for the various heat transfer regimes.

Phonons as Energy Carriers

We first consider the transport of energy in crystalline materials in which energy is passed through the crystal lattice via lattice vibrational energy. In a classical model, the entire crystal lattice may be considered in terms of an array of *mechanical oscillators* (i.e., spring–mass systems). The individual atoms oscillate with various degrees of freedom as defined by the particular lattice structure. The key feature of a classical mechanical oscillator is that the energy–distance function is *continuous* within the physical constraints of the system, which corresponds to a continuous state of vibrational energy contained within the crystal lattice structure. In contrast, the crystal lattice is correctly considered as an array of *quantum mechanical oscillators* in which the energy states are no longer continuous, but rather the vibrational energy is confined to a series of *discrete energy states* for a given wavenumber. Such a description of vibrational energy states is analogous to the familiar quantum model of electron energies, which are confined to discrete energy levels characterized by electron orbits. While one may have an instinctive feel for a quantized electron model, the imposition of a quantized model for vibrational energy is perhaps not as intuitive. Nonetheless, it is the quantum mechanical model of vibrational energy that we will impose on our crystalline structures to characterize the transfer of thermal energy. In our

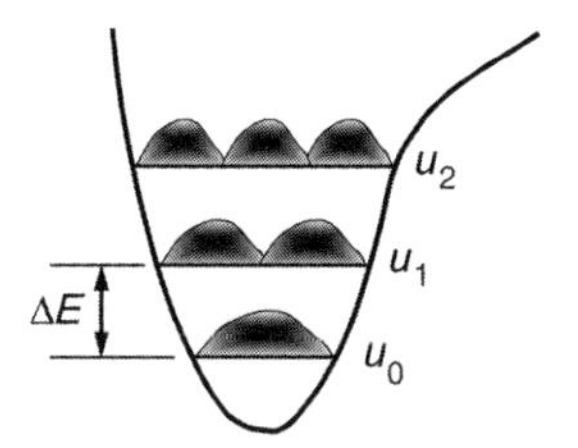

Figure 16-3 Vibrational quantum well showing discrete energy levels.

quantum mechanical model, the fundamental discrete unit of vibrational energy will be the *phonon*.

Let the vibrational frequency of the crystal structure be given by ω (s^{-1}), noting that the classical models are often well-suited for calculating the vibrational modes and the vibrational frequency. Typical vibrational frequencies in crystalline materials are in the range of 10^{12}–10^{13} Hz [10]. The vibrational energy u is then proportional to the vibrational frequency, namely, $u \propto \hbar\omega$, where the proportionality constant is the *reduced Planck constant* given as $\hbar \equiv h/2\pi$, and where h is the Planck constant given as 6.626×10^{-34} J · s. The exact values of vibrational energy are nicely conceptualized in terms of the vibrational energy well, as depicted in Figure 16-3 in which the vibrational energy levels are quantized.

The actual vibrational energy states are given for each vibrational energy level by the expression

$$u_n = \left(n + \tfrac{1}{2}\right)\hbar\omega \qquad \text{for} \qquad n = 0, 1, 2 \ldots \tag{16-3}$$

where n is the vibrational quantum number. Equation (16-3) then gives the vibrational *phonon energy*, which is seen to have discrete quantities, for example, $u_0 = \frac{1}{2}\hbar\omega$ and $u_1 = \frac{3}{2}\hbar\omega$. The lowest vibrational energy state, u_0, is considered the *ground vibrational state*. From Figure 16-3, we see that the difference in energy between any two contiguous vibrational energy states, namely, ΔE, is equal to $\hbar\omega$. For a vibrational frequency of 5×10^{12} Hz, $\Delta E = 0.0033$ eV; hence vibrational energies are orders of magnitude less than typical covalent and ionic bond strengths. We now have a quantum mechanical model for the vibrational energy states of the crystal lattice structure. Before moving on, it is useful to say a few additional words about the conceptual model depicted in Figure 16-3. If the local crystal structure is characterized by greater thermal energy, local vibrational wells are populated at energy levels above the ground state. Conceptually, an upper vibrational energy state is still characterized by the *same* vibrational frequency ω, however, the expected position of displacement of the vibrating atoms with respect to their equilibrium position is shifted such that there is greater probability of finding the atoms away from equilibrium and, therefore, on average at a greater energy state. These probability distributions of displacement are depicted schematically in Figure 16-3 for the first three energy levels.

The actual diffusive transfer of energy through a crystalline structure by phonons is now idealized by the movement of phonons (i.e., discrete waves of energy) through the crystal lattice. As will be discussed in greater detail in the following sections, the distance that the phonon wave travels between energy exchanges is considered the energy carrier mean free path. For the phonon as energy carrier, the conduction of heat is realized as the movement of these discrete phonon waves through the crystal lattice.

Debye Model of the Phonon

A convenient description of the phonon dispersion is given by the Debye model, which assumes a linear dispersion, as described below, and that the crystal waves are coherent for spacing much larger than the lattice spacing. Note also that the Debye model assumes that all phonon modes are approximated by three identical *acoustic branches*, which propagate at a constant velocity equal to the sonic velocity (i.e., speed of sound) within the crystal. As discussed in considerable detail in references 10–14, the Debye model is not a valid representation at the edge of the *Brillouin zone*, where the phonon dispersion is flat, nor for optical phonons, which are characterized by considerably higher frequencies than acoustic phonons. Having stated the limitations of the Debye model, we now introduce the linear dispersion model for our phonons as given by

$$\omega_{\text{ph}} = v_g K \tag{16-4}$$

where ω_{ph} is the *frequency of the phonon* (not the vibrational frequency of the lattice described above), v_g is the group velocity (i.e., the speed of sound in the crystal), and K is the *phonon wavenumber* (m^{-1}) given by

$$K = \frac{2\pi}{\lambda_{\text{ph}}} \tag{16-5}$$

where λ_{ph} is the characteristic *wavelength of the phonon*. One can think of the phonon wavelength as analogous to the wavelength of a photon, as both the photon and phonon represent massless discrete quanta of energy.

In electromagnetic theory, the wavelength of photons is described by a continuum; however, acoustic phonons must be sustained by the crystalline lattice and, therefore, a minimum λ_{ph} exists, and hence a maximum value of K. In the Debye model, a maximum cutoff wavenumber K_D is defined as

$$K_D = \left(6\pi^2 \eta\right)^{1/3} \tag{16-6}$$

where η is the limiting number of oscillators per unit volume ($\eta = N/V$), which may be approximated in terms of the effective lattice constant as $\eta \cong (\pi/6)(1/a^3)$. In regard to this limitation of the Debye model, the cutoff wavenumber K_D represents the edge of the Brillouin zone. The *Debye cutoff frequency* is then readily calculated from equation (16-4) as

$$\omega_D = v_g K_D = v_g \left(6\pi^2 \eta\right)^{1/3} \tag{16-7}$$

As described above, the phonon energy is given by $\hbar\omega$. The phonon energy may also be described as an equivalent thermal energy given by $k_B T$, where k_B is the Boltzmann constant equal to 1.380×10^{-23} J/K. We can then equate the phonon energy at the Debye cutoff frequency with an equivalent temperature, namely, $\hbar\omega_D = k_B \Theta_D$, which yields the *Debye temperature*:

$$\boxed{\Theta_D = \frac{\hbar\omega_D}{k_B} = \frac{\hbar v_g \left(6\pi^2 \eta\right)^{1/3}}{k_B}} \tag{16-8}$$

The group velocity, or speed of sound, for solid crystalline materials may be approximated as $v_g \cong a\sqrt{g/m}$, where a is the lattice constant, g is the spring constant of the lattice (i.e. atomic bonding force constant), and m is the mean atomic mass. For crystalline solids, $v_g \sim 10^3$ m/s, with an upper limit of about 2×10^4 m/s. For example, the average group velocity for silicon is about 6000 m/s [12], while for diamond it is about 12,000 m/s. The Debye temperature represents an upper bound on the phonon frequency, and a corresponding lower bound on the phonon wavelength, for actual temperatures above the Debye temperature. Table 16-2 lists the Debye temperatures for a range of crystalline materials, noting that we report a range of Debye temperatures for each material as taken from the various literature sources.

We are now ready to summarize the above discussion to give a characteristic wavelength for the phonon as an energy carrier, namely $\lambda_{c,\mathrm{ph}}$, based on the

TABLE 16-2 Debye Temperature of Selected Crystalline Materials

Material	Θ_D (K)	Material	Θ_D (K)
Ag	215–225	Ni	375–450
Al	394–428	Pd	274–275
Au	165–170	Pt	230–240
C (diamond)	1860–2230	Sb	200–211
Cd	120–209	Si	625–645
Cr	460–630	W	310–400
Cu	315–343	NaCl	281–321
Fe	400–470	NaBr	224–240
Ga	240–320	KCl	231–232
Ge	360–374	KBr	172–177
Mn	400–410	LiCl	422–477
Mo	380–450	RbCl	165–168

Sources: From references 12, 13, 15, 18, and 19.

Debye theory. The result is divided into two regions, corresponding to temperatures below the Debye temperature and to temperatures in excess of the Debye temperature [20, 21]. The result is summarized as

$$\lambda_{c,\mathrm{ph}} \begin{cases} \cong a & \text{for} \quad T > \Theta_D \\ \cong \dfrac{\Theta_D}{T} a & \text{for} \quad T < \Theta_D \end{cases} \tag{16-9}$$

which provides an estimate of the overall range of characteristic phonon wavelengths that extends from a few angstroms to tens of angstroms. For the materials of interest given in Tables 16-1 and 16-2, at a temperature of 300 K, the characteristic phonon wavelength ranges from a minimum of 3.8 Å for palladium to a maximum of 2.4 nm for diamond.

Electrons as Energy Carriers

We now consider the transport of energy in materials, notably metallic and semiconductor structures, in which the energy is passed through the lattice via electrons through the electron gas (i.e., Fermi gas) or in the conduction band, respectively. The electronic structure within solids is a complex topic and the reader is referred to references 10–17 for a comprehensive treatment of the topic. Our goal here is to briefly describe the physics of the electron as an energy carrier, and to define an appropriate characteristic length scale in the context of heat conduction via the electron.

For metals, we consider the free electron theory in which electrons move and carry energy through the periodic potential of the ion core, in other words through the electron gas. Hence the electrons are the primary energy carriers, and interactions between these free electrons and the ions within the potential are neglected, as is electron–electron scattering. For semiconductors, electrons are thermally excited from the valence band to the conduction band, where they serve as the primary electron energy carriers. We note that for semiconductors in particular, thermal energy may be transported simultaneously by both phonons and electrons. We define our characteristic wavelength $\lambda_{c,e}$ of the electron as an energy carrier in terms of the de Broglie wavelength, giving

$$\lambda_{c,e} = \frac{h}{m^* v} \tag{16-10}$$

where we define m^* as the effective mass of the electron, typically $\sim \frac{1}{4}-\frac{1}{2}$ of the rest mass (electron rest mass $= 9.11 \times 10^{-31}$ kg), and v as the electron thermal velocity. We now define the relevant thermal velocity for metals and for semiconductors. For metals, the Fermi velocity is appropriate, giving

$$v_f \cong \frac{\hbar}{m^*} \left(3\pi^2 \eta\right)^{1/3} \tag{16-11a}$$

For electrons in the electron gas, the Fermi velocity is on the order of 10^6 m/s, with copper giving a value of 1.6×10^6 m/s. For semiconductors, the appropriate velocity for electrons in the conductance band is

$$v \cong \left(\frac{3k_B T}{m^*}\right)^{1/2} \tag{16-11b}$$

which gives velocities on the order of 10^5 m/s. The substitution of equations (16-11) into equation (16-10) yields the characteristic wavelength of the electron as an energy carrier for metals and semiconductors, giving the following range of values

$$\lambda_{c,e} \begin{cases} \cong 1\text{–}10\ \text{Å} & \text{for metals} \\ \cong 1\text{–}10\ \text{nm} & \text{for semiconductors} \end{cases} \tag{16-12}$$

In summary, we now have introduced the concept of phonons and electrons as the actual energy carriers, noting that materials such as semiconductors may transport energy via both electrons and phonons simultaneously. In addition, we have defined a characteristic length scale λ_c for both of our energy carriers, which varies by about two orders of magnitude for dielectric crystals, metals, and semiconductors, ranging from a few angstroms to about 10 nm.

16-3 ENERGY STORAGE AND TRANSPORT

Energy Storage: Specific Heat

The energy stored within solids is partitioned within electronic and vibrational energy states. For crystalline materials, increased temperature is realized as increased internal vibrational energy, with the population of upper vibrational energy levels $u_1, u_2, \ldots,$ (see Fig. 16-3) increasing proportionately. Energy is also stored within electrons, with the Fermi energy distribution describing the total energy state. With the internal energy $u(T)$ given by the appropriate functions for phonons or electrons, the specific heat c is defined as

$$c = \frac{du(T)}{dT} \tag{16-13}$$

We will first treat the various contributions to the specific heat from lattice storage and from electron storage separately and will then combine the two for the total specific heat. Beginning with lattice storage within crystalline materials, the Debye model provides an expression for the volumetric specific heat (J/m^3 K), c_l, where we have introduced the subscript l to indicate lattice storage (i.e., vibrational storage). For temperatures well below the Debye temperature, we have

$$c_l = \frac{36\pi^4}{15}\eta k_B \left(\frac{T}{\Theta_D}\right)^3 \qquad \text{for} \qquad T < \Theta_D \tag{16-14a}$$

Equation (16-14a) is known as the T^3 law and has been observed for many solids below the Debye temperature limit. When the temperature approaches or exceeds the Debye temperature, in other words as phonons approach the boundary of the Brillouin zone, the validity of equation (16-14a) no longer holds, and the specific heat approaches a constant value, namely,

$$c_l \cong 3\eta k_B \equiv 3\overline{R} \qquad \text{for} \qquad T \gtrsim \Theta_D \tag{16-14b}$$

which is known as the Dulong–Petit law. We have introduced the ideal gas constant $\overline{R} = (N/V)k_B$. An alternative model for vibrational energy storage is the Einstein model, which treats each atom as an independent oscillator. In general, the Einstein model is better suited for amorphous materials than crystals. The Einstein model gives the specific heat in the form

$$c_E = 3\overline{R}\,\frac{\Theta_E^2}{T^2}\,\frac{e^{\Theta_E/T}}{\left(e^{\Theta_E/T} - 1\right)^2} \tag{16-15}$$

where Θ_E is the Einstein temperature. In the large temperature limit, $T \gg \Theta_E$, equation (16-15) is in perfect agreement with equation (16-14b), namely, $c_l = c_E \equiv 3\overline{R}$. In Figure 16-4 we depict the behavior of the volumetric specific heat for vibrational energy storage within crystalline solids.

We now consider the storage of energy by electrons in metals and semiconductors. Using the Fermi–Dirac energy distribution, the specific heat for electron storage is given by

$$c_e = \frac{\pi^2}{2}\eta_e k_B\left(\frac{T}{T_f}\right) \tag{16-16}$$

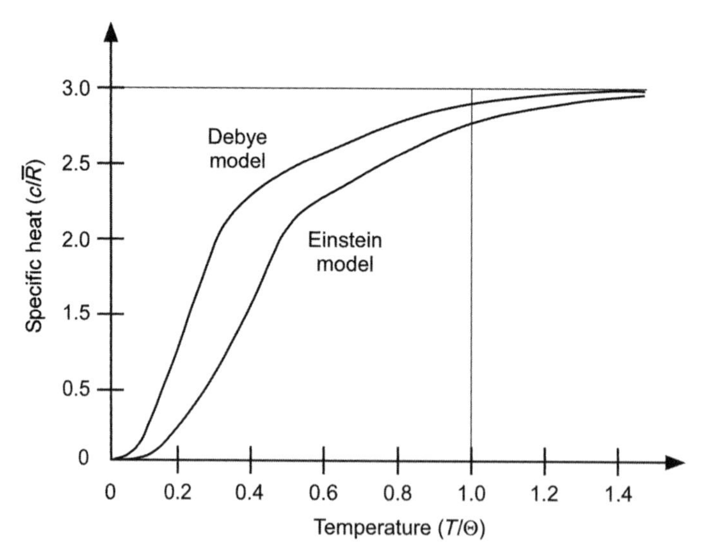

Figure 16-4 Idealized behavior of the specific heat for vibrational energy storage in crystalline materials.

where now η_e is the number density of electrons (N_e/V) rather than the number density of oscillators, and where T_f is the Fermi temperature given by $T_f = E_f/k_B$, where E_f is the Fermi energy.

With the specific heats now defined above for both the phonon and electron contributions, we can consider the total specific heat as the sum of these two contributions, giving

$$c(T) = c_l + c_e \tag{16-17}$$

where we define the individual contributions by equations (16-14) – (16-16). Equation (16-17) may be divided into different regimes depending on the temperature range and the nature of the solid. For metals at low to moderate temperatures, both the phonons and electrons contribute to the specific heat, giving

$$c(T) = A\,T^3 + \gamma_s T \qquad \text{for} \qquad T \ll \Theta_D \tag{16-18a}$$

where γ_s is the *Sommerfeld constant*, readily determined from equation (16-16), and where A is determined from equation (16-14a). In general, at very low temperatures of a few kelvins, the electron contribution tends to dominate. As discussed in references 12 and 15, theoretical values of the Sommerfeld constant generally agree with experimental data for noble and alkali metals. For transition metals that exhibit magnetism or paramagnetism, the measured value of γ_s may be an order of magnitude greater than the value predicted by theory. For semimetals such as Bi, the measured value may be an order of magnitude smaller than the theoretical values. When the temperature is near or above the Debye temperature, the specific heat is given as

$$c(T) = 3\eta\, k_B + \gamma_s T \qquad \text{for} \qquad T \gtrsim \Theta_D \tag{16-18b}$$

where we have used the asymptotic limit given by equation (16-14b) for the contribution of phonons. Equations (16-17) and (16-18) hold for metals, semimetals, and semiconductors, while for pure nonconducting materials (i.e., dielectrics), these equations simply reduce to equations (16-14).

Energy Transfer: Thermal Conductivity

We now turn our attention to perhaps our most important thermal property with respect to conduction heat transfer, namely, the thermal conductivity. In view of our above discussion, one can envision that the thermal conductivity will be treated in a similar manner to the specific heat, that is, with contributions to the thermal conductivity from phonon transport and from electron transport. This approach corresponds to the *electron–phonon transport theory* and assumes that the conduction of heat in a solid is due to the motion of free electrons and due to phonon transport via the lattice vibrational energy. Before turning our attention to the physics of phonons and electrons, we first consider energy transport in the classic case of Fourier's law. As developed in reference 10 (Section 1.6.2), for the general case of energy transport via generic particles (i.e., phonons or electrons), one may readily derive an expression for the heat flux assuming local thermodynamic equilibrium such that the energy density $u = u(T)$, which gives

$$q_x'' = -\frac{1}{3} c\, v\, \Lambda \frac{dT}{dx} \tag{16-19}$$

where c is the specific heat, v is the average particle velocity, and Λ is the energy carrier *mean free path*. No assumptions are made in equation (16-19) with regard to the nature of the energy carrier; hence the expression is universal for all energy carriers. Equation (16-19) is recognized as Fourier's law for the definition of thermal conductivity given as

$$k \equiv \tfrac{1}{3} c\, v\, \Lambda \tag{16-20}$$

We may now write equation (16-20) in the following manner:

$$k \equiv \tfrac{1}{3}\left[(c\, v\, \Lambda)_l + (c\, v\, \Lambda)_e\right] \tag{16-21}$$

where the first term in the brackets accounts for the contribution of the phonon energy transport through the lattice and the second term accounts for electron energy transport. The difficulty of equations (16-20) and (16-21) lies in the calculation of the mean free path, which is quite difficult. Rather than use these equations as a means to evaluate the thermal conductivity, they are generally used to calculate the mean free path on the basis of experimental values of thermal conductivity or other direct theoretical values, as noted in reference 11.

We begin with a discussion of thermal conductivity for insulating materials (i.e., dielectrics) in which energy transport is only via phonons. For such a case, the mean free path for phonons Λ_{ph} may be expressed using *Matthiessen's rule* in the form

$$\frac{1}{\Lambda_{\text{ph}}} \equiv \frac{1}{\Lambda_{\text{ph-ph}}} + \frac{1}{\Lambda_{\text{ph-de}}} \tag{16-22}$$

where the subscripts ph–ph and ph–de denote phonon–phonon scattering and phonon–defect scattering, respectively. Following the discussion in references 10 and 12, phonon–phonon scattering is dominant and varies inversely with temperature in the high–temperature regime. Hence to first order, thermal conductivity generally decreases with increasing temperature in the high-temperature limit, as observed in Chapter 1 in Figure 1.3. As temperature is decreased, scattering of phonons by defects dominates and becomes nearly constant. Since the specific heat in the low-temperature limit is dependent on T^3, recall equation (16-14a), the thermal conductivity likewise follows the T^3 behavior, again as observed in Figure 1-3.

The Cahill–Pohl model is a hybrid model that combines the localized oscillators of the Einstein model and the coherence of the Debye model [22, 23]. For phonon transport in materials, the Cahill–Pohl model gives the thermal conductivity in the form

$$k_{\mathrm{C-P}} = \left(\frac{\pi}{6}\right)^{1/3} k_B \eta^{2/3} \sum_i v_i \left(\frac{T}{\Theta_i}\right)^2 \int_{x=0}^{\Theta_i/T} \frac{x^3 e^x}{(e^x - 1)^2} dx \qquad \text{(16-23a)}$$

where the sum is taken over three acoustic modes (two transverse and one longitudinal) with respective group velocities v_i, and where Θ_i is the characteristic cutoff temperature for each state, as given by

$$\Theta_i = \frac{v_i}{k_B} \left(6\pi^2 \eta\right)^{1/3} \qquad \text{(16-23b)}$$

The Cahill–Pohl model provides very good agreement with experimental data over the temperature range from about 50 to 300 K for Si, SiO_2, Ge, and several other materials. For temperatures below ~50 K, discrepancies between the predicted and experimental thermal conductivity values are observed, which the authors attribute to the fact that the model does not include heat transport by phonons with long mean free paths, which are known to exist at low temperatures.

We now consider the thermal conductivity due to energy transport via electrons. In metals, we may cast the thermal conductivity in terms of the *Wiedemann–Franz law*, which gives

$$k = \sigma L_0 T \qquad \text{(16-24a)}$$

where σ is the electrical conductivity and L_0 is the Lorenz number given by

$$L_0 = \frac{\pi^2}{3} \left(\frac{k_B}{e}\right)^2 \qquad \text{(16-24b)}$$

where e is the fundamental electron charge. The theoretical value of the Lorenz number is given as $L_0 = 2.44 \times 10^{-8}\ W\Omega/\mathrm{K}^2$ [24]. Overall, equation (16-24a) shows the thermal conductivity is electron-driven in metals, although it remains an empirical formula, and therefore not so satisfying as a model for the thermal conductivity for metals and semiconductors given the additional need for electrical conductivity data.

Overall, the theoretical transport of energy via electrons remains a challenge and a continuing topic of research. To date, no comprehensive models are available. A few general comments are offered with regard to the temperature trends of thermal conductivity in metals and semiconductors. At low temperatures, the electron mean free path is controlled largely by scattering resulting from defects in the crystal lattice structure, including grain boundaries, surface defects, dislocations, and point defects. The mean free path is essentially constant in the low-temperature regime while the specific heat varies linearly with temperature, see equation (16-16), giving the thermal conductivity an approximately linear variation with temperature as well. At moderate to high temperatures, out-of-phase vibrations in the lattice can create local dipoles, which subsequently scatter electrons, although as a general rule, the thermal conductivity of metals does not display much temperature dependence in this regime. A detailed treatment of phonon scattering and the role of boundaries on thermal conductivity is given by reference 25. Additional discussion and theoretical considerations of thermal conductivity for metals, semiconductors, and dielectrics are given by references 26 and 27.

Ultimately, we are left with some theoretical models and expressions for the thermal conductivity of crystalline materials dominated by phonon transport and an empirical expression for the thermal conductivity of metals. Generally, given that the thermal conductivity of a given material is highly parameter dependent (e.g., temperature dependent), experimentally measured values become the best recourse for selection of representative values, although the above discussion can help guide the selection of appropriate values when interpolating or extrapolating over a given temperature range.

Energy Carriers: Mean Free Path

Having completed our discussion of the physics of energy carriers, including the energy carrier wavelengths, specific heat, and thermal conductivity, we are now ready to introduce our next important length scale, namely, the energy carrier *mean free path*. The energy carrier mean free path is defined as the average distance the energy carrier (e.g., phonon or electron) travels within the solid structure before transferring its excess energy to the next carrier or to the bulk structure. This concept was described above for the phonon mean free path in terms of the diffusion of energy via phonon waves through the crystalline lattice. For an electron carrier, the mean free path corresponds to electron transport through the period potential and represents the distance Λ that the electron travels before undergoing a scattering or energy transfer event. Overall, the mean free path is determined by the various physical scattering processes, including phonon–phonon, electron–electron, phonon–electron interactions, as well as scattering by defects and impurities. Because scattering is in general a function of temperature, the mean free path is also temperature dependent.

Based on our discussions above regarding equation (16-20), we now define the mean free path for a phonon in terms of the thermal conductivity, lattice specific heat, and the group velocity, namely,

$$\Lambda_{\text{ph}} = \frac{3k}{c_l\, v_g} \tag{16-25a}$$

The thermal conductivity in equation (16-25) may come from an appropriate model such as the Cahill–Pohl model or experimental values, the specific heat should come from the Debye or Einstein models using equations (16-14) and (16-15), respectively, and the group velocity is the average speed of sound for the material of interest.

TABLE 16-3 Carrier Mean Free Path for Selected Materials

Material	Carrier	Temperature (K)	Mean Free Path
Diamond	Phonon	300	100–460 nm
Si	Phonon	300	40–300 nm
Ge	Phonon	300	30–480 nm
Al-nitride	Phonon	300	30–70 nm
Al-oxide	Phonon	293	5 nm
SiO2	Phonon	273/300	97/6-8Å
NaCl	Phonon	273	67 nm
Al	Electron	298	10–40 nm
Ag	Electron	298	49–58 nm
Au	Electron	298	31–40 nm
Pt	Electron	298	12 nm
Co	Electron	273/373	21/17 nm
Cu	Electron	298	36–40 nm
La	Electron	298	41Å
Nd	Electron	298	41Å
Ce	Electron	298	53Å

Sources: From references 28–37.

For electron transport, the mean free path is given by the expression

$$\Lambda_e = \frac{3k}{c_e\, v} \tag{16-25b}$$

where the velocity is given by the Fermi velocity for a metal, equation (16-11a), and the thermal velocity for a semiconductor, equation (16-11b). The specific heat may be given by equations (16-16) or (16-17) as appropriate for metals and semiconductors, while the thermal conductivity may come from direct experimental values or through the empirical Wiedemann–Franz law given by equation (16-24).

Table 16.3 presents data on the carrier mean free paths for a range of materials and temperatures. Overall, the phonon mean free paths vary from about 1 nm to nearly 500 nm at room temperature, while the electron mean free paths vary from about 10 nm to less than 100 nm at room temperature.

16-4 LIMITATIONS OF FOURIER'S LAW AND THE FIRST REGIME OF MICROSCALE HEAT TRANSFER

We now have described in detail three length scales relevant to conduction heat transfer, namely, the characteristic dimension of the physical system (L_c), the characteristic length of the energy carrier (λ_c), which corresponds to either the phonon or the electron, and, finally, the carrier mean free path (Λ), which again

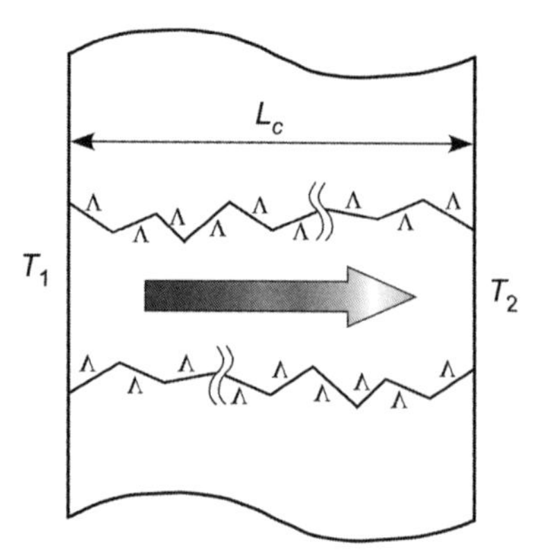

Figure 16-5 Diffusion of heat due to a temperature gradient.

can correspond to the phonon or the electron. For our energy carriers, λ_c ranges from a few angstroms to about 10 nm, while Λ ranges from about 1 nm to nearly 500 nm at room temperature and considerably longer for very low temperatures. With these three length scales in mind, we are ready to define the limitations of Fourier's law and the subsequent regimes of microscale heat transfer.

Limitations of Fourier's Law

The use of Fourier's law requires the presence of a temperature gradient, which requires a statistically valid distribution of energy carriers over the physical dimension of the problem (i.e., over the characteristic length L_c). This requires that multiple exchanges of energy between the energy carriers take place through the domain of the problem. Such a scenario is depicted in Figure 16-5, which shows the diffusion of heat as the energy carriers undergo a number of energy transfer events, each of approximate distance Λ, over the physical domain L_c, which is separated by a temperature difference $T_1 - T_2$.

The requirement of multiple energy exchanges may be interpreted as the characteristic length being much greater than the carrier mean free path, namely,

$$\boxed{L_c \gg \Lambda \qquad \text{for Fourier's law}} \tag{16-26}$$

The requirement of much larger than is generally considered satisfied by a factor of 10; hence we consider $L_c > 10\Lambda$ as our guideline for the applicability of Fourier's law. We may define the *Knudsen number* for our energy carriers as

$$Kn \equiv \frac{\Lambda}{L_c} \qquad \text{Knudsen number} \tag{16-27}$$

which represents the ratio of the energy carrier mean free path to the characteristic length. We note that in general the Knudsen number is used to compare other mean free paths to physical length scales, such as the mean free path of molecules in air, and is often used to define the limit of the continuum approximation. We may now define our limiting guideline as the $Kn \ll 1$, or simply as $Kn < 0.1$, for the validity of Fourier's law.

As an example, if we consider a thin diamond film with an average phonon mean free path of ~250 nm (see Table 16-3), we would consider Fourier's law as applicable for a film thickness down to about 2.5 μm, which we note is about $\frac{1}{40}$ the thickness of an average human hair. For a thin gold coating, with a mean free path for the electron carrier equal to ~40 nm, we would consider Fourier's law for heat conduction across the film valid for a thickness greater than about 400 nm. However, let us consider a gold nanoparticle with a diameter of 90 nm. Using equation 16-1, the corresponding characteristic length is about 15 nm, which is actually less than the mean free path of 40 nm. Accordingly, Fourier's law is not a valid means to access heat conduction through such a nanoparticle.

First Regime of Microscale Heat Transfer

We now define what may be called the *first regime of microscale heat transfer* [21], which corresponds to the following two conditions:

$$\boxed{\begin{aligned} &L_c \approx \Lambda \qquad (\text{Kn} \approx 1) \\ &L_c \gg \lambda_c \end{aligned}} \tag{16-28}$$

which together correspond to the case of a comparable characteristic length scale and mean free path, but with a characteristic length scale still much greater than the energy carrier wavelength. Let us reconsider our 90-nm gold nanoparticle discussed above, with a characteristic length about 15 nm. In consideration of equation (16-12), we consider a conservative estimate of the characteristic electron wavelength to be 1 nm. Since 15 nm is much greater than 1 nm, the gold nanoparticles would be considered to fall within our first regime of microscale heat transfer. Rather than the classical heat equation, which applies to the continuum regime as developed in Chapter 1, we now turn to the *Boltzmann transport equation* as our governing equation in the first microscale regime.

Boltzmann Transport Equation

The Boltzmann transport equation (BTE) is a statistical mechanical description of a system of particles (e.g., the particles can be atoms, electrons, phonons, or molecules) that is applicable in the absence of local thermodynamic equilibrium [38–40]. For example, when the length scale and/or temporal scale of interest is on the same order as the characteristic length scale of our particles (i.e., the mean free path) or the time scale of our particles (i.e., the relaxation time), respectively, our macroscopic or continuum models are no longer valid, and therefore only a statistical distribution of the particle system is available. The BTE provides such a model and may be written in the form

$$\frac{\partial f}{\partial t} + \hat{v} \cdot \nabla_{\hat{r}} f + \hat{a} \cdot \nabla_{\hat{v}} f = \left(\frac{\partial f}{\partial t}\right)_{\text{scat}} \tag{16-29}$$

where $f(\hat{r}, \hat{p}, t)$ is the statistical distribution of the particles giving their position, $\hat{r}$, and their momentum, $\hat{p}$, at any time t. In equation (16-29), $\hat{v}$ and $\hat{a}$ are the velocity and acceleration vectors, respectively, with the latter accounting for body forces in the three principal directions. The term on the right-hand side accounts for the rate of change of the distribution due to scattering (i.e., collisional) effects. We define the two gradient operators as

$$\nabla_{\hat{r}} = \frac{\partial}{\partial x}\hat{i} + \frac{\partial}{\partial y}\hat{j} + \frac{\partial}{\partial z}\hat{k} \tag{16-30a}$$

and

$$\nabla_{\hat{v}} = \frac{\partial}{\partial v_x}\hat{i} + \frac{\partial}{\partial v_y}\hat{j} + \frac{\partial}{\partial v_z}\hat{k} \tag{16-30b}$$

The scattering term in general may be nonlinear, making solution of the BTE very difficult. A common simplification is to use the following *relaxation time approximation* given by

$$\left(\frac{\partial f}{\partial t}\right)_{\text{scat}} = \frac{f_0 - f}{\tau(\hat{v}, \hat{p})} \tag{16-31}$$

where f_0 is the distribution function at equilibrium (e.g., Bose–Einstein distribution for phonons or Fermi–Dirac distribution for electrons), and $\tau(\hat{v}, \hat{p})$ is the relaxation time, which is itself a function of velocity and momentum. The relaxation time represents the time required to restore equilibrium to the distribution via the scattering/collision process following a perturbation. As a further approximation, we may neglect the functional dependency of the relaxation time and treat it as constant, that is, $\tau(\hat{v}, \hat{p}) \cong \tau$. With these assumptions, the BTE becomes

$$\boxed{\frac{\partial f}{\partial t} + \hat{v} \cdot \nabla_{\hat{r}} f + \hat{a} \cdot \nabla_{\hat{v}} f = \frac{f_0 - f}{\tau}} \tag{16-32}$$

Boltzmann Transport Equation and Fourier's Law

The 1-D form of the Boltzmann transport equation given by equation (16-32) can lead directly to Fourier's law using a few additional assumptions. It can be shown, see reference 10, that the energy flux vector in its most general form is given as

$$q''(\hat{r}, t) = \int v(\hat{r}, t) f(\hat{r}, \varepsilon, t) \varepsilon\, D(\varepsilon)\, d\varepsilon \tag{16-33}$$

where $D(\varepsilon)$ is the density of energy states. Considering equation (16-32), we may neglect the body forces ($\hat{a} = 0$) and make the additional assumption that $t \gg \tau$ (i.e., our time scales of interest are much greater than the relaxation time), which gives $\partial f / \partial t \approx 0$. We may also assume that $L_c \gg \Lambda$, which gives $\partial f / \partial x \approx \partial f_0 / \partial x$ and corresponds to the condition of *local thermodynamic equilibrium*. With these assumptions, equation (16-32) reduces to

$$v_x \frac{df_0}{dx} = \frac{f_0 - f}{\tau} \tag{16-34a}$$

which yields the following quasi-equilibrium solution:

$$f \cong f_0 - \tau v_x \frac{df_0}{dx} \tag{16-34b}$$

Since our equilibrium distribution f_0 is only a function of temperature under these assumptions, we can express the derivative in equation (16-34b) as

$$\frac{df_0}{dx} = \frac{df_0}{dT}\frac{dT}{dx} \tag{16-34c}$$

Substituting equations (16-34a) to (16-34c) into the 1-D form of equation (16-33) yields the following expression for the heat flux:

$$q_x'' = \int v_x \left(f_0 - \tau v_x \frac{df_0}{dT}\frac{dT}{dx} \right) \varepsilon D(\varepsilon)\, d\varepsilon \tag{16-35}$$

Equation (16-35) is readily expressed as the difference of two integrals, with the integrand of the first integral equal to $v_x f_0 \varepsilon D(\varepsilon)$. For the equilibrium distribution f_0, this first integral is identically zero. Making this simplification to equation (16-35), and pulling the term dT/dx from within the integral, gives

$$q_x'' = -\frac{dT}{dx} \int \tau v_x^2 \frac{df_0}{dT} \varepsilon D(\varepsilon)\, d\varepsilon = -k\frac{dT}{dx} \tag{16-36}$$

Equation (16-36) reduces directly to Fourier's law, as shown, for the thermal conductivity given by the following expression [10]:

$$k = \tau \int v_x^2 \frac{df_0}{dT} \varepsilon D(\varepsilon)\, d\varepsilon \equiv \frac{1}{3} c v^2 \tau \tag{16-37}$$

For integration of v_x^2, we have made the replacement with $\frac{1}{3}v^2$. Finally, we note that the relaxation time may be related to the mean free path by $\Lambda \cong v\tau$. Making this final substitution in equation (16-37), we have $k = \frac{1}{3} c v \Lambda$, which is in exact agreement with our definition of thermal conductivity as given above by equation (16-20).

In summary, within the first regime of microscale heat transfer, our governing equation becomes the Boltzmann transport equation, which represents a statistical mechanical approach to conservation of energy. Solution of the BTE remains challenging, and the reader is referred to references 10–12 and 38–40 for additional treatment and insight into the BTE.

16-5 SOLUTIONS AND APPROXIMATIONS FOR THE FIRST REGIME OF MICROSCALE HEAT TRANSFER

We present here several governing equations for conduction heat transfer within the first regime of microscale heat transfer. Our intention here is not a rigorous treatment of the topic, for which the reader is referred to the many references presented above, but rather to make the reader aware of several well-known approaches for problems in which our classical heat equation and Fourier's law have limited application.

Hyperbolic Heat Conduction Equation: Cattaneo Equation

We now consider the case for which we assume a continuum with regard to the spatial dimensions of interest; hence we assume that our characteristic length scale $L_c \gg \Lambda$ and, therefore, that any deviations from spatial equilibrium are small. For our problem, however, we are interested in temporal time scales that are on the order of our energy carrier relaxation times, that is, $t \approx \tau$. Here we consider τ an average relaxation time ($\tau \cong \Lambda/v$). For example, with an electron as the energy carrier in gold, we have a representative mean free path of 40 nm (see Table 16-3) and a Fermi velocity of about 10^6 m/s, giving an estimated relaxation time of 40×10^{-15} s, or 40 fs. For aluminum nitride, with a phonon mean free path of about 50 nm and an average group velocity (longitudinal and shear) of 8200 m/s, we have an estimated relaxation time of 6×10^{-12} s, or 6 ps. These relaxation time scales are on the order of many ultrafast laser pulses, which have pulse widths ranging from the picosecond to low femtoscale time scales. Therefore, modeling the thermal response of pulsed-laser energy deposition into materials at these time scales, but over bulk spatial scales, would be consistent with our hyperbolic heat equation regime (i.e. $t \approx \tau$ and $L_c \gg \Lambda$). We note that such laser–material interactions may involve physics well beyond conduction (e.g., material ablation, electron ejection, plasma formation, etc.), but we limit our discussion here to the diffusion of heat via conduction.

Under the conditions discussed above, insufficient time is available for the establishment of thermodynamic equilibrium and the application of the heat equation and Fourier's law is not generally applicable. An alternative form of the heat equation was developed by Carlo Cattaneo, now referred to as the *Cattaneo equation*, which for constant thermal conductivity is given by [41–43]

$$\nabla^2 T(\hat{r}, t) = \frac{1}{\alpha}\left[\frac{\partial T(\hat{r}, t)}{\partial t} + \tau \frac{\partial^2 T(\hat{r}, t)}{\partial t^2}\right] \tag{16-38a}$$

for which we list the appropriate regime as

$$L_c \gg \Lambda \qquad (Kn \ll 1) \qquad \text{and} \qquad t \approx \tau \tag{16-38b}$$

Cattaneo proposed the above equation to solve what he referred to as the *paradox of heat conduction*, by which he meant that the classical heat equation is parabolic in nature and therefore implies an infinite speed of propagation of temperature perturbations [44]. Equation (16-38) is hyperbolic in nature and has the effect of a wavelike solution (i.e., thermal wave) of the heat propagation, which is characteristic of nonequilibrium in thermodynamic transition. As noted in reference 42, like other constitutive models in engineering, modifications to Fourier's law are motivated by its deficiencies in advanced applications, for example, the case of ultrafast laser–material interactions.

Acoustically Thin Phonon Transport: Casimir Limit

As discussed in Section 16-3, the application of Fourier's law requires the presence of a temperature gradient. However, if we consider a thin film of thickness L_c, separated by some temperature difference $\Delta T = T_1 - T_2$, for the special case of $L_c \approx \Lambda$ (i.e., Kn $\approx$ 1), there is no temperature *gradient* present. Because the temperature at a point can only be defined under local thermodynamic equilibrium, a valid temperature can only be defined at points separated on average by a mean free path [45]. As such, we have no temperature gradient for the diffusion of heat and no energy carrier scattering within the domain. We now consider this scenario only for phonon transport, hence in a dielectric crystal. For such a case, the physics of phonon transport are essentially identical to the physics of photon transport (i.e., blackbody radiation), and we can define the heat transfer across our thin film by the expression [46–47]

$$q''_x = \sigma_{\text{ph}} \left[T_1^4 - T_2^4 \right] \tag{16-39a}$$

Equation (16-39a) is known as the Casimir limit for phonon transport, where σ_{ph} is the acoustic equivalence of the Stefan–Boltzmann constant for phonons, given as

$$\sigma_{\text{ph}} = \frac{\pi^2 k_B^4}{120\, h^3} \sum_i \frac{1}{v_i^2} = \frac{\pi^2 k_B^4}{40\, h^3 v^2} \tag{16-39b}$$

where the summation is over the three speeds of sound (one longitudinal and two transverse), and where we have replaced the summation in terms of the average speed of sound, namely, by $\frac{1}{3}v^2$. For diamond, we have $\sigma_{\text{ph}} \cong 50\ W/\text{m}^2 \cdot \text{K}^4$ [45].

As depicted in Figure 16-6, the Casimir limit represents what is often called the *ballistic limit* for phonon transport in that heat is transferred across the thin film phonon by phonon without any diffusion (i.e., without any scattering), but rather the phonons travel directly across the gap in a straight path much like "bullets."

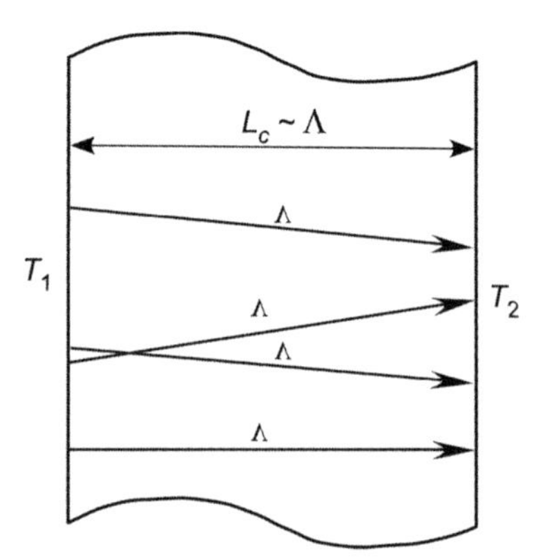

Figure 16-6 Casimir limit for ballistic transport of phonons.

Let us consider, for example, a thin diamond film of thickness 300 nm. This is clearly on the order of the mean free path for diamond phonons (see Table 16-3); hence we consider the Casimir limit to estimate the heat flux. Assuming a $\Delta T = 0.1\ K$ at 300 K, equation (16-39) predicts a heat flux of 54 kW/cm^2. Fourier's law, using the thermal conductivity of diamond as 2300 W/m · K, predicts a heat flux of 76.7 kW/cm^2, which is about 40% greater than the value predicted by the Casimir limit. Such behavior is typical, namely, that extrapolating beyond the limits of Fourier's law typically overpredicts the heat flux.

Fourier's Law and the Effective Thermal Conductivity

As detailed above, Fourier's law is limited at length scales $L_c \approx \Lambda$ (i.e., Kn $\approx$ 1). As noted in the above example, Fourier's law tends to overpredict the heat flux in this regime, which is largely due to the fact that at such small length scales, the actual thermal conductivity is reduced from the thermal conductivity of the bulk solid (i.e., k_{bulk}). For length scales in the range $L_c \lesssim \Lambda$, boundary scattering losses become important, with the effect of reducing the thermal conductivity. With this in mind, one approach is to modify Fourier's law using an *effective thermal conductivity*, giving

$$q''_x \cong -k_{\text{eff}} \frac{dT}{dx} \tag{16-40}$$

where $k_{\text{eff}} < k_{\text{bulk}}$. Several models are available for estimation of the effective thermal conductivity [12, 45, 48–51], with most casting $k_{\text{eff}} = f(\text{Kn})$, where the Knudsen number is given by equation (16-27), namely, $\text{Kn} \equiv \Lambda / L_c$. The most common expression for the effective thermal conductivity is of the form

$$\frac{k_{\text{eff}}}{k_{\text{bulk}}} = (1 + A\ \text{Kn})^{-1} \tag{16-41}$$

where the constant A is generally considered dependent on the range of Kn. Applying Matthiessen's rule for the expression of the effective mean free path in terms of the bulk mean free path and the thin-film mean free path, we have $A = 1$. This approach generally underpredicts the effective thermal conductivity [12]. For $0.1 \lesssim \text{Kn} \lesssim 1$, values of A are reported in the following range [12, 45, 49]:

$$\tfrac{1}{3} \leq A \leq \tfrac{4}{3} \qquad \text{for} \qquad 0.1 \lesssim \text{Kn} \lesssim 1 \tag{16-42}$$

Considering the example of our 300-nm diamond film at the end of the previous section, the $\text{Kn} \cong 1$. Letting $A = \frac{3}{8}$, per reference 47, we have $k_{\text{eff}} \cong 0.727 k_{\text{bulk}}$. Using the resulting effective thermal conductivity value of 1673 W/m · K in conjunction with Fourier's law, the predicted heat flux is corrected to yield nearly exact agreement with the value obtained from the Casimir limit. For larger values of Knudsen number, namely $\text{Kn} \gg 1$, equation (16-41) is generally modified to the form

$$\frac{k_{\text{eff}}}{k_{\text{bulk}}} \cong \frac{4 \ln(2\text{Kn})}{\pi \, \text{Kn}} \qquad \text{for} \qquad \text{Kn} \gg 1 \tag{16-43}$$

For Kn in the range from about 1 to 5, one should interpolate smoothly between the results of equations (16-41) and (16-43). In Figure 16-7, we depict the effective thermal conductivity as a function of Knudsen number, which provides a qualitative view of the effective thermal conductivity models. For $\text{Kn} \approx 100$, the effective thermal conductivity is approaching zero, with $k_{\text{eff}}/k_{\text{bulk}} \cong 0.067$ for $\text{Kn} = 100$, and the overall applicability of equation (16-43) becomes limited. As discussed in the next section, such a large value of Knudsen number may very well be approaching the second regime of microscale heat transfer.

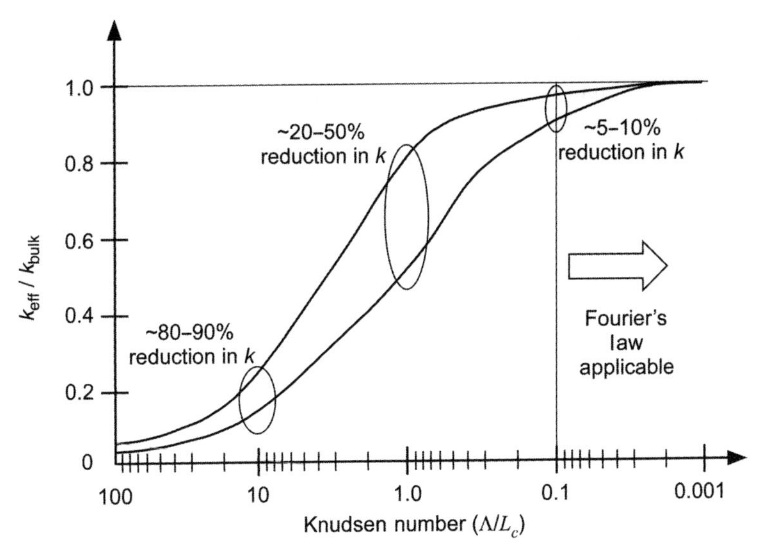

Figure 16-7 Range of effective thermal conductivity as a function of the Knudsen number.

16-6 SECOND AND THIRD REGIMES OF MICROSCALE HEAT TRANSFER

In the previous section we presented some guidelines for what is sometimes called the first regime of microscale heat transfer. This region was limited to characteristic length scales $L_c \approx \Lambda$ and/or to time scales $t \approx \tau$. We now consider the second and third regimes of microscale heat transfer [21], defining the *second regime of microscale heat transfer* as

$$L_c \tilde{<} \Lambda \qquad \text{and} \qquad L_c \approx \lambda_c \tag{16-44}$$

and the *third regime of microscale heat transfer* as

$$L_c \tilde{<} \lambda_c \tag{16-45}$$

We see that the second microscale regime is now defined by the characteristic length scale being on the order of the energy carrier wavelength. As noted in reference 21, classic size effects now hold and classical models such as the Drude and Lorentze models have some applicability. In the third microscale regime, we have the wavelength of the energy carrier, namely, our electron or phonon, on the order of the characteristic length scale. Under these conditions, quantum effects are generally dominant [52–55].

16-7 SUMMARY REMARKS

In our final chapter, we have attempted to elucidate the physics behind the transfer of energy via conduction heat transfer and to use our understanding to place the heat equation and Fourier's law in the context of electrons and phonons as energy carriers. By examining the relevant length scales for the physical heat transfer domain, the energy carrier, and the mean free paths associated with our energy carriers, we have provided what we feel are useful guidelines for understanding the limitations of our continuum approach and, importantly, to Fourier's law. Toward this goal, we have greatly condensed the extensive literature regarding the subject of microscale energy transport, and we refer the readers to the many excellent references discussed in this chapter. In addition, we refer the readers to a number of comprehensive review papers on the subject of microscale heat transfer [21, 56–59].

REFERENCES

1. Royal Society and The Royal Academy of Engineering, *Nanoscience and Nanotechnologies: Opportunities and Uncertainties*, The Royal Society, London.
2. C. M. Lieber, *MRS Bull.*, **28**, 486–491, 2003.
3. C. N. R. Rao and A. K. Cheetham, J. *Mat. Chem.* **11**, 2887–2894, 2001.

4. Y. N. Xia, P. D. Yang, Y. G. Sun, Y. Y. Wu, B. Mayers, B. Gates, Y. D. Yin, F. Kim, Y. Q. Yan, et al., *Adv. Mat.* **15**, 353–389, 2003.
5. J. T. Hu, T. W. Odom, and C. M. Lieber, *Acc. Chem. Res.* **22**, 435–445, 1999.
6. G. Decher, *Science*, **277**, 1232–1237, 1997.
7. L. H. Dubois, and R. G. Nuzzo, *Ann. Rev. Phys. Chem.* **43**, 437–463, 1992.
8. A. P. Alivisatos, *Science* **271**, 933–937, 1996.
9. C. Burda, X. B. Chen, R. Narayanan, M. A. El-Sayed, et al., *Chem. Rev.* **105**, 1025–1102, 2005.
10. C.-L. Tien, A. Majumdar, and F. M. Gerner (Eds.), *Microscale Energy Transport*, Taylor and Francis, Washington, DC, 1998.
11. G. Chen, *Nanoscale Energy Transport and Conversion*, Oxford University Press, Oxford, England, 2005.
12. Z. M. Zhang, *Nano/Miroscale Heat Transfer*, McGraw-Hill, New York, 2007.
13. M. Kaviany, *Heat Transfer Physics*, Cambridge University Press, New York, 2008.
14. C. B. Sobhan and G. P. Peterson, *Microscale and Nanoscale Heat Transfer*, CRC Press, New York, 2008.
15. N. W. Ashcroft and N. D. Mermin, *Solid State Physics*, Holt, Rinehart and Winston, New York, 1976.
16. J. Donohue, *The Structures of the Elements*, Wiley, New York, 1974.
17. W. D. Callister and D. G. Rethwisch, *Materials Science and Engineering: An Introduction*, 8th ed., Wiley, Hoboken, NJ, 2009.
18. C. Kittel, *Introduction to Solid State Physics*, 7th ed., Wiley, New York, 1996.
19. D. B. Sirdeshmukh, L. Sirdeshmukh, and K. G. Subhadra, *Micro- and Macro-Mechanical Properties of Solids*, Springer, Berlin, 2006.
20. V. L. Gurevich, *Transport in Phonon Systems*, Elsevier, New York, 1986.
21. C. L. Tien and G. Chen, *J. Heat Transfer* **116**, 799–807, 1994.
22. D. G. Cahill and R. O. Pohl, *Ann. Rev. Phys. Chem.* **39**, 93–121, 1988.
23. D. G. Cahill and R. O. Pohl, *Solid State Comm.* **70**, 927–930, 1989.
24. G. S. Kumar, G. Prasad and R. O. Pohl, *J. Mat. Sci.* **28**, 4261–4272, 1993.
25. P. D. Thacher, *Phys. Rev.* **156**, 975–988, 1967.
26. R. P. Tye, *Thermal Conductivity*, Vols. 1 and 2, Academic, London, 1969.
27. T. M. Tritt (Ed.), *Thermal Conductivity, Theory, Properties, and Applications*, Kluwer Academic, New York, 2004.
28. R. Gereth and K. Hubner, *Phys. Rev.* **134**, A235–A240, 1964.
29. H. Kanter, *Phys. Rev. B* **4**, 522–536, 1970.
30. R. Berman, *Cryogenics* **5**, 297, 1965.
31. M. Kaviany, *Principles of Heat Transfer*, Wiley, New York, 2002.
32. G. Fischer, H. Hoffmann, and J. Vancea, *Phys. Rev. B* **22**, 6065–6073, 1980.
33. C. Reale, *Appl. Phys. A: Mat. Sci. Process.*, **2**, 183–185, 1973.
34. P. Warrier and A. Teja, *Nanoscale Res. Lett.*, **6**, 247–252, 2011.
35. J. Borrajo and J. M. Heras, *Sur. Sci.* **28**, 132–144, 1971.
36. H. J. Goldsmid, *Proc. Phys. Soc. B* **67**, 360–363, 1954.
37. A. Al Shaikhi and G.P. Srivastava, *Diamond Related Mat.* **16**, 1413–1416, 2007.

38. S. Harris, *Introduction to the Theory of the Boltzmann Equation*, Dover, New York, 1999.
39. C. Cercignani, *The Boltzmann Equations and Its Applications*, Springer, New York, 1988.
40. S. Chapman and T. G. Cowling, *The Mathematical Theory of Non-uniform Gases*, 3rd ed., Cambridge University Press, London, 1990.
41. C. Cattaneo, *Comptes Rendus, French Acad. Sci.* **247**, 431–433, 1958.
42. M. N. Özişik and D. Y. Tzou, *J. Heat Transfer* **116**, 526–535, 1994.
43. A. Compte and R. Metzler, *J. Phys. A: Math. Gen.* **30**, 7277–7289, 1997.
44. A. Greven, G. Keller, and G. Warnecke (Eds.), *Entropy*, Chapter 5, Princeton University Press, Princeton, NJ, 2003.
45. A. Majumdar, *J. Heat Trans.* **115**, 7–16, 1993.
46. H. B. G. Casimir, *Physica* **5**, 495–500, 1938.
47. E. T. Swartz and R. O. Pohl, *Rev. Mod. Phy.* **61**, 605–668, 1989.
48. K. Fuchs, *Proc. Cambridge Philosophical Soc.* **24**, 100–108, 1938.
49. M. I. Flik and C. L. Tien, *J. Heat Transfer* **112**, 872–881, 1990.
50. G. Chen, *Phys. Rev. B* **57**, 14958–14953, 1998.
51. X. Zhang, H. Gu, and M. Fujii, *J. Appl. Phy.* **100**, 044325-1-6, 2006.
52. J.M. Tour, M. Kozaki, and J. M. Seminario, *J. Am. Chem. Soc.*, **120**, 8486–8493, 1998.
53. G. Chen and M. Neagu, *Appl. Phys. Lett.* **71**, 2761–2763, 1997.
54. A. Ozpineci and S. Ciraci, *Phys. Rev. B* **63**, 125415, 2001.
55. J-W. Jiang, J-S. Wang, and B. Li, *J. Appl. Phys.* **109**, 014326, 2011.
56. D. G. Cahill, K. Goodson, and A. Majumdar, *J. Heat Transfer* **124**, 223–241, 2002.
57. G. Chen and A. Shakouri, *J. Heat Transfer* **124**, 242–252, 2002.
58. D. G. Cahill, W. K. Ford, K. E. Goodson, G. D. Mahan, H. J. Maris, R. Merlin, and S. R. Phillpot, *Appl. Phys. Rev.* **93**, 793–818, 2003.
59. M. Matsumoto, *J. Thermal Sci. Tech.* **3**, 309–318, 2008.

APPENDIXES

APPENDIX I

PHYSICAL PROPERTIES

TABLE I-1 Physical Properties of Metals

		Properties at 20°C			
Metal	Melting Point, °C	ρ, $\frac{kg}{m^3}$	C, $\frac{KJ}{kg \cdot K}$	k, $\frac{W}{m \cdot K}$	α, $\frac{m^2}{s} \times 10^5$
Aluminum					
Pure	660	2,707	0.896	204	8.418
Al–Cu (Duralumin), 94–96% Al, 3–5% Cu, trace Mg		2,787	0.883	164	6.676
Beryllium	1,277	1,850	1.825	200	5.92
Bismuth	272	9,780	0.122	7.86	0.66
Cadmium	321	8,650	0.231	96.8	4.84
Copper					
Pure	1,085	8,954	0.3831	386	11.234
Aluminum bronze 95% Cu, 5% Al		8,666	0.410	83	2.330
Constantan 60% Cu, 40% Ni		8,922	0.410	22.7	0.612

(continued overleaf)

TABLE I-1 (*Continued*)

Metal	Melting Point, °C	Properties at 20°C			
		ρ, $\frac{kg}{m^3}$	C, $\frac{KJ}{kg \cdot K}$	k, $\frac{W}{m \cdot K}$	α, $\frac{m^2}{s} \times 10^5$
Iron					
Pure	1,537	7,897	0.452	73	2.034
Wrought iron, 0.5% C		7,849	0.46	59	1.626
Carbon steel					
C ≈ 0.5%		7,833	0.465	54	1.474
1.0%		7,801	0.473	43	1.172
Chrome steel					
Cr = 0%		7,897	0.452	73	2.026
1%		7,865	0.46	61	1.665
5%		7,833	0.46	40	1.110
Nickel steel					
Ni ≅ 0%		7,897	0.452	73	2.026
20%		7,933	0.46	19	0.526
Lead	328	11,373	0.130	35	2.343
Magnesium					
Pure	650	1,746	1.013	171	9.708
Mg–Al (electrolytic)		1,810	1.00	66	3.605
6–8% Al, 1–2% Zn					
Molybdenum	2,621	10,220	0.251	123	4.790
Nickel					
Pure (99.9%)	1,455	8,906	0.4459	90	2.266
Ni–Cr					
90% Ni, 10% Cr		8,666	0.444	17	0.444
80% Ni, 20% Cr		8,314	0.444	12.6	0.343
Silver					
Purest	962	10,524	0.2340	419	17.004
Pure (99.9%)		10,525	0.2340	407	16.563
Tin, pure	232	7,304	0.2265	64	3.884
Tungsten	3,387	19,350	0.1344	163	6.271
Uranium	1,133	19,070	0.116	27.6	1.25
Zinc, pure	420	7,144	0.3843	112.2	4.106

TABLE I-2 Physical Properties of Nonmetals

Material	T, °C	k, $\frac{W}{m \cdot K}$	ρ, $\frac{kg}{m^3}$	C, $\frac{KJ}{kg \cdot K}$	α, $\frac{m^2}{s} \times 10^7$
Asphalt	20–55	0.74–0.76			
Brick					
Building brick,	20	0.69	1600	0.84	5.2
common face		1.32	2000		
Carborundum	600	18.5			
brick	1400	11.1			
Chrome brick	200	2.32	3000	0.84	9.2
	900	1.99			7.9
Diatomaceous					
earth, molded	200	0.24			
and fired	870	0.31			
Fireclay brick,	500	1.04	2000	0.96	5.4
burned 1330°C	800	1.07			
Clay	30	1.3	1460	0.88	
Cement, portland	23	0.29	1500		
Coal. anthracite	30	0.26	1200–1500	1.26	
Concrete, cinder	23	0.76			
Stone 1-2-4 mix	20	1.37	1900–2300	0.88	8.2–6.8
Cotton	20	0.06	80	1.30	
Glass, window	20	0.78 (avg)	2700	0.84	3.4
Pyrex	30	1.4	2225	0.835	
Paper	30	0.011	930	1.340	
Paraffin	30	0.020	900	2.890	
Plaster, gypsum	20	0.48	1440	0.84	4.0
Rubber, vulcanized					
Soft	30	0.012	1100	2.010	
Hard	30	0.013	1190	—	
Sand	30	0.027	1515	0.800	
Stone					
Granite		1.73–3.98	2640	0.82	8–18
Limestone	100–300	1.26–1.33	2500	0.90	5.6–5.9
Marble		2.07–2.94	2500–2700	0.80	10–13.6
Sandstone	40	1.83	2160–2300	0.71	11.2–11.9
Teflon	30	0.35	2200	—	
Tissue, human skin	30	0.37	—	—	
Wood (across grain)					
Balsa	30	0.055	140		
Cypress	30	0.097	460		
Fir	23	0.11	420	2.72	0.96
Maple or oak	30	0.166	540	2.4	1.28
Yellow pine	23	0.147	640	2.8	0.82
White pine	30	0.112	430		

TABLE I-3 Physical Properties of Insulating Materials

Material	T, °C	k, $\frac{W}{m \cdot K}$	ρ, $\frac{kg}{m^3}$	C, $\frac{KJ}{kg \cdot K}$	α, $\frac{m^2}{s} \times 10^7$
Asbestos					
Loosely packed	0	0.154	470–570	0.816	3.3–4
	100	0.161			
Asbestos–cement boards	20	0.74			
Sheets	51	0.166			
Felt, 40 laminations in	38	0.057			
Balsam wool	32	0.04	35		
Board and slab					
Cellular glass	30	0.058	145	1.000	
Glass fiber, organic bonded	30	0.036	105	0.795	
Polystyrene, expanded extruded (R-12)	30	0.027	55	1.210	
Mineral fiberboard; roofing material	30	0.049	265		
Cardboard, corrugated	—	0.064			
Celotex	32	0.048			
Corkboard	30	0.043	160		
Diatomaceous earth (Sil-o-cel)	0	0.061	320		
Felt, hair	30	0.036	130–200		
Wool	30	0.052	330		
Fiber, insulating board	20	0.048	240		
Glass wool	23	0.038	24	0.7	22.6
Loose fill					
Cork, granulated	30	0.045	160		
Glass fiber, poured or blown	30	0.043	16	0.835	
Vermiculite, flakes	30	0.068	80	0.835	
Magnesia, 85%	38	0.067	270		
	150	0.074			
	204	0.080			
Rock wool, 10 lb/ft^3	32	0.040	160		
Loosely packed	150	0.067	64		
	260	0.087			
Sawdust	23	0.059			
Silica aerogel	32	0.024	140		
Wood shavings	23	0.059			

APPENDIX II

ROOTS OF TRANSCENDENTAL EQUATIONS

First Six Roots β_n of $\beta \tan \beta = c$

c	β_1	β_2	β_3	β_4	β_5	β_6
0	0	3.1416	6.2832	9.4248	12.5664	15.7080
0.001	0.0316	3.1419	6.2833	9.4249	12.5665	15.7080
0.002	0.0447	3.1422	6.2835	9.4250	12.5665	15.7081
0.004	0.0632	3.1429	6.2838	9.4252	12.5667	15.7082
0.006	0.0774	3.1435	6.2841	9.4254	12.5668	15.7083
0.008	0.0893	3.1441	6.2845	9.4256	12.5670	15.7085
0.01	0.0998	3.1448	6.2848	9.4258	12.5672	15.7086
0.02	0.1410	3.1479	6.2864	9.4269	12.5680	15.7092
0.04	0.1987	3.1543	6.2895	9.4290	12.5696	15.7105
0.06	0.2425	3.1606	6.2927	9.4311	12.5711	15.7118
0.08	0.2791	3.1668	6.2959	9.4333	12.5727	15.7131
0.1	0.3111	3.1731	6.2991	9.4354	12.5743	15.7143
0.2	0.4328	3.2039	6.3148	9.4459	12.5823	15.7207
0.3	0.5218	3.2341	6.3305	9.4565	12.5902	15.7270
0.4	0.5932	3.2636	6.3461	9.4670	12.5981	15.7334
0.5	0.6533	3.2923	6.3616	9.4775	12.6060	15.7397
0.6	0.7051	3.3204	6.3770	9.4879	12.6139	15.7460
0.7	0.7506	3.3477	6.3923	9.4983	12.6218	15.7524
0.8	0.7910	3.3744	6.4074	9.5087	12.6296	15.7587
0.9	0.8274	3.4003	6.4224	9.5190	12.6375	15.7650
1.0	0.8603	3.4256	6.4373	9.5293	12.6453	15.7713

(continues overleaf)

c	β_1	β_2	β_3	β_4	β_5	β_6
1.5	0.9882	3.5422	6.5097	9.5801	12.6841	15.8026
2.0	1.0769	3.6436	6.5783	9.6296	12.7223	15.8336
3.0	1.1925	3.8088	6.7040	9.7240	12.7966	15.8945
4.0	1.2646	3.9352	6.8140	9.8119	12.8678	15.9536
5.0	1.3138	4.0336	6.9096	9.8928	12.9352	16.0107
6.0	1.3496	4.1116	6.9924	9.9667	12.9988	16.0654
7.0	1.3766	4.1746	7.0640	10.0339	13.0584	16.1177
8.0	1.3978	4.2264	7.1263	10.0949	13.1141	16.1675
9.0	1.4149	4.2694	7.1806	10.1502	13.1660	16.2147
10.0	1.4289	4.3058	7.2281	10.2003	13.2142	16.2594
15.0	1.4729	4.4255	7.3959	10.3898	13.4078	16.4474
20.0	1.4961	4.4915	7.4954	10.5117	13.5420	16.5864
30.0	1.5202	4.5615	7.6057	10.6543	13.7085	16.7691
40.0	1.5325	4.5979	7.6647	10.7334	13.8048	16.8794
50.0	1.5400	4.6202	7.7012	10.7832	13.8666	16.9519
60.0	1.5451	4.6353	7.7259	10.8172	13.9094	17.0026
80.0	1.5514	4.6543	7.7573	10.8606	13.9644	17.0686
100.0	1.5552	4.6658	7.7764	10.8871	13.9981	17.1093
∞	1.5708	4.7124	7.8540	10.9956	14.1372	17.2788

Roots are all real if $c > 0$.

First Six Roots β_n of $\beta \cot \beta = -c$

c	β_1	β_2	β_3	β_4	β_5	β_6
−1.0	0	4.4934	7.7253	10.9041	14.0662	17.2208
−0.995	0.1224	4.4945	7.7259	10.9046	14.0666	17.2210
−0.99	0.1730	4.4956	7.7265	10.9050	14.0669	17.2213
−0.98	0.2445	4.4979	7.7278	10.9060	14.0676	17.2219
−0.97	0.2991	4.5001	7.7291	10.9069	14.0683	17.2225
−0.96	0.3450	4.5023	7.7304	10.9078	14.0690	17.2231
−0.95	0.3854	4.5045	7.7317	10.9087	14.0697	17.2237
−0.94	0.4217	4.5068	7.7330	10.9096	14.0705	17.2242
−0.93	0.4551	4.5090	7.7343	10.9105	14.0712	17.2248
−0.92	0.4860	4.5112	7.7356	10.9115	14.0719	17.2254
−0.91	0.5150	4.5134	7.7369	10.9124	14.0726	17.2260
−0.90	0.5423	4.5157	7.7382	10.9133	14.0733	17.2266
−0.85	0.6609	4.5268	7.7447	10.9179	14.0769	17.2295
−0.8	0.7593	4.5379	7.7511	10.9225	14.0804	17.2324
−0.7	0.9208	4.5601	7.7641	10.9316	14.0875	17.2382
−0.6	1.0528	4.5822	7.7770	10.9408	14.0946	17.2440
−0.5	1.1656	4.6042	7.7899	10.9499	14.1017	17.2498
−0.4	1.2644	4.6261	7.8028	10.9591	14.1088	17.2556

c	β_1	β_2	β_3	β_4	β_5	β_6
−0.3	1.3525	4.6479	7.8156	10.9682	14.1159	17.2614
−0.2	1.4320	4.6696	7.8284	10.9774	14.1230	17.2672
−0.1	1.5044	4.6911	7.8412	10.9865	14.1301	17.2730
0	1.5708	4.7124	7.8540	10.9956	14.1372	17.2788
0.1	1.6320	4.7335	7.8667	11.0047	14.1443	17.2845
0.2	1.6887	4.7544	7.8794	11.0137	14.1513	17.2903
0.3	1.7414	4.7751	7.8920	11.0228	14.1584	17.2961
0.4	1.7906	4.7956	7.9046	11.0318	14.1654	17.3019
0.5	1.8366	4.8158	7.9171	11.0409	14.1724	17.3076
0.6	1.8798	4.8358	7.9295	11.0498	14.1795	17.3134
0.7	1.9203	4.8556	7.9419	11.0588	14.1865	17.3192
0.8	1.9586	4.8751	7.9542	11.0677	14.1935	17.3249
0.9	1.9947	4.8943	7.9665	11.0767	14.2005	17.3306
1.0	2.0288	4.9132	7.9787	11.0856	14.2075	17.3364
1.5	2.1746	5.0037	8.0385	11.1296	14.2421	17.3649
2.0	2.2889	5.0870	8.0962	11.1727	14.2764	17.3932
3.0	2.4557	5.2329	8.2045	11.2560	14.3434	17.4490
4.0	2.5704	5.3540	8.3029	11.3349	14.4080	17.5034
5.0	2.6537	5.4544	8.3914	11.4086	14.4699	17.5562
6.0	2.7165	5.5378	8.4703	11.4773	14.5288	17.6072
7.0	2.7654	5.6078	8.5406	11.5408	14.5847	17.6562
8.0	2.8044	5.6669	8.6031	11.5994	14.6374	17.7032
9.0	2.8363	5.7172	8.6587	11.6532	14.6870	17.7481
10.0	2.8628	5.7606	8.7083	11.7027	14.7335	17.7908
15.0	2.9476	5.9080	8.8898	11.8959	14.9251	17.9742
20.0	2.9930	5.9921	9.0019	12.0250	15.0625	18.1136
30.0	3.0406	6.0831	9.1294	12.1807	15.2380	18.3018
40.0	3.0651	6.1311	9.1987	12.2688	15.3417	18.4180
50.0	3.0801	6.1606	9.2420	12.3247	15.4090	18.4953
60.0	3.0901	6.1805	9.2715	12.3632	15.4559	18.5497
80.0	3.1028	6.2058	9.3089	12.4124	15.5164	18.6209
100.0	3.1105	6.2211	9.3317	12.4426	15.5537	18.6650
∞	3.1416	6.2832	9.4248	12.5664	15.7080	18.8496

Roots are all real if $c > -1$.

APPENDIX III

ERROR FUNCTIONS

Numerical Values of Error Function $\mathrm{erf}(z) = \frac{2}{\sqrt{\pi}} \int_{z=0}^{z} e^{-\xi^2} d\xi$

z	erf z	z	erf z	z	erf z	z	erf z	z	erf z
0.00	0.00000	0.50	0.52049	1.00	0.84270	1.50	0.96610	2.00	0.99532
0.01	0.01128	0.51	0.52924	1.01	0.84681	1.51	0.96727	2.20	0.99814
0.02	0.02256	0.52	0.53789	1.02	0.85083	1.52	0.96841	2.40	0.99931
0.03	0.03384	0.53	0.54646	1.03	0.85478	1.53	0.96951	2.60	0.99976
0.04	0.04511	0.54	0.55493	1.04	0.85864	1.54	0.97058	2.80	0.99992
0.05	0.05637	0.55	0.56332	1.05	0.86243	1.55	0.97162	3.00	0.99998
0.06	0.06762	0.56	0.57161	1.06	0.86614	1.56	0.97262		
0.07	0.07885	0.57	0.57981	1.07	0.86977	1.57	0.97360		
0.08	0.09007	0.58	0.58792	1.08	0.87332	1.58	0.97454		
0.09	0.10128	0.59	0.59593	1.09	0.87680	1.59	0.97546		
0.10	0.11246	0.60	0.60385	1.10	0.88020	1.60	0.97634		
0.11	0.12362	0.61	0.61168	1.11	0.88353	1.61	0.97720		
0.12	0.13475	0.62	0.61941	1.12	0.88678	1.62	0.97803		
0.13	0.14586	0.63	0.62704	1.13	0.88997	1.63	0.97884		
0.14	0.15694	0.64	0.63458	1.14	0.89308	1.64	0.97962		
0.15	0.16799	0.65	0.64202	1.15	0.89612	1.65	0.98037		
0.16	0.17901	0.66	0.64937	1.16	0.89909	1.66	0.98110		
0.17	0.18999	0.67	0.65662	1.17	0.90200	1.67	0.98181		
0.18	0.20093	0.68	0.66378	1.18	0.90483	1.68	0.98249		
0.19	0.21183	0.69	0.67084	1.19	0.90760	1.69	0.98315		

z	erf z	z	erf z	z	erf z	z	erf z	z	erf z
0.20	0.22270	0.70	0.67780	1.20	0.91031	1.70	0.98379		
0.21	0.23352	0.71	0.68466	1.21	0.91295	1.71	0.98440		
0.22	0.24429	0.72	0.69143	1.22	0.91553	1.72	0.98500		
0.23	0.25502	0.73	0.69810	1.23	0.91805	1.73	0.98557		
0.24	0.26570	0.74	0.70467	1.24	0.92050	1.74	0.98613		
0.25	0.27632	0.75	0.71115	1.25	0.92290	1.75	0.98667		
0.26	0.28689	0.76	0.71753	1.26	0.92523	1.76	0.98719		
0.27	0.29741	0.77	0.72382	1.27	0.92751	1.77	0.98769		
0.28	0.30788	0.78	0.73001	1.28	0.92973	1.78	0.98817		
0.29	0.31828	0.79	0.73610	1.29	0.93189	1.79	0.98864		
0.30	0.32862	0.80	0.74210	1.30	0.93400	1.80	0.98909		
0.31	0.33890	0.81	0.74800	1.31	0.93606	1.81	0.98952		
0.32	0.34912	0.82	0.75381	1.32	0.93806	1.82	0.98994		
0.33	0.35927	0.83	0.75952	1.33	0.94001	1.83	0.99034		
0.34	0.36936	0.84	0.76514	1.34	0.94191	1.84	0.99073		
0.35	0.37938	0.85	0.77066	1.35	0.94376	1.85	0.99111		
0.36	0.38932	0.86	0.77610	1.36	0.94556	1.86	0.99147		
0.37	0.39920	0.87	0.78143	1.37	0.94731	1.87	0.99182		
0.38	0.40900	0.88	0.78668	1.38	0.94901	1.88	0.99215		
0.39	0.41873	0.89	0.79184	1.39	0.95067	1.89	0.99247		
0.40	0.42839	0.90	0.79690	1.40	0.95228	1.90	0.99279		
0.41	0.43796	0.91	0.80188	1.41	0.95385	1.91	0.99308		
0.42	0.44746	0.92	0.80676	1.42	0.95537	1.92	0.99337		
0.43	0.45688	0.93	0.81156	1.43	0.95685	1.93	0.99365		
0.44	0.46622	0.94	0.81627	1.44	0.95829	1.94	0.99392		
0.45	0.47548	0.95	0.82089	1.45	0.95969	1.95	0.99417		
0.46	0.48465	0.96	0.82542	1.46	0.96105	1.96	0.99442		
0.47	0.49374	0.97	0.82987	1.47	0.96237	1.97	0.99466		
0.48	0.50274	0.94	0.83423	1.48	0.96365	1.98	0.99489		
0.49	0.51166	0.99	0.83850	1.49	0.96489	1.99	0.99511		

The error function of argument x is defined as

$$\mathrm{erf}(x) = \frac{2}{\sqrt{\pi}} \int_{\eta=0}^{x} e^{-\eta^2}\, d\eta \tag{1}$$

and we have

$$\mathrm{erf}(\infty) = 1 \qquad \text{and} \qquad \mathrm{erf}(-x) = -\mathrm{erf}(x) \tag{2}$$

The complementary error function, $\mathrm{erfc}(x)$, is defined as

$$\mathrm{erfc}(x) = 1 - \mathrm{erf}(x) = \frac{2}{\sqrt{\pi}} \int_{\eta=x}^{\infty} e^{-\eta^2}\, d\eta \tag{3}$$

The derivatives of error function are given as

$$\frac{d}{dx}\text{erf}(x) = \frac{2}{\sqrt{\pi}}e^{-x^2}, \qquad \frac{d^2}{dx^2}\text{erf}(x) = -\frac{4}{\sqrt{\pi}}xe^{-x^2}, \ldots \tag{4}$$

The repeated integrals of error function are defined as

$$i^n\text{erfc}(x) = \int_{\eta=x}^{\infty} i^{n-1}\text{erfc}\,\eta\,d\eta \qquad n = 0, 1, 2, \ldots \tag{5a}$$

with

$$i^{-1}\text{erfc}(x) = \frac{2}{\sqrt{\pi}}e^{-x^2}, \qquad i^0\ \text{erfc}\ x = \text{erfc}\ x \tag{5b}$$

Then we have

$$i\text{erfc}(x) = \frac{1}{\sqrt{\pi}}e^{-x^2} - x\ \text{erfc}\ x \tag{6}$$

$$i^2\text{erfc}(x) = \frac{1}{4}\left[(1 + 2x^2)\text{erfc}\,x - \frac{2}{\sqrt{\pi}}xe^{-x^2}\right] \tag{7}$$

Series expansion for error function is given as

$$\text{erf}(x) = \frac{2}{\sqrt{\pi}}\sum_{n=0}^{\infty}(-1)^n\frac{x^{2n+1}}{n!(2n+1)} \tag{8}$$

For large values of x, its asymptotic expansion is

$$\text{erfc}(x) = 1 - \text{erf}(x) \cong \frac{e^{-x^2}}{\sqrt{\pi}x}\left[1 + \sum_{n=1}^{\infty}(-1)^n\frac{1.3\cdots(2n-1)}{(2x^2)^n}\right] \tag{9}$$

The error function, its derivatives, and its integrals have been tabulated [1, 2].

REFERENCES

1. M. Abramowitz and I. A. Stegun, *Handbook of Mathematical Functions*, National Bureau of Standards, Applied Mathematic Series 55, U.S. Government Printing Office, Washington, DC, 1964.
2. E. Jahnke and F. Emde, *Tables of Functions*, 2nd ed., Dover, New York, 1945.

APPENDIX IV

BESSEL FUNCTIONS

The differential equation, see equation (2-36b),

$$\frac{d^2R}{dz^2} + \frac{1}{z}\frac{dR}{dz} + \left(1 - \frac{v^2}{z^2}\right)R = 0 \tag{1}$$

is called *Bessels's differential equation of order* v. Two linearly independent solutions of this equation for all values of v are $J_v(z)$, the Bessel function of the first kind of order v, and $Y_v(z)$, the Bessel function of the second kind of order v. Thus, the general solution of equation (1) is written as [1–3]

$$R(z) = c_1 J_v(z) + c_2 Y_v(z) \tag{2}$$

The Bessel function $J_v(z)$ in series form is defined as

$$J_v(z) = \left(\frac{1}{2}z\right)^v \sum_{k=0}^{\infty} (-1)^k \frac{(\frac{1}{2}z)^{2k}}{k!\Gamma(v+k+1)} \tag{3}$$

where $\Gamma(x)$ is the gamma function.

The differential equation

$$\frac{d^2R}{dz^2} + \frac{1}{z}\frac{dR}{dz} - \left(1 + \frac{v^2}{z^2}\right)R = 0 \tag{4}$$

is called *Bessel's modified differential equation of order* v. Two linearly independent solutions of this equation for all values of v are $I_v(z)$, the modified

Bessel function of the first kind of order ν, and $K_\nu(z)$, the modified Bessel function of the second kind of order ν. Thus, the general solution of equation (4) is written as

$$R(z) = c_1 I_\nu(z) + c_2 K_\nu(z) \tag{5}$$

where $I_\nu(z)$ and $K_\nu(z)$ are real and positive when $\nu > -1$ and $z > 0$. The Bessel function $I_\nu(z)$ in series form is given by

$$I_\nu(z) = \left(\frac{1}{2}z\right)^\nu \sum_{k=0}^{\infty} \frac{(\frac{1}{2}z)^{2k}}{k!\Gamma(\nu + k + 1)} \tag{6}$$

When ν *is not zero or not a positive integer*, the general solutions (2) and (5) can be taken, respectively, in the form

$$R(z) = c_1 J_\nu(z) + c_2 J_{-\nu}(z) \tag{7a}$$

$$R(z) = c_1 I_\nu(z) + c_2 I_{-\nu}(z) \tag{7b}$$

When $\nu = n$ is a positive integer, the solutions $J_n(z)$ and $J_{-n}(z)$ are not independent; they are related by

$$J_n(z) = (-1)^n J_{-n}(z) \qquad \text{and} \qquad J_{-n}(z) = J_n(-z) \ (n = \text{integer}) \tag{8}$$

similarly, when $\nu = n$ is a positive integer, the solutions $I_n(z)$ and $I_{-n}(z)$ are not independent.

We summarize various forms of solutions of equation (1) as [2]

$$R(z) = c_1 J_\nu(z) + c_2 Y_\nu(z) \qquad \text{always} \tag{9a}$$

$$R(z) = c_1 J_\nu(z) + c_2 J_{-\nu}(z) \qquad \nu \text{ is not zero or a positive integer} \tag{9b}$$

and the solutions of equation (4) as [2]

$$R(z) = c_1 I_\nu(z) + c_2 K_\nu(z) \qquad \text{always} \tag{10a}$$

$$R(z) = c_1 I_\nu(z) + c_2 I_{-\nu}(z) \qquad \nu \text{ is not zero or positive integer} \tag{10b}$$

IV-1 GENERALIZED BESSEL EQUATION

Sometimes a given differential equation, after suitable transformation of the independent variable, yields a solution that is a linear combination of Bessel functions. A convenient way of finding out whether a given differential equation possesses a solution in terms of Bessel functions is to compare it with the *generalized*

Bessel equation developed by Douglas [4, p. 210]:

$$\frac{d^2R}{dx^2} + \left[\frac{1-2m}{x} - 2\alpha\right] \cdot \frac{dR}{dx} + \left[p^2a^2x^{2p-2} + \alpha^2 + \frac{\alpha(2m-1)}{x} + \frac{m^2 - p^2v^2}{x^2}\right] R = 0 \tag{11a}$$

and the corresponding solution of which is

$$R = x^m \cdot e^{\alpha x}[c_1 J_v(ax^p) + c_2 Y_v(ax^p)] \tag{11b}$$

where c_1 and c_2 are arbitrary constants.

For example, by comparing the differential equation

$$\frac{d^2R}{dx^2} + \frac{1}{x}\frac{dR}{dx} - \frac{\beta}{x}R = 0 \tag{12}$$

with the above generalized Bessel equation we find

$$\alpha = 0, \qquad m = 0, \qquad p = \tfrac{1}{2}, \qquad p^2v^2 = -\beta, \qquad a = 2i\sqrt{\beta}, \qquad v = 0$$

Hence, the solution of differential equation (12) is in the form

$$R = c_1 J_0(2i\sqrt{\beta x}) + c_2 Y_0(2i\sqrt{\beta x}) \tag{13a}$$

or

$$R = c_1 I_0(2\sqrt{\beta x}) + c_2 K_0(2\sqrt{\beta x}) \tag{13b}$$

which involves Bessel functions.

IV-2 LIMITING FORM FOR SMALL Z

For small values of z (i.e., $z \to 0$), the retention of the leading terms in the series results in the following approximations for the values of Bessel functions [5, p. 360]:

$$J_v(z) \cong \left(\frac{1}{2}z\right)^v \frac{1}{\Gamma(v+1)} \qquad v \neq -1, -2, -3, \ldots \tag{14a}$$

$$Y_v(z) \cong -\frac{1}{\pi}\left(\frac{2}{z}\right)^v \Gamma(v) \qquad v \neq 0 \qquad \text{and} \qquad Y_0(z) \cong \frac{2}{\pi}\ln z \tag{14b}$$

$$I_z(z) \cong \left(\frac{1}{2}z\right)^v \frac{1}{\Gamma(v+1)} \qquad v \neq -1, -2, -3, \ldots \tag{15a}$$

$$K_v(z) \cong \frac{1}{2}\left(\frac{2}{z}\right)^v \Gamma(v) \qquad v \neq 0 \quad \text{and} \quad K_0(z) \cong -\ln z \tag{15b}$$

where we note $\Gamma(1) \equiv 1$.

IV-3 LIMITING FORM FOR LARGE Z

For large values of z (i.e., $z \to \infty$) the values of Bessel functions can be approximated as [5, pp. 364, 377]

$$J_\nu(z) \cong \sqrt{\frac{2}{\pi z}} \cdot \cos\left(z - \frac{\pi}{4} - \frac{\nu\pi}{2}\right) \tag{16a}$$

$$Y_\nu(z) \cong \sqrt{\frac{2}{\pi z}} \cdot \sin\left(z - \frac{\pi}{4} - \frac{\nu\pi}{4}\right) \tag{16b}$$

$$I_\nu(z) \cong \frac{e^z}{\sqrt{2\pi z}} \quad \text{and} \quad K_\nu(z) \cong \sqrt{\frac{\pi}{2z}} \cdot e^{-z} \tag{16c}$$

IV-4 DERIVATIVES OF BESSEL FUNCTIONS

$$\frac{d}{dz}[z^\nu W_\nu(\beta z)] = \begin{cases} \beta z^\nu W_{\nu-1}(\beta z) & \text{for} \quad W \equiv J, Y, I \quad (17a) \\ -\beta z^\nu W_{\nu-1}(\beta z) & \text{for} \quad W \equiv K \quad (17b) \end{cases}$$

$$\frac{d}{dz}[z^{-\nu} W_\nu(\beta z)] = \begin{cases} -\beta z^{-\nu} W_{\nu+1}(\beta z) & \text{for} \quad W \equiv J, Y, K \\ \beta z^{-\nu} W_{\nu+1}(\beta z) & \text{for} \quad W \equiv I \end{cases} \tag{18b}$$

For example, by setting $\nu = 0$, we obtain

$$\frac{d}{dz}[W_0(\beta z)] = \begin{cases} -\beta W_1(\beta z) & \text{for} \quad W \equiv J, Y, K \quad (19a) \\ \beta W_1(\beta z) & \text{for} \quad W = I \quad (19b) \end{cases}$$

as given by reference 3 [pp. 161–163].

IV-5 INTEGRATION OF BESSEL FUNCTIONS

$$\int z^\nu W_{\nu-1}(\beta z)\, dz = \frac{1}{\beta} z^\nu W_\nu(\beta z) \quad \text{for} \quad W \equiv J, Y, I \tag{20}$$

$$\int \frac{1}{z^\nu} W_{\nu+1}(\beta z)\, dz = -\frac{1}{\beta z^\nu} W_\nu(\beta z) \quad \text{for} \quad W \equiv J, Y, K \tag{21}$$

For example, by setting $\nu = 1$ in equation (20), are obtain

$$\int z W_0(\beta z)\, dz = \frac{1}{\beta} z W_1(\beta z) \quad \text{for} \quad W \equiv J, Y, I \tag{22}$$

Infinite integrals involving Bessel functions are [1, pp. 394–395]

$$\int_{z=0}^{\infty} e^{-pz^2} z^{\nu+1} J_\nu(az)\, dz = \frac{a^\nu}{(2p)^{\nu+1}} e^{-a^2/4p} \tag{23}$$

$$\int_{z=0}^{\infty} e^{-pz^2} z J_\nu(az) J_\nu(bz)\, dz = \frac{1}{2p} e^{-(a^2+b^2)/4p} I_\nu\left(\frac{ab}{2p}\right) \tag{24}$$

The indefinite integral of the square of Bessel functions is given by [1, p. 135; 2, p. 110]

$$\int r G_\nu^2(\beta r)\, dr = \frac{1}{2} r^2 [G_\nu^2(\beta r) - G_{\nu-1}(\beta r) G_{\nu+1}(\beta r)] \tag{25a}$$

$$= \frac{1}{2} r^2 \left[G_\nu'^2(\beta r) + \left(1 - \frac{\nu^2}{\beta^2 r^2}\right) G_\nu^2(\beta r) \right] \tag{25b}$$

where $G_\nu(\beta\ r)$ is any Bessel function of the first or second kind of order ν and where prime notation designates differentiation with respect to the entire argument.

The indefinite integral of the product of two Bessel functions can be expressed in the form [6, equation 9]

$$\int r G_\nu(\beta r) \overline{G}_\nu(\beta r)\, dr = \frac{r^2}{2} \left\{ G_\nu'(\beta r) \overline{G}_\nu'(\beta r) + \left[1 - \left(\frac{\nu}{\beta r}\right)^2 \right] G_\nu(\beta r) \overline{G}_\nu(\beta r) \right\} \tag{26a}$$

or in the form [1, p. 134; 2, p. 110]

$$\int r G_\nu(\beta r) \overline{G}^\nu(\beta r)\, dr = \frac{1}{4} r^2 [2 G_\nu(\beta r) \overline{G}_\nu(\beta r) - G_{\nu-1}(\beta r) \overline{G}_{\nu+1}(\beta r) - G_{\nu+1}(\beta r) \overline{G}_{\nu-1}(\beta r)] \tag{26b}$$

where $G_\nu(\beta\ r)$ and $\overline{G}_\nu(\beta r)$ can be any Bessel function of the first or second kind and where prime notation designates differentiation with respect to the entire argument. We note that equations (25a,b) are special cases of the integrals (26a,b).

IV-6 WRONSKIAN RELATIONSHIP

The Wronskian relationship for the Bessel functions

$$J_\nu(\beta r) Y_\nu'(\beta r) - Y_\nu(\beta r) J_\nu'(\beta r) = \frac{2}{\pi \beta r} \tag{27}$$

is useful in the simplification of expressions involving Bessel functions, where prime notation designates differentiation with respect to the entire argument.

TABLE IV-1 Numerical Values of Bessel Functions

	$J_0(z)$									
z	0	0.1	0.2	0.3	0.4	0.5	0.6	0.7	0.8	0.9
0	1.0000	0.9975	0.9900	0.9776	0.9604	0.9385	0.9120	0.8812	0.8463	0.8075
1	0.7652	0.7196	0.6711	0.6201	0.5669	0.5118	0.4554	0.3980	0.3400	0.2818
2	0.2239	0.1666	0.1104	0.0555	0.0025	−0.0484	−0.0968	−0.1424	−0.1850	−0.2243
3	−0.2601	−0.2921	−0.3202	−0.3443	−0.3643	−0.3801	−0.3918	−0.3992	−0.4026	−0.4018
4	−0.3971	−0.3887	−0.3766	−0.3610	−0.3423	−0.3205	−0.2961	−0.2693	−0.2404	−0.2097
5	−0.1776	−0.1443	−0.1103	−0.0758	−0.0412	−0.0068	0.0270	0.0599	0.0917	0.1220
6	0.1506	0.1773	0.2017	0.2238	0.2433	0.2601	0.2740	0.2851	0.2931	0.2981
7	0.3001	0.2991	0.2951	0.2882	0.2786	0.2663	0.2516	0.2346	0.2154	0.1944
8	0.1717	0.1475	0.1222	0.0960	0.0692	0.0419	0.0146	−0.0125	−0.0392	−0.0653
9	−0.0903	−0.1142	−0.1367	−0.1577	−0.1768	−0.1939	−0.2090	−0.2218	−0.2323	−0.2403
10	−0.2459	−0.2490	−0.2496	−0.2477	−0.2434	−0.2366	−0.2276	−0.2164	−0.2032	−0.1881
11	−0.1712	−0.1528	−0.1330	−0.1121	−0.0902	−0.0677	−0.0446	−0.0213	0.0020	0.0250
12	0.0477	0.0697	0.0908	0.1108	0.1296	0.1469	0.1626	0.1766	0.1887	0.1988
13	0.2069	0.2129	0.2167	0.2183	0.2177	0.2150	0.2101	0.2032	0.1943	0.1836
14	0.1711	0.1570	0.1414	0.1245	0.1065	0.0875	0.0679	0.0476	0.0271	0.0064
15	−0.0142	−0.0346	−0.0544	−0.0736	−0.0919	−0.1092	−0.1253	−0.1401	−0.1533	−0.1650

When $z > 15.9$,

$$J_0(z) \simeq \sqrt{\left(\frac{2}{\pi z}\right)} \left\{ \sin(z + \frac{1}{4}\pi) + \frac{1}{8z} \sin(z - \frac{1}{4}\pi) \right\}$$

z	$J_1(z)$									
	0	0.1	0.2	0.3	0.4	0.5	0.6	0.7	0.8	0.9
0	0.0000	0.0499	0.0995	0.1483	0.1960	0.2423	0.2867	0.3290	0.3688	0.4059
1	0.4401	0.4709	0.4983	0.5220	0.5419	0.5579	0.5699	0.5778	0.5815	0.5812
2	0.5767	0.5683	0.5560	0.5399	0.5202	0.4971	0.4708	0.4416	0.4097	0.3754
3	0.3391	0.3009	0.2613	0.2207	0.1792	0.1374	0.0955	0.0538	0.0128	−0.0272
4	−0.0660	−0.1033	−0.1386	−0.1719	−0.2028	−0.2311	−0.2566	−0.2791	−0.2985	−0.3147
5	−0.3276	−0.3371	−0.3432	−0.3460	−0.3453	−0.3414	−0.3343	−0.3241	−0.3110	−0.2951
6	−0.2767	−0.2559	−0.2329	−0.2081	−0.1816	−0.1538	−0.1250	−0.0953	−0.0652	−0.0349
7	−0.0047	0.0252	0.0543	0.0826	0.1096	0.1352	0.1592	0.1813	0.2014	0.2192
8	0.2346	0.2476	0.2580	0.2657	0.2708	0.2731	0.2728	0.2697	0.2641	0.2559
9	0.2453	0.2324	0.2174	0.2004	0.1816	0.1613	0.1395	0.1166	0.0928	0.0684
10	0.0435	0.0184	−0.0066	−0.0313	−0.0555	−0.0789	−0.1012	−0.1224	−0.1422	−0.1603
11	−0.1768	−0.1913	−0.2039	−0.2143	−0.2225	−0.2284	−0.2320	−0.2333	−0.2323	−0.2290
12	−0.2234	−0.2157	−0.2060	−0.1943	−0.1807	−0.1655	−0.1487	−0.1307	−0.1114	−0.0912
13	−0.0703	−0.0489	−0.0271	−0.0052	+0.0166	0.0380	0.0590	0.0791	0.0984	0.1165
14	0.1334	0.1488	0.1626	0.1747	0.1850	0.1934	0.1999	0.2043	0.2066	0.2069
15	0.2051	0.2013	0.1955	0.1879	0.1784	0.1672	0.1544	0.1402	0.1247	0.1080

When $z > 15.9$,

$$J_1(z) \simeq \sqrt{\left(\frac{2}{\pi z}\right)} \left\{ \sin(z - \frac{1}{4}\pi) + \frac{3}{8z} \sin(z + \frac{1}{4}\pi) \right\}$$

	$Y_0(z)$									
z	0	0.1	0.2	0.3	0.4	0.5	0.6	0.7	0.8	0.9
0	$-\infty$	−1.5342	−1.0811	−0.8073	−0.6060	−0.4445	−0.3085	−0.1907	−0.0868	0.0056
1	0.0883	0.1622	0.2281	0.2865	0.3379	0.3824	0.4204	0.4520	0.4774	0.4968
2	0.5104	0.5183	0.5208	0.5181	0.5104	0.4981	0.4813	0.4605	0.4359	0.4079
3	0.3769	0.3431	0.3071	0.2691	0.2296	0.1890	0.1477	0.1061	0.0645	0.0234
4	−0.0169	−0.0561	−0.0938	−0.1296	−0.1633	−0.1947	−0.2235	−0.2494	−0.2723	−0.2921
5	−0.3085	−0.3216	−0.3313	−0.3374	−0.3402	−0.3395	−0.3354	−0.3282	−0.3177	−0.3044
6	−0.2882	−0.2694	−0.2483	−0.2251	−0.1999	−0.1732	−0.1452	−0.1162	−0.0864	−0.0563
7	−0.0259	0.0042	0.0339	0.0628	0.0907	0.1173	0.1424	0.1658	0.1872	0.2065
8	0.2235	0.2381	0.2501	0.2595	0.2662	0.2702	0.2715	0.2700	0.2659	0.2592
9	0.2499	0.2383	0.2245	0.2086	0.1907	0.1712	0.1502	0.1279	0.1045	0.0804
10	0.0557	0.0307	0.0056	−0.0193	−0.0437	−0.0675	−0.0904	−0.1122	−0.1326	−0.1516
11	−0.1688	−0.1843	−0.1977	−0.2091	−0.2183	−0.2252	−0.2299	−0.2322	−0.2322	−0.2298
12	−0.2252	−0.2184	−0.2095	−0.1986	−0.1858	−0.1712	−0.1551	−0.1375	−0.1187	−0.0989
13	−0.0782	−0.0569	−0.0352	−0.0134	0.0085	0.0301	0.0512	0.0717	0.0913	0.1099
14	0.1272	0.1431	0.1575	0.1703	0.1812	0.1903	0.1974	0.2025	0.2056	0.2065
15	0.2055	0.2023	0.1972	0.1902	0.1813	0.1706	0.1584	0.1446	0.1295	0.1132

When $z > 15.9$,

$$Y_0(z) \simeq \sqrt{\left(\frac{2}{\pi z}\right)}\left\{\sin(z - \frac{1}{4}\pi) - \frac{1}{8z}\sin(z + \frac{1}{4}\pi)\right\}$$

z	$Y_1(z)$									
	0	0.1	0.2	0.3	0.4	0.5	0.6	0.7	0.8	0.9
0	$-\infty$	−6.4590	−3.3238	−2.2931	−1.7809	−1.4715	−1.2604	−1.1032	−0.9781	−0.8731
1	−0.7812	−0.6981	−0.6211	−0.5485	−0.4791	−0.4123	−0.3476	−0.2847	−0.2237	−0.1644
2	−0.1070	−0.0517	0.0015	0.0523	0.1005	0.1459	0.1884	0.2276	0.2635	0.2959
3	0.3247	0.3496	0.3707	0.3879	0.4010	0.4102	0.4154	0.4167	0.4141	0.4078
4	0.3979	0.3846	0.3680	0.3484	0.3260	0.3010	0.2737	0.2445	0.2136	0.1812
5	0.1479	0.1137	0.0792	0.0445	0.0101	−0.0238	−0.0568	−0.0887	−0.1192	−0.1481
6	−0.1750	−0.1998	−0.2223	−0.2422	−0.2596	−0.2741	−0.2857	−0.2945	−0.3002	−0.3029
7	−0.3027	−0.2995	−0.2934	−0.2846	−0.2731	−0.2591	−0.2428	−0.2243	−0.2039	−0.1817
8	−0.1581	−0.1331	−0.1072	−0.0806	−0.0535	−0.0262	0.0011	0.0280	0.0544	0.0799
9	0.1043	0.1275	0.1491	0.1691	0.1871	0.2032	0.2171	0.2287	0.2379	0.2447
10	0.2490	0.2508	0.2502	0.2471	0.2416	0.2337	0.2236	0.2114	0.1973	0.1813
11	0.1637	0.1446	0.1243	0.1029	0.0807	0.0579	0.0348	0.0114	−0.0118	−0.0347
12	−0.0571	−0.0787	−0.0994	−0.1189	−0.1371	−0.1538	−0.1689	−0.1821	−0.1935	−0.2028
13	−0.2101	−0.2152	−0.2182	−0.2190	−0.2176	−0.2140	−0.2084	−0.2007	−0.1912	−0.1798
14	−0.1666	−0.1520	−0.1359	−0.1186	−0.1003	−0.0810	−0.0612	−0.0408	−0.0202	0.0005
15	0.0211	0.0413	0.0609	0.0799	0.0979	0.1148	0.1305	0.1447	0.1575	0.1686

When $z > 15.9$,

$$Y_1(z) \simeq \sqrt{\left(\frac{2}{\pi z}\right)}\left\{\sin(z - \frac{3}{4}\pi) + \frac{3}{8z}\sin(z - \frac{1}{4}\pi)\right\}$$

	z	$I_0(z)$									
		0	0.1	0.2	0.3	0.4	0.5	0.6	0.7	0.8	0.9
	0	1.0000	1.0025	1.0100	1.0226	1.0404	1.0635	1.0920	1.1263	1.1665	1.2130
	1	1.2661	1.3262	1.3937	1.4693	1.5534	1.6467	1.7500	1.8640	1.9896	2.1277
	2	2.2796	2.4463	2.6291	2.8296	3.0493	3.2898	3.5533	3.8417	4.1573	4.5027
	3	4.8808	5.2945	5.7472	6.2426	6.7848	7.3782	8.0277	8.7386	9.5169	10.369
$10 \times$	4	1.1302	1.2324	1.3442	1.4668	1.6010	1.7481	1.9093	2.0858	2.2794	2.4915
$10 \times$	5	2.7240	2.9789	3.2584	3.5648	3.9009	4.2695	4.6738	5.1173	5.6038	6.1377
$10 \times$	6	6.7234	7.3663	8.0718	8.8462	9.6962	10.629	11.654	12.779	14.014	15.370
$10^2 \times$	7	1.6859	1.8495	2.0292	2.2266	2.4434	2.6816	2.9433	3.2309	3.5468	3.8941
$10^2 \times$	8	4.2756	4.6950	5.1559	5.6626	6.2194	6.8316	7.5046	8.2445	9.0580	9.9524
$10^3 \times$	9	1.0936	1.2017	1.3207	1.4514	1.5953	1.7535	1.9275	2.1189	2.3294	2.5610

When $z \geq 10$,

$$I_0(z) \simeq \frac{0.3989e^z}{z^{1/2}}\left\{1 + \frac{1}{8z} + \frac{9}{128z^2} + \frac{75}{1024z^3}\right\}$$

	z	$K_0(z)$									
		0	0.1	0.2	0.3	0.4	0.5	0.6	0.7	0.8	0.9
	0	∞	2.4271	1.7527	1.3725	1.1145	0.9244	0.7775	0.6605	0.5653	0.4867
	1	0.4210	0.3656	0.3185	0.2782	0.2437	0.2138	0.1880	0.1655	0.1459	0.1288
10^{-1} ×	2	1.1389	1.0078	0.8926	0.7914	0.7022	0.6235	0.5540	0.4926	0.4382	0.3901
10^{-1} ×	3	0.3474	0.3095	0.2759	0.2461	0.2196	0.1960	0.1750	0.1563	0.1397	0.1248
10^{-2} ×	4	1.1160	0.9980	0.8927	0.7988	0.7149	0.6400	0.5730	0.5132	0.4597	0.4119
10^{-2} ×	5	0.3691	0.3308	0.2966	0.2659	0.2385	0.2139	0.1918	0.1721	0.1544	0.1386
10^{-3} ×	6	1.2440	1.1167	1.0025	0.9001	0.8083	0.7259	0.6520	0.5857	0.5262	0.4728
10^{-3} ×	7	0.4248	0.3817	0.3431	0.3084	0.2772	0.2492	0.2240	0.2014	0.1811	0.1629
10^{-4} ×	8	1.4647	1.3173	1.1849	1.0658	0.9588	0.8626	0.7761	0.6983	0.6283	0.5654
10^{-4} ×	9	0.5088	0.4579	0.4121	0.3710	0.3339	0.3006	0.2706	0.2436	0.2193	0.1975

When $z \geq 10$,

$$K_0(z) \simeq \frac{1.2533e^{-z}}{z^{1/2}} \left\{1 - \frac{1}{8z} + \frac{9}{128z^2} - \frac{75}{1024z^3}\right\}$$

	z	$I_1(z)$									
		0	0.1	0.2	0.3	0.4	0.5	0.6	0.7	0.8	0.9
	0	0	0.0501	0.1005	0.1517	0.2040	0.2579	0.3137	0.3719	0.4329	0.4971
	1	0.5652	0.6375	0.7147	0.7973	0.8861	0.9817	1.0848	1.1963	1.3172	1.4482
	2	1.5906	1.7455	1.9141	2.0978	2.2981	2.5167	2.7554	3.0161	3.3011	3.6126
	3	3.9534	4.3262	4.7343	5.1810	5.6701	6.2058	6.7927	7.4357	8.1404	8.9128
$10\times$	4	0.9759	1.0688	1.1706	1.2822	1.4046	1.5389	1.6863	1.8479	2.0253	2.2199
$10\times$	5	2.4336	2.6680	2.9254	3.2080	3.5182	3.8588	4.2328	4.6436	5.0946	5.5900
$10\times$	6	6.1342	6.7319	7.3886	8.1100	8.9026	9.7735	10.730	11.782	12.938	14.208
$10^2\times$	7	1.5604	1.7138	1.8825	2.0679	2.2717	2.4958	2.7422	3.0131	3.3110	3.6385
$10^2\times$	8	3.9987	4.3948	4.8305	5.3096	5.8366	6.4162	7.0538	7.7551	8.5266	9.3754
$10^3\times$	9	1.0309	1.1336	1.2467	1.3710	1.5079	1.6585	1.8241	2.0065	2.2071	2.4280

When $z \geq 10$,

$$I_1(z) \simeq \frac{0.3989e^z}{z^{1/2}}\left\{1 - \frac{3}{8z} - \frac{15}{128z^2} - \frac{105}{1024z^3}\right\}$$

	z	$K_1(z)$									
		0	0.1	0.2	0.3	0.4	0.5	0.6	0.7	0.8	0.9
	0	∞	9.8538	4.7760	3.0560	2.1844	1.6564	1.3028	1.0503	0.8618	0.7165
	1	0.6019	0.5098	0.4346	0.3725	0.3208	0.2774	0.2406	0.2094	0.1826	0.1597
$10^{-1}\times$	2	1.3987	1.2275	1.0790	0.9498	0.8372	0.7389	0.6528	0.5774	0.5111	0.4529
$10^{-1}\times$	3	0.4016	0.3563	0.3164	0.2812	0.2500	0.2224	0.1979	0.1763	0.1571	0.1400
$10^{-2}\times$	4	1.2484	1.1136	0.9938	0.8872	0.7923	0.7078	0.6325	0.5654	0.5055	0.4521
$10^{-2}\times$	5	0.4045	0.3619	0.3239	0.2900	0.2597	0.2326	0.2083	0.1866	0.1673	0.1499
$10^{-3}\times$	6	1.3439	1.2050	1.0805	0.9691	0.8693	0.7799	0.6998	0.6280	0.5636	0.5059
$10^{-3}\times$	7	0.4542	0.4078	0.3662	0.3288	0.2953	0.2653	0.2383	0.2141	0.1924	0.1729
$10^{-4}\times$	8	1.5537	1.3964	1.2552	1.1283	1.0143	0.9120	0.8200	0.7374	0.6631	0.5964
$10^{-4}\times$	9	0.5364	0.4825	0.4340	0.3904	0.3512	0.3160	0.2843	0.2559	0.2302	0.2072

When $z \geq 10$,

$$K_1(z) \simeq \frac{1.2533e^{-z}}{z^{1/2}} \left\{ 1 + \frac{3}{8z} - \frac{15}{128z^2} + \frac{105}{1024z^3} \right\}$$

IV-7 RECURRENCE RELATIONS

The recurrence formulas for the Bessel functions are given as [1, pp. 45 and 66; 5, p. 361]

$$W_{\nu-1}(z) + W_{\nu+1}(z) = \frac{2\nu}{2} W_\nu(z) \tag{28a}$$

$$W_{\nu-1}(z) - W_{\nu+1}(z) = 2W_\nu'(z) \tag{28b}$$

$$W_{\nu-1}(z) - \frac{\nu}{z} W_\nu(z) = W_\nu(z) \tag{28c}$$

$$-W_{\nu+1}(z) + \frac{\nu}{2} W_\nu(z) = W_\nu'(z) \tag{28d}$$

where $W = J$ or Y or any linear combination of these functions, the coefficients of which are independent of z and ν.

A systematic tabulation of various integrals involving Bessel functions is given in Ref. 7 and 8.

In Table IV-1 we present the numerical values of $J_n(z)$, $Y_n(z)$, $I_n(z)$, and $K_n(z)$ functions for $n = 0$ and 1 [2, pp. 215–221], and in Table IV-2 we present the first 10 roots of $J_n(z)$ function for $n = 0, 1, 2, 3, 5$.

Finally, in Tables IV-3 and IV-4 we present the roots of $\beta J_1(\beta) - cJ_0(\beta) = 0$ and $J_0(\beta)Y_0(c\beta) - Y_0(\beta)J_0(c\beta) = 0$, respectively [9, p. 493; 5, pp. 414–415].

TABLE IV-2 First 10 Roots of $J_n(z) = 0$, $n = 0, 1, 2, 3, 4, 5$

	J_0	J_1	J_2	J_3	J_4	J_5
1	2.4048	3.8317	5.1356	6.3802	7.5883	8.7715
2	5.5201	7.0156	8.4172	9.7610	11.0647	12.3386
3	8.6537	10.1735	11.6198	13.0152	14.3725	15.7002
4	11.7915	13.3237	14.7960	16.2235	17.6160	18.9801
5	14.9309	16.4706	17.9598	19.4094	20.8269	22.2178
6	18.0711	19.6159	21.1170	22.5827	24.0190	25.4303
7	21.2116	22.7601	24.2701	25.7482	27.1991	28.6266
8	24.3525	25.9037	27.4206	28.9084	30.3710	31.8117
9	27.4935	29.0468	30.5692	32.0649	33.5371	34.9888
10	30.6346	32.1897	33.7165	35.2187	36.6990	38.1599

TABLE IV-3 First Six Roots of $\beta J_1(\beta) - cJ_0(\beta) = 0$

c	β_1	β_2	β_3	β_4	β_5	β_6
0	0	3.8317	7.0156	10.1735	13.3237	16.4706
0.01	0.1412	3.8343	7.0170	10.1745	13.3244	16.4712
0.02	0.1995	3.8369	7.0184	10.1754	13.3252	16.4718
0.04	0.2814	3.8421	7.0213	10.1774	13.3267	16.4731
0.06	0.3438	3.8473	7.0241	10.1794	13.3282	16.4743
0.08	0.3960	3.8525	7.0270	10.1813	13.3297	16.4755
0.1	0.4417	3.8577	7.0298	10.1833	13.3312	16.4767
0.15	0.5376	3.8706	7.0369	10.1882	13.3349	16.4797
0.2	0.6170	3.8835	7.0440	10.1931	13.3387	16.4828
0.3	0.7465	3.9091	7.0582	10.2029	13.3462	16.4888
0.4	0.8516	3.9344	7.0723	10.2127	13.3537	16.4949
0.5	0.9408	3.9594	7.0864	10.2225	13.3611	16.5010
0.6	1.0184	3.9841	7.1004	10.2322	13.3686	16.5070
0.7	1.0873	4.0085	7.1143	10.2419	13.3761	16.5131
0.8	1.1490	4.0325	7.1282	10.2516	13.3835	16.5191
0.9	1.2048	4.0562	7.1421	10.2613	13.3910	16.5251
1.0	1.2558	4.0795	7.1558	10.2710	13.3984	16.5312
1.5	1.4569	4.1902	7.2233	10.3188	13.4353	16.5612
2.0	1.5994	4.2910	7.2884	10.3658	13.4719	16.5910
3.0	1.7887	4.4634	7.4103	10.4566	13.5434	16.6499
4.0	1.9081	4.6018	7.5201	10.5423	13.6125	16.7073
5.0	1.9898	4.7131	7.6177	10.6223	13.6786	16.7630
6.0	2.0490	4.8033	7.7039	10.6964	13.7414	16.8168
7.0	2.0937	4.8772	7.7797	10.7646	13.8008	16.8684
8.0	2.1286	4.9384	7.8464	10.8271	13.8566	16.9179
9.0	2.1566	4.9897	7.9051	10.8842	13.9090	16.9650
10.0	2.1795	5.0332	7.9569	10.9363	13.9580	17.0099
15.0	2.2509	5.1773	8.1422	11.1367	14.1576	17.2008
20.0	2.2880	5.2568	8.2534	11.2677	14.2983	17.3442
30.0	2.3261	5.3410	8.3771	11.4221	14.4748	17.5348
40.0	2.3455	5.3846	8.4432	11.5081	14.5774	17.6508
50.0	2.3572	5.4112	8.4840	11.5621	14.6433	17.7272
60.0	2.3651	5.4291	8.5116	11.5990	14.6889	17.7807
80.0	2.3750	5.4516	8.5466	11.6461	14.7475	17.8502
100.0	2.3809	5.4652	8.5678	11.6747	14.7834	17.8931
∞	2.4048	5.5201	8.6537	11.7915	14.9309	18.0711

Source: From Carslaw and Jaeger [9].

TABLE IV-4 First Five Roots of $J_0(\beta)Y_0(c\beta) - Y_0(\beta)J_0(c\beta) = 0$

c	β_1	β_2	β_3	β_4	β_5
1.2	15.7014	31.4126	47.1217	62.8302	78.5385
1.5	6.2702	12.5598	18.8451	25.1294	31.4133
2.0	3.1230	6.2734	9.4182	12.5614	15.7040
2.5	2.0732	4.1773	6.2754	8.3717	10.4672
3.0	1.5485	3.1291	4.7038	6.2767	7.8487
3.5	1.2339	2.5002	3.7608	5.0196	6.2776
4.0	1.0244	2.0809	3.1322	4.1816	5.2301

REFERENCES

1. G. N. Watson, *A Treatise on the Theory of Bessel Functions*, 2nd ed., Cambridge University Press, London, 1966.
2. N. W. McLachlan, *Bessel Functions for Engineers*, 2nd. ed., Oxford, Clarendon Press, London, 1961.
3. F. B. Hildebrand, *Advanced Calculus for Engineers,* Prentice-Hall, Englewood Cliffs, NJ, 1949.
4. T. K. Sherwood and C. E. Reed, *Applied Mathematics in Chemical Engineering*, McGraw-Hill, New York, 1939.
5. M. Abramowitz and I. A. Stegun, *Handbook of Mathematical Functions*, National Bureau of Standards, Applied Mathematic Series 55, U.S. Government Printing Office, Washington, DC, 1964.
6. G. Cinelli, *Int. J. Eng. Sci.* **3**, 539–559, 1965.
7. I. S. Gradshteyn and I. M. Ryzhik, *Table of Integrals, Series, and Products* (trans. from the Russian and ed. by A. Jeffrey), Academic, New York, 1965.
8. Y. L. Luke, *Integrals of Bessel Functions*, McGraw-Hill, New York, 1962.
9. H. S. Carslaw and J. G. Jaeger, *Conduction of Heat in Solids*, Oxford at the Clarendon Press, London, 1959.

APPENDIX V

NUMERICAL VALUES OF LEGENDRE POLYNOMIALS OF THE FIRST KIND

x	$P_1(x)$	$P_2(x)$	$P_3(x)$	$P_4(x)$	$P_5(x)$	$P_6(x)$	$P_7(x)$
0.00	0.0000	−0.5000	0.0000	0.3750	0.0000	−0.3125	0.0000
0.01	0.0100	−0.4998	−0.0150	0.3746	0.0187	−0.3118	−0.0219
0.02	0.0200	−0.4994	−0.0300	0.3735	0.0374	−0.3099	−0.0436
0.03	0.0300	−0.4986	−0.0449	0.3716	0.0560	−0.3066	−0.0651
0.04	0.0400	−0.4976	−0.0598	0.3690	0.0744	−0.3021	−0.0862
0.05	0.0500	−0.4962	−0.0747	0.3657	0.0927	−0.2962	−0.1069
0.06	0.0600	−0.4946	−0.0895	0.3616	0.1106	−0.2891	−0.1270
0.07	0.0700	−0.4926	−0.1041	0.3567	0.1283	−0.2808	−0.1464
0.08	0.0800	−0.4904	−0.1187	0.3512	0.1455	−0.2713	−0.1651
0.09	0.0900	−0.4878	−0.1332	0.3449	0.1624	−0.2606	−0.1828
0.10	0.1000	−0.4850	−0.1475	0.3379	0.1788	−0.2488	−0.1995
0.11	0.1100	−0.4818	−0.1617	0.3303	0.1947	−0.2360	−0.2151
0.12	0.1200	−0.4784	−0.1757	0.3219	0.2101	−0.2220	−0.2295
0.13	0.1300	−0.4746	−0.1895	0.3129	0.2248	−0.2071	−0.2427
0.14	0.1400	−0.4706	−0.2031	0.3032	0.2389	−0.1913	−0.2545
0.15	0.1500	−0.4662	−0.2166	0.2928	0.2523	−0.1746	−0.2649
0.16	0.1600	−0.4616	−0.2298	0.2819	0.2650	−0.1572	−0.2738
0.17	0.1700	−0.4566	−0.2427	0.2703	0.2769	−0.1389	−0.2812
0.18	0.1800	−0.4514	−0.2554	0.2581	0.2880	−0.1201	−0.2870
0.19	0.1900	−0.4458	−0.2679	0.2453	0.2982	−0.1006	−0.2911
0.20	0.2000	−0.4400	−0.2800	0.2320	0.3075	−0.0806	−0.2935

(continues overleaf)

x	$P_1(x)$	$P_2(x)$	$P_3(x)$	$P_4(x)$	$P_5(x)$	$P_6(x)$	$P_7(x)$
0.21	0.2100	−0.4338	−0.2918	0.2181	0.3159	−0.0601	−0.2943
0.22	0.2200	−0.4274	−0.3034	0.2037	0.3234	−0.0394	−0.2933
0.23	0.2300	−0.4206	−0.3146	0.1889	0.3299	−0.0183	−0.2906
0.24	0.2400	−0.4136	−0.3254	0.1735	0.3353	0.0029	−0.2861
0.25	0.2500	−0.4062	−0.3359	0.1577	0.3397	0.0243	−0.2799
0.26	0.2600	−0.3986	−0.3461	0.1415	0.3431	0.0456	−0.2720
0.27	0.2700	−0.3906	−0.3558	0.1249	0.3453	0.0669	−0.2625
0.28	0.2800	−0.3824	−0.3651	0.1079	0.3465	0.0879	−0.2512
0.29	0.2900	−0.3738	−0.3740	0.0906	0.3465	0.1087	−0.2384
0.30	0.3000	−0.3650	−0.3825	0.0729	0.3454	0.1292	−0.2241
0.31	0.3100	−0.3558	−0.3905	0.0550	0.3431	0.1492	−0.2082
0.32	0.3200	−0.3464	−0.3981	0.0369	0.3397	0.1686	−0.1910
0.33	0.3300	−0.3366	−0.4052	0.0185	0.3351	0.1873	−0.1724
0.34	0.3400	−0.3266	−0.4117	−0.0000	0.3294	0.2053	−0.1527
0.35	0.3500	−0.3162	−0.4178	−0.0187	0.3225	0.2225	−0.1318
0.36	0.3600	−0.3056	−0.4234	−0.0375	0.3144	0.2388	−0.1098
0.37	0.3700	−0.2946	−0.4284	−0.0564	0.3051	0.2540	−0.0870
0.38	0.3800	−0.2834	−0.4328	−0.0753	0.2948	0.2681	−0.0635
0.39	0.3900	−0.2718	−0.4367	−0.0942	0.2833	0.2810	−0.0393
0.40	0.4000	−0.2600	−0.4400	−0.1130	0.2706	0.2926	−0.0146
0.41	0.4100	−0.2478	−0.4427	−0.1317	0.2569	0.3029	0.0104
0.42	0.4200	−0.2354	−0.4448	−0.1504	0.2421	0.3118	0.0356
0.43	0.4300	−0.2226	−0.4462	−0.1688	0.2263	0.3191	0.0608
0.44	0.4400	−0.2096	−0.4470	−0.1870	0.2095	0.3249	0.0859
0.45	0.4500	−0.1962	−0.4472	−0.2050	0.1917	0.3290	0.1106
0.46	0.4600	−0.1826	−0.4467	−0.2226	0.1730	0.3314	0.1348
0.47	0.4700	−0.1686	−0.4454	−0.2399	0.1534	0.3321	0.1584
0.48	0.4800	−0.1544	−0.4435	−0.2568	0.1330	0.3310	0.1811
0.49	0.4900	−0.1398	−0.4409	−0.2732	0.1118	0.3280	0.2027
0.50	0.5000	−0.1250	−0.4375	−0.2891	0.0898	0.3232	0.2231
0.51	0.5100	−0.1098	−0.4334	−0.3044	0.0673	0.3166	0.2422
0.52	0.5200	−0.0944	−0.4258	−0.3191	0.0441	0.3080	0.2596
0.53	0.5300	−0.0786	−0.4228	−0.3332	0.0204	0.2975	0.2753
0.54	0.5400	−0.0626	−0.4163	−0.3465	−0.0037	0.2851	0.2891
0.55	0.5500	−0.0462	−0.4091	−0.3590	−0.0282	0.2708	0.3007
0.56	0.5600	−0.0296	−0.4010	−0.3707	−0.0529	0.2546	0.3102
0.57	0.5700	−0.0126	−0.3920	−0.3815	−0.0779	0.2366	0.3172
0.58	0.5800	0.0046	−0.3822	−0.3914	−0.1028	0.2168	0.3217
0.59	0.5900	0.0222	−0.3716	−0.4002	−0.1278	0.1953	0.3235
0.60	0.6000	0.0400	−0.3600	−0.4080	−0.1526	0.1721	0.3226
0.61	0.6100	0.0582	−0.3475	−0.4146	−0.1772	0.1473	0.3188
0.62	0.6200	0.0766	−0.3342	−0.4200	−0.2014	0.1211	0.3121

x	$P_1(x)$	$P_2(x)$	$P_3(x)$	$P_4(x)$	$P_5(x)$	$P_6(x)$	$P_7(x)$
0.63	0.6300	0.0954	−0.3199	−0.4242	−0.2251	0.0935	0.3023
0.64	0.6400	0.1144	−0.3046	−0.4270	−0.2482	0.0646	0.2895
0.65	0.6500	0.1338	−0.2884	−0.4284	−0.2705	0.0347	0.2737
0.66	0.6600	0.1534	−0.2713	−0.4284	−0.2919	0.0038	0.2548
0.67	0.6700	0.1734	−0.2531	−0.4268	−0.3122	−0.0278	0.2329
0.68	0.6800	0.1936	−0.2339	−0.4236	−0.3313	−0.0601	0.2081
0.69	0.6900	0.2142	−0.2137	−0.4187	−0.3490	−0.0926	0.1805
0.70	0.7000	0.2350	−0.1925	−0.4121	−0.3652	−0.1253	0.1502
0.71	0.7100	0.2562	−0.1702	−0.4036	−0.3796	−0.1578	0.1173
0.72	0.7200	0.2776	−0.1469	−0.3933	−0.3922	−0.1899	0.0822
0.73	0.7300	0.2994	−0.1225	−0.3810	−0.4026	−0.2214	0.0450
0.74	0.7400	0.3214	−0.0969	−0.3666	−0.4107	−0.2518	0.0061
0.75	0.7500	0.3438	−0.0703	−0.3501	−0.4164	−0.2808	−0.0342
0.76	0.7600	0.3664	−0.0426	−0.3314	−0.4193	−0.3081	−0.0754
0.77	0.7700	0.3894	−0.0137	−0.3104	−0.4193	−0.3333	−0.1171
0.78	0.7800	0.4126	0.0164	−0.2871	−0.4162	−0.3559	−0.1588
0.79	0.7900	0.4362	0.0476	−0.2613	−0.4097	−0.3756	−0.1999
0.80	0.8000	0.4600	0.0800	−0.2330	−0.3995	−0.3918	−0.2397
0.81	0.8100	0.4842	0.1136	−0.2021	−0.3855	−0.4041	−0.2774
0.82	0.8200	0.5086	0.1484	−0.1685	−0.3674	−0.4119	−0.3124
0.83	0.8300	0.5334	0.1845	−0.1321	−0.3449	−0.4147	−0.3437
0.84	0.8400	0.5584	0.2218	−0.0928	−0.3177	−0.4120	−0.3703
0.85	0.8500	0.5838	0.2603	−0.0506	−0.2857	−0.4030	−0.3913
0.86	0.8600	0.6094	0.3001	−0.0053	−0.2484	−0.3872	−0.4055
0.87	0.8700	0.6354	0.3413	0.0431	−0.2056	−0.3638	−0.4116
0.88	0.8800	0.6616	0.3837	0.0947	−0.1570	−0.3322	−0.4083
0.89	0.8900	0.6882	0.4274	0.1496	−0.1023	−0.2916	−0.3942
0.90	0.9000	0.7150	0.4725	0.2079	−0.0411	−0.2412	−0.3678
0.91	0.9100	0.7422	0.5189	0.2698	0.0268	−0.1802	−0.3274
0.92	0.9200	0.7696	0.5667	0.3352	0.1017	−0.1077	−0.2713
0.93	0.9300	0.7974	0.6159	0.4044	0.1842	−0.0229	−0.1975
0.94	0.9400	0.8254	0.6665	0.4773	0.2744	0.0751	−0.1040
0.95	0.9500	0.8538	0.7184	0.5541	0.3727	0.1875	0.0112
0.96	0.9600	0.8824	0.7718	0.6349	0.4796	0.3151	0.1506
0.97	0.9700	0.9114	0.8267	0.7198	0.5954	0.4590	0.3165
0.98	0.9800	0.9406	0.8830	0.8089	0.7204	0.6204	0.5115
0.99	0.9900	0.9702	0.9407	0.9022	0.8552	0.8003	0.7384
1.00	1.0000	1.0000	1.0000	1.0000	1.0000	1.0000	1.0000

Source: From W. E. Byerly, *Fourier Series and Spherical, Cylindrical, and Ellipsoidal Harmonics*, Dover, New York, 1959, pp. 280–281.

APPENDIX VI

PROPERTIES OF DELTA FUNCTIONS

The symbol $\delta(x)$, known as Dirac's delta function, is zero for every value of x except the origin $x = 0$ where it is infinite in such a way that

$$\int_{x=-\infty}^{\infty} \delta(x)\, dx = 1 \tag{1}$$

Such a definition is not meaningful in the true mathematical sense, but the theory of distributions justifies its use as well as its derivatives. Then $\delta(x)$ has the following properties:

$$\delta(x-b) = 0 \qquad \text{everywhere } x \neq b \tag{2}$$

For every continuous function $F(x)$ we write

$$\int_{x=-\infty}^{\infty} F(x)\delta(x-b)\, dx = F(b) \tag{3}$$

$$\int_{x=-\infty}^{\infty} F(x)\delta(x-0)\, dx = F(0) \tag{4}$$

$$F(x)\delta(x-b) = F(b)\delta(x-b) \tag{5}$$

VI-1 DERIVATIVES OF DELTA FUNCTION

$$\int_{x=-\infty}^{\infty} F(x)\delta'(x)\,dx = -\int_{x=-\infty}^{\infty} F'(x)\delta(x)\,dx = -F'(0) \tag{6}$$

$$\int_{x=-\infty}^{\infty} F(x)\delta''(x)\,dx = -\int_{x=-\infty}^{\infty} F'(x)\delta'(x)\,dx = -F''(0) \tag{7}$$

The delta function itself is the derivative of the unit step function $U(x)$, that is,

$$U'(x) = \delta(x) \tag{8}$$

where

$$\begin{aligned} U(x) &= 1 \qquad \text{for} \qquad x > 0 \\ &= 0 \qquad \text{for} \qquad x < 0 \end{aligned} \tag{9}$$

Note that the derivative of $U(x)$ is zero for $x < 0$, zero for $x > 0$, and undefined for $x = 0$.

INDEX

S

T

U

V

W